Nucleotide and Protein Expansions and Human Disease

Editors
Jozef Gécz, Adelaide
Grant R. Sutherland, Adelaide

108 figures, 21 in color, and 28 tables, 2003

KARGER

Basel · Freiburg · Paris · London · New York ·
Bangalore · Bangkok · Singapore · Tokyo · Sydney

Reprint of **Cytogenetic and Genome Research** (ISSN 1424–8581)
Vol. 100, No. 1–4, 2003

Cover illustration

Modified Figure 2 of J. Cleary and C. Pearson (this issue) showing genic location of various unstable repeats, the DNA, RNA, amino acids and associate disorders. The disorders covered in this single topic volume of *Cytogenetic and Genome Research* are boxed. The sequencing lanes show dideoxy manual sequencing of an interruption-stabilized FRAXA 39 repeat tract that is larger than the stability threshold of 34 repeats, $(CGG)_9AGG(CGG)_9AGG(CGG)_9AGG(CGG)_9$.

S. Karger
Medical and Scientific Publishers
Basel · Freiburg · Paris · London
New York · Bangalore · Bangkok
Singapore · Tokyo · Sydney

Drug Dosage
The authors and the publisher have exerted every effort to ensure that drug selection and dosage set forth in this text are in accord with current recommendations and practice at the time of publication. However, in view of ongoing research, changes in government regulations, and the constant flow of information relating to drug therapy and drug reactions, the reader is urged to check the package insert for each drug for any change in indications and dosage and for added warnings and precautions. This is particularly important when the recommended agent is a new and/or infrequently employed drug.

P.O. Box, CH–4009 Basel (Switzerland)
Printed in Switzerland on acid-free paper by
Reinhardt Druck, Basel
ISBN 3–8055–7621–8

KARGER

Fax +41 61 306 12 34
E-Mail karger@karger.ch
www.karger.com

Contents

Cytogenet Genome Res 100:5–6 (2003)
DOI: 10.1159/000072835

Preface

In 1991 two key events occurred that led to the recognition of DNA repeat sequence expansion as a mutational mechanism. The first of these events was the report of the (CAG) expansion in the androgen receptor gene as the genetic anomaly in Kennedy disease (spino-bulbar muscular atrophy; La Spada et al., 1991). This was a major finding but it did not have a great impact until the second event. This was the report that the fragile site at Xq27.3, also known as FRAXA or the Fragile X, was a gross expansion of a normally occurring CCG repeat sequence in the 5′-UTR of the FMR1 gene (Oberlé et al., 1991; Verkerk et al., 1991; Yu et al., 1991).

The fragile X had received much attention both scientifically and clinically prior to 1991. It was recognised to be associated with the most common form of familial mental retardation and at that time could only be diagnosed cytogenetically. Cytogenetic diagnosis was not straightforward and was inconclusive in many normal carrier females and all normal premutation carrier males.

The scientific interest in the fragile X chromosome hinged on its bizarre genetics, and there was much speculation as to what it might be at the DNA level. The fragile X syndrome of mental handicap and dysmorphic features was clearly X-linked and mapped to the fragile site yet there were normal carrier males and the penetrance of the syndrome in males was a function of the phenotype of their mothers. The behaviour of the expansion, in being unstable between generations, explained the segregation patterns (Fu et al., 1991; Yu et al., 1992) and provided an explanation for genetic anticipation (Sutherland et al., 1992).

Repeat (at that time trinucleotide repeat) expansion was soon shown to be the mutational mechanism in two high profile genetic diseases: myotonic dystrophy (Brooks et al., 1992; Fu et al., 1992; Mahadevan et al., 1992) and Huntington disease (Huntington's Disease Collaborative Research Group, 1993). Trinucleotide repeat expansion and its study virtually became a sub-specialty area of human molecular genetics.

Over the next few years a number of neurodegenerative disorders were also shown to be due to expansions of CAG repeats. Indeed the genetic classification of such disorders became of more value than the prior neurological categorisation. There are now more than fourteen such diseases that are recognised as being due to CAG repeat expansions (Cleary and Pearson, this issue).

Similarly, other fragile sites of the same type as the fragile X (rare folate-sensitive) were shown to be due to CCG expansions. At this time only one more fragile site, FRAXE, has been shown to be of clinical significance (Gécz, 2000).

Repeat expansion mutations spread beyond trinucleotides when minisatellite repeats were found to be expanded in two non-folate-sensitive fragile sites, FRA10B (Hewett et al., 1998), FRA16B (Yu et al., 1997). A 12-bp repeat was found to be expanded in a form of epilepsy (Lalioti et al., 1997).

The trinucleotides at the fragile sites were non coding as were the minisatellite expansions. The CAG repeats in the neurological disorders encoded glutamines and indeed this group of diseases has become known as "polyglutamine diseases".

Another group of diseases, due to an expanded (CCG) repeat, but in the coding region of genes in frame to encode alanine have been now recognized (e.g. OPMD, Brais et al., this issue or Amiel et al., 2003 and references therein). The DNA repeats in these genes seem to be more stable between generations and they are, as such, not dynamic mutations.

In this monograph many aspects of nucleotide expansion mutations and the proteins they affect are reviewed, including their mechanisms of expansion, molecular basis of the repeat-associated disorders, animal models, as well as clinical insights. We have also invited papers on the fragile sites (rare, common, and the forgotten) so that the monograph reviews much of what is currently known about these sites and their contribution to human disease.

While many aspects of repeat expansion mutations still remain to be discovered one of their curious features is their apparently human specific nature. Some animal models of the repeat expansion diseases have been developed but no naturally occurring disease has been found in other species.

We trust that readers will find this monograph of interest and it will become a readily available reference source for current information on fragile sites and nucleotide and protein repeat expansion mutations.

J. Gécz
G.R. Sutherland

KARGER
Fax + 41 61 306 12 34
E-mail karger@karger.ch
www.karger.com

Accessible online at:
www.karger.com/cgr

References

Amiel J, Laudier B, Attie-Bitach T, Trang H, De Pontual L, Gener B, Trochet D, Etchevers H, Ray P, Simonneau M, Vekemans M, Munnich A, Gaultier C, Lyonnet S: Polyalanine expansion and frameshift mutations of the paired-like homeobox gene PHOX2B in congenital central hypoventilation syndrome. Nature Genet 33:459–461 (2003).

Brook JD, McCurrach ME, Harley HG, Buckler AJ, Church D, Aburatani H, Hunter K, Stanton VP, Thirion JP, Hudson T, et al: Molecular basis of myotonic dystrophy: expansion of a trinucleotide (CTG) repeat at the 3′ end of a transcript encoding a protein kinase family member. Cell 68:799–808 (1992).

Cleary CD, Pearson CE: The contribution of *cis*-elements to disease-associated repeat instability: Clinical and experimental evidence. Cytogenet Genome Res 100:25–55 (2003).

Fu YH, Pizzuti A, Fenwick RG Jr, King J, Rajnarayan S, Dunne PW, Dubel J, Nasser GA, Ashizawa T, de Jong P, et al: An unstable triplet repeat in a gene related to myotonic muscular dystrophy. Science 255:1256–1258 (1992).

Fu YH, Kuhl DP, Pizzuti A, Pieretti M, Sutcliffe JS, Richards S, Verkerk AJ, Holden JJ, Fenwick RG Jr, Warren ST, et al: Variation of the CGG repeat at the fragile X site results in genetic instability: resolution of the Sherman paradox. Cell 67:1047–1058 (1991).

Gécz J: The FMR2 gene, FRAXE and non-specific X-linked mental retardation: clinical and molecular aspects. Ann hum Genet 64:95–106 (2000).

Hewett DR, Handt O, Hobson L, Mangelsdorf M, Eyre HJ, Baker E, Sutherland GR, Schuffenhauer S, Mao JI, Richards RI: FRA10B structure reveals common elements in repeat expansion and chromosomal fragile site genesis. Mol Cell 1:773–1781 (1998).

La Spada AR, Wilson EM, Lubahn DB, Harding AE, Fischbeck KH: Androgen receptor gene mutations in X-linked spinal and bulbar muscular atrophy. Nature 352:77–79 (1991).

Lalioti MD, Scott HS, Buresi C, Rossier C, Bottani A, Morris MA, Malafosse A, Antonarakis SE: Dodecamer repeat expansion in cystatin B gene in progressive myoclonus epilepsy. Nature 386:847–851 (1997).

Mahadevan M, Tsilfidis C, Sabourin L, Shutler G, Amemiya C, Jansen G, Neville C, Narang M, Barcelo J, O'Hoy K, et al: Myotonic dystrophy mutation: an unstable CTG repeat in the 3′ untranslated region of the gene. Science 255:1253–1255 (1992).

Oberle I, Rousseau F, Heitz D, Kretz C, Devys D, Hanauer A, Boue J, Bertheas MF, Mandel JL: Instability of a 550-base pair DNA segment and abnormal methylation in fragile X syndrome. Science 252:1097–1102 (1991).

Sutherland GR, Haan EA, Kremer E, Lynch M, Pritchard M, Yu S, Richards RI: Hereditary unstable DNA: a new explanation for some old genetic questions? Lancet 338:289–292 (1991).

The Huntington's Disease Collaborative Research Group: A novel gene containing a trinucleotide repeat that is expanded and unstable on Huntington's disease chromosomes. Cell 72:971–983 (1993).

Yu S, Mangelsdorf M, Hewett D, Hobson L, Baker E, Eyre HJ, Lapsys N, Le Paslier D, Doggett NA, Sutherland GR, Richards RI: Human chromosomal fragile site FRA16B is an amplified AT-rich minisatellite repeat. Cell 88:367–374 (1997).

Yu S, Mulley J, Loesch D, Turner G, Donnelly A, Gedeon A, Hillen D, Kremer E, Lynch M, Pritchard M, et al: Fragile-X syndrome: unique genetics of the heritable unstable element. Am J hum Genet 50:968–980 (1992).

Yu S, Pritchard M, Kremer E, Lynch M, Nancarrow J, Baker E, Holman K, Mulley JC, Warren ST, Schlessinger D, et al: Fragile X genotype characterized by an unstable region of DNA. Science 252:1179–1181 (1991).

Verkerk AJ, Pieretti M, Sutcliffe JS, Fu YH, Kuhl DP, Pizzuti A, Reiner O, Richards S, Victoria MF, Zhang FP, et al: Identification of a gene (FMR-1) containing a CGG repeat coincident with a breakpoint cluster region exhibiting length variation in fragile X syndrome. Cell 65:905–914 (1991).

Cytogenet Genome Res 100:7–24 (2003)
DOI: 10.1159/000072836

Trinucleotide repeat instability: a hairpin curve at the crossroads of replication, recombination, and repair

B.A. Lenzmeier[a] and C.H. Freudenreich[b]

[a] Department of Molecular Biology, Princeton University, Princeton NJ;
[b] Department of Biology, Tufts University, Medford MA (USA)

Abstract. The trinucleotide repeats that expand to cause human disease form hairpin structures in vitro that are proposed to be the major source of their genetic instability in vivo. If a replication fork is a train speeding along a track of double-stranded DNA, the trinucleotide repeats are a hairpin curve in the track. Experiments have demonstrated that the train can become derailed at the hairpin curve, resulting in significant damage to the track. Repair of the track often results in contractions and expansions of track length. In this review we introduce the in vitro evidence for why CTG/CAG and CCG/CGG repeats are inherently unstable and discuss how experiments in model organisms have implicated the replication, recombination and repair machinery as contributors to trinucleotide repeat instability in vivo.

Expansion of CTG/CAG or CCG/CGG trinucleotide repeat sequences, termed dynamic mutation (Richards and Sutherland, 1992), is associated with at least 13 different human diseases including several spinocerebellar ataxias, Huntington's disease, fragile X syndrome and myotonic dystrophy (reviewed in Ashley and Warren, 1995; Reddy and Housman, 1997; Timchenko and Caskey, 1999; Cummings and Zoghbi, 2000; Usdin and Grabczyk, 2000; Bowater and Wells, 2001; Kovtun et al., 2001; Richards, 2001). All individuals carry trinucleotide repeats at each of the disease loci. A small number of repeats, typically less than 35–40, is normal and does not cause the disease phenotype. For some of the diseases, an intermediate number of repeats is considered a premutation, with the individual not showing the disease phenotype but being susceptible to passing on a disease-causing expansion to the next generation. Expansion to the full mutation number of repeats causes the disease phenotype by altering the expression levels or biological activity of the gene(s) at that locus (reviewed in Ashley and Warren, 1995; Reddy and Housman, 1997; Timchenko and Caskey, 1999; Cummings and Zoghbi, 2000; Richards, 2001; O'Donnell and Warren, 2002). The number of repeats associated with the disease phenotype varies from locus to locus and depends on the location of the repeat tract relative to the gene coding sequence. The longest expansions to more than 2000 repeats are observed only when the trinucleotide repeat tract is located outside of the open reading frame of the disease associated gene (reviewed in Timchenko and Caskey, 1999; Usdin and Grabczyk, 2000). Interestingly, expansion of the repeat tract in an affected individual seems to occur only at the disease locus (Goellner et al., 1997). Other loci containing trinucleotide repeats are typically unaffected, suggesting the cause of the expansion is inherent to the locus itself and is not primarily due to mutation of trans-acting factors that affect the general stability of trinucleotide repeats (McMurray, 1999). Although the importance of flanking DNA cannot be ruled out entirely, a significant source of instability within the disease locus is the CTG/CAG or CCG/CGG tract itself, as high frequencies and rates of mutation occur when the tract alone is integrated at different loci in yeast (Schweitzer and Livingston, 1997; Freudenreich et al., 1998; Miret et al., 1998; White et al.,

B.A.L. is supported by postdoctoral fellowship no. PF00-013-01-GMC from the American Cancer Society and C.H.F. is supported by NIH grant no. GM63066.

Received 25 October 2002; manuscript accepted 6 January 2003.

Request reprints from Catherine H. Freudenreich
Department of Biology, Tufts University, Medford, MA 02155 (USA)
telephone: 617-627-4037; fax: 617-627-3805
e-mail: catherine.freudenreich@tufts.edu

KARGER
Fax + 41 61 306 12 34
E-mail karger@karger.ch
www.karger.com

Accessible online at:
www.karger.com/cgr

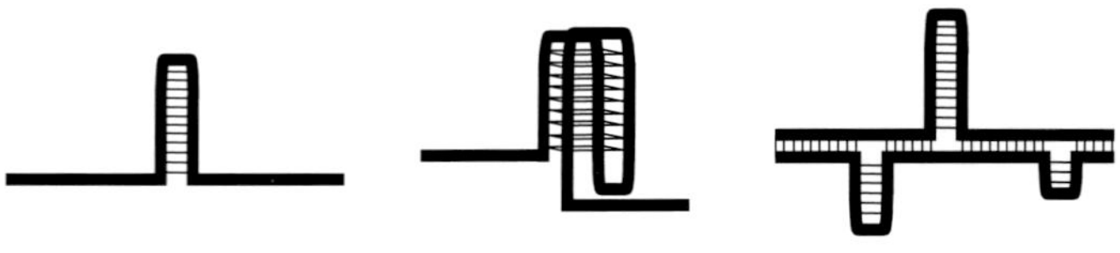

Fig. 1. Secondary structures formed by trinucleotide repeats in vitro. The stability of each single-stranded sequence relative to its base-paired partner is noted below the hairpin and quadruplex structures. For example, CTG hairpins are more stable than CAG hairpins. Slipped strand structures require double stranded DNA and the sequences shown to form slipped strand structures is noted below the structure.

1999; Balakumaran et al., 2000). Since numerous studies have also demonstrated that CTG/CAG and CCG/CGG repeats form stable secondary structures in vitro (reviewed in Mitas, 1997; Darlow and Leach, 1998; Pearson and Sinden, 1998; Sinden, 1999), these structures are proposed to be the locus-specific source of trinucleotide repeat instability in vivo. This review will focus on the in vitro and in vivo evidence for trinucleotide repeat secondary structures and how these structures wreak havoc on the replication, recombination, and repair machinery of the model organisms *E. coli* and *S. cerevisiae.*

Cis contribitions to CTG/CAG and CCG/CGG instablity

The primary sequence of a trinucleotide repeat tract determines its level of instability in vitro and in vivo. The ability of the sequence to form secondary structures, the purity of the repeat tract, the number of repeats and the chromosomal context all contribute in *cis* to the amount of instability observed for each repeat.

Structural features of trinucleotide repeat DNA

Of primary importance to trinucleotide repeat instability is the ability of the sequence to form secondary structures. Of the ten possible triplet repeat combinations only the expansion prone CTG/CAG, CCG/CGG and GAA/CTT repeats, and the rarely observed GAC/GTC repeats, form stable secondary structures in vitro (reviewed in Mitas, 1997; Darlow and Leach, 1998; Pearson and Sinden, 1998; Sinden, 1999). Systematic studies in *E. coli* and *S. cerevisiae* have confirmed the importance of primary sequence to trinucleotide repeat instability, as secondary structure forming repeats display significantly more instability in vivo than other trinucleotide repeat tracts. (Kang et al., 1995a; Ohshima et al., 1996; Shimizu et al., 1996; Spiro et al., 1999; Rolfsmeier et al., 2001).

The most common secondary structures observed for trinucleotide repeats in vitro are hairpins consisting of a single strand of DNA folding back upon itself (Fig. 1). Physical, biochemical, and theoretical approaches have clearly demonstrated a propensity for forming hairpins by single strands of CTG, CAG, CCG, CGG, GTC and GAC (Chen et al., 1995; Gacy et al., 1995; Mitas et al., 1995; Yu et al., 1995a, 1995b; Mariappan et al., 1996a, 1996b; Petruska et al., 1996; Zheng et al., 1996; Mitas, 1997; Ohshima and Wells, 1997; Yu A, et al., 1997; Yu S, et al., 1997; Gacy and McMurray, 1998; Mariappan et al., 1998a, 1998b). Stable hairpins are not observed under physiological conditions with other trinucleotide repeats or DNA sequences with a non-repetitive but otherwise identical base composition (Gacy et al., 1995; Gacy and McMurray, 1998). Hairpins formed by longer CTG/CAG and CCG/CGG tracts have melting temperatures that are above the physiological cellular temperatures of humans, yeast and bacteria, suggesting that if formed in vivo they would persist unless repaired by cellular activities (Gacy et al., 1995; Mitas et al., 1995; Yu et al., 1995a; Petruska et al., 1996; Mitas, 1997; Yu A, et al., 1997; Gacy and McMurray, 1998). CTG, CAG, CCG, and CGG hairpins of 15 repeats or more show differing degrees of stability in vitro: CGG hairpins are considerably more stable than CCG hairpins and CTG hairpins are slightly more stable than CAG hairpins (Petruska et al., 1996; Mitas, 1997; Gacy and McMurray, 1998). The importance of the strand bias of hairpin stability in vitro to the mechanisms of trinucleotide repeat instability in vivo is strongly supported by orientation-dependent instability experiments in *E. coli* and *S. cerevisiae* that are discussed below.

Although no physical evidence has been provided for trinucleotide repeat hairpin formation in vivo, genetic experiments analogous to those used to demonstrate palindrome and inverted repeat instability in bacteria and yeast are strongly suggestive of their presence. In *E. coli*, a long palindrome placed in bacteriophage λ DNA decreases plaque formation, presumably by forming secondary structures that inhibit bacteriophage λ replication (reviewed in Leach, 1994). Placement of a hairpin forming sequence at the center of the palindrome such that it would be located at the tip of the hairpin, stabilizes the palindrome and further reduces plaque size (Davison and Leach, 1994). Placement of trinucleotide repeats CTG/CAG or CCG/CGG in the center of the bacteriophage λ palindrome also significantly reduces plaque size, whereas placement of the GAC/GTC sequence that is not expanded in humans does not inhibit plaque formation (Darlow and Leach, 1995, 1998). These data strongly suggest that the trinucleotide repeats associated with expansion stabilize the tips of hairpin structures in vivo. Genetic experiments in yeast are also suggestive of hairpins forming in vivo. During yeast meiotic recombination between homologous chromosomes, any heterozygous sequences present within a heteroduplex create a mismatch that is typically removed. A palindrome placed within the mismatch, however, is not removed and causes non-Mendelian segregation events that are presumably due to the palindrome forming stable hairpin loops that escape detection and repair (Nag and Petes, 1991). When Moore et al. (1999) placed ten repeats of CGG, CCG, CTG or CAG in the same genomic location as the palindrome they escaped repair during meiosis in a manner that was similar to that of the palindrome. Importantly, non-hairpin forming trinucleotide repeats, such as AAG, CAA and AGT, were readily repaired. Also of note, CTG and CGG loops, which form more stable structures in vitro, were repaired less efficiently than their CAG and CCG counterparts in vivo (Moore et al., 1999). Although the number of trials was not enough to convey statistical significance to these differences, these data suggest that hairpin stability in vivo might correlate with what has been observed in vitro. While these experiments do not provide physical proof for hairpin formation, the clever use of genetic assays has provided the strongest evidence to date for secondary structure formation by trinucleotide repeats in vivo.

Higher order DNA structures have also been observed for the trinucleotide repeats (reviewed in Mitas, 1997; Darlow and Leach, 1998; Pearson and Sinden, 1998; Sinden, 1999). The GAA/TTC repeats associated with Friedreich's Ataxia readily form triplex structures in vitro (Mariappan et al., 1999; Sakamoto et al., 1999; LeProust et al., 2000; Vetcher et al., 2002), and there is speculation that CCG/CGG and CTG/CAG repeats might also form triplex DNA (Kuryavyi and Jovin, 1995). Quadruplex structures have also been observed in vitro with selected trinucleotide repeats. The CGG (Fry and Loeb, 1994; Chen, 1995; Kettani et al., 1995; Smith et al., 1995) and CCG (Fojtik and Vorlickova, 2001) repeat strands form tetraplexes that can be comprised of a unimolecular quartet (Fig. 1) or can be the result of two hairpins interacting. Although there is as yet no physical or genetic evidence for these structures in vivo, several observations suggest they might form. First, human proteins such as WRN helicase, human Ku and telomeric tetraplex DNA binding proteins qTBP42 and uqTBP25 preferentially recognize and modulate the stability of CGG quadruplex structures in vitro (Fry and Loeb, 1999; Uliel et al., 2000; Weisman-Shomer et al., 2000b). Second, the CGG quadruplex structures act as a block to DNA synthesis in vitro (Usdin and Woodford, 1995; Usdin, 1998; Kamath-Loeb et al., 2001). Third, this block can be alleviated by unwinding of the CGG quadruplex by addition of the human WRN protein (Fry and Loeb, 1999; Kamath-Loeb et al., 2001), a helicase whose mutation is associated with a premature aging disease that is hallmarked by genome instability (Brosh and Bohr, 2002). Together, these in vitro data raise the possibility that CGG quadruplexes might have a role in CCG/CGG instability.

Long CTG/CAG and CCG/CGG repeat tracts also have the ability to form slipped strand DNA intermediates (Pearson and Sinden, 1996; Pearson et al., 1998b). These structures, which contain simple hairpin loops on both strands, cruciform structures, or more complicated structures involving multiple hairpins (reviewed in Pearson and Sinden, 1998), are formed in vitro when the two complementary CTG/CAG or CCG/CGG strands are denatured and then allowed to re-anneal (Pearson and Sinden, 1996; Pearson et al., 1998b). An example of one of these structures is shown in Fig. 1. Slipped strand structures are recognized and bound by the human mismatch repair protein MSH2 in vitro (Pearson et al., 1997), and analyses of contraction products from interrupted CTG/CAG tracts passaged through mismatch repair deficient *E coli* are suggestive of slipped strand formation in vivo (Wells et al., 1998).

One structural feature of trinucleotide repeats in vitro that might be important for mechanisms of instability is the observation that double stranded tracts of CTG/CAG and CCG/CGG are inherently flexible and have an extended α-helical structure that is not observed in random B form DNA (Bacolla et al., 1997; Chastain and Sinden, 1998). This extended structure has been proposed to serve as a sink for superhelical density (Bacolla et al., 1997; Gellibolian et al., 1997), a feature that might have repercussions on the replication fork progression rate and the amount of mispairing observed in the tracts. It is not known if the CTG/CAG or CCG/CGG repeats are unique in this feature, as other trinucleotide repeat tracts were not examined.

A second structural feature associated with CTG/CAG and CCG/CGG trinucleotide repeats is their differing abilities to assemble into nucleosomes in vitro. While CTG/CAG repeats form very stable nucleosomes and act as highly efficient nucleosome positioning elements (Wang et al., 1994; Wang and Griffith, 1995; Godde and Wolffe, 1996), CCG/CGG repeats exclude nucleosome formation in vitro (Godde et al., 1996; Wang et al., 1996; Wang and Griffith, 1996). The role that chromatin structure plays in the mechanisms of CTG/CAG and CCG/CGG instability in vivo has not been examined directly.

Length, purity and chromosomal context

Trinucleotide repeat tract length contributes significantly to the extent of instability observed. This statement is underlined by the observation that in humans only tracts above the threshold premutation length are prone to instability and longer tracts are more likely to expand than shorter tracts (reviewed in Richards, 2001). In vitro, long CTG/CAG and CCG/CGG tracts form hairpins with dramatically longer half-lives than short repeat tracts (Gacy and McMurray, 1998). In bacteria and yeast longer tracts are more prone to frequent changes in tract length than shorter tracts (Kang et al., 1995a; Shimizu et al., 1996; Freudenreich et al., 1998; Hirst and White, 1998; Sarkar et al., 1998; Schweitzer and Livingston, 1999; White et al., 1999; Balakumaran et al., 2000; Richard et al., 2000; Rolfsmeier et al., 2001; Hashem et al., 2002). A sensitive genetic assay used to estimate the number of repeats required for expansion and contraction in yeast determined the threshold to be approximately 15 and 17 repeats respectively (Rolfsmeier et al., 2001), which correlates nicely with the number of repeats needed for stable hairpin formation in vitro (Mitas, 1997).

The purity of the trinucleotide repeat tract also contributes to the amount of instability observed. For example, interruptions in the CTG/CAG and CCG/CGG tracts found in the SCA1 and FRAXA loci suppress expansion potential in humans, and when expansions are observed they are usually accompanied by loss of the interruptions and generation of a long pure tract (reviewed in Kovtun et al., 2001). Interruptions destabilize hairpins, slipped strand structures and CGG quadruplexes in vitro (Gacy et al., 1995; Pearson et al., 1998a; Lyons-Darden and Topal, 1999b; Weisman-Shomer et al., 2000a). Interruptions also appear to decrease the frequency of tract length changes in vivo in model organisms (Kang et al., 1996; Shimizu et al., 1996; Hirst and White, 1998; Maurer et al., 1998; Jakupciak and Wells, 1999; White et al., 1999), and a systematic study into the influence of interruptions on CTG/CAG expansion rates found that an ATGATG interruption in a $(CTG)_{25}$ tract can stabilize the tract by as much as 90 fold (Rolfsmeier and Lahue, 2000). Together these data provide a

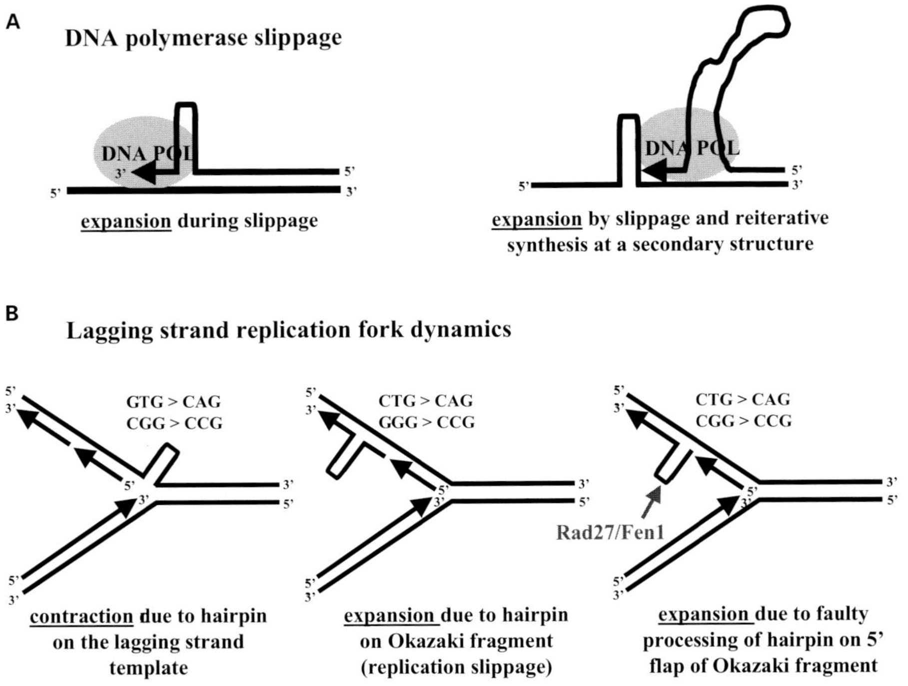

Fig. 2. Replication-based mechanisms for trinucleotide repeat instability. (**A**) DNA slippage could occur on the nascent strand to generate smaller expansions (left) or a secondary structure present in the template strand could cause slippage and reiterative synthesis of the nascent strand (right). (**B**) Lagging strand replication dynamics contribute to trinucleotide repeat instability when hairpins are not processed correctly. Contractions occur more frequently when the more stable secondary structure-forming strand (CTG or CGG) is present on the lagging strand template (left). Expansions are more likely to be observed when the more stable secondary structure-forming sequence is present on the Okazaki fragment (middle and right). Although both slippage (middle) and faulty 5′ flap processing (right) could account for the hairpin generation needed for expansions, the majority of experimental data (Fen1/Rad27 mutants) implicates the model on the right as the most likely source of generating the hairpins that lead to expansions during replication.

body of evidence that interruptions likely stabilize tracts in vivo by destabilizing secondary structures.

Finally, chromosomal context also plays a role in trinucleotide repeat instability. The direction of replication through the repeats and the distance from the nearest origin are both significant determinants to the amount and type of instability observed and will be discussed in detail below.

For further review of *cis*-elements and trinucleotide repeat instability see also "The contribution of *cis*-elements to disease-associated repeat instability: clinical and experimental evidence" by Cleary and Pearson in this issue.

The replication fork and trinucleotide repeat instability

The traditional replication slippage model

It has long been recognized that insertion/deletion errors can occur during DNA replication by slippage of the nascent and template strands with respect to each other (Fresco and Alberts, 1960; Streisinger et al., 1966). By their repetitive nature, trinucleotide repeat sequences are perfect candidates for strand slippage, and slippage models for repeat expansion were proposed (Fig. 2A) early on by several investigators (Sinden and Wells, 1992; Richards and Sutherland, 1994; McMurray, 1995). The ability of trinucleotide repeats to form secondary structures could further aid slippage. Other features of repeat expansion are also consistent with replication slippage models. Evolutionary studies of repeat loci such as FRAXA and SCA1 show that variability of the repeat length occurs principally at the 3′ end of the repeat (Eichler et al., 1994; Jodice et al., 1994; Kunst and Warren, 1994; Richards and Sutherland, 1994; Snow et al., 1994). This polarity of expansion implicates a replication dependent mechanism that might be a result of differences between leading and lagging strand replication (Trinh and Sinden, 1991; Richards and Sutherland, 1994; Oostra and Willems, 1995).

Anyone who has tried to amplify a long trinucleotide repeat by PCR knows that replication enzymes have a tough time traversing the repeat. In vitro studies have shown that both CTG/CAG and CCG/CGG repeats cause length-dependent replication pausing by human, bacterial and phage polymerases (Kang et al., 1995b; Usdin and Woodford, 1995; Ohshima and Wells, 1997; Kamath-Loeb et al., 2001; Viguera et al., 2001). Analyses of stalled products with chemical probes indicate they exist as hairpin structures (Ohshima and Wells, 1997 Viguera et al., 2001). Long expansions can occur in vitro with a variety of DNA polymerases, presumably by replication slippage. The extent of repeat expansion by slippage depends on sequence length, composition, purity, orientation, flanking sequence, temperature, and Mg^{2+} concentration, all factors that influence hairpin stability (Schlotterer and Tautz, 1992; Ji et al., 1996; Petruska et al., 1998; Lyons-Darden and Topal, 1999b; Hartenstine et al., 2000). While slippage has been proposed to be a source of CTG/CAG and CCG/CGG instability in vivo, slippage in vitro is not specific to the repeats that actually expand in vivo. All trinucleotide repeat sequences cause slippage to some extent, and AAT repeats that are not expansion-prone in vivo show a much higher propensity for slippage than the expansion-prone CTG repeats (Schlotterer and Tautz, 1992; Lyons-Darden and Topal, 1999b). These data suggest that if slippage is a mechanism for CTG/CAG and CCG/CGG expansion in vivo, it must somehow be specifically suppressed at other repeats in the genome.

Orientation effects implicate replication in repeat instability

Strong evidence that direction of replication through a CTG/CAG tract is a source of instability has come from in vivo experiments in *E.coli* and yeast. On plasmids in *E. coli* and in three locations in the yeast genome, orientation of the repeat tract with respect to direction of replication is a strong determinant of tract stability (Kang et al., 1995a; Maurer et al., 1996; Freudenreich et al., 1997; Miret et al., 1998). In both yeast and bacteria, contractions are in general more frequent than expansions. However, in all cases, the orientation where the CTG sequence was on the Okazaki fragment was less prone to contractions, and more prone to expansions. Conversely, when the CTG sequence was on the lagging strand template, contractions were highly favored over expansions. This orientation dependence of expansions and contractions is exactly that predicted from the differential stability of CTG and CAG hairpins: when the more stable CTG hairpins are located on the Okazaki fragments then expansions are more likely, and when CTG hairpins form on the lagging strand template contractions are more likely (Fig. 2B). A model in which mistakes occur primarily on the lagging strand fits with data that shows the lagging strand is more prone to mutations (Trinh and Sinden, 1991; Veaute and Fuchs, 1993). However, since length changes do occur in both tract orientations, either CAG hairpins also occasionally form on the lagging strand (on either template or Okazaki fragment) leading to contractions and expansions (respectively), or some events reflect changes on the leading strand. Leading strand rolling circle replication of both CTG/CAG and CCG/CGG repeats from the f1 phage origin in *E. coli* does result in repeat instability, primarily deletions (Iyer and Wells, 1999). Since DNA dynamics during rolling circle replication may be quite different from passage of a replication fork through double stranded DNA and eukaryotic chromatin, it is unclear whether replication defects on the leading strand play a significant role in human repeat instability.

Most recently, the orientation dependence to instability was directly examined in mammalian cells and extracts. Replication of an SV40 plasmid containing a CTG/CAG tract in vivo by primate cells in tissue culture or in vitro using human cell extracts, also demonstrated that expansion is more likely when CTG is on the Okazaki fragment (Cleary et al., 2002; Panigrahi et al., 2002). In order to recover enough material to analyze, the CTG/CAG-containing plasmid had to be passaged through bacteria after the in vivo or in vitro replication step, resulting in the necessity to subtract out a high background frequency of instability that occurred in *E. coli*. Nonetheless, the increase in both the number and size of expansions is consistent with the increases occurring during replication in the primate cells or upon incubation with mammalian proteins, and demonstrates that the orientation bias for expansion is likely to generalize to human tissues.

CCG/CGG instability has also been studied in model organisms. As with the CTG/CAG repeats, orientation in the genome has an important effect on CCG/CGG stability. During replication of plasmid DNA containing CCG/CGG repeats in *E. coli*, deletions predominate in both orientations, but are more frequent when CGG is on the lagging strand template (Shimizu et al., 1996; Hirst and White, 1998). Again, this is consistent with the idea that the more structure-prone CGG strand is forming hairpins on the lagging strand template (Fig. 2B, left). In experiments by two separate groups, long CCG/CGG tracts were integrated into haploid yeast chromosomes in both orientations and stability was observed over multiple mitotic cell divisions. In both cases, the same orientation effect was observed as in *E. coli* except that the orientation with CGG on the Okazaki fragment (Fig. 2B, middle and right) showed a significant number of expansions in yeast that were not observed in bacteria (White et al., 1999; Balakumaran et al., 2000). Recently, a third group confirmed this observation when they utilized a sensitive genetic assay to determine expansion rates of CCG/CGG tracts in vivo. Rolfsmeier et al. (2001) found that CCG/CGG tract expansions occur at rates approximately 130 and 5,200 fold higher than other trinucleotide repeat combinations that do not form secondary structures in vitro, with the highest expansion rate being detected when CGG was on the Okazaki fragment. The CCG/CGG tract expansions in yeast, which were not detected at significant levels in bacteria, might reflect the shared aspects of yeast and human replication biology, such as chromatin structure, conserved replication fork proteins, lengths of Okazaki fragments, and regulation of replication. Two experiments suggested that mitotic replication, rather than recombination, was responsible for the CCG/CGG instability observed in the yeast system. First, instability was not increased in a diploid cell or in meiosis when recombination is highly active in yeast (White et al., 1999). Second, although instability was orientation dependent, double strand break formation at the CCG/CGG tract, which would likely be healed by recombination in yeast, was not orientation dependent (Balakumaran et al., 2000).

The idea that the location of secondary structure with respect to lagging/leading and nascent/template strands is important in determining repeat stability is also supported by direct visualization of *E. coli* plasmid replication intermediates. Using two-dimensional gel electrophoresis, (Samadashwily et al., 1997) demonstrated a length and sequence-dependent slowing or pausing of the replication fork that occurs specifically within the trinucleotide repeat tract. Longer tracts caused more pausing than shorter tracts and the extent of fork pausing was dependent upon which sequence was on the lagging strand template, with the more stable structure-forming sequences CGG and CTG causing more replication pausing than the CCG and CAG counterparts. Also of note, the presence of AGG interruptions in a CGG tract somewhat alleviated the pausing, suggesting that secondary structures might be responsible for the replication fork progression defects and that the AGG interruptions present in the FRAXA locus might stabilize the repeat by facilitating fork progression through the tract. Similar experiments have been done examining fork progression through CCG/CGG repeats integrated on chromosomes in yeast. Consistent with Samadashwily et al. (1997), replication fork pausing can be detected by 2-D gel analysis with shorter tract lengths when CGG is on the lagging strand template (Lenzmeier and Zakian, unpublished data). Also of note, replication forks can arrest, or stop completely, in long CCG/CGG tracts when CGG is on the lagging strand template, suggesting that resolution of converged or collapsed replication forks might also play a role in the mechanisms of trinucleotide repeat instability (Lenzmeier and Zakian, unpublished data).

The sum of the studies correlating repeat stability with replication direction and their consistency with each other across species and chromosomal location strongly suggests that the passage of a replication fork through a trinucleotide repeat sequence of a length long enough for secondary structure formation is a likely player in repeat instability. The data from model systems suggests that the human loci that expand to cause disease are in an orientation such that the sequence most likely to form secondary structure is located on the Okazaki fragment. What remains is to directly test this model. Unfortunately, it is very difficult to determine replication direction of a human chromosomal locus. Even highly sensitive pulse-chase replication timing experiments relying on bromodeoxyuridine incorporation and PCR have been unable to determine the direction of replication through the FRAXA, FRA10B, and FRA16B loci (Hansen et al., 1997; Handt et al., 2000). An alternative strategy would be to replace an expanded trinucleotide repeat at a given locus in a human cell line with the same repeat in the opposite orientation and compare stability. T. Ashizawa, D.G. Monckton, C.T. Caskey and colleagues have successfully used single cell cloning and small pool PCR to follow CTG repeat stability in cell lines from myotonic dystrophy patients and transgenic mice (Ashizawa et al., 1996; Gomes-Pereira et al., 2001). This method could perhaps be used to address the importance of replication direction at human disease loci.

In addition to direction of replication, proximity to a replication origin has recently been shown to influence CTG/CAG stability in an SV40 DNA replication system (Cleary et al., 2002). A $(CTG)_{79}$ repeat 103 base pairs away from the origin shows a bias for expansions when CTG is on the Okazaki fragment. The same repeat tract moved just 130 base pairs further away, the typical size of an Okazaki fragment, shows a bias for deletions. The authors hypothesize that tract stability is determined by the exact location of the repeat with respect to the single-stranded template DNA of the Okazaki initiation zone, a region ~290 base pairs long. They also speculate that Okazaki fragments that terminate within a repeat are more prone to expansion. A similar analysis of substrates that move the repeat by half-Okazaki lengths so that either the 3′ or 5′ end of the fragment contains repeats would be informative, especially since the mutants defective in processing the 5′ ends of Okazaki fragments such as Fen1/Rad27 (see below) show a particularly large effect on triplet repeat expansions.

Replication in model systems versus human disease loci

How does the instability observed in model systems relate to instability in human trinucleotide diseases? One difference between yeast/bacteria and human pedigree data is the ratio of expansions to contractions. In yeast, the contraction rate for a short $(CTG)_{25}$ sequence is about three-fold higher than the expansion rate, and this difference increases as tract length increases (Rolfsmeier et al., 2001). Single sperm analysis of CAG repeat size variation at the Huntington's and SBMA loci show a bias for contractions for normal size alleles, but a dramatic bias towards expansions as allele size increases into the disease-causing range (36–51 repeats) (Zhang et al., 1994; Leeflang et al., 1995; Zhang et al., 1995). However, the bias for expansions vs. contractions can vary widely in human cells depending on the cell type (i.e. sperm vs. somatic tissues), the locus studied, and even the patient. For example, cultured lymphoblastoid cell lines from myotonic dystrophy patients show a bias for contractions (Ashizawa et al., 1996), and some DM sperm samples show a distribution suggesting both expansions and contractions, consistent with the observation that expanded DM alleles occasionally reduce in size during transmission (Shelbourne et al., 1992; O'Hoy et al., 1993; Monckton et al., 1995). Such differences in expansion/contraction ratio probably reflect subtle differences in the replication and repair of various cell types at various times of development. For example, CGG repeats at the fragile X locus undergo deletions during early development before methylation but are relatively stable in adult tissues (Malter et al., 1997), and CpG methylation stabilizes CCG/CGG repeats in *E. coli* (Nichol and Pearson, 2002). Fragile X adults also display somatic instability, with hypomethylated repeats showing more tract length heterogeneity than hypermethylated repeats, providing further evidence that methylation of the repeat contributes to stability (Wohrle et al., 1998). Since yeast DNA is naturally unmethylated, it could mimic the state of the quickly replicating and unmethylated early embryonic DNA.

One unresolved issue that is relevant to the mechanism of instability is the timing of expansions during human development. One possibility is that most expansions occur in the germline, which could involve either meiosis or the many mitotic divisions needed for gamete production. An argument for germline expansions is observation of paternal and maternal biases for transmission of expanded alleles (reviewed in La Spada, 1997). The shorter repeat expansions (typically under 100 repeats) seen in the polyglutamine repeat diseases and intermediate size FMR1 CGG repeats (40–59 repeats) tend to show greater bias for expansions when transmitted paternally (La Spada, 1997; Sullivan et al., 2002). This could be a reflection of the greater number of cell divisions that occur during spermatogenesis compared to oogenesis or a difference in the developmental program leading to mature spermatozoa, such as repair of breaks that occur during tight chromatin packaging (Kovtun and McMurray, 2001). Larger expansions to thousands of repeats that tend to be transmitted maternally could be due either to a different expansion mechanism, for example meiotic recombination in the oocyte, or selection against sperm with long repeats (Jansen et al., 1994; Malter et al., 1997). Alternatively, occurrence of expansions during early embryonic development could also explain size differences between parent and offspring, and a mouse model of Huntington's disease shows an expansion/contraction bias depending on the gender of the embryo (Jansen et al., 1994; Kovtun et al., 2000).

Somatic expansions also occur in an age and tissue-specific manner, and may have a significant influence on the progression and severity of disease (Anvret et al., 1993; Thornton et al., 1994; Wong et al., 1995; Kennedy and Shelbourne, 2000). Surprisingly, cells grown in tissue culture from both human myotonic dystrophy patients and a transgenic mouse model have shown that cells with longer CTG repeats have a selective growth advantage, perhaps explaining some of the bias towards expansion in the soma (Gomes-Pereira et al., 2001; Khajavi et al., 2001). The mouse tissue culture model also shows that instability is not correlated with cell division rates, implicating *trans* factors, such as repair, that can operate in non-dividing cells and have tissue-specific differences (Gomes-Pereira et al., 2001). Some variation in pathology of trinucleotide repeats sequences may also be due to slippage during transcription, producing transcripts with longer (or shorter) repeat sequences than the template DNA. Recently, it has been demonstrated in yeast that transcription of a $(CAG)_{98}$, $(CTG)_{98}$ or $(CTG)_{255}$ repeat but not a $(TTG)_{87}$ repeat leads to poly(A) RNA several kilobases longer than the expected transcript size (Fabre et al., 2002). In *E. coli*, transcription through a repeat tract can destabilize the repeat DNA, causing an increase in deletions (Bowater et al., 1997; Schumacher et al., 2001). At the FRAXA locus in humans, methylation of the FMR1 promoter leads to transcriptional silencing (Sutcliffe et al., 1992), thus the decrease in transcription might also contribute to stability of the tract observed in somatic cells (reviewed in Usdin and Grabczyk, 2000). In summary, the lessons learned from model systems about *cis* factors that influence repeat stability are likely to recapitulate at least some of the complexities of repeat instability in humans.

Replication proteins play a role in repeat instability

It is clear that *cis* elements such as direction of replication, proximity to an origin, repeat length, interruptions, and chromosomal location play an important role in repeat stability. However, there is also good evidence that *trans*-acting factors can play an important role in maintaining repeat stability or causing repeat instability. One example of a probable role for *trans* factors is the differences in repeat instability between organisms. In general, triplet repeats of lengths that are highly unstable in humans are relatively stable in mice, both intergenerationally and in somatic tissue. The degree of instability is also dependent on the mouse line, since some transgenic mouse models show instability that approaches that in humans, and others commonly show changes of only a few repeats, with the larger changes seen in humans, yeast and bacteria being rare or absent. In addition, the bias towards expansions vs. contractions in various organisms could reflect subtle differences in replication, repair or recombination proteins and pathways. Differences in *trans* factors between somatic and germline cells or between different tissues could also explain the noted stability differences.

One *trans*-acting factor that has a dramatic effect on triplet repeat stability is the flap exo/endonuclease protein Fen1 (Rad27 in yeast). Fen1/Rad27 plays a role in processing the 5′ ends of Okazaki fragments and in base excision repair (Harrington and Lieber, 1994; Murante et al., 1994; Reagan et al., 1995; Sommers et al., 1995; reviewed in Bambara et al., 1997; Lieber, 1997; Matsumoto, 2001). In yeast, deletion of the Fen1/Rad27 gene causes a significant destabilization of CTG/CAG and CCG/CGG tracts, with both contractions and expansions being increased (Freudenreich et al., 1998; Schweitzer and Livingston, 1998; Spiro et al., 1999; White et al., 1999). Interestingly, there is a clear bias for expansions in yeast *rad27Δ* strains even though wild-type yeast display a bias toward contractions, therefore mutation of a single protein can change the expansion/contraction bias. The increase in expansions ranged from 15–180-fold over wild type depending on repeat size, orientation and assay used (Freudenreich et al., 1998; Schweitzer and Livingston, 1998; Spiro et al., 1999; White et al., 1999). The effect of a mutation in Fen1/Rad27 is not exclusive to triplet repeats, as there is also an increase in duplication mutations between short direct repeats (Tishkoff et al., 1997), and an increase in instability of mononucleotide, dinucleotide and larger tandem repeats of 4–20 base pairs (Johnson et al., 1995; Kokoska et al., 1998). Similar effects have been observed when the 5′ exonuclease domain of *E. coli* pol I, which is homologous to Fen1/Rad27, is deleted (Morel et al., 1998).

Because of the role of Fen1/Rad27 in Okazaki fragment processing, it is expected that a defect in the nuclease will lead to unprocessed flaps (Fig. 2B, right). Thus, one logical explanation for the expansions detected in the mutant background is that this unprocessed flap persists, and eventual ligation or repair without flap removal leads to the incorporation of extra sequence (Gordenin et al., 1997; Freudenreich et al., 1998; Schweitzer and Livingston, 1998). Because single strands of repeat DNA, and especially CTG and CGG sequences, are predicted to form stable hairpins, inhibition of correct flap processing by secondary structure could be a mechanism of expansion in wild-type cells as well. This model is supported by data that show DNA flaps containing CNG repeats form secondary structures in vitro, and these structures inhibit cleavage by the Fen1 enzyme (Spiro et al., 1999; Henricksen et al., 2000). Recently, in vitro studies using substrates designed to mimic the last steps of Okazaki fragment processing showed that ligation of upstream and downstream primers containing either triplet (CTG) or tandem repeats (10 bp) generates expansion products (Henricksen et al., 2002). Importantly, there is a competition between flap removal by Fen1 and ligation by DNA ligase, with Fen1 activity decreasing as primer repeat length increases (Henricksen et al., 2002).

Mutation of many other proteins that have a role in replication have also been shown to affect repeat stability, although a deletion of Fen1/Rad27 has the largest effect on expansions to date. Two temperature-sensitive polymerization domain mutations (pol3-14, pol3-t) in the yeast DNA polymerase delta (Pol δ), thought to be the polymerase responsible for lagging strand replication, cause an increase in CTG/CAG contraction frequency (Schweitzer and Livingston, 1999; Callahan et al., submitted), and the pol3-t mutation also increases deletions of direct repeats and micro- and minisatellite DNA sequences (Tran et al., 1995; Kokoska et al., 1998). These results suggest that slowing of DNA polymerization on the lagging strand may give more opportunity for template strand secondary structure, leading to deletions. In contrast, mutation of Pol δ exonuclease activity had no effect on CAG instability, suggesting that faulty proofreading does not play a role in repeat instability (Schweitzer and Livingston, 1999). In addition to proofreading, the Pol δ exonuclease domain is thought to cooperate with Fen1/Rad27 in Okazaki fragment maturation (Jin et al., 2001).

This view comes from data demonstrating that most mutations in Pol δ exonuclease domains, but not in Pol δ polymerization domains, are synthetically lethal with mutations in Fen1/Rad27, indicating that the two activities can substitute for each other (Kokoska et al., 1998; Gary et al., 1999; Jin et al., 2001). In vitro, Fen1/Rad27 flap endonuclease cleavage is stimulated when a small 3′ flap on the adjacent Okazaki fragment is also present, suggesting that a double flap is the in vivo substrate (Kaiser et al., 1999; Kao et al., 2002). The Pol δ 3′-5′ exo is likely responsible for processing this 3′ flap during Okazaki fragment maturation. Jin et al. hypothesized that this 3′ flap processing can substitute for the absence of Fen1/Rad27, creating a ligatable nick (Jin et al., 2001). However, the absence of an effect of a Pol δ exo-mutation on trinucleotide repeat stability suggests that the Fen1/Rad27 protein can completely substitute for defective Pol δ exonuclease in terms of maintaining correct processing of triplet-containing Okazaki fragments. In contrast to Pol δ, mutations in DNA polymerase epsilon (Pol ε) have no effect on CTG/CAG tract or dinucleotide microsatellite stability (Strand et al., 1993; Schweitzer and Livingston, 1999). Recent data indicate that Pol ε and Pol δ replicate different strands in yeast (Karthikeyan et al., 2000). Since Pol δ is strongly implicated in lagging strand synthesis, it is likely that Pol ε mutants would primarily have defects in leading strand synthesis, suggesting the majority of trinucleotide repeat instability in yeast occurs on the lagging strand. In contrast to yeast, bacterial Pol III proof-reading mutants do destabilize both CGG and CAG repeats (Iyer et al., 2000). Since there is also evidence in bacteria that expansions and contractions can occur during leading strand (rolling circle) replication (Iyer and Wells, 1999), these data suggest that trinucleotide repeat instability in eukaryotes like yeast and humans might be inhibited by *cis* constraints or *trans* factors that specifically prevent secondary structures from forming during leading strand synthesis.

Besides DNA polymerases, many other accessory proteins are involved in lagging strand replication and maturation, and several studies in yeast have analyzed the effect of mutating these proteins on trinucleotide repeat stability. Mutations in DNA primase or DNA polymerase alpha (Pol α), which are responsible for synthesis of the RNA primer and the first ~25 DNA nucleotides of each Okazaki fragment, are similar to pol δ mutations in that they increase contractions (Schweitzer and Livingston, 1999; Ireland et al., 2000; Callahan et al., submitted). These results suggest that in general, events that inhibit or slow nascent strand synthesis favor repeat contractions. In contrast, a subset of proteins that function with Fen1/Rad27 to process and join Okazaki fragments increase expansions. For example, mutations in PCNA, the processivity clamp that is also involved in bringing Fen1/Rad27 and DNA ligase to the 5′ flap, increase expansions 5–11 fold, with the most dramatic increase in expansion to contraction ratio associated with the mutation that has the greatest loss of interaction with Fen1/Rad27 (Schweitzer and Livingston, 1999). Mutations in DNA ligase also increase expansions, although the magnitude of the effect varies with the exact mutation and the study (Ireland et al., 2000; Callahan et al., submitted). Interestingly, deletion of RNase HI, the yeast protein responsible for RNA primer removal, has no effect on CTG/CAG stability, perhaps because its role can be bypassed by Fen1/Rad27 (Ireland et al., 2000; Callahan et al., submitted). Similarly, mutation of the Dna2 nuclease/helicase, thought to work with Fen1/Rad27 in 5′ flap processing, results in only a very slight increase in instability, perhaps for the same reason (Ireland et al., 2000; Callahan et al., submitted). In *E. coli*, deletion of the single-stranded binding protein (SSB) leads to an increased frequency of large deletions (Rosche et al., 1996). In this case, the increased deletions may well be due to SSB preventing secondary structures from forming on the lagging strand template. In yeast, the SSB homolog, RPA, has been shown to play an important role in flap processing by mediating the sequential cleavage activities of Dna2 and Fen1 by first stimulating Dna2 flap cleavage to generate a short flap that is then processed further by Fen1 (Bae et al., 2001). This role in flap processing suggests that SSB may potentially play a role in preventing expansions by binding to the single-stranded tails of the 5′ flap and preventing hairpin structures from forming, but mutations in SSB have not been tested for their effect on triplet repeat stability in eukaryotes to date.

In summary, mutation of many replication proteins implicates defective Okazaki fragment maturation as the primary in vivo pathway to expansions, at least in yeast, whereas polymerase mutations are more likely to generate contractions. This suggests that in vivo, flap processing on the lagging strand may be more important than polymerase slippage in generating expansions. Because the expansions observed in some of these mutant backgrounds were large (mean of 17–34 repeats added in *rad27Δ* strains, mean of 23 for PCNA mutant pol30-90 (Schweitzer and Livingston, 1998; Schweitzer and Livingston, 1999; Callahan et al., submitted), this type of mechanism can easily explain both the initial expansion to disease, as well as subsequent intergenerational or somatic expansions. This does not mean, however, that individuals with trinucleotide repeat expansion diseases have a mutation in one of these genes. In fact one analysis of fifteen Huntington's disease parent/child pairs did not show any Fen1 mutations compared to controls (Otto et al., 2001). Rather, the instability phenotypes of cells mutant for the various enzymes gives insight into the type of mistakes that could happen, even in wild-type cells, to expand a sequence vulnerable to secondary structure formation. A homozygous knockout of FEN1 in mice appears to be lethal, while heterozygous mice show a mild tumor predisposition phenotype (Kucherlapati et al., 2002). Interestingly, tumors from mice heterozygous for both *Fen1*^{null} and a mutation in the adenomatous polyposis coli *(Apc)* gene, which is mutated in a majority of sporadic colorectal tumors, have a microsatellite instability phenotype that is not present in tumors from mice only heterozygous for the *Apc* mutation (Kucherlapati et al., 2002). These observations suggest that the instability phenotype in the mutant is due to loss of *Fen1*. It would be interesting to cross the *Fen1*^{null} allele into one of the mouse models with an expanded triplet repeat locus to specifically test the effect of *Fen1* haploinsufficiency on triplet repeat stability in vertebrates.

Instability by double strand break repair and recombination

Although replication and recombination models were both originally proposed to explain trinucleotide repeat instability, replication-based models were long favored to explain repeat expansion. This bias was because the original recombination models required an exchange of information between sister or homologous chromosomes that was not supported by observations in humans (Richards and Sutherland, 1992; Kremer et al., 1995; Leeflang et al., 1995; Kovtun et al., 2001). Recent studies in bacteria and yeast, however, have implicated double strand break repair and recombination via gene conversion as sources of trinucleotide repeat instability. As discussed above, there is a strong association between trinucleotide repeat instability and replication defects like slippage, faulty Okazaki fragment processing, and replication fork pausing. There is also a strong association between replication defects and double strand break formation, as replication fork arrest causes double strand breaks in *E coli.* (Michel et al., 1997) and slowing of replication with drugs that lower nucleotide pools induces fragile site formation at human loci containing CCG/CGG expansions (reviewed in Sutherland et al., 1998). These observations suggest that faulty replication of trinucleotide repeats might lead to double strand breaks. Healing of these breaks by recombination might contribute to repeat instability.

Do double strand breaks occur at trinucleotide repeats? Experiments on human cells treated with agents that slow replication have demonstrated that long CCG/CGG repeat tracts confer fragile site expression at the expanded locus (reviewed in Sutherland et al., 1998). However, expanded CTG/CAG repeats have not yet been associated with fragile site formation in human cells (Wenger et al., 1996). Since a fragile site observed in a metaphase chromosome spread could be due either to a double strand break or to lack of DNA staining due to unfinished replication, whether or not chromosomes physically break at these sites in humans is somewhat controversial. Experiments in yeast, however, have confirmed that both CTG/CAG and CCG/CGG repeats induce double strand breaks. Trinucleotide repeat-dependent physical breaks of otherwise intact yeast chromosomes can be detected by pulsed field gel electrophoresis and two different genetic assays have been used to demonstrate that trinucleotide repeats act as preferential sites of chromosome breakage (Freudenreich et al., 1998; Balakumaran et al., 2000; Callahan et al., submitted; Lenzmeier and Zakian, unpublished data). While both CTG/CAG and CCG/CGG repeats cause chromosome breakage in yeast, CCG/CGG repeats of comparable length have much higher rates of chromosome breakage than the CTG/CAG repeats, suggesting the inability to detect fragility at CTG/CAG repeats in humans might be due to issues of sensitivity (Freudenreich et al., 1998; Balakumaran et al., 2000).

Is trinucleotide repeat instability associated with double strand break repair and recombination? In yeast, CCG/CGG breakage occurs to the same extent in both the stable and unstable orientation, suggesting that repair of DNA breaks might not be the primary source of CCG/CGG instability in yeast (Balakumaran et al., 2000). In addition, CCG/CGG tracts are not unstable during meiosis, suggesting that meiotic recombination, presumably initiated by double strand breaks, is also not a major cause of CCG/CGG instability (White et al., 1999). These data suggest that CCG/CGG instability in yeast is more likely due to defects during replication rather than recombination. Originally, it was also thought that recombination was not involved in CTG/CAG instability in yeast. Deletion of RAD52, which abolishes homologous recombination and single strand annealing, had no effect on instability of CTG/CAG tracts of 25, 71, 78 or 130 repeats (Miret et al., 1997; Freudenreich et al., 1998; Miret et al., 1998; Schweitzer and Livingston, 1999). Also, mutation of RAD52 did not exacerbate or alleviate the increased instability of a 71 or 78 CTG/CAG repeat tract in a pol δ (pol3-14) mutant, further suggesting instability is replication-based (Schweitzer and Livingston, 1999). However, experiments using longer CTG/CAG tracts have implicated double strand break repair by recombination as a mechanism for trinucleotide repeat instability: CTG/CAG tracts of 130–250 repeats induce double strand breaks on yeast chromosomes and long CTG/CAG repeats expand at relatively high rates when a recombination event is initiated adjacent to the tract (Freudenreich et al., 1998). Perhaps the contribution of homologous recombination to instability of shorter CTG/CAG tracts is not as great because they break at a lower rate or pathways other than homologous recombination, such as gap repair or non-homologous end joining, preferentially repair breaks at short tracts. Further support for a role of recombination in the mechanism of expansion comes from the observation that deletion of the replication and repair enzyme Fen1/Rad27 dramatically increases the frequency of CTG/CAG expansions (Freudenreich et al., 1998; Schweitzer and Livingston, 1998; Spiro et al., 1999). *rad27Δ* cells are inviable when double strand break repair genes are also deleted (Symington, 1998; Debrauwere et al., 2001), indicating that the lesions created in *rad27Δ* cells are repaired by double-strand break repair, and suggesting that expansions may occur during the repair process. In addition, expansions to greater than twice the original tract size can be detected in *rad27Δ* cells, suggesting that some expansions cannot be explained simply by flap processing defects (Spiro et al., 1999). Several reviews have since speculated that CTG/CAG instability is due in part to double strand break repair of unprocessed CTG/CAG Okazaki fragments (Jakupciak and Wells, 2000; Usdin and Grabczyk, 2000; Kovtun et al., 2001; Richards, 2001). Recent experiments have confirmed this prediction. Using a genetic YAC-based assay where double strand breaks are detected by telomere addition rather than recombination, Callahan et al. found that mutations in replication fork proteins (like Fen1/Rad27) that destabilize CTG/CAG tracts also increase the amount of tract dependent chromosome breakage (Callahan et al., submitted). Since double strand breaks are preferentially repaired by homologous recombination in yeast (reviewed in Paques and Haber, 1999), CTG/CAG tract length changes might occur during recombinational repair of structures that have formed during lagging strand synthesis.

Support for CTG/CAG tract instability by recombination also comes from experiments examining meiotic instability. Whether on plasmids, YACs, or integrated on yeast chromosomes, CTG/CAG tracts show a higher propensity toward expansions during meiosis when compared to the same constructs replicated in mitosis (Arai et al., 1999; Cohen et al., 1999; Jankowski et al., 2000; Schweitzer et al., 2001; Jankowski and Nag, 2002). Recently, experiments addressing the mechanism of meiotic instability have implicated double strand break repair and recombination. Jankowski et al. (2000) integrated long CTG/CAG tracts into the HIS4 locus, a known meiotic recombination hotspot in yeast and a location where trinucleotide repeats likely form hairpins in vivo (Nag et al., 1989; Detloff et al., 1992; Moore et al., 1999). Length-dependent double strand breaks were detected within the CTG/CAG tract (Moore et al., 1999; Jankowski et al., 2000) and the high amount of instability associated with the tract was dependent on creation of the double strand breaks by the meiosis specific endonuclease SPO11 (Jankowski and Nag, 2002). It is important to note that instability of CTG/CAG tracts was much greater at the HIS4 locus during meiosis than at other loci examined (Cohen et al., 1999; Schweitzer et al., 2001). This difference is likely explained by the high amount of breakage and recombination that is inherent to the HIS4 locus during meiosis, as maximal instability of the human minisatellite CEB1 during meiosis was also dependent upon integration of the repeats near preferential sites of meiotic double strand breaks (Debrauwere et al., 1999). In summary, these data suggest that double strand breaks occur in CTG/CAG tracts in mitosis and meiosis, and the repair of these breaks by recombination is a likely source of instability.

Does trinucleotide repeat instability occur during double strand break repair? Paques et al. developed an elegant experimental system to directly address this question (Paques et al., 1998; Richard and Paques, 2000). In the experiment, cell viability requires recombinational repair of a double strand break on a yeast chromosome using a plasmid carrying CTG/CAG repeats as the template (Paques et al., 1998). Experiments with this system found CTG/CAG repeats are highly unstable when being copied during double strand break repair (Richard et al., 1999; Richard et al., 2000). In fact, Richard et al., (1999) estimated that a 39 repeat CTG/CAG tract is 800 times more likely to undergo contractions during gene conversion than the same repeat when it is copied via conventional replication. When 98 repeats were used in the repair template, both expansions and contractions were observed at levels that were much higher than would be expected for conventional replication (Richard et al., 2000). The threshold length for expansion of CTG/CAG repeats during gene conversion in yeast is between 40 and 97 repeats, within the premutation range for expansion of CTG/CAG repeats at the DM locus in humans (reviewed in Usdin and Grabczyk, 2000). Also consistent with what has been observed in human cells (Richards and Sutherland, 1992; Kremer et al., 1995; Leeflang et al., 1995; Kovtun et al., 2001), CTG/CAG tract length changes were only observed on the recipient DNA molecule and there was no detectable exchange of flanking information (Richard et al., 2000). These data confirm that in yeast, expansion of trinucleotide repeats occurs during double strand break repair (Fig. 3A) and that gene conversion is a valid recombination-based model for trinucleotide repeat expansions in humans.

Recombination and double strand break repair have also been implicated as a powerful tool to generate trinucleotide repeat expansions in bacteria. Initial work by Jakupciak and Wells, (1999) demonstrated that long runs of trinucleotide repeats, specifically CTG/CAG, can be transferred from a long tract on one plasmid to produce an expansion of a shorter tract on another plasmid. The expansion of the shorter tract was dependent upon the recombination gene *RecA*. Analyses of flanking DNA suggested both crossing over and gene conversion events occurred. Recombination was dependent on having the same trinucleotide repeat sequence on both plasmids, and consistent with what was observed in yeast meiosis, CCG/CGG repeats showed a much lower propensity for recombination-based instability than the CTG/CAG repeats (Jakupciak and Wells, 1999; White et al., 1999). Recent studies from the same laboratory have now demonstrated that long CTG/CAG tracts specifically stimulate both intra and intermolecular recombination in *E. coli* (Napierala et al., 2002; Pluciennik et al., 2002). Since the orientation of the tracts relative to the direction of replication fork movement influenced the amount of recombination, the authors propose that the recombinigenic nature of the CTG/CAG tracts is due to secondary structure dependent stalling of replication followed by a double strand break in the tract (Napierala et al., 2002; Pluciennik et al., 2002). Another group also proposed a replication stalling-double strand break repair model to explain CTG/CAG tract instability in bacteria. Schumacher et al. (2001) detected a significant increase in the number of expansions when a CTG/CAG tract was placed next to a *LacI* repressor complex that was previously known to stall replication (Vilette et al., 1992). Since it was already known that stalling of replication causes double strand breaks and genome instability (Bierne et al., 1997; Michel et al., 1997) and since replication forks pause in CTG/CAG tracts in bacteria (Samadashwily et al., 1997), the recent implication of recombination as a powerful mechanism for expansions in *E. coli* increases the likelihood that a source of expansions in humans is also double strand breaks followed by recombinational repair.

DNA repair and trinucleotide repeat instability

The data summarized above on the effect of replication through a trinucleotide repeat tract and fragility of triplet repeats are consistent with a model in which the repeats create a site of DNA damage that must then be dealt with and repaired by cellular machinery. Expansions could occur either from inability to repair the damage properly, for example failure to excise a flap folded into a hairpin, or they could occur during the repair process itself. The following DNA repair processes have been implicated in trinucleotide repeat instability.

Trinucleotide repeat expansion and the SbcCD and MRX complexes

If repair of hairpin and slipped structures formed during replication and recombination are a source of trinucleotide repeat expansion, then gene products involved in removal or processing of these structures should affect trinucleotide repeat

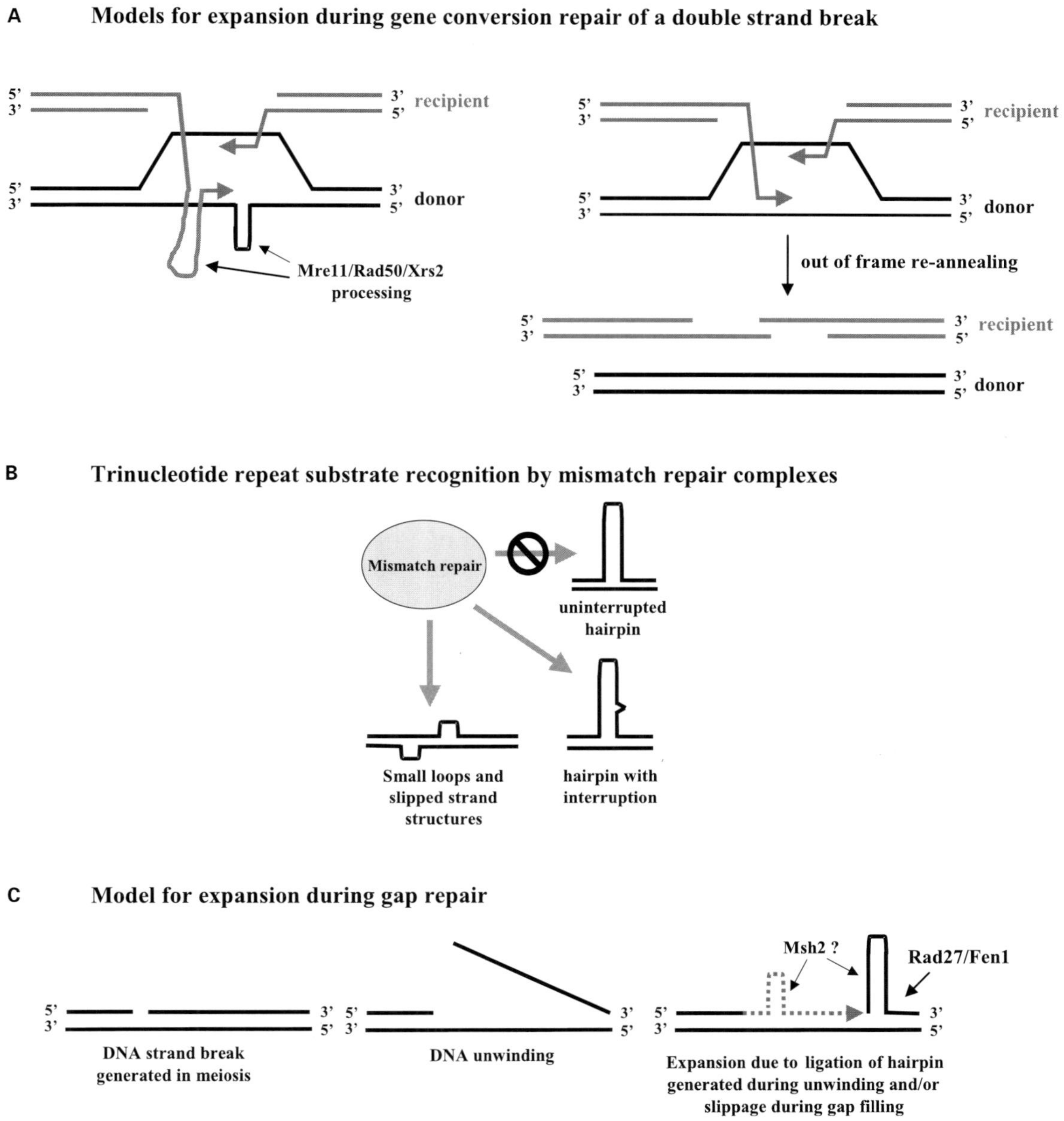

Fig. 3. DNA repair processes implicated in trinucleotide repeat instability. (**A**) Expansions can occur during gene conversion (recombinational) repair of a double strand break. Two models to explain expansion are presented: left- reiterative synthesis during strand invasion; right-strand invasion followed by out-of-frame re-annealing. For other models, including out-of-frame invasion, see the review by Richard and Paques (2000). The putative role of the Mre11/Rad50/Xrs2 complex in preventing expansions by removing potential hairpins is noted. (**B**) The DNA mismatch repair machinery recognizes and repairs some, but not all, trinucleotide repeat secondary structures. (**C**) Model for expansion during gap repair presented by Kovtun and McMurray (2001). The proposed role for Msh2 in promoting expansions by hairpin stabilization is noted. In addition, inhibition of 5′ flap processing during repair by the Fen1/Rad27 nuclease could lead to retention of the hairpin and subsequent expansion.

instability. Gene products previously implicated in the processing of palindrome and inverted repeat structures have also been associated with trinucleotide repeat instability (Sarkar et al., 1998; Richard et al., 2000). Like trinucleotide repeats, palindromes and inverted repeats that form hairpins and cruciform structures, are sites of DNA instability, and stimulate recombination in a manner that is also likely dependent upon replication (Chalker et al., 1988; Gordenin et al., 1993; Leach, 1994; Tran et al., 1995; Leach et al., 1997; Tran et al., 1997; Gordenin and Resnick, 1998; Lobachev et al., 1998; Cromie et al., 2000; Lobachev et al., 2000; Nasar et al., 2000; Pan and Leach, 2000; Bzymek and Lovett, 2001; Connelly and Leach, 2002). The gene products most implicated in palindrome and inverted repeat metabolism are the SbcCD and MRE11/RAD50/XRS2 (MRX) complexes from bacteria and yeast respectively (reviewed in Leach, 1994; Gordenin and Resnick, 1998; Connelly and Leach, 2002). These dynamic complexes are conserved throughout evolution and have been implicated in hairpin excision and processing in vitro and in vivo (Sharples and Leach, 1995; Connelly and Leach, 1996, 2002; Connelly et al., 1999; D'Amours and Jackson, 2002; Lobachev et al., 2002). Two groups have recently demonstrated that the SbcCD and MRX complexes are required to prevent expansions in *E. coli* and *S. cerevisiae*. Sarkar et al. (1998) characterized conditions that lead to extraordinarily long CTG/CAG expansions in the SURE strain of *E. coli*. Deletion of the *SbcC* gene was found to be responsible for the long expansions in the SURE strain and re-introduction of *SbcC* back into the strain alleviated the expansion phenotype, suggesting the expansions were caused by an inability of the cell to excise or process CTG/CAG hairpin structures (Sarkar et al., 1998). Experiments in yeast have implicated the MRX complex in removing structures during gene conversion of trinucleotide repeats. Richard et al. (2000) found that double strand break repair was inhibited when a long CTG/CAG tract was present in the repair template. Overexpression of the MRE11 gene in an otherwise wild type background restored double strand break repair efficiency to the levels of the control substrates, and reduced the number of expansions observed during gene conversion of a CTG/CAG tract (Richard et al., 2000). These data suggest that MRE11 facilitates gene conversion by unwinding or excising hairpins that form during recombinational repair of double strand breaks (Fig. 3A). Together, these observations from yeast and bacteria demonstrate that processing of secondary structures is likely an important step in preventing expansion of trinucleotide repeats in vivo. Since the MRX complex is also involved in checkpoint signaling after DNA damage (D'Amours and Jackson, 2002), expansion of trinucleotide repeats might also be prevented by induction of a DNA damage checkpoint. It will be of great interest to determine if checkpoint function is also required to prevent trinucleotide repeat expansion.

Mismatch repair and trinucleotide repeat instability

Mismatch repair (MMR) deficiency is associated with microsatellite instability and human diseases such as colorectal cancer (reviewed in Modrich and Lahue, 1996; Prolla, 1998; Kolodner and Marsischky, 1999). Since trinucleotide repeats form stable hairpin structures with mismatches in the stem, MMR would be predicted to prevent trinucleotide repeat expansion by recognizing and repairing hairpin structures (Gacy et al., 1995). Experiments carried out in human cells, however, have not yet supported a role for MMR deficiencies in the trinucleotide repeat expansions associated with disease. Human MMR deficient cell lines do not show increased frequencies of long expansions at the myotonic dystrophy or fragile X loci (Kramer et al., 1996). Huntington's disease patients only show instability at the HD locus, whereas patients with hereditary colorectal cancer show instability at multiple loci containing simple tandem repeats (Goellner et al., 1997). Initial genetic experiments in yeast and bacteria demonstrated that mutations in MMR components do not affect the frequency of long expansions, but instead cause frequent and incremental 1–2 repeat expansions and contractions (Jaworski et al., 1995; Miret et al., 1997, 1998; Schweitzer and Livingston, 1997; Freudenreich et al., 1998; Maurer et al., 1998; Schumacher et al., 1998; Wells et al., 1998; White et al., 1999; Parniewski et al., 2000; Schmidt et al., 2000; Hashem et al., 2002). These data suggest that small slipped structures are likely recognized by MMR components, but the longer hairpins putatively formed by trinucleotide repeats can evade detection and repair by the MMR machinery (Fig. 3B). Recent experiments, however, have demonstrated that MMR components can likely recognize and repair longer hairpin containing trinucleotide repeats, as long as the hairpin contains an interruption (Fig. 3b). Rolfsmeier et al., (2000) found that expansions in interrupted CTG/CAG tracts occurred at 3–30 fold higher rates in yeast cells deleted for various MMR components, but expansion rates of uninterrupted CTG/CAG tracts were not affected by disruption of MMR function. Although other groups had evidence that interruptions affected the amount and type of trinucleotide repeat instability observed in MMR deficient cells (Maurer et al., 1998; Wells et al., 1998), this was the first direct demonstration that MMR can function in preventing expansions. Since interruptions are known to stabilize trinucleotide repeat tracts and are often lost when a premutation length expansion occurs (reviewed in Kovtun et al., 2001), these data suggest that MMR might play a role in suppressing the initial events that would make an individual a carrier for a trinucleotide repeat expansion disease.

In contrast, experiments in transgenic mouse models suggest a different role for mismatch repair in creating or stabilizing small expansions that could initiate more dramatic instability by other mechanisms. Small CAG expansions in haploid germ cells and in somatic tissues in a Huntington's Disease exon 1 transgenic mouse model have been shown to require the MSH2 gene (Manley et al., 1999; Kovtun and McMurray, 2001). This could indicate that mistakes leading to incorporation of extra repeats may happen during MSH2-dependent repair of small loops expected to occur in trinucleotide repeat tracts, or that MSH2 plays a role in binding the hairpin and protecting it from excision repair (Manley et al., 1999; Kovtun and McMurray, 2001). Recently, van den Broek et al. (2002) showed that a deletion of Msh3, but not Msh6, blocks somatic repeat instability in a myotonic dystrophy knock-in mouse containing 84 CTG repeats. This suggests that trinucleotide repeat hairpins are probably recognized by the Msh2/Msh3 complex, which is known to act on larger loops, rather than the Msh2/Msh6 complex, which prefers base-base mismatches.

In addition to mismatch repair, which is not thought to recognize loops larger than ~ 15 nucleotides efficiently, large loop repair activities that are independent of mismatch repair have been biochemically characterized in both yeast and mammalian cell extracts (Umar et al., 1994; Corrette-Bennett et al., 1999; Littman et al., 1999; Corrette-Bennett et al., 2001). Corrette-Bennett et al. (2001) analyzed the in vivo efficiency of repair of large (38 nt) loops that were either unstructured, or able to base-pair to form a hairpin. The unstructured loop was repaired with approximately 90% efficiency, but repair efficiency was decreased by 50% for a $(CTG)_8$ hairpin loop, suggesting that triplet repeat hairpins can be repaired by this activity, but not with full efficiency. Interestingly, a hairpin with the same number of G-C base pairs as the $(CTG)_8$ loop but in a non-repeating sequence reduced repair by only 25%, suggesting that both secondary structure and the repeating sequence play a role in inhibiting the large loop repair activity. Several features of the large loop repair activity suggest that it is a good candidate for repair of trinucleotide repeat hairpins in vivo: it is not decreased by deletion of mismatch repair proteins or the Rad1 protein involved in nucleotide excision repair, neither of which significantly affect repeat stability in yeast, and it is stimulated by a 5′ nick, the expected situation if flap processing is impaired (Corrette-Bennett et al., 2001). However, the reduced activity on a CTG hairpin loop indicates that inability of this pathway to repair trinucleotide-repeat structures with full efficiency could also contribute to their propensity to expand.

DNA excision and gap repair and trinucleotide repeat expansions

In addition to mismatch repair, two types of DNA excision repair operate to remove lesions. Base excision repair removes damaged bases or abasic sites, whereas nucleotide excision repair, which has a wider substrate range, is involved in removing bulky DNA adducts, especially pyrimidine dimers and other helix-distorting lesions including DNA loops, as well as oxidized or methylated bases (reviewed in Sancar, 1994; Lindahl and Wood, 1999). Although the DNA glycosylase enzymes that normally initiate base excision repair would not be expected to recognize a lesion as large as a triplet repeat hairpin, this system could potentially be involved in repairing nicks or small gaps left after hairpin processing by another enzyme. Indeed, an abasic site analog placed within a triplet repeat template induced expansions upon in vitro primer extension (Lyons-Darden and Topal, 1999a). Similar to other in vitro slippage experiments (see above) the dependence on repeat sequence (AAT ~ ATT > CAG > CTG) did not reflect in vivo expansion propensities, however the experiment did demonstrate that DNA damage induces triplet repeat expansions in vitro. Interestingly, Fen1/Rad27 is required for long patch base excision repair in eukaryotes, raising the possibility that this enzyme may not only play a role in creating triplet hairpin lesions, but might be important for repairing them as well (Fig. 3C).

Because nucleotide excision repair (NER) removes bulkier adducts and creates a large 27 to 29 nucleotide gap (reviewed in Sancar, 1994), it seems like a better candidate for repair of TNR lesions. Indeed, deletion of UvrA, a component of the ABC excision nuclease necessary for NER in *E. coli*, further increased deletions when $(CTG)_{175}$ was the on the lagging strand template but did not effect the more stable orientation, shorter $(CTG/CAG)_{50}$ repeats or GAA/TTC and CCG/CGG repeats (Parniewski et al., 1999). The increase in deletions was dependent on transcription through the repeat, consistent with the known recruitment of NER to DNA damage generated during transcription (Parniewski et al., 1999). In contrast, a separate study recently showed a stabilization of CTG/CAG repeats in a UvrA- strain and the ability of the UvrA protein to bind to loops of 1, 2 or 17 CAG repeats in vitro (Oussatcheva et al., 2001). Initial experiments in yeast, however, showed that a deletion of the Rad1 nuclease, which is required for both the single-strand-annealing pathway of recombination and NER, does not significantly increase CTG/CAG expansions or contractions in a tract that is not transcribed (Freudenreich et al., 1998). Since the data from bacteria implicates NER in repeat instability, and since Parniewski et al. (1999) demonstrated that transcription might play a role in repeat instability via NER, it would be of great interest to determine if excision repair or even transcription-coupled repair contributes to trinucleotide repeat instability in eukaryotes.

Recent experiments in eukaryotes have implicated the filling of single stranded gaps, like those generated during nucleotide excision repair, as a mechanism for generating trinucleotide repeat expansions. Kovtun and McMurray (2001) studied the timing of expansion of the Huntington's disease CTG/CAG repeat in sperm cells of transgenic mice, and showed that small expansions of two to eight repeats were first detected in the post-meiotic haploid spermatozoa. These data indicate that neither mitotic replication nor recombination between sister chromatids or homologs were directly involved in the expansion mechanism and led the authors to propose that expansion occurred during gap repair (Kovtun and McMurray, 2001). How could gap repair lead to an expansion? Presumably a DNA strand break occurs in the CTG/CAG repeats sometime during meiosis and a hairpin forms at one of the broken DNA ends, generating a single stranded gap (Fig. 3C). Expansion could occur either by replication slippage during gap fill-in and/or by ligation of a hairpin at the broken DNA end to the newly filled-in strand (Fig. 3C). The Msh2 protein was required for this germline expansion, suggesting that either the mismatch repair machinery stabilizes hairpins during gap repair (Fig. 3C) or is somehow involved in promoting the initial gap generation step (Kovtun and McMurray, 2001). It will be interesting to determine what causes the initial lesion or strand break during meiosis and if gap repair plays a role in trinucleotide repeat instability in humans.

Conclusions

Expansion of trinucleotide repeats was originally termed "dynamic mutation" by Richards and Sutherland (1992) because the nature of the mutation is a change in copy number and the likelihood of mutation is dependent upon copy number. Little did we know at the time how truly dynamic the mechanisms of trinucleotide repeat instability would turn out to be. Experiments in model organisms like bacteria and yeast have begun to uncover the cellular processes involved in trinucleotide repeat expansions. Many of these discoveries have been useful in generating models that are consistent with what is observed in human cells. Integral to all models for trinucleotide repeat expansion is the propensity of the repeats to form stable secondary structures in vitro. Formation and stability of these structures is influenced by the length, purity, and chromosomal context of the tract. Experiments in model systems have shown that passage of a replication fork provides opportunity for secondary structure formation, and it is equally possible that a chance repair event in the neighborhood of a repeat tract could provide that opportunity. Studies of proteins whose mutation affects trinucleotide repeats implicates 5′ flap formation, which occurs during both replication and repair, as a particularly dangerous situation with a high potential to allow secondary structure formation. Slippage of the nascent or template strand followed by continued synthesis is another possible mechanism. Resolution of the putative secondary structures in vivo involves almost all aspects of DNA metabolism, and probably provides additional opportunities for changes in repeat length. Simple flap religation, excision of secondary structures, gene conversion repair of double strand breaks, mismatch repair and gap repair are all cellular processes that have been implicated in trinucleotide repeat expansion and each process influences the size of the expansion differently. For example, the small (initial) expansions that increase tract length from the normal to the premutation range are likely suppressed by things like replication fork components, mismatch repair enzymes, and the excision repair machinery. Mismatch repair enzymes also likely prevent expansion of short (and long?) interrupted repeats. Finally, the long expansions observed in humans once the premutation tract length is obtained might be generated by processing of unresolved 5′ flaps generated during replication or repair, reiterative DNA synthesis during replication or gap repair (perhaps due to fork stalling or template hairpins) or recombinational repair of a double strand break that occurred within the tract. So what is THE mechanism of trinucleotide repeat expansion? The answer, like the repeats, is quite dynamic.

Acknowledgements

We thank V. Zakian, members of the Zakian laboratory and members of the Freudenreich laboratory for helpful comments regarding the manuscript.

References

Anvret M, Ahlberg G, Grandell U, Hedberg B, Johnson K, Edstrom L: Larger expansions of the CTG repeat in muscle compared to lymphocytes from patients with myotonic dystrophy. Hum molec Genet 2:1397–1400 (1993).

Arai N, Akiyama R, Niimi N, Nakatsubo H, Inoue T: Meiotic contraction of CAG repeats in *Saccharomyces cerevisiae.* Genes Genet Syst 74:159–167 (1999).

Ashizawa T, Monckton DG, Vaishnav S, Patel BJ, Voskova A, Caskey CT: Instability of the expanded $(CTG)_n$ repeats in the myotonin protein kinase gene in cultured lymphoblastoid cell lines from patients with myotonic dystrophy. Genomics 36:47–53 (1996).

Ashley CTJ, Warren ST: Trinucleotide repeat expansion and human disease. A Rev Genet 29:703–728 (1995).

Bacolla A, Gellibolian R, Shimizu M, Amirhaeri S, Kang S, Ohshima K, Larson JE, Harvey SC, Stollar BD, Wells RD: Flexible DNA: genetically unstable CTGCAG and CGGCCG from human hereditary neuromuscular disease genes. J biol Chem 272: 16783–16792 (1997).

Bae SH, Bae KH, Kim JA, Seo YS: RPA governs endonuclease switching during processing of Okazaki fragments in eukaryotes. Nature 412:456–461 (2001).

Balakumaran BS, Freudenreich CH, Zakian VA: CGG/CCG repeats exhibit orientation-dependent instability and orientation-independent fragility in *Saccharomyces cerevisiae.* Hum molec Genet 9:93–100 (2000).

Bambara RA, Murante RS, Henricksen LA: Enzymes and reactions at the eukaryotic DNA replication fork. J biol Chem 272:4647–4650 (1997).

Bierne H, Ehrlich SD, Michel B: Deletions at stalled replication forks occur by two different pathways. EMBO J 16:3332–3340 (1997).

Bowater RP, Jaworski A, Larson JE, Parniewski P, Wells RD: Transcription increases the deletion frequency of long CTG·CAG triplet repeats from plasmids in *Escherichia coli.* Nucl Acids Res 25:2861–2868 (1997).

Bowater RP, Wells RD: The intrinsically unstable life of DNA triplet repeats associated with human hereditary disorders. Prog Nucl Acid Res Mol Biol 66:159–202 (2001).

Brosh RM Jr, Bohr VA: Roles of the Werner syndrome protein in pathways required for maintenance of genome stability. Exp Gerontol 37:491–506 (2002).

Bzymek M, Lovett ST: Instability of repetitive DNA sequences: the role of replication in multiple mechanisms. Proc natl Acad Sci, USA 98:8319–8325 (2001).

Chalker AF, Leach DR, Lloyd RG: *Escherichia coli sbcC* mutants permit stable propagation of DNA replicons containing a long palindrome. Gene 71:201–205 (1988).

Chastain PD, Sinden RR: CTG repeats associated with human genetic disease are inherently flexible. J molec Biol 275:405–411 (1998).

Chen FM: Acid-facilitated supramolecular assembly of G-quadruplexes in $d(CGG)_4$. J biol Chem 270: 23090–23096 (1995).

Chen X, Mariappan SSV, Catasti P, Ratliff R, Moyzis RK, Laayoun A, Smith SS, Bradbury EM, Gupta G: Hairpins are formed by the single DNA strands of the fragile X triplet repeats: Structure and biological implications. Proc natl Acad Sci, USA 92:5199–5203 (1995).

Cleary JD, Pearson CE: The contribution of *cis*-elements to disease-associated repeat instability: clinical and experimental evidence. Cytogenet Genome Res 100:25–55 (2003).

Cleary JD, Nichol K, Wang YH, Pearson CE: Evidence of *cis*-acting factors in replication-mediated trinucleotide repeat instability in primate cells. Nature Genet 31:37–46 (2002).

Cohen H, Sears DD, Zenvirth D, Hieter P, Simchen G: Increased instability of human CTG repeat tracts on yeast artificial chromosomes during gametogenesis. Mol cell Biol 19:4153–4158 (1999).

Connelly JC, De Leau ES, Leach DR: DNA cleavage and degradation by the SbcCD protein complex from *Escherichia coli.* Nucl Acids Res 27:1039–1046 (1999).

Connelly JC, Leach DR: The *sbcC* and *sbcD* genes of *Escherichia coli* encode a nuclease involved in palindrome inviability and genetic recombination. Genes Cells 1:285–291 (1996).

Connelly JC, Leach DR: Tethering on the brink: the evolutionarily conserved Mre11-Rad50 complex. Trends Biochem Sci 27:410–418 (2002).

Corrette-Bennett SE, Mohlman NL, Rosado Z, Miret JJ, Hess PM, Parker BO, Lahue RS: Efficient repair of large DNA loops in *Saccharomyces cerevisiae.* Nucl Acids Res 29:4134–4143 (2001).

Corrette-Bennett SE, Parker BO, Mohlman NL, Lahue RS: Correction of large mispaired DNA loops by extracts of *Saccharomyces cerevisiae.* J biol Chem 274:17605–17611 (1999).

Cromie GA Millar CB Schmidt KH and Leach DR: Palindromes as substrates for multiple pathways of recombination in *Escherichia coli.* Genetics 154: 513–522 (2000).

Cummings CJ, Zoghbi HY: Fourteen and counting: unraveling trinucleotide repeat diseases. Hum molec Genet 9:909–916 (2000).

D'amours D, Jackson SP: The Mre11 complex: at the crossroads of DNA repair and checkpoint signaling. Nature Rev Mol cell Biol 3:317–27 (2002).

Darlow J, Leach D: Secondary structures in d(CGG) and s(CCG) repeat tracts. J biol Chem 22:3–16 (1998).

Darlow JM, Leach DR: Evidence for two preferred hairpin folding patterns in d(CGG)d(CCG) repeat tracts in vivo. J molec Biol 275:17–23 (1998).

Darlow JM, Leach DR: The effects of trinucleotide repeats found in human inherited disorders on palindrome inviability in *Escherichia coli* suggest hairpin folding preferences in vivo. Genetics 141:825–832 (1995).

Davison A, Leach DR: The effects of nucleotide sequence changes on DNA secondary structure formation in *Escherichia coli* are consistent with cruciform extrusion in vivo. Genetics 137:361–368 (1994).

Debrauwere H, Buard J, Tessier J, Aubert D, Vergnaud G, Nicolas A: Meiotic instability of human minisatellite CEB1 in yeast requires DNA double-strand breaks. Nature Genet 23:367–371 (1999).

Debrauwere H, Loeillet S, Lin W, Lopes J, Nicolas A: Links between replication and recombination in *Saccharomyces cerevisiae*: a hypersensitive requirement for homologous recombination in the absence of Rad27 activity. Proc natl Acad Sci, USA 98:8263–8269 (2001).

Detloff P, White MA, Petes TD: Analysis of a gene conversion gradient at the HIS4 locus in *Saccharomyces cerevisiae.* Genetics 132:113–123 (1992).

Eichler EE, Holden JJ, Popovich BW, Reiss AL, Snow K, Thibodeau SN, Richards CS, Ward PA, Nelson DL: Length of uninterrupted CGG repeats determines instability in the FMR1 gene. Nature Genetics 8:88–94 (1994).

Fabre E, Dujon B, Richard GF: Transcription and nuclear transport of CAG/CTG trinucleotide repeats in yeast. Nucl Acids Res 30:3540–3547 (2002).

Fojtik P, Vorlickova M: The fragile X chromosome (GCC) repeat folds into a DNA tetraplex at neutral pH. Nucl Acids Res 29:4684–4690 (2001).

Fresco JR, Alberts BM: The accomodation of noncomplementary bases in helical polyribonucleotides and deoxyribonucleic acid. Proc Natl Acad Sci USA 46:311–321 (1960).
Freudenreich CH, Stavenhagen JB, Zakian VA: Stability of a CTG/CAG trinucleotide repeat in yeast is dependent on its orientation in the genome. Mol cell Biol 17:2090–2098 (1997).
Freudenreich CH, Kantrow SM, Zakian V: Expansion and length-dependent fragility of CTG repeats in yeast. Science 279:853–856 (1998).
Fry M, Loeb LA: The fragile X syndrome $d(CGG)_n$ nucleotide repeats form a stable tetrahelical structure. Proc natl Acad Sci, USA 91:4950–4954 (1994).
Fry M, Loeb LA: Human Werner syndrome DNA helicase unwinds tetrahelical structures of the fragile X syndrome repeat sequence $d(CGG)_n$. J biol Chem 274:12797–12802 (1999).
Gacy AM, Goellner G, Juranic N, Macura S, McMurray CT: Trinucleotide repeats that expand in human disease form hairpin structures in vitro. Cell 81:533–540 (1995).
Gacy AM, McMurray CT: Influence of hairpins on template reannealing at trinucleotide repeat duplexes: a model for slipped DNA. Biochemistry 37(:9426–9434 (1998).
Gary R, Park MS, Nolan JP, Cornelius HL, Kozyreva OG, Tran HT, Lobachev KS, Resnick MA, Gordenin DA: A novel role in DNA metabolism for the binding of Fen1/Rad27 to PCNA and implications for genetic risk. Mol cell Biol 19:5373–5382 (1999).
Gellibolian R, Bacolla A, Wells RD: Triplet repeat instability and DNA topology: an expansion model based on statistical mechanics. J biol Chem 272:16793–16797 (1997).
Godde JS, Wolffe AP: Nucleosome assembly on CTG triplet repeats. J biol Chem 271:15222–15229 (1996).
Godde JS, Kass SU, Hirst MC, Wolffe AP: Nucleosome assembly on methylated CGG triplet repeats in the fragile X mental retardation gene 1 promoter. J biol Chem 271:24325–24328 (1996).
Goellner GM, Tester D, Thibodeau S, Almqvist E, Goldberg YP, Hayden MR, McMurray CT: Different mechanisms underlie DNA instability in Huntington disease and colorectal cancer. Am J hum Genet 60:879–890 (1997).
Gomes-Pereira M, Fortune MT, Monckton DG: Mouse tissue culture models of unstable triplet repeats: in vitro selection for larger alleles mutational expansion bias and tissue specificity but no association with cell division rates. Hum molec Genet 10:845–854 (2001).
Gordenin DA, Kunkel TA, Resnick MA: Repeat expansion – all in a flap? Nature Genetics 16:116–118 (1997).
Gordenin DA, Lobachev KS, Degtyareva NP, Malkova AL, Perkins E, Resnick MA: Inverted DNA repeats: a source of eukaryotic genomic instability. Mol cell Biol 13:5315–5322 (1993).
Gordenin DA, Resnick MA: Yeast ARMs (DNA at-risk motifs) can reveal sources of genome instability. Mutat Res 400:45–58 (1998).
Handt O, Baker E, Dayan S, Gartler SM, Woollatt E, Richards RI, Hansen RS: Analysis of replication timing at the FRA10B and FRA16B fragile site loci. Chrom Res 8:677–688 (2000).
Hansen RS, Canfield TK, Fjeld AD, Mumm S, Laird CD, Gartler SM: A variable domain of delayed replication in FRAXA fragile X chromosomes: X inactivation-like spread of late replication. Proc natl Acad Sci, USA 94:4587–4592 (1997).
Harrington JJ, Lieber MR: The characterization of a mammalian DNA structure-specific endonucleases. EMBO J 13:1235–1246 (1994).
Hartenstine MJ, Goodman MF, Petruska J: Base stacking and even/odd behavior of hairpin loops in DNA triplet repeat slippage and expansion with DNA polymerase. J biol Chem 275:18382–18390 (2000).
Hashem VI, Rosche WA, Sinden RR: Genetic assays for measuring rates of (CAG)(CTG) repeat instability in *Escherichia coli.* Mutat Res 502:25–37 (2002).
Henricksen LA, Tom S, Liu Y, Bambara RA: Inhibition of flap endonuclease 1 by flap secondary structure and relevance to repeat sequence expansion. J biol Chem 275:16420–16427 (2000).
Henricksen LA, Veeraraghavan J, Chafin DR, Bambara RA: DNA ligase I competes with FEN1 to expand repetitive DNA sequences in vitro. J biol Chem 277:22361–22369 (2002).
Hirst MC, White PJ: Cloned human FMR1 trinucleotide repeats exhibit a length- and orientation-dependent instability suggestive of in vivo lagging strand secondary structure. Nucl Acids Res 26: 2353–2358 (1998).
Ireland MJ, Reinke SS, Livingston DM: The impact of lagging strand replication mutations on the stability of CAG repeat tracts in yeast. Genetics 155: 1657–1665 (2000).
Iyer RR, Wells RD: Expansion and deletion of triplet repeat sequences in *Escherichia coli* occur on the leading strand of DNA replication. J biol Chem 274:3865–3877 (1999).
Iyer RR, Pluciennik A, Rosche WA, Sinden RR, Wells RD: DNA polymerase III mutants enhance the expansion and deletion of triplet repeat sequences in *Escherichia coli.* J biol Chem 275:2174–2184 (2000).
Jakupciak JP, Wells RD: Genetic instabilities in (CTG·CAG) repeats occur by recombination. J biol Chem 274:23468–23479 (1999).
Jakupciak JP, Wells RD: Genetic instabilities of triplet repeat sequences by recombination. Intl Union Biochem molec Biol Life 50:355–359 (2000).
Jankowski C, Nag DK: Most meiotic CAG repeat tract-length alterations in yeast are SPO11 dependent. Mol Genet Genom 267:64–70 (2002).
Jankowski C, Nasar F, Nag DK: Meiotic instability of CAG repeat tracts occurs by double-strand break repair in yeast. Proc Natl Acad Sci 97:2134–2139 (2000).
Jansen G, Willems P, Coerwinkel M, Nillesen W, Smeets H, Vits L, Howeler C, Brunner H, Wieringa B: Gonosomal mosaicism in myotonic dystrophy patients: involvement of mitotic events in $(CTG)_n$ repeat variation and selection against extreme expansion in sperm. Am J hum Genet 54:575–585 (1994).
Jaworski A, Rosche WA, Gellibolian R, Kang S, Shimizu M, Bowater RP, Sinden RR, Wells RD: Mismatch repair in *Escherichia coli* enhances instability of $(CTG)_n$ triplet repeats from human hereditary diseases. Proc Natl Acad Sci 92:11019–11023 (1995).
Ji J, Clegg NJ, Peterson KR, Jackson AL, Laird CD, Loeb LA: in vitro expansion of GGC:GCC repeats: identification of the preferred strand of expansion. Nucl Acids Res 24:2835–2840 (1996).
Jin YH, Obert R, Burgers PM, Kunkel TA, Resnick MA, Gordenin DA: The $3' \rightarrow 5'$ exonuclease of DNA polymerase delta can substitute for the 5′ flap endonuclease Rad27/Fen1 in processing Okazaki fragments and preventing genome instability. Proc natl Acad Sci, USA 98:5122–5127 (2001).
Jodice C, Malaspina P, Persichetti F, Novelletto A, Spadaro M, Giunti P, Morocutti C, Terrenato L, Harding AE, Frontali M: Effect of trinucleotide repeat length and parental sex on phenotypic variation in spinocerebellar ataxia I. Am J hum Genet 54:959–965 (1994).
Johnson RE, Kowali GK, Prakash L, Prakash S: Requirement of the yeast RTH1 5′ to 3′ exonuclease for the stability of simple repetitive DNA. Science 269:238–240 (1995).
Kaiser MW, Lyamicheva N, Ma W, Miller C, Neri B, Fors L, Lyamichev VI: A comparison of eubacterial and archaeal structure-specific 5′- exonucleases. J biol Chem 274:21387–21394 (1999).
Kamath-Loeb AS, Loeb LA, Johansson E, Burgers PM, Fry M: Interactions between the Werner syndrome helicase and DNA polymerase delta specifically facilitate copying of tetraplex and hairpin structures of the $d(CGG)_n$ trinucleotide repeat sequence. J biol Chem 276:16439–16446 (2001).
Kang S, Jaworski A, Ohshima K, Wells RD: Expansion and deletion of CTG repeats from human disease genes are determined by the direction of replication in *E. coli.* Nature Genetics 10:213–218 (1995a).
Kang S, Ohshima K, Shimizu M, Amirhaeri S, Wells RD: Pausing of DNA synthesis in vitro at specific loci in CTG and CGG triplet repeats from human hereditary disease genes. J biol Chem 270:27014–27021 (1995b).
Kang S, Ohshima K, Jaworski A, Wells RD: CTG triplet repeats from the myotonic dystrophy gene are expanded in *Escherichia coli* distal to the replication origin as a single large event. J molec Biol 258 543–547 (1996).
Kao H-I, Henricksen LA, Liu Y, Bambara RA: Cleavage specificity of *Saccharomyces cerevisiae* flap endonuclease 1 suggests a double-flap structure as the cellular substrate. J biol Chem 277:14379–14389 (2002).
Karthikeyan R, Vonarx EJ, Straffon AF, Simon M, Faye G, Kunz BA: Evidence from mutational specificity studies that yeast DNA polymerases delta and epsilon replicate different DNA strands at an intracellular replication fork. J molec Biol 299: 405–419 (2000).
Kennedy L, Shelbourne PF: Dramatic mutation instability in HD mouse striatum: does polyglutamine load contribute to cell-specific vulnerability in Huntington's disease? Hum molec Genet 9:2539–2544 (2000).
Kettani A, Kumar RA, Patel DJ: Solution structure of a DNA quadruplex containing the fragile X syndrome triplet repeat. J molec Biol 254:638–656 (1995).
Khajavi M, Tari AM, Patel NB, Tsuji K, Siwak DR, Meistrich ML, Terry NH, Ashizawa T: "Mitotic drive" of expanded CTG repeats in myotonic dystrophy type 1 (DM1). Hum molec Genet 10:855–863 (2001).
Kokoska RJ, Stefanovic L, Tran HT, Resnick MA, Gordenin DA, Petes TD: Destabilization of yeast micro- and minisatellite DNA sequences by mutations affecting a nuclease involved in Okazaki fragment processing (rad27) and DNA polymerase d (pol3-t). Mol cell Biol 18:2779–2788 (1998).
Kolodner RD, Marsischky GT: Eukaryotic DNA mismatch repair. Curr Opin Genet Dev 9:89–96 (1999).
Kovtun I, McMurray C: Trinucleotide expansion in haploid germ cells by gap repair. Nature Genet 27:407–411 (2001).
Kovtun IV, Therneau TM, McMurray CT: Gender of the embryo contributes to CAG instability in transgenic mice containing a Huntington's disease gene. Hum molec Genet 9:2767–2775 (2000).
Kovtun IV, Goellner G, McMurray CT: Structural features of trinucleotide repeats associated with DNA expansion. Biochem Cell Biol 79:325–336 (2001).
Kramer PR, Pearson CE, Sinden RR: Stability of triplet repeats of myotonic dystrophy and fragile X loci in human mutator mismatch repair cell lines. Hum Genet 98:151–157 (1996).

Kremer B, Almqvist E, Theilmann J, Spence N, Telenius H, Goldberg YP, Hayden MR: Sex-dependent mechanisms for expansions and contractions of the CAG repeat on affected Huntington disease chromosomes. Am J hum Genet 57:343–350 (1995).

Kucherlapati M, Yang K, Kuraguchi M, Zhao J, Lia M, Heyer J, Kane MF, Fan K, Russell R, Brown AM, Kneitz B, Edelmann W, Kolodner RD, Lipkin M, Kucherlapati R: Haploinsufficiency of flap endonuclease (Fen1) leads to rapid tumor progression. Proc natl Acad Sci, USA 99:9924–9929 (2002).

Kunst CB, Warren ST: Cryptic and polar variation of the fragile X repeat could result in predisposing normal alleles. Cell 77:853–861 (1994).

Kuryavyi VV, Jovin TM: Triad-DNA: a model for trinucleotide repeats. Nature Genet 9:339–341 (1995).

La Spada AR: Trinucleotide repeat instability: Genetic features and molecular mechanisms. Brain Path 7:943–963 (1997).

Leach DR: Long DNA palindromes cruciform structures genetic instability and secondary structure repair. Bioessays 16:893–900 (1994).

Leach DR, Okely EA, Pinder DJ: Repair by recombination of DNA containing a palindromic sequence. Mol Microbiol 26:597–606 (1997).

Leeflang EP, Zhang L, Tavare S, Hubert R, Srinidhi J, MacDonald ME, Myers RH, De Young M, Wexler NS, Gusella JF, Arnheim N: Single sperm analysis of the trinucleotide repeats in the Huntington's disease gene: quantification of the mutation frequency spectrum. Hum molec Genet 4:1519–1526 (1995).

Leproust EM, Pearson CE, Sinden RR, Gao X: Unexpected formation of parallel duplex in GAA and TTC trinucleotide repeats of Friedreich's ataxia. J molec Biol 302:1063–1080 (2000).

Lieber MR: The FEN-1 family of structure-specific nucleases in eukaryotic DNA replication recombination and repair. Bioessays 19:233–240 (1997).

Lindahl T, Wood RD: Quality control by DNA repair. Science 286:1897–1905 (1999).

Littman SJ, Fang WH, Modrich P: Repair of large insertion/deletion heterologies in human nuclear extracts is directed by a 5′ single-strand break and is independent of the mismatch repair system. J biol Chem 274:7474–81 (1999).

Lobachev KS, Shor BM, Tran HT, Taylor W, Keen JD, Resnick MA, Gordenin DA: Factors affecting inverted repeat stimulation of recombination and deletion in *Saccharomyces cerevisiae.* Genetics 148:507–524 (1998).

Lobachev KS, Stenger JE, Kozyreva OG, Jurka J, Gordenin DA, Resnick MA: Inverted Alu repeats unstable in yeast are excluded from the human genome. EMBO J 19:3822–3830 (2000).

Lobachev KS, Gordenin DA, Resnick MA: The Mre11 complex is required for repair of hairpin-capped double-strand breaks and prevention of chromosome rearrangements. Cell 108:183–193 (2002).

Lyons-Darden T, Topal MD: Abasic sites induce triplet-repeat expansion during DNA replication in vitro. J biol Chem 274:25975–25978 (1999a).

Lyons-Darden T, Topal MD: Effects of temperature Mg^{2+} concentration and mismatches on triplet-repeat expansion during DNA replication in vitro. Nucl Acids Res 27:2235–2240 (1999b).

Malter HE, Iber JC, Willemsen R, De Graaff E, Tarleton JC, Leisti J, Warren ST, Oostra BA: Characterization of the full fragile X syndrome mutation in fetal gametes. Nature Genet 15:165–169 (1997).

Manley K, Shirley TL, Flaherty L, Messer A: Msh2 deficiency prevents in vivo somatic instability of the CAG repeat in Huntington disease transgenic mice. Nature Genet 23:471–473 (1999).

Mariappan SV, Catasti P, Chen X, Ratliff R, Moyzis RK, Bradbury EM, Gupta G: Solution structures of the individual single strands of the fragile X DNA triplets $(GCC)_n(GGC)_n$. Nucl Acids Res 24:784–792 (1996a).

Mariappan SV, Garcoa AE, Gupta G: Structure and dynamics of the DNA hairpins formed by tandemly repeated CTG triplets associated with myotonic dystrophy. Nucl Acids Res 24:775–783 (1996b).

Mariappan SV, Silks LA 3rd, Bradbury EM, Gupta G: Fragile X DNA triplet repeats $(GCC)_n$ form hairpins with single hydrogen-bonded cytosinecytosine mispairs at the CpG sites: isotope-edited nuclear magnetic resonance spectroscopy on $(GCC)_n$ with selective 15N4-labeled cytosine bases. J molec Biol 283:111–120 (1998a).

Mariappan SV, Silks LA 3rd, Chen X, Springer PA, Wu R, Moyzis RK, Bradbury EM, Garcia AE, Gupta G: Solution structures of the Huntington's disease DNA triplets $(CAG)_n$. J Biomol Struct Dyn 15: 723–744 (1998b).

Mariappan SV, Catasti P, Silks LA 3rd, Bradbury EM, Gupta G: The high-resolution structure of the triplex formed by the GAA/TTC triplet repeat associated with Friedreich's ataxia. J molec Biol 285: 2035–2052 (1999).

Matsumoto Y: Molecular mechanism of PCNA-dependent base excision repair. Prog Nucl Acid Res Mol Biol 68:129–138 (2001).

Maurer DJ, O'Callaghan BL, Livingston DM: Orientation dependence of trinucleotide CAG repeat instability in *Saccharomyces cerevisiae.* Mol cell Biol 16:6617–6622 (1996).

Maurer DJ, O'Callaghan BL, Livingston DM: Mapping the polarity of changes that occur in interrupted CAG repeat tracks in yeast. Mol cell Biol 18:4597–4604 (1998).

McMurray CT: Mechanisms of DNA expansion. Chromosoma 104:2–13 (1995).

McMurray CT: DNA secondary structure: A common and causative factor for expansion in human disease. Proc Natl Acad Sci 96:1823–1825 (1999).

Michel B, Erlich SD, Uzest M: DNA double-strand breaks caused by replication arrest. EMBO J 16:430–438 (1997).

Miret JJ, Pessoa-Brandao and Lahue RS: Instability of CAG and CTG trinucleotide repeats in *Saccharomyces cerevisiae.* Mol cell Biol 17:3382–3387 (1997).

Miret JJ, Pessoa-Brandao and Lahue RS: Orientation-dependent and sequence-specific expansions of CTG/CAG trinucleotide repeats in *Saccharomyces cerevisiae.* Proc Natl Acad Sci 95:12438–12443 (1998).

Mitas M: Trinucleotide repeats associated with human disease. Nucl Acids Res 25:2245–2253 (1997).

Mitas M, Yu A, Dill J, Kamp TJ, Chambers EJ, Haworth IS: Hairpin properties of single-stranded DNA containing a GC-rich triplet repeat: $(CTG)_{15}$. Nucl Acids Res 23:1050–1059 (1995).

Modrich P, Lahue R: Mismatch repair in replication fidelity genetic recombination and cancer biology. A Rev Biochem 65:101–133 (1996).

Monckton DG, Wong LJ, Ashizawa T, Caskey CT: Somatic mosaicism germline expansions germline reversions and intergenerational reductions in myotonic dystrophy males: small pool PCR analyses. Hum molec Genet 4:1–8 (1995).

Moore H, Greenwell PW, Liu C-P, Arnheim N, Petes TD: Triplet repeats form secondary structures that escape DNA repair in yeast. Proc Natl Acad Sci USA 96:1504–1509 (1999).

Morel P, Reverdy C, Michel B, Ehrlich SD, Cassuto E: The Role of SOS and flap processing in microsatellite instability in *Escherichia coli.* Proc Natl Acad Sci USA 95:10003–10008 (1998).

Murante RS, Huang L, Turchi JJ, Bambara RA: The calf 5′- to 3′-exonuclease is also an endonuclease with both activities dependent on primers annealed upstream of the point of cleavage. J biol Chem 269:1191–1196 (1994).

Nag DK, Petes TD: Seven-base-pair inverted repeats in DNA form stable hairpins in vivo in *Saccharomyces cerevisiae.* Genetics 129:669–673 (1991).

Nag DK, White MA, Petes TD: Palindromic sequences in heteroduplex DNA inhibit mismatch repair in yeast. Nature 340:318–320 (1989).

Napierala M, Parniewski P, Pluciennik A, Wells RD: Long CTGCAG repeat sequences markedly stimulate intramolecular recombination. J biol Chem 277:34087–34100 (2002).

Nasar F, Jankowski C, Nag DK: Long palindromic sequences induce double-strand breaks during meiosis in yeast. Mol cell Biol 20:3449–3458 (2000).

Nichol K, Pearson CE: CpG methylation modifies the genetic stability of cloned repeat sequences. Genome Res 12:1246–256 (2002).

O'Donnell WT, Warren ST: A decade of molecular studies of fragile X syndrome. A Rev Neurosci 25:315–338 (2002).

O'Hoy KL, Tsilfidis C, Mahadevan MS, Neville CE, Barcelo J, Hunter AG, Korneluk RG: Reduction in size of the myotonic dystrophy trinucleotide repeat mutation during transmission. Science 259:809–812 (1993).

Ohshima K, Kang S, Larson JE, Wells RD: Cloning characterization and properties of seven triplet repeat DNA sequences. J biol Chem 271:16773–16783 (1996).

Ohshima K, Wells RD: Hairpin formation during DNA synthesis primer realignment in vitro in triplet repeat sequences from human hereditary disease genes. J biol Chem 272:16798–16806 (1997).

Oostra BA, Willems PJ: A fragile gene. BioEssays 17:941–947 (1995).

Otto CJ, Almqvist E, Hayden MR, Andrew SE: The "flap" endonuclease gene FEN1 is excluded as a candidate gene implicated in the CAG repeat expansion underlying Huntington disease. Clin Genet 59:122–127 (2001).

Oussatcheva EA, Hashem VI, Zou Y, Sinden RR, Potaman VN: Involvement of the nucleotide excision repair protein UvrA in instability of CAG*CTG repeat sequences in *Escherichia coli.* J biol Chem 276:30878–30884 (2001).

Pan X, Leach DR: The roles of mutS sbcCD and recA in the propagation of TGG repeats in *Escherichia coli.* Nucl Acids Res 28:3178–31784 (2000).

Panigrahi GB, Cleary JD, Pearson CE: in vitro (CTG)*(CAG) expansions and deletions by human cell extracts. J biol Chem 277:13926–13934 (2002).

Paques F, Haber JE: Multiple pathways of recombination induced by double-strand breaks in *Saccharomyces cerevisiae.* Microbiol Mol Biol Rev 63:349–404 (1999).

Paques F, Leung WY, Haber JE: Expansions and contractions in a tandem repeat induced by double-strand break repair. Mol cell Biol 18:2045–2054 (1998).

Parniewski P, Bacolla A, Jaworski A, Wells RD: Nucleotide excision repair affects the stability of long transcribed (CTG·CAG) tracts in an orientation-dependent manner in *Escherichia coli.* Nucl Acids Res 27:616–623 (1999).

Parniewski P, Jaworski A, Wells RD, Bowater RP: Length of CTGCAG repeats determines the influence of mismatch repair on genetic instability. J molec Biol 299:865–874 (2000).

Pearson CE, Sinden RR: Alternative structures in duplex DNA formed within the trinucleotide repeats of the myotonic dystrophy and fragile X loci. Biochemistry 35:5041–5053 (1996).

Pearson CE, Sinden RR: Trinucleotide repeat DNA structures: dynamic mutations from dynamic DNA. Curr Opin Struct Biol 8:321–330 (1998).

Pearson CE, Ewel A, Acharya S, Fishel RA, Sinden RR: Human MSH2 binds to trinucleotide repeat DNA structures associated with neurodegenerative diseases. Hum molec Genet 6:1117–1123 (1997).

Pearson CE, Eichler EE, Lorenzetti D, Kramer SF, Zoghbi HY, Nelson DL, Sinden RR: Interruptions in the triplet repeats of SCA1 and FRAXA reduce the propensity and complexity of slipped strand DNA (S-DNA) formation. Biochemistry 37:2701–2708 (1998a).

Pearson CE, Wang YH, Griffith JD, Sinden RR: Structural analysis of slipped-strand DNA (S-DNA) formed in $(CTG)_n$ $(CAG)_n$ repeats from the myotonic dystrophy locus. Nucl Acids Res 26:816–823 (1998b).

Petruska J, Arnheim N, Goodman MF: Stability of intrastrand hairpin structures formed by the CAG/CTG class of DNA triplet repeats associated with neurological diseases. Nucl Acids Res 24:1992-1998 (1996).

Petruska J, Hartenstine MJ, Goodman MF: Analysis of strand slippage in DNA polymerase expansions of CAG/CTG triplet repeats associated with neurodegenerative disease. J biol Chem 273:5204–5210 (1998).

Pluciennik A, Iyer RR, Napierala M, Larson JE, Filutowicz M, Wells RD: Long CTGCAG repeats from myotonic dystrophy are preferred sites for intermolecular recombination. J biol Chem 277:34074–34086 (2002).

Prolla TA: DNA mismatch repair and cancer. Curr Opin Cell Biol 10:311–316 (1998).

Reagan MS, Pittenger C, Siede W, Friedberg EC: Characterization of a mutant strain of *Saccharomyces cerevisiae* with a deletion of the RAD27 gene, a structural homolog of the RAD2 nucleotide excision repair gene. J Bacteriol 177:364–371 (1995).

Reddy PS, Housman DE: The complex pathology of trinucleotide repeats. Curr Opin Cell Biol 9:364–372 (1997).

Richard GF, Paques F: Mini- and microsatellite expansions: the recombination connection. EMBO Rep 1:122–126 (2000).

Richard G-F, Dujon B, Haber JE: Double-strand break repair can lead to high frequencies of deletions within short CAG/CTG trinucleotide repeats. Mol gen Genet 261:871–882 (1999).

Richard GF, Goellner GM, McMurray CT, Haber JE: Recombination-induced CAG trinucleotide repeat expansions in yeast involve the MRE11-RAD50-XRS2 complex. EMBO J 19:2381–2390 (2000).

Richards RI: Dynamic mutations: a decade of unstable expanded repeats in human genetic disease. Hum molec Genet 10:2187–2194 (2001).

Richards RI, Sutherland GR: Dynamic mutations: a new class of mutations causing human disease. Cell 70:709–712 (1992).

Richards RI, Sutherland GR: Simple repeat DNA is not replicated simply. Nature Genet 6:114–116 (1994).

Rolfsmeier ML, Lahue RS: Stabilizing effects of interruptions on trinucleotide repeat expansions in *Saccharomyces cerevisiae*. Mol cell Biol 20:173–180 (2000).

Rolfsmeier ML, Dixon MJ, Lahue RS: Mismatch repair blocks expansions of interrupted trinucleotide repeats in yeast. Mol Cell 6:1501–1507 (2000).

Rolfsmeier ML, Dixon MJ, Pessoa-Brandao L, Pelletier R, Miret JJ, Lahue RS: *Cis*-elements governing trinucleotide repeat instability in *Saccharomyces cerevisiae*. Genetics 157:1569–1579 (2001).

Rosche WA, Jaworski A, Kang S, Kramer SF, Larson JE, Geidroc DP, Wells RD, Sinden RR: Single-stranded DNA-binding protein enhances the stability of CTG triplet repeats in *Escherichia coli*. J Bacteriol 178:5042–5044 (1996).

Sakamoto N, Chastain PD, Parniewski P, Ohshima K, Pandolfo M, Griffith JD, Wells RD: Sticky DNA: self-association properties of long GAATTC repeats in RRY triplex structures from Friedreich's ataxia. Mol Cell 3:465–475 (1999).

Samadashwily GM, Raca G, Mirkin SM: Trinucleotide repeats affect DNA replication in vivo. Nature Genet 17:298–304 (1997).

Sancar A: Mechanisms of DNA excision repair. Science 266:1954–1956 (1994).

Sarkar PS, Chang H-C, Boudi B, Reddy S: CTG repeats show bimodal amplification in *E. coli*. Cell 95: 531–540 (1998).

Schlotterer C, Tautz D: Slippage synthesis of simple sequence DNA. Nucl Acids Res 20:211–215 (1992).

Schmidt KH, Abbott CM, Leach DRF: Two opposing effects of mismatch repair on CTG repeat instability in *Echerichia coli*. Mol Microbiol 35:463–471 (2000).

Schumacher S, Fuchs RP, Bichara M: Expansion of CTG repeats from human disease genes is dependent upon replication mechanisms in *Escherichia coli:* the effect of long patch mismatch repair revisited. J molec Biol 279:1101–1110 (1998).

Schumacher S, Pinet I, Bichara M: Modulation of transcription reveals a new mechanism of triplet repeat instability in *Escherichia coli*. J molec Biol 307:39–49 (2001).

Schweitzer JK, Livingston DM: Destabilization of CAG trinucleotide repeat tracts by mismatch repair mutations in yeast. Hum molec Genet 6:349–355 (1997).

Schweitzer JK, Livingston DM: Expansions of CAG repeat tracts are frequent in a yeast mutant defective in Okazaki fragment maturation. Hum molec Genet 7:69–74 (1998).

Schweitzer JK, Livingston DM: The effect of DNA replication mutations on CAG Tract stability in yeast genetics. Society Am 152:953–963 (1999).

Schweitzer JK, Reinke SS, Livingston DM: Meiotic alterations in CAG repeat tracts. Genetics 159:1861–1865 (2001).

Sharples GJ, Leach DR: Structural and functional similarities between the SbcCD proteins of *Escherichia coli* and the RAD50 and MRE11 (RAD32) recombination and repair proteins of yeast. Mol Microbiol 17:1215–1217 (1995).

Shelbourne P, Winqvist R, Kunert E, Davies J, Leisti J, Thiele H, Bachmann H, Buxton J, Williamson B, Johnson K: Unstable DNA may be responsible for the incomplete penetrance of the myotonic dystrophy phenotype. Hum molec Genet 1:467–473 (1992).

Shimizu M, Gellibolian R, Oostra BA, Wells RD: Cloning, characterization and properties of plasmids containing CGG triplet repeats from the FMR-1 gene. J molec Biol 258:614–626 (1996).

Sinden RR: Human genetics '99: Trinucleotide repeats, biological implications of the DNA structures associated with disease-causing triplet repeats. Am J hum Genet 64:346–353 (1999).

Sinden RR, Wells RD: DNA structure, mutations and human genetic disease. Curr Opin Biotechnol 3:612–622 (1992).

Smith FW, Schultze P, Feigon J: Solution structures of unimolecular quadruplexes formed by oligonucleotides containing Oxytricha telomere repeats. Structure 3:997–1008 (1995).

Snow K, Tester DJ, Kruckeberg KE, Schaid DJ, Thibodeau SN: Sequence analysis of the fragile X trinucleotide repeat: implications for the origin of the fragile X mutation. Hum molec Genet 3:1543–1551 (1994).

Sommers CH, Miller EJ, Dujon B, Prakash S, Prakash L: Conditional lethality of null mutations in RTH1 that encodes the yeast counterpart of a mammalian 5′- to 3′-exonuclease required for lagging strand synthesis in reconstituted systems. J biol Chem 270:4193–4196 (1995).

Spiro C, Pelletier R, Rolfsmeir ML, Dixon MJ, Lahue RS, Gupta G, Park MS, Chen X, Mariapp SV, McMurray CT: Inhibition of FEN-1 processing by DNA secondary structure at trinucleotide repeats. Mol Cell 4:1079–1085 (1999).

Strand M, Prolla TA, Liskay RM, Petes TD: Destabilization of tracts of simple repetitive DNA in yeast by mutations affecting DNA mismatch repair. Nature 365:274–276 (1993).

Streisinger G, Okada Y, Emrich J, Newton J, Tsugita A, Terzaghi E, Inouye M: Frameshift mutations and the genetic code. This paper is dedicated to Professor Theodosius Dobzhansky on the occasion of his 66th birthday. Cold Spring Harb Symp quant Biol 31:77–84 (1966).

Sullivan AK, Crawford DC, Scott EH, Leslie ML, Sherman SL: Paternally transmitted FMR1 alleles are less stable than maternally transmitted alleles in the common and intermediate size range. Am J hum Genet 70:1532–1544 (2002).

Sutcliffe JS, Nelson DL, Zhang F, Pieretti M, Caskey CT, Saxe D, Warren ST: DNA methylation represses FMR-1 transcription in fragile X syndrome. Hum molec Genet 1:397–400 (1992).

Sutherland GR, Baker E, Richards RI: Fragile sites still breaking. Trends Genet 14:501–506 (1998).

Symington LS: Homologous recombination is required for the viability of rad27 mutants. Nucl Acids Res 26:5589–5595 (1998).

Thornton CA, Johnson K, Moxley RT 3rd: Myotonic dystrophy patients have larger CTG expansions in skeletal muscle than in leukocytes. Ann Neurol 35:104–107 (1994).

Timchenko LT, Caskey CT: Triplet repeat disorders: discussion of molecular mechanisms. Cell Mol Life Sci 55:1432–1447 (1999).

Tishkoff DX, Filosi N, Gaida GA, Kolodner RD: A novel mutation avoidance mechanism dependent on *S. cerevisiae* RAD27 is distinct from DNA mismatch repair. Cell 88:253–263 (1997).

Tran H, Degtyareva N, Gordenin D, Resnick MA: Altered replication and inverted repeats induce mismatch repair-independent recombination between highly diverged DNAs in yeast. Mol cell Biol 17:1027–1036 (1997).

Tran HT, Degtyareva NP, Koloteva NN, Sugino A, Masumoto H, Gordenin DA, Resnick MA: Replication slippage between distant short repeats in *Saccharomyces cervisiae* depends on the direction of replication and the RAD50 and RAD52 genes. Mol cell Biol 15:5607–5617 (1995).

Trinh TQ, Sinden RR: Preferential DNA secondary structure mutagenesis in the lagging strand of replication in *E. coli*. Nature 352:544–547 (1991).

Uliel L, Weisman-Shomer P, Oren-Jazan H, Newcomb T, Loeb LA, Fry M: Human Ku antigen tightly binds and stabilizes a tetrahelical form of the Fragile X syndrome $d(CGG)_n$ expanded sequence. J biol Chem 275:33134–33141 (2000).

Umar A, Boyer JC, Kunkel TA: DNA loop repair by human cell extracts. Science 266:814–816 (1994).

Usdin K: NGG-triplet repeats form similar intrastrand structures: implications for the triplet expansion diseases. Nucl Acids Res 26:4078–4085 (1998).

Usdin K, Grabczyk E: DNA repeat expansions and human disease. Cell Mol Life Sci 57:914–931 (2000).

Usdin K, Woodford KJ: CGG repeats associated with DNA instability and chromosome fragility form structures that block DNA synthesis in vitro. Nucl Acids Res 23:4202–4209 (1995).

Van Den Broek WJ, Nelen MR, Wansink DG, Coerwinkel MM, Te Riele H, Groenen PJ, Wieringa B: Somatic expansion behaviour of the $(CTG)_n$ repeat in myotonic dystrophy knock-in mice is differentially affected by Msh3 and Msh6 mismatch- repair proteins. Hum molec Genet 11:191–198 (2002).

Veaute X, Fuchs RPP: Greater susceptibility to mutations in lagging strand of DNA replication in *Escherichia coli* than in leading strand. Science 261:598–600 (1993).

Vetcher AA, Napierala M, Iyer RR, Chastain PD, Griffith JD, Wells RD: Sticky DNA a long GAA·GAA·TTC triplex that is formed intramolecularly in the sequence of intron 1 of the frataxin gene. J biol Chem 277:39217–39227 (2002).

Viguera E, Canceill D, Ehrlich SD: Replication slippage involves DNA polymerase pausing and dissociation. EMBO J 20:2587–2595 (2001).

Vilette D, Uzest M, Ehrlich SD, Michel B: DNA transcription and repressor binding affect deletion formation in *Escherichia coli* plasmids. EMBO J 11:3629–3634 (1992).

Wang Y-H, Griffith J: Expanded CTG triplet blocks from the myotonic dystrophy gene create the strongest known natural nucleosome positioning elements. Genomics 25:570–573 (1995).

Wang YH, Griffith J: Methylation of expanded CCG triplet repeat DNA from fragile X syndrome patients enhances nucleosome exclusion. J biol Chem 271:22937–22940 (1996).

Wang Y-H, Amirhaeri S, Kang S, Wells RD, Griffith JD: Preferential nucleosome assembly at DNA triplet repeats from the myotonic dystrophy gene. Science 265:669–671 (1994).

Wang YH, Gellibolian R, Shimizu M, Wells RD, Griffith J: Long CCG triplet repeat blocks exclude nucleosomes: a possible mechanism for the nature of fragile sites in chromosomes. J molec Biol 263:511–516 (1996).

Weisman-Shomer P, Cohen E, Fry M: Interruption of the fragile X syndrome expanded sequence d(CGG)(n) by interspersed d(AGG) trinucleotides diminishes the formation and stability of d(CGG)(n) tetrahelical structures. Nucl Acids Res 28:1535–1541 (2000a).

Weisman-Shomer P, Naot Y, Fry M: Tetrahelical forms of the fragile X syndrome expanded sequence d(CGG)(n) are destabilized by two heterogeneous nuclear ribonucleoprotein-related telomeric DNA-binding proteins. J biol Chem 275:2231–2238 (2000b).

Wells RD, Parniewski P, Pluciennik A, Bacolla A, Gellibolian R, Jaworski A: Small slipped register genetic instabilities in *Escherichia coli* in triplet repeat sequences associated with hereditary neurological diseases. J biol Chem 273:19532–19541 (1998).

Wenger SL, Giangreco CA, Tarleton J, Wessel HB: Inability to induce fragile sites at CTG repeats in congenital myotonic dystrophy. Am J med Gen 66:60–63 (1996).

White PJ, Borts RH, Hirst MC: Stability of the human fragile X $(CGG)_n$ triplet repeat array in *Saccharomyces cerevisiae* deficient in aspects of DNA metabolism. Mol cell Biol 19:5675–5684 (1999).

Wohrle D, Salat U, Glaser D, Mucke J, Meisel-Stosiek M, Schindler D, Vogel W, Steinbach P: Unusual mutations in high functioning fragile X males: apparent instability of expanded unmethylated CGG repeats. J med Genet 35:103–111 (1998).

Wong LJ, Ashizawa T, Monckton DG, Caskey CT, Richards CS: Somatic heterogeneity of the CTG repeat in myotonic dystrophy is age and size dependent. Am J hum Genet 56:114–122 (1995).

Yu A, Dill J, Mitas M: The purine-rich trinucleotide repeat sequences $d(CAG)_{15}$ and $d(GAC)_{15}$ form hairpins. Nucl Acids Res 23:4055–4057 (1995a).

Yu A, Dill J, Wirth SA, Huang G, Lee VH, Haworth IS, Mitas M: The trinucleotide repeat sequence $d(GTC)_{15}$ adopts a hairpin conformation. Nucl Acids Res 23:2706–2714 (1995b).

Yu A, Barron MD, Romero RM, Christy M, Gold B, Dai J, Gray DM, Haworth IS, Mitas M: At physiological pH $d(CCG)_{15}$ forms a hairpin containing protonated cytosines and a distorted helix. Biochemistry 36:3687–3699 (1997).

Yu S, Mangelsdorf M, Hewett D, Hobson L, Baker E, Eyre HJ, Lapsys N, Paslier DL, Doggett NA, Sutherland GR, Richards RI: Human chromosomal fragile site FRA16B is an amplified AT-rich minisatellite repeat. Cell 88:367–374 (1997).

Zhang L, Leeflang EP, Yu J, Arnheim N: Studying human mutations by sperm typing: instability of CAG trinucleotide repeats in the human androgen receptor gene. Nature Genet 7:531–535 (1994).

Zhang L, Fischbeck KH, Arnheim N: CAG repeat length variation in sperm from a patient with Kennedy's disease. Hum molec Genet 4:303–305 (1995).

Zheng M, Huang X, Smith GK, Yang X, Gao X: Genetically unstable CXG repeats are structurally dynamic and have a high propensity for folding. An NMR and UV spectroscopic study. J molec Biol 264:323–336 (1996).

Cytogenet Genome Res 100:25–55 (2003)
DOI: 10.1159/000072837

The contribution of *cis*-elements to disease-associated repeat instability: clinical and experimental evidence

J.D. Cleary and C.E. Pearson

Program of Genetics and Genomic Biology, The Hospital for Sick Children, and Department of Molecular and Medical Genetics, University of Toronto, Toronto, Ontario (Canada)

Abstract. Alterations in the length (instability) of gene-specific microsatellites and minisatellites are associated with at least 35 human diseases. This review will discuss the various *cis*-elements that contribute to repeat instability, primarily through examination of the most abundant disease-associated repetitive element, trinucleotide repeats. For the purpose of this review, we define *cis*-elements to include the sequence of the repeat units, the length and purity of the repeat tracts, the sequences flanking the repeat, as well as the surrounding epigenetic environment, including DNA methylation and chromatin structure. Gender-, tissue-, developmental- and locus-specific *cis*-elements in conjunction with *trans*-factors may facilitate instability through the processes of DNA replication, repair and/or recombination. Here we review the available human data that supports the involvement of *cis*-elements in repeat instability with limited reference to model systems. In diverse tissues at different developmental times and at specific loci, repetitive elements display variable levels of instability, suggesting vastly different mechanisms may be responsible for repeat instability amongst the disease loci and between various tissues.

Table of contents

Work in the Pearson laboratory is supported by the Canadian Institutes of Health Research (CIHR), Muscular Dystrophy Association USA, and the Fragile X Research Foundation of Canada. C.E.P. is a CIHR Scholar and a Canadian Genetic Disease Network Scholar. J.D.C is supported by a CIHR doctoral research award.

Received 24 December 2002; manuscript accepted 11 February 2003.

Request reprints from Christopher E. Pearson
Program of Genetics and Genomic Biology, The Hospital for Sick Children
555 University Avenue, Elm Wing 11-135
Toronto, Ontario, M5G 1X8 (Canada)
telephone: (416) 813-8256; fax: (416) 813-4931
e-mail: cepearson@genet.sickkids.on.ca

KARGER
Fax + 41 61 306 12 34
E-mail karger@karger.ch
www.karger.com

0301–0171/03/1004–0025$19.50/0

Accessible online at:
www.karger.com/cgr

Section I: The instability of repetitive elements in patients/humans

Introduction

Alterations in the length of repetitive elements occur naturally throughout the human genome. However, only in specific situations does repeat instability result in disease. In 1991, alterations in the length of trinucleotide repeats (TNRs) was identified as the disease causing mutation for X-linked spinal and bulbar muscular atrophy (SBMA), myotonic dystrophy type 1 (DM1) and fragile X syndrome (FRAXA). Numerous other diseases have subsequently been identified as resulting from repeat-specific length alterations (Tables 1 and 2). Analysis of these repetitive elements has identified DNA sequences and sequence combinations that are more prone to instability. In addition, identical repeats at different genomic loci have

Table 1. Unstable DNA repeats located in coding regions[a]

Disorder/gene disease	Repeat unit	Number of repeats			Interruption	Evidence for haplotype	Parent-of-origin effect
		Normal	Premutation	Full/disease			
Polyglutamine (CAG) tracts							
AIB-I/NCOA3							
prostate cancer risk	(CAG/CAA)n•(CTG/TTG)n	29/29 or 28/29	<29/29	<29/<29	–	ND	ND
DRPLA/DRPLA							
Haw Rver Syndrome	(CAG)n•(CTG)n	7–25	ND	49–75	–	Yes	Paternal
HD/HD							
Huntngton's disease	(CAG)n•(CTG)n	10–34	36–39	40–121	–	Yes	Paternal
HDL2/JPH3							
Huntngton's disease-like 2	(CAG)n•(CTG)n	5–18	ND	50–60	–	Yes	ND
SBMA/AR							
Kennedy's disease/ spinobulbar muscular atrophy	(CAG)n•(CTG)n	9–36	ND	38–62	–	Yes	ND
SCA1/ATX1							
spino-cerebellar ataxia 1	(CAG)n•(CTG)n	6–39	–	40–82	CAT	Yes	Paternal
SCA2/ATX2							
spino-cerebellar ataxia 2	(CAG)n•(CTG)n	13–33	–	32–200	CAA	Yes	Paternal
SCA3/ATX3							
spino-cerebellar ataxia 3	(CAG)n•(CTG)n	13–44	ND	55–84	CGG,GCC,CAG,CAA	Yes	Paternal
SCA6/CACNAIA							
spino-cerebellar ataxia 6	(CAG)n•(CTG)n	4–18	ND	20–29	–	Yes	ND
SCA7/SCA7							
spino-cerebellar ataxia 7	(CAG)n•(CTG)n	4–35	ND	37–306	–	Yes	Paternal
SCA17/TBP							
spino-cerebellar ataxia 17	(CAG)n•(CTG)n	25–42	ND	47–63	–	ND	ND
Polyalanine (GCG) tracts							
ARX/ARX							
infantile spasm syndrome	(GCG)n•(CGC)n	10–16	ND	17–23	GCA, GCT, GCC	ND	ND
CBFA1/CBFA1							
deidocranial dysplasia	(GCG)n•(CGC)n	11,17	–	27	GCA, GCT	ND	ND
FOXL2/FOXL2							
blepharophimosis/ptosis/epicanthus invensus syndrome type B	(GCG)n•(CGC)n	14	ND	10, 24	GCA, GCT, GCC	ND	ND
HOXA13/HOXA13							
hand-foot-genital syndrome	(GCG)n•(CGC)n	18	ND	26	GCA, GCT, GCC	ND	ND
HOXD13/HOXD13							
synpolydactyly	(GCG)n•(CGC)n	15	ND	22–29	GCA,GCT,GCC	ND	ND
OPDM/PABP2							
oculopharyngeal muscular dystrophy	(GCG)n•(CGC)n	6–7	ND	8–13	GCA	Yes	ND
TCFBR1/TCFBR1							
potential cancer risk association	(GCG)n•(CGC)n	6–12	ND	ND	ND	ND	ND
ZIC2/ZIC2							
holoprosencephaly	(GCG)n•(CGC)n	15	–	25	GCA, GCT, GCC	ND	ND
Other unstable repetitive elements							
CJD/PRNP							
Creutzfeldt-Jakob disease	24-bp repeat	4	ND	5–13	sequence variations	ND	ND
MED-PSACH/COMP							
multiple skeletal dysplasias	(GAC)n•(GTC)n	5	ND	4, 6, 7	–	ND	ND

[a] The relative genomic location of each repeat is listed in Fig. 2. For the listed repeat sequences the first repeat is that of the coding strand. The use of "ND" denotes insufficient data, sporadic reporting or unknown.

varying levels of instability, indicating a contribution of flanking *cis*-elements to repeat instability. For the purpose of this review, we define *cis*-elements to include the sequence of the repeat units, the length and purity of the repeat tracts, the sequences flanking the repeat as well as the surrounding epigenetic environment, including DNA methylation and chromatin structure. The first sections of this review (I and II) will focus on the analysis of patients and patient material that implicates *cis*-elements in repeat instability, while the last sections (III and IV) will focus on experimental evidence for potential *cis*-elements.

Table 2. Unstable DNA repeats located in non-coding regions[a]

Disorder/gene disease	Repeat unit	Number of repeats			Interruption	Evidence for haplotype	Parent-of-origin effect
		Normal	Premutation	Full/disease			
TNRs							
DM1/DMPK myotonic dystrophy	(CTG)n•(CAG)n	5–37	50–80	80–1000	–	Yes	Maternal
C14orf4/C14orf4 unknown	(CAG)n•(CTG)n	20–31	ND	ND	ND	ND	ND
CTG18.1 /SEF2-1 B possible bipolar disorder	(CTG)n•(CAG)n	10–37	53–250	800–2100	–	Yes	ND
ERDA1 expanded repeat domain	(CTG)n•(CAG)n	10–92	ND	ND	CAT/CAC	Yes	Yes
FRDA/FRDA Friedreich's ataxia	(GAA)n (TTC)n	6–32	33–60	200–1700	(GAG,GGA,GAA)	Yes	Maternal
SCA8/SCA8 spino-cerebellar ataxia 8	(CTG)n•(CAG)n	16–92	ND	100–127	CCG, TTG, CTA CCA,GTC,CTT	Yes	Maternal
SCA12/PPP2P2B spino-cerebellar ataxia 12	(CAG)n•(CTG)n	7–45	ND	55–78	–	ND	ND
RELN susceptible to autism	(GGC)n•(GCC)n	8–10	–	11–23	–	ND	ND
Rare fragile site-associated repeats							
Folate sensitive							
FRAXA/FMR1 fragile XA	(CGG)n•(CCG)n	6–52	59–230	230–2000	AGG	Yes	Maternal
FRAXE/FMR2 fragile XE	(CGG)n•(CCG)n	4–39	(31–61)	200–900	–	Yes	ND
FRAXF/FAM11A fragile XF	(CGG)n•(CCG)n	7–40	ND	306–1008	(GCCGTC)$_{3-4}$	Yes	ND
FRA11B/CBL2 Jacobsen syndrome	(CCG)n•(CGG)n	11	80	100–1000	–	ND	Paternal
FRA16A no disease	(CCG)n•(CGG)n	16–49	ND	1000–1900	(CCT,CTG,CCTCCA)	Yes	Maternal
Distamycin A-inducible							
FRA16B no disease	33-bp AT-rich VNTR	7–12	ND	up to 2000	sequence variations	ND	ND
BrdU inducible							
FRA10B no disease	42-bp AT-rich VNTR	1–74	ND	> 75	considerable sequence variations	Yes	ND
Other unstable repetitive elements							
DM2/ZNF9 myotonic dystrophy 2	(CCTG)n•(CAGG)$_n$	104–176	ND	75–11,000	(GCTG,TCTG)	Yes	ND
EPM1/CSTB1 myoclonus epilepsy	(CCCCGCCCCGCG)$_n$	2–3	12–17	30–75		Yes	Paternal
FSHD/FSHDIA facioscapulohumeral muscular dystrophy-1A	3.3 kb D4Z4 repeat	11–150	ND	<11	sequence variations	Yes	ND
IDDM2 diabetes mellitus type 2	(ACAGGGGT(G/C)(T/C)GGGG)n	110–150	45–109	30–44	numerous variations	Yes	Maternal
SCA10/SCA10 spino- cerebellar ataxia 10	(AATCT)•(AGATT)	10–22	ND	800–4500		Yes	ND
h-Ras VNTR/hRAS1 increased risk of ovarian cancer	28-bp VNTR 5	five common alleles	ND	rare alleles	allele not size specific	ND	ND

[a] The relative genomic location of each repeat is listed in Fig. 2. For the listed repeat sequences the first repeat is that of the coding strand. The use of "ND" denotes insufficient data, sporadic reporting or unknown.

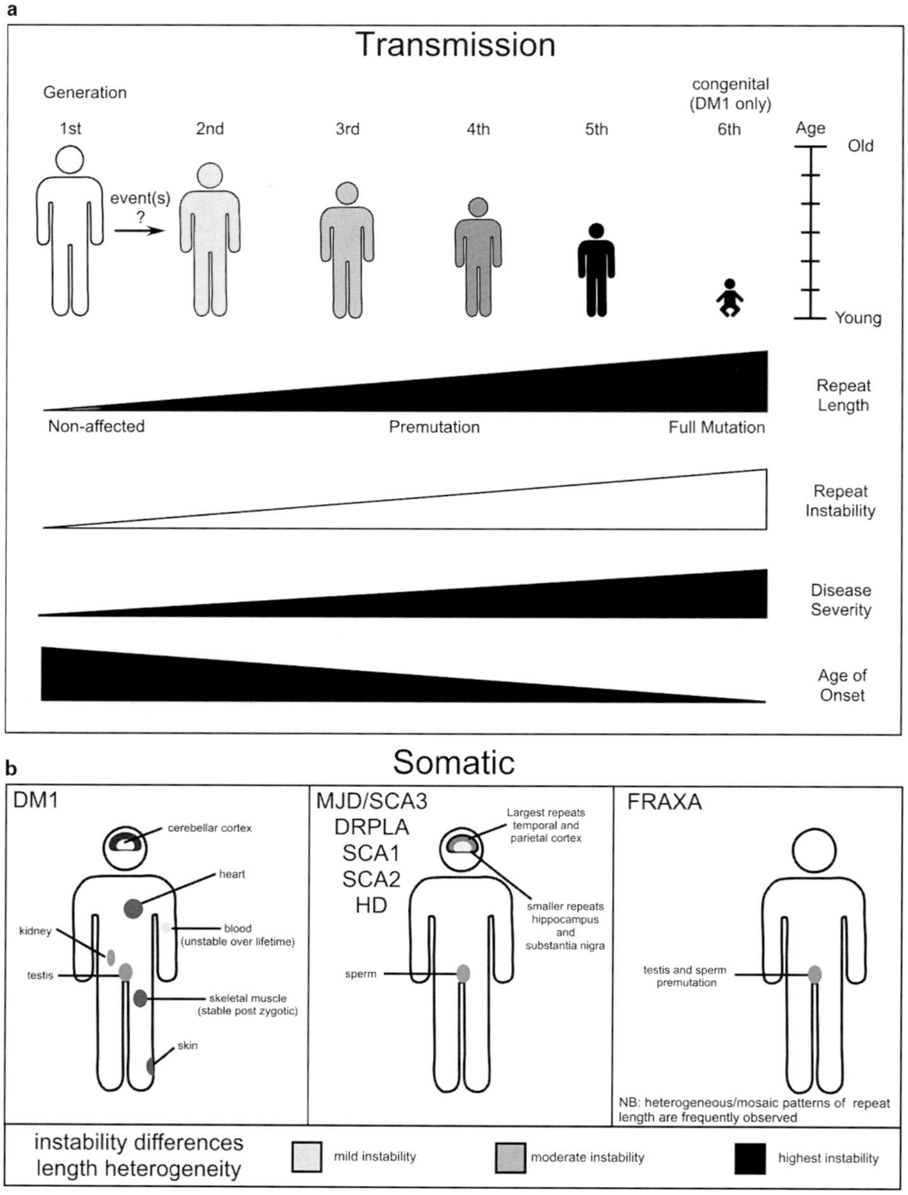

Fig. 1. Transmission versus somatic instability. (**a**) Most TNRs are subject to the phenomenon of genetic anticipation observed as an increase in the severity of symptoms (middle ramp) and/or a decreased age of onset (bottom ramp and size of male figure) as the disease progresses from generation to generation. The transmission of an unstable expanding-repeat (gradient) illustrated by shading (white to grey to black) accounts for anticipation. The transmission of extremely large repeats (2000–5000) results in the development of a congenital form of myotonic dystrophy type 1 (small fetus). The initial events in which premutation repeats develop from normal lengths are currently unknown and may occur over several generations or as the result of a single mutagenic event. (**b**) Somatic instability, differences in the instability of expanded alleles between different tissues, occurs for certain TNRs. In DM1 patients, there is considerable somatic instability, with repeat lengths in brain (except cerebellar cortex), heart and skeletal muscle larger than that observed in blood. For DM1 instability is observed over a lifetime in the blood, while in skeletal muscle it appears stable in adulthood. In FRAXA males, premutation lengths are found in the testis. In a variety of disorders, including DRPLA, SCA2, SCA3 and HD, instability varies in non-proliferative brain tissues, with smaller lengths observed in cerebellar cortex.

TNRs and other unstable repetitive elements

A variety of human diseases, ranging from neurological disorders to cancer susceptibility, are associated with repetitive element instability, with TNRs comprising the largest class. Most of the TNR-associated diseases demonstrate genetic anticipation (Fig. 1a), seen as the increase in disease severity and decrease in age of onset from one generation to next due to the transmission of an expanding TNR. The term "dynamic mutation" is used to describe the TNR mutation process, as there is a greater propensity for the expanded TNR tract to undergo further expansions. Within the general population, TNR tract length is small (~5 to 15 repeats), polymorphic and stably transmitted. For the majority of TNRs, there is a threshold of instability at ~34 repeats. Unstable TNRs and other repetitive elements can be categorized based upon their genic location (Fig. 2). Repeats in coding regions, such as the CAG repeats of Huntington's disease (HD), usually encode glutamine and have a short disease range (40–100 repeats). Repeats in non-coding regions, such as the CTG repeats of DM1, are characterized by large expansions (up to several thousand repeats) in affected individuals and intermediate expansions (40–200 repeats) that evolve into full mutations through germline expansions. Other repetitive elements from simple mononucleotide repeats to complex large macrosatellites, account for an increasing number of human diseases, ranging from diabetes susceptibility (Bennett et al., 1995), epilepsy (Virtaneva et al., 1997), prion disorders (Goldfarb et al., 1991) and cancer (Krontiris et al., 1993).

Genome-wide vs. locus-specific instability

Repeat instability can occur genome-wide with alterations at numerous loci, or it can occur locus-specifically with altera-

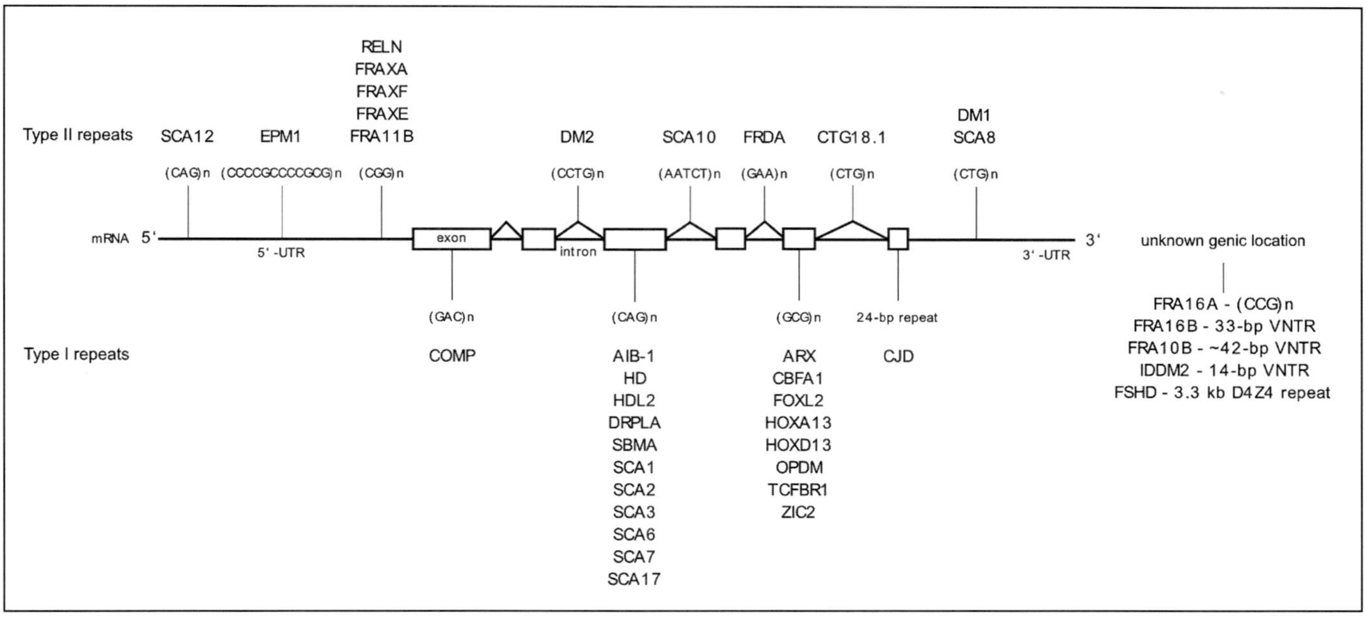

Fig. 2. Genomic location of unstable repeats. The placement of the repeats within the figure represent the general location within one of five positions (5′ UTR, exon, intron, 3′ UTR and other chromosomal locations [non-genomic or unclassified]). Repeats have been roughly grouped according to repeat sequence coding (bottom, with amino acid indicated) and non-coding (top). The RNAs for each repeat are indicated. A complete listing of repeat, reading frame and size lengths are available in Tables 1 and 2.

tions at only one locus. Genome-wide instability is observed in several diseases, generally those associated with defects in *trans*-factors including, but not limited, to sporadic colon cancers and Hereditary Non-Polyposis Colon Cancer (HNPCC-reviewed in de la Chapelle and Peltomaki, 1995). Individuals are predisposed to HNPCC by inherited heterozygous mutations in one of at least four DNA mismatch repair (MMR) genes. Tumors from individuals harboring homozygous mutations display genome-wide instability at simple tandem repeats (microsatellite instability, MSI), which manifests as the gain or loss of 1–4 repeats. Loci of similar length and/or sequence display different levels of instability, making some loci more informative for MSI (Dietmaier et al., 1997; Boland et al., 1998; Duval et al., 2002). Within these repair-deficient cells, the locus-specific differences in instability suggest a contribution of flanking sequence to repeat instability. Unlike the *trans*-factor induced instability in repair-deficient cells, TNR instability is highly locus-specific, as other, often adjacent, repetitive elements are generally unaffected. TNR tracts of similar length (Richards et al., 1996) and sequence (La Spada et al., 1992; Leeflang et al., 1995) located at different chromosomal loci also display differing levels of instability. While this suggests that locus-specific flanking *cis*-elements contribute to instability, no common flanking sequences are associated with unstable TNRs, apart from a general association with G+C rich DNA (Gourdon et al., 1997; Brock et al., 1999). Taken together, this evidence suggests that for both genome-wide and locus-specific instability, *cis*-elements within the flanking sequence contribute to repeat instability. The identity of flanking *cis*-elements that may actually "drive" instability is currently unknown.

Tissue-specific somatic instability

Tissue-specific somatic repeat instability is observed for some TNR-associated disorders (Fig. 1b). Tissue specificity may stem from either tissue-specific *trans*-factors and/or *cis*-elements, such as chromatin packaging, transcription activity or replication regulation. Repeat length heterogeneity has been observed in both proliferating and non-proliferating tissues (Anvret et al., 1993; Telenius et al., 1994; Thornton et al., 1994; Chong et al., 1995; Martorell et al., 1995; Wong et al., 1995), suggesting factors other than proliferative status contribute to somatic instability. It is currently unknown why somatic instability is observed only for certain diseases and within specific tissues.

Somatic instability has been closely examined in DM1. For the $(CTG)_n$ repeat tract of DM1, larger expansions are observed in skeletal muscle, skin and fibroblasts than in peripheral blood leukocytes from the same individual (Abeliovich et al., 1993; Wohrle et al., 1995; Peterlin et al., 1996). Samples from highly proliferative blood revealed that the repeat continues to expand over periods of 7–15 years (Martorell et al., 1995; Wong et al., 1995; Martorell et al., 1998), whereas multiple muscle samples from a given patient taken over time indicated the expanded repeat was relatively stable (Anvret et al., 1993; Thornton et al., 1994). The average repeat size in muscle tissues does not correlate well with the severity of disease symptoms and age of onset (Ashizawa et al., 1993; Thornton et al., 1994; Monckton et al., 1995; Wohrle et al., 1995). While many regions of the cerebral cortex show heterogeneous expansions larger than those observed in blood of the same patient (Sergeant et al., 2001), the cerebellar cortex consistently shows shorter lengths in DM1

(Thornton et al., 1994; Ishii et al., 1996), HD (De Rooij et al., 1995; Telenius et al., 1994), DRPLA (Aoki et al., 1996), SBMA (Tanaka et al., 1996a), SCA1 and SCA3 (Cancel et al., 1998). The difference in repeat lengths between the cerebral cortex and other tissues is remarkably greater for DM1 as compared to the milder differences at the other disease loci. For DM1 the shorter repeat lengths in the cerebrellar cortex is suggestive of large somatic contractions in this tissue. Somatic instability may contribute to the progressive nature and tissue specificity of DM1 disease symptoms.

Mixed results regarding the correlation between tissue-specific instability and proliferation rates have been obtained for other TNR loci. For HD and SCA1, tissues that undergo cell division (i.e. liver) display greater instability than those which are quiescent (i.e. brain) (Telenius et al., 1994; Chong et al., 1995). Although HD shares a similar pattern of somatic instability with SCA1, the diseases affect different cell types and tissues. In SCA1, the Purkinje cells of the cerebellum are affected (Chong et al., 1995), whereas for HD there is selective degeneration of the spiny neurons in the striatum (Strong et al., 1993). This suggests that there may be no relationship between somatic instability and tissue-specific neuropathology, and that an observed correlation, at least for these two diseases, may be merely coincidental. Tissue-specific instability correlates with cell turnover for some disorders, seen as increased instability in liver and colon of MJD and DRPLA, but not in others, as similar tissues for SBMA do not show significant instability (Ueno et al., 1995; Tanaka et al., 1996b). Tissue-specific repeat instability was also observed in DM1 transgenic mice (Gourdon, et al., 1997; Monckton, et al., 1997; Fortune, et al., 2000; van Den Broek, et al., 2002) and HD transgenic mice (Mangiarini, et al., 1997). Importantly, the levels and/or tissue specificity of instability varied considerably between different integration sites of the same sequence (Mangiarini, et al., 1997; Fortune, et al., 2000). The contribution of both *trans*-factors and *cis*-element, regulated in a tissue- and/or locus-specific manner, may account for the variable somatic instability observed for different disease loci.

Parent-of-origin effect

For many TNR loci, instability varies depending upon the sex of the transmitting parent, seen as increased instability following maternal or paternal transmission. It is unclear what sex-specific aspect(s) of allele transmission and/or embryogenesis may be responsible for this "parent-of-origin" effect. Generally in humans there is a trend towards increased instability (expansions versus deletions) following male transmission for TNR disorders such as HD, SCA1, SCA2, SCA3, SCA7, DRPLA, FRA11B and EPM1. The paternal bias is presented as: a strong increase in anticipation with paternal transmission for DRPLA (Nagafuchi et al., 1994), a strong expansion bias with dramatic changes in repeat lengths for HD (Trottier et al., 1994), and large repeat expansions compared to little or no increases for maternal transmissions for the majority of the spinocerebellar ataxias (Yvert and Mandel, 1999). This bias for instability following paternal transmission has been speculated to arise from the numerous mitotic divisions during sperm development (see below: *Instability and human germline development*).

A maternal transmission bias for instability is observed for a number of disorders, including FRAXA, DM1, FRDA, SCA8, FRA16A and IDDM2. This maternal bias is well documented for FRAXA as the full mutation/disease occurs almost exclusively by female transmission. Carrier females with premutation $(CGG)_n$ lengths have a very high probability of expansion upon transmission (Smits et al., 1992). In contrast, similar lengths observed in non-affected males never expand to full mutations upon transmission, hence the term "normal transmitting males" (Sherman et al., 1985). However these individuals can display limited instability upon transmission (Hornstra et al., 1993). In fact, even though the number of FRAXA males who reproduce is small, these individuals also transmit only premutation lengths (Willems et al., 1992; Rousseau et al., 1994). This strong sex bias may result from the combination of meiotic events in the production of oocytes (Malter et al., 1997) and the presence of contractions in the male germline (Fig. 3a). Maternal transmission can also result in a greater tendency to expand for FRDA, even for shorter maternal alleles (Pianese et al., 1997), or have significantly larger magnitude of expansions for SCA8 (–11 to +600 vs. –86 to +7; Koob et al., 1999). As oocytes undergo limited mitotic divisions in comparison to sperm, a more complex mechanism, possibly an in utero *trans*-factor, may be responsible for maternal transmission bias.

For DM1, the effect of parental gender on repeat instability varies depending upon the $(CTG)_n$ tract length. For lengths greater than 600 repeats, DM1 shows a bias for maternal transmission of exceedingly large repeat expansions (> 1000 repeats), which results in the most severe or congenital form of DM1 (CDM). However, rare instances of paternally transmitted CDM have been reported (Zeesman et al., 2002). For $(CTG)_n$ lengths of roughly 200–600 repeats, both paternal and maternal transmission show a bias for expansions (Yvert and Mandel, 1999). Interestingly, an excess of grandpaternal transmissions are observed in the 50–200 $(CTG)_n$ range (Lopez de Munain et al., 1995) and premutation alleles are more likely to expand when transmitted paternally (Martorell et al., 2001), suggestive of a paternal bias for expansion. This trimodal transmission gender-bias may in part be due to the extremely large range of repeat lengths assumed by the DM1 repeat (50–4000 repeats).

Intergenerational instablity is also observed in several, but not all, murine models of repeat instability (Gourdon, et al., 1997; Monckton, et al., 1997; Mangiarini, et al., 1997), as relatively small changes in repeat size are seen over several generations. In addition, several of these transmissions demonstrate a parent-of-origin instability bias with expansions following male transmission (Monckton, et al., 1997; Mangiarini, et al., 1997; Zhang, et al., 2002). A tendency to delete following female transmission was observed in the DM1 but not the HD transgenic mouse (Gourdon, et al., 1997; Monckton, et al., 1997). The underlying cause for the parent-of-origin effect, in mice and humans, is currently unknown but may involve certain aspects of germline development.

Instability and human germline development

The production of the maternal and paternal germlines may account for the gender-specific transmission of instability. Fe-

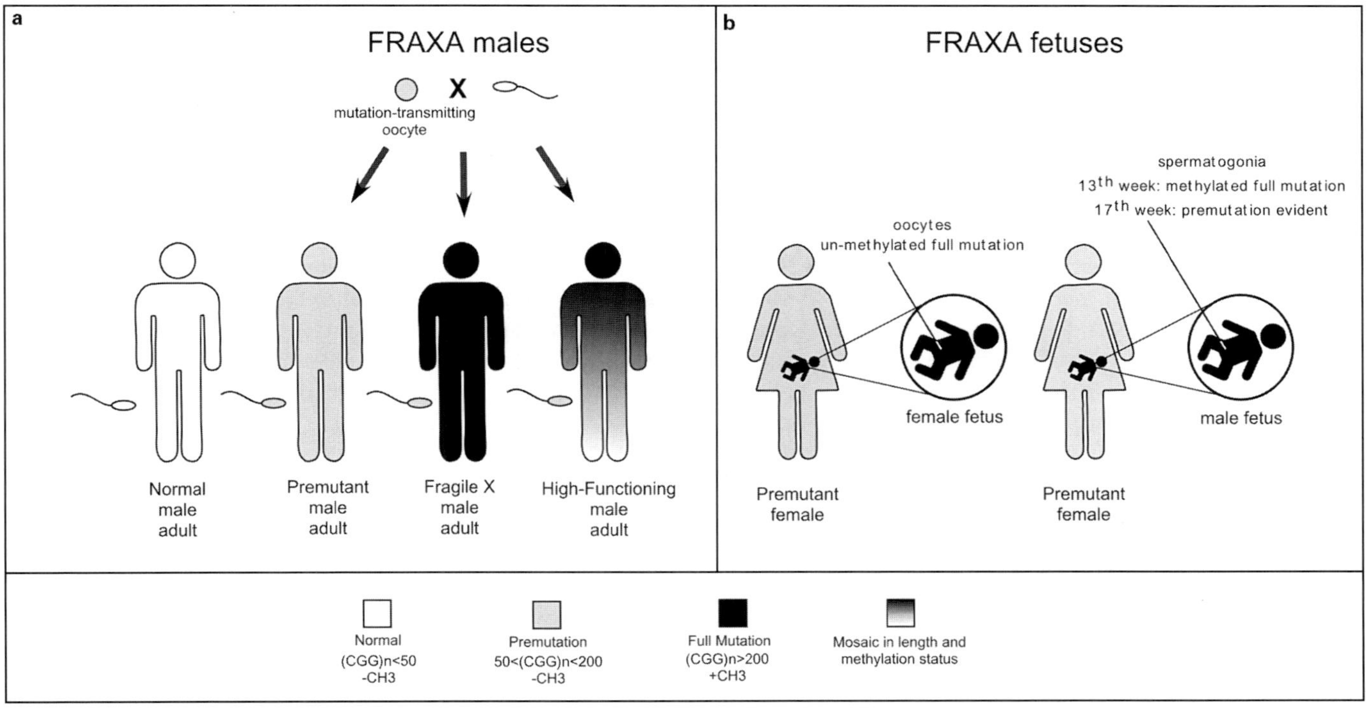

Fig. 3. Repeat instability, methylation status and fragile X syndrome. (**a**) Transmission of the FRAXA expanded repeat through the female germline (grey circle) can result in males having unmethylated premutation lengths (solid grey), methylated full mutation FRAXA males (solid black) or "high-functioning" males (gradient) which are mosaic for both aberrant methylation status and repeat length. Despite the differences in repeat length within somatic tissue, all males contain premutation length in the sperm (grey sperm). (**b**) Analysis of FRAXA fetuses reveals that the development of full mutation occurs either meiotically or during early embryogenesis. Female FRAXA fetuses contain unmethylated full mutations in oocytes. In contrast male FRAXA fetuses contain methylated full mutation in spermatogonia early in development (13th week) with both premutation and full mutation lengths evident later in development (17th week).

male germline cell production is thought to be complete before birth and requires a limited number of cell divisions (~32 in total) (Vogel and Rathenberg, 1975). In contrast, the majority of male germline cell production occurs postnatally with regular spermatogonial stem cell cycles (about every 16 days) and four divisions prior to mature sperm production (Heller and Clermont, 1964). In total, for a 30-year-old male, roughly 400 cell divisions will have occurred towards the production of sperm (Drost and Lee, 1995). As other highly proliferative tissues have differing levels of instability than sperm, the number of cell divisions may not be the sole factor influencing repeat instability (Chong et al., 1995). Differences in DNA repair, which can occur during most stages of oocyte production but decrease during spermatogenesis (for review see Ashwood-Smith and Edwards, 1996; Baarends et al., 2001; Nouspikel, 2002), may also contribute to repeat instability and the "parent-of-origin" effect. Taken together, this evidence indicates that a complex arrangement of gender-specific *cis*-elements, *trans*-factors and cellular processes (replication, repair and/or recombination) may account for the "parent-of-origin" effect.

In DM1 transgenic mice, age-dependent instability was also observed in sperm, suggesting mutations accumulate over time in spermatogenic cells (Zhang, et al., 2002). This instability depended upon the location of transgene integration, highlighting a potential *cis*-effect. In addition, enzymes associated with mismatch repair are expressed at high levels in murine spermatogonia, declining as germ cell development progresses (Richardson, et al., 2000). Recently in one mouse model (Kovtun, et al., 2000), but not all (Zhang, et al., 2002), the sex of a murine embryo was observed to influence CAG repeat instability, a difference that may be explained by *cis*-effects.

Developmental-specific instability

In humans, repeat instability may first arise during specific developmental stages (Duyao et al., 1993; Wohrle et al., 1993; Jansen et al., 1994; Telenius et al., 1994; Chong et al., 1995; Martorell et al., 1997). This timing in instability may arise from developmental specific *trans*-factors and/or *cis*-elements. Within the testis of FRAXA male fetuses, methylated full mutation lengths are observed at the 13th week (Fig. 3b), while by the 17th week premutation lengths become evident (Malter et al., 1997), with only unmethylated premutation lengths observed in the sperm and testis of adult FRAXA and premutation males (Reyniers et al., 1993, 1999; Tassone et al., 1999). This finding suggests that repeat deletion and the selective growth advantage during the development of the testis, in conjunction with methylation, contribute to the presence of premutation alleles in adult sperm. Only full mutation alleles are detected elsewhere in FRAXA fetuses, including unmethylated full mutations in female oocytes. This pattern limits the developmental window for FRAXA expansions to either meiosis (Malter et al., 1997; Moutou et al., 1997), or during the early stages (<7 days) of post-zygotic cell divisions (Reyniers et al., 1993; Wohrle et al., 1993) prior to germline separation (Malter et al., 1997; Moutou et al., 1997; Helderman-van den Enden et al., 1999). This timing could place both expansions and deletions within a narrow window in early embryogenesis in which there are waves of *de novo* global methylation (Frank et al., 1991; Migeon et al., 1991; Razin and Shemer, 1995; Reik and Dean, 2001), although the exact role of methylation in instability is currently unknown (see below: Section III, *CpG methylation*). Regardless of whether the initial expansion event was meiotic or mitotic, it is clear that some post-zygotic mitotic instability occurs, as specific tissues from a single patient show differing degrees of CGG expansions (Devys et al., 1992; Sutcliffe et al., 1992; Reyniers et al., 1999; Taylor et al., 1999).

A developmental window for instability has also been observed from the analysis of DM1 fetal samples. Tissues from a 13-week fetus showed no signs of inter-tissue repeat instability, whereas instability was observed in tissues from a 16-week fetus (Jansen et al., 1994; Wohrle et al., 1995; Martorell et al., 1997). It has been speculated that the repeat instability is silenced or absent during the differentiation that accompanies fetal development during the first trimester, but present at the early onset of the second trimester during the period of rapid fetal growth (Martorell et al., 1997). Inter-tissue differences in instability do not appear to correlate with cell division rates, suggesting elements other than proliferation rate contribute to developmental repeat instability. The very large expansions in muscle seem to occur during early development and plateau in childhood. Postnatal repeat instability in lymphocytes is observed over the lifetime of DM1 patients (Martorell et al., 1995; Martorell et al., 1998). As such, the length differences between muscle and blood repeats is smaller in younger patients than in older patients (Thornton et al., 1994; Zatz et al., 1995). No one element (methylation, replication ...) appears to account for the pattern of developmental instability, suggesting a complex multi-factorial pathway, perhaps a combination of *cis*-elements and *trans*-factors

"Reversal mutations", repeat deletions and fragile sites

Deletions of expanded disease-causing repeats, termed "reversal mutations", have been documented for several TNR disorders. For DM1, CTG repeat contractions occur in ~6 % of DM1 offspring, more frequently from paternal than maternal transmissions (Ashizawa et al., 1994). In a rare number of these transmissions, a normal length allele is transmitted from clinically affected parent to offspring who were clinically normal (Brunner et al., 1993). Other rare transmissions involve the offspring receiving a normal allele on the affected DM1 chromosomal background (Shelbourne et al., 1992), sometimes involving exchange of flanking sequence (O'Hoy et al., 1993), suggesting gene conversion may be responsible for repeat deletions. In fact, based on haplotype and patient data, gene conversion is suspected in a number of reported transmissions of deleted alleles at both DM1 (Brunner et al., 1993; Krahe et al., 1995; Tishkoff et al., 1998) and FRAXA loci (van den Ouweland et al., 1994; Brown et al., 1996; Losekoot et al., 1997). Repeat deletions in sperm have been observed for FRDA (Pianese et al., 1997), SCA8 (Moseley et al., 2000) and SBMA (Zhang et al., 1994), suggesting a male-specific factor contributing to repeat deletions. For FRAXA, methylation-associated repeat deletions in the developing testis may account for the absence of large expanded lengths and the presence of only premutation lengths in the sperm of adult patients. Rare "high-functioning" FRAXA males display somatic CGG deletions that appear to be related to reduced levels of aberrant methylation (Wohrle et al., 1998). Model systems that generally delete the repeat tract, such as bacteria and yeast, may provide clues into the *cis*-elements and *trans*-factors contributing to deletions observed in some human TNR disorders (Brook, 1993; Monckton et al., 1995; Martorell et al., 2001).

Five of the rare folate-sensitive fragile sites (FRAXA, FRAXE, FRAXF, FRA11B and FRA16A) are associated with chromosomal breakage or deletions within or around expanded $(CGG)_n$ repeats. Rare deletions limited to the FRAXA CGG repeat, as well as those including part of the repeat and some 5′ and 3′ flanking sequences, have been reported (Gecz et al., 1996; Hammond et al., 1997). Many deletions occur at certain CpG rich "hot spots" upstream of repeats (Hammond et al., 1997). Similar deletions, either within or around the repeat, are associated with the FRAXE $(CGG)_n$ tract (see Fig. 2 in Gecz, 2000). The association of $(CGG)_n$ repeats at the FRA11B locus with chromosomal deletions characteristic of Jacobsen syndrome (Jones et al., 1995) was the first demonstration of a direct link between a fragile site and chromosome breakage *in vivo*. The rare BrdU-inducible FRA10B (Hewett et al., 1998) and Distamycin-A inducible FRA16B (Yu et al., 1997) fragile sites are associated with expanded tracts of 42- and 33-bp AT-rich repeats, respectively. Recent evidence from the examination of the FRA16B AT-rich repeats (Hsu and Wang, 2002) suggests that nucleosome exclusion, a suspected *cis*-element (see below: Section III, *Nucleosomes*), may also be the underlying connection between these repetitive elements and chromosomal fragile sites.

Repeat deletions can also result in phenotypic effects, depending upon the chromosomal context of the repeat. The CAG repeat within the first exon of the androgen receptor (AR) provides an excellent example as transcriptional activity, and androgen-induced gene activation is inversely correlated with repeat length (Choong et al., 1996). Altered transcriptional activation mediated by the AR can affect many processes. Short tract lengths (≤22 repeats) result in increased risk for prostate cancer due to increased androgen-gene activation (Irvine et al., 1995; Hardy et al., 1996), while shorter tract lengths (<11 repeats) reduce activation, resulting in androgen insensitivity syndrome (AIS) (Sultan et al., 1993; Choong et al., 1996). Larger repeat tracts (>23 repeats) also reduce activation, increasing the risk of infertility (Dadze et al., 2000; von Eckardstein et al., 2001). In contrast, large expansions (> 38 repeats) in the same CAG repeat produce both AIS-like symptoms and a gain-of-function mutation, producing the SBMA neurological symptoms (La Spada et al., 1991). Disease-associated repeat deletions also affect larger repetitive elements, as deletions of the 3.3-kb D4Z4 repeat results in inappropriate gene expression and facioscapulohumeral muscular dystrophy (FSHD), with deletion size corresponding to the disease severity (Hewitt et al., 1994; Winokur et al., 1994). Even relatively minor repeat instability, such as the instability associated with polyalanine disorders (see Table 1), may result in serious consequences. Alterations by one or two GAC repeats within the COMP gene can produce two distinct forms of skeletal dysplasia (Delot et al., 1999). While similar *cis*-elements may contribute to repeat expansions and deletions, the protein and chromosomal context of the repeat tract ultimately determines disease development.

Human cell culture and repeat instability

Various cellular models of repeat instability have proved useful in revealing clues to the process of expansion-biased repeat instability. Cultured fetal or adult DM1 fibroblasts, as well as DM1 myoblasts (Abeliovich et al., 1993; Wohrle et al., 1995; Furling et al., 2001), displayed "synchronous" CTG expansions. The term "synchronous" describes the progressive expansion of the majority of the DM1 alleles by approximately the same incremental change in repeat numbers (Wohrle et al., 1995). Synchronous expansion is a mutation frequency approaching 100%, allowing for detection through Southern blotting or standard PCR conditions. EBV-transformed DM1 lymphoblastoid cells (Ashizawa et al., 1996; Khajavi et al., 2001), SV40-transformed HD transgenic mouse brain cells (Manley et al., 1999a), and cultured primary cells of a DM1 transgenic mouse (Gomes-Pereira et al., 2001) displayed expansions and/or deletions that occurred in a non-synchronous fashion. Instability in the EBV-transformed DM1 lymphoblastoid cells, which was expansion biased, could only be detected by the sensitive small-pool PCR method indicating that length changes were rare (Ashizawa et al., 1996). In contrast, EBV-transformed FRDA lymphoblastoid cells displayed frequent contractions and expansions of the GAA tract (Bidichandani et al., 1999). It is noteworthy that virally transformed cells and spontaneously transformed murine cells can display different mechanisms of replication, repair and epigenetic regulation compared to their progenitor cells (Meek et al., 1977; Oppenheim et al., 1981; Aiyar et al., 1998; Habib et al., 1999). Interestingly, it has been noted that relative to native DM1 lymphocytes, the EBV transformation of DM1 lymphoblasts can lead to considerable de-methylation of sequences flanking the expanded CTG repeat (Petronis et al., 2000). The effect of altered methylation in these cells is unknown, but could affect repeat instability (Nichol and Pearson, 2002) (see below: Section III, *CpG methylation*).

Each of the human and mouse CTG/CAG cell culture models showed expansion-biased CTG/CAG instability. Several appeared to display a selective growth advantage for cells with longer alleles (Ashizawa et al., 1996; Manley et al., 1999a; Gomes-Pereira et al., 2001; Khajavi et al., 2001). The cell lines derived from different tissues from a given mouse displayed instability that depended upon the tissue of origin, with specificities paralleling in vivo observation (Gomes-Pereira et al., 2001). As in the mice (Lia et al., 1998; Gomes-Pereira et al., 2001), instability in the cultured cells did not correlate with proliferative status. While these studies did not uncouple repeat instability from cellular proliferation, they strongly suggest that tissue-specific factors may be more critical determinants of instability than a simple association with proliferation. The tissue-specific factors might include *cis*-elements such as specific replication programs (replication origin usage, preferred sites for the initiation of Okazaki fragments, rates of fork progression, or replication timing). Another, but not mutually exclusive alternative is the role of tissue-specific *trans*-acting factors such as the activity of specific replication, repair or recombination proteins. Recent results from our laboratory have demonstrated a requirement of cellular proliferation for CTG instability in a primary DM1 fetal fibroblast line (Pearson, unpublished data). Our results would suggest that tissue-specific *cis*-elements are contributors to instability.

Cells derived from adult or fetal FRAXA tissues did not reveal any alterations in the lengths of the expanded CGG repeat, even after long-term culturing (Wohrle et al., 1993; Wohrle et al., 1995). In contrast, CGG instability was observed in proliferating cells derived from the "high-functioning" FRAXA males that display both methylation mosaicism (reduced aberrant CpG methylation) and increased length heterogeneity at the FRAXA locus (Wohrle et al., 1998; Burman et al., 1999). Most reports observed CGG deletions (Burman et al., 1999; Salat et al., 2000; Wohrle et al., 2001), while some also observed expansions (Burman et al., 1999; Glaser et al., 1999). Together, these results strongly suggest that expanded CGG tracts are destabilized by the absence of CpG methylation – an observation that agrees with the deletions to premutation lengths in testis and sperm of FRAXA patients (see above: Section I, *Developmental-specific instability*).

Evidence for cis*-elements in murine models of instability*

A number of murine model systems have been established to examine TNR instability. Evidence from these models strongly suggests flanking sequences contribute to instability. Some mouse models reproduce similar tissue specific, age-dependent and expansion-biased repeat instability, although not the large transmitted expansions associated with somatic mosaicism in DM1 patients. While larger repeats are unstable (>100 CTG/CAG, Mangiarini et al., 1997; Monckton et al., 1997), flanking genomic sequence is required for instability of moderate repeats (45–78 CTG/CAG) from DM1 (Gourdon et al., 1997), SBMA (La Spada et al., 1998) and DRPLA (Sato et al., 1999). Mice with identical inbred genetic backgrounds carrying a single insertion of a large CTG repeat with minimal human flanking sequence, but integrated at different loci, demonstrated remarkably different instabilities (from 3–10 fold difference; Zhang et al., 2002), suggesting a contribution from murine *cis*-elements. In contrast, mice containing CTG repeats with a large amount of DM1 patient-derived flanking sequence (45 kb) demonstrated similar levels of instability for all insertion locations (Gourdon et al., 1997; Seznec et al., 2000). Similar experiments in SCA7 mice, using flanking sequences, from either patient genomic or cDNA derived sources, demonstrated that marked instability occurs only in the proper chromosomal context (Libby et al., 2003). Strikingly, mice produced from an SCA7 chromosomal deletion with loss of 3′ sequences failed to be unstable, strongly suggesting the existence of flanking *cis*-elements. Differences in repeat stability have been observed for different murine strain backgrounds carrying the same sequence, suggesting strain-specific *trans* or *cis*-elements can contribute to development of repeat instability (van Den Broek et al., 2002). Taken together, this data suggests that flanking sequences, both murine and human, contribute to repeat instability.

Single-genome analysis

The examination of a single genome, either by analysis of single cells (i.e. sperm) or by small-pool PCR (spPCR), provides a powerful tool for the investigation of repetitive elements. PCR analysis of sperm has supported the presence of deletions in sperm of premutation FRAXA patients (Mornet et al., 1996; Kunst et al., 1997; Nolin et al., 1999; Crawford et al., 2000; Sullivan et al., 2002), reduced paternal penetrance of SCA8 (Moseley et al., 2000), and a role for germline mitotic division in repeat instability of HD (Leeflang et al., 1995). Germline differences in the repeat distribution of sperm between DM1 patients (Monckton et al., 1995) and spPCR analysis of sperm from DM1 mice models (Zhang et al., 2002), supports a role for *cis*-elements in repeat instability. Much of the variation observed in the repeat distribution between sperm and leukocytes of FRAXA patients detected by spPCR analysis (Crawford et al., 2000), appears to stem from individual differences. Interestingly, while both FRDA and DM1 are non-coding repeats with unmethylated repeat in somatic cells, spPCR analysis of sperm from FRDA patients shows a natural tendency to contract in vivo (Sharma et al., 2002), compared to expansions for DM1. This finding further illustrates the locus specificity of repeat instability that suggests a strong flanking sequence contribution. The length changes, estimated by sperm analysis of DM1 (Monckton et al., 1995; Martorell et al., 2000), HD (Leeflang et al., 1999), and SBMA (Zhang et al., 1994, 1995), support roles for replicative slippage and hairpin formation in repeat instability. While high levels of repeat instability can be detected by Southern blotting and standard PCR (Libby et al., 2003), spPCR reveals the magnitude of the observed changes. The spPCR technique has proved useful in detecting rare and difficult to detect events (rare larger expansions) thereby providing support for a role for *cis*-elements in repeat instability.

Chromosomal elements associated with instability

A number of chromosomal elements, such as *Alu* repeats, are frequently associated with TNRs, suggesting a possible connection to repeat instability. *Alu* repeats are an ancient mobile genetic element representing 10% of the human genome (Smit, 1999), whose insertion may account for certain human disorders (Deininger and Batzer, 1999). *Alu* repeats are responsive to changes in chromatin structure (Kim et al., 2001), have altered instability in the absence of yeast FEN1 (Lobachev et al., 2000), and are associated with chromosomal breakpoints (Rudiger et al., 1995; Kolomietz et al., 2002). Furthermore, active *Alu* repeats appear to be passed primarily through paternal germlines, in a parent-of-origin manner (Jurka et al., 2002). Interestingly, di-, tri- and tetranucleotide repeats are often located adjacent to *Alu* repeats (Economou et al., 1990; Yandava et al., 1997), with the majority of GAA repeats, including the FRDA $(GAA)_n$ repeats, associated with the 3′ end of the "A" stretch present in *Alu* repeats (Chauhan et al., 2002). The extent of GAA repeat polymorphism appears to be modulated by the length of the flanking *Alu* poly "A" stretch (Chauhan et al., 2002). In addition, $(TTA)_n$ repeats located adjacent to *Alu* repeats have a high frequency of length alterations (Zuliani and Hobbs, 1990). The unstable expanded DM1 $(CTG)_n$ repeats are also linked to an *Alu* polymorphism (Imbert et al., 1993; Neville et al., 1994). Taken together, this data suggests that while *Alu* repeats may have a common genesis with microsatellites (Arcot et al., 1995), their presence may promote instability in nearby repetitive elements. Other chromosomal elements, such as CTCF binding sites (the cellular target of the zinc finger CTCF protein), are also associated with a number of CTG/CAG repeats (Filippova et al., 2001). At the DM1 locus, these sites can form insulator elements whose methylation prevents CTCF binding and may contribute to congenital DM1. Together with the recently identified germline specific CTCF paralogue BORIS (Loukinov et al., 2002), CTCF sites represent attractive *trans*-factor interacting *cis*-elements (Fig. 4d) for developmental and sex specific instability.

Internal and external cis-*elements*

There is strong evidence in humans that *cis*-elements within and flanking the repeat contribute to instability and ultimately disease. Most of the unstable repeats are associated with CNG sequences, with longer repeat tracts more likely to undergo expansion mutations (Richards and Sutherland, 1992). The presence of tract interruptions reduces the instability of many of these TNRs. Thus, within the repeat tract, repeat sequence, tract length and purity contribute to instability representing internal *cis*-elements. However, stably transmitted repeats, considerably larger than the observed instability threshold in humans (~ 34 repeats) have been observed (Nakamoto et al., 1997). In addition, repeats of similar sequence, length and purity located at different chromosomal location have differing levels of instability (La Spada et al., 1992; Leeflang et al., 1995). Therefore, external *cis*-elements, such as sequences flanking the repeat and the surrounding epigenetic environment also contribute to the repeat instability. The association of chromosomal elements such as *Alu* repeats, CTCF sites, chromatin structure and DNA methylation with TNRs suggests that these elements may act as possible external *cis*-elements. Both external and internal *cis*-elements may participate in the formation of mutagenic DNA structures that alter the fidelity of DNA replication, repair and/or recombination. As the mutation mechanisms occurring in any one patient tissue for any disease loci is poorly understood, the next sections will examine the patient data that supports the involvement of *cis*-elements both within and flanking the repeat in instability.

Fig. 4. Mechanism of instability and candidate *cis*-elements. Various *cis*-elements may contribute to repeat instability. (**a**) The location of the repeat in the Okazaki initiation zone will determine the location of Okazaki initiation and processing, which may be required to occur within the repeat tract for instability. (**b**) The positioning of nucleosomes within and around the repeat tracts may alter the progression of the replication fork (see panel f), CTG repeats are strong nucleosome position elements, while CGG repeats

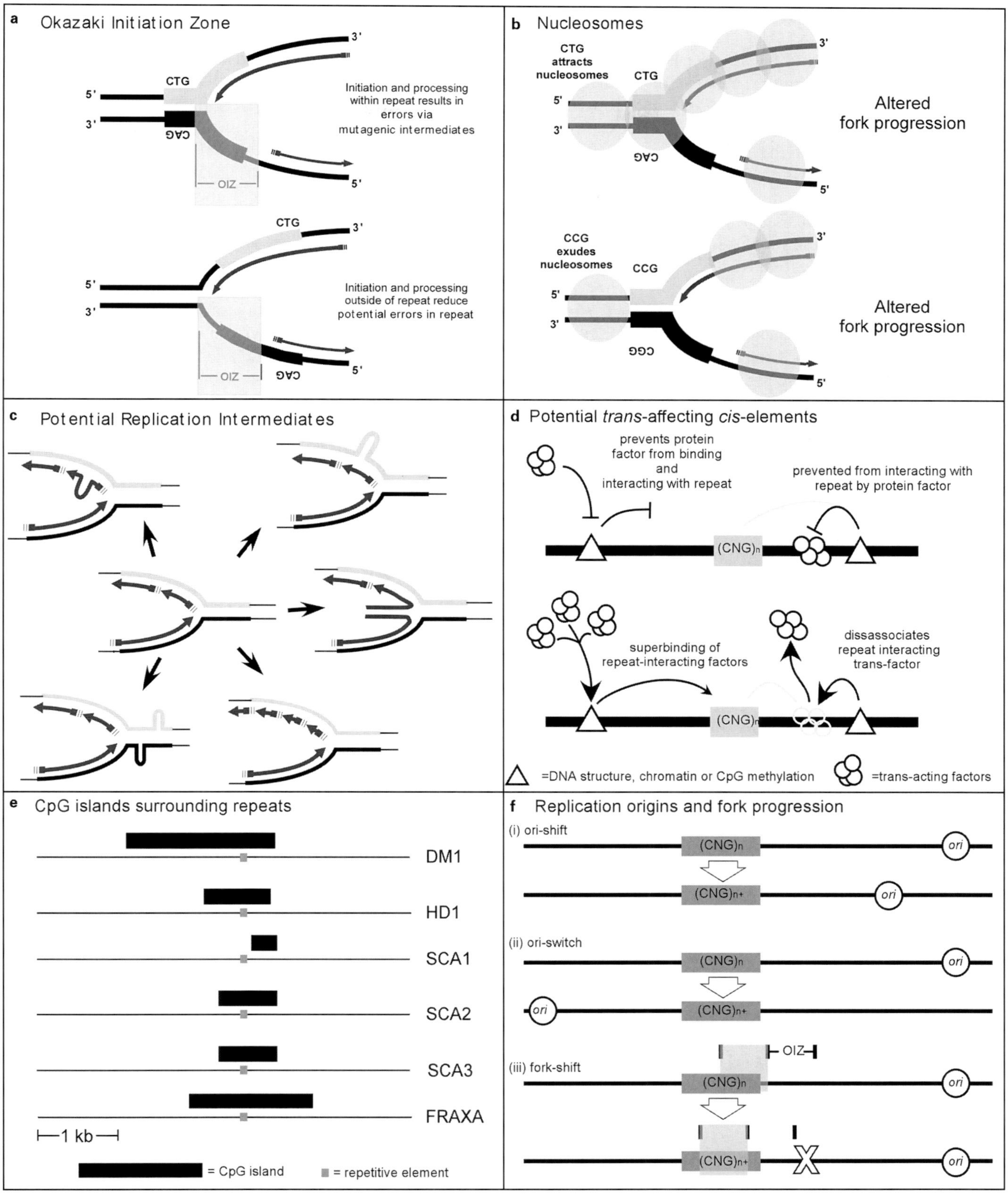

repel nucleosomes; (**c**) The formation of various mutagenic structural intermediates may occur upon replication fork stalling within the repeat tract. These products may escape repair or be incorrectly processed (see Fig. 6) thereby facilitating repeat instability; (**d**) The interaction of *cis*-elements with *trans*-factors may facilitate changes in epigenetic structure within or around the repeat (see panel f); (**e**) Unstable TNRs are located within or in proximity to CpG islands, suggesting an association between methylation and repeat instability. (**f**) Alteration in the origin of replication, either on different sides (ori switch, i); on the same side (ori shift, ii) or alterations in the progression of the replication fork by epigenetic factors. For details please see text.

Section II: Clinical evidence of *cis*-elements

Cis-elements within and flanking repetitive elements may contribute to repeat instability. The effect of repeat sequence, tract length and purity, and flanking sequence are well documented and characterized in patient samples. This section will highlight the evidence for the participation of these *cis*-elements in repeat instability through examination of clinical data with pertinent reference to model systems. The role of more complex *cis*-elements, such as unusual DNA structures and other epigenetic factors, is also strongly implicated by clinical data and will be discussed in Section III.

Repeat sequence, tract length and purity as cis-*elements*

Repeat sequence: The sequence of a repetitive element is a critical determinant of its instability, as only specific repeat nucleotide combinations are associated with disease (see Tables 1 and 2). For TNRs, the same nucleotide combination can occur in all reading frames (CNG, GCN or NGC), although interestingly, there is a preference for the $(CNG)_n$ reading frame (Astolfi et al., 2001; Katti et al., 2001). Altered instability was observed when the G+C content, but not repeat unit size or allele length of a tetranucleotide repeat, was decreased (Edelmann et al., 1996), indicating sequence specificity occurs for other repetitive elements. The sequence specificity for the disease-associated repeats may arise from the sequence-specific DNA structures formed by these repeats (see below: Section III, *Unusual DNA structures*). For the CTG/CAG TNRs, specific sequence-dependent structures are formed, such that the repeat lengths within these structures correlate with observed length changes in human tissues (see below: Section III, *Unusual DNA structures*) and Pearson et al., 2002). The ability of specific sequences to form structures, and the processing of these structures by the cell, may be the underlying cause behind the sequence preference for TNR instability.

Repeat length: The length of a TNR tract (i.e. the number of repeat units) affects its stability, as generally TNR tracts greater than 35–40 repeats become genetically unstable. The expansion of short repetitive elements causing disease appears limited to humans (Sutherland and Richards, 1995). For the majority of repeat loci, a polymorphic distribution of repeat lengths is present in the non-affected population (Tables 1 and 2). These non-affected lengths rarely display instability during transmission. However, the presence of larger non-affected alleles is speculated to give rise to premutations and ultimately full mutation lengths. For example, 97% of the general population has $(CAG)_n$ lengths at the SCA7 locus between 10–13 repeats, with the majority consisting of $(CAG)_{10}$ (Gouw et al., 1998). While repeats in the large normal range of 20–35 are extremely rare (<0.5%), several examples exist of de novo expansions to full mutation from this range (Stevanin et al., 1998). Given the fitness restraints against TNR expansions, the existence of a pool of large non-affected alleles may account for the historical persistence of TNRs within humans. For other TNR loci, a paucity of alleles between the high-end normal and low-end disease exists, suggestive of a highly unstable length range prone to expansion or of distinct facultative expansion events.

A number of terms (pre-, proto-mutation, intermediate alleles) are used to describe the repeat length range between non-affected and unstable disease-associated for TNR loci – most notably DM1, FRAXA and HD. For DM1, the term premutation describes lengths between high-end non-affected and low-end disease (38–49 repeats), which are highly unstable and frequently expand following paternal transmission (Yamagata et al., 1994; Martorell et al., 2001). DM1 protomutation lengths (50–80 repeats) can also expand upon transmission and may eventually cause the development of mild symptoms but are asymptomatic at the time of assessment (Barcelo et al., 1993). For Huntington's, the term "mutable normal allele" or "intermediate allele (IA)" are used for large normal alleles (27–35 repeats), while "HD allele with reduced penetrance" describes an allele range (36–39 repeats) that does not show an overall bias for expansions but in which de novo mutation may arise (HD Genetic Testing Working Group, 1998). A larger length range (60–200 repeats) is termed premutation for FRAXA, in which only maternal transmission results in expansions (see above: Section I, *Parent-of-origin effect*). For FRAXA, the term "grey zone" describes the length range (34 to 55 repeats; Zhong et al., 1996) in which interrupted repeats are stably transmitted, whereas for "pure" repeat tracts the allele is unstably transmitted. As all four terms loosely describe similar situations, the terms are often used interchangeably.

For the polyalanine tract repeats, or $(GCG)_n$ repeats, instability and disease symptoms are observed at lengths far below the stability threshold of the majority of TNRs. Relatively minor changes in repeat length can result in disease, as expansion to less than 30 repeats results in synpolydactyly (Goodman et al., 1997), cleidocranial dysplasia (Mundlos et al., 1997) and oculopharyngeal muscular dystrophy (OPMD; (Brais et al., 1998) (see Table 1). Expansion of the $(GAC)_5$ repeats within the human COMP gene to six repeats results in one type of skeletal dysplasia, while expansion to seven or deletion to four repeats results in a more severe type of dysplasia. Disease development at these loci may relate more to the protein context of the repeat. While some of these loci do not display somatic instability (Nakamoto et al., 2002), disease progression does occur at these loci at repeat lengths far below the stability threshold of most disease-associated TNRs. The small changes in repeat number at these loci causing disease development are compatible with replicative slippage as the underlying mutation mechanism.

Length dependent instability has also been observed for larger repetitive elements, such as the CEB1, B6.7 minisatellites (Jeffreys et al., 1994) or the 4.7-kb RS447 repeat (Okada et al., 2002). However, unlike TNRs, no threshold length for instability exists for minisatellites, as even five CEB1 units are associated with instability (Buard et al., 1998). While an association exists between tract length and instability for numerous loci of differing repeat size (micro, mini and macrosatellites) this association is not exclusive. TNR tracts (>50 repeats) larger than the instability length threshold (~34 repeats) can be found at several human loci which display no intergenerational instability, including the ERDA1 locus (Nakamoto et al., 1997). This phenomenon extends to larger repeat units, as the MS32 minisatellite (29 bp repeat) does not show an increased mutation rate with longer tract lengths (Jeffreys et al., 1994). Thus, repeat tract length is an important, but not exclusive, contributor to repeat instability.

The polarity effect: Repeat length alterations have been observed to occur preferentially on one side of the repeat tract (Eichler et al., 1994; Richards and Sutherland, 1994), suggesting a "polarity" effect. A bias for 3′ end length changes was observed through analysis of di- and trinucleotide repeats replicated in mammalian cells (Suzuki et al., 2001), as well as the human MS205 minisatellite locus (Armour et al., 1993; He et al., 1999). For several TNRs, including the FRAXA $(CGG)_n$, length alterations also appear biased toward the 3′ end of the repeat (Eichler et al., 1994). In yeast, a 3′-end polarity was also observed for TNRs, but this was altered in an Okazaki processing mutant (rad27), or absent in a mismatch repair mutant (msh2) (Maurer et al., 1998). For the CAG repeats of SCA2, variation was mostly observed at the 5′ end of the repeat tract (Choudhry et al., 2001). Replicative slippage within the repeat tract on the lagging strand template is thought to generate the polarized instability. As this effect roughly stems from the location of the repeat relative to the advancing replication fork, the differences in the polarity between the various repeat loci may relate to the location of the nearest replication origin.

Repeat purity: The presence of interruptions within an otherwise homogeneous (pure) tract of repeats has been associated with increased repeat stability. "Pure" TNR tracts are more likely to expand following transmission than interrupted repeats of equal size. Interestingly, the sequence of the interruptions appears to differ between those found in humans with those induced by *E. coli* (Pearson and Sinden, 1998, Pearson unpublished data), suggesting species-specific preferences for interruption formation. In humans, a number of the TNR loci are associated with repeat interruptions (see Tables 1 and 2), including FRAXA (Eichler et al., 1994), SCA1 (Chong et al., 1995), SCA2 (Choudhry et al., 2001) and FRDA (Cossee et al., 1997; Montermini et al., 1997). For the FRAXA $(CGG)_n$ repeats, the absence of AGG interruptions correlates with unstable alleles for the "grey zone" lengths (Nolin et al., 1996). Even the presence or absence of a single interruption within a high-end normal allele can significantly alter the stability of a repeat tract. SCA1 $(CAG)_n$ tracts as large as 39 repeats with a single CAT interruption are somatically stable, whereas alleles with 40 "pure" repeats are unstable (Chong et al., 1995). Thus, for the SCA1 repeats, the presence of a single base substitution (CAG to CAT) appears to prevent repeat instability. The stabilizing effect of TNR interruptions suggests that the loss of an interruption may be an important mutational event in the generation of unstable alleles. Interruption loss may be the first step in the long road to instability but, to date, the loss of an interruption leading to disease has yet to be detected.

The stabilizing effect of interruptions is observed for other repetitive elements such as dinucleotides (Weber, 1990), although not universally. There are several disease-associated expanded repeats which remain interrupted following expansion, including SCA3, SCA8 and SCA17. The expanded disease-associated CAG repeats of SCA3/MJD maintain a 5′ AGG interruption encoding an essential lysine residue (Kawaguchi et al., 1994), suggesting protein context as important for maintaining interruptions. In SCA8, the expanded $(CAG)_n$ repeats appear to be more susceptible to interruption-generating point mutations (Moseley et al., 2000), which are duplicated in a cassette-like mechanism. Repeat duplication has been reported for de novo expansions of the SCA17 $(CAG)_n$ repeat (Koide et al., 1999). Both the SCA17 $(CAG)_n$ repeat and the interrupting CAA repeats encode glutamine, such that the repeat length necessary for disease development can be reached by addition of an interrupted repeat. This contrasts with the stabilization of CAG repeats with CAA interruptions and the effect of interruption loss at the SCA2 locus (Choudhry et al., 2001). Generally, minisatellites are not overtly influenced by repeat heterogeneity. The expanded CEB1 minisatellites are heavily intermingled with interruptions, suggesting that repeat instability may actually generate heterogeneity (Buard et al., 1998). Differences in the underlying mutation mechanism (slippage/structure formation vs. duplication) occurring at different repetitive elements, as well as the repeats protein context, may explain the differential effect of repeat interruptions.

The genetic stabilizing effect of repeat interruptions may stem from their ability to disrupt the formation of unusual DNA structures (see below: Section III, *Unusual DNA structures*) by repetitive sequences. For the SCA1 and FRAXA repeats, the effect of tract purity on the propensity to form slipped structures correlates nicely with its effect on genetic instability and disease (Pearson et al., 1998). Similarly the presence of interruptions appears to disrupt the formation of quadruplex DNA by CGG repeats (Weisman-Shomer et al., 2000). For FRDA, the presence of GGA interruptions within the GAA repeat tract was demonstrated to inhibit the formation of triplex and "sticky" DNA (Fig. 5) by introducing destabilizing base–base mismatches (Sakamoto et al., 2001). Interruptions also stabilize repeats in yeast and in some but not all systems (Petes et al., 1997) their stabilizing effect was diminished in the absence of mismatch repair (Rolfsmeier et al., 2000). In humans however, the stabilizing effects of repeat interruptions may not require a functional DNA mismatch repair system (Bacon et al., 2000, Sturzeneker et al., 2000). Thus, the presence of tract interruptions appears to block or limit the formation of mutagenic intermediates, but for those that do form, correct repair may be facilitated by the presence of interruptions.

Contiguous repeat tracts

Contiguous repeats are found within, or immediately flanking, numerous disease-associated unstable repeat tracts. For the closely spaced FRAXA and FRAXE fragile-site associated TNRs, the instability of each repeat appears connected. Larger repeats, minor instability and deletions at the FRAXE locus are all connected with the expanded FRAXA repeats (Brown et al., 1997; Murray et al., 1997), suggesting a commonality in the mechanism of instability for these two loci. A similarly close connection between repeat loci can be observed between *Alu* repeats and a number of the unstable repetitive elements (Economou et al., 1990; Chauhan et al., 2002). For the DM1 locus a series of 5′ *Alu* repeats are closely linked to the

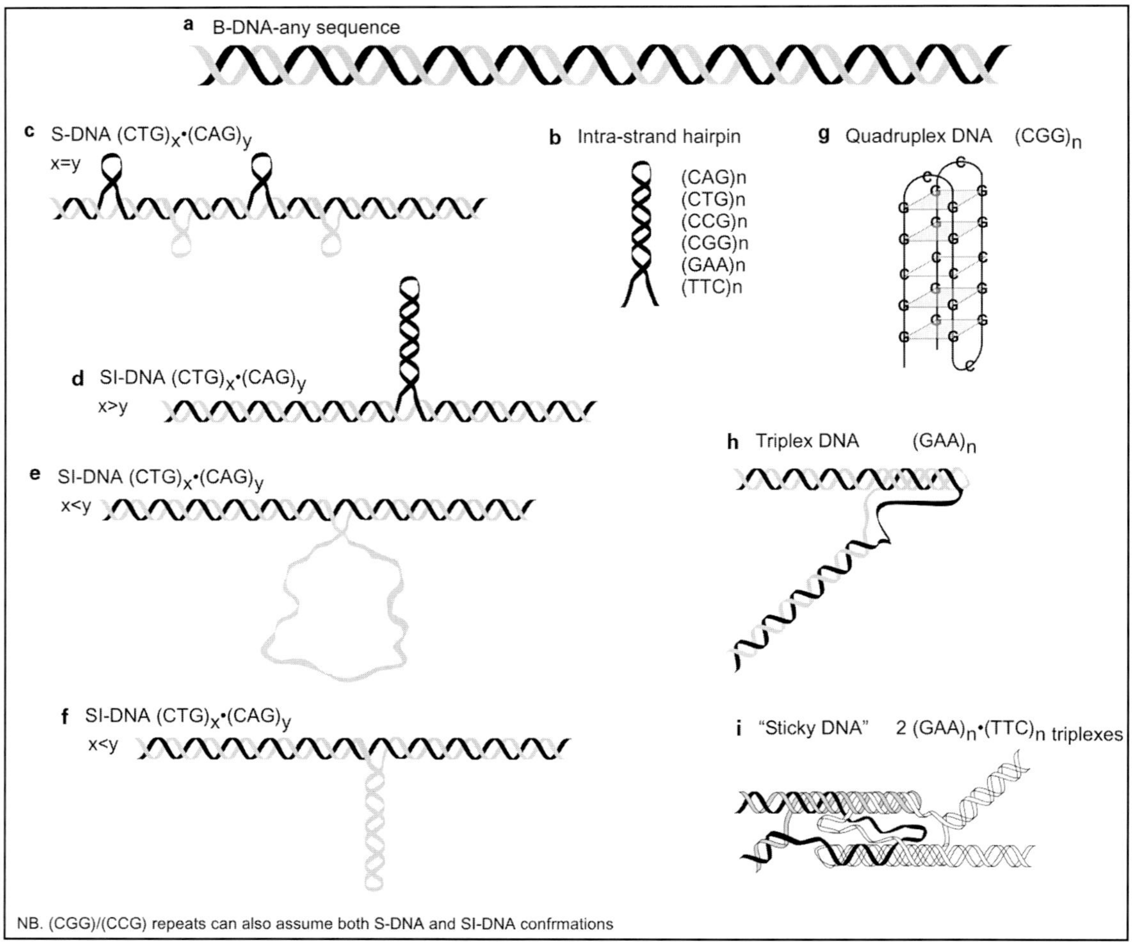

Fig. 5. Unusual DNA structure formed by TNRs. TNRs can form unusual DNA structures that may serve as mutagenic intermediates during cellular processes. (**a**) Normal B-DNA. (**b**) Intra-strand hairpins. (**c**) Slipped homoduplex S-DNAs are formed between two strands having the same number of repeats, and consist of short slip-outs (1–10 repeats) which vary in number and position along the tract. (**d–f**) Heteroduplex slipped intermediates SI-DNAs formed between two strands having different numbers of repeats. Most of the excess repeats slipped-out, in SI-DNAs, at preferred locations, forming defined three-way slip-out junctions. Slipped-out CAG and slipped-out CTG repeats were predominantly in the random-coil (**e**) and hairpin conformations (**f**), respectively. (**g**) Quadruplex structures are formed by single stranded repeats folding up into four-stranded structures and can be formed by $(CGG)_n$ repeats. (**h**) Intra-molecular triplex DNA can be formed by polypurine/polypyrimidine sequences, such as GAA repeats. (**i**) Two intramolecular triplexes may stably associate to form "sticky DNA".

expanded $(CTG)_n$ repeat (see below: Section II, *Haplotype and sequence elements flanking the unstable repeats*) (Imbert et al., 1993; Neville et al., 1994). The association between a contiguous repeat and expanded TNRs is also observed for the polymorphic $(GGC)_{11-17}$ repeat located ~1 kb 3′ of the SBMA $(CAG)_n$ repeats (Sleddens et al., 1993). In a number of populations (Irvine et al., 1995; Correa-Cerro et al., 1999; Lund et al., 2000; Lund et al., 2001; Chang et al., 2002), specific $(GGC)_n$ lengths are associated with the expanded repeat, such as (GGC)18 in Scandinavian populations (Lund et al., 2000). Similarly, a polymorphic $(CCG)_{1-2}$ is associated with the expanded $(CAG)_n$ repeats of SCA2 (Mizushima et al., 1999), although the contiguous repeat remains unchanged despite instability in the expanded CAG repeat tract. At the HD locus, several alleles of a polymorphic $(CCG)_n$ contiguous repeat are associated with large or expanded HD $(CAG)_n$ repeats (Rubinsztein et al., 1993). Interspersed between the HD disease $(CAG)_n$ tract and the nearby $(CCG)_n$ tract are CAA and CCA repeats respectively, which may have arisen from degeneracy. Minor changes in these variants (CAA to CAG and CCA to CCG) resulting in longer unstable repeats, have been documented for one family (Goldberg et al., 1995). Due to the rarity of such events, "interruption loss" does not appear to be a common process and therefore not a general mechanism for repeat instability at the HD locus. Ultimately, the smaller size of many contiguous repeat tracts may shield them from substantial instability. Contiguous repeats appear to be associated with and may even participate as *cis*-elements in the instability of disease-associated repeats.

Haplotype and sequence elements flanking the unstable repeats

Specific combinations of chromosomal markers (polymorphisms, restriction sites, microsatellites and/or insertions/deletions) are often associated with expanded TNR tracts. These haplotype markers, or tightly linked *cis*-elements, may ultimately contribute to instability. Alternatively, the initial mutational event generating the instability may have occurred on a specific chromosomal background. As many disease-associated unstable repeats are identified in small isolated populations, marker linkage may also result from limited marker availability (founder effect). The association of specific chromosomal backgrounds with repeat instability for a number of disorders suggests a contribution from *cis*-elements. Only recently has research begun to identify potential *cis*-elements, such as replication dynamics (Gourdon et al., 1997; Cleary et al., 2002; Zhang et al., 2002; Libby et al., 2003) that may explain the contribution of flanking sequences to instability.

One specific chromosomal background associated with numerous markers gives rise to repeat expansions at the DM1 locus (Imbert et al., 1993; Neville et al., 1994). Strikingly, genetic analysis of nine different flanking polymorphisms, spanning 30 kb and encompassing the unstable $(CTG)_n$, revealed only one combination of markers (haplotype) out of 2^9 potential combinations to be in complete allelic association with expanded repeats (Imbert et al., 1993; Neville et al., 1994). This haplotype includes a 5′ variable *Alu* repeat in which five Alu's are associated with $(CTG)_5$ and $(CTG)_{>19}$ including all but one reported case of disease chromosome, in contrast to the two Alu's associated with $(CTG)_{11-14}$ repeats. Only one case of a disease-length $(CTG)_n$ repeat from a non-DM1 haplotype has been reported, although this event was speculated to have arisen from gene conversion (Krahe et al., 1995). These data strongly suggest the expanded repeat occurs on one specific chromosomal background, with possibly several infrequent associated backgrounds (Goldman et al., 1995). The region covered by the DM1 haplotype is large (~30 kb) and contains several potential epigenetic factors, such as a CpG island and large insertion/deletion polymorphisms that could act as *cis*-elements. While the close association of the expanded repeat with the DM1 haplotype markers suggests that *cis*-elements within this region may contribute to instability, other *cis*-elements outside this region may also participate.

There is also a distinct connection with a specific chromosomal background for the $(CAG)_n$ repeats of SCA3/MJD. The first non-repeat base at the 3′ end of the repeat tract, normally G, was found to be C in 91% of expanded alleles (Igarashi et al., 1996). This terminal base is also C in 55% of non-affected individuals in the high-end repeat length range (27–40 repeats) (Limprasert et al., 1996). The combination of the expanded repeat with the C allele and a normal repeat with the G allele is more unstable than all other possible combinations, suggesting conversion as a mechanism for MJD instability (Igarashi et al., 1996). A worldwide population study has linked the majority of SCA3 expanded repeats to one intragenic three marker haplotype (Gaspar et al., 2001), although variation is observed in flanking microsatellites located further from repeat tract. However within a smaller population study, the same haplotype was not associated with larger normal alleles (Maciel et al., 1999). The sharing of one intragenic haplotype within the majority of worldwide SCA3/MJD families suggests the *cis*-elements that contribute to instability may be located within this intragenic region.

A limited set of haplotypes rather than a single haplotype is associated with most unstable repeat loci. At the HD $(CAG)_n$ locus, a flanking polymorphic $(CCG)_n$ repeat ranges from 7–12 repeats within the normal population (Rubinsztein et al., 1993), with $(CCG)_{<7}$ associated with large non-affected and full mutations alleles (Andrew et al., 1993; Almqvist et al., 1995). A survey of five polymorphic markers, including restriction sites, insertions/deletions and the $(CCG)_n$ repeat, spanning 150 kb of the HD locus revealed two haplotypes associated with the majority of expanded HD $(CCG)_n$ repeat tracts (Squitieri et al., 1994). On these HD haplotypes, normal and intermediate alleles (27–35 repeats) have larger average repeat lengths, with the majority of de novo cases of HD arising on these haplotypes (Maat-Kievit et al., 2001). A combination of *cis*-elements, not all associated with polymorphic markers, might contribute to HD instability thereby accounting for the multiple haplotypes associated with disease alleles.

The expanded CGG repeats at the FRAXA locus are linked to haplotypes different from those associated with non-affected alleles. Analysis of the markers flanking the FRAXA locus suggests that the transition from normal to premutation lengths (>40 repeats) occurs on one haplotype background (Chiurazzi et al., 1996). The same analysis indicates the majority of full mutation FRAXA alleles are associated with three major haplotypes, with the remaining alleles associated with rare haplotypes (for review see Chiurazzi et al., 1996). In addition, using a single nucleotide polymorphism linked to the expanded allele, Gunter et al. (1998) suggested that the *cis*-elements responsible for the instability should be located within 19.6 kb of the $(CGG)_n$ repeats. Analysis of the FRAXA haplotypes and the pattern of interruptions has lead to speculation that two mutation pathways exist for repeat expansion (Eichler et al., 1996). One pathway involves loss of AGG interruptions on multiple haplotypes, subdivided into sporadic interruption loss and frequent, recurrent interruption loss. These two subgroups may account for the diversity of haplotypes associated with FRAXA. In contrast the second pathway involves a single haplotype that maintains the two AGG interruptions yet remains associated with FRAXA. Haplotype association with expanded FRAXA repeats is dependent upon the populations in which linkage is performed (Crawford et al., 2002), and may be due to founder effects (Edelmann et al., 1996). The existence of multiple mechanisms leading to instability, each of which may or may not be haplotype associated, increases the complexity in detecting and identifying *cis*-elements at the FRAXA locus.

The association between specific chromosomal backgrounds and the expanded TNR tract varies greatly between disorders. For some disorders this association can be highly specific, as with the one haplotype associated with expanded SCA2 repeats across diverse ethnic populations (Takano et al., 1998; Choudhry et al., 2001; van de Warrenburg et al., 2002). Rather than a specific haplotype, a subset of haplotypes may be associated with the expanded repeat, as occurs with FRDA (Cossee et al., 1997), DRPLA (Yanagisawa et al., 1996) and SBMA (Lund et al., 2001). For FRDA, one major haplotype and numerous related haplotypes are associated not only with expanded repeat but also with high-end large non-affected alleles (Cossee et al., 1997). Interestingly, the frequency of these high-end non-affected alleles correlates well with the incidence of FRDA in several different populations (Labuda et al., 2000), suggesting a predisposing haplotype that serves as a reservoir for larger expansions. A number of disorders, including SCA1 (Jodice et al., 1993), SCA6 (Dichgans et al., 1999), SCA7 (Stevanin et al., 1999), and OPMD (Grewal et al., 1999; Scacheri et al., 1999; Blumen et al., 2000) are only associated with haplotypes within specific populations. In contrast, some disorders show no association with a particular haplotype, as with SCA8 (Juvonen et al., 2000; Moseley et al., 2000). Finally, a number of disorders are sufficiently rare as to prevent an accurate analysis of haplotype association (listed as "ND" in Tables 1 and 2), such as SCA10, SCA12, SCA17 and IDDM2. The diversity of linkage between chromosomal markers and the various TNRs suggests that flanking *cis*-elements may contribute to only a subset of TNR instabilities, and/or these elements may not be readily associated with identifiable chromosomal markers.

Other genomic repeats

The association between flanking sequences and repeat instability is observed for other repetitive elements. A single base transversion (G→C) 48 bp upstream of the MS32 minisatellite repeat is strongly associated with repeat instability. MS32 repeats with the transversion show reduced population variance and instability in sperm, although no effect on the rate or level of somatic instability was observed (Monckton et al., 1994). The unstable minisatellite of EPM1 is associated with one marker in 96% of EPM1-bearing chromosomes, although analysis of additional markers revealed no striking linkage (Lafreniere et al., 1997), except in a Finnish population (Virtaneva et al., 1997). The large DZ4Z repeats of FSHD are associated with one of two alleles of a distal 10-kb polymorphic segment (van Geel et al., 2002). Thus the association between chromosomal backgrounds and repeat instability extends beyond TNRs to other repetitive elements, a general contribution of flanking *cis*-elements to repeat instability is suggested. As mutation mechanisms likely differ between the various repetitive elements at different loci, the manner of the contribution by flanking *cis*-elements to repeat instability may differ.

Section III: Candidates for *cis*-elements in repeat instability

Cis-elements within and flanking repetitive elements contribute to repeat instability. Complex candidate *cis*-elements, such as nucleosomes, CpG methylation, and unusual DNA structures have been suggested to contribute to repeat instability. This section will highlight the participation of these *cis*-elements in repeat instability through examination of evidence from model systems, as well as patient-derived data.

Nucleosomes

The length threshold for TNR instability (30–40 repeats = 90–120 bp) approximates the length of DNA in a nucleosome (146 bp; Wolffe, 1994) – making nucleosomes and chromatin packaging an attractive candidate *cis*-element that may contribute to repeat instability. Alterations in chromatin packaging govern a number of processes including transcription, replication, repair and recombination that have been implicated in TNR instability. In a manner similar to TNR instability, chromatin packaging is regulated in a tissue-specific and developmental window-specific manner (Felsenfeld, 1996; Gribnau et al., 2000). Some repetitive sequences, such as $((A/T)_3NN(G/C)_3NN)_n$ (Sadowski, 1971), $(TATAAACGCC)_n$ (Widlund et al., 1999) and the triplet repeats $(CTG)_n$ and $(GGA)_n$ favor nucleosome assembly. In contrast, repetitive sequences such as telomeric $(TTAGGG)_n$ repeats (Cacchione et al., 1997), $(TTGA)_n$ (Cao et al., 1998), $(CGG)_n$ (Wang and Griffith, 1996) and $poly(A)_n$ (Shimizu et al., 2000; Suter et al., 2000) disfavor nucleosome assembly. The sequence preference for nucleosome assembly may relate to the metabolic processing (replication, repair or recombination) of these sequences. While there is strong support for a connection between repeats and nucleosome formation, the exact nature of the relationship to repeat instability is currently unknown.

There has been extensive analysis of the connection between TNRs (mostly $(CTG)_n$ and $(CGG)_n$) and their propensity for nucleosome assembly (reviewed in Griffith and Wang, 1998). $(CTG)_n$ repeats are considered the strongest known nucleosome positioning elements (Wang and Griffith, 1995). Nucleosome assembly on $(CAG)_n$ tract is disrupted by the presence of CAT interruptions (Pearson and Wang, unpublished), a phenomenon which may correlate with the genetic stabilizing effects of repeat interruptions in the CAG tracts of SCA1 and SCA2 (Chung et al., 1993; Chong et al., 1995; Pulst et al., 1996; Choudhry et al., 2001). In vivo, increased protection from DNaseI cleavage (Otten and Tapscott, 1995), a more compact chromatin state, and the loss of a nearby hypersensitive site are associated with the expanded DM1 $(CTG)_n$ repeat – suggestive of alternations in the chromatin structure surrounding the expanded repeat tract. In addition, *Alu* repeats, which are frequently proximal to unstable tandem repeats (Economou et al., 1990; Mahadevan et al., 1992; Chauhan et al., 2002), are known to influence nucleosome positioning, an effect which may extend to neighboring sequences (Englander and Howard, 1995). Furthermore, boundary elements closely connected with chromatin structure flank the DM1 locus and $(CTG)_n$ repeats (Filippova et al., 2001). Flanking DNA sequence, specifically the haplotype (including the *Alu* polymorphism) associated with the disease-associated DM1 repeat, may alter the chromatin environment at the DM1 repeats, thereby increasing the repeat tract's susceptibility to instability. Alternatively, the presence of an altered chromatin environment around expanded repeats may represent the effect of an increased number of repeats.

Chromosomal fragile sites appear cytogenetically to be regions of altered chromatin structure, which have been ascribed to regions containing nicks, gaps or under-replicated sequences. Unstable $(CGG)_n$ repeats are responsible for at least five rare fragile sites (FRAXA, FRAXE, FRAXF, FRA16A and FRA11B) suggestive of a connection between altered chromatin states and $(CGG)_n$ repeats. $(CGG)_n$ tracts greater than 50 repeats demonstrate strong nucleosomal exclusion (Wang et al., 1996), an observation that might explain the cytogenetic chromosomal fragile sites. CpG methylation of the expanded CGG repeats further enhances nucleosome exclusion two-fold beyond the inherent exclusion due to sequence alone (Godde et al., 1996; Wang and Griffith, 1996). Nevertheless, CGG expansions but not methylation, is sufficient for fragile site expression (Smeets et al., 1995). For FRAXA cells in vivo, regions proximal to the fragile X site are sensitive to DNAseI digestion (Nolin et al., 1988) and resistant to restriction digestion in regions flanking the CGG tract (Eberhart and Warren, 1996). Together, these results support an altered chromatin status of the FRAXA locus. The association between $(CGG)_n$ repeats and chromosomal breakpoints around the FRA11B locus (Jones et al., 2000) is the first demonstration of a direct link between fragile sites and chromosome breakage in vivo. In contrast to expanded CGG repeats, the expanded CTG repeats at the DM1 locus are not associated with and cannot be induced to form fragile sites (Werntges et al., 1986; Jalal et al., 1993). Surprisingly none of the cloned common fragile sites (FRA3B, FRA7G and FRA7H) have been associated with known repeats (reviewed in Sutherland and Richards, 1999). Taken together, this evidence indicates that some, but not all of expanded $(CGG)_n$ repeats can exclude chromatin packaging, which may be the basis for their association as chromosomal fragile sites.

If nucleosomes are actively contributing to, and not merely an effect of, repeat instability, they may act by modulating the fidelity of DNA replication, repair or recombination. Certain types of repair, such as nucleotide-excision repair, are strongly influenced by the presence of nucleosomes (for review see Lieber, 1999; Green and Almouzni, 2002). Recombination processes, such as site-specific immunoglobin V(D)J recombination, are also strongly regulated by chromatin structure (Lieber et al., 1987; McMurry and Krangel, 2000; Roth and Roth, 2000). Similar to disease-associated TNR instability, V(D)J recombination occurs in a locus-, tissue-, sequence- and development-specific manner, and is associated with alternative DNA structures (Lieber et al., 1987; McMurry and Krangel, 2000). Chromatin structure can control DNA replication initiation by regulating the accessibility of initiation factors to replication origins (for review see DePamphilis, 2000), or histone acetylation (Iizuka and Stillman, 1999). Although the exact nature of the connection between chromatin structure and site-specific replication initiation is unknown, chromatin remodelling represents an excellent mechanism for controlling DNA replication without the sequence specificity observed in lower organisms. Chromatin structure may influence, in a manner similar to replication, other cellular processes such as transcription, repair and/or recombination.

CpG methylation

In humans, DNA methylation occurs primarily, but not exclusively, at CpG sites (Woodcock et al., 1997; Dodge et al., 2002) or CpG islands, which are stretches of DNA greater than 200 nt that are enriched in 5′-CpG-3′ dinucleotides above that expected for its G+C content. Methylation of these CpG sites, like TNR instability, occurs in a tissue- and development-specific manner (Shiota et al., 2002). Alterations in chromosomal methylation status have been associated with tumour specific microsatellite instability (Herman et al., 1998; Toyota et al., 1999), as well as the genomic instability of ICF patients (Immunodeficiency, Centromeric instability and Facial abnormalities) (Xu et al., 1999). Hypomethylation, induced by either drug treatment (Haaf, 1995) or loss of DNA methyltransferases (Jeanpierre et al., 1993; Chen et al., 1998), also induces genomic instability. Although there is no general association amongst all repetitive elements and methylatable CpG sites, the most unstable TNRs are flanked by, or embedded in, CpG islands (Fig. 4e and Gourdon et al., 1997; Brock et al., 1999). In humans, DNA methylation affects numerous cellular DNA processes, including repeat instability, making it an excellent *cis*-element candidate.

The connection between CpG methylation and repeat instability is best characterized for the FRAXA $(CGG)_n$ repeat (Fig. 3). The distinct large expansions in FRAXA individuals are typically methylated and somatically stable. However, a subset of cases, the "high-functioning" FRAXA patients, display both heterogeneous lengths of $(CGG)_n$ expansions (full expansions to normal lengths) and reduced aberrant CpG methylation (Fig. 3a) (Hagerman et al., 1994), with unmethylated small expansions and methylated large expansions. The lack of aberrant methylation in these individuals is region specific, as the surrounding *Alu* repeats and other repetitive elements on X and Y-chromosomes remain methylated (Burman et al., 1999). Cultured fibroblasts from "high-functioning" FRAXA individuals also exhibit methylation-dependent somatic instability, with a bias for deletions (Wohrle et al., 1993; Burman et al., 1999; Salat et al., 2000; Wohrle et al., 2001). This contrasts with the stable methylated $(CGG)_n$ repeats in fibroblasts of typical FRAXA patients (Wohrle et al., 1998). Similarly, in FRAXA males, while most tissues contain methylated full expansions, the testes and sperm contain only unmethylated premutations (Reyniers et al., 1993; Malter et al., 1997; Tassone et al., 1999). In FRAXA cells, the hypermethylated repeat demonstrates similar patterns of discrete repeat lengths within different fetal tissues (Wohrle et al., 1993). Conversely, the unmethylated repeats are present as smears of repeat lengths with extensive changes. This association with methylation status suggests that somatic instability for FRAXA may arise as the result of the selective deletion of unmethylated $(CGG)_n$ repeats.

Early human embryogenesis is a period of considerable change in terms of both DNA methylation and repeat stability. Cells undergo massive waves of de novo methylation and demethylation during early embryogenesis (6–8 weeks) (Razin and Shemer, 1995; Reik and Dean, 2001), roughly at the same time that FRAXA repeat instability is postulated to occur (Devys et al., 1992; Reyniers et al., 1999; Tassone et al., 1999). Changes in the methylation status of the gametes and developing embryo are regulated differently between the male and female germ cells (Trasler et al., 1990; Bestor and Tycko, 1996; Bestor, 1998; Mertineit et al., 1998; Trasler, 1998), suggestive of a possible source for the FRAXA parent-of-origin effect. For the normal FRAXA locus, the methylation pattern is established by at least six weeks of development (8–21 weeks post-conception) (Luo et al., 1993). The lack of methylation of the expanded CGG repeat in oocytes (Malter et al., 1997) suggests that expansion may occur prior to abnormal methylation (Sutcliffe et al., 1992). In the testis of a FRAXA male fetus, premutation alleles become evident as early as the 17th week of development (Malter et al., 1997). These alleles in adult testis and sperm are unmethylated (Reyniers et al., 1993; Tassone et al., 1999), while full mutations remain methylated, suggesting methylation may participate in the deletion process. Limited post-zygotic mitotic instability, observed as tissue-specific differences in expansion size within the same patient (Devys et al., 1992; Sutcliffe et al., 1992; Reyniers et al., 1999; Taylor et al., 1999), suggests that instability may also occur after the methylation pattern is established. In addition, methylation-independent instability has been observed in human-mouse hybrids (Burman et al., 1999) in which differentiation appears to trigger instability in methylated expanded FRAXA repeats. It is possible, given that the exact mechanism of DNA methylation upon instability is unknown, that methylation may only influence instability under specific conditions, such as early embryogenesis.

The effects of DNA methylation may be multifaceted and extend to regions beyond methylatable sites (Hodges-Garcia and Hagerman, 1992; Zacharias, 1993; Nichol and Pearson, 2002). Recently, our laboratory has examined the association of CpG methylation with the stability of various repetitive elements, including mono-, di-, tri-, penta- and minisatellite sequences (Nichol and Pearson, 2002). In general, methylation appears to reduce repeat instability, observed as a lower frequency and magnitude of deletion events, although a minority of repeat sequences were destabilized or unaffected by methylation. The stabilizing effect of CpG methylation extended to unmethylated repeats sequences, such as $(CTG)_n$, $(CA)_n$ and $(CT)_n$ repeats, suggesting methylation of flanking sequence may affect repeat instability. It is possible that, in addition to chromosomal context, the methylation of flanking sequences can determine the differential susceptibility to instability for various repeat loci (Dietmaier et al., 1997; Boland et al., 1998; Duval et al., 2002). Methylation stabilization of repeat instability was observed in bacteria and primate cells, two evolutionarily divergent systems that have different protein machinery for methylation, replication, repair and chromatin packaging. Accordingly, the effect of methylation on repeat instability may be transmitted through a component common to both systems, the methylation status of the DNA.

The methylation status of DNA may alter the ability of nearby repeats to form mutagenic intermediates, similar to the ability of flanking sequence to strongly influence the extrusion of various DNA structures (including cruciforms, Z-DNA and triplex DNA) at other DNA sequences (Sullivan and Lilley, 1986). Methylated DNA has an increased melting temperature (Zacharias, 1993), which may affect both the rate and fidelity of DNA polymerase synthesis. Methylation is also closely linked to chromatin structure, as the hypermethylated 5′ end of FMR1 is known to be associated locally with histone deacetylation and chromatin remodeling (Coffee et al., 1999). It has been suggested that methylation modifies genetic stability through the action of *trans*-factors (Bird and Wolffe, 1999) sensitive to the methylation status of DNA (Muller-Hartmann et al., 2000). Thus, the effects of methylation on repeat instability may arise from the combined efforts of both *cis*-elements and *trans*-factors (Fig. 4d) in altering DNA metabolism at or around the repetitive sequences.

Unusual DNA structures

Many repetitive elements, including TNRs, possess the ability to form unusual DNA structures (Fig. 5) that may contribute to repeat instability. In vitro, a number of TNRs form intra-strand hairpins (Fig. 5b) (Gacy et al., 1995; Mitas et al., 1995; Yu et al., 1995; Darlow and Leach, 1998). Hairpins are critical components of the more complex slipped-strand DNA structure. Long disease-relevant TNR tracts form slipped-structures which display unique structural features, including slip-out junctions seen as multiple or individual slip-outs which can assume an intra-strand hairpin or random coil conformation (Pearson et al., 2002). Many of these structures are not formed by shorter repeat lengths. For TNRs, both instability and the propensity for slipped-structure formation are sensitive to repeat tract length and purity (Pearson and Sinden, 1996, 1998; Pearson et al., 1998). The ability of large TNRs to form unusual DNA structures appears intrinsically linked to their instability.

There are two types of slipped DNAs that TNRs can form (Fig. 5): (1) homoduplex slipped DNAs (S-DNA), formed between two strands having the same number of repeats; and (2) heteroduplex slipped intermediate DNAs (SI-DNA), formed between two strands having different numbers of repeats. S-DNAs (Fig. 5c) are composed predominantly of short slip-outs (1–10 repeats), which vary in number and position along the tract (Pearson et al., 2002). The pattern of S-DNAs differs for the same $(CTG)_n$ lengths generated from the DM1 or SCA1 loci (Pearson et al., 1997, 1998), supporting a contribution for flanking sequence in S-DNA formation. The number of repeats contained within the slip-outs of S-DNAs (1–10 repeats) are similar to the majority of intra-tissue repeat length changes observed in non-proliferating neural brain tissues of HD, DRPLA and SCA3 patients (Pearson et al., 2002). S-DNAs may represent the mutagenic intermediates of error-prone DNA metabolism (repair or recombination) in these quiescent cells.

In contrast to S-DNAs, SI-DNAs are generated from complementary repeat sequences differing in repeat number $(CTG)_x/(CAG)_y$. This difference results in a limited set of isomers with unique slipped-out regions at specific locations (off center of the shorter repeat tract-see Pearson et al., 2002). Excess repeats extrude from the duplex at three-way slip-out junctions. Slipped structures containing an excess of CAG repeats assume a mixture of single-stranded random coil structures (Fig. 5e) as well as intrastrand hairpins (Fig. 5f). In contrast, slipped structures containing an excess of CTG repeats assume only distinct intrastrand hairpins (Fig. 5d). Unlike S-DNAs, flanking sequence does not appear to affect the formation of SI-DNAs (Pearson et al., 1997, 1998). The size of the slip-out in SI-DNAs can be as great as the total difference between the repeat tracts on either strand. This length range is similar to the large meiotic or somatic jumps in repeat length (20–500 repeats) observed in patients, suggesting a role for SI-DNAs in replication-mediated instability. SI-DNAs may be formed by replication slippage (Pearson et al., 1997), during unequal crossover recombination between sister chromatids (Smith, 1976), or homologous recombination (Richard et al., 2000) (Fig. 6c). The correlation of slip-out size in SI-DNAs with observed tract length changes in proliferating tissues, and slip-out size of S-DNAs with length changes in non-proliferating tissues, suggest that they may act as mutagenic intermediates in proliferative and non-proliferative processes of instability respectively.

Quadruplex DNA is another unusual structure formed by TNRs, as well as, other G-rich repetitive elements. These structures are formed by single-stranded repeats folding up into four-stranded structures (Fig. 5g). Quadruplex DNA has been observed in human telomeric repeats (Parkinson et al., 2002), EPM1 repeats (Saha and Usdin, 2001), and in vivo by antibody staining in nuclei of certain protozoans (Schaffitzel et al., 2001). Some repeats, such as FRAXA and FRAXE $(CGG)_n$, are capable of forming both quadruplexes and hairpins (reviewed in Darlow and Leach, 1998). In addition, quadruplexes formed by $(CGG)_n$ repeats are inhibited by interruptions (Weisman-Shomer et al., 2000). Like slipped-structures, quadruplex DNA may act as a *cis*-element through perturbation of DNA replication, repair and/or recombination processes.

The $(GAA)_n$ repeats of Freidreich's ataxia (FRDA) are associated with the formation of hairpins (Heidenfelder et al., 2002), parallel DNA (LeProust et al., 2000), and intermolecular triplexes (Fig. 5h), two of which may stably associate to form the unusual DNA structure termed "sticky DNA" (Fig. 5i) (Ohshima et al., 1999). Sticky DNA formation occurs in a length-dependent manner (Sakamoto et al., 1999). The presence of several GGA interruptions (more than 20%) within long GAA tracts disrupts structure formation, enhances genetic stability and ultimately interferes with the ability of the structures to inhibit transcription (Sakamoto et al., 1999). The unique intermolecular structural properties of the $(GAA)_n$ tracts may be related to the recessive nature of FRDA, a characteristic unique to this TNR-associated disease.

The formation, recognition and processing of unusual DNA structures may occur during numerous biological processes. The presence of single-stranded DNA, formed during lagging-strand replication, transcription, recombination, DNA metabolism and DNA "breathing" (Fig. 6), may facilitate formation of these structures. In fact, long stretches (up to 10 kb) of single-stranded DNA have been detected in both replicating (Bjursell et al., 1979) and non-replicating human cells (Henson, 1978), as well as in increasing quantity within non-proliferating tissues of older mice (Price et al., 1971; Chetsanga et al., 1975; Nakanishi et al., 1979). The biophysical features of unusual DNA structures may critically determine whether they are recognized and correctly processed by replication, repair, or recombination proteins (Fig. 4d). Proteins such as SSB and MSH2 are known to associate with slipped-structures, both preferentially binding CAG slip-outs over CTG slip-outs (Pearson et al., 1997). Short hairpins of CTG or CAG repeats escape repair in yeast, although it is unknown if they do so in mammalian cells (Moore et al., 1999). The formation and biophysical nature of unusual DNA structures may determine if they act as *cis*-elements, as well as how they interact, correctly or aberrantly, with *trans*-factors.

Section IV: Mechanisms of instability

Despite the growing number of repeat-associated disorders, little is currently known about the underlying mechanisms of repeat instability. Repeat instability has been associated with both the processes of genome duplication (replication) and genome maintenance (repair or recombination) (Fig. 6). The observation of instability in non-proliferative tissues (brain and/or gonads) of HD, SCA1 and DM1 patients (Telenius et al., 1994; Thornton et al., 1994; Chong et al., 1995; Ishii et al., 1996) and in the haploid germ cells of SCA1 and HD transgenic mice (Kaytor et al., 1997; Kovtun and McMurray, 2001) suggests that metabolic processes independent of cell proliferation contribute to repeat instability. Differences in somatic CTG instability were observed between various tissues (Lia et al., 1998) and in proliferating cells lines derived from different tissues of a DM1 transgenic mouse (Gomes-Pereira et al., 2001) suggesting instability may not be strictly dependent upon cell division rates. The repeat instability in DM1 proliferative tissues (Martorell et al., 1998) and during the rapid cellular growth of embryogenesis supports duplication-mediated instability. Our laboratory has recently demonstrated that cellular proliferation is required for CTG instability in primary DM1 fetal fibroblasts (Pearson, unpublished data), suggesting a role for DNA replication or proliferation-associated metabolic processes (repair and recombination).

Processes implicated in repeat instability include replication slippage, the direction of replication fork progression, lagging strand errors, Okazaki fragment processing, mismatch repair, gap-filling, double-strand break repair and recombination. It is very likely that multiple processes contribute to repeat instability in patients, such that different mechanisms participate at different loci, in different tissues, and/or during different developmental stages. A particular mechanism implicated by any one model system may reflect the instability in only one tissue, at only one locus, or even during a specific developmental stage. Therefore, in turn, each of these biological processes may be influenced by *cis*-elements within, as well as flanking the repeat tract, in a similar tissue-, locus-, and developmental-specific manner, resulting in the highly specific repeat instability observed in humans.

Replication

The occurrence of repeat instability in highly proliferative tissues for some TNR-disorders is supportive of a connection between repeat stability and replication. For DM1, instability occurs in, but is not limited to, proliferating cells (Martorell et al., 1995, 1998; Wong et al., 1995). It also occurs during periods of high cell proliferation, such as embryogenesis and fetal development (Martorell et al., 1995; Wohrle et al., 1995), and may cease or slow following differentiation and muscle cell

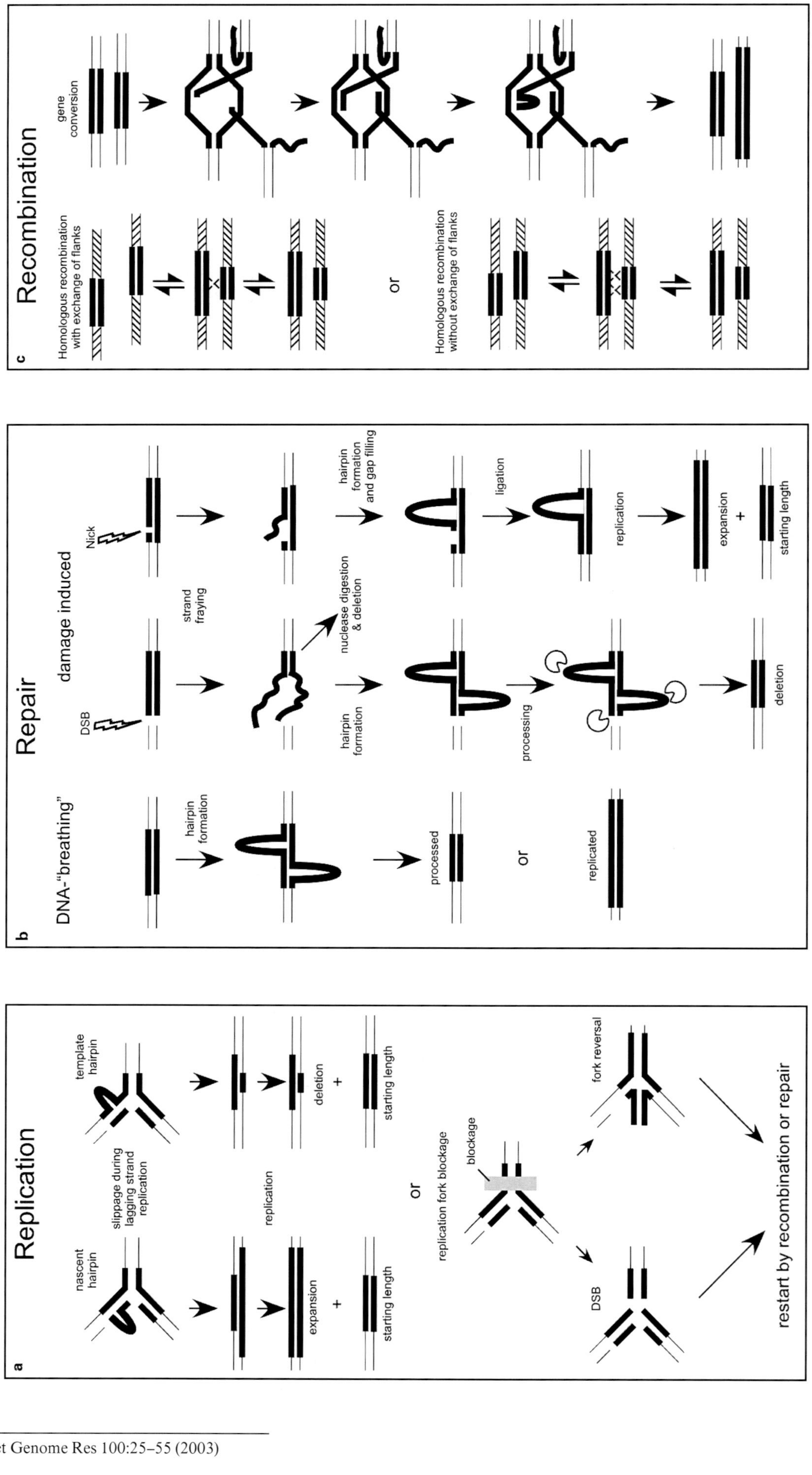
a
Replication
nascent hairpin
slippage during lagging strand replication
template hairpin
replication
expansion
+
starting length
deletion
+
starting length
or
replication fork blockage
blockage
DSB
fork reversal
restart by recombination or repair
b
DNA-“breathing”
hairpin formation
processed
or
replicated
Repair
damage induced
DSB
strand fraying
hairpin formation
nuclease digestion & deletion
processing
deletion
Nick
hairpin formation and gap filling
ligation
replication
expansion
+
starting length
c
Recombination
Homologous recombination with exchange of flanks
or
Homologous recombination without exchange of flanks
gene conversion

quiescence (Anvret et al., 1993; Thornton et al., 1994; Zatz et al., 1995). Larger expansions in tumors of DM1 patients relative to their non-neoplastic tissue suggest that expansions may also occur during acquired cell proliferation (Kinoshita et al., 2002; Osani et al., 2000). Bacterial (Kang et al., 1995; Samadashwily et al., 1997) and yeast (Freudenreich et al., 1997; Miret et al., 1998; Schweitzer and Livingston, 1998) models of TNR instability support a role for DNA replication, in that the direction of replication through the TNR tract affects its stability. Proliferation also occurs in the majority of cell-culture models in which instability was observed (see above: Section I, *Human cell culture and repeat instablilty*). Strand slippage and/or structure formation (see above: Section III, *Unusual DNA structures*) within the repeat tract during replication is suspected to contribute to instability (Fig. 6a). In this manner, instability may be influenced by the sequence of the repeat (CTG vs. CAG) comprising the leading or lagging strand, and its propensity to form mutagenic intermediates (Fig. 4c).

Some models of somatic TNR instability involve aberrant processing of Okazaki fragments initiated within the repeat tract (Gordenin et al., 1997; Spiro et al., 1999; Henricksen et al., 2000). Repeat instability may require the sequence-specific process of either Okazaki initiation (RNA priming and pol$\alpha \rightarrow$ polδ switching) and/or Okazaki processing (RNA removal, gap filling and ligation) to occur within the repeat tract. This hypothesis is supported by the observation that the TNR length stability threshold (~ 30–40 repeats = 90–120 bp) is close to the length of an Okazaki fragment (~ 135–145 bp) (Anderson and DePamphilis, 1979; Hay and DePamphilis, 1982). In addition, work with yeast rad27/FEN1, an enzyme that processes the RNA primers of Okazaki fragments (Waga et al., 1994; Bambara et al., 1997), indicates that instability involves Okazaki fragment processing within the repeat (Freudenreich et al., 1998; Schweitzer and Livingston, 1999). Biochemically, FEN1 is unable to recognize and process hairpin substrates composed of $(CTG)_n$ or $(CAG)_n$ repeats (Spiro et al., 1999; Henricksen et al., 2000). Pausing associated with polα/polδ switching (Nethanel and Kaufmann, 1990; Waga et al., 1994) may allow for slippage of either template or nascent strands (Fig. 6a), with the sequence at this location influencing the probability and identity of the mutagenic structures generated. A number of factors may influence Okazaki initiation, including sequences in the non-repeating flanking regions, differences in density of preferred Okazaki priming hotspots (3′-PuT-5′; Waga and Stillman, 1994) at either side of the repeat tract, chromatin structure, and methylation status.

Fig. 6. Replication, repair and recombination. (**a**) Replication mediated instability may occur via replicative slippage during lagging strand synthesis resulting in hairpin formation (other structures are possible). The location of the hairpin (template vs. nascent) and possibly the sequence of the hairpin may determine if and how cellular proteins recognize and repair these structures. Replication across the hairpin structure results in expansion for nascent hairpins (left pathway) and deletions for template hairpins (right pathway). Alternatively the formation of unusual DNA structure by repeats may stall the advancing replication fork (bottom pathway), resulting in double-strand breaks (DSB) or fork reversal. Replication restart may occur in these situations through repair or recombination (see middle and right panel). (**b**) Repair-mediated instability may occur following the formation of slipped-structure DNAs by breathing and aberrant processing, DSBs or nicks within or around the repeat. The resulting repeat end may be more susceptible: to nuclease degradation (middle path); to end-fraying leading to increased single-strand degradation of the repeat (middle path); to end-fraying allowing for the formation of unusual structures, which are recognized and processed (left path); hairpin formation and reiterative synthesis (right path). (**c**) Recombination-mediated instability may occur independent or in conjunction with processes of replication and/or repair. Homologous recombination between allelic repeat tracts may occur with the exchange of flanking sequence (top left) or without exchange of flanks (bottom left). Gene conversion (right) has been postulated to occur following DSB or replication fork blockage. Single-strands from the broken repeat invade the sister tract, allowing for completion/restart of synthesis. During this period of synthesis hairpin structures or reiterative synthesis may occur, prompting repeat length changes. The resulting products are then ligated, allowing for replication to restart. Note: Owing to the commonality between these pathways, considerable overlap may occur so that the presented pathways are simplified to reduce confusion.

It is possible that the replication fork may become blocked as it passes through the triplet repeat tract (Freudenreich et al., 1997; Samadashwily et al., 1997; Pelletier et al., 2003). Replication pausing at TNRs has been observed both in vitro (Ohshima and Sinden, 1998), using a number of polymerases including human pol β (Kang et al., 1995) and in vivo, in a length-dependent manner in yeast and bacteria (Samadashwily et al., 1997; Pelletier et al., 2003). This blockage may be mediated by a variety of mutagenic intermediates containing slipped-structures, formed either between the nascent and template strands or between the unreplicated template strands (Ohshima and Wells, 1997; Samadashwily et al., 1997). Elimination of the blockage and subsequent resumption of fork progression may occur by a number of mechanisms, including replication fork reversal (Flores et al., 2001), double-strand break-mediated recovery (Haber, 1999; Michel, 2000; Saintigny et al., 2001), and/or recombination (Pearson and Sinden, 1996; Richard and Paques, 2000) (Fig. 6). These attempts to restart replication within a repetitive sequence may facilitate repeat instability.

Using a primate model system, we recently demonstrated that replication-mediated instability for a premutation length DM1 $(CTG)_n$ repeat is dependent, in part, upon the location of the repeat (Fig. 4a) within the single stranded Okazaki initiation zone (Cleary et al., 2002). Factors that influence the selection of sites for initiation, and ultimately processing of Okazaki fragments, will affect the frequency with which Okazaki fragments will contain TNR repeats, as well as the number of repeat units at their termini. Based in part upon this evidence, it has been suggested (Cleary et al., 2002; Mirkin and Smirnova, 2002) that disease-associated repeat instability may be the result of three possible scenarios (Fig. 4f): (1) Replication origin switch, in which alterations in replication origin usage on either side of the repeat tract occur between diseased and non-affected chromosomes. This concept is supported by the differential stability of various repeat tracts depending upon the direction of replication (Kang et al., 1995; Freudenreich et al., 1997; Panigrahi et al., 2002). (2) Replication origin shift, in which usage of the replication origin on one side of the repeat tract shifts between diseased and non-affected chromosomes. It is possible that the *Alu* polymorphism (insertion of three Alu's) associated with the expanded DM1 allele (Neville et al., 1994) could alter the location of replication origins but not the direction of replication. (3) Replication fork shift, in which the progress of the advancing replication fork shifts relative to the repeat, due to the presence of an epigenetic element, such as chromatin packaging around the repeat. Such modification could be mediated by CTCF sites on either side of the DM1 repeat tract (Filippova et al., 2001). All three mechanisms are supported by our primate-based work (Cleary et al., 2002). These mechanisms may each occur in different individuals, tissues, loci, and/or developmental stages. Not surprisingly, many of the elements known to affect Okazaki initiation and replication fork progression, such as chromatin structure and unusual DNA structures, are also candidate *cis*-elements for repeat instability.

Transcription

As all disease-associated repeats are transcribed, the process of transcription is a potential *cis*-element influencing repeat instability. The induction of transcription through TNRs in bacterial model systems (Bowater et al., 1997; Schumacher et al., 2001) results in increased repeat instability, favoring deletion events. This effect of transcription is both sequence- and length-dependent, occurring only when the transcribed strand is also the lagging strand template. In addition, both tri- and dinucleotide repeats cloned into transcription units show altered stability (Jaworski et al., 1989; Shimizu et al., 1996). Collisions between transcription and replication forks have been shown to result in replication fork pausing (Liu and Alberts, 1995; Elias-Arnanz and Salas, 1997), increased recombination (McGlynn and Lloyd, 2000), and repeat instability (Schumacher et al., 2001). In *E. coli*, transcription of mononucleotide repeats blocks their subsequent replication (Krasilnikova et al., 1998). Transcription may influence repeat stability either through the collision of replication and transcription machineries, by promoting the formation of mutagenic structural intermediates, or through aberrant transcription-coupled repair (Fig. 4c).

Repair

DNA repair, which may occur in conjunction with or independent of DNA replication, has also been implicated in repeat instability (Sinden, 2001). For some TNR loci, the lack of association between somatic repeat instability and tissue proliferative capacity suggests that instability might occur by metabolic processes independent of DNA replication, such as maintenance repair. Such repair processes may include proteins typically associated with replication, i.e. MMR proteins (see below). Error-prone repair of nicks, gaps and double-strand breaks may lead to TNR instability. In yeast, TNRs may be susceptible to double-strand break (DSB) formation through processes such as polymerase slippage or endonuclytic cleavage of secondary structures (Freudenreich et al., 1998; Cohen et al., 1999; Jankowski et al., 2000). In yeast the formation of DSBs within or proximal to the repeats is length dependent (Freudenreich et al., 1998). It has been suggested that subsequent repair of these breaks can lead to instability of the repeat tract (Jankowski et al., 2000; Richard et al., 2000). The use of a donor sequence containing an expanded CTG repeat to repair a DSB, lead to the integration and expansion of the TNR tract (Richard et al., 2000). However, in that study the initial DSB substrate did not harbour any CTG/CAG repeats. Taken together, this evidence suggests a role for DNA repair either in association with DNA replication, or alternatively as a form of maintenance repair, that occurs independently, but not exclusive of, replication in repeat instability.

The processing of errors incurred during DNA replication may facilitate repeat instability, through either escaped repair or incorrect processing (Fig. 6b). However, conflicting conclusions regarding the role of DNA repair in repeat instability have been reported in various model systems. These results include decreased (Jaworski et al., 1995) and increased (Schumacher et al., 1998) levels of deletion events in bacteria, no effect on instability (Miret et al., 1997; Miret et al., 1998; Rolfsmeier et al., 2000) and destabilization (Schweitzer and Livingston, 1997) in yeast systems. A requirement for the participation of MSH2 (Manley et al., 1999b; Kovtun and McMurray, 2001; Saavouret et al., 2003) and MSH3 (van Den Broek et al., 2002) in CTG/CAG instability has been provided by transgenic mice. Recent results from a DM1-transgenic mouse revealed a requirement for MSH2 for expansions but not deletions of the CTG/CAG repeat (Savouret et al., 2003). The differences in the role of MSH2 in CTG/CAG instability between the DM1 (Savouret et al., 2003) and HD (Manley et al., 1999b; Kovtun and McMurray, 2001; van Den Broek et al., 2002) mice may be due to mouse strain-specific differences (van Den Broek et al., 2002). The requirement of MMR for TNR instability may be similar to its requirement for site-specific hypremutation of the immunoglobin genes (Cascalho et al., 1998). HNPCC tumors which are deficient in MMR display genome-wide instability at simple tandem repeats, with minor variations observed in TNRs at the DM1, HD and FRAXA loci (Kramer et al., 1996; Goellner et al., 1997). However, the lengths examined were less than the TNR stability threshold (~ 34 repeats). Thus, defects in human MMR are capable of influencing the stability of simple repeats and possibly TNRs. While an MMR defect may not cause the initial repeat expansion event, MMR may contribute, at least for mice and possibly humans, to the low levels of instability observed in the expanded repeats of somatic and germ tissues.

Cis-elements which influence repeat instability have also been shown to influence repair. Flanking sequence is known to influence the structure and biophysical stability of repair substrates, such as base-base mismatches (Fazakerley et al., 1986; Werntges et al., 1986; Fagan et al., 1996). Similarly, S-DNAs differ for the same $(CTG)_n$ lengths generated from the DM1 or SCA1 loci (Pearson et al., 1997; Pearson et al., 1998), also supporting a contribution for flanking sequence in S-DNA formation. The efficiency of the repair process itself can also be influenced in a structure-specific manner by sequences flanking the damaged site, as demonstrated for MMR and base-excision repair in bacteria (Seibert et al., 2002). Genomic context contributes to both efficiency and strand-specificity of transcription-coupled repair (Feng et al., 2002). Other candidate *cis*-elements, such as chromatin structure (see above: Section III, *Nucleosomes*), are also important components in determining repair efficiency (Li et al., 1999). *Cis*-elements may contribute to repeat instability by altering the efficiency or specificity of repair of mutagenic structural intermediates formed at the repeat tract, in conjunction with genome replication, or separately during genome maintenance.

Recombination

Recombination, in the form of either gene conversion or crossing over (Fig. 6c), has been suggested as an alternative mechanism for the generation of repeat instability. Based on patient data, gene conversion has been suggested to account for the mix of chromosomal markers between normal and premutation alleles in some FRAXA cases (van den Ouweland et al., 1994), as well as the loss of the FRAXA full mutation (Losekoot et al., 1997). Similar gene conversion events resulting in an apparent deletion to non-affected lengths of $(CGG)_n$ repeats may account for the very rare revertants that occur in premutation FRAXA carrier offspring (Brown et al., 1996). Rare events suggestive of recombination (gene conversion) have also been reported for DM1 as; deleted alleles have been found on the DM1 risk haplotype (Brunner et al., 1993), expanded DM1 alleles occur on the DM1 non-risk haplotype, and linkage disequilibrium is seen between haplotype variations (Krahe et al., 1995). Recombination is also suspected in the majority of the polyalanine expansions, as these repeats appear to expand by duplicating segments of the repeat tract, including interruptions (Goodman et al., 1997; Brais et al., 1998; Nakamoto et al., 2002). The presence of interruptions in the $(GCG)_n$ repeats would also be expected to reduce the chance of replicative slippage, thereby implicating other methods of repeat expansion (Goodman et al., 1997). Evidence for recombination has been observed through germline analysis of human minisatellites, suggesting gene conversion as a source of instability (Jeffreys et al., 1994; Buard et al., 1998). While there is evidence suggesting recombination may participate in rare instances of repeat instability, a defined role has yet to be determined.

Both yeast and bacterial models suggest a potential role for recombination (reviewed in Jakupciak and Wells, 1999) in repeat instability. Work in bacteria has shown the repair-associated processes such as inter- and intra-strand recombination and gene conversion can strongly influence the expansions of CTG repeats (Jakupciak and Wells, 1999; Jakupciak and Wells, 2000; Napierala et al., 2002; Pluciennik et al., 2002). Such processes may correlate to the rare deletion events suggestive of recombination observed in various human TNR disorders. Owing to the difficulty in detecting and analyzing repeat-specific recombination events within the human genome, model systems are an essential tool for analysis of the extent of the involvement of recombination in repeat instability.

Recombination, like DNA repair, may act in concert with genome duplication (DNA replication). The formation of unusual DNA structures by repetitive elements has been proposed to stall replication forks, which can result in strand breakage (Fig. 6a). Recombination may be one possible mechanism for repair of these stalled replication forks (Fig. 6c). Moreover, *cis*-elements such as chromatin packaging and CpG methylation, which affect DNA replication and repair, may also alter recombination and contribute to repeat instability. In a similar manner, recombination may participate in repeat instability in a locus-, tissue- or development-specific manner.

Conclusions

The increasing number of human disorders caused by the instability of repetitive elements raises the need to better understand the mechanisms underlying this serious mutation. The large amount of patient-derived data, combined with data from model systems, supports *cis*-elements as contributing factors in repeat instability. In diverse tissues at different developmental times and at specific loci, repetitive elements display variable levels of instability, suggesting vastly different mechanisms may be responsible for repeat instability amongst the disease loci. A particular mechanism implicated by any one model system may reflect the instability at only one locus, in only one tissue, or during a specific developmental stage. Given the complex nature of dynamic mutations, it is likely that both *cis*-elements and *trans*-factors act in concert to produce these tissue-, development- and locus-specific instabilities. Based on evidence in humans, it is clear that *cis*-elements within the repeat tract, such as sequence, tract length and purity contribute to instability. The involvement of *cis*-elements flanking the repeat, including methylation, nucleosome structures and other epigenetic elements, is strongly supported by evidence from humans, as well as the results of several model systems. The discovery of the identity of these flanking *cis*-elements may help facilitate the understanding of the mechanism behind repeat instability, and perhaps lead to future avenues of disease prevention.

Acknowledgements

Owing to the page constraints, we wish to apologize to any individuals whose work is not completely referenced. The authors wish to thank the members of the Pearson laboratory for their input and support.

References

Abeliovich D, Lerer I, Pashut-Lavon I, Shmueli E, Raas-Rothschild A, Frydman M: Negative expansion of the myotonic dystrophy unstable sequence. Am J hum Genet 52:1175–1181 (1993).

Aiyar A, Tyree C, Sugden B: The plasmid replicon of EBV consists of multiple *cis*-acting elements that facilitate DNA synthesis by the cell and a viral maintenance element. EMBO J 17:6394–6403 (1998).

Almqvist E, Spence N, Nichol K, Andrew SE, Vesa J, Peltonen L, Anvret M, Goto J, Kanazawa I, Goldberg YP, et al: Ancestral differences in the distribution of the delta 2642 glutamic acid polymorphism is associated with varying CAG repeat lengths on normal chromosomes: insights into the genetic evolution of Huntington disease. Hum molec Genet 4:207–214 (1995).

Anderson S, DePamphilis ML: Metabolism of Okazaki fragments during simian virus 40 DNA replication. J biol Chem 254:11495–11504 (1979).

Andrew SE, Goldberg YP, Kremer B, Telenius H, Theilmann J, Adam S, Starr E, Squitieri F, Lin B, Kalchman MA, et al: The relationship between trinucleotide (CAG) repeat length and clinical features of Huntington's disease. Nature Genet 4:398–403 (1993).

Anvret M, Ahlberg G, Grandell U, Hedberg B, Johnson K, Edstrom L: Larger expansions of the CTG repeat in muscle compared to lymphocytes from patients with myotonic dystrophy. Hum molec Genet 2:1397–1400 (1993).

Aoki M, Abe K, Tobita M, Kameya T, Watanabe M, Itoyama Y: Reduction of CAG expansions in cerebellar cortex and spinal cord of DRPLA. Clin Genet 50:199–201 (1996).
Arcot SS, Wang Z, Weber JL, Deininger PL, Batzer MA: *Alu* repeats: a source for the genesis of primate microsatellites. Genomics 29:136–144 (1995).
Armour JA, Harris PC, Jeffreys AJ: Allelic diversity at minisatellite MS205 (D16S309): evidence for polarized variability. Hum molec Genet 2:1137–1145 (1993).
Ashizawa T, Dubel JR, Harati Y: Somatic instability of CTG repeat in myotonic dystrophy. Neurology 43:2674–2678 (1993).
Ashizawa T, Anvret M, Baiget M, Barcelo JM, Brunner H, Cobo AM, Dallapiccola B, Fenwick RG Jr, Grandell U, Harley H, et al: Characteristics of intergenerational contractions of the CTG repeat in myotonic dystrophy. Am J hum Genet 54:414–423 (1994).
Ashizawa T, Monckton DG, Vaishnav S, Patel BJ, Voskova A, Caskey CT: Instability of the expanded $(CTG)_n$ repeats in the myotonin protein kinase gene in cultured lymphoblastoid cell lines from patients with myotonic dystrophy. Genomics 36:47–53 (1996).
Ashwood-Smith MJ, Edwards RG: DNA repair by oocytes. Mol Hum Reprod 2:46–51 (1996).
Astolfi P, Bellizzi D, Losso MA, Sgaramella V: Triplet repeats over-expanded in neuromuscular diseases are under-represented in mammalian DNA: a survey of models. Brain Res Bull 56:265–271 (2001).
Baarends WM, van der Laan R, Grootegoed JA: DNA repair mechanisms and gametogenesis. Reprod 121:31–39 (2001).
Bacon AL, Farrington SM, Dunlop MG: Sequence interruptions confer differential stability at microsatellite alleles in mismatch repair-deficient cells. Hum molec Genet 9:2707–2713 (2000).
Bambara RA, Murante RS, Henricksen LA: Enzymes and reactions at the eukaryotic DNA replication fork. J biol Chem 272:4647–4650 (1997).
Barcelo JM, Mahadevan MS, Tsilfidis C, MacKenzie AE, Korneluk RG: Intergenerational stability of the myotonic dystrophy protomutation. Hum molec Genet 2:705–709 (1993).
Bennett ST, Lucassen AM, Gough SC, Powell EE, Undlien DE, Pritchard LE, Merriman ME, Kawaguchi Y, Dronsfield MJ, Pociot F, et al: Susceptibility to human type 1 diabetes at IDDM2 is determined by tandem repeat variation at the insulin gene minisatellite locus. Nature Genet 9:284–292 (1995).
Bestor TH: Cytosine methylation and the unequal developmental potentials of the oocyte and sperm genomes. Am J hum Genet 62:1269–1273 (1998).
Bestor TH, Tycko B: Creation of genomic methylation patterns. Nature Genet 12:363–367 (1996).
Bidichandani SI, Purandare SM, Taylor EE, Gumin G, Machkhas H, Harati Y, Gibbs RA, Ashizawa T, Patel PI: Somatic sequence variation at the Friedreich ataxia locus includes complete contraction of the expanded GAA triplet repeat, significant length variation in serially passaged lymphoblasts and enhanced mutagenesis in the flanking sequence. Hum molec Genet 8:2425–2436 (1999).
Bird AP, Wolffe AP: Methylation-induced repression – belts, braces and chromatin. Cell 99:451–454 (1999).
Bjursell G, Gussander E, Lindahl T: Long regions of single-stranded DNA in human cells. Nature 280:420–423 (1979).
Blumen SC, Korczyn AD, Lavoie H, Medynski S, Chapman J, Asherov A, Nisipeanu P, Inzelberg R, Carasso RL, Bouchard JP, Tome FM, Rouleau GA, Brais B: Oculopharyngeal MD among Bukhara Jews is due to a founder $(GCG)_9$ mutation in the PABP2 gene. Neurology 55:1267–1270 (2000).
Boland CR, Thibodeau SN, Hamilton SR, Sidransky D, Eshleman JR, Burt RW, Meltzer SJ, Rodriguez-Bigas MA, Fodde R, Ranzani GN, Srivastava S: A National Cancer Institute Workshop on Microsatellite Instability for cancer detection and familial predisposition: development of international criteria for the determination of microsatellite instability in colorectal cancer. Cancer Res 58:5248–5257 (1998).
Bowater RP, Jaworski A, Larson JE, Parniewski P, Wells RD: Transcription increases the deletion frequency of long CTGCAG triplet repeats from plasmids in *Escherichia coli.* Nucl Acids Res 25:2861–2868 (1997).
Brais B, Bouchard JP, Xie YG, Rochefort DL, Chretien N, Tome FM, Lafreniere RG, Rommens JM, Uyama E, Nohira O, Blumen S, Korczyn AD, Heutink P, Mathieu J, Duranceau A, Codere F, Fardeau M, Rouleau GA, Korcyn AD: Short GCG expansions in the PABP2 gene cause oculopharyngeal muscular dystrophy. Nature Genet 18:164–167 (1998).
Brock GJ, Anderson NH, Monckton DG: *Cis*-acting modifiers of expanded CAG/CTG triplet repeat expandability: associations with flanking GC content and proximity to CpG islands. Hum molec Genet 8:1061–1067 (1999).
Brook JD: Retreat of the triplet repeat? Nature Genet 3:279–281 (1993).
Brown TC, Tarleton JC, Go RC, Longshore JW, Descartes M: Instability of the FMR2 trinucleotide repeat region associated with expanded FMR1 alleles. Am J med Genet 73:447–455 (1997).
Brown WT, Houck GE Jr, Ding X, Zhong N, Nolin S, Glicksman A, Dobkin C, Jenkins EC: Reverse mutations in the fragile X syndrome. Am J med Genet 64:287–292 (1996).
Brunner H, Jansen G, Nillesen W, Nelen MR, de Die CE, Howeler CJ, van Oost BA, Wieringa B, Ropers HH, Smeets HJ: Brief report: reverse mutation in myotonic dystrophy. New Engl J Med 328:476–480 (1993).
Buard J, Bourdet A, Yardley J, Dubrova Y, Jeffreys AJ: Influences of array size and homogeneity on minisatellite mutation. EMBO J 17:3495–3502 (1998).
Burman RW, Popovich BW, Jacky PB, Turker MS: Fully expanded FMR1 CGG repeats exhibit a length- and differentiation-dependent instability in cell hybrids that is independent of DNA methylation. Hum molec Genet 8:2293–2302 (1999).
Cacchione S, Cerone MA, Savino M: *In vitro* low propensity to form nucleosomes of four telomeric sequences. FEBS Lett 400:37–41 (1997).
Cancel G, Gourfinkel-An I, Stevanin G, Didierjean O, Abbas N, Hirsch E, Agid Y, Brice A: Somatic mosaicism of the CAG repeat expansion in spinocerebellar ataxia type 3/Machado-Joseph disease. Hum Mutat 11:23–27 (1998).
Cao H, Widlund HR, Simonsson T, Kubista M: TGGA repeats impair nucleosome formation. J molec Biol 281:253–260 (1998).
Cascalho M, Wong J, Steinberg C, Wabl M: Mismatch repair co-opted by hypremutation. Science 279:1207–1210 (1998).
Chang BL, Zheng SL, Hawkins GA, Isaacs SD, Wiley KE, Turner A, Carpten JD, Bleecker ER, Walsh PC, Trent JM, Meyers DA, Isaacs WB, Xu J: Polymorphic GGC repeats in the androgen receptor gene are associated with hereditary and sporadic prostate cancer risk. Hum Genet 110:122–129 (2002).
Chauhan C, Dash D, Grover D, Rajamani J, Mukerji M: Origin and instability of GAA repeats: insights from *Alu* elements. J Biomol Struct Dyn 20:253–264 (2002).
Chen RZ, Pettersson U, Beard C, Jackson-Grusby L, Jaenisch R: DNA hypomethylation leads to elevated mutation rates. Nature 395:89–93 (1998).
Chetsanga CJ, Boyd V, Peterson L, Rushlow K: Single-stranded regions in DNA of old mice. Nature 253:130–131 (1975).
Chiurazzi P, Macpherson J, Sherman S, Neri G: Significance of linkage disequilibrium between the fragile X locus and its flanking markers. Am J med Genet 64:203–208 (1996).
Chong SS, McCall AE, Cota J, Subramony SH, Orr HT, Hughes MR, Zoghbi HY: Gametic and somatic tissue-specific heterogeneity of the expanded SCA1 CAG repeat in spinocerebellar ataxia type 1. Nature Genet 10:344–350 (1995).
Choong CS, Kemppainen JA, Zhou ZX, Wilson EM: Reduced androgen receptor gene expression with first exon CAG repeat expansion. Mol Endocrinol 10:1527–1535 (1996).
Choudhry S, Mukerji M, Srivastava AK, Jain S, Brahmachari SK: CAG repeat instability at SCA2 locus: anchoring CAA interruptions and linked single nucleotide polymorphisms. Hum molec Genet 10:2437–2446 (2001).
Chung MY, Ranum LP, Duvick LA, Servadio A, Zoghbi HY, Orr HT: Evidence for a mechanism predisposing to intergenerational CAG repeat instability in spinocerebellar ataxia type I. Nature Genet 5:254–258 (1993).
Cleary JD, Nichol K, Wang YH, Pearson CE: Evidence of *cis*-acting factors in replication-mediated trinucleotide repeat instability in primate cells. Nature Genet 31:37–46 (2002).
Coffee B, Zhang F, Warren ST, Reines D: Acetylated histones are associated with FMR1 in normal but not fragile X-syndrome cells. Nature Genet 22:98–101 (1999).
Cohen H, Sears DD, Zenvirth D, Hieter P, Simchen G: Increased instability of human CTG repeat tracts on yeast artificial chromosomes during gametogenesis. Mol Cell Biol 19:4153–4158 (1999).
Correa-Cerro L, Wohr G, Haussler J, Berthon P, Drelon E, Mangin P, Fournier G, Cussenot O, Kraus P, Just W, Paiss T, Cantu JM, Vogel W: $(CAG)_nCAA$ and GGN repeats in the human androgen receptor gene are not associated with prostate cancer in a French-German population. Eur J hum Genet 7:357–362 (1999).
Cossee M, Schmitt M, Campuzano V, Reutenauer L, Moutou C, Mandel JL, Koenig M: Evolution of the Friedreich's ataxia trinucleotide repeat expansion: founder effect and premutations. Proc natl Acad Sci, USA 94:7452–7457 (1997).
Crawford DC, Meadows KL, Newman JL, Taft LF, Scott E, Leslie M, Shubek L, Holmgreen P, Yeargin-Allsopp M, Boyle C, Sherman SL: Prevalence of the fragile X syndrome in African-Americans. Am J med Genet 110:226–233 (2002).
Crawford DC, Wilson B, Sherman SL: Factors involved in the initial mutation of the fragile X CGG repeat as determined by sperm small pool PCR. Hum molec Genet 9:2909–2918 (2000).
Dadze S, Wieland C, Jakubiczka S, Funke K, Schroder E, Royer-Pokora B, Willers R, Wieacker PF: The size of the CAG repeat in exon 1 of the androgen receptor gene shows no significant relationship to impaired spermatogenesis in an infertile Caucasoid sample of German origin. Mol Hum Reprod 6:207–214 (2000).
Darlow JM, Leach DR: Secondary structures in d(CGG) and d(CCG) repeat tracts J molec Biol 275: 3–16 (1998).
de la Chapelle A, Peltomaki P: Genetics of hereditary colon cancer A Rev Genet 29:329–348 (1995).
De Rooij KE, De Koning Gans PAM, Roos RAC, Van Ommen G-JB, Den Dunnen JT: Somatic expansion of the $(CAG)_n$ repeat in Huntington disease brains. Hum Genet 95:270–274 (1995).
Deininger PL, Batzer MA: *Alu* repeats and human disease. Mol Genet Metab 67:183–193 (1999).

Delot E, King LM, Briggs MD, Wilcox WR, Cohn DH: Trinucleotide expansion mutations in the cartilage oligomeric matrix protein (COMP) gene. Hum molec Genet 8:123–128 (1999).

DePamphilis ML: Review: nuclear structure and DNA replication. J Struct Biol 129:186–197 (2000).

Devys D, Biancalana V, Rousseau F, Boue J, Mandel JL, Oberle I: Analysis of full fragile X mutations in fetal tissues and monozygotic twins indicate that abnormal methylation and somatic heterogeneity are established early in development. Am J med Genet 43:208–216 (1992).

Dichgans M, Schols L, Herzog J, Stevanin G, Weirich-Schwaiger H, Rouleau G, Burk K, Klockgether T, Zuhlke C, Laccone F, Riess O, Gasser T: Spinocerebellar ataxia type 6: evidence for a strong founder effect among German families. Neurology 52:849–851 (1999).

Dietmaier W, Wallinger S, Bocker T, Kullmann F, Fishel R, Ruschoff J: Diagnostic microsatellite instability: definition and correlation with mismatch repair protein expression. Cancer Res 57:4749–4756 (1997).

Dodge JE, Ramsahoye BH, Wo ZG, Okano M, Li E: *De novo* methylation of MMLV provirus in embryonic stem cells: CpG versus non-CpG methylation. Gene 289:41–48 (2002).

Drost JB, Lee WR: Biological basis of germline mutation: comparisons of spontaneous germline mutation rates among Drosophila, mouse and human Environ Mol Mutagen 25:48–64 (1995).

Duval A, Reperant M, Hamelin R: Comparative analysis of mutation frequency of coding and non coding short mononucleotide repeats in mismatch repair deficient colorectal cancers. Oncogene 21:8062–8066 (2002).

Duyao M, Ambrose C, Myers R, Novelletto A, Persichetti F, Frontali M, Folstein S, Ross C, Franz M, Abbott M, et al: Trinucleotide repeat length, instability and age of onset in Huntington's disease. Nature Genet 4:387–392 (1993).

Eberhart DE, Warren ST: Nuclease sensitivity of permeabilized cells confirms altered chromatin formation at the fragile X locus. Somat Cell molec Genet 22:435–441 (1996).

Economou EP, Bergen AW, Warren AC, Antonarakis SE: The polydeoxyadenylate tract of *Alu* repetitive elements is polymorphic in the human genome. Proc natl Acad Sci, USA 87:2951–2954 (1990).

Edelmann W, Cohen PE, Kane M, Lau K, Morrow B, Bennett S, Umar A, Kunkel T, Cattoretti G, Chaganti R, Pollard JW, Kolodner RD, Kucherlapati R: Meiotic pachytene arrest in MLH1-deficient mice. Cell 85:1125–1134 (1996).

Eichler EE, Holden JJ, Popovich BW, Reiss AL, Snow K, Thibodeau SN, Richards CS, Ward PA, Nelson DL: Length of uninterrupted CGG repeats determines instability in the FMR1 gene. Nature Genet 8:88–94 (1994).

Eichler EE, Macpherson JN, Murray A, Jacobs PA, Chakravarti A, Nelson DL: Haplotype and interspersion analysis of the FMR1 CGG repeat identifies two different mutational pathways for the origin of the fragile X syndrome. Hum molec Genet 5:319–330 (1996).

Elias-Arnanz M, Salas M: Bacteriophage phi29 DNA replication arrest caused by codirectional collisions with the transcription machinery. EMBO J 16:5775–5783 (1997).

Englander EW, Howard BH: Nucleosome positioning by human *Alu* elements in chromatin. J biol Chem 270:10091–10096 (1995).

Fagan PA, Fabrega C, Eritja R, Goodman MF, Wemmer DE: NMR study of the conformation of the 2-aminopurine:cytosine mismatch in DNA. Biochemistry 35:4026–4033 (1996).

Fazakerley GV, Quignard E, Woisard A, Guschlbauer W, van der Marel GA, van Boom JH, Jones M, Radman M: Structures of mismatched base pairs in DNA and their recognition by the *Escherichia coli* mismatch repair system. EMBO J 5:3697–3703 (1986).

Felsenfeld G: Chromatin unfolds. Cell 86:13–19 (1996).

Feng Z, Hu W, Komissarova E, Pao A, Hung MC, Adair GM, Tang MS: Transcription-coupled DNA repair is genomic context-dependent. J biol Chem 277:12777–12783 (2002).

Filippova GN, Thienes CP, Penn BH, Cho DH, Hu YJ, Moore JM, Klesert TR, Lobanenkov VV, Tapscott SJ: CTCF-binding sites flank CTG/CAG repeats and form a methylation-sensitive insulator at the DM1 locus. Nature Genet 28:335–343 (2001).

Flores MJ, Bierne H, Ehrlich SD, Michel B: Impairment of lagging strand synthesis triggers the formation of a RuvABC substrate at replication forks. EMBO J 20:619–629 (2001).

Fortune MT, Vassilopoulos C, Coolbaugh MI, Siciliano MJ, Monckton DG: Dramatic expansion-biased age-dependent tissue-specific somatic mosaicism in a transgenic mouse model of triplet repeat instability. Hum molec Genet 9:439–445 (2000).

Frank D, Keshet I, Shani M, Levine A, Razin A, Cedar H: Demethylation of CpG islands in embryonic cells. Nature 351:239–241 (1991).

Freudenreich CH, Stavenhagen JB, Zakian VA: Stability of a CTG/CAG trinucleotide repeat in yeast is dependent on its orientation in the genome. Mol Cell Biol 17:2090–2098 (1997).

Freudenreich CH, Kantrow SM, Zakian VA: Expansion and length-dependent fragility of CTG repeats in yeast. Science 279:853–856 (1998).

Furling D, Lemieux D, Taneja K, Puymirat J: Decreased levels of myotonic dystrophy protein kinase (DMPK) and delayed differentiation in human myotonic dystrophy myoblasts. Neuromuscul Disord 11:728–735 (2001).

Gacy AM, Goellner G, Juranic N, Macura S, McMurray CT: Trinucleotide repeats that expand in human disease form hairpin structures *in vitro*. Cell 81:533–540 (1995).

Gaspar C, Lopes-Cendes I, Hayes S, Goto J, Arvidsson K, Dias A, Silveira I, Maciel P, Coutinho P, Lima M, Zhou YX, Soong BW, Watanabe M, Giunti P, Stevanin G, Riess O, Sasaki H, Hsieh M, Nicholson GA, Brunt E, Higgins JJ, Lauritzen M, Tranebjaerg L, Volpini V, Wood N, Ranum L, Tsuji S, Brice A, Sequeiros J, Rouleau GA: Ancestral origins of the Machado-Joseph disease mutation: a worldwide haplotype study. Am J hum Genet 68:523–528 (2001).

Gecz J: FMR3 is a novel gene associated with FRAXE CpG island and transcriptionally silent in FRAXE full mutations. J med Genet 37:782–784 (2000).

Gecz J, Gedeon AK, Sutherland GR, Mulley JC: Identification of the gene FMR2 associated with FRAXE mental retardation. Nature Genet 13:105–108 (1996).

Glaser D, Wohrle D, Salat U, Vogel W, Steinbach P: Mitotic behavior of expanded CGG repeats studied on cultured cells: further evidence for methylation-mediated triplet repeat stability in fragile X syndrome. Am J med Genet 84:226–228 (1999).

Godde JS, Kass SU, Hirst MC, Wolffe AP: Nucleosome assembly on methylated CGG triplet repeats in the fragile X mental retardation gene 1 promoter. J biol Chem 271:24325–24328 (1996).

Goellner GM, Tester D, Thibodeau S, Almqvist E, Goldberg YP, Hayden MR, McMurray CT: Different mechanisms underlie DNA instability in Huntington disease and colorectal cancer. Am J hum Genet 60:879–890 (1997).

Goldberg YP, McMurray TC, Zeisler J, Almqvist E, Sillence D, Richards F, Gacy AM, Buchanan J, Telenius H, Hayden MR: Increased instability of intermediate alleles in families with sporadic Huntington disease compared to similar sized intermediate alleles in the general population. Hum molec Genet 4:1911–1918 (1995).

Goldfarb LG, Brown P, McCombie WR, Goldgaber D, Swergold GD, Wills PR, Cervenakova L, Baron H, Gibbs CJ Jr, Gajdusek DC: Transmissible familial Creutzfeldt-Jakob disease associated with five seven and eight extra octapeptide coding repeats in the PRNP gene. Proc natl Acad Sci, USA 88:10926–10930 (1991).

Goldman A, Ramsay M, Jenkins T: New founder haplotypes at the myotonic dystrophy locus in southern Africa. Am J hum Genet 56:1373–1378 (1995).

Gomes-Pereira M, Fortune MT, Monckton DG: Mouse tissue culture models of unstable triplet repeats: *in vitro* selection for larger alleles mutational expansion bias and tissue specificity but no association with cell division rates. Hum mol Genet 10:845–854 (2001).

Goodman FR, Mundlos S, Muragaki Y, Donnai D, Giovannucci-Uzielli ML, Lapi E, Majewski F, McGaughran J, McKeown C, Reardon W, Upton J, Winter RM, Olsen BR, Scambler PJ: Synpolydactyly phenotypes correlate with size of expansions in HOXD13 polyalanine tract. Proc natl Acad Sci, USA 94:7458–7463 (1997).

Gordenin DA, Kunkel TA, Resnick MA: Repeat expansion – all in a flap? [news] Nature Genet 16:116–118 (1997).

Gourdon G, Dessen P, Lia AS, Junien C, Hofmann-Radvanyi H: Intriguing association between disease associated unstable trinucleotide repeat and CpG island. Ann Genet 40:73–77 (1997).

Gourdon G, Radvanyi F, Lia AS, Duros C, Blanche M, Abitbol M, Junien C, Hofmann-Radvanyi H: Moderate intergenerational and somatic instability of a 55-CTG repeat in transgenic mice. Nature Genet 15:190–192 (1997).

Gouw LG, Castaneda MA, McKenna CK, Digre KB, Pulst SM, Perlman S, Lee MS, Gomez C, Fischbeck K, Gagnon D, Storey E, Bird T, Jeri FR, Ptacek LJ: Analysis of the dynamic mutation in the SCA7 gene shows marked parental effects on CAG repeat transmission. Hum molec Genet 7:525–532 (1998).

Green CM, Almouzni G: When repair meets chromatin. First in series on chromatin dynamics. EMBO Rep 3:28–33 (2002).

Grewal RP, Karkera JD, Grewal RK, Detera-Wadleigh SD: Mutation analysis of oculopharyngeal muscular dystrophy in Hispanic American families. Arch Neurol 56:1378–1381 (1999).

Gribnau J, Diderich K, Pruzina S, Calzolari R, Fraser P: Intergenic transcription and developmental remodeling of chromatin subdomains in the human beta-globin locus. Mol Cell 5:377–386 (2000).

Griffith JD, Wang Y-H: Nucleosome analysis and disease of chromatin structure, in Wells RD, Warren ST (eds): Genetic Instabilities and Hereditary Neurological Diseases, pp 677–690 (Academic Press, San Diego 1998).

Gunter C, Paradee W, Crawford DC, Meadows KA, Newman J, Kunst CB, Nelson DL, Schwartz C, Murray A, Macpherson JN, Sherman SL, Warren ST: Re-examination of factors associated with expansion of CGG repeats using a single nucleotide polymorphism in FMR1. Hum molec Genet 7:1935–1946 (1998).

Haaf T: The effects of 5-azacytidine and 5-azadeoxycytidine on chromosome structure and function: implications for methylation-associated cellular processes. Pharmacol Ther 65:19–46 (1995).

Haber JE: DNA recombination: the replication connection. Trends Biochem Sci 24:271–275 (1999).

Habib M, Fares F, Bourgeois CA, Bella C, Bernardino J, Hernandez-Blazquez F, de Capoa A, Niveleau A: DNA global hypomethylation in EBV-transformed interphase nuclei. Exp Cell Res 249:46–53 (1999).
Hagerman RJ, Hull CE, Safanda JF, Carpenter I, Staley LW, O'Connor RA, Seydel C, Mazzocco MM, Snow K, Thibodeau SN, et al: High functioning fragile X males: demonstration of an unmethylated fully expanded FMR-1 mutation associated with protein expression. Am J med Genet 51:298–308 (1994).
Hammond LS, Macias MM, Tarleton JC, Shashidhar Pai G: Fragile X syndrome and deletions in FMR1: new case and review of the literature. Am J med Genet 72:430–434 (1997).
Hardy DO, Scher HI, Bogenreider T, Sabbatini P, Zhang ZF, Nanus DM, Catterall JF: Androgen receptor CAG repeat lengths in prostate cancer: correlation with age of onset. J Clin Endocrinol Metab 81:4400–4405 (1996).
Hashida H, Goto J, Kurisaki H, Mizusawa H, Kanazawa I: Brain regional differences in the expansion of a CAG repeat in the spinocerebellar ataxias: dentatorubral-pallidoluysian atrophy, Machado-Joseph disease and spinocerebellar ataxia type 1. Ann Neurol 41:505–511 (1997).
Hay RT, DePamphilis ML: Initiation of SV40 DNA replication *in vivo*: location and structure of 5′ ends of DNA synthesized in the ori region. Cell 28:767–779 (1982).
He Q, Cederberg H, Armour JA, May CA, Rannug U: *Cis*-regulation of inter-allelic exchanges in mutation at human minisatellite MS205 in yeast. Gene 232:143–153 (1999).
Heidenfelder BL, Makhof AM, Topal MD: Hairpin formation in Friedreich's Ataxia triplet-repeat expansion. J biol Chem 18:18 (2002).
Helderman-van den Enden AT, Maaswinkel-Mooij PD, Hoogendoorn E, Willemsen R, Maat-Kievit JA, Losekoot M, Oostra BA: Monozygotic twin brothers with the fragile X syndrome: different CGG repeats and different mental capacities. J med Genet 36:253–257 (1999).
Heller CG, Clermont Y: Kinetics of germinal epithelium in man. Recent Prog Hormone Res 20:545–575 (1964).
Henricksen LA, Tom S, Liu Y, Bambara RA: Inhibition of flap endonuclease 1 by flap secondary structure and relevance to repeat sequence expansion. J biol Chem 275: 16420–16427 (2000).
Henson P: The presence of single-stranded regions in mammalian DNA. J molec Biol 119:487–506 (1978).
Herman JG, Umar A, Polyak K, Graff JR, Ahuja N, Issa JP, Markowitz S, Willson JK, Hamilton SR, Kinzler KW, Kane MF, Kolodner RD, Vogelstein B, Kunkel TA, Baylin SB: Incidence and functional consequences of hMLH1 promoter hypermethylation in colorectal carcinoma. Proc natl Acad Sci, USA 95:6870–6875 (1998).
Hewett DR, Handt O, Hobson L, Mangelsdorf M, Eyre HJ, Baker E, Sutherland GR, Schuffenhauer S, Mao JI, Richards RI: FRA10B structure reveals common elements in repeat expansion and chromosomal fragile site genesis. Mol Cell 1:773–781 (1998).
Hewitt JE, Lyle R, Clark LN, Valleley EM, Wright TJ, Wijmenga C, van Deutekom JC, Francis F, Sharpe PT, Hofker M, et al: Analysis of the tandem repeat locus D4Z4 associated with facioscapulohumeral muscular dystrophy. Hum molec Genet 3:1287–1295 (1994).
Hodges-Garcia Y, Hagerman PJ: Cytosine methylation can induce local distortions in the structure of duplex DNA. Biochemistry 31:7595–7599 (1992).
Hornstra IK, Nelson DL, Warren ST, Yang TP: High resolution methylation analysis of the FMR1 gene trinucleotide repeat region in fragile X syndrome. Hum molec Genet 2:1659–1665 (1993).
Hsu YY, Wang YH: Human fragile site FRA16B DNA excludes nucleosomes in the presence of distamycin. J biol Chem 277:17315–17319 (2002).
Huntington Disease Genetic Testing Working Group: ACMG/ASHG Statement-Laboratory Guidelines for Huntington disease genetic testing. Am J hum Genet 62:1243–1247 (1998).
Igarashi S, Takiyama Y, Cancel G, Rogaeva EA, Sasaki H, Wakisaka A, Zhou YX, Takano H, Endo K, Sanpei K, Oyake M, Tanaka H, Stevanin G, Abbas N, Durr A, Rogaev EI, Sherrington R, Tsuda T, Ikeda M, Cassa E, Nishizawa M, Benomar A, Julien J, Weissenbach J, Tsuji S, et al: Intergenerational instability of the CAG repeat of the gene for Machado-Joseph disease (MJD1) is affected by the genotype of the normal chromosome: implications for the molecular mechanisms of the instability of the CAG repeat. Hum Mol Genet 5:923–932 (1996).
Iizuka M, Stillman B: Histone acetyltransferase HBO1 interacts with the ORC1 subunit of the human initiator protein. J biol Chem 274:23027–23034 (1999).
Imbert G Kretz C Johnson K and Mandel JL: Origin of the expansion mutation in myotonic dystrophy. Nature Genet 4:72–76 (1993).
Irvine RA, Yu MC, Ross RK, Coetzee GA: The CAG and GGC microsatellites of the androgen receptor gene are in linkage disequilibrium in men with prostate cancer. Cancer Res 55:1937–1940 (1995).
Ishii S, Nishio T, Sunohara N, Yoshihara T, Takemura K, Hikiji K, Tsujino S, Sakuragawa N: Small increase in triplet repeat length of cerebellum from patients with myotonic dystrophy. Hum Genet 98:138–140 (1996).
Jakupciak JP, Wells RD: Genetic instabilities in (CTGCAG) repeats occur by recombination. J biol Chem 274:23468–23479 (1999).
Jakupciak JP, Wells RD: Gene conversion (recombination) mediates expansions of CTG(CAG repeats. J biol Chem 275:40003–40013 (2000).
Jalal SM, Lindor NM, Michels VV, Buckley DD, Hoppe DA, Sarkar G, Dewald GW: Absence of chromosome fragility at 19q13.3 in patients with myotonic dystrophy. Am J med Genet 46: 441–443 (1993).
Jankowski C, Nasar F, Nag DK: Meiotic instability of CAG repeat tracts occurs by double-strand break repair in yeast. Proc natl Acad Sci, USA 97:2134–2139 (2000).
Jansen G, Willems P, Coerwinkel M, Nillesen W, Smeets H, Vits L, Howeler C, Brunner H, Wieringa B: Gonosomal mosaicism in myotonic dystrophy patients: involvement of mitotic events in $(CTG)_n$ repeat variation and selection against extreme expansion in sperm. Am J hum Genet 54:575–585 (1994).
Jaworski A, Blaho JA, Larson JE, Shimizu M, Wells RD: Tetracycline promoter mutations decrease non-B DNA structural transitions, negative linking differences and deletions in recombinant plasmids in *Escherichia coli*. J molec Biol 207:513–526 (1989).
Jaworski A, Rosche WA, Gellibolian R, Kang S, Shimizu M, Bowater RP, Sinden RR, Wells RD: Mismatch repair in *Escherichia coli* enhances instability of $(CTG)_n$ triplet repeats from human hereditary diseases. Proc natl Acad Sci, USA 92:11019–11023 (1995).
Jeanpierre M, Turleau C, Aurias A, Prieur M, Ledeist F, Fischer A, Viegas-Pequignot E: An embryonic-like methylation pattern of classical satellite DNA is observed in ICF syndrome. Hum molec Genet 2:731–735 (1993).
Jeffreys AJ, Tamaki K, MacLeod A, Monckton DG, Neil DL, Armour JA: Complex gene conversion events in germline mutation at human minisatellites. Nature Genet 6:136–145 (1994).
Jodice C, Frontali M, Persichetti F, Novelletto A, Pandolfo M, Spadaro M, Giunti P, Schinaia G, Lulli P, Malaspina P, et al: The gene for spinal cerebellar ataxia 1 (SCA1) is flanked by two closely linked highly polymorphic microsatellite loci. Hum molec Genet 2:1383–1387 (1993).
Jones C, Penny L, Mattina T, Yu S, Baker E, Voullaire L, Langdon WY, Sutherland GR, Richards RI, Tunnacliffe A: Association of a chromosome deletion syndrome with a fragile site within the proto-oncogene CBL2. Nature 376:145–149 (1995).
Jones C, Mullenbach R, Grossfeld P, Auer R, Favier R, Chien K, James M, Tunnacliffe A, Cotter F: Co-localisation of CCG repeats and chromosome deletion breakpoints in Jacobsen syndrome: evidence for a common mechanism of chromosome breakage. Hum molec Genet 9:1201–1208 (2000).
Jurka J, Krnjajic M, Kapitonov VV, Stenger JE, Kokhanyy O: Active *Alu* elements are passed primarily through paternal germlines. Theor Popul Biol 61:519–530 (2002).
Juvonen V, Hietala M, Paivarinta M, Rantamaki M, Hakamies L, Kaakkola S, Vierimaa O, Penttinen M, Savontaus ML: Clinical and genetic findings in Finnish ataxia patients with the spinocerebellar ataxia 8 repeat expansion. Ann Neurol 48:354–361 (2000).
Kang S, Jaworski A, Ohshima K, Wells RD: Expansion and deletion of CTG repeats from human disease genes are determined by the direction of replication in *E. coli*. Nature Genet 10:213–218 (1995).
Kang S, Ohshima K, Shimizu M, Amirhaeri S, Wells RD: Pausing of DNA synthesis *in vitro* at specific loci in CTG and CGG triplet repeats from human hereditary disease genes. J biol Chem 270:27014–27021 (1995).
Katti MV, Ranjekar PK, Gupta VS: Differential distribution of simple sequence repeats in eukaryotic genome sequences. Mol Biol Evol 18:1161–1167 (2001).
Kawaguchi Y, Okamoto T, Taniwaki M, Aizawa M, Inoue M, Katayama S, Kawakami H, Nakamura S, Nishimura M, Akiguchi I, et al: CAG expansions in a novel gene for Machado-Joseph disease at chromosome 14q32.1. Nature Genet 8: 221–228 (1994).
Kaytor MD, Burright EN, Duvick LA, Zoghbi HY, Orr HT: Increased trinucleotide repeat instability with advanced maternal age. Hum molec Genet 6:2135–2139 (1997).
Khajavi M, Tari AM, Patel NB, Tsuji K, Siwak DR. Meistrich ML, Terry NH, Ashizawa T: 'Mitotic drive' of expanded CTG repeats in myotonic dystrophy type 1 (DM1). Hum molec Genet 10:855–863 (2001).
Kim C, Rubin CM, Schmid CW: Genome-wide chromatin remodeling modulates the *Alu* heat shock response. Gene 276:127–133 (2001).
Kinboshita M, Ossani R, Kikkawa M, Adachi A, Ohtake T, Komori T, Hashimoto K, Itoyama S, Mitarai T, Hirose K: A patient with myotonic dystrophy type 1 (DM1) accompanied by laryngeal and renal cell carcinomas has a small CTG triplet repeat expansion but no somatic instability in normal tissues. Inter Med 41:312–318 (2002).
Koide R, Kobayashi S, Shimohata T, Ikeuchi T, Maruyama M, Saito M, Yamada M, Takahashi H, Tsuji S: A neurological disease caused by an expanded CAG trinucleotide repeat in the TATA-binding protein gene: a new polyglutamine disease? Hum molec Genet 8:2047–2053 (1999).
Kolomietz E, Meyn MS, Pandita A, Squire JA: The role of *Alu* repeat clusters as mediators of recurrent chromosomal aberrations in tumors. Genes Chrom Cancer 35:97–112 (2002).
Koob MD, Moseley ML, Schut LJ, Benzow KA, Bird TD, Day JW, Ranum LP: An untranslated CTG expansion causes a novel form of spinocerebellar ataxia (SCA8). Nature Genet 21: 379–384 (1999).

Kovtun IV, McMurray CT: Trinucleotide expansion in haploid germ cells by gap repair. Nature Genet 27:407–411 (2001).

Kovtun IV, Therneau TM, McMurray CT: Gender of the embryo contributes to CAG instability in transgenic mice containing a Huntington's disease gene. Hum molec Genet 9:2767–2775 (2000).

Krahe R, Eckhart M, Ogunniyi AO, Osuntokun BO, Siciliano MJ, Ashizawa T: *De novo* myotonic dystrophy mutation in a Nigerian kindred. Am J hum Genet 56:1067–1074 (1995).

Kramer PR, Pearson CE, Sinden RR: Stability of triplet repeats of myotonic dystrophy and fragile X loci in human mutator mismatch repair cell lines. Hum Genet 98:151–157 (1996).

Krasilnikova MM, Samadashwily GM, Krasilnikov AS, Mirkin SM: Transcription through a simple DNA repeat blocks replication elongation. EMBO J 17:5095–5102 (1998).

Krontiris TG, Devlin B, Karp DD, Robert NJ, Risch N: An association between the risk of cancer and mutations in the HRAS1 minisatellite locus. New Engl J Med 329:517–523 (1993).

Kunst CB, Leeflang EP, Iber JC, Arnheim N, Warren ST: The effect of FMR1 CGG repeat interruptions on mutation frequency as measured by sperm typing. J med Genet 34:6276–6231 (1997).

La Spada AR, Wilson EM, Lubahn DB, Harding AE, Fischbeck KH: Androgen receptor gene mutations in X-linked spinal and bulbar muscular atrophy. Nature 352:77–79 (1991).

La Spada AR, Roling DB, Harding AE, Warner CL, Spiegel R, Hausmanowa-Petrusewicz I, Yee WC, Fischbeck KH: Meiotic stability and genotype-phenotype correlation of the trinucleotide repeat in X-linked spinal and bulbar muscular atrophy. Nature Genet 2:301–304 (1992).

La Spada AR, Peterson KR, Meadows SA, McClain ME, Jeng G, Chmelar RS, Haugen HA, Chen K, Singer MJ, Moore D, Trask BJ, Fischbeck KH, Clegg CH, McKnight GS: Androgen receptor YAC transgenic mice carrying CAG 45 alleles show trinucleotide repeat instability. Hum molec Genet 7:959–967 (1998).

Labuda M, Labuda D, Miranda C, Poirier J, Soong BW, Barucha NE, Pandolfo M: Unique origin and specific ethnic distribution of the Friedreich ataxia GAA expansion. Neurology 54:2322–2324 (2000).

Lafreniere RG, Rochefort DL, Chretien N, Rommens JM, Cochius JI, Kalviainen R, Nousiainen U, Patry G, Farrell K, Soderfeldt B, Federico A, Hale BR, Cossio OH, Sorensen T, Pouliot MA, Kmiec T, Uldall P, Janszky J, Pranzatelli MR, Andermann F, Andermann E, Rouleau GA: Unstable insertion in the 5′ flanking region of the cystatin B gene is the most common mutation in progressive myoclonus epilepsy type 1 EPM1. Nature Genet 15:298–302 (1997).

Leeflang EP, Zhang L, Tavare S, Hubert R, Srinidhi, J MacDonald ME, Myers RH, de Young M, Wexler NS, Gusella JF, et al: Single sperm analysis of the trinucleotide repeats in the Huntington's disease gene: quantification of the mutation frequency spectrum. Hum molec Genet 4:1519–1526 (1995).

Leeflang EP, Tavare S, Marjoram P, Neal CO, Srinidhi J, MacFarlane H, MacDonald ME, Gusella JF, de Young M, Wexler NS, Arnheim N: Analysis of germline mutation spectra at the Huntington's disease locus supports a mitotic mutation mechanism. Hum molec Genet 8:173–183 (1999).

LeProust EM, Pearson CE, Sinden RR, Gao X: Unexpected formation of parallel duplex in GAA and TTC trinucleotide repeats of Friedreich's ataxia. J molec Biol 302:1063–1080 (2000).

Li S, Livingstone-Zatchej M, Gupta R, Meijer M, Thoma F, Smerdon MJ: Nucleotide excision repair in a constitutive and inducible gene of a yeast minichromosome in intact cells. Nucl Acids Res 27:3610–3620 (1999).

Lia AS, Seznec H, Hofmann-Radvanyi H, Radvanyi F, Duros C, Saquet C, Blanche M, Junien C, Gourdon G: Somatic instability of the CTG repeat in mice transgenic for the myotonic dystrophy region is age dependent but not correlated to the relative intertissue transcription levels and proliferative capacities. Hum molec Genet 7:1285–1291 (1998).

Libby RT, Monckton DG, Fu YH, Martinez RA, McAbney JP, Lau R, Einum DD, Nichol K, Ware CB, Ptacek LJ, Pearson CE, La Spada AR: Genomic context drives SCA7 CAG repeat instability while expressed SCA7 cDNAs are intergenerationally and somatically stable in transgenic mice. Hum molec Genet 12:41–50 (2003).

Lieber MR: The biochemistry and biological significance of nonhomologous DNA end joining: an essential repair process in multicellular eukaryotes. Genes Cells 4:77–85 (1999).

Lieber MR, Hesse JE, Mizuuchi K, Gellert M: Developmental stage specificity of the lymphoid V(D)J recombination activity. Genes Dev 1:751–761 (1987).

Limprasert P, Nouri N, Heyman RA, Nopparatana C, Kamonsilp M, Deininger PL, Keats BJ: Analysis of CAG repeat of the Machado-Joseph gene in human chimpanzee and monkey populations: a variant nucleotide is associated with the number of CAG repeats. Hum molec Genet 5:207–213 (1996).

Liu B, Alberts BM: Head-on collision between a DNA replication apparatus and RNA polymerase transcription complex. Science 267:1131–1137 (1995).

Lobachev KS, Stenger JE, Kozyreva OG, Jurka J, Gordenin DA, Resnick MA: Inverted *Alu* repeats unstable in yeast are excluded from the human genome. EMBO J 19:3822–3830 (2000).

Lopez de Munain A, Cobo AM, Poza JJ, Navarrete D, Martorell L, Palau F, Emparanza JI, Baiget M: Influence of the sex of the transmitting grandparent in congenital myotonic dystrophy. J med Genet 32:689–691 (1995).

Losekoot M, Hoogendoorn E, Olmer R, Jansen CC, Oosterwijk JC, van den Ouweland AM, Halley DJ, Warren ST, Willemsen R, Oostra BA, Bakker E: Prenatal diagnosis of the fragile X syndrome: loss of mutation owing to a double recombinant or gene conversion event at the FMR1 locus. J med Genet 34:924–926 (1997).

Loukinov DI, Pugacheva E, Vatolin S, Pack SD, Moon H, Chernukhin I, Mannan P, Larsson E, Kanduri C, Vostrov AA, Cui H, Niemitz EL, Rasko JE, Docquier FM, Kistler M, Breen JJ, Zhuang Z, Quitschke WW, Renkawitz R, Klenova EM, Feinberg AP, Ohlsson R, Morse HC 3rd, Lobanenkov VV: BORIS a novel male germ-line-specific protein associated with epigenetic reprogramming events shares the same 11-zinc-finger domain with CTCF the insulator protein involved in reading imprinting marks in the soma. Proc natl Acad Sci, USA 99:6806–6811 (2002).

Lund A, Udd B, Juvonen V, Andersen PM, Cederquist K, Ronnevi LO, Sistonen P, Sorensen SA, Tranebjaerg L, Wallgren-Pettersson C, Savontaus ML: Founder effect in spinal and bulbar muscular atrophy (SBMA) in Scandinavia. Eur J hum Genet 8:631–636 (2000).

Lund A, Udd B, Juvonen V, Andersen PM, Cederquist K, Davis M, Gellera C, Kolmel C, Ronnevi LO, Sperfeld AD, Sorensen SA, Tranebjaerg L, Van Maldergem L, Watanabe M, Weber M, Yeung L, Savontaus ML: Multiple founder effects in spinal and bulbar muscular atrophy (SBMA Kennedy disease) around the world. Eur J hum Genet 9:431–436 (2001).

Luo S, Robinson JC, Reiss AL, Migeon BR: DNA methylation of the fragile X locus in somatic and germ cells during fetal development: relevance to the fragile X syndrome and X inactivation. Somat Cell molec Genet 19:393–404 (1993).

Maat-Kievit A, Helderman-van den Enden P, Losekoot M, de Knijff P, Belfroid R, Vegter-van der Vlis M, Roos R, Breuning M: Using a roster and haplotyping is useful in risk assessment for persons with intermediate and reduced penetrance alleles in Huntington disease. Am J med Genet 105:737–744 (2001).

Maciel P, Gaspar C, Guimaraes L, Goto J, Lopes-Cendes I, Hayes S, Arvidsson K, Dias A, Sequeiros J, Sousa A, Rouleau GA: Study of three intragenic polymorphisms in the Machado-Joseph disease gene (MJD1) in relation to genetic instability of the $(CAG)_n$ tract. Eur J hum Genet 7:147–156 (1999).

Mahadevan M, Tsilfidis C, Sabourin L, Shutler G, Amemiya C, Jansen G, Neville C, Narang M, Barcelo J, O'Hoy K, et al: Myotonic dystrophy mutation: an unstable CTG repeat in the 3′ untranslated region of the gene. Science 255:1253–1255 (1992).

Malter HE, Iber JC, Willemsen R, de Graaff E, Tarleton JC, Leisti J, Warren ST, Oostra BA: Characterization of the full fragile X syndrome mutation in fetal gametes. Nature Genet 15:165–169 (1997).

Mangiarini L, Sathasivam K, Mahal A, Mott R, Seller M, Bates GP: Instability of highly expanded CAG repeats in mice transgenic for the Huntington's disease mutation. Nature Genet 15:197–200 (1997).

Manley K, Pugh J, Messer A: Instability of the CAG repeat in immortalized fibroblast cell cultures from Huntington's disease transgenic mice. Brain Res 835:74–79 (1999a).

Manley K, Shirley TL, Flaherty L, Messer A: Msh2 deficiency prevents *in vivo* somatic instability of the CAG repeat in Huntington disease transgenic mice. Nature Genet 23:471–473 (1999b).

Martorell L, Martinez JM, Carey N, Johnson K, Baiget M: Comparison of CTG repeat length expansion and clinical progression of myotonic dystrophy over a five year period. J med Genet 32:593–596 (1995).

Martorell L, Johnson K, Boucher CA, Baiget M: Somatic instability of the myotonic dystrophy $(CTG)_n$ repeat during human fetal development. Hum molec Genet 6:877–880 (1997).

Martorell L, Monckton DG, Gamez J, Johnson KJ, Gich I, de Munain AL, Baiget M: Progression of somatic CTG repeat length heterogeneity in the blood cells of myotonic dystrophy patients. Hum molec Genet 7:307–312 (1998).

Martorell L, Monckton DG, Gamez J, Baiget M: Complex patterns of male germline instability and somatic mosaicism in myotonic dystrophy type 1. Eur J hum Genet 8:423–430 (2000).

Martorell L, Monckton DG, Sanchez A, Lopez De Munain A, Baiget M: Frequency and stability of the myotonic dystrophy type 1 premutation. Neurology 56:328–335 (2001).

Matsuura T, Yamagata T, Burgess DL, Rasmussen A, Grewal RP, Watase K, Khajavi M, McCall AE, Davis CF, Zu L, Achari M, Pulst SM, Alonso E, Noebels JL, Nelson DL, Zoghbi HY, Ashizawa T: Large expansion of the ATTCT pentanucleotide repeat in spinocerebellar ataxia type 10. Nature Genet 26:191–194 (2000).

Maurer DJ, O'Callaghan BL, Livingston DM: Mapping the polarity of changes that occur in interrupted CAG repeat tracts in yeast. Mol Cell Biol 18:4597–4604 (1998).

McGlynn P, Lloyd RG: Modulation of RNA polymerase by (p)ppGpp reveals a RecG-dependent mechanism for replication fork progression. Cell 101:35–45 (2000).

McMurry MT, Krangel MS: A role for histone acetylation in the developmental regulation of VDJ recombination. Science 287:495–498 (2000).

Meek RL, Bowman PD, Daniel CW: Establishment of mouse embryo cells *in vitro*. Relationship of DNA synthesis, senescence and malignant transformation. Exp Cell Res 107:277–284 (1977).

Mertineit CJ, Yoder A, Taketo T, Laird DW, Trasler JM, Bestor TH: Sex-specific exons control DNA methyltransferase in mammalian germ cells. Development 125:889–897 (1998).
Michel B: Replication fork arrest and DNA recombination. Trends Biochem Sci 25:173–178 (2000).
Migeon BR, Holland MM, Driscoll DJ, Robinson JC: Programmed demethylation in CpG islands during human fetal development. Somat Cell Mol Genet 17:159168 (1991).
Miret J, Pessoa-Brandao JL, Lahue RS: Instability of CAG and CTG trinucleotide repeats in *Saccharomyces cerevisiae*. Mol Cell Biol 17:3382–3387 (1997).
Miret JJ, Pessoa-Brandao L, Lahue RS: Orientation-dependent and sequence-specific expansions of CTG/CAG trinucleotide repeats in *Saccharomyces cerevisiae*. Proc natl Acad Sci, USA 95:12438–12443 (1998).
Mirkin SM, Smirnova EV: Positioned to expand. Nature Genet 31:5–6 (2002).
Mitas M, Yu A, Dill J, Kamp TJ, Chambers EJ, Haworth IS: Hairpin properties of single-stranded DNA containing a GC-rich triplet repeat: $(CTG)_{15}$. Nucl Acids Res 23:1050–1059 (1995).
Mizushima K, Watanabe M, Kondo I, Okamoto K, Shizuka M, Abe K, Aoki M, Shoji M: Analysis of spinocerebellar ataxia type 2 gene and haplotype analysis: (CCG)1–2 polymorphism and contribution to founder effect. J med Genet 36:112–114 (1999).
Monckton DG, Neumann R, Guram T, Fretwell N, Tamaki K, MacLeod A, Jeffreys AJ: Minisatellite mutation rate variation associated with a flanking DNA sequence polymorphism. Nature Genet 8:162–170 (1994).
Monckton DG, Wong LJ, Ashizawa T, Caskey CT: Somatic mosaicism germline expansions, germline reversions and intergenerational reductions in myotonic dystrophy males: small pool PCR analyses. Hum molec Genet 4: 1–8 (1995).
Monckton DG, Coolbaugh MI, Ashizawa KT, Siciliano MJ, Caskey CT: Hypermutable myotonic dystrophy CTG repeats in transgenic mice. Nature Genet 15:193–196 (1997).
Montermini L, Andermann E, Labuda M, Richter A, Pandolfo M, Cavalcanti F, Pianese L, Iodice L, Farina G, Monticelli A, Turano M, Filla A, De Michele G, Cocozza S: The Friedreich ataxia GAA triplet repeat: premutation and normal alleles. Hum molec Genet 6:1261–1266 (1997).
Moore H, Greenwell PW, Liu CP, Arnheim N, Petes TD: Triplet repeats form secondary structures that escape DNA repair in yeast. Proc natl Acad Sci, USA 96:1504–1509 (1999).
Mornet E, Chateau C, Hirst MC, Thepot F, Taillandier A, Cibois O, Serre JL: Analysis of germline variation at the FMR1 CGG repeat shows variation in the normal-premutated borderline range. Hum mol Genet 5:821–825 (1996).
Moseley ML, Schut LJ, Bird TD, Koob MD, Day JW, Ranum LP: SCA8 CTG repeat: en masse contractions in sperm and intergenerational sequence changes may play a role in reduced penetrance. Hum molec Genet 9:2125–2130 (2000).
Moutou C, Vincent MC, Biancalana V, Mandel JL: Transition from premutation to full mutation in fragile X syndrome is likely to be prezygotic. Hum molec Genet 6:971–979 (1997).
Muller-Hartmann H, Deissler H, Naumann F, Schmitz B, Schroer J, Doerfler W: The human 20-kDa 5′-(CGG)(n)-3′-binding protein is targeted to the nucleus and affects the activity of the FMR1 promoter. J biol Chem 275:6447–6452 (2000).
Mundlos S, Otto F, Mundlos C, Mulliken JB, Aylsworth AS, Albright S, Lindhout D, Cole WG, Henn W, Knoll JH, Owen MJ, Mertelsmann R, Zabel BU, Olsen BR: Mutations involving the transcription factor CBFA1 cause cleidocranial dysplasia. Cell 89:773–779 (1997).
Murray A, Macpherson JN, Pound MC, Sharrock A, Youings SA, Dennis NR, McKechnie N, Linehan P, Morton NE, Jacobs PA: The role of size sequence and haplotype in the stability of FRAXA and FRAXE alleles during transmission. Hum molec Genet 6:173–184 (1997).
Nagafuchi S, Yanagisawa H, Sato K, Shirayama T, Ohsaki E, Bundo M, Takeda T, Tadokoro K, Kondo I, Murayama N, et al: Dentatorubral and pallidoluysian atrophy expansion of an unstable CAG trinucleotide on chromosome 12p. Nature Genet 6:14–18 (1994).
Nakamoto M, Nakano S, Kawashima S, Ihara M, Nishimura Y, Shinde A, Kakizuka A: Unequal crossing-over in unique PABP2 mutations in Japanese patients: a possible cause of oculopharyngeal muscular dystrophy. Arch Neurol 59:474–477 (2002).
Nakamoto M, Takebayashi H, Kawaguchi Y, Narumiya S, Taniwaki M, Nakamura Y, Ishikawa Y, Akiguchi I, Kimura J, Kakizuka A: A CAG/CTG expansion in the normal population. Nature Genet 17:385–386 (1997).
Nakanishi K, Shima A, Fukuda M, Fujita S: Age associated increase of single-stranded regions in the DNA of mouse brain and liver cells. Mech Ageing Dev 10:273–281 (1979).
Napierala M, Parniewski P, Pluciennik A, Wells RD: Long CTG*CAG repeat sequences markedly stimulate intramolecular recombination. J biol Chem 3:3 (2002).
Nethanel T, Kaufmann G: Two DNA polymerases may be required for synthesis of the lagging DNA strand of simian virus 40. J Virol 64:5912–5918 (1990).
Neville CE, Mahadevan MS, Barcelo JM, Korneluk RG: High resolution genetic analysis suggests one ancestral predisposing haplotype for the origin of the myotonic dystrophy mutation. Hum molec Genet 3:45–51 (1994).
Nichol K, Pearson CE: CpG methylation modifies the genetic stability of cloned repeat sequences. Genome Res 12:1246–1256 (2002).
Nolin SL, Houck GE Jr, Gargano AD, Blumstein H, Dobkin CS, Brown WT: FMR1 CGG-repeat instability in single sperm and lymphocytes of fragile-X premutation males. Am J hum Genet 65:680–688 (1999).
Nolin SL, Jenkins EC, Brown WT, Dobkin CS: In situ nick translation of the fragile X region. Am J med Genet 30:443–450 (1988).
Nolin SL, Lewis FA 3rd, Ye LL, Houck GE Jr, Glicksman AE, Limprasert P, Li SY, Zhong N, Ashley AE, Feingold E, Sherman SL, Brown WT: Familial transmission of the FMR1 CGG repeat. Am J hum Genet 59:1252–1261 (1996).
Nouspikel T, Hanawalt PC: DNA repair in terminally differentiated cells. DNA Repair 1:59–75 (2002).
O'Hoy K, Tsilfidis LC, Mahadevan MS, Neville CE, Barcelo J, Hunter AG, Korneluk RG: Reduction in size of the myotonic dystrophy trinucleotide repeat mutation during transmission. Science 259:809–812 (1993).
Ohshima K, Sakamoto N, Labuda M, Poirier J, Moseley ML, Montermini L, Ranum LP, Wells RD, Pandolfo M: A nonpathogenic GAAGGA repeat in the Friedreich gene: implications for pathogenesis. Neurology 53:1854–1857 (1999).
Ohshima K, Sinden RR: *In vitro* DNA synthesis of triplet repeat sequences, in Wells RB, Warren ST (eds): Genetic Instabilities and Hereditary Neurological Diseases, pp 717–735 (Academic Press, San Diego 1998).
Ohshima K, Wells RD: Hairpin formation during DNA synthesis primer realignment *in vitro* in triplet repeat sequences from human hereditary disease genes. J biol Chem 272:16798–16806 (1997).
Okada T, Gondo Y, Goto J, Kanazawa I, Hadano S, Ikeda J E: Unstable transmission of the RS447 human megasatellite tandem repetitive sequence that contains the USP17 deubiquitinating enzyme gene. Hum Genet 110:302–313 (2002).
Oppenheim A, Shlomai Z, Ben-Bassat H: Initiation points for cellular deoxyribonucleic acid replication in human lymphoid cells converted by Epstein-Barr virus. Mol Cell Biol 1:753–762 (1981).
Osani R, Kinoshita M, Hirose K, Homma T, Kawabata I: CTG triplet repeat expansion in a laryngeal carcinonoma from a patient with myotonic dystrophy. Muscle Nerve 23:804–806 (2000).
Otten AD, Tapscott SJ: Triplet repeat expansion in myotonic dystrophy alters the adjacent chromatin structure. Proc natl Acad Sci, USA 92:5465–5469 (1995).
Panigrahi GB, Cleary JD, Pearson CE: *In vitro* (CTG)*(CAG) expansions and deletions by human cell extracts. J biol Chem 277:13926–13934 (2002).
Parkinson GN, Lee MP, Neidle S: Crystal structure of parallel quadruplexes from human telomeric DNA. Nature 417:876–880 (2002).
Pearson CE, Sinden RR: Alternative structures in duplex DNA formed within the trinucleotide repeats of the myotonic dystrophy and fragile X loci. Biochemistry 35:5041–5053 (1996).
Pearson CE, Sinden RR: Slipped Strand DNA Dynamic Mutations and Human Disease in Genetic Instabilities and Hereditary Neurological Diseases, Wells RD, Warren ST (eds), pp 585–621 (Academic Press, New York 1998).
Pearson CE, Sinden RR: Trinucleotide repeat DNA structures: dynamic mutations from dynamic DNA. Curr Opin Struct Biol 8:321–330 (1998).
Pearson CE, Ewel A, Acharya S, Fishel RA, Sinden R R: Human MSH2 binds to trinucleotide repeat DNA structures associated with neurodegenerative diseases. Hum Mol Genet 6:1117–1123 (1997).
Pearson CE, Eichler EE, Lorenzetti D, Kramer SF, Zoghbi HY, Nelson DL, Sinden RR: Interruptions in the triplet repeats of SCA1 and FRAXA reduce the propensity and complexity of slipped strand DNA (S-DNA) formation. Biochemistry 37:2701–2708 (1998).
Pearson CE, Tam M, Wang Y-H, Montgomery SE, Dar A, Cleary JD Nichol K: Slipped-strand DNAs formed by long (CAG)(CTG) repeats: slipped-out repeats and slip-out junctions. Nucl Acids Res 30:4534–4547 (2002).
Pelletier R, Krasilnikova MM, Samadashwily GM, Lahue R, Mirkin SM: Replication and expansion of trinucleotide repeats in yeast. Mol Cell Biol 23:1349–1357 (2003).
Petes TD, Greenwell PW, Dominska M: Stabilization of microsatellite sequences by variant repeats in the yeast *Saccharmoyces cerevisiae*. Genet 146: 491–498 (1997).
Peterlin B, Logar N, Zidar J: CTG repeat analysis in lymphocytes, muscles and fibroblasts in patients with myotonic dystrophy. Pflugers Arch 431:R199–200 (1996).
Petronis A, Vincent JB, Surh LC, Ashizawa T, Kennedy JL: Polyglutamine-containing proteins in schizophrenia: an effect of lymphoblastoid cells? Mol Psychiatry 5:234–236 (2000).
Pianese L, Cavalcanti F, De Michele G, Filla A, Campanella G, Calabrese O, Castaldo I, Monticelli A, Cocozza S: The effect of parental gender on the GAA dynamic mutation in the FRDA gene. Am J hum Genet 60:460–463 (1997).
Pluciennik A, Iyer RR, Napierala M, Larson JE, Filutowicz M, Wells RD: Long CTG*CAG repeats from myotonic dystrophy are preferred sites for intermolecular recombination. J biol Chem 26:26 (2002).

Price GB, Modak SP, Makinodan T: Age-associated changes in the DNA of mouse tissue. Science 171:917–920 (1971).
Pulst SM, Nechiporuk A, Nechiporuk T, Gispert S, Chen XN, Lopes-Cendes I, Pearlman S, Starkman S, Orozco-Diaz G, Lunkes A, DeJong P, Rouleau GA, Auburger G, Korenberg JR, Figueroa C, Sahba S: Moderate expansion of a normally biallelic trinucleotide repeat in spinocerebellar ataxia type 2. Nature Genet 14:269–276 (1996).
Razin A, Shemer R: DNA methylation in early development. Hum molec Genet 4:1751–1755 (1995).
Reik W, Dean W: DNA methylation and mammalian epigenetics. Electrophoresis 22:2838–2843 (2001).
Reyniers E, Vits L, De Boulle K, Van Roy B, Van Velzen D, de Graaff E, Verkerk AJ, Jorens HZ, Darby JK, Oostra B, et al: The full mutation in the FMR-1 gene of male fragile X patients is absent in their sperm. Natue Genet 4: 143–146 (1993).
Reyniers E, Martin JJ, Cras P, Van Marck E, Handig I, Jorens HZ, Oostra BA, Kooy RF, Willems PJ: Postmortem examination of two fragile X brothers with an FMR1 full mutation. Am J med Genet 84:245–249 (1999).
Richard GF, Goellner GM, McMurray CT, Haber JE: Recombination-induced CAG trinucleotide repeat expansions in yeast involve the MRE11-RAD50-XRS2 complex. EMBO J 19:2381–2390 (2000).
Richard GF, Paques F: Mini- and microsatellite expansions: the recombination connection. EMBO Rep 1:122–126 (2000).
Richards RI, Crawford J, Narahara K, Mangelsdorf M, Friend K, Staples A, Denton M, Easteal S, Hori TA, Kondo I, Jenkins T, Goldman A, Panich V, Ferakova E, Sutherland GR: Dynamic mutation loci: allele distributions in different populations. Ann Hum Genet 60:391–400 (1996).
Richards RI, Sutherland GR: Dynamic mutations: a new class of mutations causing human disease. Cell 70:709–712 (1992).
Richards RI, Sutherland GR: Simple repeat DNA is not replicated simply. [news] Nature Genet 6:114–116 (1994).
Richardson LL, Pedigo C, Handel AM: Expression of deoxyribonucleic acid repair enzymes during spermatogenesis in mice. Biol Reprod 62:789–796 (2000).
Rolfsmeier ML, Dixon MJ, Lahue RS: Mismatch repair blocks expansions of interrupted trinucleotide repeats in yeast. Mol Cell 6:1501–1507 (2000).
Roth DB, Roth SY: Unequal access: regulating V(D)J recombination through chromatin remodeling. Cell 103:699–702 (2000).
Rousseau F, Heitz D, Tarleton J, MacPherson J, Malmgren H, Dahl N, Barnicoat A, Mathew C, Mornet E, Tejada I, et al: A multicenter study on genotype-phenotype correlations in the fragile X syndrome using direct diagnosis with probe StB123: the first 2253 cases. Am J hum Genet 55:225–237 (1994).
Rubinsztein DC, Barton DE, Davison BC, Ferguson-Smith MA: Analysis of the huntingtin gene reveals a trinucleotide-length polymorphism in the region of the gene that contains two CCG-rich stretches and a correlation between decreased age of onset of Huntington's disease and CAG repeat number. Hum molec Genet 2:1713–1715 (1993).
Rudiger NS, Gregersen N, Kielland-Brandt MC: One short well conserved region of Alu-sequences is involved in human gene rearrangements and has homology with prokaryotic chi. Nucl Acids Res 23:256–260 (1995).
Sadowski PD: Bacteriophage T7 endonuclease I. Properties of the enzyme purified from T7 phage-infected *Escherichia coli B.* J biol Chem 246:209–216 (1971).
Saha T, Usdin K: Tetraplex formation by the progressive myoclonus epilepsy type-1 repeat: implications for instability in the repeat expansion diseases. FEBS Lett 491:184–187 (2001).
Saintigny Y, Delacote F, Vares G, Petitot F, Lambert S, Averbeck D, Lopez BS: Characterization of homologous recombination induced by replication inhibition in mammalian cells. EMBO J 20:3861–3870 (2001).
Sakamoto N, Chastain PD, Parniewski P, Ohshima K, Pandolfo M, Griffith JD, Wells RD: Sticky DNA: self-association properties of long GAATTC repeats in RRY triplex structures from Friedreich's ataxia. Mol Cell 3:465–75 (1999).
Sakamoto N, Larson JE, Iyer RR, Montermini L, Pandolfo M, Wells RD: GGA*TCC-interrupted triplets in long GAA*TTC repeats inhibit the formation of triplex and sticky DNA structures, alleviate transcription inhibition, and reduce genetic instabilities. J biol Chem 276:27178–27187 (2001).
Salat U, Bardoni B, Wohrle D, Steinbach P: Increase of FMRP expression raised levels of FMR1 mRNA and clonal selection in proliferating cells with unmethylated fragile X repeat expansions: a clue to the sex bias in the transmission of full mutations? J med Genet 37:842–850 (2000).
Samadashwily GM, Raca G, Mirkin SM: Trinucleotide repeats affect DNA replication *in vivo.* Nature Genet 17:298–304 (1997).
Sato T, Oyake M, Nakamura K, Nakao K, Fukusima Y, Onodera O, Igarashi S, Takano H, Kikugawa K, Ishida Y, Shimohata T, Koide R, Ikeuchi T, Tanaka H, Futamura N, Matsumura R, Takayanagi T, Tanaka F, Sobue G, Komure O, Takahashi M, Sano A, Ichikawa Y, Goto J, Kanazawa I, et al: Transgenic mice harboring a full-length human mutant DRPLA gene exhibit age-dependent intergenerational and somatic instabilities of CAG repeats comparable with those in DRPLA patients. Hum mol Genet 8:99–106 (1999).
Scacheri PC, Garcia C, Hebert R, Hoffman EP: Unique PABP2 mutations in "Cajuns" suggest multiple founders of oculopharyngeal muscular dystrophy in populations with French ancestry. Am J med Genet 86: 477–481 (1999).
Schaffitzel C, Berger I, Postberg J, Hanes J, Lipps HJ, Pluckthun A: *In vitro* generated antibodies specific for telomeric guanine-quadruplex DNA react with *Stylonychia lemnae* macronuclei. Proc natl Acad Sci, USA 98:8572–8577 (2001).
Schumacher S, Fuchs RP, Bichara M: Expansion of CTG repeats from human disease genes is dependent upon replication mechanisms in *Escherichia coli*: the effect of long patch mismatch repair revisited. J molec Biol 279:1101–1110 (1998).
Schumacher S, Pinet I, Bichara M: Modulation of transcription reveals a new mechanism of triplet repeat instability in *Escherichia coli.* J molec Biol 307:39–49 (2001).
Schweitzer JK, Livingston DM: Destabilization of CAG trinucleotide repeat tracts by mismatch repair mutations in yeast. Hum molec Genet 6:349–355 (1997).
Schweitzer JK, Livingston DM: Expansions of CAG repeat tracts are frequent in a yeast mutant defective in Okazaki fragment maturation. Hum molec Genet 7:69–74 (1998).
Schweitzer JK, Livingston DM: The effect of DNA replication mutations on CAG tract stability in yeast. Genetics 152:953–963 (1999).
Sergeant N, Sablonniere B, Schraen-Maschke S, Ghestem A, Maurage CA: Dysregulation of human brain microtubule-associated tau mRNA maturation in myotonic dystrophy type 1. Hum molec Genet 10:2143–2155 (2001).
Seibert E, Ross JB, Osman R: Role of DNA flexibility in sequence-dependent activity of uracil DNA glycosylase. Biochemistry 41:10976–10984 (2002).
Seznec H, Lia-Baldini AS, Duros C, Fouquet C, Lacroix C, Hofmann-Radvanyi H, Junien C, Gourdon G: Transgenic mice carrying large human genomic sequences with expanded CTG repeat mimic closely the DM CTG repeat intergenerational and somatic instability. Hum molec Genet 9:1185–1194 (2000).
Sharma R, Bhatti S, Gomez M, Clark RM, Murray C, Ashizawa T, Bidichandani SI: The GAA triplet-repeat sequence in Friedreich ataxia shows a high level of somatic instability *in vivo* with a significant predilection for large contractions. Hum molec Genet 11:2175–2187 (2002).
Shelbourne P, Winqvist R, Kunert E, Davies J, Leisti J, Thiele H, Bachmann H, Buxton J, Williamson B, Johnson K: Unstable DNA may be responsible for the incomplete penetrance of the myotonic dystrophy phenotype. Hum molec Genet 1:467–473 (1992).
Sherman SL, Jacobs PA, Morton NE, Froster-Iskenius U, Howard-Peebles PN, Nielsen KB, Partington MW, Sutherland GR, Turner G, Watson M: Further segregation analysis of the fragile X syndrome with special reference to transmitting males. Hum Genet 69:289–299 (1985).
Shimizu M, Gellibolian R, Oostra BA, Wells RD: Cloning, characterization and properties of plasmids containing CGG triplet repeats from the FMR-1 gene. J molec Biol 258:614–626 (1996).
Shimizu M, Mori T, Sakurai T, Shindo H: Destabilization of nucleosomes by an unusual DNA conformation adopted by poly(dA) small middle dotpoly(dT) tracts *in vivo.* EMBO J 19:3358–3365 (2000).
Shiota K, Kogo Y, Ohgane J, Imamura T, Urano A, Nishino K, Tanaka S, Hattori N: Epigenetic marks by DNA methylation specific to stem, germ and somatic cells in mice. Genes Cells 7:961–969 (2002).
Sinden RR: Neurodegenerative diseases. Origins of instability. Nature 411:757–758 (2001).
Sleddens HF, Oostra BA, Brinkmann AO, Trapman J: Trinucleotide (GGN) repeat polymorphism in the human androgen receptor (AR) gene. Hum molec Genet 2:493 (1993).
Smeets HJ, Smits AP, Verheij CE, Theelen JP, Willemsen R, van de Burgt I, Hoogeveen AT, Oosterwijk JC, Oostra BA: Normal phenotype in two brothers with a full FMR1 mutation. Hum molec Genet 4:2103–2108 (1995).
Smit AF: Interspersed repeats and other mementos of transposable elements in mammalian genomes. Curr Opin Genet Dev 9:657–663 (1999).
Smith GP: Evolution of repeated DNA sequences by unequal crossover. Science 191:528–535 (1976).
Smits A, Smeets D, Dreesen J, Hamel B, de Haan A, van Oost B: Parental origin of the fra(X) gene is a major determinant of the cytogenetic expression and the CGG repeat length in female carriers. Am J med Genet 43:261–267 (1992).
Spiro C, Pelletier R, Rolfsmeier ML, Dixon MJ, Lahue RS, Gupta G, Park MS, Chen X, Mariappan SV, McMurray CT: Inhibition of FEN-1 processing by DNA secondary structure at trinucleotide repeats. Mol Cell 4:1079–1085 (1999).
Squitieri F, Andrew SE, Goldberg YP, Kremer B, Spence N, Zeisler J, Nichol K, Theilmann J, Greenberg J, Goto J, et al: DNA haplotype analysis of Huntington disease reveals clues to the origins and mechanisms of CAG expansion and reasons for geographic variations of prevalence. Hum mol Genet 3:2103–2114 (1994).
Stevanin G, David G, Durr A, Giunti P, Benomar A, Abada-Bendib M, Lee MS, Agid Y, Brice A: Multiple origins of the spinocerebellar ataxia 7 (SCA7) mutation revealed by linkage disequilibrium studies with closely flanking markers including an intragenic polymorphism (G3145TG/A3145TG). Eur J hum Genet 7:889–896 (1999).

Stevanin G, Giunti P, Belal GD, Durr A, Ruberg M, Wood N, Brice A: *De novo* expansion of intermediate alleles in spinocerebellar ataxia 7. Hum molec Genet 7:1809–1813 (1998).
Strong TV, Tagle DA, Valdes JM, Elmer LW, Boehm K, Swaroop M, Kaatz KW, Collins FS, Albin RL: Widespread expression of the human and rat Huntington's disease gene in brain and nonneural tissues. Nature Genet 5:259–265 (1993).
Sturzeneker R, Bevilacqua RAU, Haddad LA, Simpson AJG, Pena SDJ: Microsatellite instability in tumors as a model to study the process of microsatellite mutations. Hum molec Genet 9:347–352 (2000).
Sullivan AK, Crawford DC, Scott EH, Leslie ML, Sherman SL: Paternally transmitted FMR1 alleles are less stable than maternally transmitted alleles in the common and intermediate size range. Am J hum Genet 70:1532–1544 (2002).
Sullivan KM, Lilley DM: A dominant influence of flanking sequences on a local structural transition in DNA. Cell 47:817–827 (1986).
Sultan C, Lumbroso S, Poujol N, Belon C, Boudon C, Lobaccaro JM: Mutations of androgen receptor gene in androgen insensitivity syndromes. J Steroid Biochem Mol Biol 46:519–530 (1993).
Sutcliffe JS, Nelson DL, Zhang F, Pieretti M Caskey CT, Saxe D, Warren ST: DNA methylation represses FMR-1 transcription in fragile X syndrome. Hum molec Genet 1:397–400 (1992).
Suter B, Schnappauf G, Thoma F: Poly(dAdT) sequences exist as rigid DNA structures in nucleosome-free yeast promoters *in vivo*. Nucl Acids Res 28:4083–4089 (2000).
Sutherland GR, Richards RI: Simple tandem DNA repeats and human genetic disease. Proc natl Acad Sci, USA 92:3636–3641 (1995).
Sutherland GR, Richards RI: Fragile sites-cytogenetic similarity with molecular diversity. Am J hum Genet 64:354–359 (1999).
Suzuki A, Maruno A, Tahira T, Hayashi K: Polar alteration of short tandem repeats (STRs) in mammalian cells. Mutat Res 474:159–168 (2001).
Takano H, Cancel G, Ikeuchi T, Lorenzetti D, Mawad R, Stevanin G, Didierjean O, Durr A, Oyake M, Shimohata T, Sasaki R, Koide R, Igarashi S, Hayashi S, Takiyama Y, Nishizawa M, Tanaka H, Zoghbi H, Brice A, Tsuji S: Close associations between prevalences of dominantly inherited spinocerebellar ataxias with CAG-repeat expansions and frequencies of large normal CAG alleles in Japanese and Caucasian populations. Am J hum Genet 63:1060–1066 (1998).
Tanaka F, Doyu M, Ito Y, Matsumoto M, Mitsuma T, Abe K, Aoki M, Itoyama Y, Fischbeck KH, Sobue G: Founder effect in spinal and bulbar muscular atrophy (SBMA). Hum molec Genet 5:1253–1257 (1996a).
Tanaka F, Sobue G, Doyu M, Ito Y, Yamamoto M,, Shimada N, Yamamoto K, Riku S, Hshizume Y, Mitsuma T: Differential pattern in tissue-specific somatic mosaicism of expanded CAG trinucleotide repeats in dentatorubral-pallidoluysian atrophy, Machado-Joseph disease and X-linked recessive spinal and bulbar muscular atrophy. J Neurol Sci 135:43–50 (1996b).
Tassone F, Hagerman RJ, Gane LW, Taylor AK: Strong similarities of the FMR1 mutation in multiple tissues: postmortem studies of a male with a full mutation and a male carrier of a premutation. Am J med Genet 84:240–244 (1999).
Taylor AK, Tassone F, Dyer PN, Hersch SM, Harris JB, Greenough WT, Hagerman RJ: Tissue heterogeneity of the FMR1 mutation in a high-functioning male with fragile X syndrome. Am J med Genet 84:233–239 (1999).
Telenius H, Kremer B, Goldberg YP, Theilmann J, Andrew SE, Zeisler J,, Adam S, Greenberg C, Ives EJ, Clarke LA, et al: Somatic and gonadal mosaicism of the Huntington disease gene CAG repeat in brain and sperm. Nature Genet 6:409–414 (1994).
Thornton CA, Johnson K, Moxley RT 3rd: Myotonic dystrophy patients have larger CTG expansions in skeletal muscle than in leukocytes. Ann Neurol 35:104–107 (1994).
Tishkoff SA, Goldman A, Calafell F, Speed WC, Deinard AS, Bonne-Tamir B, Kidd JR, Pakstis AJ, Jenkins T, Kidd KK: A global haplotype analysis of the myotonic dystrophy locus: implications for the evolution of modern humans and for the origin of myotonic dystrophy mutations. Am J hum Genet 62:1389–1402 (1998).
Toyota M, Ahuja N, Ohe-Toyota M, Herman JG, Baylin SB, Issa JP: CpG island methylator phenotype in colorectal cancer Proc natl Acad Sci, USA 96:8681–8686 (1999).
Trasler JM: Origin and roles of genomic methylation patterns in male germ cells. Semin Cell Dev Biol 9:467–474 (1998).
Trasler JM, Hake LE, Johnson PA, Alcivar AA, Millette CF, Hecht NB: DNA methylation and demethylation events during meiotic prophase in the mouse testis. Mol Cell Biol 10:1828–1834 (1990).
Trottier Y, Biancalana V, Mandel JL: Instability of CAG repeats in Huntington's disease: relation to parental transmission and age of onset. J Med Genet 31:377–382 (1994).
Ueno S, Kondoh K, Kotani Y, Komure O, Kuno S, Kawai J, Hazama F, Sano A: Somatic mosaicism of CAG repeat in dentatorubral-pallidoluysian atrophy (DRPLA). Hum molec Genet 4:663–666 (1995).
van de Warrenburg BP, Sinke RJ, Verschuuren-Bemelmans CC, Scheffer H, Brunt ER, Ippel PF, Maat-Kievit JA, Dooijes D, Notermans NC, Lindhout D, Knoers NV, Kremer HP: Spinocerebellar ataxias in the Netherlands: prevalence and age at onset variance analysis. Neurology 58: 702–708 (2002).
van Den Broek WJ, Nelen MR, Wansink DG, Coerwinkel MM, te Riele H, Groenen PJ, Wieringa B: Somatic expansion behaviour of the (CTG)(n) repeat in myotonic dystrophy knock-in mice is differentially affected by Msh3 and Msh6 mismatch-repair proteins, Hum molec Genet 11:191–198 (2002).
van den Ouweland AM, Deelen WH, Kunst CB, Uzielli ML, Nelson DL, Warren ST, Oostra BA, Halley DJ: Loss of mutation at the FMR1 locus through multiple exchanges between maternal X chromosomes. Hum Mol Genet 3:1823–1827 (1994).
van Geel M, Dickson MC, Beck AF, Bolland DJ, Frants RR, van der Maarel SM, de Jong PJ, Hewitt JE: Genomic analysis of human chromosome 10q and 4q telomeres suggests a common origin. Genomics 79:210–217 (2002).
Virtaneva K, D'Amato E, Miao J, Koskiniemi M, Norio R, Avanzini G, Franceschetti S, Michelucci R, Tassinari CA, Omer S, Pennacchio LA, Myers RM, Dieguez-Lucena JL, Krahe R, de la Chapelle A, Lehesjoki AE: Unstable minisatellite expansion causing recessively inherited myoclonus epilepsy EPM1. Nature Genet 15:393–396 (1997).
Vogel F, Rathenberg R: Spontaneous mutation in man. Adv Hum Genet 5:223–318 (1975).
von Eckardstein S, Syska A, Gromoll J, Kamischke A, Simoni M, Nieschlag E: Inverse correlation between sperm concentration and number of androgen receptor CAG repeats in normal men. J Clin Endocrinol Metab 86:2585–2590 (2001).
Waga S, Stillman B: Anatomy of a DNA replication fork revealed by reconstitution of SV40 DNA replication *in vitro*. Nature 369:207–212 (1994).
Waga S, Bauer G, Stillman B: Reconstitution of complete SV40 DNA replication with purified replication factors. J biol Chem 269:10923–10934 (1994).
Wang YH, Griffith J: Expanded CTG triplet blocks from the myotonic dystrophy gene create the strongest known natural nucleosome positioning elements. Genomics 25:570–573 (1995).
Wang YH, Griffith J: Methylation of expanded CCG triplet repeat DNA from fragile X syndrome patients enhances nucleosome exclusion. J biol Chem 271:22937–22940 (1996).
Wang Y, Gellibolian HR, Shimizu M, Wells RD, Griffith J: Long CCG triplet repeat blocks exclude nucleosomes: a possible mechanism for the nature of fragile sites in chromosomes. J molec Biol 263:511–516 (1996).
Weber JL: Informativeness of human (dC-dA)n(dG-dT)n polymorphisms. Genomics 7:524–530 (1990).
Weisman-Shomer P, Cohen E, Fry M: Interruption of the fragile X syndrome expanded sequence d(CGG)(n) by interspersed d(AGG) trinucleotides diminishes the formation and stability of d(CGG)(n) tetrahelical structures. Nucl Acids Res 28:1535–1541 (2000).
Werntges H, Steger G, Riesner D, Fritz HJ: Mismatches in DNA double strands: thermodynamic parameters and their correlation to repair efficiencies. Nucl Acids Res 14:3773–3790 (1986).
Widlund HR, Kuduvalli PN, Bengtsson M, Cao H, Tullius TD, Kubista M: Nucleosome structural features and intrinsic properties of the TATAAACGCC repeat sequence. J biol Chem 274: 31847–31852 (1999).
Willems PJ, Van Roy B, De Boulle K, Vits L, Reyniers E, Beck O, Dumon JE, Verkerk A, Oostra B: Segregation of the fragile X mutation from an affected male to his normal daughter. Hum molec Genet 1:511–515 (1992).
Winokur ST, Bengtsson U, Feddersen J, Mathews KD, Weiffenbach B, Bailey H, Markovich RP, Murray JC, Wasmuth JJ, Altherr MR, et al: The DNA rearrangement associated with facioscapulohumeral muscular dystrophy involves a heterochromatin-associated repetitive element: implications for a role of chromatin structure in the pathogenesis of the disease. Chrom Res 2:225–234 (1994).
Wohrle D, Hennig I, Vogel W, Steinbach P: Mitotic stability of fragile X mutations in differentiated cells indicates early post-conceptional trinucleotide repeat expansion. Nature Genet 4:140–142 (1993).
Wohrle D, Kennerknecht I, Wolf M, Enders H, Schwemmle S, Steinbach P: Heterogeneity of DM kinase repeat expansion in different fetal tissues and further expansion during cell proliferation *in vitro*: evidence for a casual involvement of methyl-directed DNA mismatch repair in triplet repeat stability. Hum molec Genet 4:1147–1153 (1995).
Wohrle D, Salat U, Glaser D, Mucke J, Meisel-Stosiek M, Schindler D, Vogel W, Steinbach P: Unusual mutations in high functioning fragile X males: apparent instability of expanded unmethylated CGG repeats. J Med Genet 35:103–111 (1998).
Wohrle D, Salat U, Hameister H, Vogel W, Steinbach P: Demethylation reactivation and destabilization of human fragile X full-mutation alleles in mouse embryocarcinoma cells. Am J hum Genet 69:504–515 (2001).
Wolffe AP: Transcription: in tune with the histones. Cell 77:13–16 (1994).
Wong LJ, Ashizawa T, Monckton DG, Caskey CT, Richards CS: Somatic heterogeneity of the CTG repeat in myotonic dystrophy is age and size dependent. Am J hum Genet 56:114–122 (1995).

Woodcock DM, Lawler CB, Linsenmeyer ME, Doherty JP, Warren WD: Asymmetric methylation in the hypermethylated CpG promoter region of the human L1 retrotransposon. J biol Chem 272:7810–7816 (1997).

Xu GL, Bestor TH, Bourc'his D, Hsieh CL, Tommerup N, Bugge M, Hulten M, Qu X, Russo JJ, Viegas-Pequignot E: Chromosome instability and immunodeficiency syndrome caused by mutations in a DNA methyltransferase gene. Nature 402:187–191 (1999).

Yamagata H, Miki T, Yamanaka N, Takemoto Y, Kanda F, Takahashi K, Inui T, Kinoshita M, Nakagawa M, Higuchi I, et al: Characteristics of dynamic mutation in Japanese myotonic dystrophy. Jpn J hum Genet 39:327–335 (1994).

Yanagisawa H, Fujii K, Nagafuchi S, Nakahori Y, Nakagome Y, Akane A, Nakamura M, Sano A, Komure O, Kondo I, Jin DK, Sorensen SA, Potter NT, Young SR, Nakamura K, Nukina N, Nagao Y, Tadokoro K, Okuyama T, Miyashita T, Inoue T, Kanazawa I, Yamada M: A unique origin and multistep process for the generation of expanded DRPLA triplet repeats. Hum molec Genet 5:373–379 (1996).

Yandava CN, Gastier JM, Pulido JC, Brody T, Sheffield V, Murray J, Buetow K, Duyk GM: Characterization of *Alu* repeats that are associated with trinucleotide and tetranucleotide repeat microsatellites. Genome Res 7:716–724 (1997).

Yu A, Dill J, Mitas M: The purine-rich trinucleotide repeat sequences d(CAG)15 and d(GAC)15 form hairpins. Nucl Acids Res 23:4055–4057 (1995).

Yu S, Mangelsdorf M, Hewett D, Hobson L, Baker E, Eyre HJ, Lapsys N, Le Paslier D, Doggett NA, Sutherland GR, Richards RI: Human chromosomal fragile site FRA16B is an amplified AT-rich minisatellite repeat. Cell 88:367–374 (1997).

Yvert G, Mandel JL: Variation on a trinucleotide theme. Nature Med 5:383–384 (1999).

Zacharias W: Methylation of cytosine influences the DNA structure. Exs 64:27–38 (1993).

Zatz M, Passos-Bueno MR, Cerqueira A, Marie SK, Vainzof M, Pavanello RC: Analysis of the CTG repeat in skeletal muscle of young and adult myotonic dystrophy patients: when does the expansion occur? Hum molec Genet 4:401–406 (1995).

Zeesman S, Carson N, Whelan DT: Paternal transmission of the congenital form of myotonic dystrophy type 1: a new case and review of the literature. Am J med Genet 107:222–226 (2002).

Zhang L, Fischbeck KH, Arnheim N: CAG repeat length variation in sperm from a patient with Kennedy's disease. Hum mol Genet 4:303–305 (1995).

Zhang L, Leeflang EP, Yu J, Arnheim N: Studying human mutations by sperm typing: instability of CAG trinucleotide repeats in the human androgen receptor gene. Nature Genet 7:531–535 (1994).

Zhang Y, Monckton DG, Siciliano MJ, Connor TH, Meistrich ML: Age and insertion site dependence of repeat number instability of a human DM1 transgene in individual mouse sperm. Hum molec Genet 11:791–798 (2002).

Zhong N, Ju W, Pietrofesa J, Wang D, Dobkin C, Brown WT: Fragile X "gray zone" alleles: AGG patterns expansion risks and associated haplotypes. Am J med Genet 64:261–265 (1996).

Zuliani G, Hobbs HH: A high frequency of length polymorphisms in repeated sequences adjacent to *Alu* sequences. Am J hum Genet 46:963–969 (1990).

Cytogenet Genome Res 100:56–64 (2003)
DOI: 10.1159/000072838

Reactivation of silenced genes and transcriptional therapy

P. Chiurazzi and G. Neri

Institute of Medical Genetics, Catholic University, Rome (Italy)

Abstract. The purpose of this review is to discuss the potential role of "transcriptional therapy" to modulate the expression of target genes in order to treat monogenic as well as multifactorial disorders. In vitro and in vivo experiments with DNA demethylating and histone hyperacetylating drugs are currently performed in several laboratories on a variety of genes. In attempting to place these results into perspective, we divided the target genes into four major categories: (1) single genes with a hypermethylated CpG island; (2) single genes without a CpG island; (3) groups of genes silenced by aberrant DNA methylation; and (4) groups of genes silenced by lack of histone acetylation. We discuss the latest advances in the field of chromatin regulation and, in particular, the role of histone methylation and that of RNA interference in gene silencing. We can expect that in the future regulation of transcription will become an effective treatment for several genetic conditions.

Reactivation of silent or poorly expressed genes represents a potential tool to treat monogenic as well as multifactorial disorders, such as cancer. The results of experimental studies in vitro and in animal models, while promising, raise several questions about specificity and potential genome-wide effects of nonspecific molecules. Although a basic knowledge of the regulation of transcription is rapidly accumulating, the general picture remains incomplete, and a thorough understanding of chromatin remodeling is necessary before any therapeutic trial in humans can take place.

The central role previously assigned to DNA methylation is being revised and should be integrated into a more complex network of chromatin modifications (Fig. 1).

In this review, genes that hold the potential of being targeted by reactivating drugs were divided into four categories (Table 1): (1) genes with an abnormally hypermethylated CpG island that can be reactivated with DNA demethylating agents (e.g., 5-azadeoxycytidine and procainamide); (2) genes lacking a CpG island that can be silenced usually by histone deacetylation and whose expression can be increased with hyperacetylating drugs (e.g., butyrate and its analogs); (3) groups of genes with a CpG island that can be inactivated by inappropriate genome-wide DNA methylation, as has been observed in several types of cancer; and (4) groups of genes that can be silenced because of inadequate histone acetylation, as recently discovered in some neurodegenerative disorders.

We shall also stress that "transcriptional therapy" using simple, small-molecule drugs (Table 2) has a potential genome-wide effect on DNA methylation or histone acetylation and may therefore be more useful in treating multifactorial conditions, such as those discussed below under global dysregulation in cancer and neurodegenerative disorders. However, these drugs should also be tried for treating monogenic disorders in animal models before concentrating our efforts on artificial transcription factors, such as those envisaged by Reik et al. (2002). Lastly, we discuss the possibility that it might be possible to reactivate or suppress the transcription of specific target genes with the help of RNA interference techniques.

The authors' work quoted in this review was supported by FRAXA Foundation, Conquer Fragile X Foundation, MIUR Italia, Fondazione Telethon, and Sigma-Tau.

Received 22 November 2002; revision accepted 18 February 2003.

Request reprints from Dr. Giovanni Neri
Istituto di Genetica Medica, Università Cattolica
Largo F. Vito 1, 00168 Rome (Italy)
telephone: +39-06-3054449; fax: +39-06-3050031
e-mail: gneri@rm.unicatt.it

KARGER
Fax + 41 61 306 12 34
E-mail karger@karger.ch
www.karger.com

0301–0171/03/1004–0056$19.50/0

Accessible online at:
www.karger.com/cgr

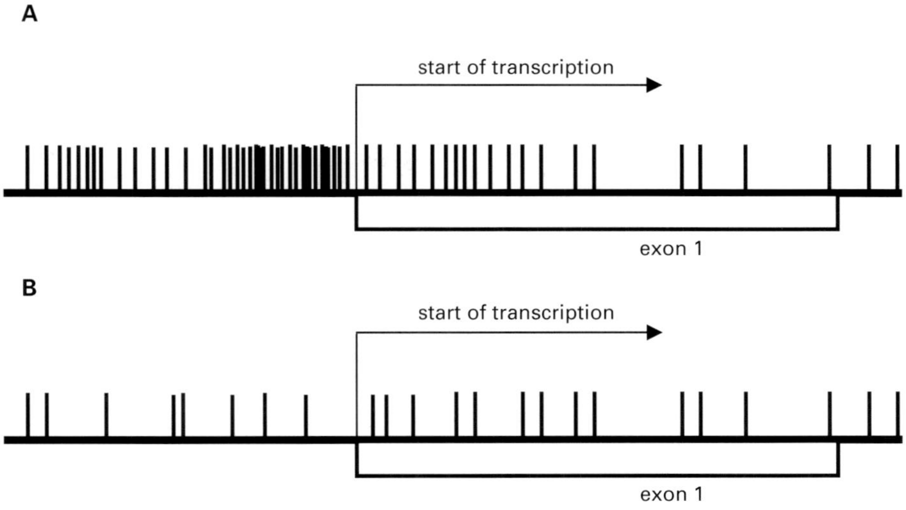

Fig. 1. CpG sites density in different types of promoters: (**A**) promoter with CpG island, usually associated with "housekeeping" genes; (**B**) promoter without a CpG island, sometimes found in tissue-specific genes. Each CpG site is indicated by a vertical bar on the thick line representing the DNA sequence. Start of transcription is indicated by a the arrow pointing to the right, and exon 1 is indicated by a box.

Table 1. Some genes or groups of genes that have been targeted in reactivation experiments

	Single target	Multiple targets
Genes with a CpG island (methylated DNA when silenced)	Genes with a CGG expansion: FMR1 (FRAXA), FAM11A (FRAXF)	Cancer genes (tumor-suppressor genes): MLH1, CDKN2A, TIMP3, PTGS2, etc.
	Imprinted genes: SNRPN gene at the Prader-Willi imprinting center (PW-IC)	Genes on the inactive X chromosome
Genes without a CpG island (no DNA methylation when silenced)	Partially silenced genes: gamma-globin	Genes downregulated in poly-glutamine (poly-Q) diseases: Huntington, spinocerebellar ataxias (SCAs), etc.
	Active genes (potentiation): ALDPL1, SMN2 (via SR proteins), CFTR (F508) ?	

Table 2. Some drugs that have been employed in reactivation experiments

Drug class	Examples
DNA demethylating drugs (DNA methyltransferase inhibitors)	Azacytidine (azaC) 5-Azadeoxycytidine (5-azadC) Procainamide Hydralazine
Histone hyperacetylating drugs (HDAC inhibitors)	Sodium butyrate 4-Phenylbutyrate (4-PBA) Trichostatin A (TSA) Suberoylanilide hydroxamic acid (SAHA)
Other drugs (action unknown)	Hydroxyurea

DNA methylation

Cytosine methylation in position 5 of the pyrimidine ring is a common modification in DNA of mammals and other vertebrates, which usually is associated with the loss of transcription and heterochromatin formation (Bestor and Tycko, 1996). The methyl group is added to CpG sites by at least three DNA methyltransferases: DNMT1, DNMT3A, and DNMT3B (Bestor, 2000). The latter two are also expressed during early development (when a de novo wave of methylation takes place) and cooperate with DNMT1 to ensure maintenance of the methylation pattern during DNA replication (Rhee et al., 2000, 2002; Liang et al., 2002; Robert et al., 2003). Although commonly associated with transcriptional regulation, cytosine methylation mainly appears to direct silencing of transposons and endogenous retroviruses (Bestor and Tycko, 1996; Yoder et al., 1997; Walsh and Bestor, 1999; Liang et al., 2002). Gene promoters are almost always devoid of cytosine methylation, even when the gene is inactive, independently of their type, with or without a CpG-rich "island," as shown in Fig. 1.

CpG islands seem to owe their high cytosine content to the fact that transcription and/or replication factors "protect" DNA from methylation by binding promoters of "housekeeping" genes during development (Antequera and Bird, 1999). Methylcytosines can easily mutate by deamination to thymines, which can escape repair, whereas deamination of cytosine to uracil is more readily corrected. CpG islands are only methylated in promoters of imprinted genes or in genes located on the inactive X chromosome. In these cases the corresponding genes are physiologically inactive (i.e., "silent") and reside in heterochromatic domains.

Reactivation of genes with a methylated CpG island

Table 1 lists genes that may become a target of "transcriptional therapy," i.e., whose reactivation or increased expression might be beneficial in certain genetic conditions. The first class of targets that we discuss are single genes with a CpG island. Such genes can be pathologically silenced because of the presence of an expanded sequence (e.g., genes associated with the FRAXA, FRAXE, and FRAXF fragile sites).

The fragile X syndrome of mental retardation is due to the pathological silencing of the FMR1 gene located at Xq27.3, due

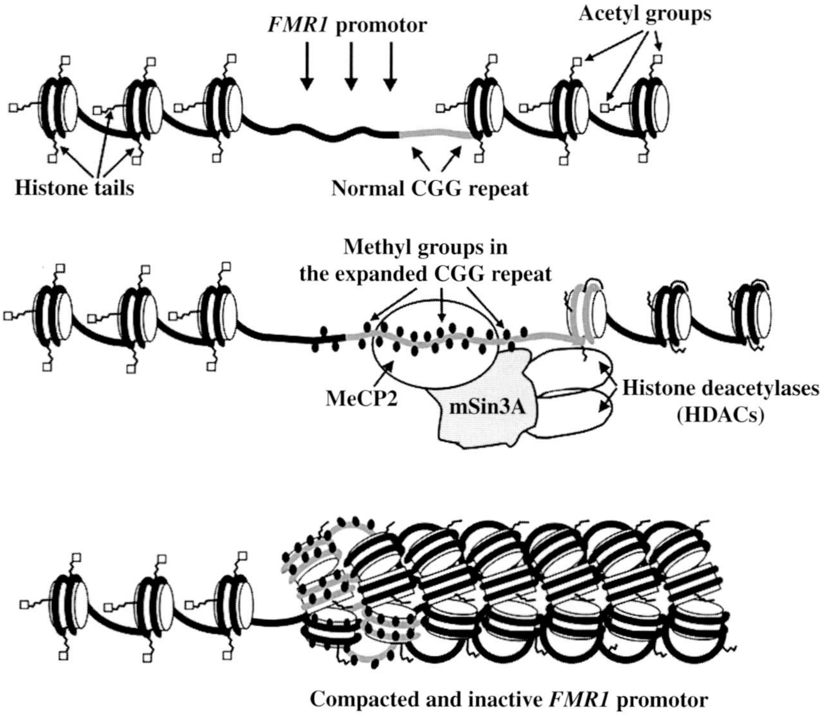

Fig. 2. Schematic representation of chromatin structures in the FMR1 promoter region and the CGG repeat. In cells with a normal number of CGG repeats, the promoter region is accessible for transcription factors, and the gene is active (upper panel). When the CGG repeat is expanded beyond the threshold of ~200 repeats, cytosines in the repeat and in the adjacent promoter become methylated, and a methyl-DNA binding protein such as MeCP2, is supposed to attach to the site (middle panel). MeCP2 recruits one (or more) chromatin-remodeling complex(es) that cause histone deacetylation and eventually heterochromatin formation (lower panel).

to the abnormal expansion of a CGG repeat located in its promoter overlapping a CpG island. Upon expansion of the repeat (> 200 repeats), the entire promoter region becomes methylated and transcription stops (Chiurazzi and Neri, 2001). The clinical phenotype of the syndrome is due to the lack of FMRP protein, which can be produced only when the FMR1 gene promoter is unmethylated. The expanded CGG sequence is probably treated by the cell as a potentially dangerous repeated element (Bestor and Tycko, 1996) and is therefore methylated to achieve transcriptional silencing. The fragile site FRAXA, which gives the name to the syndrome, actually coincides with the CGG expansion in patients. Two other similar fragile sites, FRAXE and FRAXF, are located at Xq28, close to FRAXA. These two fragile sites are also due the expansion and hypermethylation of a CGG repeat, and the corresponding genes, FMR2 and FAM11A, become inactivated, just like FMR1 in the fragile X syndrome.

5-Azadeoxycytidine (5-azadC) is a nucleotide analog that can be incorporated into DNA and, as shown in Table 2, becomes a potent and irreversible inhibitor of DNA methyltransferases (Jackson-Grusby et al., 1997). We have shown that in vitro treatment with 5-azadC of fragile X cells leads to reactivation of FMR1 transcription (Chiurazzi et al., 1998), after causing passive DNA demethylation (Pietrobono et al., 2002). Furthermore, we recently proved that treating cell lines of FRAXF individuals with 5-azadC also causes transcriptional reactivation of the FAM11A gene (Shaw et al., 2002).

Deletion or uniparental disomy of an imprinted gene can also cause gene silencing. In fact, loss of the only functional copy of the SNRPN and neighboring genes, expressed only from the paternal chromosome 15, causes Prader-Willi syndrome. In this case, Saitoh and Wada (2000) were able to achieve reactivation of the silenced copy of SNRPN in cell lines of Prader-Willi patients after treatment with 5-azadC.

Histone acetylation

DNA methylation represents just one step in a cascade of events that leads to heterochromatin formation and gene inactivation (Kornberg and Lorch, 2002).

Although cytosine methylation can directly inhibit the binding of certain transcription factors, methylated DNA is recognized by different methyl-DNA binding proteins (such as MeCP2 and MBD2), which eventually recruit other enzymes capable of removing acetyl groups from histones (Bird and Wolffe, 1999; Ng and Bird, 1999). These histone deacetylases (HDACs) remove acetyl groups from lysine residues in the *N*-terminal tail of histones H3 and H4. Then, repressor proteins, such as the ISWI ATPase, bind deacetylated histone tails and initiate chromatin condensation (i.e. heterochromatin formation) (Corona et al., 2001).

Conversely, histone tails can be acetylated by histone acetyltransferases (HATs), and acetylated histones are usually associated with transcriptionally active chromatin (Grunstein, 1997). Acetylated histone tails are recognized by the so-called "bromodomain" motif (Kouzarides, 1999) that is found in many transcriptional activators, such as GCN5, P/CAF, CBP (CREB-binding protein), TAF250, and others. These activator proteins, which often have acetyltransferase activity themselves, are supposed to "open" the local chromatin architecture and effectively allow transcription to proceed. This does not necessarily mean that all genes in the region will start transcribing, but they have the potential to do so.

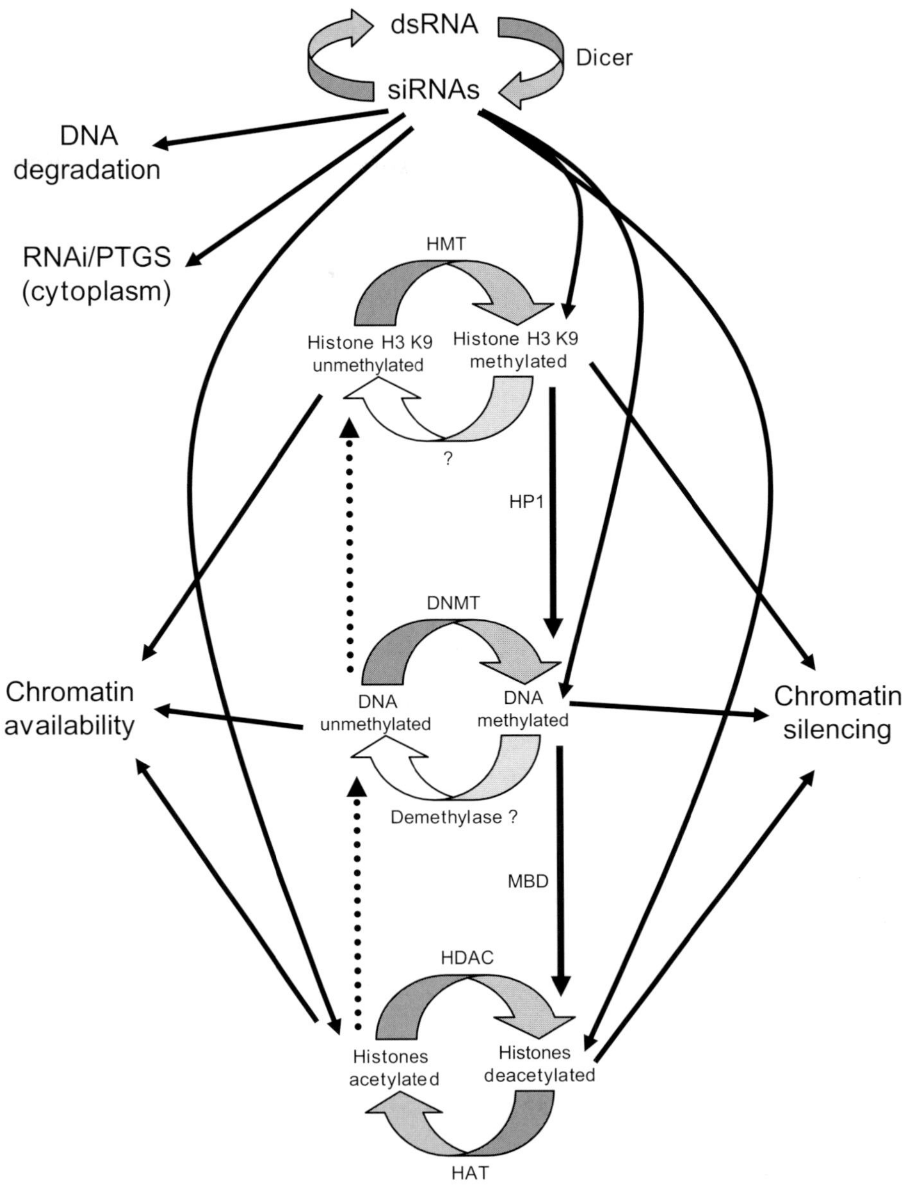

Fig. 3. Flowchart of interactions between histone acetylation, histone methylation, and DNA methylation. All of these epigenetic marks may be imposed on chromatin by short RNAs, identical to the short interfering RNAs (siRNA) responsible for post-transcriptional gene silencing (PTGS)/RNA interference (RNAi). The resulting epigenotype is then read by chromatin-activating or chromatin-silencing complexes, which mediate local remodeling and formation of euchromatin or heterochromatin, respectively. Abbreviations: dsRNA = double-stranded RNAs; siRNA = short interfering RNAs; HMT = histone methyltransferases; HP1 = heterochromatin protein 1; DNMT = DNA methyltransferases; MBD = methyl-binding domain proteins; HDAC = histone deacetylases; HAT = histone acetyltransferases.

In the case of the fragile X syndrome, we suppose that a series of events occur as depicted in Fig. 2. If DNA methylation directs histone deacetylation (middle panel) and this, in turn, leads to heterochromatin formation and gene silencing (lower panel), then it should be possible to interfere with this process by treating cells with HDAC inhibitors (Table 2). Coffee et al. (1999) showed that histones associated with the FMR1 promoter in fragile X cells are deacetylated, while we tested the effect of sodium butyrate, 4-phenylbutyrate (4-PBA), and TSA as reactivating agents for the inactive FMR1 gene. We found that HDAC inhibitors alone, although capable of reacetylating histones associated with the FMR1 promoter, caused minimal reactivation of the transcription of the gene (Chiurazzi et al., 1999). However, when sodium butyrate or 4-PBA were employed after pretreating the cells with 5-azadC, we observed a strong synergistic effect and detected a 2- to 4-fold increase in response, compared to 5-azadC alone (Chiurazzi et al., 1999). If one considers Fig. 2, it is apparent that histone hyperacetylation alone may not be sufficient for reactivation until DNA methylation is in place, favoring heterochromatin formation.

DNA methylation could drive heterochromatin formation through other pathways independent of histone deacetylation. Very recently, it was discovered that all three DNA methyltransferases interact directly with histone deacetylases HDAC1 and HDAC2 (Robertson et al., 2000; Rountree et al., 2000; Bachman et al., 2001; Fuks et al., 2001), inducing heterochromatin formation and repressing transcription, bypassing cytosine methylation (Burgers et al., 2002).

Figure 3 summarizes in its lowest part the relationship between DNA methylation and histone acetylation. Although it is now accepted that DNA methylation causes histone deacetylation, it is still not clear if histone acetylation can cause local DNA demethylation (Selker, 1998; Burgers et al., 2002) and if there exists any active DNA demethylase at all (Kress et al.,

2001). In any event, whatever modification may be found on the DNA primary sequence (e.g., cytosine methylation) or on the associated histones (lysine or arginine acetylation, serine phosphorylation, and other changes), this "epigenetic code" is read by several different chromatin-remodeling proteins. Complexes of such proteins eventually determine the balance between chromatin activation ("availability") or inactivation ("silencing") (Strahl and Allis, 2000).

(Re)activation of genes without a CpG island

Genes without a CpG island in their promoter cannot be efficiently silenced by methylation, but their transcription correlates with levels of histone acetylation in their promoters. Although the majority of such genes (Table 1) are either partially active (e.g., the SMN2 gene) or physiologically inactive (e.g., gamma-globin in adult bone marrow erythroblasts), there are good reasons to try and increase their transcription.

Gamma-globin is the model which has been studied most, because of the potential benefits of increasing fetal hemoglobin levels in patients with either beta-thalassemia or sickle-cell disease, converting these severe conditions into the much milder hereditary persistence of fetal hemoglobin. Both 5-azadC and HDAC inhibitors have been tried in vitro and in vivo (Lowrey and Nienhuis, 1993; Atweh et al., 1999; Gabbianelli et al., 2000), and encouraging clinical results were obtained when arginine butyrate was administered with a pulsed regimen. In fact, all HDAC inhibitors are strong inhibitors of cell-cycle progression (Davis et al., 2000), and continued treatment can be detrimental, unless these compounds are employed as anticancer drugs (Jung, 2001). So, dosage and timing of administration must be finely tuned.

Another gene that might be "boosted" by a "transcriptional therapy" is ALDPL1, the autosomal homolog of the XALD gene encoding a peroxisomal half-transporter which is mutated or deleted in adrenoleukodystrophy (ALD), a severe X-linked disorder with impaired very-long-chain fatty acids (VLCFA) metabolism. Kemp et al. (1998) employed 4-PBA both in vitro (cell lines from patients with ALD) and in vivo (as dietary supplementation to knockout [KO] mice). In vitro they observed a 3-fold increase in the amount of ALD-related protein, following an increased transcription of the ALDPL1 mRNA, and in vivo a substantial reduction of VLCFA in KO mice (Kemp et al., 1998).

A third condition in which sodium butyrate was used to boost a silent gene is spinal muscular atrophy (SMA). SMA is due in most cases to the deletion of the SMN1 gene. SMN2, the centromeric copy of SMN1, is very poorly active, and its transcript lacks exon 7. Treating cell lines of SMA patients with sodium butyrate, Chang et al. (2001) obtained an increase in the levels of the full-length transcript of the SMN2 gene, which includes exon 7. They showed that the effect on splicing was probably due to an increased expression of SR (Ser-Arg) proteins that play an essential role in pre-mRNA splicing. When they fed sodium butyrate to SMA-like mice, there was an amelioration of symptoms. When the drug was administered to pregnant mice bearing affected fetuses, prevention of the congenital severe SMA-phenotype was observed (Chang et al., 2001). We also confirmed an increase in full-length SMN2 transcripts after in vitro treatment with 4-PBA of SMA fibroblasts (Tiziano et al., 2002).

Finally, it has been reported that 4-PBA was successfully employed in vitro and recently also in vivo to treat cystic fibrosis due to the ΔF508 mutation. In fact, transport of the CFTR-ΔF508 mutant protein to the cell membrane is impaired. Initial reports suggested that 4-PBA increases synthesis of the mutant CFTR transcript, increasing the percentage of CFTR protein which reaches the cell membrane. Now it appears that 4-PBA may not be increasing the expression of the CFTR gene but, rather, stabilizing the CFTR-ΔF508 protein (Andersson and Roomans, 2000; Zeitlin et al., 2002), just as it has been suggested for another mutant protein, α_1-antitrypsin Z (Burrows et al., 2000).

Note on "transcriptional therapy" with simple drugs

All of the compounds listed in Table 2 (i.e., DNA demethylating drugs and HDAC inhibitors) obviously do not discriminate between chromatin regions and have a genome-wide effect. A "transcriptional therapy" based on simple drugs may therefore be more useful in treating genetic disorders due to global dysregulation of DNA methylation or histone acetylation, which will be discussed in the next two sections of this review.

A major concern on the use of simple compounds for reactivating single genes, such as FMR1 or SMN2, arises because the indiscriminate reactivation of a multitude of other genes may have catastrophic effects. However, experiments reported by Suzuki et al. (2002) show that fewer than 1% of 10,000 genes increased their transcription level in response to 5-azadC treatment and half that number increased transcription after TSA treatment (Suzuki et al., 2002). Thus, the number of responding genes is not as large as has been feared, possibly because most genes have a demethylated promoter anyhow (with the exception of imprinted genes and those on the inactive X).

Side effects would be completely avoided if artificial transcription factors designed for binding only the promoter of a specific gene were used (see reviews by Tupler et al. [2001] and Reik et al. [2002] for more details). Zhang et al. (2000) and Pabo et al. (2001), for example, describe the design and use of synthetic zinc finger transcription factors, which act only on specific genes. However, this approach requires the introduction of either expression vectors or recombinant proteins into cells. Small-molecule drugs do not present such limitations, and, in our opinion, it would be worth exploring whether a concentration range exists in which a drug would cause preferential reactivation of a given target gene.

Global dysregulation and aberrant methylation: cancer genes

Two recent reviews clearly summarize our present understanding of the epigenetic changes that occur in cancer cells (Baylin et al., 2001; Jones and Baylin, 2002). An increasing amount of evidence shows that aberrant hypermethylation of CpG islands associated with promoters is responsible for inappropriate transcriptional silencing of almost every known tumor-suppressor gene. Thus, promoter hypermethylation is at least as common as gene mutation or deletion as cause of cancer (Jones and Baylin, 2002). Promoter methylation seems to be an early event in the pathogenetic cascade leading to cancer, it involves several genes at the same time, and it appears to be related to aging (Ahuja et al., 1998). Hypermethylation of CpG islands has been demonstrated in the apparently normal mucosa surrounding colon cancers, suggesting that an epigenetic "methylator" phenotype may predispose to further cancer progression (Toyota et al., 1999), just as a "mutator" phenotype does when DNA mismatch repair genes are mutated. Sometimes, the two events coincide, if a tumor-suppressor gene crucial for DNA repair becomes inactive after abnormal methylation of its promoter, as often happens for MLH1, whose loss of function is the most common cause of microsatellite instability in colon, endometrial, and gastric cancer (Herman et al., 1998).

It is worth noting that epigenetic changes such as promoter hypermethylation, unlike DNA mutations or deletions, are potentially reversible. This plasticity can be extremely advantageous for cancer progression, allowing the same cell population to lose and then regain some important properties. For example, the ability to grow in aggregates can be lost in cancer cells upon inactivation of the E-cadherin gene (CDH1), facilitating local invasion and metastasis (Graff et al., 2000). The same property can be usefully regained to allow growth of metastatic tumors, if methylation in the CDH1 promoter is lost and re-expression of E-cadherin occurs.

Although 5-azacytidine and 5-azadeoxycytidine had been synthesized in the 1970s as potential anticancer drugs (Cihak, 1974), Laird et al. (1995) were among the first to test in a murine model the hypothesis that DNA hypomethylation might be effective in treating cancer. Cameron et al. (1999) then showed that tumor-suppressor genes silenced by hypermethylation (such as MLH1, TIMP3, and CDKN2A) could be reactivated by treating cells with 5-azadC. These authors also found that a combined treatment including a DNA demethylating drug and the histone deacetylase inhibitor TSA was more effective than 5-azadC alone. These results were concordant with our findings, as we also observed a synergistic effect of 5-azadC and the HDAC inhibitors sodium butyrate or 4-PBA on the reactivation of the FMR1 gene in fragile X cells (Chiurazzi et al., 1999). Furthermore, as Cameron et al. (1999) did, we found that HDAC inhibitors alone cannot efficiently reactivate genes with a hypermethylated promoter.

However, we must note these drugs have a strong antiproliferative effect, induce differentiation (Davis et al., 2000; McGrath-Morrow and Stahl, 2000), and are increasingly appreciated as anticancer agents (Jung, 2001). It is likely that these effects are mediated by increased expression of genes without a CpG island or at least with an unmethylated CpG island.

Suzuki et al. (2002) followed up on the observations of Cameron et al. (1999) and searched for tumor-suppressor genes silenced by promoter hypermethylation. They used a microarray approach to screen over 10,000 genes in human colorectal cancer cell lines treated with 5-azadC alone, 5-azadC plus TSA, or TSA alone. Out of 10,814 genes tested, 51 showed increased expression with 5-azadC (usually reinforced by TSA), and 23 were upregulated by TSA treatment alone (Suzuki et al., 2002). These latter genes had either no CpG island in their promoter or a completely unmethylated CpG island, as expected. On the contrary, the 51 genes reactivated by 5-azadC often had a methylated CpG island in their promoter. With this approach, Suzuki et al. (2002) were able to identify a new family of genes abnormally inactivated in colon cancer. These results underscore the importance of profiling promoter methylation to characterize human cancers. Methylation analysis will eventually facilitate early diagnosis by testing serum or even sputum DNA from patients with different types of tumors (Palmisano et al., 2000; Baylin et al., 2001).

Recently, Baylin et al. (2001) started a phase I trial with 5-azadC and 4-PBA for all tumor types, aiming at lowering the dose of each drug because of their synergistic action. If such a combined treatment, at dosages below the toxicity threshold, will be found beneficial, this will fuel new interest into the potential application of gene reactivation in treating tumors.

Global dysregulation without methylation: neurodegenerative disorders

New, exciting results suggest that a number of neurodegenerative disorders due to polyglutamine tract expansions, such as Huntington disease, may be caused by a global transcriptional dysregulation secondary to histone deacetylation (Bates, 2001). The first clue that pointed to this common pathogenetic pathway, shared by at least nine different neurodegenerative disorders, was provided by the discovery of the presence of CBP (CREB-binding protein) in the insoluble aggregates found in the nucleus of affected neurons. CBP is a protein with an acetyltransferase domain and is important for the functioning of the transcription factor CREB, whose heterozygous inactivation causes the Rubinstein-Taybi syndrome as well as a general decrease in histone acetylation (Murata et al., 2001). Steffan et al. (2001) have now shown that a fragment of huntingtin with 93 glutamines or a peptide consisting of 48 glutamines, capable of causing neurodegeneration and early death when expressed in the fruit fly, bind and inhibit the acetyltransferase domain of CBP, p300 and p300/CBP-associated factor (P/CAF), even in solution. Thus, sequestration in nuclear aggregates is not strictly required for inhibition of acetyltransferase activity, and the mere presence of polyglutamine tracts leads to an overall reduction of histone acetylation levels (Bates, 2001). Steffan et al. (2001) went on to demonstrate that the HDAC inhibitors sodium butyrate and suberoylanilide hydroxamic acid (SAHA) can restore wild-type levels of acetylated histones H3 and H4 in cell culture and can arrest ongoing progressive neuronal degeneration in vivo. Shortly after these results were published, another group used DNA microarray analysis to examine the effect of expanded polyglutamine tracts in yeast. They observed a significant defect in transcription, with an expression profile similar to that observed in yeast strains mutated for components of the histone acetyltransferase complex SAGA, and found that TSA could antagonize the observed transcriptional dysregulation (Hughes et al., 2001). Given these encouraging results, testing the effect of HDAC inhibitors in mouse models should be considered a high priority (Bates, 2001). Curiously enough, it has been reported that feeding wild-type *Drosophila* with 4-PBA can significantly increase lifespan, without diminution of locomotor vigor, resistance to stress, or reproductive ability (Kang et al., 2002). These flies, as expected, show a global increase in histone acetylation.

Histone methylation

Evidence suggests that DNA methylation is not the first step in the process leading to gene inactivation, but rather a secondary change with a reinforcing role. Organisms such as *Drosophila* do not methylate their DNA but are still capable of silencing genes and of achieving dosage compensation of their sex chromosomes. As indicated above in the section describing the role of histone acetylation, DNA methyltransferases can bypass cytosine methylation and recruit histone deacetylases (Robertson et al., 2000; Rountree et al., 2000; Fuks et al., 2001; Bachman et al., 2001). But what directs DNA methyltransferases to their target sites?

Jenuwein and Allis (2001), in their review on "translating the histone code," remind us that transcriptional activators bind acetylated lysines of histone tails by a so-called "bromodomain," while repressor complexes (such as NuRD) recognize target chromatin with a "chromodomain" that binds methylated lysines on histones. A number of histone methyltransferases (HMTs) have been characterized, each specific for a given lysine residue. Methylation of lysine 9 (K9) of histone H3 is associated with silenced chromatin, whereas methylation of histone H3 lysine 4 (K4) is found in active chromatin (Spotswood and Turner, 2002).

Now, work of Tamaru and Selker (2001) in *Neurospora crassa* shows that DNA methylation apparently depends on K9 methylation of histone H3. Furthermore, it has been shown that murine Dnmt3a co-localizes with the heterochromatin protein HP1, which binds to methylated histones through its conserved chromodomain (Bachman et al., 2001). The new data place DNA methylation in a relatively central position in our scheme (Fig. 3) and seem to suggest that histone methylation directs DNA methylation (Burgers et al., 2002). Xin et al. (2001) looked for patterns of histone H3 methylation in the Prader-Willi imprinting center overlapping the SNRPN promoter and indeed found that the inactive maternal allele is enriched in K9 methylated histone H3, whereas the active paternal chromosome is associated with K4 methylated histone H3. Coffee et al. (2002) found a similar pattern of histone methylation in the FMR1 promoter of fragile X patients: in these patients' cells there is a decrease of K4 methylation and a marked increase of K9 methylation of histone H3. Even after treatment with 5-azadC, the "mark" of K9 methylation is not entirely lost, and this may explain why DNA is apparently methylated again after treatment is discontinued (Coffee et al., 2002; Pietrobono et al., 2002).

However, experiments performed in cancer cell lines demonstrate a rapid reduction of K9 methylation following transcriptional reactivation of silenced genes after treatment with 5-azadC (Fahrner et al., 2002; Kondo et al., 2003). Furthermore, Fuks et al. (2003) recently reported that the methyl-DNA binding protein MeCP2 associates in vivo with histone methyltransferases, resulting in K9 methylation, thereby indicating that MeCP2 can reinforce transcriptional repression by linking DNA methylation to both histone deacetylation and histone H3 K9 methylation. At present, it seems that neither DNA methylation nor histone K9 methylation directs the other, but both processes cooperate in a finely tuned interplay leading to chromatin inactivation (Johnson et al., 2002).

Latest advances: RNA-guided silencing in the nucleus

In Fig. 3 we also propose a higher level of control: on top of histone methylation, short interfering RNAs (siRNAs) may play a critical role. Short interfering RNAs are small duplex RNA molecules (19 bp long) with a 2-bp 3′ overhang, which have long been known to plant geneticists as mediators of post-transcriptional gene silencing (PTGS). PTGS, also known as "RNA interference" (RNAi), consists of degradation of mRNAs homologous to the siRNAs (Matzke et al., 2001). RNAi is also a formidable tool for knocking down genes in vitro and in vivo (McManus and Sharp, 2002).

siRNAs are produced by digestion of longer double-stranded RNA molecules (dsRNAs) mediated by a ribonuclease III (RNase III) enzyme, known in *Drosophila* as Dicer, and appear to direct specific recognition of mRNA sequences which will be cut by a different ribonuclease complex, RISC (RNA-induced silencing complex). However, RNAi is not the only effect of siRNAs (also known as micro/miRNAs); in fact, these small RNAs are also found in the nucleus, where they cause RNA-directed DNA methylation (RdDM) or even programmed DNA degradation (Matzke et al., 2001). siRNAs (in this case also called shRNAs, for short heterochromatic RNAs) are able to induce heterochromatinization of homologous DNA stretches by directing DNA methylation and/or other chromatin modifications (even in organisms that do not methylate their DNA). One such modification is the methylation of histone H3 K9 residue; mutant strains of *Schizosaccharomyces pombe* deleted for genes of the RNAi machinery (including Dicer) lose histone H3 K9 methylation, accumulate aberrant complementary transcripts from centromeric heterochromatic repeats, and display impaired centromere function (Reinhart and Bartel, 2002; Volpe et al., 2002). It is worth noting that Volpe et al. (2002) observed transcriptional reactivation of silenced transgenes in yeast centromeres, indicating that RNA-directed heterochromatin formation is essential for silencing "parasitic" DNA sequences, such as transposable elements and endogenous retroviruses, which often produce dsRNAs during replication (Matzke et al., 2001). Hall et al. (2002), following up on the results of Volpe et al. (2002), showed that insertion of a centromere-homologous repeat (cenH) is sufficient for heterochromatin formation at an ectopic site. Hall et al. (2002) also showed that the RNAi machinery is necessary for heterochromatin formation, but is dispensable for its subsequent inheritance, which is maintained by histone H3 K9 methylation and recruitment of Swi6/HP1. These recent discoveries suggest that cells have an elegant RNA-based system for discriminating between chromatin domains that should stay active and genomic "chunks" that must be transcriptionally silenced (Jenuwein, 2002). Authentic promoter-driven transcription will ensure that abundant mRNA molecules are produced and exported to the cytoplasm. Conversely, "spurious" bidirectional transcription (usually associated with clusters of repeated sequences) will greatly favor the formation of dsRNA and the nuclear accumulation of shRNAs which induce heterochromatin formation, irrespective of the actual sequence of the repeated DNA. Although much work remains to be done in order to characterize naturally occurring short RNAs (Matzke et al., 2001), antisense transcripts (Lehner et al., 2002), and proteins interacting with short RNAs, it is clear that RNA plays a critical role in directing the epigenetic status of the genome (Jenuwein, 2002).

Conclusions

Basic research on the mechanisms of transcriptional regulation has shown that the chain of events leading to heterochromatin formation includes RNA interference, histone H3 K9 methylation, DNA methylation, and histone deacetylation (Fig. 3). In organisms that do not methylate their DNA, such as *Drosophila*, DNA methyltransferases favor heterochromatin formation only by recruiting other repressor proteins, such as histone deacetylases (Burgers et al., 2002).

Our present knowledge suggests a number of potential ways of interfering with chromatin inactivation. Eventually, a "transcriptional therapy" may become available for all those genetic disorders that are not due to structural DNA mutations but, rather, the mis-regulated expression of genes. Reik et al. (2002) discuss the possibility of designing artificial transcription factors that will recognize promoters of specific target genes. One could also speculate on the possibility of interacting with the RNA interference machinery in order to achieve selective reactivation of silenced genes. However attractive in principle, these options would require introduction of expression vectors or proteins into cells, presenting the same problems of gene therapy.

In contrast, in this review we considered the possibility of employing small-molecule drugs, such as those listed in Table 2, to obtain reactivation of an inactive gene or increase transcription of an already active gene. The unspecific (re)activation of multiple genes may be minimized by optimizing dosage and combination of drugs. A "second generation" of reactivating compounds causing DNA demethylation or histone acetylation, with fewer or no toxic effects, may soon become available and will hopefully turn "transcriptional therapy" into a clinical reality.

References

Ahuja N, Li Q, Mohan AL, Baylin SB, Issa JP: Aging and DNA methylation in colorectal mucosa and cancer. Cancer Res 58:5489–5494 (1998).

Andersson C, Roomans GM: Activation of deltaF508 CFTR in a cystic fibrosis respiratory epithelial cell line by 4-phenylbutyrate, genistein and CPX. Eur Respir J 15:937–941 (2000).

Antequera F, Bird A: CpG islands as genomic footprints of promoters that are associated with replication origins. Curr Biol 9:R661–667 (1999).

Atweh GF, Sutton M, Nassif I, Boosalis V, Dover GJ, Wallenstein S, Wright E, McMahon L, Stamatoyannopoulos G, Faller DV, Perrine SP: Sustained induction of fetal hemoglobin by pulse butyrate therapy in sickle cell disease. Blood 93:1790–1797 (1999).

Bachman KE, Rountree MR, Baylin SB: Dnmt3a and Dnmt3b are transcriptional repressors that exhibit unique localization properties to heterochromatin. J biol Chem 276:32282–32287 (2001).

Bates GP: Exploiting expression. Nature 413:691–694 (2001).

Baylin SB, Esteller M, Rountree MR, Bachman KE, Schuebel K, Herman JG: Aberrant patterns of DNA methylation, chromatin formation and gene expression in cancer. Hum molec Genet 10:687–692 (2001).

Bestor TH: The DNA methyltransferases of mammals. Hum molec Genet 9:2395–2402 (2000).

Bestor TH, Tycko B: Creation of genomic methylation patterns. Nature Genet 12:363–367 (1996).

Bird A, Wolffe AP: Methylation-induced repression–belts, braces and chromatin. Cell 99:451–454 (1999).

Burgers WA, Fuks F, Kouzarides T: DNA methyltransferases get connected to chromatin. Trends Genet 18:275–277 (2002).

Burrows JAJ, Willis LK, Perlmutter DH: Chemical chaperones mediate increased secretion of mutant α_1-antitrypsin (α_1-AT) Z: a potential pharmacological strategy for prevention of liver injury and emphysema in α_1-AT deficiency. Proc natl Acad Sci, USA 97:1796–1801 (2000).

Cameron EE, Bachman KE, Myohanen S, Herman JG, Baylin SB: Synergy of demethylation and histone deacetylase inhibition in the re-expression of genes silenced in cancer. Nature Genet 21:103–107 (1999).

Chang JG, Hsieh-Li HM, Jong YJ, Wang NM, Tsai CH, Li H: Treatment of spinal muscular atrophy by sodium butyrate. Proc natl Acad Sci, USA 98:9808–9813 (2001).

Chiurazzi P, Neri G: Pharmacological reactivation of inactive genes: the fragile X experience. Brain Res Bull 56:383–387 (2001).

Chiurazzi P, Pomponi MG, Pietrobono R, Bakker CE, Neri G, Oostra BA: Synergistic effect of histone hyperacetylation and DNA demethylation in the reactivation of the FMR1 gene. Hum molec Genet 8:2317–2323 (1999).

Chiurazzi P, Pomponi MG, Willemsen R, Oostra BA, Neri G: In vitro reactivation of the FMR1 gene involved in fragile X syndrome. Hum molec Genet 7:109–113 (1998).

Cihak A: Biological effects of 5-azacytidine in eukaryotes. Oncology 30:405–422 (1974).

Coffee B, Zhang F, Ceman S, Warren S, Reines D: Histone modifications depict an aberrantly heterochromatinized FMR1 gene in fragile X syndrome. Am J hum Genet 71:923–932 (2002).

Coffee B, Zhang F, Warren S, Reines D: Acetylated histones are associated with FMR1 in normal but not fragile X syndrome cells. Nature Genet 22:98–101 (1999).

Corona DF, Clapier CR, Becker PB, Tamkun JW: Modulation of ISWI function by site-specific histone acetylation. EMBO Rep 3:242–247 (2001).

Davis T, Kennedy C, Chiew YE, Clarke CL, deFazio A: Histone deacetylase inhibitors decrease proliferation and modulate cell cycle gene expression in normal mammary epithelial cells. Clin Cancer Res 6:4334–4342 (2000).

Fahrner JA, Eguchi S, Herman JG, Baylin SB: Dependence of histone modifications and gene expression on DNA hypermethylation in cancer. Cancer Res 62:7213–7218 (2002).

Fuks F, Burgers WA, Godin N, Kasai M, Kouzarides T: Dnmt3a binds deacetylases and is recruited by a sequence-specific repressor to silence transcription. EMBO J 20:2536-2544 (2001).

Fuks F, Hurd PJ, Wolf D, Nan X, Bird AP, Kouzarides T: The methyl-CpG-binding protein MeCP2 links DNA methylation to histone methylation. J biol Chem 278:4035–4040 (2003).

Gabbianelli M, Testa U, Massa A, Pelosi E, Sposi NM, Riccioni R, Luchetti L, Peschle C: Hemoglobin switching in unicellular erythroid culture of sibling erythroid burst-forming units: kit ligand induces a dose-dependent fetal hemoglobin reactivation potentiated by sodium butyrate. Blood 95:3555–3561 (2000).

Graff JR, Gabrielson E,Fujii H, Baylin SB, Herman JG: Methylation patterns of the E-cadherin 5′ CpG island are unstable and reflect the dynamic, heterogeneous loss of E-cadherin expression during metastatic progression. J biol Chem 275:2727–2732 (2000).

Grunstein M: Histone acetylation in chromatin structure and transcription. Nature 389:349–352 (1997).

Hall IM, Shankaranarayana GD, Noma K, Ayoub N, Cohen A, Grewal SIS: Establishment and maintenance of a heterochromatin domain. Science 297:2232–2237 (2002).

Herman JG, Umar A, Polyak K, Graff JR, Ahuja N, Issa JPJ, Markowitz S, Willson JKV, Hamilton SR, Kinzler KW, Kane MF, Kolodner RD, Vogelstein B, Kunkel TA, Baylin SB: Incidence and functional consequences of hMLH1 promoter hypermethylation in colorectal carcinoma. Proc natl Acad Sci, USA 95:6870–6875 (1998).

Hughes RE, Lo RS, Davis C, Strand AD, Neal CL, Olson JM, Fields S: Altered transcription in yeast expressing expanded polyglutamine. Proc natl Acad Sci, USA 98:13201–13206 (2001).

Jackson-Grusby L, Laird PW, Magge SN, Moeller BJ, Jaenisch R: Mutagenicity of 5-aza-2′-deoxycytidine is mediated by the mammalian DNA methyltransferase. Proc natl Acad Sci, USA 94:4681–4685 (1997).

Jenuwein T: An RNA-guide pathway for the epigenome. Science 297:2215–2218 (2002).

Jenuwein T, Allis CD: Translating the histone code. Science 293:1074–1080 (2001).

Johnson LM, Cao X, Jacobsen SE: Interplay between two epigenetic marks: DNA methylation and histone H3 lysine 9 methylation. Curr Biol 12:1360–1367 (2002).

Jones PA, Baylin SB: The fundamental role of epigenetic events in cancer. Nature Genet Rev 3:415–428 (2002).

Jung M: Inhibitors of histone deacetylase as new anticancer agents. Curr Med Chem 8:1505–1511 (2001).

Kang HL, Benzer S, Min KT: Life extension in *Drosophila* by feeding a drug. Proc natl Acad Sci, USA 99:838–843 (2002).

Kemp S, Wei HM, Lu JF, Braiterman LT, McGuinness MC, Moser AB, Watkins PA, Smith KD: Gene redundancy and pharmacological gene therapy: implications for X-linked adrenoleukodystrophy. Nature Med 4:1261–1268 (1998).

Kondo Y, Shen LL, Issa JPJ: Critical role of histone methylation in tumor suppressor gene silencing in colorectal cancer. Mol Cell Biol 23:206–215 (2003).

Kornberg RD, Lorch Y: Chromatin and transcription: where do we go from here? Curr Op Genet Dev 12:249–251 (2002).

Kouzarides T: Histone acetylases and deacetylases in cell proliferation. Curr Op Genet Dev 9:40–48 (1999).

Kress C, Thomassin H, Grange T: Local DNA demethylation in vertebrates: how could it be performed and targeted ? FEBS Lett 494:135–140 (2001).

Laird PW, Jackson-Grusby L, Fazeli A, Dickinson SL, Jung WE, Li E, Weinberg RA, Jaenisch R: Suppression of intestinal neoplasia by DNA hypomethylation. Cell 81:197–205 (1995).

Lehner B, Williams G, Campbell RD, Sanderson CM: Antisense transcripts in the human genome. Trends Genet 18:63–65 (2002).

Liang G, Chan MF, Tomigahara Y, Tsai YC, Gonzales FA, Li E, Laird PW, Jones PA: Cooperativity between DNA methyltransferases in the maintenance methylation of repetitive elements. Mol Cell Biol 22:480–491 (2002).

Lowrey CH, Nienhuis AW: Treatment with azacitidine of patients with end stage beta-thalassemia. New Engl J Med 329:845–848 (1993).

Matzke M, Matzke AJM, Kooter JM: RNA: guiding gene silencing. Science 293:1080–1083 (2001).

McGrath-Morrow SA, Stahl JL: G_1 phase growth arrest and induction of p21 in IB3-1 cells treated with 4-sodium phenylbutyrate. J Pharm exp Ther 294:941–947 (2000).

McManus MT, Sharp PA: Gene silencing in mammals by small interfering RNAs. Nature Rev Genet 3:737–747 (2002).

Murata T, Kurokawa R, Krones A, Tatsumi K, Ishii M, Taki T, Masuno M, Ohashi H, Yanagisawa M, Rosenfeld MG, Glass CK, Hayashi Y: Defect of histone acetyltransferase activity of the nuclear transcriptional coactivator CBP in Rubinstein-Taybi syndrome. Hum molec Genet 10:1071–1076 (2001).

Ng HH, Bird A: DNA methylation and chromatin modification. Curr Op Genet Dev 9:158–163 (1999).

Pabo CO, Peisach E, Grant RA: Design and selection of novel cys2his2 zinc finger proteins. A Rev Biochem 70:313–340 (2001).

Palmisano WA, Divine KK, Saccomanno G, Gilliland FD, Baylin SB, Herman JG, Belinsky SA: Predicting lung cancer by detecting aberrant promoter methylation in sputum. Cancer Res 60:5954–5958 (2000).

Pietrobono R, Pomponi MG, Tabolacci E, Oostra B, Chiurazzi P, Neri G: Quantitative analysis of DNA demethylation and transcriptional reactivation of the FMR1 gene in fragile X cells treated with 5-azadeoxycytidine. Nucl Acids Res 30:3278–3285 (2002).

Reik A, Gregory PD, Urnov FD: Biotechnologies and therapeutics: chromatin as target. Curr Op Genet Dev 12:233–242 (2002).

Reinhart BJ, Bartel DP: Small RNAs correspond to centromere heterochromatic repeats. Science 297:1831 (2002).

Rhee I, Bachman KE, Park BH, Jair KW, Yen RW, Schuebel KE, Cui H, Feinberg AP, Lengauer C, Kinzler KW, Baylin SB, Vogelstein B: DNMT1 and DNMT3B cooperate to silence genes in human cancer cells. Nature 416:552–556 (2002).

Rhee I, Jair KW, Yen RW, Lengauer C, Herman JG, Kinzler KW, Vogelstein B, Baylin SB, Schuebel KE: CpG methylation is maintained in human cancer cells lacking DNMT1. Nature 404:1003–1007 (2000).

Robert MF, Morin S, Beaulieu N, Gauthier F, Chute IC, Barsalou A, MacLeod AR: DNMT1 is required to maintain CpG methylation and aberrant gene silencing in human cancer cells. Nature Genet 33:61–65 (2003).

Robertson KD, Ait-Si-Ali S, Yokochi T, Wade PA, Jones PL, Wolffe AP: DNMT1 forms a complex with Rb, E2F1 and HDAC1 and represses transcription from E2F-responsive promoters. Nature Genet 25:338–342 (2000).

Rountree MR, Bachman KE, Baylin SB: DNMT1 binds HDAC2 and a new co-repressor, DMAP1, to form a complex at replication foci. Nature Genet 25:269–277 (2000).

Saitoh S, Wada T: Parent-of-origin specific histone acetylation and reactivation of a key imprinted gene locus in Prader-Willi syndrome. Am J Hum Genet 66:1958–1962 (2000).

Selker EU: Trichostatin A causes selective loss of DNA methylation in *Neurospora*. Proc natl Acad Sci, USA 95:9430–9435 (1998).

Shaw MA, Chiurazzi P, Romain DR, Neri G, Gecz: A novel gene, FAM11A, associated with the FRAXF CpG island is transcriptionally silent in FRAXF full mutation. Eur J hum Genet 10:767–772 J (2002).

Spotswood HT, Turner BM: An increasingly complex code. J clin Invest 110:577–582 (2002).

Steffan JS, Bodai L, Pallos J, Poelman M, McCampbell A, Apostol BL, Kazantsev A, Schmidt E, Zhu YZ, Greenwald M, Kurokawa R, Housman DE, Jackson GR, Marsh JL, Thompson LM: Histone deacetylase inhibitors arrest polyglutamine-dependent neurodegeneration in *Drosophila*. Nature 413:739–743 (2001).

Strahl BD, Allis CD: The language of covalent histone modifications. Nature 403:41–45 (2000).

Suzuki H, Gabrielson E, Chen W, Anbazhagan R, van Engeland M, Weijenberg MP, Herman JG, Baylin SB: A genomic screen for genes upregulated by demethylation and histone deacetylase inhibition in human colorectal cancer. Nature Genet 31:141–149 (2002).

Tamaru H, Selker EU: A histone H3 methyltransferase controls DNA methylation in *Neurospora crassa*. Nature 414:277–283 (2001).

Tiziano FD, Andreassi C, Angelozzi C, Vitali T, De Vincenzi E, Boninsegna A, Villanova M, Neri G, Brahe C: Upregulation of *SMN2* gene expression by 4-phenylbutyrate: relevance for a therapeutic approach to spinal muscular atrophy. Am J hum Genet 71(Suppl):587 (2002).

Toyota M, Ahuja N, Ohe-Toyota M, Herman JB, Baylin SB, Issa JP: CpG island methylator phenotype in colorectal cancer. Proc natl Acad Sci, USA 96:8681–8686 (1999).

Tupler R, Perini G, Green MR: Expressing the human genome. Nature 409:832–833 (2001).

Volpe TA, Kidner C, Hall IM, Teng G, Grewal SIS, Martienssen RA: Regulation of heterochromatic silencing and histone H3 lysine-9 methylation by RNAi. Science 297:1833–1837 (2002).

Walsh CP, Bestor TH: Cytosine methylation and mammalian development. Genes Dev 13:26–34 (1999).

Xin Z, Allis CD, Wagstaff J: Parent-specific complementary patterns of histone H3 lysine 9 and H3 lysine 4 methylation at the Prader-Willi syndrome imprinting center. Am J hum Genet 69:1389–1394 (2001).

Yoder JA, Walsh CP, Bestor TH: Cytosine methylation and the ecology of intragenomic parasites. Trends Genet 13:335–340 (1997).

Zeitlin PL, Diener-West M, Rubenstein RC, Boyle MP, Lee CK, Brass-Ernst L: Evidence of CFTR function in cystic fibrosis after systemic administration of 4-phenylbutyrate. Mol Ther 6:119–126 (2002).

Zhang L, Spratt SK, Liu Q, Johnstone B, Qi H, Raschke EE, Jamieson AC, Rebar EJ, Wolffe AP, Case CC: Synthetic zinc finger transcription factor action at an endogenous chromosomal site: activation of the human erythropoietin gene. J biol Chem 275: 33850–33860 (2000).

Cytogenet Genome Res 100:65–76 (2003)
DOI: 10.1159/000072839

Transcription defects induced by repeat expansion: fragile X syndrome, FRAXE mental retardation, progressive myoclonus epilepsy type 1, and Friedreich ataxia

E. Greene, V. Handa, D. Kumari and K. Usdin

Section on Genomic Structure and Function, Laboratory of Molecular and Cellular Biology, National Institute of Diabetes, Digestive and Kidney Diseases, National Institutes of Health, Bethesda, MD (USA)

Abstract. Fragile X mental retardation syndrome, FRAXE mental retardation, Progressive myoclonus epilepsy Type I, and Friedreich ataxia are members of a larger group of genetic disorders known as the Repeat Expansion Diseases. Unlike other members of this group, these four disorders all result from a primary defect in the initiation or elongation of transcription. In this review, we discuss current models for the relationship between the expanded repeat and the disease symptoms.

The repeat expansion diseases are a group of genetic disorders caused by an increase in the number of repeats (expansion) in a specific tandem array. Unaffected individuals have a number of repeats below a threshold value, the exact number depending on the disease. Expansion of this repeat beyond this threshold value is associated with disease symptoms. Nineteen repeat expansion diseases have been identified to date including 16 trinucleotide expansions (Ranum and Day, 2002), and one tetranucleotide expansion, a pentanucleotide expansion, and a dodecamer expansion (Hashem and Sinden, 2002). The exact pathological consequences of expansion depend on the location of the repeat in the affected gene, and in some instances also on the sequence or other properties of the repeat. Where expansion occurs in an open reading frame, the connection between expansion and disease symptoms is relatively straightforward. For example, many of the repeat expansion diseases result from expansion of the repeat CAG·CTG in the coding sequence of the affected gene. This results in an increase in the number of contiguous glutamine residues in the protein product of the gene. It is this polyglutamine tract that results in the characteristic neuropathology seen in these diseases (see Taylor and Fischbeck, 2002 for a recent review). These diseases are all dominantly inherited neurodegenerative disorders and include Huntington disease (HD1), and a number of spinocerebellar ataxias (see Fig. 1).

Where the repeat is located outside of the open reading frame, the connection between expansion and disease symptoms is less clear. This group of diseases can be divided into two categories (see Fig. 1): the dominantly inherited disorders that may result, at least in part, from a toxic gain-of-function of the affected RNA. This group includes the mytonic dystrophies, DM1 and DM2 (see Ranum and Day, 2002 for a recent review). The second category contains diseases that are recessive or X-linked and that result from insufficiency of the protein product of the affected gene. Diseases in this category are fragile X mental retardation syndrome, FRAXE mental retardation, Progressive myoclonus epilepsy type 1, and Friedreich ataxia. This review will focus on the molecular basis of disease pathology in these four disorders.

Supported by the National Institute of Diabetes, Digestive and Kidney Diseases, NIH, Bethesda MD.

Received 2 October 2002; revision accepted 6 February 2003.

Request reprints from Karen Usdin, Building 8, Room 202
National Institutes of Health, 8 Center Drive MSC 0830
Bethesda MD 20892-0830 (USA); telephone: 301-496-2189
fax: 301-402-0053; e-mail: ku@helix.nih.gov

KARGER Fax + 41 61 306 12 34
E-mail karger@karger.ch
www.karger.com

Accessible online at:
www.karger.com/cgr

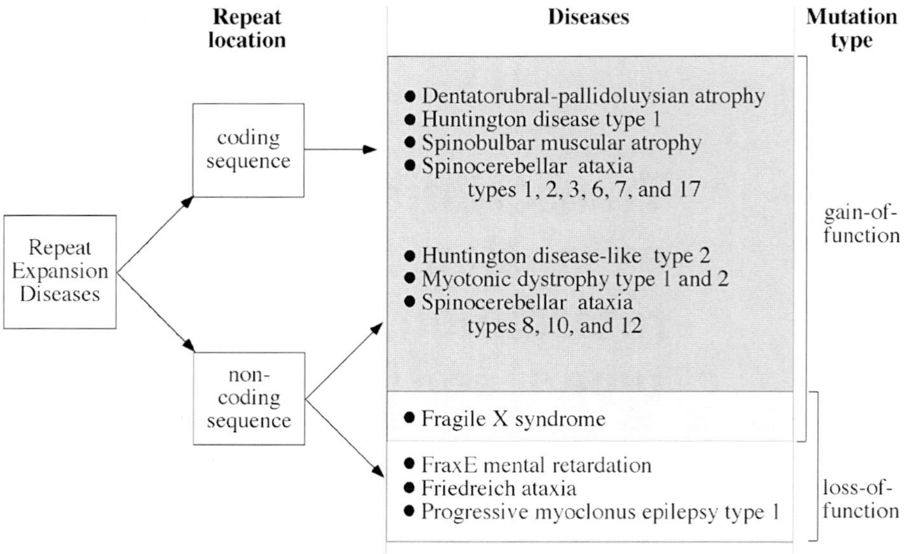

Fig. 1. Classification of the currently known repeat expansion diseases. The dark grey box contains those diseases in which expansion produces a gain-of-function, the open box those diseases in which expansion results in a loss-of-function. FXS is shown in the region of overlap of the two boxes because while the symptoms seen in individuals with the full mutation are due to a loss-of-function, the symptoms seen in premutation carriers are probably due to a toxic gain-of-function.

Fragile X syndrome: epigenetic effects on the initiation of transcription.

Background

Fragile X syndrome (FXS; MIM 309550) is the most common form of inherited mental retardation, affecting one in 4,500 males and one in 9,000 females (Warren and Sherman, 2001). FXS most often manifests as moderate to severe cognitive difficulties and developmental delay. It is frequently accompanied by impaired communication and gaze avoidance, social anxiety, attention deficit hyperactivity disorder, seizures, sleep disturbances, connective tissue abnormalities, scoliosis, macroorchidism, and mild facial anomalies. FXS results from mutations in the fragile X mental retardation 1 (FMR1) gene located at Xq27.3 (Fu et al., 1991; Verkerk et al., 1991). In many FXS patients FMR1 mRNA is absent or its levels are significantly reduced. In all patients there is a deficit of the functional FMR1 gene product (FMRP) (Pieretti et al., 1991). The symptoms seen in individuals with different intragenic point mutations (De Boulle et al., 1993; Lugenbeel et al., 1995), demonstrate that fragile X syndrome results from a loss of functional FMRP. In fact, an inverse linear effect of FMRP levels is seen on the Full Scale IQ (FSIQ) and all FSIQ-adjusted Wechsler subtest and summary scores of affected individuals (Loesch et al., 2002).

FMRP is an RNA binding protein that is highly expressed in brain and testes, the two organs most affected by its absence. FMRP normally binds about 4% of all brain mRNAs (Ashley et al., 1993a). It recognizes RNA that forms folded hairpins or tetraplexes which are four-stranded structures stabilized primarily by G-quartets that are held together by Hoogsteen bonds (Darnell et al., 2001; Schaeffer et al., 2001). FMRP shuttles between the nucleus and cytoplasm (Feng et al., 1997b). However, it resides predominantly in the cytoplasm in association with actively translating polysomes (Feng et al., 1997a; Khandjian et al., 1998), where it acts to regulate the translation of the mRNA to which it binds (Brown et al., 2001). Absence of FMRP in hippocampal neurons results in an immature dendritic spine morphology (Irwin et al., 2000), and a delay in establishing synaptic connections (Braun and Segal, 2000). FMRP has two autosomal paralogs, FXR1P and FXR2P, with which it interacts (Siomi et al., 1995; Zhang et al., 1995). These three proteins have a similar but not identical tissue distribution. Since absence of FMRP leads to symptoms of FXS despite the presence of FXR1P and FXR2P, the functions of these proteins presumably do not completely overlap. In Drosophila which has only one gene belonging to the FMR1/FXR family (Wan et al., 2000), loss of gene activity has a number of effects that parallel those seen in individuals with FXS. This includes circadian rhythm disturbances (Dockendorff et al., 2002; Inoue et al., 2002; Morales et al., 2002), and defects in neuronal morphology (Morales et al., 2002). An important clue to the role of human FMRP in translation control is that the Drosophila protein has been shown to interact with components of the RNA interference (RNAi) machinery (Caudy et al., 2002; Ishizuka et al., 2002). RNAi is a group of processes resulting in homology-dependent gene silencing triggered by double-stranded RNA (dsRNA) (Nishikura, 2001).

Most FXS mutations involve expansion of a CGG·CCG-tract in the 5′ UTR of the FMR1 gene (Ashley et al., 1993b), as illustrated in Fig. 2. These expansions are also associated with the appearance of a folate-sensitive fragile site coincident with the repeat that gives this disease its name. The molecular basis of this fragility is not known. Unaffected individuals have fewer than 60 repeats. Premutation carriers have 60–200 repeats and female carriers of the premutation are at risk of transmission of a much larger allele to their offspring. Affected individuals have more than 200 and can have >2,000 repeats. Premutation carriers generally do not exhibit the symptoms of FXS although mild cognitive difficulties and social anxiety may be present. However, these individuals do have symptoms not seen in fully affected individuals, such as a predisposition to premature ovarian failure (POF) (Allingham-Hawkins et al., 1999; Uzielli et al., 1999) and ataxia and cerebellar tremor in older males (Leehey et al., 2002). These males also have neuronal nuclear inclusions (Greco et al., 2002), reminiscent of patients with a polyglutamine repeat expansion disorder.

The apparently paradoxical effects of repeat expansion

While many individuals with full mutation alleles show a significant deficit of FMR1 mRNA, premutation carriers can have as much as seven times more FMR1 mRNA than individuals with normal alleles (Tassone et al., 2000a, 2000b, 2001; Kenneson et al., 2001). Current evidence suggests that both the elevated levels of FMR1 RNA in premutation carriers and the decrease in fully affected individuals result from an effect on the initiation of transcription, rather than on mRNA stability (Tassone et al., 2000b). Therefore, understanding the regulation of the FMR1 promoter and how expansion affects this regulation is fundamental to our understanding of the disease pathology, and is important if attempts to correct or compensate for the FXS mutation are to be successful.

The normal regulation of the FMR1 promoter

The FMR1 promoter is GC-rich, and lacks both a functional TATA- and Initiator sequence (Kumari and Usdin, 2001). It contains putative binding sites for a number of transcription factors; however only four of these sites are bound in the cells from unaffected individuals (Drouin et al., 1997; Schwemmle et al., 1997). Two of these regions bind the transcription factors USF1/USF2 and NRF-1 in brain extracts (Kumari and Usdin, 2001). A point mutation in both the USF1/USF2 and NRF-1 binding sites reduces promoter activity to background levels in transient transfection assays. The remaining two in vivo protein-binding sites are GC boxes. These sites are bound by members of the Sp family of transcription factors. No evidence of Sp1 binding was seen in vitro, but Sp3 binding was observed on a sub-fragment of the promoter in PC12 cells (Kumari and Usdin, 2001). The significance of this binding remains to be determined.

Upregulation of FMR1 transcription in premutation carriers

The reason for the increased levels of FMR1 transcript in premutation carriers is unclear. Since FMR1 mRNA levels are not elevated in an individual who carries an FMR1 point mutation that negatively affects FMRP function (Kenneson et al., 2001), it seems unlikely that cells of premutation carriers are directly responding to absent or malfunctioning FMRP. It has been suggested that long CGG·CCG-repeat tracts may increase transcription by altering promoter conformation (Raca et al., 2000), perhaps by virtue of the tendency of long FXS repeats to exclude nucleosomes (Wang et al., 1996), thereby making the promoter region more accessible to transcription factors. It is also possible that the expanded repeats can bind more of a specific transcriptional activator.

Premutation carriers have FMRP levels very close to normal, and significantly higher than individuals with full mutations. Thus it seems likely that an increase in FMR1 mRNA, rather than decreased FMRP, is responsible for the increased incidence of POF and cerebellar ataxia in this group of individuals. Analysis of FMR1 mRNA levels in individuals with these

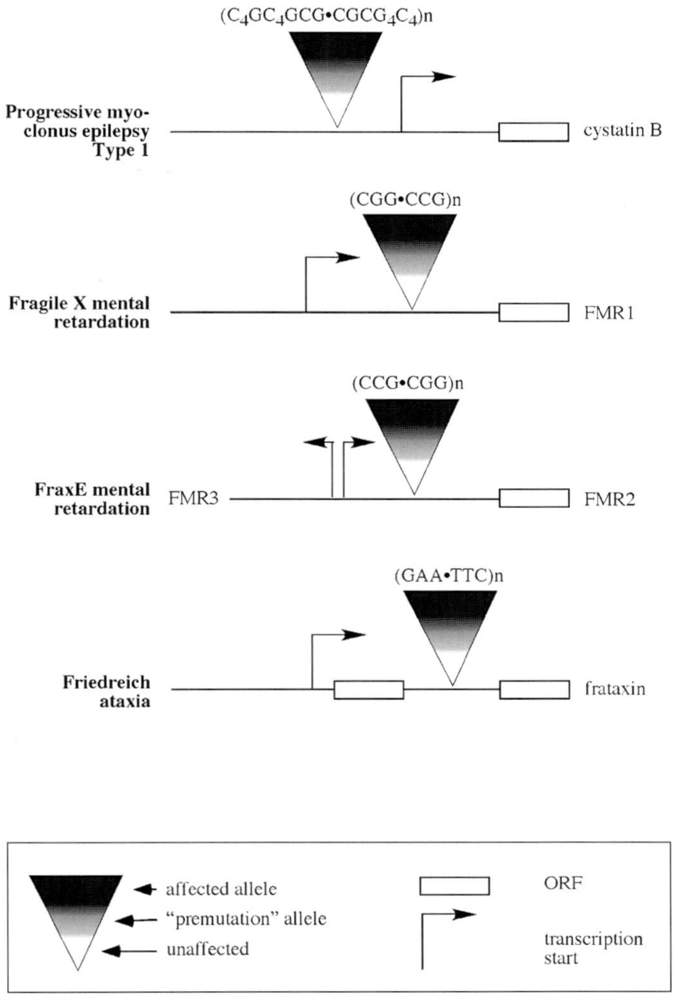

Fig. 2. The location of the repeat tract relative to the transcription start site and open reading frames in the 5′ ends of the genes affected in FXS, FRAXE MR, EPM1 and FRDA. The diagram is not to scale. The repeat numbers in normal, premutation, and fully affected alleles in FXS, FRAXE MR, EPM1 and FRDA are illustrated by the white, grey and black areas of each triangle.

symptoms will be important in verifying this hypothesis. While the gain-of-function in premutation carriers may simply be due to the elevated levels of RNA, it is likely that the long CGG-tract within this RNA is significant. We have shown that CGG-RNA, like its DNA counterpart, forms hairpins as shown in Fig. 3. (Saha and Usdin, manuscript in preparation). One cytoplasmic effect of large CGG-hairpins might be on the protein kinase, PKR. PKR causes apoptosis and is activated by dsRNA (see Tan and Katze, 1999 for a review). The CGG-hairpin could also act to down-regulate genes containing a CGG-repeat tract by a mechanism involving RNAi. Alternatively, by analogy to DM1, some of the CGG-RNA could be retained in the nucleus where it may sequester proteins needed for the normal regulation of other genes (see review by Ranum and Day, 2000). This protein binding in the nucleus may also lead to the neuronal inclusions seen in male carriers with cerebellar degeneration (Greco et al., 2002).

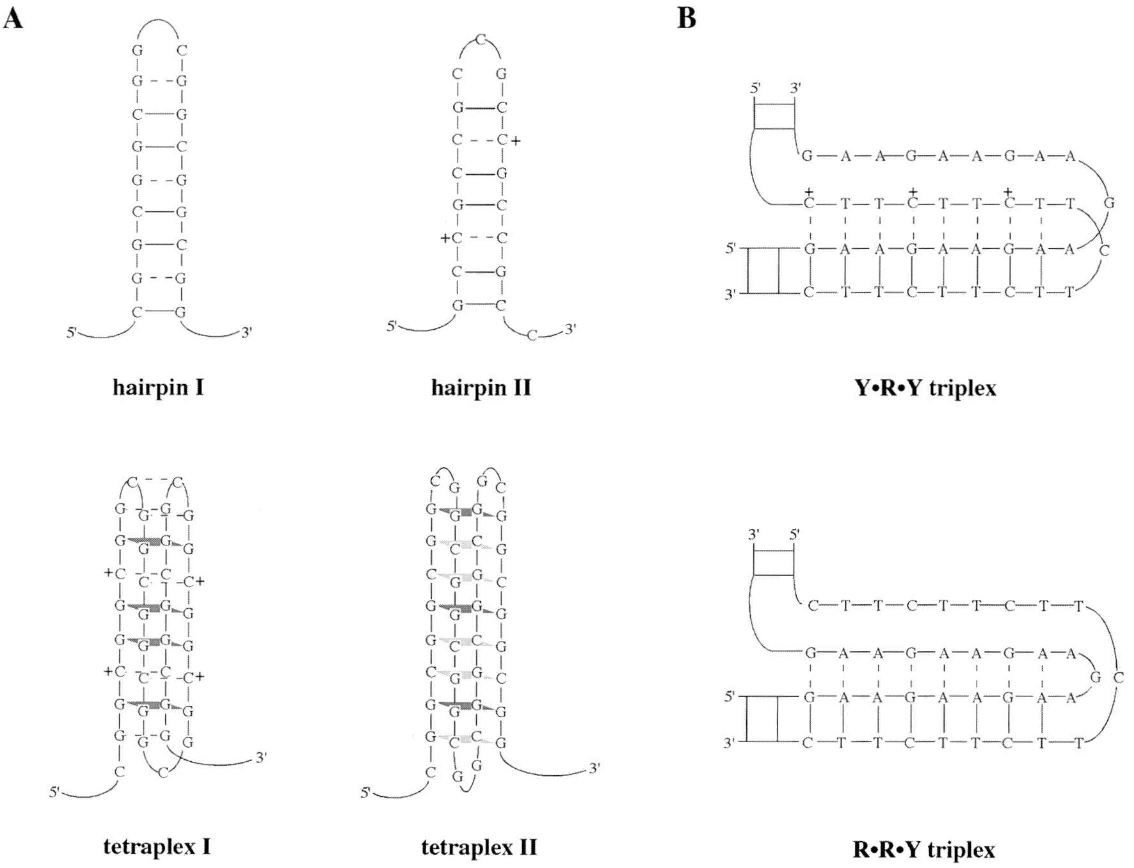

Fig. 3. Secondary structures formed by CGG·CCG (**A**) and GAA·TTC repeats (**B**). For clarity these structures are drawn with only a small number of repeats. It is not intended to imply that structures containing this small number of repeats would in fact be stable. Not all possible structures formed by these repeats are shown. The Watson-Crick base pairs are shown as solid lines connecting the bases, non-Watson-Crick base pairs are shown with dotted lines. Protonated bases are indicated by + signs. G4-tetrads are shown by dark gray parallelograms, and the G·C·G·C-tetrads by light gray ones. Hairpin I is formed by the CGG strand and has a mixture of Watson-Crick G·C pairs, and Hoogsteen G·G pairs (Mitas et al., 1995; Nadel et al., 1995; Usdin, 1998), while hairpin II is formed by CCG strand and contains a mixture of Watson-Crick G·C pairs and C·C+ mispairs (Mariappan et al., 1996, 1998; Yu et al., 1997). Tetraplex I is shown containing protonated cytosines since this structure is stabilized by low pH. (Fry and Loeb, 1994; Usdin and Woodford, 1995; Usdin, 1998; Patel et al., 2000). These cytosines may be involved in C·C+ pairs or C4-tetrads (Patel et al., 2000). Tetraplex II is based on a projected structure of uninterrupted CGG-repeats derived from studies on the association of two hairpins containing the sequence d(G-C-G-G-T3-G-C-G-G) (Kettani et al., 1995). The EPM1 repeat forms a very similar array of structures (Pataskar et al., 2001a, 2001b; Saha and Usdin, 2001).

Since the FMR1 mRNA in premutation carriers is no less stable than that of unaffected individuals, the fact that FMRP levels are close to normal despite the elevated levels of FMR1 mRNA likely reflects a decrease in the translation of the transcript. This decrease may be due to stalling of 40S ribosomal subunits that is seen in the repeat tract (Feng et al., 1995), perhaps because of hairpin formation by the expanded repeats.

Promoter down-regulation in individuals with full mutations

In most instances, the expanded repeat and adjacent CpG-rich promoter of full-mutation alleles are heavily methylated (Pieretti et al., 1991; Sutcliffe et al., 1992; Hornstra et al., 1993). The expanded repeat may trigger DNA methylation directly. Many of the secondary structures formed by the FXS repeat contain C·C mispairs (see Fig. 3). Such mispairs are good targets for methylation by the human DNA methyltransferase, perhaps because it is easier to flip the target cytosine into the active site of the enzyme (Laayoun and Smith, 1995). Consistent with this idea is the fact that the hairpins formed by the CCG-rich strand are methylated in vitro much more efficiently than a Watson-Crick duplex (Chen et al., 1995; Chen et al., 1998). Such secondary structures may thus facilitate methylation of the repeat (Chen et al., 1995, 1998), which could then spread to the adjacent promoter (Lindsay and Adams, 1996). The methylated cytosines would then be bound by methyl-binding proteins that would then recruit other proteins such as histone deacetylases and histone methyltransferases, leading to the establishment of transcriptionally silent chromatin (Fuks et al., 2002).

However, it is also possible that DNA methylation is a secondary consequence of other epigenetic modifications triggered by expansion. For example in *Neurospora crassa* DNA methylation is triggered by methylation of histone H3 at lysine 9 (Tamaru and Selker, 2001). It is known that mutations in the mammalian SWI2/SNF-2-like chromatin-remodeling proteins lead to DNA demethylation (Jeddeloh et al., 1999; Dennis et al.,

2001). In addition, during X chromosome inactivation, histone methylation and gene silencing occur long before DNA methylation can be detected (Heard et al., 2001). Thus DNA methylation in humans may also be dependent on K-9 methylation of histone H3. The trigger for histone methylation is not known. One intriguing possibility is that in the case of the FMR1 gene, it may be the CGG-RNA hairpins. In yeast, mutations in genes in the RNAi pathway lead to loss of gene silencing (Volpe et al., 2002), and in plants, transgene silencing by dsRNA is associated with methylation of the region homologous to the dsRNA (Mette et al., 2000; Jones et al., 2001). Given the many parallels between yeast, plants and mammals in various aspects of RNAi, it may be that dsRNA is also a signal for histone methylation in mammals (Allshire, 2002).

A role for DNA methylation in silencing of the FMR1 promoter is supported by the observation that in vitro methylation of the FMR1 promoter leads to a significant drop in promoter activity in transient transfection assays (Sandberg and Schalling, 1997). In addition, treatment of fully methylated cell lines from FXS patients with 5-aza-2′-deoxycytidine (5-aza-dC), an inhibitor of DNA methyltransferases, leads to a decrease in promoter methylation that is accompanied by an increase in FMR1 expression (Chiurazzi et al., 1998; Coffee et al., 1999, 2002; Pietrobono et al., 2002).

DNA methylation can affect promoter activity by directly blocking the binding of important transcription factors that are methylation-sensitive. Alternatively, the condensed chromatin associated with methylated DNA may prevent access to the promoter by the transcription machinery (Razin, 1998). Chromatin condensation is facilitated by deacetylation of the tails of histones H3 and H4. On the FMR1 promoter in cells from FXS patients both histone H3 and H4 are hypoacetylated (Coffee et al., 1999). However, while trichostatin A (TSA), a histone deacetylase inhibitor, increases the amount of acetylated histone H4 on the FMR1 promoter, it has little if any effect on the acetylation status of FMR1 H3 histones (Coffee et al., 1999). This is correlated with the continued presence of condensed chromatin in the promoter and failure to reactivate the gene. This is presumably because the K9-methylation of histone H3 blocks acetylation at that position (Coffee et al., 2002). In addition to the effect of chromatin modification on gene inactivation, DNA methylation of the NRF-1 site directly blocks transcription factor binding. This significantly reduces promoter activity (Kumari and Usdin, 2001), and suggests that DNA demethylation may be necessary to fully reactivate the FMR1 gene. The key differences between the promoter in unaffected and affected individuals are illustrated in Fig. 4.

While most individuals with the full mutation show heavy methylation of the FMR1 promoter, a number of individuals have been described that do not (Smeets et al., 1995; Burman et al., 1999). This suggests the existence of a class of genetic modifiers that ameliorates FXS symptoms by affecting the likelihood of promoter methylation.

In addition, some individuals with fully methylated alleles still have mRNA levels similar to those seen in unaffected individuals (Tassone et al., 2000a, 2000b, 2001). If one takes into account the effect of increasing repeat number on transcription in premutation carriers, these individuals probably still have a significant decrease in promoter activity relative to an unmethylated allele of the same size. Since the NRF-1 site is frequently still methylated in these individuals, the residual transcription may be due to binding of transcription factors that are not highly methylation-sensitive, such as USF1/USF2, whose binding site in any event often remains unmethylated (Pietrobono et al., 2002). In vivo DNA footprinting of the FMR1 promoter in these individuals should help resolve this issue.

Reversing the epigenetic modifications of the promoter

In spite of significant FMR1 transcription from some full mutation alleles, FMRP levels are often very low in these individuals (Tassone et al., 2000a, 2000b, 2001). The same is true of alleles inactivated by methylation and subsequently treated with 5-aza-dC (Coffee et al., 2002). This reduced translation may due in part to the aforementioned effect of hairpin formation in the RNA that blocks ribosome scanning. Such a defect in translation makes a strategy aimed at ameliorating the symptoms of FXS by reactivating the FMR1 gene more problematic, since even if the effects of methylation could be alleviated, the problem of translation of the expanded message may still remain. However, the observation that some individuals with unmethylated full mutations, make significant amounts of FMRP and do not suffer apparent cognitive difficulties (Smeets et al., 1995), suggests not only a potential additional class of genetic modifiers of disease severity, but also that a gene reactivation strategy may benefit at least a subset of individuals with methylated full mutation alleles.

FRAXE mental retardation: Another case of methylation-induced transcriptional silencing

Background

FRAXE mental retardation (FRAXE MR; MIM 309548) is the most common form of so-called nonspecific X-linked mental retardation (MRX) accounting for the symptoms seen in about two-thirds of families with MRX thus far identified (Gecz, 2000a). The mental retardation in affected individuals, which is generally mild, may be accompanied by an assortment of other features including nasal abnormalities, irregular teeth, mild facial hypoplasia, and speech delay. As in FXS this disorder is associated with the appearance of a folate-sensitive fragile site coincident with a CGG·CCG-repeat expansion. The expanded repeat is located in the 5′ UTR of a gene, FMR2, as illustrated in Fig. 2. FMR2 is located 600 kb downstream of the FMR1 gene at Xq28. In this instance the CCG-rich strand is the non-template strand (Gecz et al., 1996; Gu et al., 1996). The FMR2 gene encodes a protein that localizes to the nucleus (Hillman and Gecz, 2001), and is highly expressed in the hippocampus (Chakrabarti et al., 1998). This protein has significant homology to three proline- and serine-rich autosomal genes: AF4, LAF4, and AF5Q31. AF4 and LAF4 have transcriptional transactivation activity (Hillman and Gecz, 2001), and by extension FMR2 is hypothesized to be a transcriptional activator.

As in FXS, expansion in the FMR2 gene is also accompanied by hypermethylation and a deficit of FMR2 mRNA. *Fmr2*

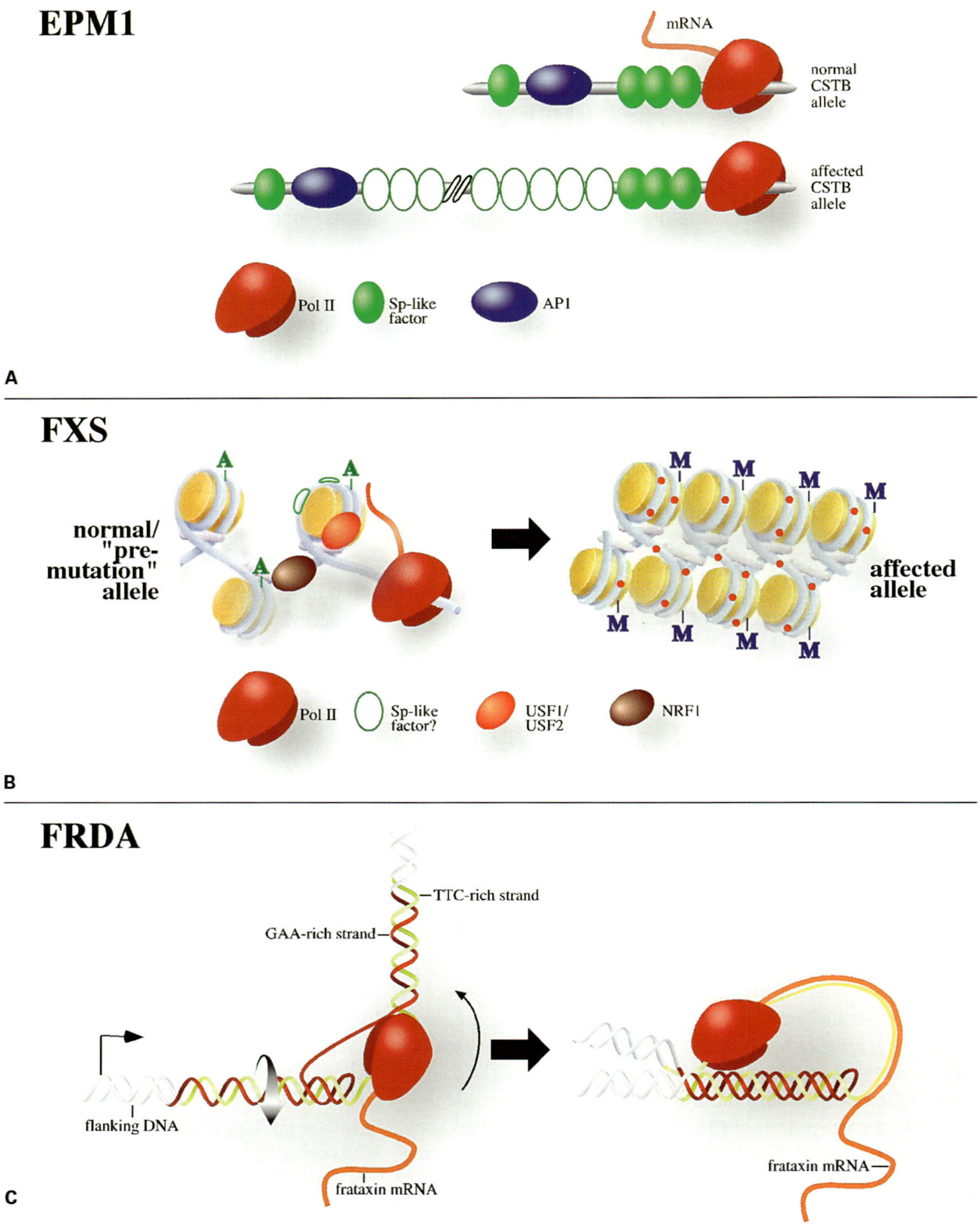

Fig. 4. Models for the transcription defects in FXS, FRAXE MR, EPM1 and FRDA. (**A**) Diagrammatic representation of the model for the transcription defect in EPM1 proposed by Lalioti et. al. (1999). Generic Sp-like factors binding to GC boxes and a generic AP1 factor are shown in green and blue respectively. In normal alleles AP1 and, for example, Sp1 may be close enough to each other and to the basal transcription machinery to facilitate a synergistic activation of the promoter as reported for other promoters (Banks et al., 1998; Kaluzova et al., 2001). In affected alleles the distance between AP1 and the basal transcription machinery and adjacent GC boxes is increased, leading to a decrease in promoter activity. (**B**) Model for the effect of expansion on transcription in FXS. In the normal or premutation allele, the promoter is unmethylated and is associated with acetylated histones (indicated by the letter A). The chromatin conformation is thus open and the DNA is accessible to the transcription factors USF1/USF2 and NRF-1 that are important for transcriptional activity in transient transfection assays. In vivo the two GC boxes located between the USF and NRF1 binding sites are also occupied and thus may be important for promoter regulation in situ. The outlines of generic Sp factors that bind the GC boxes are shown in green. The histones become deacetylated and the unmethylated repeats are a target for histone methyltransferases resulting in K9-methylation of histone H3 (indicated by the letter M). This facilitates the recruitment of proteins such as HP1 that have an affinity for histone tails methylated at this position. This leads to the establishment of transcriptionally silent heterochromatin that prevents access of transcription factors to the promoter. DNA methyltransferases are recruited resulting in promoter methylation (indicated by the red dots). (**C**). Model for the effect of expansion on transcription elongation in FRDA. During transcription of the GAA·TTC repeat tract, RNA polymerase temporarily covers parts of the template strand shown in yellow, leav-

knockout mice have a delay-dependent conditioned fear deficit that is accompanied by an increase in long-term potentiation (LTP) (Gu et al., 2002). The impaired learning and memory seen in these mice supports the idea that an effect on FMR2 is responsible for the symptoms of FRAXE MR. However, some individuals with the FRAXE mutation and low FMR2 levels seem to be relatively unaffected (Murgia et al., 1996; Gecz et al., 1997; Lo Nigro et al., 2000). It has been suggested that the absence of an apparent effect of FMR2 silencing in these cases results from balancing effects of other more prominent determinants of cognitive ability. Somatic mosaicism resulting in shorter repeat lengths or less extensive methylation in the neural tissue of these individuals is also a possibility (Gecz, 2000a).

In addition the FMR2 5′ UTR overlaps the putative promoter of a gene designated FMR3 that is transcribed in the opposite direction (Gecz, 2000b). Transcription of this gene is also decreased on expansion of the CGG·CCG-tract. No open reading frames have yet been identified for this transcript, and its role, if any, in the pathology of FRAXE MR remains to be determined. Like FMR1, mutations in the FMR2/FMR3 region are also associated with premature ovarian failure. However, since the mutations in this instance are microdeletions involving the region around the repeat rather than its expansion (Murray et al., 1999), the mechanisms are likely to be different. It is not known if elevated FMR2 levels occur in FRAXE premutation carriers. Moreover, the transcribed strand contains CCG-repeats that form less stable secondary structures at physiologically reasonable pHs than do CGG-repeats. It is thus unclear whether premutation carriers will suffer significant negative consequences of repeat expansion, as FXS premutation carriers do.

FMR2 and FMR3 promoter silencing

While FMR2 gene regulation has not been studied as extensively as that of FMR1, it seems likely that FMR2 transcriptional silencing mechanisms will be similar to those seen in FXS. The start of FMR2 transcription is embedded in a short CCG·CGG tract that presumably is methylated on repeat expansion. Transcription is thought to be particularly sensitive to methylation in the "preinitiation" region and this promoter may thus be extremely sensitive to its effects. Given its location just 24 bp upstream of the start of transcription, the CGG·CCG-tract is presumably an integral part of the FMR3 promoter. Expansion may affect FMR3 activity by some combination of promoter disruption, as proposed for EPM1 (see below), and an epigenetic effect on transcription initiation, as in FXS.

ing the GAA-rich non-template strand, shown in red, unpaired. As the polymerase proceeds through the repeat tract this unpaired strand folds back and hydrogen bonds to the GAA·TTC duplex that has already reannealed behind the polymerase. This results in the formation of an R·R·Y triplex. The polymerase becomes trapped at the triplex·duplex junction leading to the formation of truncated transcripts and sequestration of the polymerase on the triplex-containing templates.

Progressive myoclonus epilepsy type 1: A promoter disruption problem?

Background

Progressive myoclonic epilepsy type 1 or Unverricht-Lundborg disease (EPM1, MIM 254800) is one of more than 12 different progressive myoclonic epilepsies. This rare but debilitating group of disorders also includes progressive myoclonus epilepsy type 2 (Lafora disease; EPM2, MIM 254780), myoclonus epilepsy with ragged red fibers (MERRF; MIM 545000), neuraminidase deficiency (MIM 256550), and some of the neuronal ceroid lipofuscinoses. Like the other members of this group, EPM1 is characterized by marked progressive photosensitive myoclonus, and epilepsy. Symptoms may also include progressive cerebellar ataxia and slow intellectual decline. Polyspike electroencephalographic (EEG) activity is typically seen on photic stimulation. Histologically, the brain shows degenerative changes without the accumulation of inclusion bodies. The disease typically has an age of onset between 6 and 16 years of age and shows a variable rate of progression. Until recently, early death was not uncommon, most frequently from pneumonia. It is less common today, presumably because of antibiotics and the availability of better anticonvulsive medications. EPM1 is inherited in an autosomal recessive fashion, and while less common in other parts of the world, has an incidence of about one in 20,000 in Finland and some parts of North Africa (Lalioti et al., 1997).

EPM1 is caused by mutations in the cystatin B (CSTB) gene located at 21q22.3 (Pennacchio et al., 1996; Lafreniere et al., 1997). Cystatin B is a member of the cystatin family of cysteine proteinase inhibitors. It is expressed in almost all forebrain neurons, but not in glia. Highest levels of cystatin B expression are seen in the hippocampus. A role for cystatin B in epilepsy is suggested by the fact that in rats kindling-induced recurring seizures results in increased *Cstb* expression (D'Amato et al., 2000). Further mice with *Cstb* disruptions exhibit myoclonic seizures, progressive ataxia, and cerebellar pathology (Shannon et al., 2002). They also show apoptotic loss of neurons in the hippocampal formation, the entorhinal cortex and cerebellum. Gliosis that is seen in older mice is consistent with neuronal atrophy being an important consequence of cystatin-B deficiency. In addition *Cstb*-deficient mice show elevated levels of a number of genes thought to be involved in increased proteolysis, apoptosis and glial activation (Lieuallen et al., 2001).

Cystatin B binds the cathepsins B, H, L and S in vitro (Kinne et al., 2002). Lymphoblasts of individuals with EPM1 show elevated activities of cathepsin, B, L, and S but for reasons that are still unclear, not cathepsin H (Kinne et al., 2002). Elevated protease activity is consistent with a model in which cystatin B acts to prevent apotosis by binding cysteine proteases (D'Amato et al., 2000). However, cystatin B does not colocalize with the cathepsins in many cell types (Riccio et al., 2001). Rather it is found in a complex with a number of structural proteins, such as NF-L and β-spectrin, that are thought to be involved in the regulation of cytoskeletal functions (Di Giaimo et al., 2002). The connection between this complex and the cathepsins thus remains to be resolved.

More than 84% of cases of EPM1 result from expansions of a tandem array of the dodecamer C4GC4GCG·CGCG4CG4 in the CSTB promoter 66 and 77 bp upstream of the two transcription start sites, and 175 bp upstream of the translation initiation codon as illustrated in Fig. 2. The repeat is present at 2–3 copies in unaffected individuals, 12–17 copies in premutation carriers and 30–150 copies in EPM1 patients (Lalioti et al., 1997). Expansion causes a severe reduction in CSTB mRNA in some, but perhaps not all, cell types (Pennacchio et al., 1996; Lafreniere et al., 1997; Lalioti et al., 1997). A smaller but still significant decrease is seen in premutation carriers (Alakurtti et al., 2000). Since individuals with point mutations in CSTB (Pennacchio et al., 1996; Lafreniere et al., 1997; Virtaneva et al., 1997), and *Cstb* knockout mice (Lieuallen et al., 2001), have a similar phenotype to individuals with repeat expansions, it is likely that EPM1 is caused by cystatin B insufficiency.

The effect of repeat expansion on CSTB promoter activity

The regulation of the CSTB gene is not well understood. The minimal region necessary for full CSTB promoter activity and the transcription factors that bind it are still controversial (Lalioti et al., 1999; Alakurtti et al., 2000). The region upstream of the first transcription start site in unaffected individuals contains a number of transcription factor binding sites including those for AP1 and the Sp family (GC boxes). In addition to the single GC box upstream of the repeats, each of the EPM1 repeats contains a GC box. The AP1 sites and GC boxes are bound in vitro by purified AP1 and Sp1, as illustrated in Fig. 4. A 2–4 fold reduction in promoter activity was seen for constructs containing 50 copies of the dodecamer repeat in some, but not all cell lines tested (Lalioti et al., 1999). Replacement of the repeat unit with similar sized pieces of heterologous DNA caused a comparable reduction in promoter activity. This finding suggests that expansion affects promoter activity by increasing the distance between an important transcription factor, perhaps AP1, bound to a site 5′ of the repeat, and the basal transcriptional machinery at the transcriptional start site 3′ of the repeat (Lalioti et al., 1999), as illustrated in Fig. 4.

However, the negative effect of the repeats on promoter activity in these experiments is much smaller than the reduction in CSTB transcription seen in affected individuals. Moreover, this effect was only seen in certain cell lines. With a different promoter construct, a much larger and more consistent inhibition of transcription was seen with only 19 repeats (Alakurtti et al., 2000). Unfortunately, the effect of substituting heterologous DNA for the repeat tract was not tested. It thus remains to be seen whether aberrant spacing accounts for the entire transcription defect in EPM1.

A number of other factors may also play a role. For example, repeat expansion leads to an increase in the number of GC boxes close to the start of transcription. If Sp1 acts as a repressor on this promoter, as it does in some sequence contexts, or if in vivo the GC boxes bind other members of the Sp family, such as the repressors TIEG1 and TIEG2 (Cook et al., 1999), expansion may contribute to the transcription deficit by increasing the number of targets for repressor binding. Another possible effect of expansion on CSTB promoter activity may be the formation of an abnormal chromatin structure (Alakurtti et al., 2000). The EPM1 repeat resembles the FXS and FRAXE repeat in that it is 100% G + C, with the top strand being predominantly G-rich. Many of the biochemical properties of the repeats may thus also be similar. The repeats might exclude nucleosomes, as demonstrated for the FXS repeats (Wang et al., 1996). In addition, the individual strands of the EPM1 repeats form a variety of secondary structures including tetraplexes, i-tetraplexes and hairpins (Pataskar et al., 2001a; Pataskar et al., 2001b; Saha and Usdin, 2001), similar to those seen in the FXS repeat (see Fig. 3). Since these structures contain C·C mispairs, they may be hotspots for direct DNA methylation, as proposed for FXS (Laayoun and Smith, 1995). While the *Hpa*II sites more than 70 bp from the repeats are known to be unmethylated in EPM1 (Lalioti et al., 1999), the methylation status of the repeat is not known. However, in this case, since the repeat is part of the promoter, methylation of the repeat alone may still reduce transcription. For example, methylation of GC boxes blocks binding of some members of the Sp family of transcription factors (Mancini et al., 1999; Cao et al., 2000). Studies of the chromatin structure, methylation status of the repeat and the occupancy of the putative protein binding sites in the promoter in the cells of unaffected and affected individuals may go a long way to resolving the details of transcriptional silencing in this disorder.

Friedreich ataxia: a transcription elongation defect

Background

Friedreich ataxia (FRDA, MIM 229300), is an autosomal recessive neurodegenerative disease. It has a frequency of one to two per 50,000 individuals, making it the most common inherited ataxia. The disease is characterized by progressive gait and limb ataxia, sensory loss, muscle weakness, dysarthria, absent lower limb reflexes, and loss of the sense of position and vibration. Increased ventricular wall thickness is observed in all patients, with most developing a life-threatening hypertrophic cardiomyopathy. Diabetes mellitus is seen in 10% of patients with an additional 20% exhibiting carbohydrate intolerance. FRDA patients begin to exhibit symptoms in their early childhood and can be wheelchair-bound by their late teens or early 20s. The disease progresses relentlessly, often leading to early death, most often from cardiac failure. FRDA is caused by mutations in the frataxin gene on chromosome 9q13 (Campuzano et al., 1996).

Frataxin is an essential protein of 210 amino acids. It is nuclear encoded but localizes to the mitochondria. It is highly conserved from purple bacteria to mammals, but does not show significant homology to any known proteins. In general, its expression is highest in those tissues that are rich in mitochondria such as cardiomyocytes, skeletal muscle and large sensory neurons. The precise role of frataxin in the mitochondria is still hotly debated. It has been suggested that frataxin is involved in mitochondrial iron transport, iron binding/sequestration, protection against oxidative stress, or the biogenesis of Iron Sulfur Clusters (ISCs) (Reviewed in Pandolfo, 2001). Distinguishing among these alternatives is not easy since it is difficult to separate early effects of frataxin deficiency from its secondary consequences. However, most of the recent data supports a role for this protein in the maturation of ISCs (Foury, 1999; Huynen et al., 2001; Puccio et al., 2001; Chen et al., 2002; Muhlenhoff et al., 2002).

All FRDA patients have an expansion of a GAA·TTC-repeat in the first intron of the frataxin gene as illustrated in Fig. 2. Ninety-eight percent of individuals with FRDA have expansions on both alleles. The remaining patients are heterozygous for a GAA·TTC-repeat expansion and a point mutation. No patients are seen with point mutations in both alleles, suggesting that frataxin is an essential protein. This is supported by the finding that frataxin-null mice die in utero (Cossee et al., 2000). Unaffected individuals have 7–38 GAA·TTC-repeats, while affected individuals have from 66 to more than 1,700 repeats on each allele (Bidichandani et al., 1998). The severity of the disease symptoms is correlated with the length of the shortest GAA repeat while the age of onset is inversely proportional to the length (Durr et al., 1996; Filla et al., 1996; Isnard et al., 1997). Milder symptoms or later onset is often associated with interruptions of the GAA·TTC-tract (Ohshima et al., 1999; McDaniel et al., 2001), suggesting that the disease results from some intrinsic property of the repeat.

FRDA symptoms result from a defect in transcription elongation

RT-PCR and RNAse protection studies of RNA from the lymphoblasts of FRDA patients homozygous for the GAA·TTC expansion revealed a reduction in frataxin RNA levels consistent with a transcription or RNA maturation defect (Campuzano et al., 1996; Bidichandani et al., 1998). A similar decrease in RNA yield has been reported in vitro and in transient transfection assays (Grabczyk and Usdin, 1997, 2000a, 2000b; Ohshima et al., 1998). Since no evidence of aberrantly spliced gene products has been found (Bidichandani et al., 1998), the effect of the repeats is probably exerted at the level of transcription. No enhancer activity was detected when the repeat and its flanking sequence was cloned into the SV40 promoter upstream of a luciferase reporter construct, although it is possible that some tissue-specific enhancer activity may be found in those cells most sensitive to the frataxin deficit (Bidichandani et al., 1998). In vitro transcription studies showed no effect of repeat number on the rate of transcription initiation, suggesting that the mRNA deficiency may reflect a problem in transcription elongation (Grabczyk and Usdin, 2000b).

Molecular models for the elongation defect

A number of models have been proposed to explain the repeat-induced transcription deficit. All involve some sort of triplex, based on the propensity of sequences with a purine (R) and pyrimidine (Y) strand asymmetry to form such structures. The FRDA repeat forms Y·R·Y triplexes (Hanvey et al., 1988; Gacy et al., 1998) as well as R·R·Y triplexes (Grabczyk and Usdin, 2000b) as shown in Fig. 3. A variation on the R·R·Y theme known as "sticky DNA", once thought to involve R·R·Y triplex formation between two DNA molecules, is now believed to involve triplex formation between two different repeat tracts on the same molecule (Vetcher et al., 2002), analogous to the interaction of two different R·Y tracts to form Y·R·Y triplexes (Hampel et al., 1994; Lee et al., 1995).

In one DNA triplex-based model, an R·R·Y triplex is generated during transcription (Grabczyk and Usdin, 1997, 2000b, 2000a). In other models the triplex preexists in the chromosome and may or may not involve a duplex formed between the free pyrimidine-rich strand and the new transcript (Ohshima et al., 1998). In vitro transcription studies using T7 RNA polymerase on templates containing 88 GAA·TTC-repeats show no evidence of a block to transcription at the beginning of the repeat tract. Rather, truncated transcripts accumulate at the distal end of the repeat. This is inconsistent with triplexes containing RNA as the third strand or a preexisting triplex (Grabczyk and Usdin, 2000a, 2000b). Templates containing the repeat affect transcription in trans, indicating that the polymerases remain trapped on the template. While supercoiled templates are more effective at reducing transcription, the transcription deficit can still be seen on linear templates. This indicates that the energy required for triplex formation can be derived entirely from the transcription process.

These data are most consistent with a model in which a triplex is formed dynamically during transcription as illustrated in Fig. 4. According to this model, an R·R·Y triplex forms between the repeats that have already reannealed behind the polymerase and the free purine strand resulting from the occlusion of the template strand by the transcribing polymerase. The polymerase then becomes trapped at the triplex:duplex junction, as illustrated in Fig. 4. Support for this model comes from the observation that an RNA:DNA hybrid can be detected on the unpaired region of the template. In addition, the transcription deficit can be alleviated by the addition of a single stranded oligonucleotide containing TTC-repeats. The oligonucleotide anneals to the GAA-rich strand, thus blocking triplex formation (Grabczyk and Usdin, 2000a). This model is attractive because it does not require the existence of a preformed structure. The formation of a preexisting triplex in mammalian cells would require large regions of the chromosome to be unpaired and free of single-stranded binding protein. In contrast, the transcription-driven triplex forms readily, even on linear templates, presumably because the RNA polymerase facilitates unpairing and the wave of supercoil energy that accompanies the enzyme can be utilized to drive triplex formation.

It remains to be seen which, if any, of these models best describes what occurs on the much longer GAA·TTC tracts in the cells of patients with FRDA. Intramolecular YRY-triplex formation also blocks transcription (Sarkar and Brahmachari, 1992; Grabczyk and Fishman, 1995; Grabczyk and Usdin, 2000b). Since the triplex-forming sequence can be anywhere within the primary transcript for it to have an effect on transcription elongation, the potential for triplex-mediated transcription elongation defects in other genes is large.

Conclusions

Unlike the other repeat expansion diseases, the symptoms seen in full mutation carriers of the diseases described above all result, at least in part, from the direct effects of expansion on transcription. These effects range from what may be a permanent disruption of the promoter, in the case of EPM1, to the induction of a cascade of epigenetic modifications in FXS and FRAXE MR that reversibly blocks transcription initiation, and a dynamic block to transcription elongation in FRDA. In addition to symptoms related to loss-of-function of the affect protein in individuals with the full mutation, FXS expansions in premutation carriers may also cause RNA toxicity. In this respect FXS may resemble the dominantly inherited repeat expansion diseases such as DM1 and DM2 where the expanded repeat is transcribed but not translated.

While the other repeat expansion diseases have different proximal effects, recent evidence points to one of the downstream consequences of expansion in all of these diseases being at the level of RNA regulation. In these instances the effect may be a very general one. In the case of diseases, where the affected gene is transcribed but not translated, this effect could be on the distribution of proteins involved in transcription and RNA splicing, PKR activation, or via RNAi, while in the case of the polyglutamine diseases it may be via an effect on the interaction of transcription initiator activators with the general transcription machinery (Philips et al., 1998; Savkur et al., 2001; Buj-Bello et al., 2002; Taylor and Fischbeck, 2002).

Acknowledgements

The authors would like to thank Martine Usdin for her careful reading of the manuscript and helpful suggestions, and Marina Grasso for her input on FRAXE MR.

References

Alakurtti K, Virtaneva K, Joensuu T, Palvimo JJ, Lehesjoki AE: Characterization of the cystatin B gene promoter harboring the dodecamer repeat expanded in progressive myoclonus epilepsy, EPM1. Gene 242:65–73 (2000).

Allingham-Hawkins DJ, Babul-Hirji R, Chitayat D, Holden JJ, Yang KT, Lee C, Hudson R, Gorwill H, Nolin SL, Glicksman A, et al: Fragile X premutation is a significant risk factor for premature ovarian failure: the International Collaborative POF in Fragile X study – preliminary data. Am J med Genet 83: 322–325 (1999).

Allshire R: Molecular Biology: RNAi and heterochromatin – a hushed-up affair. Science 297:1818–1819 (2002).

Ashley CT Jr., Wilkinson KD, Reines D, Warren ST: FMR1 protein: conserved RNP family domains and selective RNA binding. Science 262:563–566 (1993a).

Ashley CT, Sutcliffe JS, Kunst CB, Leiner HA, Eichler EE, Nelson DL, Warren ST: Human and murine FMR-1: alternative splicing and translational initiation downstream of the CGG-repeat. Nature Genet 4: 244–251 (1993b).

Banks EB, Crish JF, Welter JF, Eckert RL: Characterization of human involucrin promoter distal regulatory region transcriptional activator elements-a role for Sp1 and AP1 binding sites. Biochem J 331:61–68 (1998).

Bidichandani SI, Ashizawa T, Patel PI: The GAA triplet-repeat expansion in Friedreich ataxia interferes with transcription and may be associated with an unusual DNA structure. Am J hum Genet 62:111–121 (1998).

Braun K, Segal M: FMRP involvement in formation of synapses among cultured hippocampal neurons. Cereb Cortex 10:1045–1052 (2000).

Brown V, Jin P, Ceman S, Darnell JC, O'Donnell WT, Tenenbaum SA, Jin X, Feng Y, Wilkinson KD, Keene JD, et al: Microarray identification of FMRP-associated brain mRNAs and altered mRNA translational profiles in fragile X syndrome. Cell 107:477–487 (2001).

Buj-Bello A, Furling D, Tronchere H, Laporte J, Lerouge T, Butler-Browne GS, Mandel JL: Muscle-specific alternative splicing of myotubularin-related 1 gene is impaired in DM1 muscle cells. Hum molec Genet 11:2297–2307 (2002).

Burman RW, Yates PA, Green LD, Jacky PB, Turker MS, Popovich BW: Hypomethylation of an expanded FMR1 allele is not associated with a global DNA methylation defect. Am J hum Genet 65:1375–1386 (1999).

Campuzano V, Montermini L, Molto MD, Pianese L, Cossee M, Cavalcanti F, Monros E, Rodius F, Duclos F, Monticelli A, et al: Friedreich's ataxia: autosomal recessive disease caused by an intronic GAA triplet repeat expansion. Science 271:1423–1427 (1996).

Cao YX, Jean JC, Williams MC: Cytosine methylation of an Sp1 site contributes to organ-specific and cell-specific regulation of expression of the lung epithelial gene t1alpha. Biochem J 350:883–890 (2000).

Caudy AA, Myers M, Hannon GJ, Hammond SM: Fragile X-related protein and VIG associate with the RNA interference machinery. Genes Dev 16: 2491–2496 (2002).

Chakrabarti L, Bristulf J, Foss GS, Davies KE: Expression of the murine homologue of FMR2 in mouse brain and during development. Hum molec Genet 7:441–448 (1998).

Chen OS, Hemenway S, Kaplan J: Inhibition of Fe-S cluster biosynthesis decreases mitochondrial iron export: Evidence that Yfh1p affects Fe-S cluster synthesis. Proc natl Acad Sci, USA (2002).

Chen X, Mariappan SV, Catasti P, Ratliff R, Moyzis RK, Laayoun A, Smith SS, Bradbury EM, Gupta G: Hairpins are formed by the single DNA strands of the fragile X triplet repeats: structure and biological implications. Proc natl Acad Sci, USA 92: 5199–5203 (1995).

Chen X, Mariappan SV, Moyzis RK, Bradbury EM, Gupta G: Hairpin induced slippage and hypermethylation of the fragile X DNA triplets. J Biomol Struct Dyn 15:745–756 (1998).

Chiurazzi P, Pomponi MG, Willemsen R, Oostra BA, Neri G: In vitro reactivation of the FMR1 gene involved in fragile X syndrome. Hum molec Genet 7:109–113 (1998).

Coffee B, Zhang F, Warren ST, Reines D: Acetylated histones are associated with FMR1 in normal but not fragile X-syndrome cells. Nature Genet 22:98–101 (1999).

Coffee B, Zhang F, Ceman S, Warren ST, Reines D: Histone modifications depict an aberrantly heterochromatinized FMR1 gene in fragile X syndrome. Am J hum Genet 71:(2002).

Cook T, Gebelein B, Belal M, Mesa K, Urrutia R: Three conserved transcriptional repressor domains are a defining feature of the TIEG subfamily of Sp1-like zinc finger proteins. J biol Chem 274:29500–29504 (1999).

Cossee M, Puccio H, Gansmuller A, Koutnikova H, Dierich A, LeMeur M, Fischbeck K, Dolle P, Koenig M: Inactivation of the Friedreich ataxia mouse gene leads to early embryonic lethality without iron accumulation. Hum molec Genet 9:1219–1226 (2000).

D'Amato E, Kokaia Z, Nanobashvili A, Reeben M, Lehesjoki AE, Saarma M, Lindvall O: Seizures induce widespread upregulation of cystatin B, the gene mutated in progressive myoclonus epilepsy, in rat forebrain neurons. Eur J Neurosci 12:1687–1695 (2000).

Darnell JC, Jensen KB, Jin P, Brown V, Warren ST, Darnell RB: Fragile X mental retardation protein targets G quartet mRNAs important for neuronal function. Cell 107:489–499 (2001).

De Boulle K, Verkerk AJ, Reyniers E, Vits L, Hendrickx J, Van Roy B, Van den Bos F, de Graaff E, Oostra BA, Willems PJ: A point mutation in the FMR-1 gene associated with fragile X mental retardation. Nature Genet 3:31–35 (1993).

Dennis K, Fan T, Geiman T, Yan Q, Muegge K: Lsh, a member of the SNF2 family, is required for genome-wide methylation. Genes Dev 15:2940–2944 (2001).

Di Giaimo R, Riccio M, Santi S, Galeotti C, Ambrosetti DC, Melli M: New insights into the molecular basis of progressive myoclonus epilepsy: a multiprotein complex with cystatin B. Hum molec Genet 11:2941–2950 (2002).

Dockendorff TC, Su HS, McBride SM, Yang Z, Choi CH, Siwicki KK, Sehgal A, Jongens TA: Drosophila lacking dfmr1 activity show defects in circadian output and fail to maintain courtship interest. Neuron 34:973–984 (2002).

Drouin R, Angers M, Dallaire N, Rose TM, Khandjian W, Rousseau F: Structural and functional characterization of the human FMR1 promoter reveals similarities with the hnRNP-A2 promoter region. Hum molec Genet 6:2051–2060 (1997).

Durr A, Cossee M, Agid Y, Campuzano V, Mignard C, Penet C, Mandel JL, Brice A, Koenig M: Clinical and genetic abnormalities in patients with Friedreich's ataxia. New Engl J Med 335:1169–1175 (1996).

Feng Y, Zhang F, Lokey LK, Chastain JL, Lakkis L, Eberhart D, Warren ST: Translational suppression by trinucleotide repeat expansion at FMR1. Science 268:731–734 (1995).

Feng Y, Absher D, Eberhart DE, Brown V, Malter HE, Warren ST: FMRP associates with polyribosomes as an mRNP, and the I304N mutation of severe fragile X syndrome abolishes this association. Mol Cell 1:109–118 (1997a).

Feng Y, Gutekunst CA, Eberhart DE, Yi H, Warren ST, Hersch SM: Fragile X mental retardation protein: nucleocytoplasmic shuttling and association with somatodendritic ribosomes. J Neurosci 17:1539–1547 (1997b).
Filla A, De Michele G, Cavalcanti F, Pianese L, Monticelli A, Campanella G, Cocozza S: The relationship between trinucleotide (GAA) repeat length and clinical features in Friedreich ataxia. Am J hum Genet 59:554–560 (1996).
Foury F: Low iron concentration and aconitase deficiency in a yeast frataxin homologue deficient strain. FEBS Lett 456:281–284 (1999).
Fry M, Loeb LA: The fragile X syndrome d(CGG)n nucleotide repeats form a stable tetrahelical structure. Proc natl Acad Sci, USA 91:4950–4954 (1994).
Fu YH, Kuhl DP, Pizzuti A, Pieretti M, Sutcliffe JS, Richards S, Verkerk AJ, Holden JJ, Fenwick RG Jr., Warren ST, et al: Variation of the CGG repeat at the fragile X site results in genetic instability: resolution of the Sherman paradox. Cell 67:1047–1058 (1991).
Fuks F, Hurd PJ, Wolf D, Nan X, Bird AP, Kouzarides T: The methyl-CpG-binding protein MeCP2 links DNA methylation to histone methylation. J biol Chem (2002).
Gacy AM, Goellner GM, Spiro C, Chen X, Gupta G, Bradbury EM, Dyer RB, Mikesell MJ, Yao JZ, Johnson AJ, et al: GAA instability in Friedreich's Ataxia shares a common, DNA-directed and intraallelic mechanism with other trinucleotide diseases. Mol Cell 1:583–593 (1998).
Gecz J: The FMR2 gene, FRAXE and non-specific X-linked mental retardation: clinical and molecular aspects. Ann hum Genet 64:95–106 (2000a).
Gecz J: FMR3 is a novel gene associated with FRAXE CpG island and transcriptionally silent in FRAXE full mutations. J med Genet 37:782–784 (2000b).
Gecz J, Gedeon AK, Sutherland GR, Mulley JC: Identification of the gene FMR2, associated with FRAXE mental retardation. Nature Genet 13: 105–108 (1996).
Gecz J, Oostra BA, Hockey A, Carbonell P, Turner G, Haan EA, Sutherland GR, Mulley JC: FMR2 expression in families with FRAXE mental retardation. Hum molec Genet 6:435–441 (1997).
Grabczyk E, Fishman MC: A long purine-pyrimidine homopolymer acts as a transcriptional diode. J biol Chem 270:1791–1797 (1995).
Grabczyk E, Usdin K: Length dependent transcription attenuation in the Friedreich's ataxia triplet expansion mutation (GAA) via triple helix formation. FASEB J 11:1998 (1997).
Grabczyk E, Usdin K: Alleviating transcript insufficiency caused by Friedreich's ataxia triplet repeats. Nucl Acids Res 28:4930–4937 (2000a).
Grabczyk E, Usdin K: The GAA*TTC triplet repeat expanded in Friedreich's ataxia impedes transcription elongation by T7 RNA polymerase in a length and supercoil dependent manner. Nucl Acids Res 28:2815–2822 (2000b).
Greco CM, Hagerman RJ, Tassone F, Chudley AE, Del Bigio MR, Jacquemont S, Leehey M, Hagerman PJ: Neuronal intranuclear inclusions in a new cerebellar tremor/ataxia syndrome among fragile X carriers. Brain 125:1760–1771 (2002).
Gu Y, Shen Y, Gibbs RA, Nelson DL: Identification of FMR2, a novel gene associated with the FRAXE CCG repeat and CpG island. Nature Genet 13: 109–113 (1996).
Gu Y, McIlwain KL, Weeber EJ, Yamagata T, Xu B, Antalffy BA, Reyes C, Yuva-Paylor L, Armstrong D, Zoghbi H, et al: Impaired conditioned fear and enhanced long-term potentiation in *Fmr2* knockout mice. J Neurosci 22:2753–2763 (2002).
Hampel KJ, Ashley C, Lee JS: Kilobase-range communication between polypurine, polypyrimidine tracts in linear plasmids mediated by triplex formation: a braided knot between two linear duplexes. Biochemistry 33:5674–5681 (1994).
Hanvey JC, Shimizu M, Wells RD: Intramolecular DNA triplexes in supercoiled plasmids. Proc natl Acad Sci, USA 85:6292–6296 (1988).
Hashem VI, Sinden RR: Chemotherapeutically induced deletion of expanded triplet repeats. Mutat Res 508:107–119 (2002).
Heard E, Rougeulle C, Arnaud D, Avner P, Allis CD, Spector DL: Methylation of histone H3 at Lys-9 is an early mark on the X chromosome during X inactivation. Cell 107:727–738 (2001).
Hillman MA, Gecz J: Fragile XE-associated familial mental retardation protein 2 (FMR2) acts as a potent transcription activator. J Hum Genet 46:251–259 (2001).
Hornstra IK, Nelson DL, Warren ST, Yang TP: High resolution methylation analysis of the FMR1 gene trinucleotide repeat region in fragile X syndrome. Hum molec Genet 2:1659–1665 (1993).
Huynen MA, Snel B, Bork P, Gibson TJ: The phylogenetic distribution of frataxin indicates a role in iron-sulfur cluster protein assembly. Hum molec Genet 10:2463–2468 (2001).
Inoue S, Shimoda M, Nishinokubi I, Siomi MC, Okamura M, Nakamura A, Kobayashi S, Ishida N, Siomi H: A role for the Drosophila fragile X-related gene in circadian output. Curr Biol 12:1331–1335 (2002).
Irwin SA, Galvez R, Greenough WT: Dendritic spine structural anomalies in fragile-X mental retardation syndrome. Cereb Cortex 10:1038–1044 (2000).
Ishizuka A, Siomi MC, Siomi H: A Drosophila fragile X protein interacts with components of RNAi and ribosomal proteins. Genes Dev 16:2497–2508 (2002).
Isnard R, Kalotka H, Durr A, Cossee M, Schmitt M, Pousset F, Thomas D, Brice A, Koenig M, Komajda M: Correlation between left ventricular hypertrophy and GAA trinucleotide repeat length in Friedreich's ataxia. Circulation 95:2247–2249 (1997).
Jeddeloh JA, Stokes TL, Richards EJ: Maintenance of genomic methylation requires a SWI2/SNF2-like protein. Nature Genet 22:94–97 (1999).
Jones L, Ratcliff F, Baulcombe DC: RNA-directed transcriptional gene silencing in plants can be inherited independently of the RNA trigger and requires Met1 for maintenance. Curr Biol 11:747–757 (2001).
Kaluzova M, Pastorekova S, Svastova E, Pastorek J, Stanbridge EJ, Kaluz S: Characterization of the MN/CA 9 promoter proximal region: a role for specificity protein (SP) and activator protein 1 (AP1) factors. Biochem J 359:669–677 (2001).
Kenneson A, Zhang F, Hagedorn CH, Warren ST: Reduced FMRP and increased FMR1 transcription is proportionally associated with CGG repeat number in intermediate-length and premutation carriers. Hum molec Genet 10:1449–1454 (2001).
Kettani A, Kumar RA, Patel DJ: Solution structure of a DNA quadruplex containing the fragile X syndrome triplet repeat. J molec Biol 254:638–656 (1995).
Khandjian EW, Bardoni B, Corbin F, Sittler A, Giroux S, Heitz D, Tremblay S, Pinset C, Montarras D, Rousseau F, et al: Novel isoforms of the fragile X related protein FXR1P are expressed during myogenesis. Hum molec Genet 7:2121–2128 (1998).
Kinne R, Saukko P, Jarvinen M, Lehesjoki AE: Reduced cystatin B activity correlates with enhanced cathepsin activity in progressive myoclonus epilepsy. Ann Med 34:380–385 (2002).
Kumari D, Usdin K: Interaction of the transcription factors USF1, USF2, and alpha -Pal/Nrf-1 with the FMR1 promoter. Implications for Fragile X mental retardation syndrome. J biol Chem 276:4357–4364 (2001).
Laayoun A, Smith SS: Methylation of slipped duplexes, snapbacks and cruciforms by human DNA(cytosine-5)methyltransferase. Nucl Acids Res 23:1584–1589 (1995).
Lafreniere RG, Rochefort DL, Chretien N, Rommens JM, Cochius JI, Kalviainen R, Nousiainen U, Patry G, Farrell K, Soderfeldt B, et al: Unstable insertion in the 5′ flanking region of the cystatin B gene is the most common mutation in progressive myoclonus epilepsy type 1, EPM1. Nature Genet 15:298–302 (1997).
Lalioti MD, Scott HS, Buresi C, Rossier C, Bottani A, Morris MA, Malafosse A, Antonarakis SE: Dodecamer repeat expansion in cystatin B gene in progressive myoclonus epilepsy. Nature 386:847–851 (1997).
Lalioti MD, Scott HS, Antonarakis SE: Altered spacing of promoter elements due to the dodecamer repeat expansion contributes to reduced expression of the cystatin B gene in EPM1. Hum molec Genet 8:1791–1798 (1999).
Lee JS, Ashley C, Hampel KJ, Bradley R, Scraba DG: A stable interaction between separated pyrimidine.purine tracts in circular DNA. J molec Biol 252:283–288 (1995).
Leehey MA, Hagerman RJ, Landau WM, Grigsby J, Tassone F, Hagerman PJ: Tremor/ataxia syndrome in fragile X carrier males. Mov Disord 17:744–745 (2002).
Lieuallen K, Pennacchio LA, Park M, Myers RM, Lennon GG: Cystatin B-deficient mice have increased expression of apoptosis and glial activation genes. Hum molec Genet 10:1867–1871 (2001).
Lindsay H, Adams RL: Spreading of methylation along DNA. Biochem J 320:473–478 (1996).
Lo Nigro C, Faravelli F, Cavani S, Perroni L, Novello P, Vitali M, Bricarelli FD, Grasso M: FRAXE mutation in a mentally retarded subject and in his phenotypically normal twin brother. Eur J hum Genet 8:157–162 (2000).
Loesch DZ, Huggins RM, Bui QM, Epstein JL, Taylor AK, Hagerman RJ: Effect of the deficits of fragile X mental retardation protein on cognitive status of fragile X males and females assessed by robust pedigree analysis. J Dev Behav Pediatr 23:416–423 (2002).
Lugenbeel KA, Peier AM, Carson NL, Chudley AE, Nelson DL: Intragenic loss of function mutations demonstrate the primary role of FMR1 in fragile X syndrome. Nature Genet 10:483–485 (1995).
Mancini DN, Singh SM, Archer TK, Rodenhiser DI: Site-specific DNA methylation in the neurofibromatosis (NF1) promoter interferes with binding of CREB and SP1 transcription factors. Oncogene 18:4108–4119 (1999).
Mariappan SV, Catasti P, Chen X, Ratliff R, Moyzis RK, Bradbury EM, Gupta G: Solution structures of the individual single strands of the fragile X DNA triplets $(GCC)_n.(GGC)_n$. Nucl Acids Res 24:784–792 (1996).
Mariappan SV, Silks LA, 3rd, Bradbury EM, Gupta G: Fragile X DNA triplet repeats, $(GCC)_n$, form hairpins with single hydrogen-bonded cytosine.cytosine mispairs at the CpG sites: isotope-edited nuclear magnetic resonance spectroscopy on $(GCC)_n$ with selective 15N4-labeled cytosine bases. J molec Biol 283:111–120 (1998).
McDaniel DO, Keats B, Vedanarayanan VV, Subramony SH: Sequence variation in GAA repeat expansions may cause differential phenotype display in Friedreich's ataxia. Mov Disord 16:1153–1158 (2001).

Mette MF, Aufsatz W, van der Winden J, Matzke MA, Matzke AJ: Transcriptional silencing and promoter methylation triggered by double-stranded RNA. EMBO J 19:5194–5201 (2000).

Mitas M, Yu A, Dill J, Haworth IS: The trinucleotide repeat sequence $d(CGG)_{15}$ forms a heat-stable hairpin containing Gsyn. Ganti base pairs. Biochemistry 34:12803–12811 (1995).

Morales J, Hiesinger PR, Schroeder AJ, Kume K, Verstreken P, Jackson FR, Nelson DL, Hassan BA: Drosophila fragile X protein, DFXR, regulates neuronal morphology and function in the brain. Neuron 34:961–972 (2002).

Muhlenhoff U, Richhardt N, Ristow M, Kispal G, Lill R: The yeast frataxin homolog Yfh1p plays a specific role in the maturation of cellular Fe/S proteins. Hum molec Genet 11:2025–2036 (2002).

Murgia A, Polli R, Vinanzi C, Salis M, Drigo P, Artifoni L, Zacchello F: Amplification of the Xq28 FRAXE repeats: extreme phenotype variability? Am J med Genet 64:441–444 (1996).

Murray A, Webb J, Dennis N, Conway G, Morton N: Microdeletions in FMR2 may be a significant cause of premature ovarian failure. J med Genet 36:767–770 (1999).

Nadel Y, Weisman-Shomer P, Fry M: The fragile X syndrome single strand $d(CGG)_n$ nucleotide repeats readily fold back to form unimolecular hairpin structures. J biol Chem 270:28970–28977 (1995).

Nishikura K: A short primer on RNAi: RNA-directed RNA polymerase acts as a key catalyst. Cell 107:415–418 (2001).

Ohshima K, Montermini L, Wells RD, Pandolfo M: Inhibitory effects of expanded GAA.TTC triplet repeats from intron I of the Friedreich ataxia gene on transcription and replication *in vivo*. J biol Chem 273:14588–14595 (1998).

Ohshima K, Sakamoto N, Labuda M, Poirier J, Moseley ML, Montermini L, Ranum LP, Wells RD, Pandolfo M: A nonpathogenic GAAGGA repeat in the Friedreich gene: implications for pathogenesis. Neurology 53:1854–1857 (1999).

Pandolfo M: Molecular basis of Friedreich ataxia. Mov Disord 16:815–821 (2001).

Pataskar SS, Dash D, Brahmachari SK: Intramolecular i-motif structure at acidic pH for progressive myoclonus epilepsy (EPM1) repeat $d(CCCCGCCCCG\text{-}CG)_n$. J Biomol Struct Dyn 19:307–313 (2001a).

Pataskar SS, Dash D, Brahmachari SK: Progressive myoclonus epilepsy [EPM1] repeat $d(CCCCGCC\text{-}CCGCG)_n$ forms folded hairpin structures at physiological pH. J Biomol Struct Dyn 19:293–305 (2001b).

Patel PK, Bhavesh NS, Hosur RV: Cation-dependent conformational switches in d-TGGCGGC containing two triplet repeats of fragile X syndrome: NMR observations. Biochem biophys Res Commun 278:833–838 (2000).

Pennacchio LA, Lehesjoki AE, Stone NE, Willour VL, Virtaneva K, Miao J, D'Amato E, Ramirez L, Faham M, Koskiniemi M, et al: Mutations in the gene encoding cystatin B in progressive myoclonus epilepsy (EPM1). Science 271:1731–1734 (1996).

Philips AV, Timchenko LT, Cooper TA: Disruption of splicing regulated by a CUG-binding protein in myotonic dystrophy. Science 280:737–741 (1998).

Pieretti M, Zhang FP, Fu YH, Warren ST, Oostra BA, Caskey CT, Nelson DL: Absence of expression of the FMR-1 gene in fragile X syndrome. Cell 66:817–822 (1991).

Pietrobono R, Pomponi MG, Tabolacci E, Oostra B, Chiurazzi P, Neri G: Quantitative analysis of DNA demethylation and transcriptional reactivation of the FMR1 gene in fragile X cells treated with 5-azadeoxycytidine. Nucl Acids Res 30:3278–3285 (2002).

Puccio H, Simon D, Cossee M, Criqui-Filipe P, Tiziano F, Melki J, Hindelang C, Matyas R, Rustin P, Koenig M: Mouse models for Friedreich ataxia exhibit cardiomyopathy, sensory nerve defect and Fe-S enzyme deficiency followed by intramitochondrial iron deposits. Nature Genet 27:181–186 (2001).

Raca G, Siyanova E, Mirkin S: Janus effects of premutation size CGG repeats on gene expression. Am J hum Genet 67:2046 (2000).

Ranum LP, Day JW: Dominantly inherited, non-coding microsatellite expansion disorders. Curr Opin Genet Dev 12:266–271 (2002).

Razin A: CpG methylation, chromatin structure and gene silencing – a three-way connection. EMBO J 17:4905–4908 (1998).

Riccio M, Di Giaimo R, Pianetti S, Palmieri PP, Melli M, Santi S: Nuclear localization of cystatin B, the cathepsin inhibitor implicated in myoclonus epilepsy (EPM1). Exp Cell Res 262:84–94 (2001).

Saha T, Usdin K: Tetraplex formation by the progressive myoclonus epilepsy type-1 repeat: implications for instability in the repeat expansion diseases. FEBS Lett 491:184–187 (2001).

Sandberg G, Schalling M: Effect of in vitro promoter methylation and CGG repeat expansion on FMR-1 expression. Nucl Acids Res 25:2883–2887 (1997).

Sarkar PS, Brahmachari SK: Intramolecular triplex potential sequence within a gene down regulates its expression in vivo. Nucl Acids Res 20:5713–5718 (1992).

Savkur RS, Philips AV, Cooper TA: Aberrant regulation of insulin receptor alternative splicing is associated with insulin resistance in myotonic dystrophy. Nature Genet 29:40–47 (2001).

Schaeffer C, Bardoni B, Mandel JL, Ehresmann B, Ehresmann C, Moine H: The fragile X mental retardation protein binds specifically to its mRNA via a purine quartet motif. EMBO J 20:4803–4813 (2001).

Schwemmle S, de Graaff E, Deissler H, Glaser D, Wohrle D, Kennerknecht I, Just W, Oostra BA, Dorfler W, Vogel W, et al: Characterization of FMR1 promoter elements by in vivo-footprinting analysis. Am J hum Genet 60:1354–1362 (1997).

Shannon P, Pennacchio LA, Houseweart MK, Minassian BA, Myers RM: Neuropathological changes in a mouse model of progressive myoclonus epilepsy: cystatin B deficiency and Unverricht-Lundborg disease. J Neuropathol Exp Neurol 61:1085–1091 (2002).

Siomi MC, Siomi H, Sauer WH, Srinivasan S, Nussbaum RL, Dreyfuss G: FXR1, an autosomal homolog of the fragile X mental retardation gene. EMBO J 14:2401–2408 (1995).

Smeets HJ, Smits AP, Verheij CE, Theelen JP, Willemsen R, van de Burgt I, Hoogeveen AT, Oosterwijk JC, Oostra BA: Normal phenotype in two brothers with a full FMR1 mutation. Hum molec Genet 4:2103–2108 (1995).

Sutcliffe JS, Nelson DL, Zhang F, Pieretti M, Caskey CT, Saxe D, Warren ST: DNA methylation represses FMR-1 transcription in fragile X syndrome. Hum molec Genet 1:397–400 (1992).

Tamaru H, Selker EU: A histone H3 methyltransferase controls DNA methylation in *Neurospora crassa*. Nature 414:277–283 (2001).

Tan SL, Katze MG: The emerging role of the interferon-induced PKR protein kinase as an apoptotic effector: a new face of death? J Interferon Cytokine Res 19:543–554 (1999).

Tassone F, Hagerman RJ, Loesch DZ, Lachiewicz A, Taylor AK, Hagerman PJ: Fragile X males with unmethylated, full mutation trinucleotide repeat expansions have elevated levels of FMR1 messenger RNA. Am J med Genet 94:232–236 (2000a).

Tassone F, Hagerman RJ, Taylor AK, Gane LW, Godfrey TE, Hagerman PJ: Elevated levels of FMR1 mRNA in carrier males: a new mechanism of involvement in the fragile-X syndrome. Am J hum Genet 66:6–15 (2000b).

Tassone F, Hagerman RJ, Taylor AK, Hagerman PJ: A majority of fragile X males with methylated, full mutation alleles have significant levels of FMR1 messenger RNA. J med Genet 38:453–456 (2001).

Taylor JP, Fischbeck KH: Altered acetylation in polyglutamine disease: an opportunity for therapeutic intervention? Trends Mol Med 8:195–197 (2002).

Usdin K: NGG-triplet repeats form similar intrastrand structures: implications for the triplet expansion diseases. Nucl Acids Res 26:4078–4085 (1998).

Usdin K, Woodford KJ: CGG repeats associated with DNA instability and chromosome fragility form structures that block DNA synthesis *in vitro*. Nucl Acids Res 23:4202–4209 (1995).

Uzielli ML, Guarducci S, Lapi E, Cecconi A, Ricci U, Ricotti G, Biondi C, Scarselli B, Vieri F, Scarnato P, et al: Premature ovarian failure (POF) and fragile X premutation females: from POF to to fragile X carrier identification, from fragile X carrier diagnosis to POF association data. Am J med Genet 84:300–303 (1999).

Verkerk AJ, Pieretti M, Sutcliffe JS, Fu YH, Kuhl DP, Pizzuti A, Reiner O, Richards S, Victoria MF, Zhang FP, et al: Identification of a gene (FMR-1) containing a CGG repeat coincident with a breakpoint cluster region exhibiting length variation in fragile X syndrome. Cell 65:905–914 (1991).

Vetcher AA, Napierala M, Iyer RR, Chastain PD, Griffith JD, Wells RD: Sticky DNA, a long GAA5G-AA5TTC triplex which is formed intramolecularly, in the sequence of intron 1 of the Frataxin gene. J biol Chem (2002).

Virtaneva K, D'Amato E, Miao J, Koskiniemi M, Norio R, Avanzini G, Franceschetti S, Michelucci R, Tassinari CA, Omer S, et al: Unstable minisatellite expansion causing recessively inherited myoclonus epilepsy, EPM1. Nature Genet 15:393–396 (1997).

Volpe TA, Kidner C, Hall IM, Teng G, Grewal SI, Martienssen RA: Regulation of heterochromatic silencing and histone H3 lysine-9 methylation by RNAi. Science 297:1833–1837 (2002).

Wan L, Dockendorff TC, Jongens TA, Dreyfuss G: Characterization of dFMR1, a *Drosophila melanogaster* homolog of the fragile X mental retardation protein. Mol Cell Biol 20:8536–8547 (2000).

Wang YH, Gellibolian R, Shimizu M, Wells RD, Griffith J: Long CCG triplet repeat blocks exclude nucleosomes: a possible mechanism for the nature of fragile sites in chromosomes. J molec Biol 263:511–516 (1996).

Warren ST, Sherman SL: The fragile X syndrome, in Scriver CR, Beaudet AL, Sly WS, Valle D (eds): The Metabolic and Molecular Basis of Inherited Disease, Vol 8, pp 1257–1289 (McGraw-Hill, New York 2001).

Yu A, Barron MD, Romero RM, Christy M, Gold B, Dai J, Gray DM, Haworth IS, Mitas M: At physiological pH, $d(CCG)_{15}$ forms a hairpin containing protonated cytosines and a distorted helix. Biochemistry 36:3687–3699 (1997).

Zhang Y, O'Connor JP, Siomi MC, Srinivasan S, Dutra A, Nussbaum RL, Dreyfuss G: The fragile X mental retardation syndrome protein interacts with novel homologs FXR1 and FXR2. EMBO J 14:5358–5366 (1995).

Cytogenet Genome Res 100:77–84 (2003)
DOI: 10.1159/000072840

Rare fragile sites

G.R. Sutherland

Department of Cytogenetics and Molecular Genetics, Women's and Children's Hospital, and Department of Paediatrics, University of Adelaide, Adelaide (Australia)

Abstract. Rare folate-sensitive fragile sites are the archetypal trinucleotide repeats. Although the CAG repeat in the androgen receptor, associated with spinobulbar muscular atrophy, was the first to be published in 1991, it was the publication in the same year of the molecular basis of fragile X that focused much attention on trinucleotide repeat expansion as a mutational mechanism. A number of rare fragile sites have had their repeat elements characterised since that time. The so-called "folate-sensitive" fragile sites are likely to be all CCG repeat expansions similar to the fragile X. The folate insensitive fragile sites have more complex longer repeat elements. Only two rare fragile sites (FRAXA and FRAXE) are of unequivocal clinical significance in that they are associated with intellectual disability.

Fragile sites on human chromosomes were first described in 1965 by Dekaban et al. In the early years of human cytogenetics these were rather enigmatic phenomena that were seen by some laboratories and not by others. It was not until 1977 that it was demonstrated that the type of culture medium in which lymphocytes were cultured to prepare chromosomes was a critical factor in determining whether fragile sites were seen (Sutherland, 1977). It was subsequently shown that folic acid and thymidine needed to be absent from the culture medium for fragile sites to manifest themselves (Sutherland, 1979).

Some rare fragile sites were found not to be sensitive to folic acid but inducible by a range of other compounds and a classification of the fragile sites based upon their conditions for cytogenetic induction was developed (Table 1). Little that was radically new about fragile sites emerged until the cloning of the fragile X in 1991 (Yu et al., 1991, Oberlé et al., 1991, Verkerk et al., 1991) and since that time several fragile sites have been characterised at the molecular level. Much however remains to be learned about the molecular basis of fragile sites, in particular how the molecular lesions at these loci result in the chromosomal manifestation of the fragile site.

Supported by National Health and Medical Research Council of Australia.

Received 17 February 2003; manuscript accepted 27 February 2003.

Request reprints from Grant R. Sutherland
Department of Cytogenetics and Molecular Genetics
Women's and Children's Hospital, Adelaide 5006 (Australia)
telephone: 618-8161-7284; fax: 618-8161-7342
e-mail: grant.sutherland@adelaide.edu.au

Cytogenetics

A fragile site is a region of chromatin that fails to compact normally for mitosis. When this happens the chromosome is attenuated at that point and fragile, both when prepared for microscopic examination and in vivo as evidenced by configurations that must arise from breakage at the fragile site and nondisjunction of the acentric fragments (Figs. 1–3). The fragile site is almost never seen in 100% of metaphases from an individual who has it, the proportion varying from only a few percent up to 50–60%. An exception is FRA16B which can be seen in close to 100% of metaphases when induced by berenil (Schmid et al., 1986). While fragile sites can be seen on banded chromosomes they are easier to see and more distinctive when seen on unbanded chromosomes (Sutherland and Hecht, 1985).

Most studies of fragile sites have been on chromosomes prepared from short-term (3–4 day) cultures of lymphocytes. The lymphocytes must be cultured under specific conditions for the

KARGER
Fax + 41 61 306 12 34
E-mail karger@karger.ch
www.karger.com

0301–0171/03/1004–0077$19.50/0

Accessible online at:
www.karger.com/cgr

Table 1. All rare fragile sites

Group	Gene symbol	Chromosome band	Reference[a]
Folate sensitive			
	FRA1M	1p21.3	Baker and Sutherland, 1991
	FRA2L	2p11.2	Schuffenhauer et al., 1996
	FRA2A	2q11.2	Ford and Madan, 1973
			Lejeune et al., 1968
			Quack et al., 1978
			Sutherland and Hecht, 1985
	FRA2B	2q13	Ford and Madan, 1973
			Lejeune et al., 1968
			Quack et al., 1978
			Sutherland and Hecht, 1985
	FRA2K	2q22.3	Mattei et al., 1987
	FRA5G	5q35	Sutherland and Hecht, 1985
			Howell et al., 1990
	FRA6A	6p23	Sutherland et al., 1983
	FRA7A	7p11.2	Sutherland and Hecht, 1985
	FRA8A	8q22.3	Quack et al., 1978
	FRA9A	9p21	Guichaoua et al., 1982
			Sutherland et al., 1983
	FRA9B	9q32	Petit et al., 1986
			Sutherland et al., 1983
	FRA10A	10q23.3	Dutrillaux et al., 1985
			Giraud et al., 1976
			Sutherland, 1979
	FRA11A	11q13.3	Quack et al., 1978
			Sutherland, 1979
	FRA11B	11q23.3	Sutherland et al., 1983
			Voullaire et al., 1987
	FRA12A	12q13.1	Giraud et al., 1976
			Quack et al., 1978
			Sutherland and Hinton, 1981
	FRA12D	12q24.13	Amarose et al., 1987
			Sutherland and Baker, 1993
	FRA15B[b]	15q13	Karadeniz et al., 2003
	FRA16A	16p13.11	Callen et al., 1989
			Day et al., 1967
			Sutherland, 1979
			Nancarrow et al., 1994
Folate sensitive			
	FRA19B	19p13	Chodirker et al., 1987
			Tommerup et al., 1985
	FRA20A	20p11.23	Shabtai and Halbrecht, 1981
			Sutherland, 1979
	FRA22A	22q13.1	Kahkonen et al., 1989,
			Webb and Thake, 1984
	FRAXA	Xq27.3	Lubs, 1969,
			Sutherland, 1977
			Yu et al., 1991
			Oberle et al., 1991
			Verkerk et al., 1991
	FRAXE	Xq28	Sutherland and Baker, 1992,
			Oberle et al., 1992
			Knight et al., 1993
			Shaw et al., 2002
	FRAXF	Xq28	Parrish et al., 1994
Distamycin A inducible			
BrdU inducible			
	FRA16B	16q22.1	Callen et al., 1988
			Magenis et al., 1970
			Schmid et al., 1980
			Simmers et al., 1987
			Yu et al., 1997
	FRA17A	17p12	Court Brown et al., 1966
			Sutherland et al., 1984
Not BrdU inducible			
	FRA8E	8q24.1	Ochi et al., 1988
			Takahashi et al., 1988
	FRA11I	11p15.1	Hori et al., 1988
			Takahashi et al., 1988
	FRA16E	16p12.1	Hori et al., 1988
BrdU requiring			
	FRA10B	10q25.2	Scheres and Hustinx, 1980
			Sutherland et al., 1980
			Hewett, 1998
	FRA12C	12q24.2	Sutherland et al., 1985
			Voiculescu et al., 1988

[a] References are to papers that either first specifically identified the fragile site or characterised it.
[b] Provisional, single report only, type unclear.

fragile sites to be expressed. Detailed protocols are given in Sutherland (1991). For the folate-sensitive fragile sites these conditions include absence of folic acid or presence of antifolates such as methotrexate; and low concentrations of thymidine and its analogs BrdU, BrdC, or presence of inhibitors of the enzyme thymidylate synthase (such as FudR); or high concentrations of thymidine (but not its analogs). Any of these inducing conditions need to be present for at least 24 h before chromosome preparation. All of these inducing conditions appear to lead to a relative deficiency of thymidine triphosphate at the time of DNA synthesis (Fig. 4). The failure of high concentrations of BrdU to induce these fragile sites is unexplained, a possibility that the enol form of this compound is incorporated into the DNA under such conditions and inhibits fragile site expression is a barely satisfactory explanation.

The Distamycin A inducible fragile sites are induced by specific agents which bind the minor groove of the DNA helix. These agents include Hoechst 33258, netropsin and berenil. The group 2A fragile sites are also induced by BrdU (but not thymidine). The group 2B fragile sites, which have only been described in the Japanese, are not inducible by BrdU. The group 3 fragile sites are only inducible by BrdU (and BrdC).

An interesting observation that remains unexplained is that when there are two rare folate sensitive fragile sites on chromosomes in one cell, or even two on one chromosome, these tend to "associate" in metaphase chromosome preparations (Sutherland and Baker, 2003). They give a somewhat similar appearance to acrocentric chromosome satellite stalk associations.

Genetics

The fragile X (FRAXA) is the rare fragile site whose genetics has been best studied (Sherman et al., 1984). Its segregation patterns departed greatly from regular Mendelian inheritance. Its most striking features were:

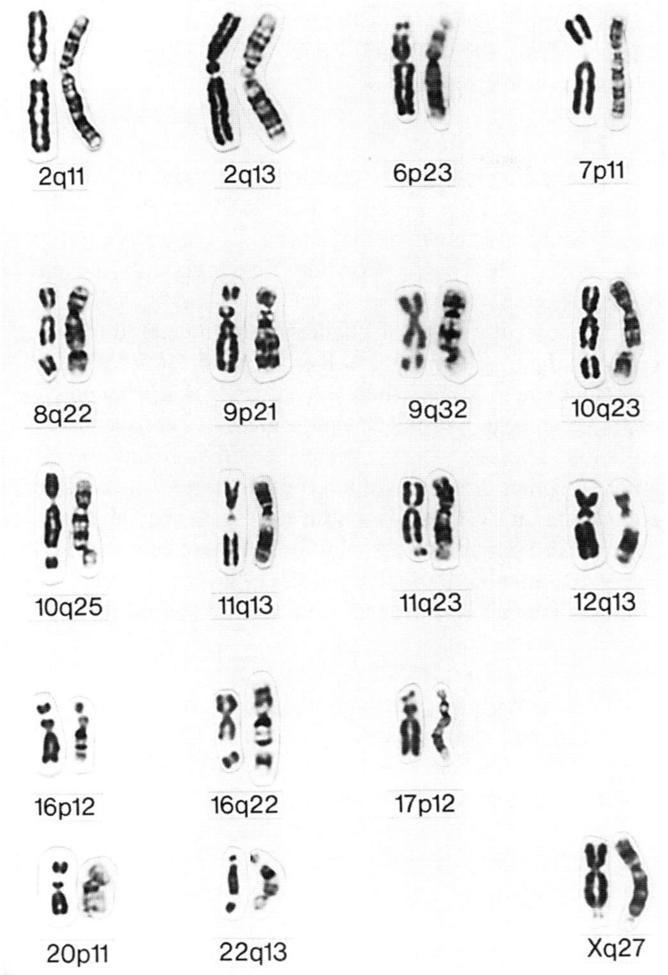

Fig. 1. An assortment of rare fragile sites. The chromosome on the left of each pair is Giemsa stained (unbanded) and the one on the right is G-banded. (Reproduced from Gardner and Sutherland, 1996).

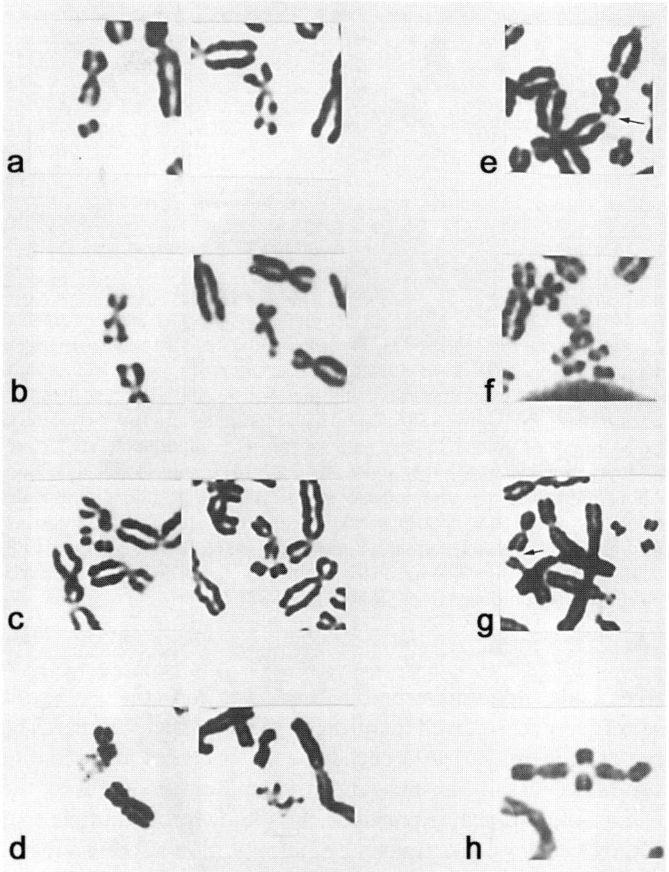

Fig. 2. Forms of expression manifested by fragile sites exemplified by FRA16B. (**a**) Isochromatid gap; (**b**) single chromatid gap; (**c**) duplication of chromatids distal to the fragile site arising from the mechanism in Fig. 3; (**d**) despiralisation of chromatin distal to the fragile site; (**e**) deletion of material distal to the fragile site (arrow); (**f**) quadruplication of chromatin distal to the fragile site; (**g**) deletion (arrow) and acentric fragments; (**h**) quadriradial figure. (Reproduced from Sutherland and Hecht 1985).

- Males could be asymptomatic carriers (transmitting males), a novel finding for an X-linked disease.
- The penetrance of the cytogenetic presence of the fragile site and the associated mental retardation syndrome was incomplete and depended upon the phenotype of the mother. Mothers who were intellectually affected showed complete penetrance in their offspring, whereas penetrance in the offspring of unaffected mothers was about 75%.
- The penetrance of mental retardation in the offspring of the mothers and daughters of transmitting males was different, a phenomenon that became known as the Sherman paradox.

The segregation of the rare autosomal fragile sites was studied by Sherman and Sutherland (1986). For the folate-sensitive fragile sites they found that there was a sex difference in penetrance (defined as the cytogenetic detection of the fragile site): when the fragile sites were transmitted by women they were fully penetrant but when transmitted by males they were only 50% penetrant. The great majority of probands in folate-sensi-

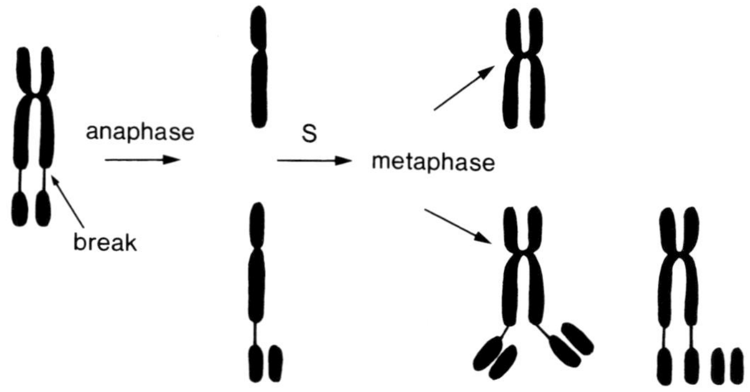

Fig. 3. Mechanism of production of triradial figures and deleted chromosomes by breakage at the fragile site and nondisjunction of an acentric fragment during mitotic anaphase. (Reproduced from Sutherland, 1983).

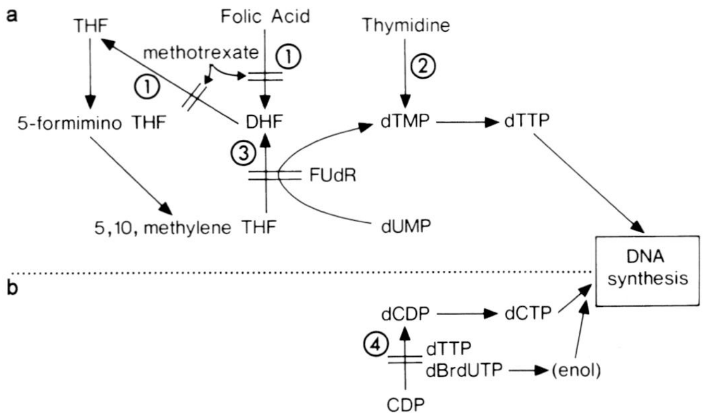

Fig. 4. Areas of nucleotide metabolism involved in folate-sensitive fragile site expression. (**a**) Pathway affected by restriction of folic acid and thymidine or induction by inhibitors of folate metabolism or inhibitors of thymidylate synthase; (**b**) pathway affected by high thymidine and BrdU concentrations. Enzymes crucial to this pathway are (1) dihydrofolate reductase, (2) thymidine kinase, (3) thymidylate synthase, (4) ribonucleotide reductase. Abbreviations: BrdU, bromodeoxyuridine; CDP, cytidine diphosphate; dCDP, deoxycytidine diphosphate; DHF dihydrofolate; FUdR, flurodeoxyuridine; dTMP, deoxythymidine monophosphate (thymidylate); dTTP, deoxythymidine triphosphate; THF, tetrahydrofolate; UMP, uridine monophosphate. (Reproduced from Sutherland, 1988).

tive fragile site families had, perhaps as a consequence of this sex difference, received their fragile site from their mother. The reasons for this sex difference have never been addressed, but sex differences in transmission are characteristic of most of the trinucleotide repeat expansions. It is known for example that FRAXA males never transmit a full mutation to their daughters and it may be that the tendency for folate-sensitive fragile site repeat expansions to regress in size during spermatogenesis is a general phenomenon. Presumably the offspring of rare folate-sensitive fragile site males receive the fragile site-carrying chromosome 50% of the time, but on about half of these occasions the size of its CCG repeat will have been reduced to the point where a fragile site is not manifested cytogenetically. This can only be tested extensively when more of this group of fragile sites has been characterised at the molecular level and their transmission studied. However, in three transmissions from a FRA16A male there were no discernible changes in CCG repeat copy number (Nancarrow et al., 1994).

FRA16B appears to follow a normal mendelian segregation pattern. Although FRA10B appeared to show some of the features of the rare folate-sensitive fragile site segregation patterns the differences from normal Mendelian segregation did not reach statistical significance (Sherman and Sutherland, 1986).

Molecular genetics

Only four autosomal and three X-linked rare fragile sites have been characterised at the molecular level. These are listed and some of their properties are shown in Table 2. All of the isolated folate-sensitive fragile sites have been shown to be expansions of naturally occurring CCG repeats that are polymorphic in length and it is reasonable to hypothesise that all of this group of fragile sites will have a similar molecular basis. The other two, FRA10B and FRA16B, are expansions of AT-rich minisatellite repeats.

FRAXA

The first fragile site to be characterised at the molecular level was FRAXA. Interest had focused on this fragile site since it was the only one known at that time to be clinically significant because of its close association with fragile X syndrome and its bizarre genetics.

The molecular bases of fragile X syndrome, and to a lesser extent of the fragile site, are well established. At the position of the fragile site in Xq27.3 there is a naturally occurring polymorphic CCG repeat. Normal chromosomes have between five and around 55 copies of this trinucleotide with a modal number of 29 or 30 copies depending upon the population studied. There are occasional CCT motifs within the repeat sequence and the patterns and population distributions of these have been extensively documented by Eichler et al. (1996). An increase in the number of copies of the repeat from around 55 to about 230 (ie. an increase of about 200 copies above the modal number of about 30 copies) results in a premutation.

Males with a premutation are clinically normal although a proportion of them will develop a tremor/ataxia syndrome in late-middle to old age (Leehey et al., 2003). These males are cytogenetically normal, not showing the fragile X.

Females with a premutation are clinically normal but some may have a subtle neuropsychiatric impairment (Steyaert et al., 2003) and others are predisposed to premature ovarian failure. Some premutation females show the fragile X in a very small proportion of their X chromosomes. It is likely that those X chromosomes that manifest the fragile site are hypermethylated inactive X chromosomes since CpG methylation of the repeat appears to be a requirement for fragile site expression.

When premutation males pass their fragile X chromosome to their daughters the repeat mostly does not change much in copy number, and any changes are mostly decreases. Their daughters are of course all premutation carriers. When full mutation males reproduce (a fairly rare event) and pass on their fragile X chromosome, the full mutation always regresses to a premutation size. Indeed sperm studies show only premutation sized fragile X alleles to be present (Reyniers et al., 1993).

When fragile X carrier (pre or full mutation) females transmit their fragile X chromosome it frequently increases in copy number, the probability of increase being a function of their copy number (Table 3). Sons with full mutations will have fragile X syndrome; daughters with such mutations have a high probability (~60%) of intellectual impairment and some of the physical features of fragile X syndrome.

Full mutation fragile X chromosomes have more than ~230 copies of the repeat, and the repeat itself and surrounding DNA is heavily CpG methylated. The repeat is within the 5′ untranslated region of the FMR1 gene which is transcriptionally silenced, and this is the pathogenetic mechanism which results in fragile X syndrome. Paradoxically, when a premutation is present the gene is overexpressed but mRNA with increased copies of the repeat is progressively ineffective in production of FMR1 protein (Tassone et al., 2000).

Table 2. Features of the molecularly characterised rare fragile sites

Fragile site	Type[a]	Repeat motif	Gene involved	Reference
FRA16A	FS	CCG (complex)	? Yes	Nancarrow et al., 1994 Nancarrow, 1998
FRA11B	FS	CCG (perfect)	CBL2	Jones et al., 1995
FRA10B	BrdU	42 bp AT-rich minisatellite	No	Yu et al., 1997
FRA16B	Distamycin A	33 bp AT-rich minisatellite	No	Hewett et al., 1998
FRAXA	FS	CCG (complex)	FMR1	Yu et al., 1991 Oberle et al., 1991
FRAXE	FS	CCG (perfect)	FMR2/FMR3	Gecz et al., 1996 Gu et al., 1996 Gecz, 2000a
FRAXF	FS	CCG (complex)	FAM11A	Shaw et al., 2002

[a] FS: folate sensitive.

FRAXE

This fragile site was discovered when individuals with a fragile X chromosome were found not to have an expansion of the FRAXA CCG repeat (Sutherland and Baker, 1992). FRAXE was soon shown to be an expansion of a CCG repeat (Knight et al., 1993). This CCG repeat, unlike the FRAXA and FRA16A repeats, does not have imperfections and ranges in size from 4–39 copies on normal chromosomes with modal sizes of 16 or 18 repeats depending upon the population studied (Zhong et al., 1996).

The repeat was shown to be at the 5′ end of the FMR2 gene that is transcriptionally silenced by full mutations, consequent to CpG methylation of the repeat and surrounding DNA (Gecz et al., 1996; Gu et al., 1996). This was the first gene for non-specific X-linked mental retardation to be isolated (Gecz, 2000a) and has been shown to be a potent transcription activator (Hillman and Gecz, 2001). Gecz (2000b) has shown that there is another transcript, FMR3, originating from the FRAXE CpG island that is also silenced in full mutation males. The function of FMR3 is unknown.

FRAXF

This fragile site was also identified after cytogenetically detected fragile Xs were shown not to have the molecular findings of FRAXA or FRAXE (Hirst et al., 1993). Cytogenetically this fragile site can be distinguished from the two more proximal fragile sites on good quality preparations. FRAXA and FRAXE are indistinguishable cytogenetically, both lying more distal on the chromosome than the common fragile site FRAXD.

This fragile site appears to be clinically harmless even in full mutation males. It is an expansion of the CCG copy number of a complex repeat reported as $(GCCGTC)_{3 \text{ or } 4}$ $(GCC)_n$ where mostly n = 6 or 8 but n values of up to 29 have been seen (Parrish et al., 1994). Recently it has been shown that the fragile site is within the 5′ UTR of a gene, and that this gene is silenced in full mutation males (Shaw et al., 2002). The gene, FAM11A, is highly conserved in mouse and both species have a transcribed pseudogene that is identical at the putative amino acid level.

Table 3. Risk of fragile X premutation expanding to a full mutation on female transmission is a function of CCG repeat copy number, data corrected for ascertainment[a]

Repeat size	No. of Transmissions	% full mutations
55–59	27	3.7
60–69	113	5.3
70–79	90	31
80–89	140	58
90–99	111	80
100–199	197	98.5

[a] Data from Nolin et al. (2003).

FRA16A

This was the first rare autosomal fragile site to be characterised (Nancarrow et al., 1994). It is an expansion of a CCG repeat that is 392 bp from the 5′ end of a large intron-less expressed sequence, which is conserved in hamster and mouse; the gene has not been further characterised (Nancarrow, 1998).

The CCG repeat has been characterised in a number of population groups (Nancarrow et al., 1995). A typical allele series has the structure $(CCG)_6$–$(CCT)_2$–$(CCG)_{6–9}$–(CTG)$(CCG)_{6–11}$ (CCTCCA). The longest uninterrupted stretches of perfect $(CCG)_n$ were seen in the Europeans. Fragile site carriers had CCG amplifications within the range 3.0–5.7 kb.

FRA11B

This is the only other autosomal rare folate-sensitive fragile site to be characterised (Jones et al., 1995). It is another expansion of a perfect CCG repeat, in this instance in the 5′ UTR of the CBL2 oncogene. FRA11B has attracted significant attention since some patients with a chromosomal deletion disorder, Jacobsen syndrome, have their deletion breakpoint at or close to this fragile site. These patients have inherited the deleted chromosome from mothers who have this fragile site on the deleted chromosome. However, the majority of Jacobsen syndrome patients however do not have their breakpoint at this location, but more distally on 11q (Michaelis et al., 1998) and presumably do not have mothers with FRA11B.

FRA16B

This is the most common of the rare fragile sites and its properties are well detailed by Felbor et al. (2003) in this monograph. The fragile site has been characterised and shown to be a massive expansion of a 33-bp AT rich minisatellite sequence. Interestingly this sequence has been reported to be 35 bp in length in a Japanese carrier of this fragile site (Yamauchi et al., 2000). This fragile site does not appear to affect any gene and homozygotes have been described who are normal (Hocking et al., 1999).

FRA10B

This is another reasonably frequent rare fragile site that is induced only by BrdU (and the related compound BrdC). It is also apparently harmless and normal homozygotes have been described. It is carried by 1 in 40 Australians.

At the molecular level it is due to an expansion of a 42-bp AT rich minisatellite repeat that does not appear to affect any gene (Hewett et al., 1998).

Clinical significance

Only FRAXA and FRAXE are unequivocally of clinical significance in that they are associated with fragile X syndrome and FRAXE non-specific X-linked mental retardation respectively.

FRA11B has been associated with the chromosome deletion disorder known as Jacobsen syndrome (Jones et al., 1995). It is clear that a minority of patients with this very rare disorder have their chromosome deletion breakpoint at or very close to the fragile site. In the majority of patients the breakpoint is located more distally. Only a small number (4 or 5) of Jacobsen patients have been shown to have a mother with FRA11B. At this time it is not possible to estimate the risk of a FRA11B carrier having a Jacobsen syndrome child, but it must be relatively small.

Some population studies have reported a higher incidence of autosomal rare fragile sites in mentally retarded groups than in controls. These studies are difficult to interpret since the controls are often not well matched and the numbers of fragile site carriers in either group is relatively small making statistical significance marginal at best. Furthermore, when autosomal fragile sites in the mentally retarded are followed within families, they do not segregate with mental retardation. A detailed review of the clinical significance of fragile sites has recently been published (Sutherland and Baker, 2000).

Replication timing

Laird (1988) originally proposed that the fragile X (FRAXA) would be late replicating and he turned out to be correct, but almost certainly for the wrong reasons. Numerous studies have shown that FRAXA, FRAXE and some of the common fragile sites (reviewed by Arlt et al., 2003) are later replicating than the same locus without either the repeat expansion (the rare sites) or the presence of specific fragile site inducers (the common sites). Data on FRA10B and FRA16B were less clear (Handt et al., 2000) and indeed these authors concluded that while "... delayed replication may play a role in fragile site expression – replication delay may not need to occur at the expanded repeat region itself ... to be permissive for fragility."

Fragile site expression

While much is known about the DNA sequences at fragile sites the mechanisms whereby these sequences give rise to the discontinuity in chromosome structure that is the visible manifestation of the fragile site remains conjectural. Almost certainly the fragile site results from a failure of chromatin compaction for normal mitosis. Many of the factors associated with induction of fragile sites, both chemical in terms of inducing agents, and physical in terms of timing of the action of such agents and of DNA replication, point to DNA at the fragile site not being completely replicated by the time a cell enters metaphase. Such under-replicated, perhaps even single-stranded, DNA is unlikely to package as would normal completely replicated DNA.

Alternatively, it has been shown that expanded CCG repeats, especially when CpG methylated, exclude nucleosomes thus providing a mechanism for packaging failure (Wang and Griffith, 1996). Furthermore, in vitro studies of the 33-bp sequence amplified in FRA16B suggest that the sequence itself, particularly in the presence of distamycin A, also excludes nucleosomes and the distamycin A effect is not limited to single nucleosomes but also to nucleosome arrays (Hsu and Wang, 2002). This could explain why FRA16B can be spontaneously expressed but that distamycin A greatly enhances its expression.

References

Amarose AP, Huttenlocher PR, Sprudzs RM, Laitsch TJ, Pettanati MJ: A heritable fragile 12q24.13 segregating in a family with the fragile X chromosome. Hum Genet 75:4–6 (1987).

Arlt MF, Casper AM, Glover TW: Common Fragile Sites. Cytogenet Genome Res 100:92–100 (2003).

Baker E, Sutherland GR: A new folate sensitive fragile site at 1p21.3. J med Genet 28:356–357 (1991).

Callen DF, Hyland VJ, Baker EG, Fratini A, Simmers RN, Mulley JC, Sutherland GR: Fine mapping of gene probes and anonymous DNA fragments to the long arm of chromosome 16. Genomics 2:144–153 (1988).

Callen DF, Hyland VJ, Baker EG, Fratini A, Gedeon AK, Mulley JC, Fernandez KE, Breuning MH, Sutherland GR: Mapping the short arm of human chromosome 16. Genomics 4:348–354 (1989).

Chodirker BN, Chudley AE, Ray M, Wickstrom DE, Riordan DL: Fragile 19p13 in a family with mental illness. Clin Genet 31:1–6 (1987).

Court Brown WM, Buckton KE, Jacobs PA, Tough IM, Kuenssberg EV, Knox JD: Chromosome studies on adults. Eugenics Laboratory Memoirs, Vol XLII. (Cambridge University Press, New York 1966).

Day EJ, Marshall R, MacDonald PA, Davidson WM: Deleted chromosome 18 with paternal mosaicism. Lancet 2:1307 (1967).

Dekaban A: Persisting clone of cells with an abnormal chromosome in a woman previously irradiated. J Nuclear Med 6:740–746 (1965).

Dutrillaux B, Muleris M, Prieur M: Exact localization of several fragile sites remains uncertain. The example of fra(10) sensitive to folate. Annls Génét 28:161–163 (1985).

Eichler EE, Macpherson JN, Murray A, Jacobs PA, Chakravarti A, Nelson DL: Haplotype and interspersion analysis of the FMR1 CGG repeat identifies two different mutational pathways for the origin of the fragile X syndrome. Hum molec Genet 5:319–330 (1996).

Felbor U, Feichtinger W, Schmid M: The rare human fragile site FRA16B. Cytogenet Genome Res 100: 85–88 (2003).

Ford CE, Madan K: Chromosome Identification, pp 98–103 (Academic Press, New York 1973).

Gardner RJM, Sutherland GR: Chromosome abnormalities and genetic counseling. Oxford Monographs on Medical Genetics No. 29 (Oxford University Press, New York 1996).

Gecz J: The FMR2 gene, FRAXE and non-specific X-linked mental retardation: clinical and molecular aspects. Ann hum Genet 64:95–106 (2000a).

Gecz J: FMR3 is a novel gene associated with FRAXE CpG island and transcriptionally silent in FRAXE full mutations. J med Genet 37: 782–784 (2000b).

Gecz J, Gedeon AK, Sutherland GR, Mulley JC: Identification of the gene FMR2 associated with FRAXE mental retardation. Nature Genet 13: 105–108 (1996).

Giraud F, Ayme S, Mattei JF, Mattei MG: Constitutional chromosomal breakage. Hum Genet 34: 125–136 (1976).

Gu Y, Shen Y, Gibbs RA, Nelson DL: Identification of FMR2, a novel gene associated with the FRAXE CCG repeat and CpG island. Nature Genet 13: 109–113 (1996).

Guichaoua M, Mattei MG, Mattei JF, Giraud F: Genetic aspects of autosomal fragile sites. Apropos of 40 cases. J Genet Hum 30:183–197 (1982).

Handt O, Baker E, Dayan S, Gartler SM, Woollatt E, Richards RI, Hansen RS: Analysis of replication timing at the *FRA10B* and *FRA16B* fragile site loci. Chrom Res 8:677–688 (2000).

Hewett DR, Handt O, Hobson L, Mangelsdorf M, Eyre H, Baker E, Sutherland GR, Schuffenhauer S, Mao J, Richards RI: FRA10B structure reveals common elements in repeat expansion and chromosomal fragile site genesis. Mol Cell 1:773–781 (1998).

Hillman MA, Gecz J: Fragile XE-associated familial mental retardation protein 2 (FMR2) acts as a potent transcription activator. J hum Genet 46: 251–259 (2001).

Hirst MC, Barnicoat A, Flynn G, Wang Q, Daker M, Buckle VJ, Davies KE, Bobrow M: The identification of a third fragile site, FRAXF, in Xq27→q28 distal to both FRAXA and FRAXE. Hum molec Genet 2: 197–200 (1993).

Hocking T, Feichtinger W, Schmid M, Haan EA, Baker E, Sutherland GR: Homozygotes for FRA16B are normal. Chrom Res 7:553–556 (1999).

Hori T, Takahashi E, Ishihara T, Minamihisamatsu M, Kaneko Y, Murata M: Distamycin A-inducible fragile sites and cancer proneness. Cancer Gene Cytogenet 34:177–187 (1988).

Howell RT, McDermott A, Evans JL: A new apparently folate sensitive fragile site, 5q35. J med Genet 27:527–528 (1990).

Hsu YY, Wang Y-H: Human fragile site FRA16B DNA excludes nucleosomes in the presence of distamycin. J biol Chem 277:17315–17319 (2002).

Jones C, Penny L, Mattina T, Yu S, Baker E, Voullaire L, Langdon WY, Sutherland GR, Richards RI, Tunnacliffe A: Association of a chromosome deletion syndrome with a fragile site within the proto-oncogene CBL2. Nature 376:145–149 (1995).

Kahkonen M, Tengstrom C, Alitalo T, Matilainen R, Kaski M, Airaksinen E: Population cytogenetics of folate-sensitive fragile sites. II. Autosomal rare fragile sites. Hum Genet 82:3–8 (1989).

Karadeniz NN, Tunca Y, Imirzalioglu N: New heritable fragile site at 15q13 in both members of a non-consanguineous couple. Am J med Genet 118A: 290–292 (2003).

Knight SJL, Flannery AV, Hirst MC, Campbell L, Christodoulou Z, Phelps SR, Pointon J, Middleton-Price HR, Barnicoat A, Pembrey ME, Holland J, Oostra BA, Bobrow M, Davies KE: Trinucleotide repeat amplification and hypermethylation of a CpG island in *FRAXE* mental retardation. Cell 74:127–134 (1993).

Laird CD: Fragile X mutation proposed to block complete reactivation in females of an inactive X chromosome. Am J med Genet 30:693–696 (1988).

Leehey MA, Munhoz RP, Lang AE, Brunberg JA, Grigsby,J, Greco C, Jacquemont S, Tassone F, Lozano AM, Hagerman PJ, Hagerman RJ: The fragile X permutation presenting as essential tremor. Arch Neurol 60:117–121 (2003).

Lejeune J, Dutrillaux B, Lafourcade J, Berger R, Abonyi D, Rethore MO: Selective endoreduplication of the long arm of the 2 chromosome in a woman and her daughter. CR Acad Sci Hebd Seances Acad Sci D 266:24–26 (1968).

Lubs HA: A marker X chromosome. Am J hum Genet 21:231–244 (1969).

Magenis RE, Hecht F, Lovrien EW: Heritable fragile site on chromosome 16: probable localization of haptoglobin locus in man. Science 170:85–87 (1970).

Mattei MG, Philip N, Pellissier MC, Piquet C, Mattei JF, Giraud F: A new rare fragile site on human chromosome 2. Cytogenet Cell Genet 46:658 (1987).

Michaelis RC, Velagaleti GV, Jones C, Pivnick EK, Phelan MC, Boyd E, Tarleton J, Wilroy RS, Tunnacliffe A, Tharapel AT: Most Jacobsen syndrome deletion breakpoints occur distal to FRA11B. Am J med Genet 76:222–228 (1998).

Nancarrow J: Studies of fragile sites on human chromosome 16. PhD Thesis, University of Adelaide (1998).

Nancarrow JK, Kremer E, Holman K, Eyre H, Doggett NA, Le Paslier D, Callen DF, Sutherland GR, Richards RI: Implications of FRA16A structure for the mechanism of chromosomal fragile site genesis. Science 264:1938–1941 (1994).

Nancarrow JK, Holman K, Mangelsdorf M, Hori T, Denton M, Sutherland GR, Richards RI: Molecular basis of $p(CCG)_n$ repeat instability at the *FRA16A* fragile site locus. Hum molec Genet 4:367–372 (1995).

Nolin SL, Brown T, Glicksman A, Houck E, Gargano AD, Sullivan A, Biancalana V, Brondum-Nielsen K, Hjalgrim H, Holinski-Feder E, Kooy F, Longshore J, Macpherson J, Mandel JL, Matthijs G, Rousseau F, Steinbach P, Vaisanen ML, von Koskull H, Sherman S: Expansion of the fragile X CGG repeat in females with premutation or intermediate alleles. Am J hum Genet 72:454–464 (2003).

Oberle I, Rousseau F, Heitz D, Kretz C, Devys D, Hanauer A, Boue J, Bertheas MF, Mandel JL: Instability of a 550-base pair DNA segment and abnormal methylation in fragile X syndrome. Science 252:1097–1102 (1991).

Oberle I, Boue J, Croquette MF, Voelckel MA, Mattei MG, Mandel JL: Three families with high expression of a fragile site at Xq27.3, lack of anomalies at the FMR-1 CpG island, and no clear phenotypic association. Am J med Genet 43:224–231 (1992).

Ochi H, Watanabe S, Yamamoto H: New heritable fragile site on chromosome 8 induced by distamycin A. Jpn J Cancer Res 79:145–177 (1988).

Parrish JE, Oostra BA, Verkerk AJMH, Richards CS, Reynolds J, Spikes AS, Shaffer LG, Nelson DL: Isolation of a GCC repeat showing expansion in FRAXF, a fragile site distal to FRAXA and FRAXE. Nature Genet 8:229–235 (1994).

Petit P, Fryns JP, van den Berghe H, Hecht F: Population cytogenetics of autosomal fragile sites. Clin Genet 29:96–100 (1986).

Quack B, Nantois Y, Mottet J, Noel B: Constitutional stereotyped gap in human chromosomes. J Genet Hum 26:55–67 (1978).

Reyniers E, Vits L, DeBoulle K, Van Roy B, Van Velzen D, de Graaff E, Verkerk AJ, Jorens HZ, Darby JK, Oostra B et al: The full mutation in the FMR-1 gene of male fragile X patients is absent in their sperm. Nature Genet 4:107–108 (1993).

Scheres JMJC, Hustinx TWJ: Heritable fragile sites and lymphocyte culture medium containing BrdU. Am J hum Genet 32:628–629 (1980).

Schmid M, Klett C, Niederhofer A: Demonstration of a heritable fragile site in human chromosome 16 with distamycin A. Cytogenet Cell Genet 28:87–94 (1980).

Schmid M, Feichtinger W, Jessberger A, Kohler J, Lange R: The fragile site (16)(q22). I. Induction by AT-specific DNA-ligands and population frequency. Hum Genet 74:67–73 (1986).

Schuffenhauer S, Lederer G, Murken J: A heritable folate-sensitive fragile site on chromosome 2p11.2 (FRA2L). Chrom Res 4:252–254 (1996).

Shabtai F, Halbrecht I: Fragile sites on human chromosomes. Clin Genet 19:536 (1981).

Shaw MA, Chiurazzi P, Romain DR, Neri G, Gecz J: A novel gene, FAM11A, associated with the FRAXF CpG island is transcriptionally silent in FRAXF full mutation. Eur J hum Genet 10:767–772 (2002).

Sherman SL, Morton NE, Jacobs PA, Turner G: The marker (X) syndrome: a cytogenetic and genetic analysis. Ann hum Genet 48:21–37 (1984).

Sherman SL, Sutherland GR: Segregation analysis of rare autosomal fragile sites. Hum Genet 72:123–128 (1986).

Simmers RN, Sutherland GR, West A, Richards RI: Fragile sites at 16q22 are not at the breakpoint of the chromosomal rearrangement in AMMoL. Science 236:92–94 (1987).

Steyaert J, Legius E, Borghgraef M, Fryns J-P: A distinct neurocognitive phenotype in female fragile-X premutation carriers assessed with visual attention tasks. Am J med Genet 116A:44–51 (2003).

Sutherland GR: Fragile sites on human chromosomes: Demonstration of their dependence on the type of tissue culture medium. Science 197:265–266 (1977).
Sutherland GR: Heritable fragile sites on human chromosomes. I. Factors affecting expression in lymphocyte culture. Am J hum Genet 31:125–135 (1979).
Sutherland GR: The fragile X chromosome. Intl Rev Cytol 81:107–143 (1983).
Sutherland GR: The role of nucleotides in human fragile site expression. Mutat Res 200:207–213 (1988).
Sutherland GR: Chromosomal fragile sites. Genet Anal Techn Appl 8:161–167 (1991).
Sutherland GR, Baker E: Characterisation of a new rare fragile site easily confused with the fragile X. Hum molec Genet 1:111–113 (1992).
Sutherland GR, Baker E: Unusual behaviour of a human autosome having two rare folate sensitive fragile sites. Annls Génét 36:159–162 (1993).
Sutherland GR, Baker E: The clinical significance of fragile sites on human chromosomes. Clin Genet 58:157–161 (2000).
Sutherland GR, Baker E: Forgotten fragile sites and related phenomena. Cytogenet Genome Res 100: 89–91 (2003).
Sutherland GR, Hecht F: Fragile Sites on Human Chromosomes. (Oxford University Press, New York 1985).
Sutherland GR, Hinton L: Heritable fragile sites on human chromosomes. VI, Characterization of the fragile site at 12q13. Hum Genet 57:217–219 (1981).
Sutherland GR, Baker E, Seshadri R: Heritable fragile sites on human chromosomes. V. A new class of fragile site requiring BrdU for expression. Am J hum Genet 32:542–548 (1980).
Sutherland GR, Jacky PB, Baker E, Manuel A: Heritable fragile sites on human chromosomes. X. New folate sensitive fragile sites: 6p23, 9p21, 9q32, 11q23. Am J hum Genet 35:432–437 (1983).
Sutherland GR, Jacky PB, Baker E: Heritable fragile sites on human chromosomes. XI. Factors affecting expression of fragile sites at 10q25, 16q22 and 17p12. Am J hum Gene 36:110–122 (1984).
Sutherland GR, Parslow MI, Baker E: New classes of fragile sites induced by 5-azacytidine and BrdU. Hum Genet 69:233–237 (1985).
Takahashi E, Kaneko Y, Ishihara T, Minamihisamatsu M, Murata M, Hori T: A new rare distamycin A-inducible fragile site, fra(11)(p15.1), found in two acute nonlymphocytic leukemia (ANLL) patients with t(7;11)(p15 → p13;p15). Hum Genet 80:124–126 (1988).
Tassone F, Hagerman RJ, Taylor AK, Gane LW, Godfrey TE, Hagerman PJ: Elevated levels of *FMR1* mRNA in carrier males: A new mechanism of involvement in the Fragile-X syndrome. Am J hum Genet 66:6–15 (2000).
Tommerup N, Nielsen J, Mikkelsen M: A folate sensitive heritable fragile site at 19p13. Clin Genet 27:510–514 (1985).
Verkerk AJMH, Pieretti M, Sutcliffe JS, Fu Y-H, Kuhl DPA, Pizzuti A, Reiner O, Richards S, Victoria MF, Zhang F, Eussen BE, van Ommen G-JB, Blonden LAJ, Riggins GJ, Chastain JL, Kunst CB, Galjaard H, Caskey CT, Nelson DL, Oostra BA, Warren ST: Identification of a gene *(FMR-1)* containing a CGG repeat coincident with a breakpoint cluster region exhibiting length variation in fragile X syndrome. Cell 65:905–914 (1991).
Voiculescu I, Hausmann C, Wolff G, Back E: A BrdU-requiring fragile site on chromosome 12. Hum Genet 78:183–185 (1988).
Voullaire LE, Webb GC, Leversha MA: Chromosome deletion at 11q23 in an abnormal child from a family with inherited fragility at 11q23. Hum Genet 76:202–204 (1987).
Wang YH, Griffith J: Methylation of expanded CCG triplet repeat DNA from fragile X syndrome patients enhances nucleosome exclusion. J biol Chem 271:22937–22940 (1996).
Webb T, Thake A: Fragile 22q13 segregating in a family. Clin Genet 26:125–128 (1984).
Yamauchi M, Tsuji S, Mita K, Saito T, Morimyo M: A novel minisatellite repeat expansion identified at *FRA16B* in a Japanese carrier. Genes Genet Syst 75:149–154 (2000).
Yu S, Pritchard M, Kremer E, Lynch M, Nancarrow J, Baker E, Holman K, Mulley JC, Warren ST, Schlessinger D, Sutherland GR, Richards RI: Fragile X genotype characterised by an unstable region of DNA. Science 252:1179–1181 (1991).
Yu S, Mangelsdorf M, Hewett D, Hobson L, Baker E, Eyre H, Lapsys N, Le Paslier D, Doggett NA, Sutherland GR, Richards RI: Human chromosomal fragile site FRA16B is an amplified AT-rich minisatellite repeat. Cell 88:367–374 (1997).
Zhong N, Ju W, Curley D, Wang D, Pietrofesa J, Wu G, Shen Y, Pang C, Poon P, Liu X, Gou S, Kajanoja E, Ryynanen M, Dobkin C, Brown WT: A survey of FRAXE allele sizes in three populations. Am J med Genet 64:415–419 (1996).

Cytogenet Genome Res 100:85–88 (2003)
DOI: 10.1159/000072841

The rare human fragile site 16B

U. Felbor, W. Feichtinger and M. Schmid

Institut für Humangenetik, Biozentrum, Würzburg (Germany)

Abstract. The rare human fragile site 16B (FRA16B) has been found to occur spontaneously. Its expression in lymphocyte cultures can also be induced or greatly enhanced by addition of chemicals which are known to bind to AT-rich DNA regions. Following optimal treatment with 150 µg/ml berenil 24 h prior to fixation, the heterozygote frequency of FRA16B is found to be about 5% in populations of European descent. Thus, FRA16B represents the most common of the rare fragile sites. Consistent with cytogenetic observations, the molecular characterization of FRA16B revealed that it is an amplified 33-base pair AT-rich minisatellite repeat. These interindividually variable, extremely large repeat expansions of 15–70 kb in size do not seem to interfere with the expression of genes essential for human development since heterozygotes and homozygotes for FRA16B are normal.

FRA16B was originally described as a fully penetrant chromosomal marker which helped to localize the gene for the α-polypeptide chain of haptoglobin on chromosome 16 (Magenis et al., 1970). Thereafter, this fragile site was observed in a variable proportion of lymphocyte metaphases from further cases (referenced in Schmid et al., 1980) and classified as non-folate-sensitive (Sutherland, 1979a). Intriguingly, AT base pair-specific, nonintercalative DNA-ligands such as distamycin A and berenil were found to induce FRA16B in individuals that do not express FRA16B spontaneously and to increase the expression of FRA16B in spontaneous cases (Schmid et al., 1980, 1986; Sutherland et al., 1984). Besides, the thymidine analogue bromodeoxyuridine (BrdU) was shown to induce FRA16B (Croci, 1983; Sanfilippo, 1983; Sutherland et al., 1984). Taken together, these observations led to the assumption that FRA16B might be a highly repetitive AT-rich DNA sequence whose chromosomal condensation is inhibited by AT-binding compounds during mitosis. Indeed, cloning of FRA16B one and a half decades later revealed that it consists of up to 2000 copies of a 33-base pair AT-rich minisatellite repeat (Yu et al., 1997). Interfamilial differences regarding the cytogenetic expression of FRA16B were consequently attributed to different repeat copy numbers. In addition, in vitro experiments with distamycin A which binds externally to AT base pairs in the minor groove of double-stranded DNA demonstrated altered rotational positioning of nucleosome assembled DNA (Brown and Fox, 1996) and nucleosome destabilization in FRA16B DNA (Hsu and Wang, 2002).

Review of the morphology

In the original description of spontaneous FRA16B, an isochromatid break at the junction of the middle and distal thirds in 16q was reported. 2–12 copies of the distal portion of the long arm were found in some lymphocyte metaphase spreads (Magenis et al., 1970). Such chromosomal breakage leading to non-staining gaps of both chromatids and, in a subset of metaphases, to deleted chromosomes 16, acentric fragments and multiradial figures was thereafter observed in conventional cultures of spontaneous FRA16B by several other investigators (referenced in Schmid et al., 1986).

Experimental induction of FRA16B essentially produces an analogous picture. Distinct non-staining gaps of at least one to five times the size of the chromatids at one-third the distance from the distal end of the long arm were seen in all FRA16B cases (Fig. 1b). Misalignments of the terminal acentric fragment with the rest of chromosome 16 and multiradial chromosome figures (Fig. 1a, c) were frequently noted (Schmid et al., 1980, 1986). However, certain probands only showed a tiny constriction without misalignment of the chromatids upon treatment which suggested interfamilial variability. This category may represent a subgroup of weakly inducible fragile sites due to low copy numbers of the 33-base pair AT-rich minisatellite repeat which are below the threshold of cytogenetic detection. However, it clearly does not meet the cytogenetic criteria mentioned above.

U.F. is supported by the Emmy Noether program of the Deutsche Forschungsgemeinschaft (FE432/6-1).

Received 11 November 2002; manuscript accepted 4 December 2002.

Request reprints from M. Schmid, Institut für Humangenetik
Biozentrum, Am Hubland, D–97074 Würzburg (Germany)
telephone: (+) 49-931-888 4077; fax: (+) 49-931-888 4058
e-mail: m.schmid@biozentrum.uni-wuerzburg.de

KARGER Fax + 41 61 306 12 34
E-mail karger@karger.ch
www.karger.com

0301–0171/03/1004–0085$19.50/0

Accessible online at:
www.karger.com/cgr

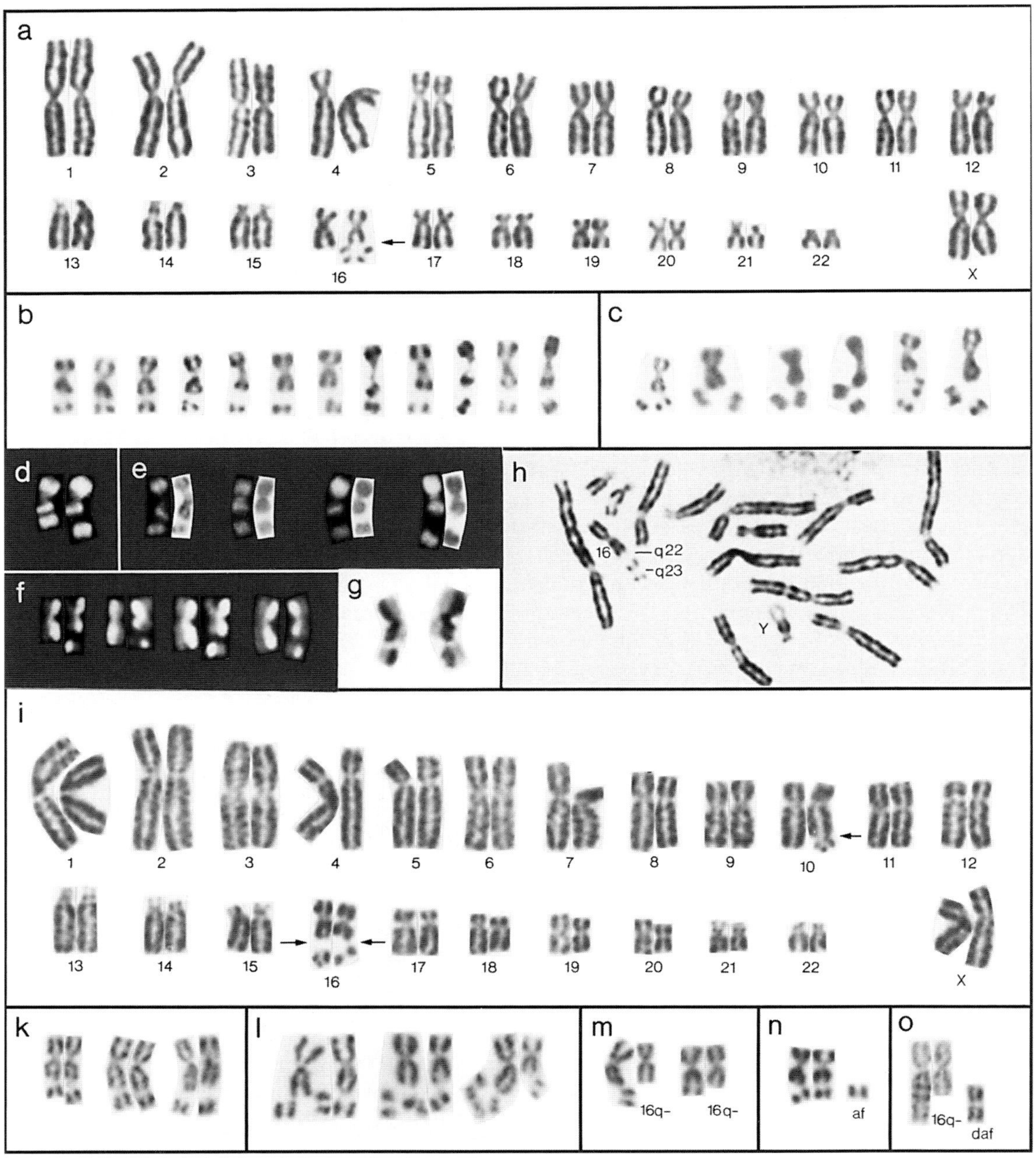

Fig. 1. Morphology of FRA16B in lymphocyte metaphase spreads. (**a**) Giemsa-stained female karyotype showing a distamycin A-induced multiradial figure at FRA16B in heterozygous form (arrow). (**b**) Distinct non-staining gaps or breaks and (**c**) triradial figures of chemically induced G-banded chromosomes 16. (**d**, **e**) Berenil-induced FRA16B stained with acridine orange and Giemsa. (**f**, **g**) SCEs within inducible FRA16B after treatment with berenil and BrdU. (**h**) Partial metaphase from a culture treated with both berenil and aphidicolin to demonstrate FRA16B at 16q22 as well as the common fragile site 16D at 16q23 in the same chromosome. The Y chromosome is undercondensed due to treatment with berenil. (**i**) Giemsa-stained karyotype of a healthy female homozygous for FRA16B (arrows) and heterozygous for FRA10B (arrow). Her lymphocyte cultures were treated with berenil. (**k**) Non-staining gaps, (**l**) multiradial figures, (**m**) deleted chromosome 16, (**n**) an acentric fragment (af) and (**o**) a double acentric fragment (daf) found in the same proband.

Banding characteristics

Spontaneous FRA16B was first localized in band 16q21 by Giraud et al. (1976) using R-banding. Later, (Magenis and Chamberlin, 1979) mapped FRA16B to 16q22.00 using G-bands. Sutherland (1979b) assigned FRA16B more precisely to 16q22.1. Furthermore, induced FRA16B could be mapped to 16q22 although Q-banded metaphases from distamycin A-treated cultures are of poor quality (Schmid et al., 1980). Since the quality of metaphases in berenil-treated cultures is far better, this compound was later chosen for mapping studies of induced FRA16B (Fig. 1d, e, h) (Schmid et al., 1986). C-banded preparations were important since they proved that the fragile site was always induced on the same homologue of chromosome 16 (Schmid et al., 1980).

Induction with AT base pair-specific DNA-ligands

The experimental induction of FRA16B using the AT-binding oligopeptide antibiotic distamycin A was first performed by Schmid et al. (1980). Since distamycin A is stable for approximately 24 hours in aqueous solutions, it was initially added to lymphocyte cultures 24 hours prior to fixation at a final concentration of 100 μg/ml. Later, it was shown that the percentage of induced FRA16B doubled if distamycin A was added 48 hours prior to fixation (Schmid et al., 1986). Apart from distamycin A, other AT-specific substances such as netropsin, Hoechst 33248, D287/170, methyl green, DAPI, and berenil are also able to induce FRA16B or to increase its expression in spontaneous cases (Sutherland et al., 1984; Schmid et al., 1986). Thorough dose response experiments identified optimal treatment conditions under which the highest percentage of chemically induced FRA16B is achieved (Table 1). The aromatic diamidine berenil (4,4′-diazoamino-dibenzamidine diaceturate) proved to have several advantages: 1) it induces the highest levels of FRA16B, 2) it has no negative effect on the quality of metaphase preparations in the optimum concentration range, 3) it is relatively inexpensive. Consequently, berenil is recommended for detection of FRA16B in population studies (Schmid et al., 1986).

Table 1. Frequency of metaphases with chemically induced FRA16B under optimal lymphocyte culture conditions (Schmid et al., 1986)

AT-binding compound	Optimal treatment time (in hours)	Optimal concentration (in μg/ml)	Frequency of FRA16B (%)
Distamycin A	48	100	50-60
Netropsin	36	200	55
DAPI	24	30	90
Berenil	24	150	90

Population frequency

The determination of the population frequency critically depends on optimal cell culture conditions. Prior to the discovery that FRA16B can easily be detected using 150 μg/ml berenil 24 hours before fixation, FRA16B was thought to be rather rare, ranging from 1:2,150 (Schwanitz, 1976) over 1:4,100 (Hecht and Kaiser-McCaw, 1979) to 1:30,000 (Sørensen et al., 1979). Inappropriate treatment of lymphocyte cultures with low dose BrdU yielded heterozygote frequencies ranging from about 1:100 to 1:40 (Sanfilippo, 1983). Similarly, distamycin A added 24 h instead of 48 h prior to harvest induced FRA16B in about 1 in 90 individuals (Sutherland, 1985). The addition of 150 μg/ml berenil 24 h prior to cell fixation demonstrated that as many as one in 20 Germans and Australians are heterozygous for FRA16B (Schmid et al., 1986; Müller et al., 1992; Hocking et al., 1999).

Evolutionary aspects

Thus far, an undercondensed region in 33% of distamycin A-treated lymphocyte metaphases was found in the gorilla at 17q21, the chromosome region corresponding to 16q22 in humans (Schmid et al., 1981). However, lymphocyte cultures from only one animal were analyzed and the undercondensed band in one of the homologous chromosomes 17 did not represent a true gap. Thus, the frequency and morphology of this distamycin A-sensitive region needs to be analyzed in a larger number of animals.

Segregation analyses

The original Portland pedigree described by Magenis et al. (1970) remains the largest pedigree analyzed: Lymphocyte cultures of 127 living family members revealed 31 affected individuals. The mode of inheritance was classified as autosomal dominant with complete penetrance. In order to analyze whether the sex of the transmitting parent influences FRA16B expression, Sherman and Sutherland (1986) evaluated five previously published pedigrees, including the large Portland pedigree, and five families from Adelaide. Their results suggested that the expression of distamycin A-inducible FRA16B did not depend on the sex of the carrier parent but segregated as a fully penetrant, simple codominant trait.

In an attempt to add more family data, Müller et al. (1992) presented a second calculation including ten newly ascertained families and two published by Schmid et al. (1980). While autosomal dominant inheritance was confirmed and no preferential transmission to male or female offspring was found, significantly more mothers than fathers seemed to be FRA16B-positive. However, taken into consideration that the sample size of both analyses was rather small, future studies will have to clarify whether sex-specific factors play a role in transmission of FRA16B.

Molecular characterization

Positional cloning of FRA16B revealed that it consists of up to 2000 copies of a complex polymorphic minisatellite repeat (Yu et al., 1997). Normal chromosomes contain 7–12 copies of a 33-base pair consensus sequence, p(ATATATTATATATTA-TATCTAATAATATAT$^{C}/_{A}$TA)$_n$, which itself is subject to deletions of up to seven base pairs at the C-terminus giving rise to a complex polymorphism. Instability at FRA16B results in extraordinarily large expansions of 15–70 kb in size when compared to trinucleotide repeat expansions, the molecular basis of folate-sensitive fragile sites. Thus, it was demonstrated that repeat motifs of varying length and composition are subject to dynamic mutations.

The molecular characterization of the sequence composition of FRA16B also allowed in vitro studies on the effect of inducing agents. It could be demonstrated that increasing numbers of 33-base pair repeats intensify the effect of distamycin A on destabilization of nucleosome formation (Hsu and Wang, 2002). This finding alludes to a common molecular mechanism for the formation of fragile sites since nucleosome exclusion has been observed in DNA fragments containing CCG repeats found in folate-sensitive fragile sites (Godde et al., 1996; Wang et al., 1996).

Clinical impact of FRA16B

There has been much speculation on FRA16B predisposing to chromosome rearrangement since FRA16B was found in parents of patients with de novo structural chromosome abnormalities (Sørensen et al., 1979). Furthermore, BrdU treatment of lymphocyte cultures revealed an exceptionally high frequency of sister chromatid exchanges (Fig. 1f, g) which supported the view that FRA16B is a chromosome site with properties of high somatic recombination (Schmid et al., 1987). However, it has been demonstrated that a specific leukemic breakpoint is, in fact, located proximal to FRA16B (Simmers et al., 1987). Most importantly, individuals heterozygous and homozygous for FRA16B are generally normal (Fig. 1i–o) (Hocking et al., 1999). In this context it is noteworthy to mention that two of the three healthy homozygotes published to date were also heterozygous for FRA10B (Fig. 1i). In addition, another female homozygote for FRA16B also proved to be heterozygous for FRA17A (Schmid et al., unpublished data) supporting the view that these non-folate-sensitive fragile sites are apparently without clinical significance.

Interstitial deletions involving 16q22 have been associated with developmental delay and multiple relatively mild dysmorphisms (Fujiwara et al., 1992; Callen et al., 1993). We have observed deleted chromosomes 16 due to spontaneous FRA16B in up to 7% of lymphocyte metaphases (Schmid et al., unpublished data). Other tissues have thus far not been assessed. However, earlier studies demonstrated that spontaneous FRA16B is expressed in lymphocytes and bone marrow, but explicitly not in fibroblasts (Sutherland, 1979b). In addition, the frequency of induced FRA16B was twenty-fold less in fibroblast cultures when compared to lymphocytes (Sutherland et al., 1984). Therefore, it can be speculated that mosaic 16q– due to FRA16B is likely to occur only in a very small percentage of cells and may be subject to negative selection. In addition, it is even conceivable that FRA16B-derived 16q– is a cell culture artefact after stimulation with phytohemagglutinine and does not reflect the in vivo situation. Furthermore, the potential clinical impact of FRA16B-derived mosaic 16q– would be difficult to determine due to the fact that many FRA16B patients ascertained in cytogenetic clinics have an underlying chromosomal abnormality such as Turner syndrome. However, given the fact that homozygous FRA16B patients with 16q– in a subset of lymphocyte metaphases are healthy (Hocking et al., 1999), it seems likely that FRA16B-derived mosaic 16q– does not lead to a major overt phenotype.

Recently, a dominant form of cleft palate with microstomia and micrognathia was reported to cosegregate with FRA16B in two families (Bettex et al., 1998; McKenzie et al., 2002), and the authors of the present article are aware of a further family (G. Gillessen-Kaesbach, personal communication). Thus, it seems likely that FRA16B largely remains the "minor chromosome variation" that serves as an "additional tool for use in gene localization" as already stated by Magenis in 1970.

References

Bettex M, Graf B, Winkler B, Gerber-Huber S: Oropalatal dysplasia Bettex-Graf – a new syndrome. Eur J Pediatr Surg 8:4–8 (1998).

Brown PM, Fox KR: Minor groove binding ligands alter the rotational positioning of DNA fragments on nucleosome core particles. J molec Biol 262: 671–685 (1996).

Callen DF, Eyre H, Lane S, Shen Y, Hansmann I, Spinner N, Zackai E, McDonald-McGinn D, Schuffenhauer S, Wauters J, van Thienen M-N, van Roy B, Sutherland GR, Haan EA: High resolution mapping of interstitial long arm deletions of chromosome 16: relationship to phenotype. J med Genet 30:828–832 (1993).

Croci G: BrdU-sensitive fragile site on long arm of chromosome 16. Am J hum Genet 35:530–533 (1983).

Fujiwara M, Yoshimoto T, Morita Y, Kamada M: Intersitial deletion of chromosome 16q: 16q22 is critical for 16q– syndrome. Am J med Genet 43:561–564 (1992).

Giraud F, Ayme S, Mattei JF, Mattei MG: Constitutional chromosomal breakage. Hum Genet 34: 125–136 (1976).

Godde JS, Kass SU, Hirst MC, Wolffe AP: Nucleosome assembly on methylated CGG triplet repeats in the fragile X mental retardation gene 1 promoter. J biol Chem 271:24325–24328 (1996).

Hecht F, Kaiser-McCaw B: The importance of being a fragile site. Am J hum Genet 31:223–225 (1979).

Hocking T, Feichtinger W, Schmid M, Haan EA, Baker E, Sutherland GR: Homozygotes for FRA16B are normal. Chrom Res 7:553–556 (1999).

Hsu YY, Wang Y-H: Human fragile site FRA16B DNA excludes nucleosomes in the presence of distamycin. J biol Chem 277:17315–17319 (2002).

Magenis RE, Hecht F, Lovrien EW: Heritable fragile site on chromosome 16: probable localization of haptoglobin locus in man. Science 170:85–87 (1970).

Magenis RE, Chamberlin J: Location of fragile site: 16q22.00. Cytogenet Cell Genet 25:180 (1979).

McKenzie F, Turner A, Withers S, Dalzell P, McGlynn M, Kirk EPE: Dominant inheritance of cleft palate, microstomia and micrognathia – possible linkage to the fragile site at 16q22 (FRA16B). Clin Dysmorphol 11:237–241 (2002).

Müller B, Feichtinger W, Bonaïti-Pellié C, Schmid M: Fragile site (16)(q22). III. Segregation analysis. Hum Genet 89:612–614 (1992).

Sanfilippo S, Neri G, Tedeschi B, Carlo-Stella N, Triolo O, Serra A: Chromosomal fragile sites: preliminary data of a population survey. Clin Genet 24:295 (1983).

Schmid M, Klett C, Niederhofer A: Demonstration of a heritable fragile site in human chromosome 16 with distamycin A. Cytogenet Cell Genet 28:87–94 (1980).

Schmid M, Poppen A, Engel W: The effects of distamycin A on gorilla-, chimpanzee- and orangutan-lymphocyte cultures. Cytogenet Cell Genet 30:211–221 (1981).

Schmid M, Feichtinger W, Jessberger A, Köhler J, Lange R: The fragile site (16)(q22). I. Induction by AT-specific DNA-ligands and population frequency. Hum Genet 74:67–73 (1986).

Schmid M, Feichtinger W, Haaf T: The fragile site (16)(q22). II. Sister chromatid exchanges. Hum Genet 76:365–368 (1987).

Schwanitz: Die Normvarianten menschlicher Chromosomen (Perimed, Erlangen 1976).

Sherman SL, Sutherland GR: Segregation analysis of rare autosomal fragile sites. Hum Genet 72:123–128 (1986).

Simmers RN, Sutherland GR, West A, Richards RI: Fragile sites at 16q22 are not at the breakpoint of the chromosomal rearrangement in AMMoL. Science 236:92–94 (1987).

Sørensen K, Nielsen J, Holm V, Haahr J: Fragile site long arm chromosome 16. Hum Genet 48:131–134 (1979).

Sutherland GR: Heritable fragile sites on human chromosomes I. Factors affecting expression in lymphocyte culture. Am J hum Genet 31:125–135 (1979a).

Sutherland GR: Heritable fragile sites on human chromosomes II. Distribution, phenotypic effects, and cytogenetics. Am J hum Genet 31:136–148 (1979b).

Sutherland GR: Heritable fragile sites on human chromosomes. XII. Population cytogenetics. Ann hum Genet 49:153–161 (1985).

Sutherland GR, Jacky PB, Baker EG: Heritable fragile sites on human chromosomes. XI. Factors affecting expression of fragile sites at 10q25, 16q22, and 17p12. Am J hum Genet 36:110–122 (1984).

Wang Y-H, Gellibolian R, Shimizu M, Wells RD, Griffith J: Long CCG triplet repeat blocks exclude nucleosomes: a possible mechanism for the nature of fragile sites in chromosomes. J molec Biol 263:511–516 (1996).

Yu S, Mangelsdorf M, Hewett D, Hobson L, Baker E, Eyre HJ, Lapsys N, Le Paslier D, Doggett NA, Sutherland GR, Richards RI: Human chromosomal fragile site FRA16B is an amplified AT-rich minisatellite repeat. Cell 88:367–374 (1997).

Cytogenet Genome Res 100:89–91 (2003)
DOI: 10.1159/000072842

Forgotten fragile sites and related phenomena

G.R. Sutherland[a, b] and E. Baker[a, b]

[a] Department of Cytogenetics and Molecular Genetics, Women's and Children's Hospital, Adelaide (Australia);
[b] Department of Paediatrics, University of Adelaide, Adelaide (Australia); and
[c] Department of Research and Development, US LABS, Irvine, CA (USA)

Abstract. Fragile sites have been recognised since 1965. Extensive molecular characterisation of them has occurred in the last decade. Yet there are a number of interesting phenomena relating to fragile sites and similar lesions on chromosomes that have not been subject to recent published research. The aim of this short review is to stimulate interest in some of these aspects of fragile sites in the hope that a more complete picture of their nature and properties will emerge.

Since the cloning of the first fragile site, FRAXA, in 1991, there has been a lot of activity in the characterisation of both rare and common fragile sites. Of the rare folate-sensitive fragile sites, the three on the X chromosome and two on the autosomes (Nancarrow et al., 1984; Jones et al., 1995) have been cloned, as have six of the aphidicolin-type common fragile sites (see Arlt et al., 2003, in this issue). The two most frequent of the non-folate-sensitive rare fragile sites (FRA10B and FRA16B) have been cloned and characterised (Yu et al., 1997; Hewett et al., 1998).

In spite of all this activity, there are a number of fragile sites and related cytogenetic phenomena that have been in the literature for a long time but have received scant attention. Groups of common fragile sites (Table 1) specifically induced by 5-bromodeoxyuridine (BrdU) or 5-azacytidine were described in 1985 (Sutherland et al., 1985). Some comparative studies in other genomes were published shortly afterward (Schmid et al., 1985; Djalali et al., 1990), but there has effectively been nothing in the literature about these fragile sites for 10 years. It should be relatively easy to characterise these fragile sites at the DNA level, given the availability of the human genome sequence and clone resources.

Table 1. Common fragile sites induced by 5-bromodeoxyuridine (BrdU) or 5-azacytidine

BrdU		5-Azacytidine	
Gene symbol	Band	Gene symbol	Band
FRA4B	4q12	FRA1J	1q12
FRA5A	5p13	FRA1H	1q42
FRA5B	5q15	FRA9F	9q12
FRA6D	6q13	FRA19A	19q13
FRA9C	9p21		
FRA10C	10q21		
FRA13B	13q21		

Three rare fragile sites (FRA8E, FRA11I, and FRA16E) inducible by distamycin A but not by BrdU were found in the Japanese population more than a decade ago (Hori et al., 1988; Takahashi et al., 1988). These fragile sites form their own category in a classification of fragile sites. None of these has been characterised at the molecular level, where they could be expected to have novel sequence characteristics. No other population or family studies have been conducted looking at these fragile sites to determine whether they are specific to the Japanese or to study their segregation patterns.

Received 12 February 2003; manuscript accepted 13 February 2003.

Request reprints from Dr. Grant R. Sutherland
Department of Cytogenetics and Molecular Genetics
Women's and Children's Hospital
Adelaide SA 5006 (Australia)
telephone: +61-8-8161-7111; fax: +61-8-8161-7342
e-mail: grant.sutherland@adelaide.edu.au

KARGER
Fax + 41 61 306 12 34
E-mail karger@karger.ch
www.karger.com

0301–0171/03/1004–0089$19.50/0

Accessible online at:
www.karger.com/cgr

Fig. 1. Stickiness and associations of fragile sites. (**a**, **b**) Partial metaphase spreads showing a viral modification site at 11q23.1. Note the lesion "sticking" to other chromosomes (arrows). Reproduced from Sutherland et al. (1988). (**c**) Three partial metaphases showing a chromosome 12 with two rare fragile sites adopting a U-shaped configuration to allow the fragile regions to associate. Reproduced from Sutherland and Baker (1993).

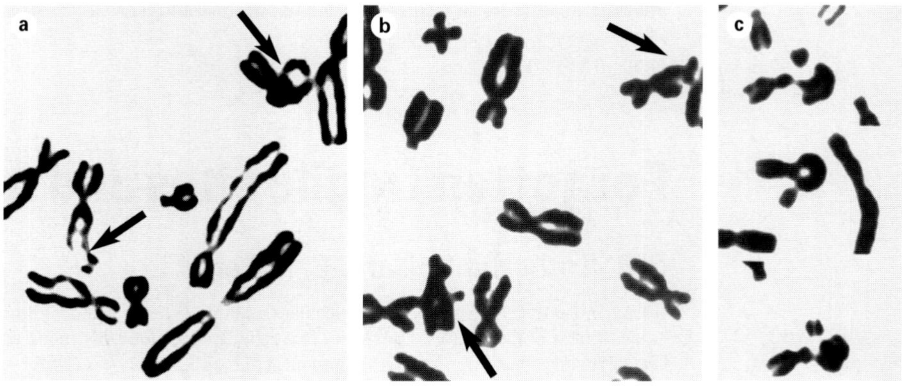

A viral modification site at 11q23.1 having a similar cytogenetic appearance to a fragile site was described in EBV-transformed lymphocytes in 1988 (Sutherland et al., 1988) and re-examined in 1992 (Seki et al., 1992b). This lesion appeared sporadically but was further inducible with BrdU and *n*-butyric acid. Seki et al. (1992b) also showed that sister chromatid exchanges (SCEs) occurred preferentially at this site, as has been shown for other rare and common fragile sites. Again, there has been nothing in the literature on this viral lesion for a decade, apart from a couple of reports of achromatic gaps in Burkitt lymphoma cell lines (e.g., Jox et al., 1997). It is of interest that the viral modification site at 11q23.1 was "sticky" in that the area of the lesion tended to associate with other chromosomes (Fig. 1a, b).

The phenomenon of "stickiness" has also been noted in cells in which more than one fragile site is present, even to the extent that a single chromosome 12 with two rare folate-sensitive fragile sites on the long arm would adopt a U-shaped appearance so that the fragile sites could associate with each other (Fig. 1c; Sutherland and Baker, 1993). There is only speculation about the reasons for such an association. We have also observed, but not studied systematically, that common fragile sites, both on homologous and nonhomologous chromosomes, occasionally associate to form quadriradial figures. We note that a similar appearance was documented in the original description of aphidicolin-type common fragile sites (Glover et al., 1984; Fig. 3b).

There appears to be an excess of individuals with more than one rare fragile site in their genome, a phenomenon called "double ascertainment" (Seki et al., 1992a). The fragile sites in the one individual need not necessarily be of the same type, but can be. For example, the rare FRA12D has only been recorded in the presence of other rare folate-sensitive fragile sites. We have seen rare fragile sites of different types on the same chromosome (FRA10A with FRA10B; Sutherland et al., 1982), and one individual with FRAXA, FRA10B, and FRA16B as an example of "triple ascertainment." The reasons for double ascertainment of rare fragile sites remain obscure. It is possible, but unlikely, that they could be artefactual in that once one fragile site is found, more detailed examination of that individual's chromosomes reveals another fragile site. The possibility does exist, however, that there could be some biological basis for double ascertainment. Could some *trans*-acting factor that causes one fragile site-associated repeat sequence to expand affect more than one such sequence and result in the generation of more than one fragile site? Should this occur, the fragile sites would be expected to be rapidly separated by random segregation within a small number of generations, or more slowly when two fragile sites were on the same chromosome (eg. Sutherland et al., 1982). It is of interest that amongst Japanese SCA6 patients, there is an increase in long SCA8 alleles and *vice versa* (Izumi et al., 2003), but it is not clear whether this is due to a common origin for the expansions or ascertainment bias because SCA8 expansion has a role in SCA6 pathology or vice versa.

References

Arlt MF, Casper AM, Glover TW: Common fragile sites. Cytogenet Genome Res 100:92–100 (2003).

Djalali M, Adolph S, Steinbach P, Winking H, Hameister H: Fragile sites induced by 5-azacytidine and 5-azadeoxycytidine in the murine genome. Hereditas 112:77–81 (1990).

Glover TW, Berger C, Coyle J, Echo B: DNA polymerase alpha inhibition by aphidicolin induces gaps and breaks at common fragile sites in human chromosomes. Hum Genet 67:136–142 (1984).

Hewett DR, Handt O, Hobson L, Mangelsdorf M, Eyre H, Baker E, Sutherland GR, Schuffenhauer S, Mao J, Richards RI: FRA10B structure reveals common elements in repeat expansion and chromosomal fragile site genesis. Mol Cell 1:773–781 (1998).

Hori T, Takahashi E, Ishihara T, Minamihisamatsu M, Kaneko Y, Murata M: Distamycin A-inducible fragile sites and cancer proneness. Cancer Genet Cytogenet 34:177–187 (1988).

Izumi Y, Maruyama H, Oda M, Morino H, Okada T, Ito H, Sasaki I, Tanaka H, Komure O, Udaka F, Nakamura S, Kawakami H: *SCA8* repeat expansion: large CTA/CTG repeat alleles are more common in ataxic patients, including those with SCA6. Am J hum Genet 72:704–709 (2003).

Jones C, Penny L, Mattina T, Yu S, Baker E, Voullaire L, Langdon WY, Sutherland GR, Richards RI, Tunnacliffe A: Association of a chromosome deletion syndrome with a fragile site within the proto-oncogene CBL2. Nature 376:145–149 (1995).

Jox A, Rohen C, Belge G, Bartnitzke S, Pawlita M, Diehl V, Bullerdiek J, Wolf J: Integration of Epstein-Barr virus in Burkitt's lymphoma cells leads to a region of enhanced chromosome instability. Annls Oncol 8(Suppl 2):131–135 (1997).

Nancarrow JK, Kremer E, Holman K, Eyre H, Doggett NA, Le Paslier D, Callen DF, Sutherland GR, Richards RI: Implications of FRA16A structure for the mechanism of chromosomal fragile site genesis. Science 264:1938–1941 (1994).

Schmid M, Ott G, Haaf T, Scheres JM: Evolutionary conservation of fragile sites induced by 5-azacytidine and 5-azadeoxycytidine in man, gorilla, and chimpanzee. Hum Genet 71:342–350 (1985).

Seki N, Sumiya H, Shimazaki J, Toyama Y, Takahashi E, Murata M, Hori T: Chromosome abnormalities and rare fragile sites detected in azoospermia patients. Jpn J hum Genet 37:215–222 (1992a).

Seki N, Tsuji H, Takahashi E, Yamauchi M, Saito T, Hashimoto T, Yamamoto K, Hori T: Induction of a BrdU-enhanceable fragile site-like lesion and sister chromatid exchanges at 11q23.1 in EBV-transformed lymphoblastoid cell lines. Cytogenet Cell Genet 61:95–98 (1992b).

Sutherland GR, Baker E: Unusual behaviour of a human autosome having two rare folate sensitive fragile sites. Annls Génét 36:159–162 (1993).

Sutherland GR, Parslow MI, Baker E: New classes of fragile sites induced by 5-azacytidine and BrdU. Hum Genet 69:233–237 (1985).

Sutherland GR, Baker E, Callen DF: A BrdU enhanceable fragile site or viral modification site at 11q23.1 in lymphoblastoid cultures. Cytogenet Cell Genet 47:201–203 (1988).

Sutherland GR, Baker E, Mulley JC: Genetic length of a human chromosomal segment measured by recombination between two fragile sites. Science 217:373–374 (1982).

Takahashi E, Kaneko Y, Ishihara T, Minamihisamatsu M, Murata M, Hori T: A new rare distamycin A-inducible fragile site, fra(11)(p15.1), found in two acute nonlymphocytic leukemia (ANLL) patients with t(7;11)(p15-p13;p15). Hum Genet 80:124–126 (1988).

Yu S, Mangelsdorf M, Hewett D, Hobson L, Baker E, Eyre H, Lapsys N, Le Paslier D, Doggett NA, Sutherland GR, Richards RI: Human chromosomal fragile site FRA16B is an amplified AT-rich minisatellite repeat. Cell 88:367–374 (1997).

Cytogenet Genome Res 100:92–100 (2003)
DOI: 10.1159/000072843

Common fragile sites

M.F. Arlt, A.M. Casper and T.W. Glover

Department of Human Genetics, University of Michigan, Ann Arbor MI (USA)

Abstract. Aphidicolin-induced common fragile sites are site-specific gaps or breaks seen on metaphase chromosomes after partial inhibition of DNA synthesis. These fragile sites were first recognized during the early studies of the fragile X syndrome and are induced by the same conditions of folate or thymidylate stress used to induce the fragile X site. Common fragile sites are normally stable in cultured human cells. However, following induction with replication inhibitors, they display a number of characteristics of unstable and highly recombinogenic DNA. From the many studies that have cloned and characterized fragile sites, it is now known that these sites extend over large regions, are associated with genes, exhibit late or delayed replication, and contain regions of high flexibility but are otherwise unremarkable in sequence. Studies showing that fragile sites and their associated genes are frequently deleted or rearranged in cancer cells have clearly demonstrated their importance in genome instability in tumorigenesis. Yet until recently, very little was known about the molecular mechanisms involved in their stability. Recent findings showing that the key checkpoint genes ATR and BRCA1 are critical for genome stability at fragile sites have shed new light on these mechanisms and on the biological significance of common fragile sites.

During the early cytogenetic investigations of the fragile X syndrome, we and others noted that certain chromosomal loci, such as 3p14.2, 16q23 and Xp22.3, repeatedly displayed gaps and breaks, just like the fragile X site, but in all individuals. Thus, we coined the term "common fragile sites" to describe them (Glover et al., 1984). Common fragile sites are now defined as site-specific gaps or breaks seen on metaphase chromosomes after partial inhibition of DNA synthesis (Fig. 1). These lesions were seen under the same conditions of folate or thymidylate stress used to induce cytogenetic expression of the fragile X (FRAXA). In further studies, it was found that these fragile sites are also induced by aphidicolin, an inhibitor of DNA polymerase (Ikegami et al., 1978; Glover et al., 1984). Fragile sites are now categorized into two main classes. "Rare" fragile sites, like the fragile X site, are seen in few individuals, are associated with CGG or AT repeats, and segregate in a Mendelian manner (Kremer et al., 1991; Sutherland et al., 1998). "Common" fragile sites are seen in all individuals and, therefore, represent a component of normal chromosome structure. This review focuses on the aphidicolin-induced common fragile sites, which comprise, by far, the largest group. Common fragile sites induced by BrdU or 5-azacytidine are reviewed elsewhere (Sutherland et al., 1998).

The exact number of aphidicolin-induced common fragile sites that exist is a matter of interpretation, with 75 such sites reported in GDB. The greater the stress placed on DNA replication, the more such fragile sites are observed, until replication stops altogether. However, not all fragile sites are expressed at the same frequency, and a small number of sites in the genome are most prone to form lesions under stress conditions (Glover et al., 1984; Yunis and Soreng, 1984). Gaps and breaks at just 20 fragile sites represent over 80% of all lesions observed in lymphocytes following treatment with low doses of aphidicolin (Glover et al., 1984). FRA3B at 3p14.2 stands out as the most "fragile" site in the genome and can be induced to break or

Supported by NIH grant CA43222 (T.W.G.) and an NSF predoctoral fellowship to A.M.C.

Received 13 January 2003; manuscript accepted 17 January 2003.

Request reprints from Thomas W. Glover, Ph.D.
Department of Human Genetics, 4909 Buhl, Box 0618
1241 E. Catherine Street, University of Michigan
Ann Arbor MI 48109-0618 (USA)
telephone: (734) 763-5222; fax: (734) 763-3784
e-mail: glover@umich.edu

KARGER
Fax + 41 61 306 12 34
E-mail karger@karger.ch
www.karger.com

Accessible online at:
www.karger.com/cgr

form gaps in the majority of treated cells. Other highly expressed fragile sites include those at 16q23 (FRA16D), 6q26 (FRA6E), 7q32.3 (FRA7H), and Xp22.3 (FRAXB) (Glover et al., 1984).

Fragile sites are unstable in vitro

Common fragile sites are normally stable in cultured human cells. However, when cells are cultured under conditions that partially inhibit DNA synthesis but allow cell division, common fragile sites are expressed and display a number of characteristics of unstable and highly recombinogenic DNA. Following induction, these sites are "hot spots" for sister-chromatid exchanges (SCEs) (Glover and Stein, 1987) and show a high rate of translocations and deletions in somatic cell hybrid systems (Glover and Stein, 1988; Wang et al., 1997). They are preferred sites of integration of transfected plasmid DNA (Rassool et al., 1991) and may be favored targets for DNA virus integration in vivo (Popescu and DiPaolo, 1989; Smith et al., 1992; Wilke et al., 1996; Mishmar et al., 1998; Thorland et al., 2000). Fragile sites may also play a pivotal role in some gene amplification events, both in vitro (Kuo et al., 1994; Coquelle et al., 1997) and in tumor cells (Ciullo et al., 2002; Hellman et al., 2002) by initiating a breakage-fusion-bridge cycle, presumably by providing recurring double-strand breaks at the boundaries of amplicons. All of these characteristics are consistent with the hypothesis that DNA strand breaks or recombination are associated with fragile site induction, either as a primary or secondary event.

Cloning fragile site regions and demonstration of instability in tumor cells

Based on the behavior of common fragile sites in vitro and their locations, many investigators suggested early on that fragile sites may play a mechanistic role in chromosome breakage and rearrangements involved in cancer (Hecht and Sutherland, 1984; Yunis and Soreng, 1984; Hecht, 1988). These associations were based largely on statistical correlations of cancer chromosome breakpoints and fragile sites at the level of banded chromosomes and assumed that instability at fragile sites can lead to chromosome breakage or recombination in vivo, as they do in vitro. To test this hypothesis, and to further study the molecular nature of common fragile sites, it was necessary to clone individual fragile sites.

The FRA3B fragile site was first targeted for cloning because it is the most frequently seen fragile site on human chromosomes and can be induced in the majority of treated cells. In addition, it maps to a region of 3p known to be associated with deletions in a number of solid tumors and a t(3;8) translocation breakpoint segregating in a family with hereditary renal cell carcinoma (RCC) (Cohen et al., 1979). A number of groups identified sequences at the FRA3B site using positional cloning approaches (Boldog et al., 1994, 1997; Wilke et al., 1994, 1996; Paradee et al., 1995, 1996; Rassool et al., 1996). For example, our group mapped genomic clones to 3p14.2 and used them in

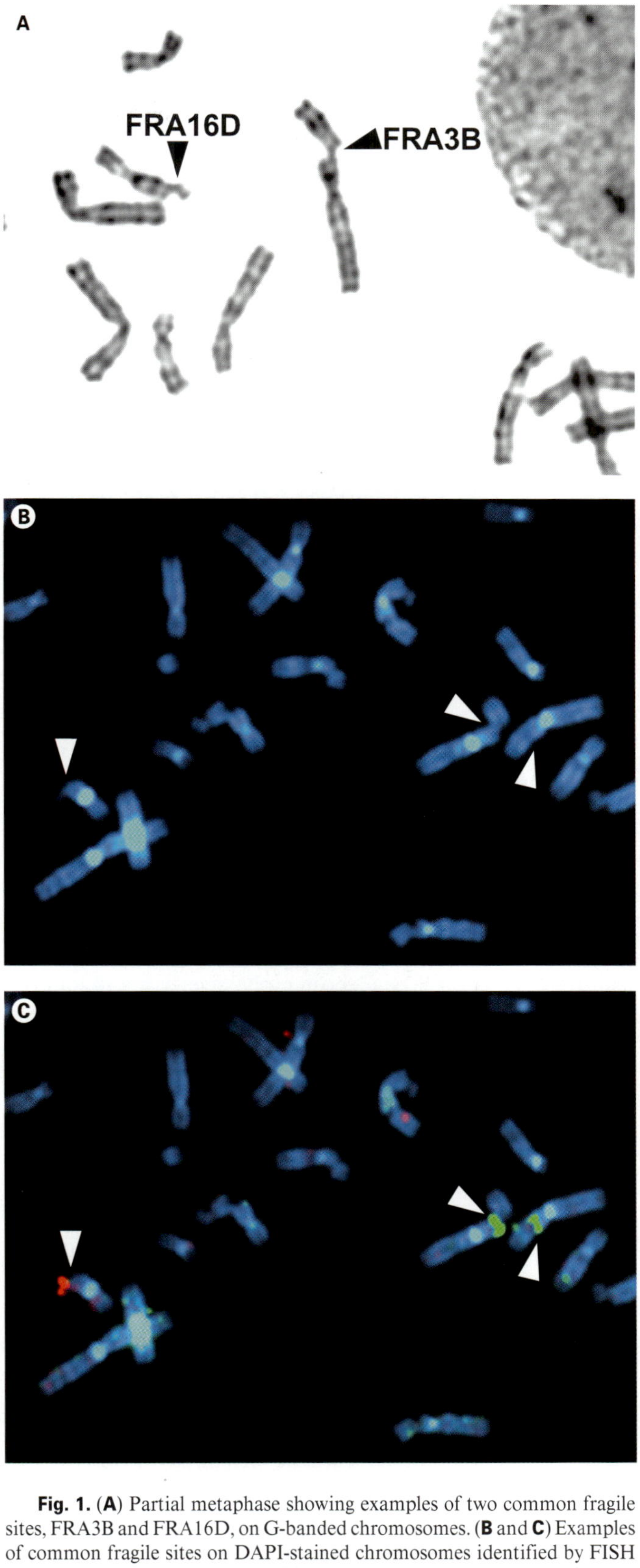

Fig. 1. (**A**) Partial metaphase showing examples of two common fragile sites, FRA3B and FRA16D, on G-banded chromosomes. (**B** and **C**) Examples of common fragile sites on DAPI-stained chromosomes identified by FISH with fragile site-specific probes. HCT116 cells treated with aphidicolin were hybridized with YAC 850A6 (green) to FRA3B with BAC 264L1 (red) to FRA16D (arrows). One homolog of each chromosome expresses the fragile site. Reprinted with modifications from Casper et al. (2002) with permission from Elsevier Science.

FISH analyses on cells expressing the fragile site to identify clones spanning FRA3B (Wilke et al., 1994). We found that FRA3B extends over a broad region, which we now know is over 400 kb (Boldog et al., 1994). No CGG or other di- or trinucleotide repeats were found in this region, suggesting that the molecular mechanism responsible for this and other common fragile sites differ from those with simple repeat expansions. Analysis of aphidicolin-induced deletion and translocation breakpoints in somatic cell hybrids and of transfected pSV2neo-plasmid integration sites has also confirmed that they, indeed, occurred within the fragile site (Glover and Stein, 1988; Rassool et al., 1991; Wang et al., 1997). From our sequence analysis of the FRA3B region, we also discovered a human papilloma virus type 16 (HPV-16) integration site, whose flanking sequence in a cervical carcinoma was reported (Wilke et al., 1996), a finding that led to more recent findings of frequent HPV integrations at fragile sites in cervical tumor cells by others (Thorland et al., 2000).

FRA3B lies within the FHIT gene – an unstable tumor suppressor gene

Based on RDA analysis of tumors, which showed homozygous deletion of genomic sequences within the 3p14.2 region, Ohta et al. (1996) used exon trapping to identify an exon of what is now known as the FHIT gene. FHIT is large, spanning an estimated ~900 kb of 3p14.2, but it encodes only a 1.1-kb transcript. Comparing the map of the FRA3B region to that of FHIT, FRA3B was found to map within the large FHIT gene in a region extending at least from intron 4 through much of the very large intron 5. FHIT encodes a 147-amino acid protein with diadenosine tri- and tetraphosphate hydrolase activity (Barnes et al., 1996). Diadenosine phosphates are ubiquitous small molecules produced from ATP during tRNA synthesis and consist of two adenosine molecules bridged by three or four phosphate groups. The biological function of these molecules is not clearly understood.

Numerous studies have demonstrated that FHIT is frequently rearranged and displays intralocus deletions and aberrant transcripts in a wide variety of tumor cells, including breast, ovarian, cervical, gastric, esophageal, and others (Wilke et al., 1994; Ohta et al., 1996; Druck et al., 1997; Michael et al., 1997; Huebner et al., 1998; Mimori et al., 1999; Fang et al., 2000). The genomic rearrangements are usually one or more large deletions of tens to hundreds of kilobases directly within the fragile site region. Multiple, sometimes overlapping, deletions have been noted within a single tumor (Michael et al., 1997). Ongoing instability in the FRA3B region is suggested by the presence of multiple, variable deletions within FRA3B homologs in tumor cell lines (Corbin et al., 2002). Translocations with breakpoints in FRA3B have also been observed (Geurts et al., 1997; Gemmill et al., 1998; Fang et al., 2000). Based on these findings, FHIT was proposed to be a tumor suppressor gene. While this conclusion has been somewhat controversial, recent functional studies lend support to this hypothesis (Huebner et al., 1998; Sozzi et al., 1998; Dumon et al., 2001). Overexpression of FHIT has been shown to inhibit growth of some tumor cells both in vitro and in vivo and may function in an apoptotic pathway (Siprashvili et al., 1997). In addition, mice with inactivation of *Fhit* show a striking increase in NMBA-induced gastric tumors (Zanesi et al., 2001).

Other fragile sites also show rearrangements and deletions in tumor cells

It has been hypothesized that the frequent deletions within FHIT are a result of instability at FRA3B in tumor cell backgrounds. Selection of cells with loss of function of FHIT may then play a role in tumor progression. This raises the question of whether other fragile sites associated with genes are unstable in cancer cells. Six fragile sites have now been cloned, and these also show large intra-locus deletions in various tumor cells, indicating that deletions are not unique to FRA3B or the FHIT gene (Mishmar et al., 1998; Huang et al., 1999; Krummel et al., 2000; Paige et al., 2000; Ried et al., 2000; Arlt et al., 2002b; Morelli et al., 2002). FRA3B and FRA16D lie within the large FHIT and WWOX genes, respectively (Ohta et al., 1996; Ried et al., 2000). In addition, as might be expected given the large size of fragile sites, most are associated with known genes. FRA7G spans several genes including the caveolin genes, CAV1 and CAV2, both of which have reported tumor-suppressor function, and is adjacent to the MET oncogene (Engelman et al., 1998). FRAXB spans a number of genes, including STS, but no genes with tumor-suppressor activity (Arlt et al., 2002b). FRA6F spans at least five known genes, including a gene associated with schizophrenia (Morelli et al., 2002). No genes have yet been found in the FRA7H region (Fig. 2).

FRA16D is particularly interesting in that it lies within the WWOX gene, a ww-domain containing oxidoreductase activity. Like FHIT, WWOX extends over 1 Mb but has a small, 2.2-kb transcript (Ried et al., 2000). WWOX is a candidate tumor-suppressor gene by virtue of finding large, intragenic, homozygous deletions and point mutations in ovarian, gastric, and other cancer cells and its ability to suppress growth of mammary tumors in mice (Mangelsdorf et al., 2000; Paige et al., 2000, 2001; Ried et al., 2000; Bednarek et al., 2001). In addition, a common translocation breakpoint in multiple myelomas is located in WWOX and FRA16D (Krummel et al., 2000).

It is clear from numerous studies that all common fragile sites studied frequently contain large, intra-locus deletions or translocations in tumor cells (Huebner and Croce, 2001). Therefore, genes at fragile sites, such WWOX and FHIT, are at a high risk for deletions or chromosomal rearrangements. The type of mutations seen in these genes in tumors, large deletions, and translocations, supports the involvement of the associated fragile sites in the mutations. A number of tumors have been identified with deletions in more than one fragile site, suggesting a common mechanism (Arlt et al., 2002b). Why the fragile site regions are unstable in these tumor cells, but not normal cells, has been unknown. As explained below, we suggest that these deletions are "signatures" of stalled replication forks caused by mutations or perturbations of the S-phase and G_2/M checkpoints or associated repair genes during tumor progression.

Mechanisms of instability

Based on the fact that fragile sites are induced by agents that retard or inhibit DNA replication or repair, it was proposed many years ago that fragile sites are associated with unreplicated DNA that can give rise to DNA strand breaks (Glover et al., 1984; Glover and Stein, 1987). DNA repair mechanisms were also suggested to play a role since caffeine, a suspected inhibitor of G_2 repair, greatly increases the expression of common fragile sites (Glover et al., 1986; Schlegel and Pardee, 1986), and SCEs are observed at the majority of expressed fragile sites (Glover and Stein, 1987).

From cloning studies, we know that common fragile sites extend over hundreds of kilobases, with gaps and breaks occurring throughout the regions. However, sequence analyses of cloned fragile site regions have not readily revealed why the sites are unstable. All fragile sites studied to date are composed of AT-rich sequences (Boldog et al., 1997; Mishmar et al., 1998; Ried et al., 2000; Arlt et al., 2002b). There are no simple repeat sequences, such as di- or trinucleotide repeats, that can explain the fragility. A detailed analysis of the FRA7H region by Mishmar et al. (1998) identified a high number of what were termed "flexibility peaks". These are short regions of a few hundred nucleotides that display high local variation in twist angle. While such peaks are dispersed throughout the genome, they are found at a high frequency in fragile site regions and may contribute to instability in the regions. Analysis of FRA3B, FRAXB, FRA6F, and FRA16D has shown that they all contain a high number of these flexibility peaks (Mishmar et al., 1998, 1998; Mimori et al., 1999; Ried et al., 2000; Morelli et al., 2002). It may be relevant that flexibility peak regions have recently been reported to be enriched in known mammalian origins of replication (Toledo et al., 2000).

Fragile sites are late replicating

Laird et al. (1987) first hypothesized that fragile sites, in particular the fragile X site (FRAXA), are late replicating. In addition to FRAXA (Subramanian et al., 1996; Hansen et al., 1997), replication timing has been studied at two other rare fragile sites, FRA10B and FRA16B (Handt et al., 2000) and two common fragile sites. LeBeau et al. (1998) used interphase FISH techniques to show that the sequences from within FRA3B are late replicating, as had previously been shown for the Fragile X site (FRAXA) (Subramanian et al., 1996; Hansen et al., 1997). Following addition of aphidicolin, the regions replicate even later, with indications that, in some cells, the regions are unreplicated in G_2. The fact that fragile sites are seen on metaphase chromosomes suggests that these fragile sites lesions can escape G_2/M checkpoint controls. Late replication of FRA3B was confirmed by Wang et al. (1999), who further suggested asynchronous replication of the homologous alleles.

The only other fragile site so far studied for replication timing is FRA7H. Using in situ hybridization on interphase nuclei, Hellman et al. (2000) showed that the replication of this region is initiated relatively early, before 30% of S-phase is completed.

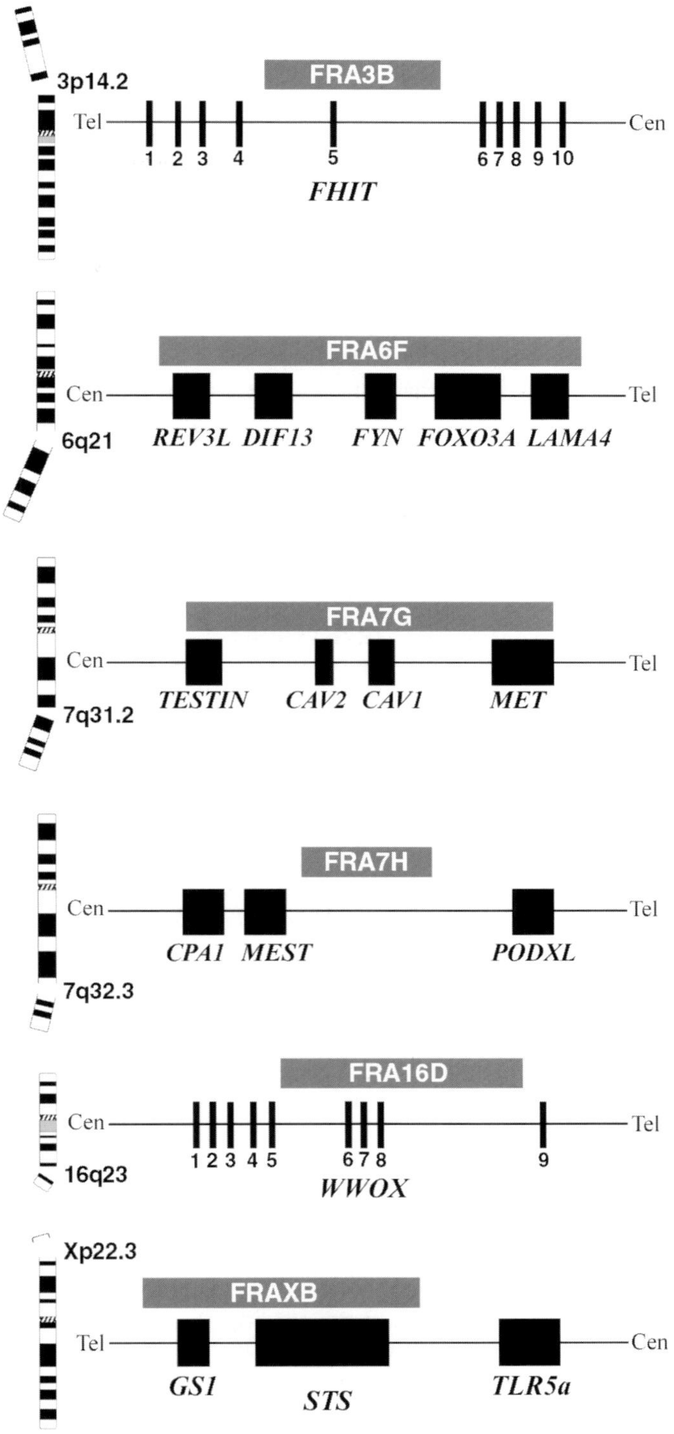

Fig. 2. Maps of cloned common fragile sites and their associated genes. Chromosomal locations of individual fragile sites are indicated on the ideograms at left. Gray horizontal bars indicate positions of fragile sites. Genes are indicated by black boxes with the gene symbol underneath. Individual exons of FHIT and WWOX are numbered.

However, a more detailed analysis showed significant differences in the replication of adjacent clones under normal growth conditions and upon aphidicolin treatment, with some clones replicating late in S-phase. This pattern significantly differed from that of two non-fragile regions, which showed coordinated replication under both conditions. These results indicate that aphidicolin is enhancing an existing difference in the replication timing along the FRA7H region. Taken together, these results suggest that fragile site regions contain late replicating DNA, perhaps juxtaposed with early replicating regions, that is particularly sensitive to further delay following partial replication inhibition. Such late or delayed replication likely contributes to instability at fragile sites, which suggests an underlying basis in sequence organization and replication origins (Toledo et al., 2000). Further studies of replication timing and dynamics at additional fragile sites will lead to a better understanding of the normal replication of these regions and why they are particularly sensitive to inhibitors of replication.

Genetic control: The ATR and BRCA1 checkpoint pathways are central to fragile site stability

As mentioned above, the mechanism for induction of all fragile sites appears to involve interruptions of normal DNA synthesis. DNA repair mechanisms may play a role since fragile sites are likely associated with DNA strand breaks or daughter-strand gaps and because caffeine increases the number of cells expressing fragile sites (Glover et al., 1984; Glover and Stein, 1987). Based on these observations, we hypothesized that analysis of fragile sites in cells with mutations in cell cycle checkpoint or repair genes that respond to DNA DS breaks could provide insight into the mechanisms of fragile site expression. However, we did not see any change in fragile site expression in cells from patients with ataxia telangiectasia (AT), Nijmegen Breakage syndrome (NBS), Bloom syndrome, or Cockayne syndrome. In addition, we did not see any significant effect of mutations in NHEJ pathway genes (XRCC4, XRCC5 and PRKDC) in rodent cells (Glover, unpublished). These results were, for a time, difficult to explain but are now clearly consistent with new findings and a model of common fragile sites.

Recent findings in the field of cell cycle checkpoints, together with earlier clues, led to the consideration of checkpoint pathways in the mechanism of fragile site instability. Studies performed over ten years ago showed that caffeine greatly increases fragile site expression (Glover et al., 1986). For years, caffeine has been known to affect the mutagenicity of a variety of agents. Until recently, the exact mechanism was unknown. However, caffeine and the related compound, 2-aminopurine (2-AP), have recently been shown to inhibit the cell cycle checkpoint kinases ATM and ATR (Sarkaria et al., 1999; Dimitrova and Gilbert, 2000). We have recently shown that 2-AP has almost identical effects to caffeine in enhancing fragile site expression (Casper et al., 2002), thus, strongly implicating ATM or ATR in fragile site stability.

ATM and ATR are key checkpoint proteins of the intra-S phase and G_2/M checkpoint pathways. Recent findings have shown that ATM and ATR act in distinct but partially overlapping pathways in response to specific types of DNA damage during cellular replication. They activate the G_2/M checkpoint via phosphorylation of CHK1, CHK2, BRCA1, and other targets (Abraham, 2001; Durocher and Jackson 2001; Shiloh, 2001). CHK1 and CHK2 regulate the CDC25A kinase and WEE1, an inhibitor of the CDC2/cyclin B complex, and, thus, progression into mitosis. S-phase checkpoint activation by ATM involves phosphorylation of CHK2 and Nbs1 to trigger two signaling cascades (Falck et al., 2002). The exact mechanisms of ATR regulation of the S-phase checkpoint are far less clear but appear to involve stabilization of existing replicons and inhibition of late firing origins (Dimitrova and Gilbert, 2000).

ATM has been extensively studied in cell lines derived from ataxia telangiectasia (AT) patients and is known to be activated by DNA double-strand breaks (Pincheira et al., 2001). Thus, AT cells are sensitive to ionizing irradiation but not UV light or stalled replication forks. We have investigated the role of ATM in fragile site stability using a series of AT cell lines. Fragile site expression was unchanged in cells with ATM deficiency, suggesting that the role of the ATM/CHK2 pathway in maintaining stability at fragile sites is negligible (Casper et al., 2002). Thus, it is unlikely that DS breaks are the primary cause of fragile site expression, although they must sometimes occur as a secondary event, as evidenced by deletions and rearrangements at these sites in some tumor cells.

Compared to ATM, much less is known about ATR, since deficiency is lethal to cells (Brown and Baltimore, 2000). As opposed to responding to DNA DS breaks, like ATM, ATR acts in a parallel pathway in response to stalled replication forks. In the past few years, overexpression of dominant-negative alleles of ATR (Cliby et al., 1998; Nghiem et al., 2001), or cell lines with inactivated ATR alleles (Cortez et al., 2001), have been used to show that ATR is required for checkpoint responses after treatment of cells with UV light and agents that block replication fork progression, such as hydroxyurea and aphidicolin (Nghiem et al., 2001), and more recently, hypoxia (Hammond et al., 2002). A major outcome of ATR deficiency was severe chromosome fragmentation following treatment of cells with high doses of aphidicolin or hydroxyurea, whereas similar experiments with ATM-deficient cells showed no chromosome fragmentation under these conditions (Nghiem et al., 2001). This fragmentation may represent an extreme example of the type of damage seen first at common fragile sites. We investigated the role of ATR in maintaining stability at fragile sites, using three independent approaches. These are the use of cells stably transfected with a dox-inducible dominant negative ATR construct (ATR-kd), inhibition of ATR using RNAi approaches, and cells with cre-lox mediated deletion of ATR. Results from these experiments clearly demonstrate that inactivation of ATR greatly increases both spontaneous and induced gaps and breaks at common fragile sites. Furthermore, ATR deficiency alone, without addition of replication inhibitors, induces breaks and gaps at fragile sites. This finding supports the concept that some level of replication stalling occurs normally at fragile sites. This major molecular pathway involving ATR is the first to have been associated with the maintenance of stability at fragile sites, and these results provide a basis for understanding the mechanism of fragile site instability and the consequences of stalled replication in mammalian cells. These observations are particularly important in cancer cells, where checkpoint pathways are likely to be disrupted, resulting in increased instability at fragile sites.

BRCA1 is a downstream target of ATR phosphorylation in response to DNA damage and has been shown to be central to G_2/M checkpoint control (Yarden et al., 2002) and homology-directed DNA repair (Moynahan et al., 2001). Recently it has been shown that after treatment with ionizing radiation, BRCA1 serine 1423 phosphorylation by ATM is required for G_2/M arrest (Xu et al., 2001), while serine 1387 phosphorylation is necessary for S-phase arrest (Xu et al., 2002). In addition, it has been shown that these same residues are phosphorylated by ATR in response to stalled replication forks (Tibbetts et al., 2000), suggesting a similar mechanism is activated through ATR as with ATM, but in response to different forms of DNA damage. In studies of the role of BRCA1 in fragile site instability, Turner et al. (2002) and Arlt et al. (2002a) have shown that HCC1937, a breast cancer cell line deficient in BRCA1 activity, demonstrates an elevated level of total gaps and breaks at specific fragile sites when treated with aphidicolin. This increase in fragile site expression is complemented to near wild type levels when wild type BRCA1 expression is restored in these cells (Arlt et al., 2002a). Thus, although these tumor cells undoubtedly have a variety of mutations in tumor-related genes, wt-BRCA1 alone complements the enhanced fragile site expression.

Other targets of ATR include CHK1, p53, and histone H2AX (Tibbetts et al., 1999; Liu et al., 2000; Ward and Chen, 2001; Hammond et al., 2002; Nghiem et al., 2002). These proteins may be important in chromatin conformation or recruitment of repair enzymes at stalled forks. The possible role of these proteins in fragile site instability has not been explored, but it is highly likely that at least CHK1 deficiency will result in increased fragile site instability, as CHK1 is not only a target of ATR but also is known to be phosphorylated by BRCA1 (Yarden et al., 2002).

Evolutionary conservation of fragile sites

Common fragile sites appear to be conserved in all mammalian species examined to date, including the Old and New World primates, horse, cow, pig, dog, cat, Indian mole rat, deer mouse, and laboratory mouse (Elder and Robinson, 1989; Stone et al., 1991, 1993; Yang and Long, 1993; McAllister and Greenbaum, 1997; Loeb et al., 2001). In all cases, the fragile sites are induced by folate stress or aphidicolin, as in humans. It has been shown in the mouse and in primates that some specific fragile sites are orthologous to those found in humans (Glover et al., 1998; Loeb et al., 2001; Shiraishi et al., 2001; Krummel et al., 2002). Furthermore, fragile sites may also have counterparts in yeast. Deficiency of the ATR homologue Mec1 in *S. cerevisiae* has recently been shown to result in non-random chromosome breakage within "replication slow zones" (RSZ) (Cha and Kleckner, 2002). These RSZs are approximately 10 kb in size and occur, on average, every 60 kb on chromosome III. Although different in size and distribution from mammalian fragile sites, Cha and Kleckner suggest that these RSZs may be functionally analogous. The fact that both Mec1 (Cha and Kleckner, 2002) and ATR (Casper et al., 2002) deficiencies lead to similar chromosomal events in yeast and mammals, respectively, strengthens this hypothesis. The conservation of fragile site regions suggests that these sites are maintained for a particular role, such as monitoring replication events during S-phase.

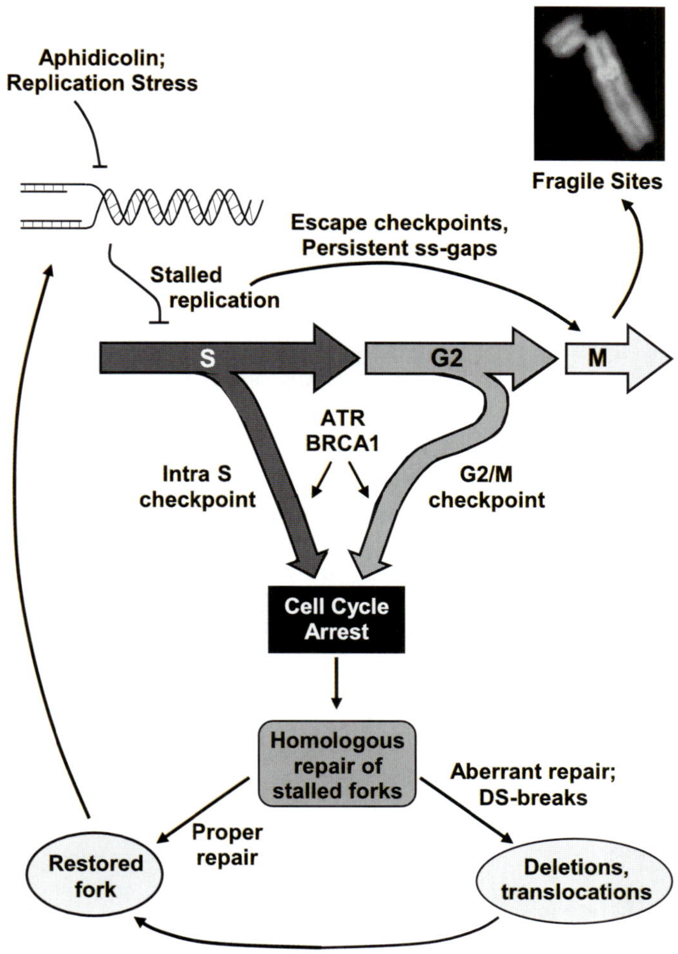

Fig. 3. Model of common fragile site expression. See text for details. Reprinted with modifications from Casper et al. (2002) with permission from Elsevier Science.

A model for common fragile sites

Our model to explain instability at common fragile sites is based on previous findings from our laboratory and others and our recent data implicating ATR and BRCA1 in the maintenance of fragile site stability (Fig. 3). We propose that fragile sites represent single-stranded (unreplicated) regions caused by stalled or collapsed replication forks following partial inhibition of DNA replication. Fragile site regions are significantly more sensitive to replication fork arrest and delayed replication than other regions in the genome. The reasons for this sensitivity are as yet unknown but are likely related to sequence or replication dynamics. A related possibility is that fragile sites may be among the latest regions to complete replication, with some replication occurring during G_2. Replication fork stalling at fragile sites is recognized by the S-phase and/or G_2/M checkpoints in which ATR and BRCA1 play key roles. However, some proportion of these lesions must escape checkpoint controls, as evidenced by the appearance of fragile sites on metaphase chromosomes. Reinitiation of replication from stalled or collapsed forks is thought to usually involve DNA homologous strand repair (HR), and the previously observed frequency of sister chromatid exchanges at fragile sites supports the suggestion that HR is employed at fragile site lesions. Rearrangements at fragile sites could result from occasional double-strand breaks or illegitimate recombination. Thus, the deletions seen in tumor cells are proposed to arise from faulty or unequal recombinational repair or from mutations or perturbations in the S-phase or G_2/M checkpoints pathways that respond to stalled replication forks. In this way, deletions at fragile sites in tumors may be "signatures" of stalled replication forks.

Summary

It has long been known that fragile sites represent regions of the genome that are particularly sensitive to genotoxic stress that interferes with DNA replication. From the many studies that have cloned and characterized fragile sites, it is now known that these sites extend over large regions, are associated with genes, may complete replication late and contain stretches of high flexibility, but are otherwise unremarkable in sequence. Studies showing that fragile sites, and associated genes, are frequently deleted or rearranged in cancer cells have clearly demonstrated their importance in genome instability in tumorigenesis. Yet until recently, very little was known about the molecular mechanisms involved in their stability. The finding that ATR and BRCA1 are critical for genome stability at fragile sites represents the first major molecular pathway to be associated with control of fragile site expression. We believe that the study of fragile sites takes on new significance based on these findings. These genes are of particular interest due to their critical role in cellular checkpoint responses to stalled replication forks and genome integrity in both normal and cancer cells, and this pathway will provide a basis for further understanding the mechanism of fragile site instability.

Acknowledgements

We thank C. Gaffney for assistance in preparation of this manuscript.

References

Abraham RT: Cell cycle checkpoint signaling through the ATM and ATR kinases. Genes Dev 15:2177–2196 (2001).

Arlt MF, Durkin SG, Casper AM, Kastan MB, Glover TW: BRCA1 deficient cells have increased fragile site expression that is complemented by wild type BRCA1. Am J hum Genet 71:181S (2002a).

Arlt MF, Miller DE, Beer DG, Glover TW: Molecular characterization of FRAXB and comparative common fragile site instability in cancer cells. Genes Chrom Cancer 33:82–92 (2002b).

Barnes LD, Garrison PN, Siprashvili Z, Guranowski A, Robinson AK, Ingram SW, Croce CM, Ohta M, Huebner K: Fhit, a putative tumor suppressor in humans, is a dinucleoside 5′,5‴-P1,P3-triphosphate hydrolase. Biochemistry 35:11529–11535 (1996).

Bednarek A, Keck-Waggoner CL, Daniel RL, Laflin KJ, Bergsagel PL, Kiguchi K, Brenner AW, Aldaz CM: WWOX, the FRA16D gene, behaves as a suppressor of tumor growth. Cancer Res 61:8068–8073 (2001).

Boldog F, Gemmill R, West J, Robinson M, Li E: Chromosome 3p14 homozygous deletions and sequence analysis of FRA3B. Hum molec Genet 6:193–203 (1997).

Boldog FL, Waggoner B, Glover TW, Chumakov I, Le Paslier D, Cohen D, Gemmill RM, Drabkin HA: Integrated YAC contig containing the 3p14.2 hereditary renal carcinoma 3;8 translocation breakpoint and the fragile site FRA3B. Genes Chrom Cancer 11:216–221 (1994).

Brown EJ, Baltimore D: ATR disruption leads to chromosomal fragmentation and early embryonic lethality. Genes Dev 14:397–402 (2000).

Casper AM, Nghiem P, Arlt MF, Glover TW: ATR regulates fragile site stability. Cell 111:779–789 (2002).

Cha RS, Kleckner N: ATR homolog Mec1 promotes fork progression, thus averting breaks in replication slow zones. Science 297:602–606 (2002).

Ciullo M, Debily MA, Rozier L, Autiero M, Billault A, Mayau V, El Marhomy S, Guardiola J, Bernheim A, Coullin P, Piatier-Tonneau D, Debatisse M: Initiation of the breakage-fusion-bridge mechanism through common fragile site activation in human breast cancer cells: the model of PIP gene duplication from a break at FRA7I. Hum molec Genet 11:2887–2894 (2002).

Cliby WA, Roberts CJ, Cimprich KA, Stringer CM, Lamb JR, Schreiber SL, Friend SH: Overexpression of a kinase-inactive ATR protein causes sensitivity to DNA-damaging agents and defects in cell cycle checkpoints. EMBO J 17:159–169 (1998).

Cohen AJ, Li FP, Berg S, Marchetto DJ, Tsai S, Jacobs SC, Brown RS: Hereditary renal-cell carcinoma associated with a chromosomal translocation. New Engl J Med 301:592–595 (1979).

Coquelle A, Pipiras E, Toledo F, Buttin G, Debatisse M: Expression of fragile sites triggers intrachromosomal mammalian gene amplification and sets boundaries to early amplicons. Cell 89:215–225 (1997).

Corbin S, Neilly ME, Espinosa III R, Davis EM, McKeithan TW, Le Beau MM: Identification of unstable sequences within the common fragile site at 3p14.2: implications for the mechanism of deletions within fragile histidine triad gene/common fragile site at 3p14.2 in tumors. Cancer Res 62:3477–3484 (2002).

Cortez D, Guntuku S, Qin J, Elledge SJ: ATR and ATRIP: partners in checkpoint signaling. Science 294:1713–1716 (2001).

Dimitrova DS, Gilbert DM: Temporally coordinated assembly and disassembly of replication factories in the absence of DNA synthesis. Nature Cell Biol 2:686–694 (2000).

Druck T, Hadaczek P, Fu T-B, Ohta M, Siprashvili Z, Baffa R, Negrini M, Kastury K, Veronese ML, Rosen D, Rothstein J, McCue P, Cotticelli MG, Inoue H, Croce CM, Huebner K: Structure and expression of the human FHIT gene in normal and tumor cells. Cancer Res 57:504–512 (1997).

Dumon KR, Ishii H, Fong LYY, Zanesi N, Fidanza V, Mancini R, Vecchione A, Baffa R, Trapasso F, During MJ, Huebner K, Croce CM: FHIT gene therapy prevents tumor development in Fhit-deficient mice. Proc natl Acad Sci, USA 98:3346–3351 (2001).

Durocher D, Jackson SP: DNA-PK, ATM and ATR as sensors of DNA damage: variations on a theme? Curr Opin Cell Biol 13:225–231 (2001).

Elder FFB, Robinson TJ: Rodent fragile sites: are they conserved? Evidence from mouse and rat. Chromosoma 97:459–464 (1989).

Engelman JA, Zhang XL, Lisanti MP: Genes encoding human caveolin-1 and -2 are co-localized to the D7S522 locus (7q31.1), a known fragile site (FRA7G) that is frequently deleted in human cancers. FEBS Lett 436:403–410 (1998).

Falck J, Petrini JHJ, Williams BR, Lukas J, Bartek J: The DNA damage-dependent intra-S phase checkpoint is regulated by parallel pathways. Nature Genet 30:290–294 (2002).

Fang JM, Arlt MF, Burgess AC, Beer DG, Glover TW: Translocation breakpoints in FHIT and FRA3B in both homologs of chromosome 3 in an esophageal adenocarcinoma. Genes Chrom Cancer 30:292–298 (2000).

Gemmill RM, West JD, Boldog F, Tanaka N, Robinson LJ, Smith DI, Li F, Drabkin HA: The hereditary renal cell carcinoma 3;8 translocation fuses FHIT to a patched-related gene, TRC8. Proc natl Acad Sci, USA 95:9572–9577 (1998).

Geurts JMW, Schoenmakers EFPM, Roijer E, Stenman G, Van de Ven WJM: Expression of reciprocal hybrid transcripts of HMGIC and FHIT in a pleomorphic adenoma of the parotid gland. Cancer Res 57:13–17 (1997).

Glover TW, Stein CK: Induction of sister chromatid exchanges at common fragile sites. Am J hum Genet 41:882–890 (1987).

Glover TW, Stein CK: Chromosome breakage and recombination at fragile sites. Am J hum Genet 43:265–273 (1988).

Glover TW, Berger C, Coyle J, Echo B: DNA polymerase a inhibition by aphidicolin induces gaps and breaks at common fragile sites in human chromosomes. Hum Genet 67:136–142 (1984).

Glover TW, Coyle-Morris J, Morgan R: Fragile sites: overview, occurrence in acute nonlymphocytic leukemia and effects of caffeine on expression. Cancer Genet Cytogenetics 19:141–150 (1986).

Glover TW, Hoge AW, Miller DE, Ascara-Wilke JE, Adam AN, Dagenais SL, Wilke CM, Dierick HA, Beer DG: The murine *Fhit* gene is highly similar to its human orthologue and maps to a common fragile site region. Cancer Res 58:3409–3414 (1998).

Hammond EM, Denko NC, Dorie MJ, Abraham RT, Giaccia AJ: Hypoxia links ATR and p53 through replication arrest. Mol Cell Biol 22:1834–1843 (2002).

Handt O, Baker E, Dayan S, Gartler SM, Woollatt E, Richards RI, Hansen RS: Analysis of replication timing at the FRA10B and FRA16B fragile site loci. Chrom Res 8:677–688 (2000).

Hansen RS, Canfield TK, Fjeld AD, Mumm S, Laird CD, Gartler SM: A variable domain of delayed replication in FRAXA fragile X chromosomes: X inactivation-like spread of late replication. Proc natl Acad Sci, USA 94:4587–4592 (1997).

Hecht F: Fragile sites, cancer chromosome breakpoints, and oncogenes all cluster in light G-bands. Cancer Genet Cytogenet 31:17–24 (1988).

Hecht F, Sutherland GR: Fragile sites and cancer breakpoints. Cancer Genet Cytogenet 12:179–181 (1984).

Hellman A, Rahat A, Scherer SW, Darvasi A, Tsui L-C, Kerem S: Replication delay along FRA7H, a common fragile site on human chromosome 7, leads to chromosomal instability. Mol Cell Biol 20:4420–4427 (2000).

Hellman A, Zlotorynski E, Scherer SW, Cheung J, Vincent JB, Smith DI, Trakhtenbrot L, Kerem B: A role for common fragile site induction in amplification of human oncogenes. Cancer Cell 1:89–97 (2002).

Huang H, Reed CP, Mordi A, Lomberk G, Wang L, Shridhar V, Hartmann L, Jenkins R, Smith DI: Frequent deletions within FRA7G at 7q31.2 in invasive epithelial ovarian cancer. Genes Chrom Cancer 24:48–55 (1999).

Huebner K, Croce CM: FRA3B and other common fragile sites: the weakest links. Nature Rev Cancer 1:214–221 (2001).

Huebner K, Druck T, Siprashvili Z, Croce CM, Kovatich A, McCue PA: The role of deletions at the FRA3B/FHIT locus in carcinogenesis. Recent Res Cancer Res 154:200–215 (1998).

Ikegami S, Taguchi T, Ohashi M, Oguro M, Nagano H, Mano Y: Aphidicolin prevents mitotic cell division by interfering with the activity of DNA polymerase-alpha. Nature 275:458–460 (1978).

Kremer EJ, Pritchard M, Lynch M, Yu S, Holman K, Baker E, Warren ST, Schlessinger D, Sutherland GR, Richards RI: Mapping of DNA instability at the fragile X to a trinucleotide repeat sequence p(CCG)n. Science 252:1711–1714 (1991).

Krummel KA, Denison SR, Calhoun E, Phillips LA, Smith DI: The common fragile site FRA16D and its associated gene WWOX are highly conserved in the mouse at Fra8E1. Genes Chrom Cancer 34:154–167 (2002).

Krummel KA, Roberts LR, Kawakami M, Glover TW, Smith DI: The characterization of the common fragile site FRA16D and its involvement in multiple myeloma translocations. Genomics 69:37–46 (2000).

Kuo MT, Vyas RC, Jiang LX, Hittelman WN: Chromosome breakage at a major fragile site associated with P-glycoprotein gene amplification in multidrug-resistant CHO cells. Mol Cell Biol 14:5202–5211 (1994).

Laird C, Jaffe E, Karpsen G, Lamb M, Nelson R: Fragile sites in human chromosomes as regions of late-replicating DNA. Trends in Genetics 3:274–281 (1987).

Le Beau MM, Rassool FV, Neilly ME, Espinosa III R, Glover TW, Smith DI, McKeithan TW: Replication of a common fragile site, FRA3B, occurs late in S phase and is delayed further upon induction: implications for the mechanism of fragile site induction. Hum Mol Genet 7:755–761 (1998).

Liu Q, Guntuku S, Cui X-S, Matsuoka S, Cortez D, Tamai K, Luo G, Carattini-Rivera S, DeMayo F, Bradley A, Donehower LA, Elledge SJ: Chk1 is an essential kinase that is regulated by Atr and required for the G_2/M DNA damage checkpoint. Genes Dev 14:1448–1459 (2000).

Loeb AM, Arlt MF, Morrish TA, Glover TW: Conservation of FRA3B and FRAXB in the great apes. Am J hum Genet 69:329S (2001).

Mangelsdorf M, Ried K, Woollatt E, Dayan S, Eyre H, Finnis M, Hobson L, Nancarrow J, Venter D, Baker E, Richards RI: Chromosomal fragile site FRA16D and DNA instability in cancer. Cancer Res 60:1683–1689 (2000).

McAllister BF, Greenbaum IF: How common are common fragile sites: variation of aphidicolin-induced chromosomal fragile sites in a population of the deer mouse *(Peromyscus maniculatus)*. Hum Genet 100:182–188 (1997).

Michael D, Beer DG, Wilke CW, Miller DE, Glover TW: Frequent deletions of FHIT and FRA3B in Barrett's metaplasia and esophageal adenocarcinomas. Oncogene 15:1653–1659 (1997).

Mimori K, Druck T, Inoue H, Alder H, Berk L, Mori M, Huebner K, Croce CM: Cancer-specific chromosome alterations in the constitutive fragile region FRA3B. Proc natl Acad Sci, USA 96:7456–7461 (1999).

Mishmar D, Mandel-Gutfreund Y, Margalit H, Rahat A, Kerem B: Common fragile sites: G-band characteristics within an R-band. Am J hum Genet 64:908–910 (1999).

Mishmar D, Rahat A, Scherer SW, Nyakatura G, Hinzmann B, Kohwi Y, Mandel-Gutfroint Y, Lee JR, Drescher B, Sas DE, Margalit H, Platzer M, Weiss A, Tsui LC, Rosenthal A, Kerem B: Molecular characterization of a common fragile site (FRA7H) on human chromosome 7 by the cloning of a simian virus 40 integration site. Proc natl Acad Sci, USA 95:8141–8146 (1998).

Morelli C, Karayianni E, Magnanini C, Mungall AJ, Thorland E, Negrini M, Smith DI, Barbanti-Brodano G: Cloning and characterization of the common fragile site FRA6F harboring a replicative senescence gene and frequently deleted in human tumors. Oncogene 21:7266–7276 (2002).

Moynahan ME, Pierce AJ, Jasin M: BRCA2 is required for homology-directed repair of chromosomal breaks. Molecular Cell 7:263–272 (2001).

Nghiem P, Park PK, Kim Y-s, Vaziri C, Schreiber SL: ATR inhibition selectively sensitizes G_1 checkpoint-deficient cells to lethal premature chromatin condensation. Proc natl Acad Sci, USA 98:9092–9097 (2001).

Nghiem P, Park PK, Sim Y-s, Desai BN, Schreiber SL: ATR is not required for p53 activation but synergizes with p53 in the replication checkpoint. J biol Chem 277:4428–4434 (2002).

Ohta M, Inoue H, Cotticelli MG, Kastury K, Baffa R, Palazzo J, Siprashvili Z, Mori M, McCue P, Druck T, Croce CM, Huebner K: The FHIT gene, spanning the chromosome 3p14.2 fragile site and renal carcinoma-associated t(3;8) breakpoint, is abnormal in digestive tract cancers. Cell 84:587–597 (1996).

Paige AJW, Taylor KJ, Stewart A, Sgouros JG, Gabra H, Sellar GC, Smyth JF, Porteous DJ, Watson JEV: A 700-kb physical map of a region of 16q23.2 homozygously deleted in multiple cancers and spanning the common fragile site FRA16D. Cancer Res 60:1690–1697 (2000).

Paige AJW, Taylor KJ, Taylor C, Hillier SG, Farrington S, Scott D, Porteous DJ, Smyth JF, Gabra H, Watson JEV: WWOX: a candidate tumor suppressor gene involved in multiple tumor types. Proc natl Acad Sci, USA 98:11417–11422 (2001).

Paradee W, Mullins C, He Z, Glover T, Wilke C, Opalka B, Schutte J, Smith DI: Precise localization of aphidicolin-induced breakpoints on the short arm of human chromosome 3. Genomics 27:358–361 (1995).

Paradee W, Wilke CM, Wang L, Shridhar R, Mullins CM, Hoge A, Glover TW, Smith DI: A 350-kb cosmid contig in 3p14.2 that crosses the t(3;8) hereditary renal cell carcinoma translocation breakpoint and 17 aphidicolin-induced FRA3B breakpoints. Genomics 35:87–93 (1996).

Pincheira J, Bravo M, Navarrete MH, Marcelain K, López-Sáez JF, de la Torre C: Ataxia telangiectasia: G2 checkpoint and chromosomal damage in proliferating lymphocytes. Mutagenesis 16:419–422 (2001).

Popescu NC, DiPaolo JA: Preferential sites for viral integration on mammalian genome. Cancer Genet Cytogenet 42:157–171 (1989).

Rassool FV, McKeithan TW, Neilly ME, van Melle E, Espinosa III R, Le Beau MM: Preferential integration of marker DNA into the chromosomal fragile site at 3p14: an approach to cloning fragile sites. Proc natl Acad Sci, USA 88:6657–6661 (1991).

Rassool FV, Le Beau MM, Shen ML, Neilly ME, Espinosa III R, Ong ST, Boldog R, Drabkin H, McCarroll R, McKeithan TW: Direct cloning of DNA sequences from the common fragile site region at chromosome band 3p14.2. Genomics 35:109–117 (1996).

Ried K, Finnis M, Hobson L, Mangelsdorf M, Sayan S, Nancarrow JK, Woollatt E, Kremmidiotis G, Gardner A, Venter D, Baker E, Richards RI: Common chromosomal fragile site FRA16D sequence: identification of the FOR gene spanning FRA16D and homozygous deletions and translocation breakpoints in cancer cells. Hum molec Genet 9:1651–1663 (2000).

Sarkaria JN, Busby EC, Tibbetts RS, Roos P, Taya Y, Karnitz LM, Abraham RT: Inhibition of ATM and ATR kinase activities by the radiosensitizing agent, caffeine. Cancer Res 59:4375–4382 (1999).

Schlegel R, Pardee AB: Caffeine-induced uncoupling of mitosis from the completion of DNA replication in mammalian cells. Science 232:1264–1266 (1986).

Shiloh Y: ATM and ATR: networking cellular responses to DNA damage. Curr Opin Genet Dev 11:71–77 (2001).

Shiraishi T, Druck T, Mimori K, Flomenberg J, Berk L, Alder H, Miller W, Huebner K, Croce CM: Sequence conservation at human and mouse orthologous common fragile regions, FRA3B/FHIT and *Fra14A2/Fhit*. Proc natl Acad Sci, USA 98:5722–5727 (2001).

Siprashvili Z, Sozzi G, Barnes LD, McCue P, Robinson AK, Eryomin V, Sard L, Tagliabue E, Greco A, Fusetti L, Schwartz G, Pierotti MC, Croce CM, Huebner K: Replacement of Fhit in cancer cells suppresses tumorigenicity. Proc natl Acad Sci, USA 94:13771–13776 (1997).

Smith PP, Friedman C, Bryant EM, McDougall JK: Viral integration and fragile sites in human papillomavirus-immortalized human keratinocyte cell lines. Genes Chrom Cancer 5:150–157 (1992).

Sozzi G, Huebner K, Croce CM: FHIT in human cancer. Adv Cancer Res 74:141–166 (1998).

Stone DM, Jacky PB, Hancock DD, Prieur DJ: Chromosomal fragile site expression in dogs. Am J med Genet 40:214–222 (1991).

Stone DM, Stephens KE, Doles J: Folate-sensitive and aphidicolin-inducible fragile sites are expressed in the genome of the domestic cat. Cancer Genet Cytogenet 65:130–134 (1993).

Subramanian PS, Nelson DL, Chinault AC: Large domains of apparent delayed replication timing associated with triplet repeat expansion at FRAXA and FRAXE. Am J hum Genet 59:407–416 (1996).

Sutherland GR, Baker E, Richards RI: Fragile sites still breaking. Trend Genet 14:501–506 (1998).

Thorland EC, Myers SL, Persing DH, Sarkar G, McGovern RM, Gostout BS, Smith DI: Human papillomavirus type 16 integrations in cervical tumors frequently occur in common fragile sites. Cancer Res 60:5916–5921 (2000).

Tibbetts RS, Brumbaugh KM, Williams JM, Sarkaria JN, Cliby WA, Shieh S-Y, Taya Y, Prives C, Abraham RT: A role for ATR in the DNA damage-induced phosphorylation of p53. Genes Dev 13:152–157 (1999).

Tibbetts RS, Cortez D, Brumbaugh KM, Scully R, Livingston D, Elledge SJ, Abraham RT: Functional interactions between BRCA1 and the checkpoint kinase ATR during genotoxic stress. Genes Dev 14:2989–3002 (2000).

Toledo F, Coquelle A, Svetlova E, Debatisse M: Enhanced flexibility and aphidicolin-induced DNA breaks near mammalian replication origins: implications for replicon mapping and chromosome fragility. Nucl Acids Res 28:4805–4813 (2000).

Turner BC, Ottey M, Zimonjic DB, Potoczek M, Hauck WW, Pequignot E, Keck-Waggoner CL, Devignani C, Aldaz CM, McCue PA, Palazzo J, Huebner K, Popescu NC: The fragile histidine triad/common chromosome fragile site 3B locus and repair-deficient cancers. Cancer Res 62:4054–4060 (2002).

Wang L, Darling J, Zhang J-S, Huang H, Liu W, Smith DI: Allele-specific late replication and fragility of the most active common fragile site, FRA3B. Hum molec Genet 8:431–437 (1999).

Wang L, Paradee W, Mullins C, Shridhar R, Rosati R, Wilke CM, Glover TW, Smith DI: Aphidicolin-induced FRA3B breakpoints cluster in two distinct regions. Genomics 41:485–488 (1997).

Ward IM, Chen J: Histone H2AX is phosphorylated in an ATR-dependent manner in response to replicational stress. J Biol Chem 276:47759–47762 (2001).

Wilke CM, Guo SW, Hall BK, Boldog F, Gemmill RM, Chandrasekharappa SC, Barcroft CL, Drabkin HA, Glover TW: Multicolor FISH mapping of YAC clones in 3p14 and identification of a YAC spanning both FRA3B and the t(3;8) associated with hereditary renal cell carcinoma. Genomics 22:319–326 (1994).

Wilke CM, Hall BK, Hoge A, Paradee W, Smith DI, Glover TW: FRA3B extends over a broad region and contains a spontaneous HPV16 integration site: direct evidence for the coincidence of viral integration sites and fragile sites. Hum molec Genet 5:187–195 (1996).

Xu B, Kim S-T, Kastan MB: Involvement of *Brca1* in S-phase and G2-phase checkpoints after ionizing irradiation. Mol Cell Biol 21:3445–3450 (2001).

Xu B, O'Donnell AH, Kim ST, Kastan MB: Phosphorylation of serine 1387 in Brca1 is specifically required for the Atm-mediated S-phase checkpoint after ionizing irradiation. Cancer Res 62:4588–4591 (2002).

Yang MY, Long SE: Folate sensitive common fragile sites in chromosomes of the domestic pig *(Sus scrofa)*. Res Vet Sci 55:231–235 (1993).

Yarden RI, Pardo-Reoyo S, Sgagias M, Cowan KH, Brody LC: BRCA1 regulates the G_2/M checkpoint by activating Chk1 kinase upon DNA damage. Nature Genet 30:285–289 (2002).

Yunis JJ, Soreng AL: Constitutive fragile sites and cancer. Science 226:1199–1204 (1984).

Zanesi N, Fidanza V, Fong LY, Mancini R, Druck T, Valtieri M, Rudiger T, McCue PA, Croce CM, Huebner K: The tumour spectrum in *Fhit*-deficient mice. Proc natl Acad Sci, USA 98:10250–10255 (2001).

Cytogenet Genome Res 100:101–110 (2003)
DOI: 10.1159/000072844

WWOX, the common chromosomal fragile site, FRA16D, cancer gene

J.H. Ludes-Meyers,[a] A.K. Bednarek,[b] N.C. Popescu,[c] M. Bedford[a] and C.M. Aldaz[a]

[a] Department of Carcinogenesis, The University of Texas M.D. Anderson Cancer Center, Science Park-Research Division, Smithville TX (USA); [b] IPh&B Department of Medical Biochemistry, Medical University of Lodz, Lodz (Poland); [c] Laboratory of Experimental Carcinogenesis, National Cancer Institute, Bethesda MD (USA)

Abstract. Gross chromosomal rearrangements and aneuploidy are among the most common somatic genomic abnormalities that occur during cancer initiation and progression, in particular in human solid tumor carcinogenesis. The loss of large chromosomal regions as consequence of gross rearrangements (e.g. deletions, monosomies, unbalanced translocations and mitotic recombination) have been traditionally associated with the existence of tumor suppressor genes within the areas affected by the loss of genetic material. The long arm of chromosome 16 was identified as being frequently associated with structural abnormalities in multiple neoplasias, that led us to focus attention on the detailed genetic dissection of this region resulting in the cloning of the putative tumor suppressor gene, WWOX (WW domain containing Oxidoreductase). Interestingly, the WWOX gene resides in the very same region as that of the common chromosomal fragile site 16D (FRA16D). The WWOX gene encodes a protein that contains two WW domains, involved in protein-protein interactions, and a short chain dehydrogenase (SDR) domain, possibly involved in sex-steroid metabolism. We have identified the WWOX WW domain ligand as the PPXY motif confirming the biochemical activity of this domain. WWOX normally resides in the Golgi and we will demonstrate that Golgi localization requires an intact SDR. Inactivation of the WWOX gene during tumorigenesis can occur by homozygous deletions and possibly mutation, however, aberrantly spliced forms of WWOX mRNA have been observed even when one allele is still intact. The aberrantly spliced mRNAs have deletions of the exons that encode the SDR and these WWOX protein isoforms display abnormal intracellular localization to the nucleus possibly functioning as dominant negative inhibitors of full length WWOX. Thus, generation of aberrant transcripts of WWOX may represent a novel mechanism to functionally inactivate WWOX without genomic alteration of the remaining allele. In this article we will review the cloning and identification of WWOX as the target of FRA16D. In addition, we will discuss the possible biochemical functions of WWOX and present evidence that ectopic WWOX expression inhibits tumor growth.

Chromosome 16q arm LOH in cancer led to the cloning of WWOX

For several years our laboratory has been interested in defining the progression of chromosomal and allelotypic abnormalities in human breast cancer (Aldaz et al., 1995; Brenner and Aldaz, 1995; Charpentier and Aldaz, 2002; Chen et al., 1996). In early studies with human samples, in order to better understand the timing for presentation of allelic losses in breast carcinogenesis, we performed a comparative allelotypic study of pre-invasive stages of breast carcinoma versus invasive ductal carcinomas. We observed that chromosome 7p, 16q and 17p17q were frequently affected by LOH and allelic imbalances at pre-invasive stages of breast cancer progression (Aldaz et al., 1995). Other studies at the time and later also reached very similar conclusions in particular implicating chromosome arms 16q and 17p17q as early anomalies (Lakhani et al., 1995; Radford et al., 1995; Fujii et al., 1996; O'Connell et al., 1998). Losses affecting 16q and ch17 were observed early in ductal hyperplastic stages of pre-malignant tumor progression (Lakhani et al., 1995). All these studies also helped to settle the notion that atypical hyper-

Received 6 January 2003; manuscript accepted 17 March 2003.

Request reprints from Dr. C. Marcelo Aldaz, MD, PhD
Department of Carcinogenesis
The University of Texas M.D. Anderson Cancer Center
Science Park-Research Division, P.O. Box 389
Smithville TX 78957 (USA); telephone: 512-237-2403
fax: 512-237-2475; e-mail: maldaz@odin.mdacc.tmc.edu

KARGER
Fax + 41 61 306 12 34
E-mail karger@karger.ch
www.karger.com

0301–0171/03/1004–0101$19.50/0

Accessible online at:
www.karger.com/cgr

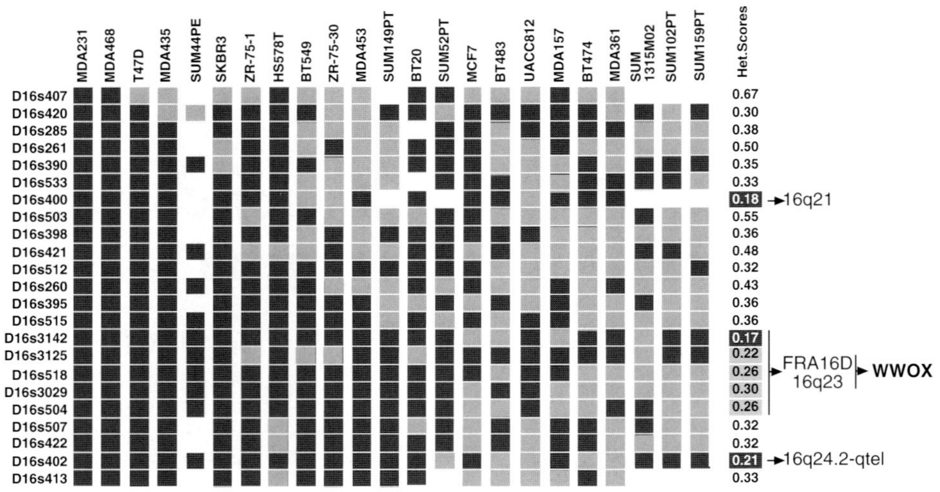

Fig. 1. High resolution allelotype of chromosome 16 in breast cancer cell lines using highly polymorphic microsatellite markers. Markers arranged in mapping and linkage order as previously described (Chen et al., 1996). Dark gray blocks indicate that a single allele was observed at the corresponding locus (i.e. hemi or homozygosity), light gray blocks indicate heterozygosity is preserved. Blank blocks were not determined. Numbers at right represent the heterozygosity scores calculated at each specific locus from analyzing this breast cancer panel. Several breast cancer lines showed hemizygosity affecting all or most of the chromosome 16q arm. Areas of high LOH mapped to 16q21, 16q23 and 16q24.2. The most commonly affected area (16q23) was determined to be FRA16D and to contain the WWOX gene. The SUM breast cancer cell lines were a generous gift from Dr. Stephen Ethier (Department of Radiation Oncology, The University of Michigan Health System).

plasias and ductal carcinoma in situ were true precursor lesions of invasive breast cancer, as previously proposed by Page and Dupont (1990). The information gained from these allelotypic studies was more recently further confirmed with other methodologies such as comparative genomic hybridization (CGH) (Lu et al., 1998; Buerger et al., 1999; Gong et al., 2001).

Interestingly, by means of conventional cytogenetics the long arm of chromosome 16 had already been suggested several years ago as a site for the occurrence of non-random primary structural abnormalities in breast cancer (Dutrillaux et al., 1990; Pandis et al., 1992). In particular, 16q was shown to participate systematically in nonrandom translocations with chromosome 1 and to show frequent deletions (Pandis et al., 1992). In order to refine the loci of interest we generated a high-resolution deletion map of ch16q in ductal carcinoma in situ, utilizing microdissected material from paraffin embedded samples (Chen et al., 1996). We identified three distinctive areas with high LOH. Two areas were previously identified by other groups 16q21 and 16q24.2→qter (Cleton-Jansen et al., 1994). However, the third was the most frequently affected area in our study spanning microsatellite markers D16S515–D16S504 with the highest incidence of LOH at D16S518 and we estimated this subregion lay in 16q23.3→q24.1 spanning approximately 2–3 Mb (Chen et al., 1996).

For the most proximal region, 16q22.1, it was proposed that the target of the losses could be the gene CADH1 (E-cadherin), a cellular adhesion molecule. However, only lobular breast carcinomas showed mutations affecting this gene and no evidence of mutations or homozygous losses were reported for the more common ductal carcinomas (Berx et al., 1995). Several other studies ruled out various candidate genes mapping more distally in ch16q as possible tumor suppressor genes (Crawford et al., 1999; Savino et al., 1999).

Further stimulating the search of the target gene(s) for LOH, was the fact that 16q and in particular the same sub-region we identified (16q23.3→q24.1) were also reported as affected by LOH in numerous other epithelial neoplasias such as: prostate cancer (Cher et al., 1995; Suzuki et al., 1996; Elo et al., 1999; Li et al., 1999), ovarian cancer (Iwabuchi et al., 1995; Kawakami et al., 1999), hepatocellular carcinomas (Nishida et al., 1992) and in Wilm's tumor (Skotnicka-Klonowicz et al., 2000). Interestingly, in many of these tumor types the LOH of this specific region of 16q associated with a worse prognosis and lower survival (Elo et al., 1999; Li et al., 1999; Skotnicka-Klonowicz et al., 2000).

In our laboratory, we then launched a classical positional cloning project attempting to identify the gene(s) targeted by the high frequency of LOH observed at 16q23.3→q24.1 (the most highly affected area in our view). In one of our first studies, we performed a 16q allelotypic analysis using highly polymorphic markers (>0.70 heterozygosity scores) on a panel of breast cancer lines attempting to identify regions of homozygous loss that could help us narrow down the region of interest (Bednarek et al., 2000). We basically confirmed the observations of our previous study identifying three distinctive regions in which the heterozygosity scores for various adjacent markers was drastically reduced from >0.70 to <0.17–0.21 indicating that very likely those regions represented areas of hemizygous loss rather than homozygosity. However, we failed to identify regions of homozygous loss. Nevertheless, we observed that many of the cell lines appeared to display hemizygosity affecting loss of the complete or most of the 16q arm (Fig. 1). We observed similar anomalies in the analysis of normal-tumor matched DCIS samples (Chen et al., 1996). This type of aberration is typical and usually the consequence of mitotic recombination events or loss of a complete chromosome arm as was

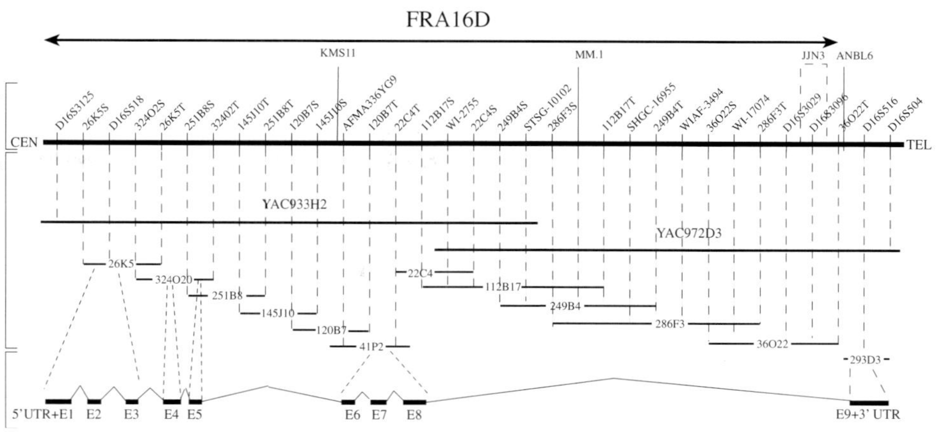

Fig. 2. Physical map of the 16q23.3→q24.1 chromosome region containing WWOX. (**a**) STS markers are ordered from centromere (left) to telomere (right). The relative distance between STS markers is arbitrary. The nomenclature used for the STSs derived from the ends of specific BACs indicates first the BAC clone address followed by S or T (i.e. Sp6 or T7 vector sequencing primers). KMS11, MM.1, JJN3 and ANBL6 represent the location of four multiple myeloma translocation breakpoints t(14;16). (**b**) The isolated BACs and YACs used for this physical map are represented by horizontal solid black lanes (Bednarek et al., 2000). The size of the clones are not to scale. (average size of a BAC 150 kb). (**c**) Black rectangular boxes represent WWOX exons as numbered. The approximate position of each exon in each corresponding BAC is illustrated.

suggested to occur frequently in 16q and possibly the main reason for LOH in this autosome (Gupta et al., 1997).

We identified only later that a breast cancer cell strain (1739) (obtained from the SW Medical Center, Dallas, gift of Dr. Lauren Gollahon) displayed a homozygous loss between STS markers 249B4S and D16S3029 (Bednarek et al., 2000). We used shotgun sequencing of this region and sequenced approximately 400,000 bp, including a continuous sequence of ~100,000 bp reported to GenBank (accession no. AF179633). We also followed a second approach, a solution hybridization cDNA capture method (Futreal et al., 1994). After sequencing and analyzing over 35 candidate cDNAs plus all candidate ESTs mapping to the region of interest we identified only one cDNA with evidence of exon-intron structure. We proceeded then to clone the complete cDNA and characterize the candidate gene (Bednarek et al., 2000) (Fig. 2).

The WWOX gene encodes a 414 amino acid protein that was predicted to contain two distinct protein domains identified using the PROSITE database. The WWOX amino terminal 88 amino acids show high sequence conservation to the WW domain family of proteins and the carboxy terminal 326 amino acids are homologous to the short-chain dehydrogenase/reductase (SDR) superfamily (Fig. 3). Therefore, we named the gene WWOX for WW domain containing Oxidoreductase.

```
          10        20        30        40
MAALRYAGLDDTDSEDELPPGWEERTTKDGWVYYANHTEEKTQWEHPKTG
          60        70        80        90
KRKRVAGDLPYGWEQETDENGQVFFVDHINKRTTYLDPRLAFTVDDNPTK
          110       120       130       140
PTTRQRYDGSTTAMEILQGRDFTGKVVVVTGANSGIGFETAKSFALHGAH
          160       170       180       190
VILACRNMARASEAVSRILEEWHKAKVEAMTLDLALLRSVQHFAEAFKAK
          210       220       230       240
NVPLHVLVCNAATFALPWSLTKDGLETTFQVNHLGHFYLVQLLQDVLCRS
          260       270       280       290
APARVIVVSSESHRFTDINDSLGKLDFSRLSPTKNDYWAMLAYNRSKLCN
                                          P
          310       320       330       340
ILFSNELHRRLSPRGVTSNAVHPGNMMYSNIHRSWWVYTLLFTLARPFTK
          360       370       380       390
SMQQGAATTVYCAAVPELEGLGGMYFNNCCRCMPSPEAQSEETARTLWAL
          410
SERLIQERLGSQSG
```

Fig. 3. WWOX amino acid sequence. WW domains are boxed and conserved tryptophans and prolines shown in bold. Note that in the second WW domain one tryptophan is replaced by a tyrosine (conservative change). The short chain dehydrogenase/reductase domain is underlined and the conserved motif **YXXXK** and **S**, characteristic of a substrate binding site are highlighted by black boxes. The motif **GXXXGXG**, typical of a coenzyme binding site (NADH or NADPH) is shown in bold italics. Leucine 291 is highlighted and the mutation to proline observed in an esophageal tumor is shown below the wildtype amino acid sequence.

WWOX, target of the second most common chromosomal fragile site, FRA16D

Mammalian chromosomes display regions of fragility where some are spontaneous and directly evident and others are disclosed upon treatment with specific chemicals or under specific culture conditions. These non-random regions prone to breakage are known as chromosomal fragile sites. They are specific loci susceptible to the spontaneous or induced occurrence of gaps, breaks, rearrangement and usually they are highly recombinogenic. There are two known classes of chromosomal fragile sites "rare" and "common". The "rare" or heritable fragile sites are transmitted in a Mendelian codominant fashion, e.g. the FRAXA, and FRAXE associated with the mental retardation fragile X syndrome or FRA11B associated with Jacobsen syn-

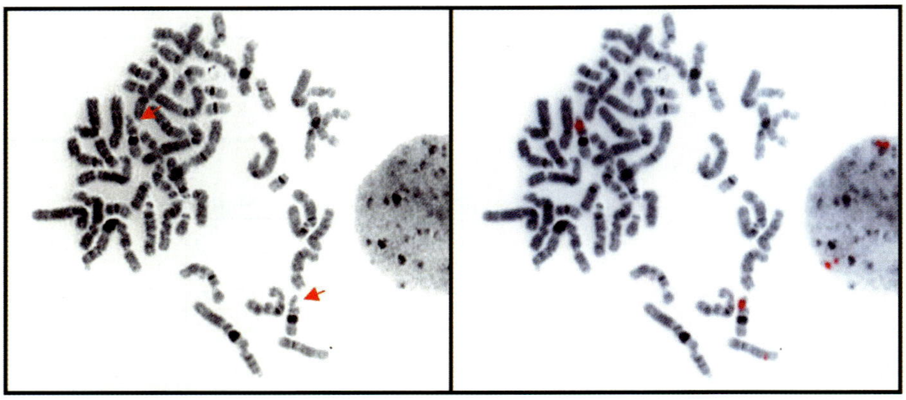

Fig. 4. WWOX is the target FRA16D. Left panel: Representative normal human G-Banded metaphase plate of cells treated with aphidicolin in order to expose constitutive fragile sites. Arrows point to both chromosomes 16 with chromatid breaks at the 16q23 region (FRA16D). Right panel: same metaphase after chromosome in situ hybridization using as probe YAC933H2 (see Fig. 2). Red dots represent the region of hybridization right on top of the FRA16D region.

drome. Rare fragile sites occur in less than 5% of the population and the mechanistic basis of their occurrence is usually associated with expansions of trinucleotide and minisatellite repeats. It was shown that they are not an in vitro artifact since they were clearly demonstrated to also occur in vivo (reviewed in Smith et al., 1998; Sutherland et al., 1998; Richards 2001). The "common" or constitutive chromosomal fragile sites are found in all individuals and are basically a constant chromosome feature. Most of these common sites become evident by exposure of cells to inhibitors of DNA replication such as aphidicolin (Glover et al., 1984). This compound inhibits DNA polymerases alpha and delta. A minority of common fragile sites can also be induced by exposure to 5-azacytidine or BrdU. The expression of common fragile sites can be further enhanced by exposure to caffeine. Approximately 80 common or constitutive chromosomal fragile sites have been described. However, by far the most frequent common fragile sites are predominantly FRA3B at 3p14.2 and FRA16D (16q23.3→q24.1) and with much less frequency FRA6E (6q26), FRA7H (7q32) and FRAXB (Xp22) (Smith et al., 1998).

The WWOX cloned gene spanned a huge genomic region of 1,113,822 bp in size (according to the human genome sequence data base) in the region 16q23.3→q24.1. It is composed of nine exons (Fig. 2). Interestingly even though the coding region itself was not large, introns 5 and 8 were very large. Intron 8 alone spans 779,639 bp. In addition, we detected from our sequencing and physical map of the region that both introns 5 and 8 harbored the sequences of previously reported translocation breakpoints in multiple myeloma. Within intron 8 we identified the translocation breakpoints MM.1, JJN3 and ANBL6 according to reported GenBank sequences (Fig. 2). These three multiple myeloma translocation breakpoints (and a fourth one KMS11 in intron 5) participated in translocations t(14;16)(q32;q23) (Chesi et al., 1998). Intron 8 also displayed repetitive sequences of various family types and we detected the presence of the sequence of a pseudogene for ribosomal protein S3 (Bednarek et al., 2000). These various unusual genomic characteristics plus the true physical mapping generated, led us to speculate that perhaps we had cloned the gene spanning the FRA16D. In fact, studies on the same region, based in part on the identification of STS markers of LOH, previously reported by us (Chen et al., 1996), were almost simultaneously reported confirming that this region was indeed that of FRA16D (Mangelsdorf et al., 2000; Paige et al., 2000). In those reports, however, they could not identify any particular gene but they observed that various tumor lines (ovary, lung, colon and gastric carcinoma) displayed homozygous deletions and instability in that specific area. Soon after our original report (Bednarek et al., 2000), Reid et al. (2000) cloned the same WWOX gene and named it FOR (fragile site FRA16D oxidoreductase). We also independently confirmed that WWOX mapped to FRA16D by hybridizing the YAC clone containing most of the coding region of WWOX to metaphase plates of cells treated with aphidicolin (Fig. 4). A more recent detailed physical map of the region by Krummel et al. (2000) also agreed with the conclusion that this region is indeed that of the FRA16D.

WWOX as a cancer-associated gene

A potential role for chromosomal fragile sites in cancer was suggested several years ago by Yunis and Soreng (1984). Since then the issue has been fairly controversial. The chromosomal regions spanned by common fragile sites have demonstrated associations with cancer. FRA3B occurs within a region deleted in multiple cancers, including balanced translocations affecting this locus in familial renal cancer t(3:8) (p14.2;q24). The gene spanning FRA3B was cloned and named FHIT (Fragile Histidine Triad protein) (Ohta et al., 1996). FRA6E occurs within a region frequently affected in ovarian and gastric cancer and FRA7H shows frequent loss in prostate, breast and pancreatic cancers (Smith et al., 1998).

It is hypothesized that the fragile sites predispose the chromosomes to breakage and since they contain highly recombinogenic sequences generate rearrangement and this constitutes the basis of the common alterations found in these various regions in cancer (Smith et al., 1998). A corollary of this hypothesis would be that such common fragility would affect the function of genes encoded in those chromosome regions and as a consequence alteration or inactivation of such genes may be of importance in cancer development or progression.

However, the main question that remains to be answered is: Are abnormalities affecting common fragile sites and their resident genes "Cause or Consequence" of tumor progression? One could argue, as was criticized for the case of FHIT, that abnormalities affecting this gene could be likely the consequence of FHIT residing in a region commonly affected by chromosomal rearrangements, rather than being the inactivation of such gene related to the origin of the cancer (Le Beau et al., 1998). Genes like FHIT, and for that matter, all the other genes residing in common fragile site regions, could just simply be innocent bystanders.

It is interesting to note that the findings of genomic and transcriptional abnormalities affecting WWOX the FRA16D gene, as discussed in previous and following sections are very similar to findings observed with FHIT, the FRA3B gene. Both genes span huge genomic regions, both genes are affected by translocations and homozygous deletions in particular affecting intron regions. In most cases of hemizygous deletion point mutations were usually not found in the remaining allele. Both genes expressed multiple aberrantly spliced forms in tumors not found in normal samples.

However, compelling evidence has accumulated indicating that FHIT also behaves as a suppressor gene when ectopically expressed. Several cases of homozygous deletions were demonstrated in various cancer lines. *Fhit* KO mice are more susceptible to cancer than wild type mice. Fhit protein is reduced or lost in many cancer types. (reviewed in Huebner and Croce, 2001; Richards 2001).

We hypothesize contrary to the bystander notion, that it is precisely the location of WWOX within a fragile site region that makes it a prime candidate to be affected by genomic rearrangements during carcinogenesis and tumor progression. It is also possible that the role in facilitating tumor progression is even more important than its putative role in the initiation of neoplasia. We can also not ignore a potential role for WWOX haploinsufficiency in the carcinogenesis process.

It is also interesting to note that there is a remarkable sequence conservation at the human and mouse orthologous common fragile site regions both for human FRA3B/FHIT with mouse *Fra14A2/Fhit* (Shiraishi et al., 2001) and more recently reported also for human FRA16D/WWOX and mouse *Fra8ei/Wwox* (Krummel et al., 2002). It is really intriguing that the sequence conservation in both cases is not limited only to the coding regions but it is found throughout very long intron regions as well. There has to be some evolutionary explanation (perhaps regulatory sequences) for preserving homology so well in non-coding DNA sequences.

WWOX, a potential tumor suppressor gene?

We determined that the area spanned by WWOX is affected by hemizygous loss in the vast majority of breast cancer cell lines and primary tumors and detected one case of homozygous deletion (Bednarek et al., 2000). As also mentioned, others have reported homozygous deletions in various cancer lines in the region spanned by WWOX. However, as was the case with the deletion that we found in breast cancer most deletions fell in intron regions of the gene. Initially, we were cautious not to call WWOX a tumor suppressor gene because when we investigated the cell lines that were hemizygous for WWOX we failed

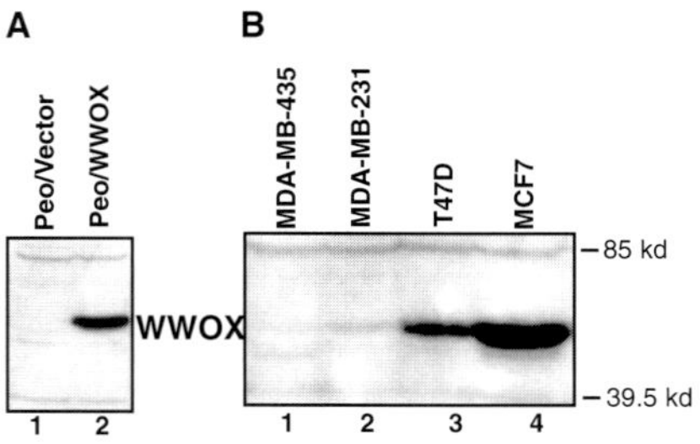

Fig. 5. WWOX protein expression in cancer cells. (**A**) Western blot analysis was used to determine the specificity of the WWOX antiserum. 50 ug of total cell protein from the cell lines, Peo/vector, (lane 1) and Peo/WWOX (lane 2). was separated by 10% SDS-PAGE and transferred to PVDF membrane. WWOX proteins were immunodetected using the anti-WWOX antiserum (1:1000) and detected using chemiluminescent detection (ECLplus, Amersham). Clearly, an ~46 kDa protein was detected in Peo/WWOX cell extracts that is undetectable in Peo/vector, having homozygous WWOX deletion, cell extracts. (**B**) We determined WWOX protein expression in the breast cancer cell lines MCF7, T47D, MDA-MB-231and MDA-MB-435 by Western blotting using the WWOX specific antiserum. 50 µg of total protein from each cell line was analyzed as described in (a). MCF7 and T47D were observed to have very prominent band at ~46 kDa while MDA-MB-231 and MDA-MB-435 had a barely visible to undetectable band. The band observed in MCF7 and T47D was shown to comigrate with the exogenously expressed WWOX band observed in Peo/WWOX.

to detect any mutation in the coding exons of the remaining allele. We detected a series of polymorphism in the WWOX coding region in those lines that were heterozygous e.g. MCF7 cells. We, however, observed variable levels of expression across breast cancer lines with some lines not expressing WWOX and others overexpressing WWOX as detected by Northern and RT-PCR analyses (Bednarek et al., 2000).

We have recently generated WWOX specific antiserum to be used to determine WWOX protein levels in normal and cancer cells. The antiserum was raised using a GST fusion to WWOX amino acid residues 12–94 containing both of the WW domains. Specific immunodetection of the WWOX protein was demonstrated by Western blotting of total cellular extracts from the ovarian cancer cell lines, Peo/vector and Peo/WWOX. (Fig. 5a). The ovarian cancer cell line, Peo, has a homozygous deletion of the WWOX gene and does not produce full-length WWOX (the original Peo cell line with homozygous WWOX deletion was kindly provided by Drs. Hanni Gabra and Adam Paige, Imperial Cancer Research Fund, Edinburgh, UK) (Paige et al., 2001). Peo cells were stably transfected with a WWOX expression vector creating the cell line Peo/WWOX to generate a positive control for WWOX protein expression. A protein of ~46 kDa, the predicted size of WWOX, was clearly evident in the extracts from Peo/WWOX positive control cell extracts that was absent in the Peo/vector negative control cell extracts.

We are currently using this antiserum to measure WWOX protein levels in cancer cell lines and in normal tissues. Figure 5b shows the results of a Western analysis of total protein extracts from breast cancer cell lines. The WWOX protein levels correlated quite well with our previous Northern and real-time PCR analyses of WWOX gene expression in these cell lines (Bednarek et al., 2000, 2001). Interestingly, WWOX expression appears to be inversely correlated with the aggres-

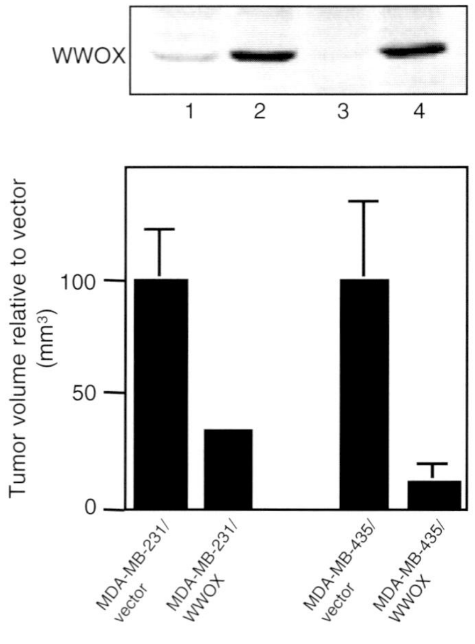

Fig. 6. WWOX tumor growth inhibition. Top panel: The breast cancer cell lines MDA MB 231 and MDA MB 435 were infected with the recombinant retroviruses expressing a WWOX cDNA or having no insert (Bednarek et al., 2001). Pools of stably expressing cell lines were selected for G418 resistance and analyzed for WWOX expression by western blotting. Cell lines infected with empty vector (MDA MB 231/vector and MDA MB 435/vector, lanes 1 and 3, respectively) had low to undetectable endogenous WWOX expression as observed previously (see Fig. 5). Cell lines infected with WWOX expressing vector (MDA MB 231/WWOX and MDA MB 435/WWOX, lanes 2 and 4, respectively) had high levels of WWOX expression. Bottom panel: Tumor growth of each cell line was observed by injecting 1 × 10^6 cells into the thoracic mammary fat pad of BALB/c athymic mice (Bednarek et al., 2001). The average tumor volume after 35 days of tumor growth is shown (± SEM). Tumor volume of the stably expressing WWOX cell lines is plotted relative to the tumor volume of the empty vector transfected cell lines. Average tumor volumes were; MDA MB 231/vector, 302 ± 71 mm^3; MDA MB 231/WWOX, 103 ± 11mm^3; MDA MB 435/vector, 664 ± 239 mm^3; MDA MB 435/WWOX, 81 ± 48 mm^3. In addition, it was observed that WWOX expressing cell lines had differences in tumor latency and kinetics of tumor growth (Bednarek et al., 2001).

siveness of these cell lines as determined by xenograph tumor growth assays. We detected very low or basically undetectable WWOX expression in the highly tumorigenic cell lines, MDA-MB-231 and MDA-MB-435 compared to the less aggressive cell lines MCF7 and T47D. It is also intriguing that high WWOX protein levels correlated with estrogen receptor (ER) status in these cell lines as well. MCF7 and T47D having high levels of WWOX are ER positive while MDA-MB-231 and MDA-MB-435 having very low levels of WWOX are ER negative. We are currently measuring WWOX protein levels in a panel of primary breast tumors to determine whether WWOX expression correlates with prognostic factors such as ER status.

It seems paradoxical that a putative tumor suppressor would be expressed at higher levels in some cancer cell lines when compared to normal cells as we observed (Bednarek et al., 2001). In some cases tumor suppressor genes have been observed to be highly expressed in tumors under certain circumstances (e.g. p16). Presently it is not clear what the functional significance of varying levels of WWOX expression may represent.

It was interesting that WWOX mRNA and protein expression was undetectable in the highly tumorigenic breast cancer cell line MDA-MB-435 even though it is hemizygous for the WWOX allele. We demonstrated that the lack of expression was not the result of a methylation phenomenon, since expression did not increase upon treatment with 5-aza-2′-deoxycytidine, an inhibitor of CpG methylation. Furthermore, we also did not find any signs of methylation in the CpG island around the promoter and first exon of WWOX when investigated by bisulfite DNA sequencing on the various cell lines (Bednarek et al., 2001). Therefore loss of WWOX expression in cancer cells is not due to transcriptional silencing by DNA methylation but is likely the result of alterations in transcription factor activity that regulates the WWOX promoter. We have recently cloned the WWOX promoter sequence to determine the mechanisms of WWOX gene transcriptional regulation.

Supporting a role of WWOX in affecting cancer growth other investigators detected homozygous losses in the same region, and more recently, but more importantly several of these deletions were confirmed to be affecting WWOX exons in ovarian and other cancer cell lines (Paige et al., 2001). Such genomic deletions were effectively knocking out the enzymatic portions of the WWOX coding region. More recently, it was observed that WWOX's LOH was complemented with a mutation in a single case of esophageal carcinoma (Kuroki et al., 2002). The mutation resulted in a Leucine to Proline substitution at amino acid 291. Although the consequence of this mutation on WWOX function is not known it is interesting to note that this amino acid is in close proximity to the predicted active site of the SDR and may affect catalytic activity (see Fig. 3).

The mouse homologue of WWOX was cloned and a further link to cancer was suggested since, it was reported that Wwox could play a role in apoptosis and may interact with p53 and it was proposed that Wwox may enhance the cytotoxicity of TNF (Chang et al., 2001). However, the nature of such putative interaction with p53 was not clearly demonstrated and requires further investigation.

It is clear that the WWOX gene is a frequent target for genomic instability affecting WWOX gene expression in multiple cancers making it a potential suppressor of tumor growth. To test this hypothesis we determined the effect of WWOX expression on in vivo tumor growth of the highly tumorigenic breast cancer cell lines, MDA-MB-231 and MDA-MB-435, (Bednarek et al., 2001) which express barely detectable and undetectable levels of WWOX, respectively (Fig. 5B). Cell lines stably expressing WWOX (Fig. 6, top panel) were injected into the mammary fat pad of nude mice and observed for 35 days. Figure 6 shows the dramatic effect on tumor growth of the WWOX expressing cell lines compared to the vector control cell lines. This is the strongest reported evidence demonstrating that WWOX behaves as a tumor suppressor. In addition, we demonstrated that ectopic WWOX expression strongly inhibited anchorage-independent growth in soft agar of breast cancer lines MDA-MB-435 and T47D (Bednarek et al., 2001). Mouse models having genetically modified WWOX alleles will more clearly define the role WWOX may play in tumorigenesis.

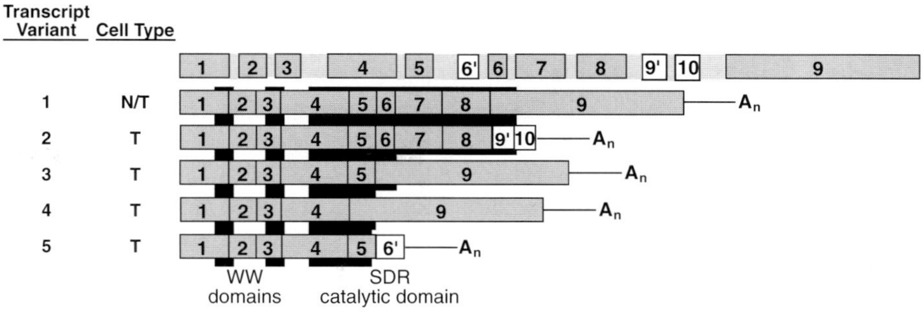

Fig. 7. Schematic representation of WWOX aberrantly spliced mRNAs. The WWOX gene is represented schematically at the top. Numbered boxes indicate exons encoded by the WWOX gene. Exons shaded grey are included in the full length WWOX transcript (transcript variant 1) that is expressed in normal and tumor cells. The alternative exons 6′, 9′ and 10 observed in aberrantly spliced transcripts are shown as unshaded boxes. Aberrantly spliced transcripts, transcript variants 2–5, are shown below the full length WWOX transcript, and have been observed only in cancer cell lines and cancer tissues. The transcript variants shown are numbered according to NCBI RefSeq. GenBank accession numbers AF211943, variant 1; AF211943, AF227526, variant 2; AF395123, variant 3; AF395124, variant 4; AF2119443, AF227528, variant 5. The black areas indicate the protein coding regions for the WW domains and the SDR catalytic domain. Transcript variants 4 and 5 have out-of-frame deletions while transcript variant 3 has an in-frame deletion. N-normal tissue; T-tumor tissue or cancer cell lines; A_n indicates polyadenylation. Transcript variants 1, 3 and 4 (Bednarek et al., 2001); transcript variants 2 and 5 (Ried et al., 2000).

WWOX aberrant transcripts

Several WWOX transcripts have been detected in normal and cancer cells (Fig. 7). The full length WWOX mRNA (transcript variant 1; Fig. 7 and Fig. 2) is the only transcript expressed in normal tissues and encodes a full length WWOX protein containing the WW domains and the SDR domain. In tumors the most commonly expressed transcripts are full length WWOX mRNA and four aberrantly spliced mRNAs (transcript variants 2–4). The transcript variants 2–4 will be referred to as aberrantly spliced transcripts based on the observations that they are only expressed in cancer tissues and have not been observed in normal tissues. In our initial studies we detected the common occurrence of aberrantly spliced WWOX transcripts with deletions of exons 6–8 and 5–8 in various carcinoma cell lines, multiple myeloma cell lines and primary breast tumors (transcript variants 3 and 4 in Fig. 7). Subsequent studies have reported the existence of additional aberrant transcripts (transcript variants 2 and 5) (Driouch et al., 2002; Ried et al., 2000). More importantly, these aberrant mRNA forms were not detected in normal tissues (Ried et al., 2000; Bednarek et al., 2001; Paige et al., 2001; Driouch et al., 2002). Furthermore, by using primers specific to the novel exon-exon junctions generated by the aberrant splicing, we determined that approximately 33% of primary breast carcinomas expressed the aberrant delta6–8 WWOX mRNA form (transcript variant 3, Bednarek et al., 2001). Other groups have found independently the common expression of aberrant WWOX spliced forms in cancer samples, in particular in breast, ovarian and esophageal cancers (Croce et al., 1999; Paige et al., 2001; Driouch et al., 2002; Kuroki et al., 2002). It is of particular interest that in breast cancer we and others have observed aberrantly spliced transcripts produced in cells that do not have genomic deletions of WWOX exons. Thus WWOX aberrant forms can be generated by alternative splicing of WWOX mRNA as well as by exon deletions of the WWOX allele in cancer.

The possibility exists that WWOX aberrantly spliced transcripts are nonfunctional mRNAs. Splicing abnormalities occur frequently in cancer, perhaps due to increased cellular proliferation or other unknown abnormalities in the splicing machinery. So it is not uncommon to find spliced variants or abnormally spliced genes in tumors cells, e.g. estrogen receptor. Furthermore, one could hypothesize that the putative splicing abnormalities would more often affect genes that have very long transcripts such as WWOX. It remains to be determined whether the aberrantly spliced WWOX transcripts are utilized to produce their encoded protein isoforms. We are currently analyzing WWOX protein expression in several tumor types, and since the anti-WWOX antibody recognizes epitopes within the WW domains it will also detect proteins encoded by the aberrantly spliced mRNAs if expressed.

An additional WWOX mRNA (transcript 2) has been described that has the first eight exons spliced to two alternate exons (9′ and 10, Ried et al., 2000). This transcript codes for a protein similar to full length WWOX encoded by transcript variant 1 containing the WW domains and an intact SDR domain but with a different carboxy terminal amino acid sequence. We have positively identified transcript variant 1 as the mRNA encoding full length WWOX protein in MCF-7 breast cancer cells by virtue of the difference in the carboxy terminal amino acid sequence using MALDI-TOF mass spectroscopy peptide sequencing of immunoprecipitated WWOX protein (Ludes-Meyers et al, unpublished observation).

In summary, the data shows that the WWOX allele is a target for disruption in multiple tumor types and WWOX expression can be altered by various mechanisms including homozygous deletion of exons, transcriptional dysregulation and aberrant splicing.

Biochemical features of the WWOX protein

WWOX is a 414 amino acid protein. We named this gene WWOX because it contains two WW-domains at the amino terminus coupled to a region with high homology to the short-chain dehydrogenase/reductase family of enzymes (SDR) (Fig. 3). WW motifs are known to be involved in protein-protein interactions. Protein-protein interactions are essential in transmission of signal-transduction information. There are a small number of modular protein-protein recognition domains. Proline-rich motifs are often targeted for protein-protein interactions, probably due to the unique structural properties of these regions. Domains that recognize proline-rich motifs include SH3, EVH1 and WW domains. The highly compact WW domains (35–45aa in length) are characterized by the presence of a pair of conserved tryptophans (i.e.W). These two "signature" W residues are spaced approximately 20–22 aa apart and play a fundamental role in the structure and function of the domain. WW domains recognize proline-containing sequences, and they have a very diverse sequence preference. Based on such binding preference there are basically four groups of WW domains. Two are major and more common, Groups I and II and two less common, Groups III and IV. Group I binds the minimal core consensus Pro–Pro–X–Tyr (PPXY). Examples of WW domain containing proteins from Group I include YAP65, dystrophin and NEDD4. The proteins containing the specific proline-rich motifs constitute specific ligands, for instance the ligand of the WW domain of dystrophin is beta-distroglycan which contains the (PPXY) motif. Group II WW domain containing proteins include the formin binding proteins and FE65 that typically bind the Pro–Pro–Leu–Pro (PPLP) motif. Group III WW domains select polyproline motifs flanked by Arg or Lys and Group IV WW domains have preference for phospho–Ser–Pro or phospho–Thr–Pro containing ligands. An example of this last group of WW domain containing proteins is Pin-1. This protein can regulate early mitotic events by interaction with various mitosis-specific phosphoproteins such as CdC25C phosphatase (reviewed in Sudol and Hunter, 2000). We are in the process of determining the in vitro and in vivo biochemical activity of the WWOX WW domains. In preliminary studies we have observed that the first WW domain specifically interacts with the PPXY ligand defining this WWOX domain as a Group I WW domain (Ludes-Meyers et al., manuscript in preparation). These studies confirm that the WWOX WW domains have specific ligand binding activity and studies are ongoing to identify in vivo protein partners involved in WWOX tumor growth inhibition.

As reported, the central portion of WWOX contains very high homology to the SDR family of enzymes. Based on the high expression of WWOX in hormonally regulated tissues (human testis, prostate and ovary) and its amino acid sequence homology to specific oxidoreductases, we postulated that WWOX may be an enzyme involved in sex-steroid metabolism (Bednarek et al., 2000). According to Kallberg et al., based on structural analyses of human SDRs and comparison across species, they detected the occurrence of 63 different SDR enzymes in humans, reduced to truly 58 after elimination of possible isozymes (Kallberg et al., 2002). Of these, 46 enzymes are of the classical type and 17 of the extended type (these are related to mostly sugar metabolism). WWOX structure is the archetypal representative of one of four separate clusters of classical SDRs (Kallberg et al., 2002). The other three clusters are represented by human Hep27, FVT1 and 17beta-HSD3. To date only the molecular function of 17beta-HSD3 has been defined (involved in tissue-specific testosterone synthesis). Interestingly, the other three clusters represented by WWOX, Hep27 and FVT1 all demonstrated some link to cancer. Hep27 was cloned from a hepatocellular carcinoma and is a nuclear protein related to growth control (Gabrielli et al., 1995). FVT1 was identified as a target for translocation in non-Hodgkin lymphomas showing overexpression in certain T cell malignancies and indicating some role in tumorigenic processes (Rimokh et al., 1993). As mentioned, we hypothesized that WWOX is related to sex-steroid metabolism and according to Kallberg and coworkers this is further supported not only by its tissue distribution and structural features but also by two other human SDRs found in WWOX's cluster, CGI-82 and PAN2. CGI-82 is abundantly expressed in prostate (Lin et al., 2001) and PAN2 displays sequence similarities to other hydroxysteroid dehydrogenases (Kallberg et al., 2002).

To date the only known link to a biological function of WWOX is its subcellular localization to the Golgi apparatus. This was demonstrated by transient expression of a GFP-fusion to full length WWOX followed by visualization of green fluorescence using confocal microscopy. Cells expressing the GFP-WWOX fusion protein clearly had perinuclear green fluorescence that colocalized only with a Golgi specific marker but not with other organelle specific markers (e.g. Mitotracker). Interestingly, GFP-fusions with proteins encoded by two aberrantly spliced transcripts, delta 6–8 and delta 5–8, showed exclusively nuclear fluorescence (Bednarek et al., 2001). A common feature of the predicted proteins encoded by WWOX delta forms is deletion of the predicted enzymatic domain of the full-length protein while retaining the WW domains. Interestingly, a nuclear localization signal has been identified in the mouse Wwox protein to reside between the two WW domains (Chang et al., 2001). Hypothetically, these proteins will be inactive enzymatically and could act as dominant-negative competitors of the normal WWOX protein.

The WWOX SDR is required for localization to the Golgi

The observation that full length WWOX resides in the Golgi apparatus, while proteins encoded by aberrantly spliced transcripts localize to the nucleus, suggests that amino acids within the SDR domain are involved in directing WWOX to the Golgi. Chang et al. (2001) reported a 65 amino acid residue sequence as a mitochondrial targeting signal located between Wwox amino acids 209–273. In previous studies we observed colocalization of GFP-WWOX with a Golgi specific marker (Golgi 58 K protein) but did not observe colocalization of GFP-WWOX with a mitochondria-specific dye (MitoTracker Red CMXRos). It is not clear why there is a discrepancy between these two observations. To more precisely define the WWOX Golgi targeting sequences we have used site directed mutagene-

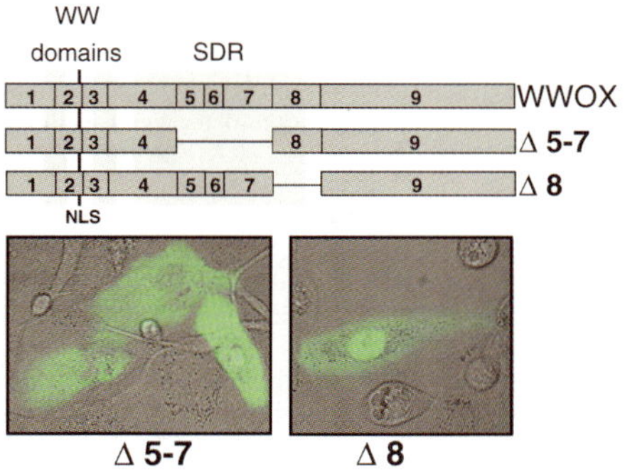

Fig. 8. The WWOX catalytic domain is sufficient for Golgi localization. To determine the domains required for WWOX Golgi localization we generated deletion mutants of WWOX in the context of the WWOX-GFP fusion protein. These mutants were transiently expressed in the immortal mammary epithelial cell line, MCF10F, and visualized using confocal microscopy. The WWOX protein and the deletion mutants are shown schematically (top panel). Two mutants were analyzed. Δ5–7 has an in-frame fusion between the coding sequence of exon 4 and exon 8 deleting the coding sequences of exons 5,6 and 7. Δ8 has an in-frame fusion between the coding sequence of exon 7 and exon 9 deleting exon 8. The catalytic active site is contained within the exon 8 coding sequence (see Fig. 7a). The confocal images show cross sections through the cytoplasmic and nuclear regions of the cells. NLS, putative nuclear localization signal.

sis to make deletions of the SDR domain while keeping the WW domains intact (Fig. 8). Mutant delta5–7 has the protein coding sequences of exons 5, 6 and 7 deleted resulting in an in-frame fusion of exons 1–4 to exons 8 and 9. Mutant delta8 has exon 8 deleted resulting in an in-frame fusion of exons 1–7 to exon 9. Cells expressing either of the mutants show similar diffuse cytoplasmic and nuclear fluorescence with striking exclusion from the nucleoli (Fig. 8). Clearly proper subcellular localization is affected by disruption of the SDR domain. Analysis of single amino acid changes within the full length WWOX protein will be required to clearly define the amino acids required for proper subcellular localization. We have also generated a GFP fusion with the SDR domain alone by deleting the WW domains. Surprisingly this mutant was observed to have perinuclear fluorescence similar to full length WWOX (J.H. Ludes-Meyers, data not shown). Although these studies do not precisely define the amino acid residues required for WWOX subcellular localization we conclude that the SDR domain is necessary and sufficient for Golgi localization.

Since the WW domains are dispensable for Golgi localization it is reasonable to postulate that they are involved in tumor growth inhibition through specific protein-protein interactions in the Golgi. Recently, a Drosophila gene *salvador* was cloned that encodes a protein containing two WW domains and is involved in cell cycle regulation and apoptosis. Most importantly, the human ortholog of salvador (SAV1) was observed to be mutated in three cancer cell lines (Tapon et al., 2002). This provides another example, together with WWOX, of a WW domain containing protein that has been linked to cancer.

Conclusions

In summary, we have cloned a candidate cancer gene mapping to region 16q23.3→q24.1. This gene is the target of FRA16D and codes for the protein WWOX. WWOX contains two WW domains and a short-chain dehydrogenase/reductase enzyme domain that is probably involved in sex-steroid hormone metabolism.

We hypothesize that the normal cellular function of WWOX may be affected by various different mechanisms at the genomic and transcriptional level. The single allele losses observed in various tumor types and in particular breast cancer, by itself may be of importance and perhaps sufficient in leading to phenotypic abnormalities, as was observed in cases of tumor suppression inactivation due to haploinsufficiency. Our data indicate that WWOX behaves as a potent suppressor of tumor growth and suggest that abnormalities affecting this gene at the genomic and transcriptional level could be relevant in carcinogenesis. The development of mouse strains with targeted inactivation of the *Wwox* allele will help define the role that WWOX may play in tumorigenesis. In addition, WWOX specific reagents are now available to determine the clinical relevance of WWOX abnormalities at the genomic and transcription levels.

References

Aldaz CM, Chen T, Sahin A, Cunningham J, Bondy M: Comparative allelotype of in situ and invasive human breast cancer: high frequency of microsatellite instability in lobular breast carcinomas. Cancer Res 55:3976–3981 (1995).

Bednarek AK, Keck-Waggoner CL, Daniel RL, Laflin KJ, Bergsagel PL, Kiguchi K, Brenner AJ, Aldaz CM: WWOX, the FRA16D gene, behaves as a suppressor of tumor growth. Cancer Res 61:8068–8073 (2001).

Bednarek AK, Laflin KJ, Daniel RL, Liao Q, Hawkins KA, Aldaz CM: WWOX, a novel WW domain-containing protein mapping to human chromosome 16q23.3→q24.1, a region frequently affected in breast cancer. Cancer Res 60:2140–2145 (2000).

Berx G, Cleton-Jansen A, Nollet F, de Leeuw W, van de Vijver M, Cornelisse C, van Roy F E-cadherin is a tumour/invasion suppressor gene mutated in human lobular breast cancers. EMBO J 14:6107–6115 (1995).

Brenner AJ, Aldaz CM: Chromosome 9p allelic loss and p16/CDKN2 in breast cancer and evidence of p16 inactivation in immortal breast epithelial cells. Cancer Res 55:2892–2895 (1995).

Buerger H, Otterbach F, Simon R, Poremba C, Diallo R, Decker T, Riethdorf L, Brinkschmidt C, Dockhorn-Dworniczak B, Boecker W: Comparative genomic hybridization of ductal carcinoma in situ of the breast-evidence of multiple genetic pathways. J Pathol 187:396–402 (1999).

Chang NS, Pratt N, Heath J, Schultz L, Sleve D, Carey GB, Zevotek N: Hyaluronidase induction of a WW domain-containing oxidoreductase that enhances tumor necrosis factor cytotoxicity. J biol Chem 276:3361–3370 (2001).

Charpentier A, Aldaz CM: The molecular basis of breast carcinogenesis, in Coleman WB, Tsongalis GJ (eds): Molecular Basis of Human Cancer: Genomic Instability and Molecular Mutation in Neoplastic Transformation, pp 347–363 (Humana Press, Totowa 2002).

Chen T, Sahin A, Aldaz C: Deletion map of chromosome 16q in ductal carcinoma in situ of the breast: refining a putative tumor suppressor gene region. Cancer Res 56:5605–5609 (1996).

Cher ML, Ito T, Weidner N, Carroll PR, Jensen RH: Mapping of regions of physical deletion on chromosome 16q in prostate cancer cells by fluorescence in situ hybridization (FISH). J Urol 153:249–254 (1995).

Chesi M, Bergsagel PL, Shonukan OO, Martelli ML, Brents LA, Chen T, Schrock E, Ried T, Kuehl WM: Frequent dysregulation of the *c-maf* proto-oncogene at 16q23 by translocation to an Ig locus in multiple myeloma. Blood 91:4457–4463 (1998).

Cleton-Jansen A, Moerland E, Kuipers-Dijkshoorn N, Callen D, Sutherland G, Hansen B, Devilee P, Cornelisse C: At least two different regions are involved in allelic imbalance on chromosome arm 16q in breast cancer. Genes Chrom Cancer 9:101–107 (1994).

Crawford J, Ianzano L, Savino M, Whitmore S, Cleton-Jansen AM, Settasatian C, d'apolito M, Seshadri R, Pronk JC, Auerbach AD, Verlander PC, Mathew CG, Tipping AJ, Doggett NA, Zelante L, Callen DF, Savoia A: The PISSLRE gene: structure, exon skipping, and exclusion as tumor suppressor in breast cancer. Genomics 56:90–97 (1999).

Croce CM, Sozzi G, Huebner K: Role of FHIT in human cancer. J Clin Oncol 17:1618–1624 (1999).

Driouch K, Prydz H, Monese R, Johansen H, Lidereau R, Frengen E: Alternative transcripts of the candidate tumor suppressor gene, WWOX, are expressed at high levels in human breast tumors. Oncogene 21:1832–1840 (2002).

Dutrillaux B, Gerbault-Seureau M, Zafrani B: Characterization of chromosomal anomalies in human breast cancer. A comparison of 30 paradiploid cases with few chromosome changes. Cancer Genet Cytogenet 49:203–217 (1990).

Elo JP, Harkonen P, Kyllonen AP, Lukkarinen O, Vihko P: Three independently deleted regions at chromosome arm 16q in human prostate cancer: allelic loss at 16q24.1 → q24.2 is associated with aggressive behaviour of the disease, recurrent growth, poor differentiation of the tumour and poor prognosis for the patient. Br J Cancer 79:156–160 (1999).

Fujii H, Szumel R, Marsh C, Zhou W, Gabrielson E: Genetic progression, histological grade, and allelic loss in ductal carcinoma in situ of the breast. Cancer Res 56:5260–5265 (1996).

Futreal PA, Cochran C, Rosenthal J, Miki Y, Swenson J, Hobbs M, Bennett LM, Haugen-Strano A, Marks J, Barrett JC, et al: Isolation of a diverged homeobox gene, MOX1, from the BRCA1 region on 17q21 by solution hybrid capture. Hum molec Genet 3:1359–1364 (1994).

Gabrielli F, Donadel G, Bensi G, Heguy A, Melli M: A nuclear protein, synthesized in growth-arrested human hepatoblastoma cells, is a novel member of the short-chain alcohol dehydrogenase family. Eur J Biochem 232:473–477 (1995).

Glover TW, Berger C, Coyle J, Echo B: DNA polymerase alpha inhibition by aphidicolin induces gap and breaks at common fragile sites in human chromosomes. Hum Genet 67:136–142 (1984).

Gong G, DeVries S, Chew KL, Cha I, Ljung BM, Waldman FM: Genetic changes in paired atypical and usual ductal hyperplasia of the breast by comparative genomic hybridization. Clin Cancer Res 7:2410–2414 (2001).

Gupta PK, Sahota A, Boyadjiev SA, Bye S, Shao C, O'Neill JP, Hunter TC, Albertini RJ, Stambrook PJ, Tischfield JA: High frequency in vivo loss of heterozygosity is primarily a consequence of mitotic recombination. Cancer Res 57:1188–1193 (1997).

Huebner K, Croce CM: FRA3B and other common fragile sites: the weakest links. Nature Rev Cancer 1:214–221 (2001).

Iwabuchi H, Sakamoto M, Sakunaga H, Ma YY, Carcangiu ML, Pinkel D, Yang-Feng TL, Gray JW: Genetic analysis of benign, low-grade, and high-grade ovarian tumors. Cancer Res 55:6172–6180 (1995).

Kallberg Y, Oppermann U, Jornvall H, Persson B: Short-chain dehydrogenase/reductase (SDR) relationships: A large family with eight clusters common to human, animal, and plant genomes. Protein Sci 11:636–641 (2002).

Kawakami M, Staub J, Cliby W, Hartmann L, Smith DI, Shridhar V: Involvement of H-cadherin (CDH13) on 16q in the region of frequent deletion in ovarian cancer. Int J Oncol 15:715–720 (1999).

Krummel KA, Denison SR, Calhoun E, Phillips LA, Smith DI: The common fragile site FRA16D and its associated gene WWOX are highly conserved in the mouse at Fra8E1. Genes Chrom Cancer 34:154–167 (2002).

Krummel KA, Roberts LR, Kawakami M, Glover TW, Smith DI: The characterization of the common fragile site FRA16D and its involvement in multiple myeloma translocations. Genomics 69:37–46 (2000).

Kuroki T, Trapasso F, Shiraishi T, Alder H, Mimori K, Mori M, Croce CM: Genetic alterations of the tumor suppressor gene WWOX in esophageal squamous cell carcinoma. Cancer Res 62:2258–2260 (2002).

Lakhani SR, Collins N, Sloane JP, Stratton MR: Loss of heterozygosity in lobular carcinoma in situ of the breast. J clin Pathol: Mol Pathol 48:M74–M78 (1995).

Le Beau MM, Drabkin H, Glover TW, Gemmill R, Rassool FV, McKeithan TW, Smith DI: An FHIT tumor suppressor gene? Genes Chrom Cancer 21:281–289 (1998).

Li C, Berx G, Larsson C, Auer G, Aspenblad U, Pan Y, Sundelin B, Ekman P, Nordenskjold M, van Roy F, Bergerheim US: Distinct deleted regions on chromosome segment 16q23 → q24 associated with metastases in prostate cancer. Genes Chrom Cancer 24:175–182 (1999).

Lin B, White JT, Ferguson C, Wang S, Vessella R, Bumgarner R, True LD, Hood L, Nelson PS: Prostate short-chain dehydrogenase reductase 1 (PSDR1): a new member of the short-chain steroid dehydrogenase/reductase family highly expressed in normal and neoplastic prostate epithelium. Cancer Res 61:1611–1618 (2001).

Lu YJ, Osin P, Lakhani SR, Di Palma S, Gusterson BA, Shipley JM: Comparative genomic hybridization analysis of lobular carcinoma in situ and atypical lobular hyperplasia and potential roles for gains and losses of genetic material in breast neoplasia. Cancer Res 58:4721–4727 (1998).

Mangelsdorf M, Ried K, Woollatt E, Dayan S, Eyre H, Finnis M, Hobson L, Nancarrow J, Venter D, Baker E, Richards RI: Chromosomal fragile site FRA16D and DNA instability in cancer. Cancer Res 60:1683–1689 (2000).

Nishida N, Fukuda Y, Kokuryu H, Sadamoto T, Isowa G, Honda K, Yamaoka Y, Ikenaga M, Imura H, Ishizaki K: Accumulation of allelic loss on arms of chromosomes 13q, 16q and 17p in the advanced stages of human hepatocellular carcinoma. Int J Cancer 51:862–868 (1992).

O'Connell P, Pekkel V, Fuqua SA, Osborne CK, Clark GM, Allred DC: Analysis of loss of heterozygosity in 399 premalignant breast lesions at 15 genetic loci. J natl Cancer Inst 90:697–703 (1998).

Ohta M, Inoue H, Cotticelli MG, Kastury K, Baffa R, Palazzo J, Siprashvili Z, Mori M, McCue P, Druck T, et al: The FHIT gene, spanning the chromosome 3p14.2 fragile site and renal carcinoma-associated t(3;8) breakpoint, is abnormal in digestive tract cancers. Cell 84:587–597 (1996).

Page DL, Dupont WD: Anatomic markers of human premalignancy and risk of breast cancer. Cancer 66:1326–1335 (1990).

Paige AJ, Taylor KJ, Stewart A, Sgouros JG, Gabra H, Sellar GC, Smyth JF, Porteous DJ, Watson JE: A 700-kb physical map of a region of 16q23.2 homozygously deleted in multiple cancers and spanning the common fragile site FRA16D. Cancer Res 60:1690–1697 (2000).

Paige AJ, Taylor KJ, Taylor C, Hillier SG, Farrington S, Scott D, Porteous DJ, Smyth JF, Gabra H, Watson JE: WWOX: a candidate tumor suppressor gene involved in multiple tumor types. Proc natl Acad Sci, USA 98:11417–11422 (2001).

Pandis N, Heim S, Bardi G, Idvall I, Mandahl N, Mitelman F: Whole-arm t(1;16) and i(1q) as sole anomalies identify gain of 1q as a primary chromosomal abnormality in breast cancer. Genes Chrom Cancer 5:235–238 (1992).

Radford D, Fair K, Phillips N, Ritter J, Steinbrueck T, Holt M, Donis-Keller H: Allelotyping of ductal carcinoma in situ of the breast: deletion of loci on 8p, 13q, 16q, 17p and 17q. Cancer Res 55:3399–3405 (1995).

Richards RI: Fragile and unstable chromosomes in cancer: causes and consequences. Trends Genet 17:339–345 (2001).

Ried K, Finnis M, Hobson L, Mangelsdorf M, Dayan S, Nancarrow JK, Woollatt E, Kremmidiotis G, Gardner A, Venter D, Baker E, Richards RI: Common chromosomal fragile site FRA16D sequence: identification of the FOR gene spanning FRA16D and homozygous deletions and translocation breakpoints in cancer cells. Hum molec Genet 9:1651–1663 (2000).

Rimokh R, Gadoux M, Bertheas MF, Berger F, Garoscio M, Deleage G, Germain D, Magaud JP FVT-1, a novel human transcription unit affected by variant translocation t(2;18)(p11;q21) of follicular lymphoma. Blood 81:136–142 (1993).

Savino M, d'Apolito M, Centra M, van Beerendonk HM, Cleton-Jansen AM, Whitmore SA, Crawford J, Callen DF, Zelante L, Savoia A: Characterization of copine VII, a new member of the copine family, and its exclusion as a candidate in sporadic breast cancers with loss of heterozygosity at 16q24.3. Genomics 61:219–226 (1999).

Shiraishi T, Druck T, Mimori K, Flomenberg J, Berk L, Alder H, Miller W, Huebner K, Croce CM: Sequence conservation at human and mouse orthologous common fragile regions, FRA3B/FHIT and Fra14A2/Fhit. Proc natl Acad Sci, USA 98:5722–5727 (2001).

Skotnicka-Klonowicz G, Rieske P, Bartkowiak J, Szymik-Kantorowicz S, Daszkiewicz P, Debiec-Rychter M: 16q heterozygosity loss in Wilms' tumour in children and its clinical importance. Eur J Surg Oncol 26:61–66 (2000).

Smith DI, Huang H, Wang L: Common fragile sites and cancer (Review). Int J Oncol 12:187–196 (1998).

Sudol M, Hunter T: NeW wrinkles for an old domain. Cell 103:1001–1004 (2000).

Sutherland GR, Baker E, Richards RI: Fragile sites still breaking. Trends Genet 14:501–506 (1998).

Suzuki H, Komiya A, Emi M, Kuramochi H, Shiraishi T, Yatani R, Shimazaki J: Three distinct commonly deleted regions of chromosome arm 16q in human primary and metastatic prostate cancers. Genes Chrom Cancer 17:225–233 (1996).

Tapon N, Harvey KF, Bell DW, Wahrer DC, Schiripo TA, Haber DA, Hariharan IK: salvador Promotes both cell cycle exit and apoptosis in Drosophila and is mutated in human cancer cell lines. Cell 110:467–478 (2002).

Yunis JJ, Soreng AL: Constitutive fragile sites and cancer. Science 226:1199–1204 (1984).

Cytogenet Genome Res 100:111–123 (2003)
DOI: 10.1159/000072845

Understanding fragile X syndrome: insights from animal models

C.E. Bakker and B.A. Oostra

Department of Clinical Genetics, Erasmus MC, Rotterdam (The Netherlands)

Abstract. The fragile X mental retardation syndrome is caused by large methylated expansions of a CGG repeat in the FMR1 gene leading to the loss of expression of FMRP, an RNA-binding protein. FMRP is proposed to act as a regulator of mRNA transport or translation that plays a role in synaptic maturation and function. To study the physiological function of the FMR1 protein, mouse and Drosophila models have been developed. The loss-of-function mouse model shows slightly enlarged testes, a subtle behavioral phenotype, and discrete anomalies of dendrite spines similar to those observed in brains of patients. Studies in Drosophila indicate that FXMR plays an important role in synaptogenesis and axonal arborization, which may underlie the observed deficits in flight ability and circadian behavior of *fxr* mutant flies. The relevance of these studies to our understanding of fragile X syndrome is discussed.

The identification of a gene involved in human disease is just a first step on the long road to understanding the biological basis of the disease. To unravel the biological mechanisms involved in normal and abnormal functioning of a disease gene, a model system might be of great help.

Due to the growing number of genome projects, it has become clear that there is an amazing degree of conservation among genes that participate in similar pathways across a wide spectrum of species. Together with the technical ability to introduce disease-causing mutations in an organism, this high degree of conservation makes it possible to use simple organisms to study human disease (reviewed by van Heyningen, 1997). Recently, yeast was shown to be a simple model for cellular aging, which may provide clues to mammalian aging events (Frohlich and Madeo, 2001). Yeast is also used as a model organism to study determinants involved in DNA repeat instability in general (Shimizu et al., 1996; Wells, 1996; White et al., 1999; Jin and Warren, 2000).

Genes involved in development and differentiation are widely studied in the nematode worm *Caenorhabditis elegans* and the fruit fly *Drosophila melanogaster*. The recent development of new technologies, like RNA interference, makes it possible to use non-mammalian vertebrates like the frog *Xenopus laevis* and the zebra fish *Danio rerio*, to study genes involved in early development of the vertebrate.

Although lower vertebrate organisms and also worms and flies can serve as good models, there are many reasons to use a mammalian system when specific details and aspects need to be studied. The laboratory mouse is, for practical and technical reasons, the mammal of choice in functional studies.

In this review we focus on fragile X syndrome with specific emphasis on the two different model organisms for this syndrome, the fruit fly *Drosophila melanogaster* and the mouse.

The human fragile X syndrome

The fragile X syndrome

The FMR1 protein (FMRP) is an RNA binding protein, highly expressed in brain, whose absence or mutation leads to the fragile X syndrome. The fragile X syndrome is an X-linked dominant disorder and the most frequent cause of inherited

Research on animal models for fragile X syndrome in the authors laboratory is supported by the National Institutes of Health (NIH 5RO1 HD 38038), HFSP (Human Frontiers Science Program) and by a grant from IOP Genomics.

Received 9 October 2002; manuscript accepted 27 November 2002.

Request reprints from Prof. Dr B.A. Oostra, Department of Clinical Genetics
Erasmus MC, P.O. Box 1738, 3000 DR Rotterdam (The Netherlands)
telephone: 31-10-4087198; fax: 31-10-4089489
e-mail: Oostra@kgen.fgg.eur.nl

KARGER Fax + 41 61 306 12 34
E-mail karger@karger.ch
www.karger.com

Accessible online at:
www.karger.com/cgr

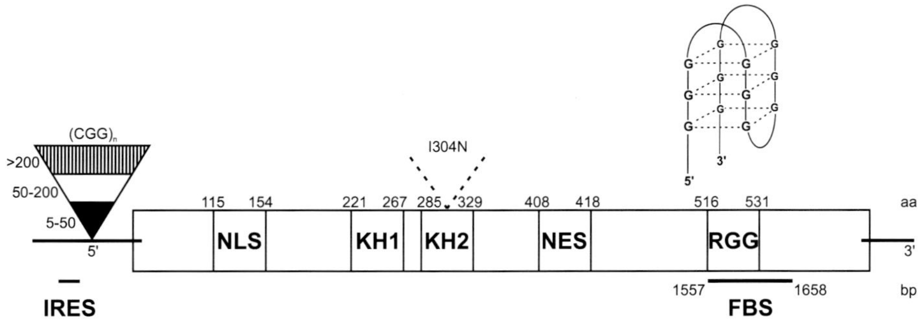

Fig. 1. Schematic representation of the FMR1 gene showing the known domains: IRES, internal ribosome entry site; NLS, nuclear location signal; KH, K-protein homology, domain; NES, nuclear export signal; RGG box, Arg-Gly-Gly triplet; FBS, FMRP binding site. In addition, the I304N missense mutation found in a single, severely affected patient is indicated. In the 5′-UTR the different repeat classes are shown. The G-quartet structure present in the mRNA is depicted above the RGG box for which it encodes and to which it binds. The numbering on top of the ORF is from the amino acids, and the numbering below the ORF is from the mRNA sequence.

mental retardation (1 in 4000 males and 1 in 6000 females) (Turner et al., 1996; De Vries et al., 1997). Mental retardation and developmental delay are the most significant clinical features of fragile X syndrome (Hagerman, 2002).

In males the mental retardation can vary from mild to severe (De Vries et al., 1993). Unlike most X-linked disorders, females can also be affected. 50–70% of females carrying a disease-causing mutation also show mental impairment, although generally less severe compared to males with a disease-causing mutation (Rousseau et al., 1991a, 1991b; Smits et al., 1994; Taylor et al., 1994; De Vries et al., 1996b).

Patients may show behavioral abnormalities such as hyperactivity, decreased attention span and a number of autistic features (Cohen et al., 1991; Fisch, 1993). In approximately 20% of fragile X males epileptic seizures are observed (Hagerman, 2002). Male patients show physical characteristics such as a long face and large everted ears and hyperextensible joints. Macroorchidism (enlargement of the testes) is seen in most adult patients and develops mostly during and after puberty (De Vries et al., 1998).

The fragile X mutation

The syndrome is mostly associated with an unstable expansion of a CGG repeat located in the 5′ UTR of the fragile X mental retardation gene (FMR1) followed by an abnormal methylation pattern that most of the time leads to a transcription silencing of the gene (Oberlé et al., 1991; Verkerk et al., 1991; Sutcliffe et al., 1992). In humans the alleles are divided into three groups: the normal, the premutation and the full mutation alleles. Normal alleles range between 5 and 50 repeat units. The most common alleles have 29 or 30 triplets (Fu et al., 1991). These alleles are stably transmitted to offspring and there is normal FMRP expression.

Premutation alleles range between 50 and 200 triplets (Fig. 1). These repeat sizes can change upon transmission. Both expansions and contractions occur, depending on the length of the premutation. An increasing length of the premutation increases the chance of expansions. Premutations do show instability upon transmission to next generations, and they do show up to normal levels of FMRP (Devys et al., 1993; Feng et al., 1995a, 1995b; Tassone et al., 2000). Therefore premutations do not result in the fragile X phenotype.

Studies in yeast and *E. coli* suggest that direction of replication, genetic background (including repair systems), transcription and growth conditions have an influence on repeat instability (Shimizu et al., 1996; Wells, 1996; White et al., 1999).

Upon female transmission the CGG repeat can expand above the threshold of 200 CGG triplets, called full mutations (Kremer et al., 1991; Oberlé et al., 1991). Full mutations coincide with the methylation of the CGG repeat and the surrounding CpG island, including the promoter region (Hansen et al., 1992; Hornstra et al., 1993). This methylation blocks the transcription of FMR1, and thus results in the absence of FMRP (Pieretti et al., 1991; Verheij et al., 1993). Due to the absence of FMRP, these full mutations are causing the fragile X phenotype. The observation that full mutation males with an unmethylated promoter are not affected provided evidence that the repeat expansion alone is not enough to cause the fragile X phenotype (McConkie-Rosell et al., 1993; Hagerman et al., 1994; Smeets et al., 1995; De Vries et al., 1996a; Wohrle et al., 1996, 1998; Taylor et al., 1999).

Rare atypical cases of fragile X syndrome, which are not associated with amplification of the trinucleotide repeat, but with deletions and single point mutations, have also been reported (Gedeon et al., 1992; Wöhrle et al., 1992; De Boulle et al., 1993; Meijer et al., 1994; Hirst et al., 1995; Mila et al., 2000).

The FMR1 protein

Protein studies showed that FMRP is expressed predominantly in the cytoplasm of cells (Devys et al., 1993; Verheij et al., 1993), but in in vitro experiments some isoforms are found in the nucleus (Eberhart et al., 1996; Fridell et al., 1996; Sittler et al., 1996; Willemsen et al., 1996). High levels of FMRP expression were found in neurons of the brain, (early) spermatogonia in the testis and in connective tissues, particularly in epidermal layers which are actively dividing (Devys et al., 1993).

FMRP contains RNA binding motifs including two KH domains and an RGG box and indeed FMRP is binding RNA homopolymers and mRNAs in vitro (Fig. 1) (Ashley et al., 1993a; Siomi et al., 1993, 1994; Brown et al., 1998; Adinolfi et al., 1999). It binds to 4% of human fetal brain mRNA, to myelin basic protein mRNA and a specific binding was demonstrated to its own mRNA (Ashley et al., 1993a; Brown et al., 1998; Schaeffer et al., 2001). Further, RNA candidates composing a very heterogeneous family have been recently characterized (Brown et al., 2001; Sung et al., 2000). FMRP interaction with its own mRNA occurs in a region, which may form a particular structure called the purine quartet (Darnell et al., 2001; Schaeffer et al., 2001). There are two close homologs of FMRP (FXR1P, and FXR2P) in human cells, and the three proteins are conserved in vertebrates (Kirkpatrick et al., 2001), whereas a single gene is found in Drosophila (Wan et al., 2000). The FMRP homologues that can form complexes with FMRP; the proteins have been supposed to have functions that overlap but are clearly distinct. For example, the pattern of expression in testis, as well as the subnuclear distribution is different (Tamanini et al., 1997, 1999). FMRP was found associated with ribosomes in the cell body and in dendrites (Eberhart et al., 1996; Khandjian et al., 1996; Tamanini et al., 1996; Corbin et al., 1997; Feng et al., 1997; Weiler et al., 1997). The presence of an NLS and an NES (Fig. 1), suggests that FMRP may function as a shuttle for mRNA export from nucleus to cytoplasm, the association with polysomes suggests a role of FMRP in modulating stability and/or translation of its target mRNAs in the cell body and also at the synapses. Human mRNAs with altered polysomal profile have been identified probing microarrays using mRNA isolated from polyribosomes of a human fragile X lymphoblastoid cell line (Brown et al., 2001). Two different groups have recently shown that FMRP functions as a repressor of translation in vitro (Laggerbauer et al., 2001; Li et al., 2001).

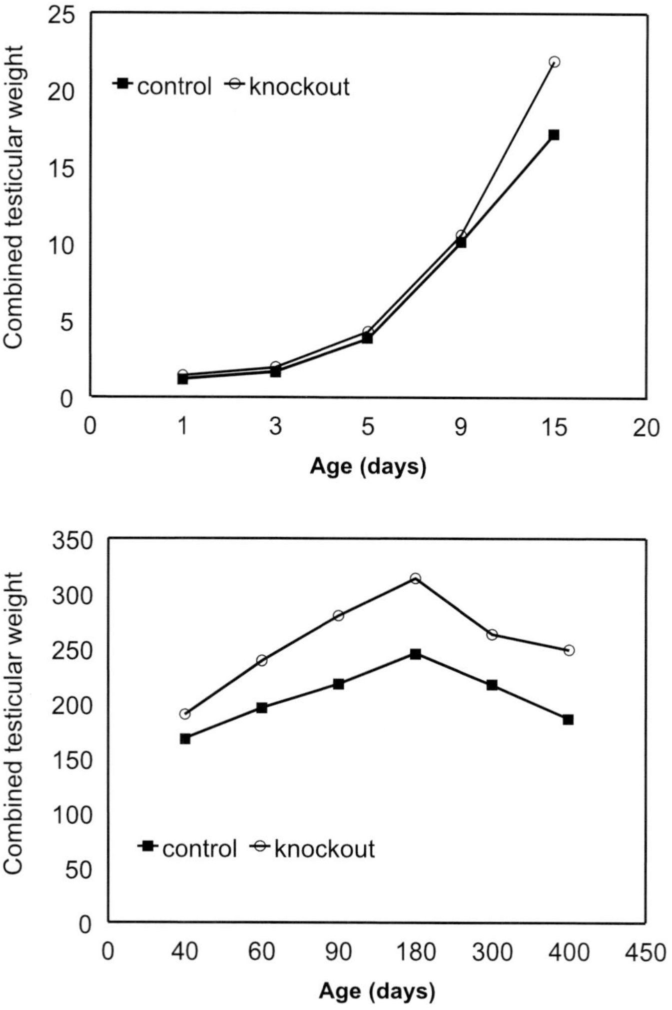

Fig. 2. Combined testicular weight of knockout (○) and control (■) mice at different ages.

Mouse models

The fragile X knockout mouse

The FMR1 gene is highly conserved among species (Verkerk et al., 1991), although the size of the human CGG repeat is longer than in other mammals (Deelen et al., 1994). Like in humans, the mouse *Fmr1* gene is also located on the X chromosome and dependent on the mouse strain. The mouse CGG repeat contains 9–11 CGG units. The murine homologue Fmrp shows 97% identity in amino acid sequence compared to the human FMRP (Ashley et al., 1993b). The expression level and the location of the murine Fmrp in different tissues is very similar to the level and location seen in humans (Abitbol et al., 1993; Bächner et al., 1993; Devys et al., 1993; Hinds et al., 1993; Bakker et al., 2000a; De Diego Otero et al., 2000). This makes the mouse a good organism to use as a model for the fragile X syndrome.

To mimic the fragile X syndrome in the mouse, several options are open. In humans, the mechanism of the disease is the growing CGG repeat, followed by methylation and inactivation of the FMR1 gene. In the mouse, it is not known whether the endogenous *Fmr1* CGG repeat can grow, and in this way result in inactivation of *Fmr1*. Also it is technically very difficult to manipulate long, pure CGG repeats for transgenic purposes. Since the fragile X phenotype in the human is caused by the absence of FMRP, the first attempt to make a mouse model was to disrupt the murine *Fmr1* gene. With this disruption the gene was inactivated leading to the absence of Fmrp in the mouse. The procedure and results of this attempt to mimic the human fragile X phenotype in the mouse are described in detail by Bakker et al. (1994).

Similar to male human patients, *Fmr1* knockout mice develop progressive macroorchidism with time (Fig. 2) (Bakker et al., 1994; Kooy et al., 1996; Slegtenhorst-Eegdeman et al., 1998). Cognitive function and behavioral profile of the knockout mice are consistent with an abnormal neurological function in line with findings in human fragile X patients (Bakker et al., 1994; Kooy et al., 1996; Paradee et al., 1999; Musumeci et al., 2000; Peier et al., 2000; Van Dam et al., 2000; Mineur et al., 2002; Nielsen et al., 2002). As the knockout mice lack Fmrp and show abnormalities comparable to human fragile X patients, it is a potentially valid model to provide insight into the physiological function of FMR1 and the pathogenesis of the fragile X syndrome.

Macroorchidism

Macroorchidism found in the *Fmr1* knockout mouse has been studied in more detail by Slegtenhorst-Eegdeman et al. (1998). In general the size of the testis is mainly determined by the number of Sertoli cells that support the proliferation and differentiation of the germ cells (Sharpe, 1993). In the knockout mice, it appeared that the proliferative activity of the Sertoli cells was significantly higher, resulting in an increase in spermatogenic cell number and testicular weight. Follicle-stimulating hormone (FSH) plays an important, but not essential, role in Sertoli cell proliferation (Kumar et al., 1997). Sertoli cells are most sensitive to the mitogenic activity of FSH at the end of the fetal period and shortly after birth (Orth, 1982). The circulating FSH level was measured in knockout mice, but not found to be elevated compared to that in wild type littermates. This observation is similar in human fragile X patients showing no evidence for an increased FSH level (Nielsen et al., 1982; Moore et al., 1991). Determination of the level of FSH receptor mRNA in the testis showed a slight increase, which, however, was not significant (Slegtenhorst-Eegdeman et al., 1998). These findings suggest that elements of the FSH signal transduction pathway are not involved in development of macroorchidism in the fragile X syndrome.

Absence of FMRP in developing germ cells as a primary cause of development of macroorchidism is not very likely, since in affected human male fetuses FMRP expression could be detected in the primordial germ cells (Malter et al., 1997). The primary cause of the increased testis size in *Fmr1* knockout mice may be found in Sertoli cells. Sertoli cells of the wild type neonatal mouse show a high expression of *Fmrp* at postnatal day three. This expression disappears at postnatal day 14, suggesting a function for *Fmrp* in Sertoli cells during early postnatal life (Bakker et al., 2000a). In knockout mice, and also in human fragile X patients, this high early postnatal expression is not present (Bakker et al., 2000a). This might indicate that the absence of Fmrp during this postnatal period leads to dysregulation of Sertoli cell maturation and proliferation, and consequently to development of macroorchidism in adult life.

Neuroanatomy

The neuroanatomy of the *Fmr1* knockout mouse has been studied in more detail using in vivo high-resolution magnetic resonance imaging (MRI) (Kooy et al., 1999). MRI studies in human fragile X males show a decrease in size in the posterior vermis (Reiss et al., 1991a), but *Fmr1* knockout mice do not show differences in this area compared to wild type littermates (Kooy et al., 1999). Along with the vermis hypoplasia, a 25–35% increase in size of the 4th ventricle has been reported in human patients (Reiss et al., 1988, 1991a, 1991b), but knockout mice did not show these differences (Kooy et al., 1999). The caudate nucleus, together with the lenticular and thalamic nucleus, forms the subcortical grey and allows the cerebral cortex to affect behavior. The caudate nucleus is significantly larger in volume in fragile X male patients (Reiss et al., 1995). In *Fmr1* knockout mice, it was not possible to identify each nucleus individually, but no evidence was found for an increase in subcortical grey volume (Kooy et al., 1999).

Using 3D-reconstruction, total brain volume of knockout mice was compared with the total brain volume of wild type mice. No significant differences were found (Kooy et al., 1999), in contrast to the human situation where fragile X brains appeared larger than control brains.

In conclusion, no evidence was found for differences in neuroanatomy between *Fmr1* knockout mice and wild type mice in those brain areas where human fragile X patients show differences with non-fragile X individuals. However, the structural differences in neuroanatomy between human fragile X patients and controls may have been overestimated in the limited number of previous studies (Jakala et al., 1997). The basis for studying vermis size was the correlation between vermis size and autistic behavior. It was found that vermis size is reduced not only in patients with autism, but also in a percentage of patients suffering from neurogenetic syndromes without autistic features (Schaefer et al., 1996).

It may well be that the mouse model does simply not resemble the human disease with respect to brain structure, although the same neuropathologic abnormalities of dendritic spines have been reported in both human and mouse (Hinton et al., 1991; Comery et al., 1997; Irwin et al., 2001).

Recently, altered sizes and distributions of hippocampal intra- and infrapyramidal mossy fiber (IIPMF) terminal fields in *Fmr1* knockout mice were described (Ivanco and Greenough, 2002; Mineur et al., 2002). It has been shown that these hippocampal fibers are involved in spatial learning tasks (Crusio and Schwegler, 1987; Schwegler et al., 1990; Schwegler and Crusio, 1995), but the precise consequences of these alterations in knockout mice are still unclear.

Mean corpuscular haemoglobin

Biochemical abnormalities in human fragile X patients are limited to a lowered cyclic AMP production (Berry-Kravis and Huttenlocher, 1992), and an increased mean corpuscular hemoglobin (MCH) level (Langenbeck et al., 1984). The difference in MCH in human patients was small, but statistically significant. The authors argued that the increase in MCH level was a result of a minor disturbance in the folate metabolism of fragile X patients. The MCH level in erythrocytes from *Fmr1* knockout and wild type mice were measured, but no difference was found (Reyniers et al., 1996). It is therefore unlikely that Fmrp has an effect on MCH.

Behavioral tests

Along with the cognitive dysfunction accounting for deficits in short-term memory and visual-spatial disabilities (Maes et al., 1994), a complex of behavioral abnormalities is seen in human fragile X patients. These abnormalities include social avoidance, hyperactivity, abnormal response to sensory stimuli, and repetitive, stereotypic autistic behavior (Brown et al., 1986; Fisch, 1992; Reiss and Freund, 1992). The *Fmr1* knockout mouse is used as a model to study the cognitive dysfunction and behavioral abnormalities.

Extensive behavioral tests have been performed to study cognitive functions of the knockout mice (Bakker et al., 1994; Kooy et al., 1996; D'Hooge et al., 1997; Paradee et al., 1999; Dobkin et al., 2000; Peier et al., 2000; Van Dam et al., 2000;

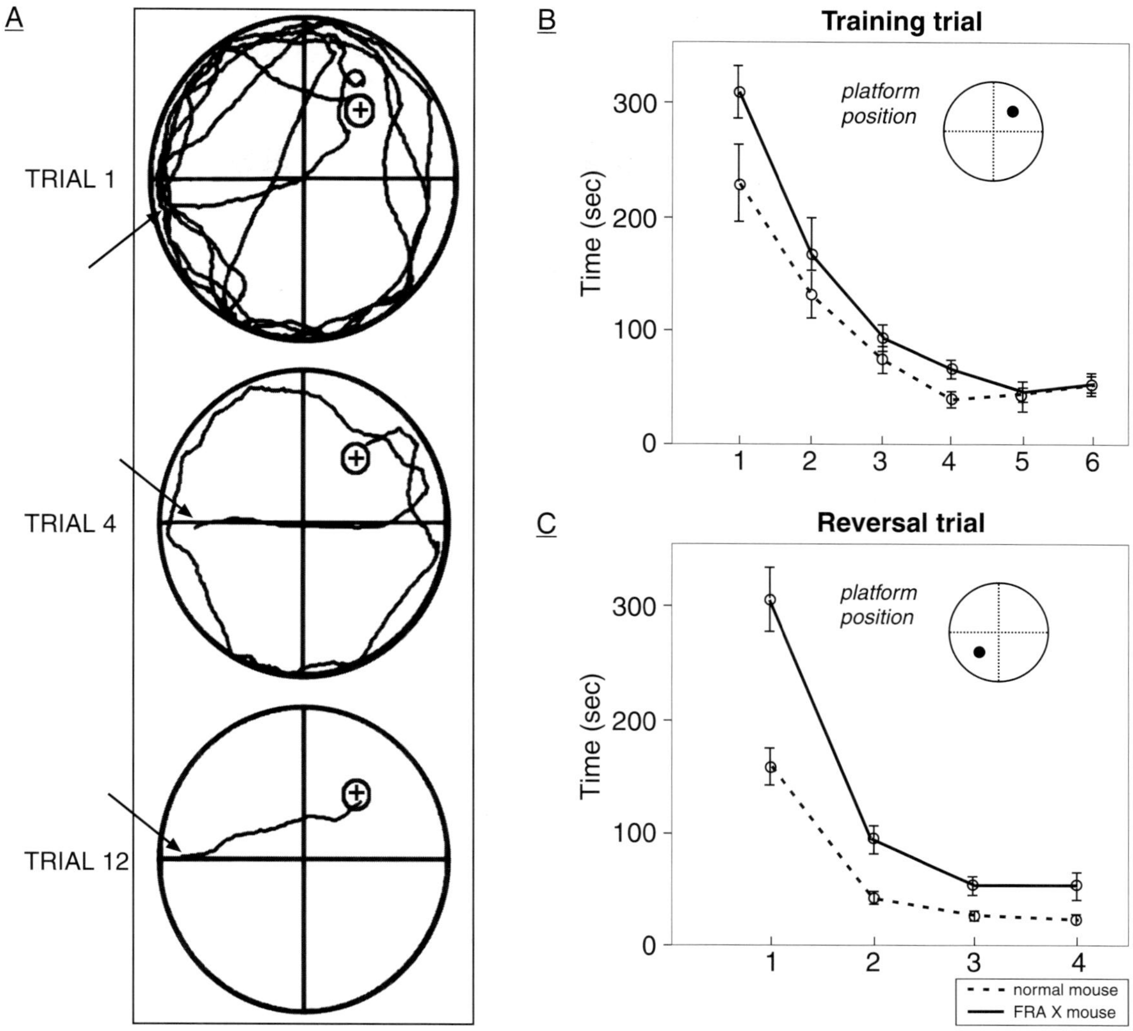

Fig. 3. Morris water maze. (**A**) Swimming trajectories of a control mouse during the first swims of three training trials of a hidden-platform water maze training. The starting position is indicated by a black arrow. The hidden platform (circle with cross) is hidden at the center of the northeast quadrant. (**B**) Escape latency to find a hidden platform: training trials. (**C**) Escape latency during reverse trials. Filled circles: knockout mice; open circles: control littermates.

Mineur et al., 2002). A mouse behavioral test that measures spatial learning and memory is the Morris water maze task (Fig. 3) (Morris et al., 1982; Morris, 1984). This task is highly dependent on hippocampal function. Since FMRP expression is high in hippocampus, but absent in *Fmr1* knockout mice, these mice were tested in the Morris water maze to analyze their hippocampal function. The knockout mice showed an impaired performance during learning and reversal trails, but the differences are rather subtle and highly dependent on the genetic background of the mouse strain used (Kooy et al., 1996; D'Hooge et al., 1997; Paradee et al., 1999; Dobkin et al., 2000; Peier et al., 2000; Van Dam et al., 2000; Mineur et al., 2002). In addition, tests measuring exploration, hyperactivity, anxiety, conditioned fear and aggression were performed (Bakker et al., 1994; Kooy et al., 1996; D'Hooge et al., 1997; Paradee et al., 1999; Dobkin et al., 2000; Peier et al., 2000; Van Dam et al., 2000; Mineur et al., 2002). It is clear from these studies that *Fmr1* knockout mice show increased activity and exploratory behavior, but no differences were seen in conditioned fear or aggression compared to control mice. Finally, tests measuring anxiety give conflicting results probably due to differences in experimental setting or interpretation (Peier et al., 2000; Mineur et al., 2002). A robust difference in performance between *Fmr1* knockout mice and controls is observed in the acoustic startle reflex (Chen and Toth, 2001; Nielsen et al., 2002).

The results of the behavioral tests illustrate the complexity of the effects of Fmrp deficiency on brain function. It has become clear that only extensive behavioral validation of the *Fmr1* knockout mouse will benefit the use of the model in pathophysiological and therapeutic studies.

Long-term potentiation and long-term depression

Malperformance of the Morris water maze task is indicative for dysfunction of the hippocampus (Morris et al., 1982). Stud-

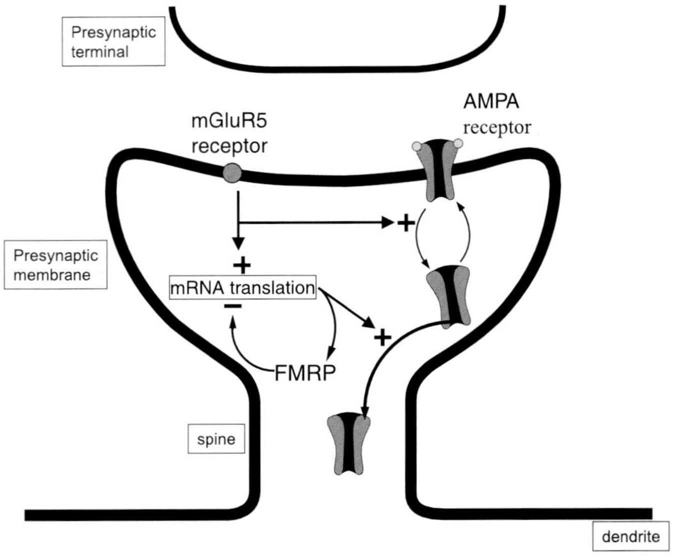

Fig. 4. One expression mechanism for LTD of synaptic transmission is the internalization of AMPA receptors. Activation of mGluR5 (metabotropic glutamate receptor) stimulates the internalization of AMPA receptors and induces local mRNA translation. This translation results in new protein synthesis which stimulates the permanent internalization of AMPA receptors. The general protein synthesis is proposed to be negatively regulated by FMRP which is also synthesized in response to mGluR activation. Therefore, in the absence of FMRP, protein synthesis is not controlled and internalization is increased and subsequently LTD magnitude is increased. This figure is adapted from (Huber et al., 2002).

ies on other, unrelated knockout mouse models have shown that also long-term potentiation (LTP) is involved in malperformance of the Morris water maze task (Grant et al., 1992; Silva et al., 1992a, 1992b; Bach et al., 1995; Sakimura et al., 1995; Wu et al., 1995). The physiological significance of LTP is still unknown, but it is generally assumed that there is a link between learning and memory, synaptic plasticity, and LTP. LTP is an input-specific and persistent increase in the strength of synaptic connections that is the major model mechanism for information storage in the brain (Bliss and Collingridge, 1993; Danysz et al., 1995; Asztely and Gustafsson, 1996; McHugh et al., 1996; Tsien, 2000; Lu et al., 2001). Electrophysiological studies in CA1 hippocampal slices from *Fmr1* knockout mice and wild type mice were performed to investigate the involvement of *Fmr1* in early-phase LTP, but no differences were found (Godfraind et al., 1996). Also short-term potentiation was similar in this study. Additionally, *Fmr1* expression in brain after seizures was monitored, to investigate whether *Fmr1* is involved in the later stages of LTP (LTP3) as an immediate early gene. No alterations in the amount of *Fmr1* mRNA could be detected in total brain or hippocampus up to 150 min after the first seizures, making it unlikely that *Fmr1* is an immediate early gene involved in LTP3 (Godfraind et al., 1996).

There is evidence suggesting that some mechanisms of synaptic plasticity are common to both hippocampus and cortex (Aniksztejn and Ben-Ari, 1991; Pelletier and Hablitz, 1996). Li et al. (2002) studied LTP in the cortex of mice and, although no changes in hippocampal LTP were found, studies on cortical LTP showed a significant reduction in knockout mice.

Another study examining late-phase long-term potentiation, the protein synthesis-dependent form of long-term potentiation, did also show no differences between knockout and wild type mice (Paradee et al., 1999). These results suggest that Fmrp may not have an influence on long-term potentiation in the hippocampus or that any influence is too subtle to be detected by the techniques used.

A second form of hippocampal synaptic plasticity, which is protein synthesis-dependent, is long-term depression (LTD) triggered by activation of group 1 metabotropic glutamate receptors (mGluR) (Huber et al., 2000, 2001; Snyder et al., 2001). This type of LTD is protein dependent and is not observed when mRNA translation is inhibited (Huber et al., 2000; Snyder et al., 2001). Since it is known that FMRP is synthesized in response to mGluR activation by glutamate (Weiler and Greenough, 1999), the involvement of FMRP in hippocampal LTD has been investigated (Fig. 4) (Huber et al., 2002). In *Fmr1* knockout mice, the hippocampal LTD is selectively enhanced compared to wild type mice. This finding is in agreement with the function of FMRP as an inhibitor of translation (Laggerbauer et al., 2001; Li et al., 2001; Zhang et al., 2001). In the absence of Fmrp, translation of mRNAs is not inhibited resulting in enhanced hippocampal LTD. The hypothesis is that this enhanced hippocampal LTD could interfere with the formation and maintenance of strong synapses required for normal brain function (Huber et al., 2002). It is also suggested that the enhanced activity- and mGluR-dependent synapse turnover is underlying the slowed development of dendrite spines in the cerebellar cortex (Nimchinsky et al., 2001) and thereby affecting the normal function of the cerebral cortex (Huber et al., 2002).

Synaptic plasticity

Both FMR1 mRNA and FMRP are present in dendrites, and the expression level is increased by activation of metabotropic glutamate receptors (Weiler et al., 1997; Weiler and Greenough, 1999) linking FMRP to synaptic function. Glutamate receptors, such as NMDA and AMPA receptors, are localized in the brain and play a critical role in learning and memory (Bliss and Collingridge, 1993; Danysz et al., 1995; Asztely and Gustafsson, 1996; McHugh et al., 1996; Tsien, 2000; Lu et al., 2001). For the AMPA receptors, it has recently been found that they have an important function in synaptic plasticity (Musleh et al., 1997; Nayak et al., 1998; Zamanillo et al., 1999). Together with the reduced LTP in cortical slices, it was of interest to study the expression of GluR1, an AMPA receptor subunit, in the cortex of *Fmr1* knockout mice. It was shown that cortical GluR1 expression is reduced in the absence of Fmrp, probably leading to depressed cortical synaptic plasticity (Li et al., 2002).

In both human fragile X patients (Rudelli et al., 1985; Hinton et al., 1991; Irwin et al., 2001) and *Fmr1* knockout mice (Comery et al., 1997), abnormalities in dendrite spines have been described (Fig. 5). Contrary to the pathological observations in patients, these anomalies appear transiently in the mouse model, limited to the first weeks of life. The spines are unusually long, resembling immature spine morphology, and are increased in density. Absence of FMRP may be involved in a defect in spine maturation and pruning (Comery et al., 1997). Spine motility is believed to represent postsynaptic participation in synapse formation (Dailey and Smith, 1996). An abnormality in spine motility early in postnatal life could affect synaptogenesis, and therefore it is of interest to determine spine motility in the absence of Fmrp. Nimchinsky et al. (2001) showed that dendrite spines in vivo in the intact brain of *Fmr1* knockout mice were abnormal early in postnatal life. These abnormalities in the somatosensory cortex were most pronounced during the period of greatest synaptogenesis in that region (White et al., 1997), which therefore supports the hypothesis that Fmrp plays a role in synaptogenesis in the normal brain. This early disturbance of synaptogenesis could be reflected at that developmental time point by abnormalities in spine morphology and density and later by cognitive defects resulting from improperly established connections (Nimchinsky et al., 2001). In contrast, in vitro studies (Steward et al., 1998; Nimchinsky et al., 2001) do not detect the spine abnormalities found in the in vivo study performed by Nimchinsky et al., probably due to differences in methods used (Nimchinsky et al., 2001).

Protein synthesis occurs near synapses, and particularly during development polyribosomal aggregates are formed. Polyribosomal aggregates in spines increase during experience-dependent synaptogenesis (Greenough et al., 1987). Some protein synthesis appears to be regulated directly by synaptic activity. This synthesis can be monitored in synaptoneurosomes. Synaptoneurosomes are preparations highly enriched in pinched-off, resealed presynaptic processes attached to resealed postsynaptic processes that retain normal function of neurotransmitter release. When synaptoneurosomes are stimulated with glutamate, mRNA is rapidly taken up into polyribosomal aggregates, and proteins are synthesized (Greenough et al., 2001). One of these proteins is Fmrp. Greenough et al. (2001) reported that the protein synthesis in response to glutamate is dramatically reduced in synaptoneurosomes derived from the cerebral cortex of *Fmr1* knockout mice. This is in contrast to reports of others describing a function of Fmrp in inhibition of mRNA translation. Absence of Fmrp, as is the case in *Fmr1* knockout mice, would then result in non-inhibited mRNA translation (Laggerbauer et al., 2001; Li et al., 2001; Zhang et al., 2001). Nevertheless, these studies suggest that Fmrp is involved in the process of a rapid localized translational response to synaptic stimulation in the cortex.

In conclusion, all data obtained with studies on dendrite spines point in the direction of an impairment of mechanisms that promote synapse maturation and pruning in the *Fmr1* knockout mouse. Fmrp plays a role in the neural maturation process, and its absence evolves in cognitive and behavioral impairments as observed in the fragile X syndrome.

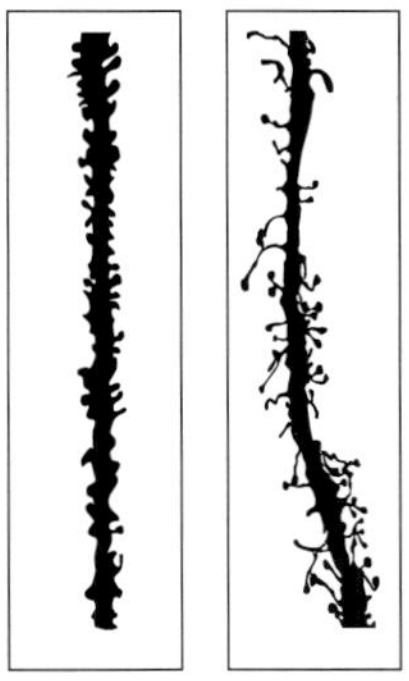

Fig. 5. Spines along the dendrite of a control and a knockout mouse. In the knockout mouse there is increasing spine density and longer, thinner spines.

Seizures

More than 50% of prepuberal fragile X patients show typical EEG abnormalities, whereas seizures are observed in more than 20% of the patients (Musumeci et al., 1999). Since *Fmr1* knockout mice have never been reported to have spontaneous epileptic seizures, acoustic stimulation was used to provoke audiogenic seizures in the knockout and wild type mice. The susceptibility to audiogenic seizures (AGS) is greater in knockout mice compared to wild type mice. These results indicate that absence of Fmrp results in an increase in cortical excitability in knockout mice (Musumeci et al., 2000).

In general, fragile X patients show hyperactivity to auditory, visual, tactile and olfactory stimuli (Merenstein et al., 1996; Hagerman et al., 1999; Miller et al., 1999; Hagerman, 2002). Chen and Toth were particularly interested in the sensory responsiveness of the *Fmr1* knockout mouse. They tested the knockout mice for susceptibility of audiogenic seizures, as did Mucumeci et al, and found an age dependent increased AGS in knockout mice from week 10 onwards (Chen and Toth, 2001). Chemical convulsants were used to test whether this seizure susceptibility is restricted to the auditory system or is a more general phenomenon. No difference between knockout and wild type mice in seizure susceptibility was found, indicating that knockout mice do have a local, presumably auditory system hyperexcitability (Chen and Toth, 2001). This auditory system hyperexcitability was confirmed in a study of the auditory startle response in *Fmr1* knockout and control mice of different genetic backgrounds. The knockout mice exhibited increased auditory startle response amplitudes to low intensity stimuli and decreased responses to high intensity stimuli (Nielsen et al., 2002). Although the decreased responses to high intensity stimuli are difficult to explain it might be that there is an abnormality in secondary brain regions that modulates the primary startle response (Nielsen et al., 2002). The responsiveness of *Fmr1* knockout mice to auditory stimuli is consistent with the sensory hypersensitivity of fragile X patients.

Since epileptic seizures induce neuronal expression of the immediate-early gene product c-Fos (Morgan and Curran, 1991) some mice that had experienced an audiogenic seizure were analyzed for localization of c-Fos expression. The c-Fos expression pattern denotes the involvement of specific brain structures in seizure propagation. In this way the neuronal network involved in audiogenic seizures in the knockout mice could be traced from auditory stimulus to the final motor component (Chen and Toth, 2001). Based on this study the authors hypothesized that absence of Fmrp in the auditory system of knockout mice results in perturbations in developmental/activity-dependent neuroplasticity, leading to an increased susceptibility to loud, audible sound (Chen and Toth, 2001).

As mentioned before, FMRP is expressed in synapses in response to activation of glutamate receptors by glutamate (Weiler et al., 1997; Weiler and Greenough, 1999) and via a negative inhibition pathway FMRP suppresses translation of mRNA (Laggerbauer et al., 2001; Li et al., 2001; Huber et al., 2002). Studies on mGluR-triggered protein synthesis in the hippocampus suggest that without inhibition of translation, for example performed by FMRP, the threshold for synaptic potentiation is reduced (Raymond et al., 2000) and epileptic activity can be triggered (Merlin et al., 1998; Wong et al., 1999). This finding might explain the hyperreactivity seen in the absence of FMRP in the fragile X syndrome (Huber et al., 2002).

Reintroduction of FMRP

Using the *Fmr1* knockout mouse as a model, the FMR1 cDNA and thus also FMRP could be reintroduced in an attempt to rescue the phenotype (Bakker et al., 2000b). In this study the expression of the protein is constitutive, and in all cell types, but there is no rescue of the phenotype. The timing of expression, the level of expression per cell and the type of cells in which FMRP is expressed seems to be essential for a rescue (Bakker et al., 2000b).

Peier et al. (2000) used a yeast artificial chromosome (YAC) containing the entire human FMR1 gene with extensive amounts of flanking sequence for insertion into the genome of the *Fmr1* knockout mouse. The transgenic protein was expressed 13–17 times higher than the endogenous Fmrp. While the *Fmr1* knockout mouse in behavioral tests displays less anxiety-related responses with increased exploratory behavior compared to wild type mice, the YAC-containing *Fmr1* knockout mice showed opposing behavioral response and additional abnormal behaviors compared to wild type mice. These observations indicate that overexpression of FMRP is harmful to the animal, and this finding has significant implications for gene therapy for the fragile X syndrome (Peier et al., 2000).

Taking into consideration the problems with reintroduction of the FMR1 gene, a more appropriate approach could be the use of pharmacological agents to reactivate the *Fmr1* gene in patients with a methylated full mutation. Since methylation of the CGG repeat and of the *Fmr1* promoter region are the inactivating cause of the fragile X syndrome, demethylation could be a way to activate FMR1 and obtain FMRP expression. Studies with demethylating agents in cultured cells from patients have been performed and show promising results, although the agents used are very toxic (Chiurazzi et al., 1998, 1999). To test this type of re-expression of Fmrp *in vivo*, an animal model containing a long repeat and showing methylation of the FMR1 promoter region, would be of much importance.

The CGG repeat mouse

A mouse model, mimicking several aspects of the fragile X syndrome in considerable detail, would be of great value not only to a demethylation therapy, but also to shed light on the mechanisms involved in FMR1 CGG repeat instability and methylation.

Previously, a transgenic mouse model was made containing a fusion gene of a long human repeat of 81 CGG units and the LacZ reporter gene. Upon transmission to the next generation, this repeat remained stable (Bontekoe et al., 1997). Even when human sequences normally surrounding the CGG repeat were included in the transgene, no instability was seen (Lavedan et al., 1997, 1998). Thus, nearby *cis*-acting elements seem not to be involved in the control of, or sufficient to induce, instability of the repeat. Therefore, placing the long repeat in its endogenous environment may provide the factors needed for repeat instability.

A mouse with a long CGG repeat at the position of the endogenous short mouse CGG repeat has been created (Bontekoe et al., 2001). The initial length of the long repeat of human origin was 98 repeats. This repeat is in the human premutation range, showing instability upon maternal transmission to the next generation in the human. In the mouse, this repeat shows mild instability, and stepwise has reached up till now a length of 108 repeats (Bontekoe et al., 2001). Creating successive generations of mice may finally lead to a full mutation repeat, initiating methylation and inactivation of the *Fmr1* gene. This mouse model will make the study of repeat instability in its mouse endogenous environment possible.

Other mouse models containing other trinucleotide repeats involved in human diseases have been generated. Instability was found in a number of them, some are discussed in this special issue.

A knockout mouse for Fxr1 and Fxr2

For the homologes of FMR1, FXR1 and FXR2, no disease causing mutations have been described yet. The absence of Fmrp in the *Fmr1* knockout mouse does not influence the expression of Fxr1p or Fxr2p markedly (Bakker et al., 2000a). Knockout mice for *Fxr1* and *Fxr2* may reveal a phenotype and could shed light on the functions of Fxr1p and Fxr2p.

The generation of a knockout mouse for *Fxr1* has not been successful until now. The first indications are that mice homozygous for the *Fxr1* knockout allele die shortly after birth (H. Siomi, personal communication). Since *Fxr1* is highly expressed in muscle tissue, and absence of Fxr1p would lead to malfunction of vital organs a logical explanation would be that heart or lung functions are impaired. A conditional knockout mouse for *Fxr1*, in which the gene can be turned off at a desired moment, or in a desired organ, is under development.

Generation of a knockout mouse for *Fxr2* has been successful, and has led to mice being vital and fertile (Bontekoe et al., 2002). The *Fxr2* knockout mice have been subjected to the same behavioral test battery, as have been the *Fmr1* knockout mice. *Fxr2* knockout mice show behavioral abnormalities compared to wild type mice. Some of these behaviors resemble those of the *Fmr1* knockout mouse and several other of these are opposing. *Fmr1* and *Fxr2* knockout mice are hyperactive in the open-field test and impaired in the rotarod test. On the other hand, *Fxr2*, but not *Fmr1*, knockout mice show poor performance during training trials of the Morris water maze task, and are impaired in the context-dependent conditioned fear test (Bontekoe et al., 2002). It has to be taken into account that, although the behavioral tests are performed in the same laboratory, they were not performed simultaneously and the genetic background of the two knockout models was different. Still, the results with the *Fxr2* knockout mice are indicative of a function of Fxr2p in those parts of the brain involved in behavioral responses (Bontekoe et al., 2002).

Generation of *Fmr1/Fxr2* double-knockout mice can provide information about possible interaction of Fmrp and Fxr2p, and whether or not these proteins show partial or complete functional complementation. Studies with the double-knockout mice are in progress (unpublished result).

Drosophila models

The dfxr gene

Drosophila is the second model organism for fragile X syndrome that has only been recently introduced. The single fly FMR homologue, dFXR also named dFMR, shares extensive amino acid sequence identity with the human FXR genes including several key domains (Wan et al., 2000). The protein product, dFXR, is a single protein with a size of 85 kDa and is an RNA-binding protein capable of interacting with itself and the mammalian counterparts, thus implying some degree of functional conservation. Expression studies indicate that dFXR is ubiquitously expressed throughout development and is localized to cytoplasm. In addition, overexpression of dFXR in the eye and wing indicated that high doses of dFXR can cause cell death (Wan et al., 2000). dFXR, like its mammalian homologues, is expressed in brain neurons but not in glia (Zhang et al., 2001; Morales et al., 2002).

Loss of function mutants

Drosophila with a loss-of-function d*fxr* allele has been studied by several groups (Zhang et al., 2001; Dockendorff et al., 2002; Morales et al., 2002). The d*fxr* mutations do not affect the viability of mutant adult flies although a semi-lethal phenotype was observed by Morales et al. This semi-lethal phenotype may be caused by an unidentified second-site mutation. Null mutations flies display peripheral (neuromuscular junction, NMJ) and central (optic lobe) synaptic transmission defects (Zhang et al., 2001). A relatively modest increase was reported in the number of arboreal branches at the NMJ and, probably consequently, a similar increase in the number of peripheral synaptic boutons.

Neuronal overexpression of dFXR caused a decrease of bouton number, without affecting the number of branches. The latter phenotype had been previously observed for hypomorphic futsch mutants. The futsch homologue, MAP-1B, was also recently identified as one of the many targets of FMRP (Brown et al., 2001). The dFXR protein was shown to repress expression of the futsch protein (Zhang et al., 2001). Zhang et al. further showed that normal synaptic structure and function are restored in a double mutant *fxr* null/*futsch* hypomorph, and that *futsch* expression is enhanced in d*fxr* null flies. In vitro, *futsch* mRNA binds to the dFXR protein, like the human and mouse homologue MAP1B mRNA to FMRP (Brown et al., 2001). The authors suggest that the upregulation of Futsch, in d*fxr* mutants is sufficient to explain the defects observed in d*fxr* mutants (Zhang et al., 2001). This is a surprising finding given the potentially large number of target RNAs of the mammalian FMRP protein. It would be interesting to determine whether other candidate targets are regulated in the same way as futsch and how FMRP affects their translation initiation or elongation at the molecular level. This suggestion of Zhang et al. is in contrast with the finding that the introduction of a *futsch* loss-of-function mutant into the d*fmr1* mutant background did not rescue the circadian phenotype (see below). It will be interesting to see whether similar effects on MAP1B expression can be documented in the knockout mouse model.

Behavioral defects

Fragile X patients show a number of abnormal behaviors including sleep problems (Hagerman, 1996; Gould et al., 2000). Two groups have studied whether a loss-of-function mutation of d*fxr* affects circadian behavior (Dockendorff et al., 2002; Inoue et al., 2002). Under constant darkness, a lack of d*fxr* expression causes arrhythmic locomotor activity and erratic patterns of locomotor activity, but in light:dark cycles, their behavioral rhythms appear normal. On the other hand overexpression of d*fxr* leads to a lengthened period. In addition, the clock-controlled eclosion rhythm is abnormal in d*fxr*-deficient flies. Molecular analysis fails to reveal any defects in the expression of clock components. Morphological analysis of neurons required for normal circadian behavior reveals subtle abnormalities, suggesting that defects in axonal pathfinding or synapse formation may cause the observed behavioral defects. These results suggest that dFXR plays a critical role in the circadian output pathway regulating locomotor activity in Drosophila. A role for human FMRP has been postulated in neuronal development. Proteins that affect neuronal development or function might be expected to affect circadian rhythms. Sleep problems have been observed in fragile X patients, but no studies have been performed on the circadian clock in fragile X patients.

Analysis of courtship behavior in the d*fxr* mutants has revealed a defect in overall courtship activity (Dockendorff et al., 2002). The mutants initiate normal numbers of courtship attempts, but are unable to maintain courtship interest. Analyses of adult brain structure reveal no apparent defects in mutant mushroom bodies, but show abnormal axon branching and an overgrowth of terminal arborizations in individual neurons required for circadian behavior.

Anatomical analyses in human, mouse and Drosophila suggest that regulation of dendrite growth and/or synapse formation may be affected (Comery et al., 1997; Zhang et al., 2001; Dockendorff et al., 2002). These abnormalities in the neurons might be the basis for the behavior phenotypes in the flies.

Conclusion

The mouse model of *Fmr1* deficiency shows slightly enlarged testes, a subtle behavioral phenotype, and discrete anomalies of dendrite spines similar to those observed in brains of patients. This animal model has been and will be very useful in studies concerning the role of FMRP in brain. The similarities between the human and fly mutant phenotypes and the shared biochemical properties of the two proteins suggest that studies of d*fxr* will be a useful model for identifying physiological pathways and substrates affected by the FMR1 gene.

The establishment of animal models provides a new avenue to dissect multiple physiological pathways affected by the mutations causing the fragile X syndrome.

Acknowledgements

We wish to thank our colleagues of the fragile X research group for fruitful collaboration and discussions.

References

Abitbol M, Menini C, Delezoide AL, Rhyner T, Vekemans M, Mallet J: Nucleus basalis magnocellularis and hippocampus are the major sites of FMR-1 expression in the human fetal brain. Nature Genet 4:147–153 (1993).

Adinolfi S, Bagni C, Musco G, Gibson T, Mazzarella L, Pastore A: Dissecting FMR1, the protein responsible for fragile X syndrome, in its structural and functional domains. RNA 5:1248–1258 (1999).

Aniksztejn L, Ben-Ari Y: Novel form of long-term potentiation produced by a K+ channel blocker in the hippocampus. Nature 349:67–69 (1991).

Ashley C Jr, Wilkinson KD, Reines D, Warren ST: FMR1 protein: conserved RNP family domains and selective RNA binding. Science 262:563–568 (1993a).

Ashley CT, Sutcliffe JS, Kunst CB, Leiner HA, Eichler EE, Nelson DL, Warren ST: Human and murine FMR-1: alternative splicing and translational initiation downstream of the CGG-repeat. Nature Genet 4:244–251 (1993b).

Asztely F, Gustafsson B: Ionotropic glutamate receptors: Their possible role in the expression of hippocampal synaptic plasticity. Mol Neurobiol 12:1–11 (1996).

Bach ME, Hawkins RD, Osman M, Kandel ER, Mayford M: Impairment of spatial but not contextual memory in CaMKII mutant mice with a selective loss of hippocampal LTP in the range of the theta frequency. Cell 81:905–915 (1995).

Bächner D, Manca A, Steinbach P, Wöhrle D, Just W, Vogel W, Hameister H, Poustka A: Enhanced expression of the murine FMR1 gene during germ cell proliferation suggests a special function in both the male and the female gonad. Hum molec Genet 2:2043–2050 (1993).

Bakker CE, de Diego Otero Y, Bontekoe C, Raghoe P, Luteijn T, Hoogeveen AT, Oostra BA, Willemsen R: Immunocytochemical and biochemical characterization of FMRP, FXR1P, and FXR2P in the mouse. Exp Cell Res 258:162–170 (2000a).

Bakker CE, Kooy RF, D'Hooge R, Tamanini F, Willemsen R, Nieuwenhuizen I, De Vries BBA, Reyniers E, Hoogeveen AT, Willems PJ, De Deyn PP, Oostra BA: Introduction of an FMR1 transgene in the fragile X knockout mouse. Neurosci Res Comm 26:265–277 (2000b).

Bakker CE, Verheij C, Willemsen R, Vanderhelm R, Oerlemans F, Vermey M, Bygrave A, Hoogeveen AT, Oostra BA, Reyniers E, Deboulle K, Dhooge R, Cras P, Van Velzen D, Nagels G, Martin JJ, Dedeyn PP, Darby JK, Willems PJ: Fmr1 knockout mice: A model to study fragile X mental retardation. Cell 78:23–33 (1994).

Berry-Kravis E, Huttenlocher PR: Cyclic AMP metabolism in fragile X syndrome. Ann Neurol 31:22–26 (1992).

Bliss TV, Collingridge GL: A synaptic model of memory: long-term potentiation in the hippocampus. Nature 361:31–39 (1993).

Bontekoe CJ, Bakker CE, Nieuwenhuizen IM, van Der Linde H, Lans H, de Lange D, Hirst MC, Oostra BA: Instability of a (CGG) (98) repeat in the Fmr1 promoter. Hum molec Genet 10:1693–1699 (2001).

Bontekoe CJ, McIlwain KL, Nieuwenhuizen IM, Yuva-Paylor LA, Nellis A, Willemsen R, Fang Z, Kirkpatrick L, Bakker CE, McAninch R, Cheng NC, Merriweather M, Hoogeveen AT, Nelson D, Paylor R, Oostra BA: Knockout mouse model for Fxr2: a model for mental retardation. Hum molec Genet 11:487–498 (2002).

Bontekoe CJM, de Graaff E, Nieuwenhuizen IM, Willemsen R, Oostra BA: FMR1 premutation allele is stable in mice. Eur J hum Genet 5:293–298 (1997).

Brown V, Jin P, Ceman S, Darnell JC, O'Donnell WT, Tenenbaum SA, Jin X, Feng Y, Wilkinson KD, Keene JD, Darnell RB, Warren ST: Microarray identification of FMRP-associated brain mRNAs and altered mRNA translational profiles in fragile X syndrome. Cell 107:477–487 (2001).

Brown V, Small K, Lakkis L, Feng Y, Gunter C, Wilkinson KD, Warren ST: Purified recombinant Fmrp exhibits selective RNA binding as an intrinsic property of the fragile X mental retardation protein. J biol Chem 273:15521–15527 (1998).

Brown WT, Jenkins EC, Cohen IL, Fish GS, Wolf-Schein EG, Gross A, Waterhouse L, Fein D, Mason-Brothers A, Ritvo E, Rittenberg BA, Bentley W, Castells S: Fragile X and autism: A multicenter survey. Am J med Genet 23:341–352 (1986).

Chen L, Toth M: Fragile X mice develop sensory hyperreactivity to auditory stimuli. Neuroscience 103: 1043–1050 (2001).

Chiurazzi P, Pomponi MG, Pietrobono R, Bakker CE, Neri G, Oostra BA: Synergistic effect of histone hyperacetylation and DNA demethylation in the reactivation of the FMR1 gene. Hum molec Genet 8:2317–2323 (1999).

Chiurazzi P, Pomponi MG, Willemsen R, Oostra BA, Neri G: In vitro reactivation of the FMR1 gene involved in fragile X syndrome. Hum molec Genet 7:109–113 (1998).

Cohen IL, Vietze PM, Sudhalter V, Jenkins EC, Brown WT: Effects of age and communication level on eye contact in fragile X males and non-fragile X autistic males. Am J med Genet 38:498–502 (1991).

Comery TA, Harris JB, Willems PJ, Oostra BA, Irwin SA, Weiler IJ, Greenough WT: Abnormal dendritic spines in fragile X knockout mice: Maturation and pruning deficits. Proc natl Acad Sci, USA 94: 5401–5404 (1997).

Corbin F, Bouillon M, Fortin A, Morin S, Rousseau F, Khandjian EW: The fragile X mental retardation protein is associated with poly(A) (+) mRNA in actively translating polyribosomes. Hum molec Genet 6:1465–1472 (1997).

Crusio WE, Schwegler H: Hippocampal mossy fiber distribution co-varies with open-field habituation in the mouse. Behav Brain Res 26:153–158 (1987).

Dailey ME, Smith SJ: The dynamics of dendritic structure in developing hippocampal slices. J Neurosci 16:2983–2994 (1996).

Danysz W, Zajaczkowski W, Parsons CG: Modulation of learning processes by ionotropic glutamate receptor ligands. Behav Pharmacol 6:455–474 (1995).

Darnell JC, Jensen KB, Jin P, Brown V, Warren ST, Darnell RB: Fragile X mental retardation protein targets G quartet mRNAs important for neuronal function. Cell 107:489–499 (2001).

De Boulle K, Verkerk AJ, Reyniers E, Vits L, Hendrickx J, Van Roy B, Van den Bos F, de Graaff E, Oostra BA, Willems PJ: A point mutation in the FMR-1 gene associated with fragile X mental retardation. Nature Genet 3:31–35 (1993).

De Diego Otero Y, Bakker CE, Raghoe P, Severijnen LWFM, Hoogeveen A, Oostra BA, Willemsen R: Immunocytochemical characterization of FMRP, FXR1P and FXR2P during embryonic development in the mouse. Gene Funct Dis 1:28–37 (2000).

De Vries BB, van den Ouweland AM, Mohkamsing S, Duivenvoorden HJ, Mol E, Gelsema K, van Rijn M, Halley DJ, Sandkuijl LA, Oostra BA, Tibben A, Niermeijer MF: Screening and diagnosis for the fragile X syndrome among the mentally retarded: an epidemiological and psychological survey. Am J hum Genet 61:660–667 (1997).

De Vries BB, Halley DJ, Oostra BA, Niermeijer MF: The fragile X syndrome. J med Genet 35:579–589 (1998).

De Vries BB, Jansen CA, Duits AA, Verheij C, Willemsen R, Van Hemel JO, Van den Ouweland, AM, Niermeijer MF, Oostra BA, Halley DJ: Variable FMR1 gene methylation leads to variable phenotype in 3 males from one fragile X family. J med Genet 33:1007–1010 (1996a).

De Vries BB, Wiegers AM, De Graaff E, Verkerk AJ, Van Hemel JO, Halley DJ, Fryns JP, Curfs LM, Niermeijer MF, Oostra BA: Mental status and fragile X expression in relation to FMR-1 gene mutation. Eur J hum Genet 1:72–79 (1993).

De Vries LB, Wiegers AM, Smits AP, Mohkamsing S, Duivenvoorden HJ, Fryns J-P, Curfs LM, Halley DJ, Oostra BA, van den Ouweland AM, Niermeijer MF: Mental status of females with an FMR1 gene full mutation. Am J hum Genet 58:1025–1032 (1996b).

Deelen W, Bakker C, Halley D, Oostra BA: Conservation of CGG region in FMR1 gene in mammals. Am J med Genet 51:513–516 (1994).

Devys D, Lutz Y, Rouyer N, Bellocq JP, Mandel JL: The FMR-1 protein is cytoplasmic, most abundant in neurons and appears normal in carriers of a fragile X premutation. Nature Genet 4:335–340 (1993).

D'Hooge R, Nagels G, Franck F, Bakker CE, Reyniers E, Storm K, Kooy RF, Oostra BA, Willems PJ, Dedeyn PP: Mildly impaired water maze performance in male *Fmr1* knockout mice. Neuroscience 76:367–376 (1997).

Dobkin C, Rabe A, Dumas R, El Idrissi A, Haubenstock H, Ted Brown W: *Fmr1* knockout mouse has a distinctive strain-specific learning impairment. Neuroscience 100:423–429 (2000).

Dockendorff TC, Su HS, McBride SMJ, Yang Z, Choi CH, Siwicki KK, Sehgal A, Jongens TA: Drosophila lacking dfmr1 activity show defects in circadian output and fail to maintain courtship interest. Neuron 34:973–984 (2002).

Eberhart DE, Malter HE, Feng Y, Warren ST: The fragile X mental retardation protein is a ribosonucleoprotein containing both nuclear localization and nuclear export signals. Hum molec Genet 5:1083–1091 (1996).
Feng Y, Absher D, Eberhart DE, Brown V, Malter HE, Warren ST: FMRP associates with polyribosomes as an mRNP, and the I304N mutation of severe fragile X syndrome abolishes this association. Mol Cell 1:109–118 (1997).
Feng Y, Lakkis D, Warren ST: Quantitative comparison of FMR1 gene expression in normal and premutation alleles. Am J hum Genet 56:106–113 (1995a).
Feng Y, Zhang FP, Lokey LK, Chastain JL, Lakkis L, Eberhart D, Warren ST: Translational suppression by trinucleotide repeat expansion at FMR1. Science 268:731–734 (1995b).
Fisch GS: Is autism associated with the fragile X syndrome? Am J med Genet 43:78–86 (1992).
Fisch GS: What is associated with the fragile X syndrome?. Am J med Genet 48:112–121 (1993).
Fridell RA, Benson RE, Hua J, Bogerd HP, Cullen BR: A nuclear role for the fragile X mental retardation protein. EMBO J 15:5408–5414 (1996).
Frohlich KU, Madeo F: Apoptosis in yeast: a new model for aging research. Exp Gerontol 37:27–31 (2001).
Fu YH, Kuhl DP, Pizzuti A, Pieretti M, Sutcliffe JS, Richards S, Verkerk AJ, Holden JJ, Fenwick R, Jr., Warren ST, Oostra BA, Nelson DL, Caskey CT: Variation of the CGG repeat at the fragile X site results in genetic instability: resolution of the Sherman paradox. Cell 67:1047–1058 (1991).
Gedeon AK, Baker E, Robinson H, Partington MW, Gross B, Manca A, Korn B, Poustka A, Yu S, Sutherland GR, Mulley JC: Fragile X syndrome without CCG amplification has an FMR1 deletion. Nature Genet 1:341–344 (1992).
Godfraind JM, Reyniers E, Deboulle K, Dhooge R, Dedeyn PP, Bakker CE, Oostra BA, Kooy RF, Willems PJ: Long-term potentiation in the hippocampus of fragile X knockout mice. Am J med Genet 64:246–251 (1996).
Gould EL, Loesch DZ, Martin MJ, Hagerman RJ, Armstrong SM, Huggins RM: Melatonin profiles and sleep characteristics in boys with fragile X syndrome: a preliminary study. Am J med Genet 95:307–315 (2000).
Grant SG, O'Dell TJ, Karl KA, Stein PL, Soriano P, Kandel ER: Impaired long-term potentiation, spatial learning, and hippocampal development in fyn mutant mice. Science 258:1903–1910 (1992).
Greenough WT, Black JE, Wallace CS: Experience and brain development. Child Dev 58:539–559 (1987).
Greenough WT, Klintsova AY, Irwin SA, Galvez R, Bates KE, Weiler IJ: Synaptic regulation of protein synthesis and the fragile X protein. Proc natl Acad Sci, USA 98:7101–7106 (2001).
Hagerman R: Physical and behavioral phenotype, in Hagerman RJ, Cronister A (eds): Fragile-X Syndrome: Diagnosis, Treatment and Research, pp 3–87 (The Johns Hopkins University Press, Baltimore 1996).
Hagerman RJ: The physical and behavioral phenotype, in Hagerman RJ, Hagerman PJ (eds): Fragile-X Syndrome: Diagnosis, Treatment and Research, pp 3–109 (The Johns Hopkins University Press, Baltimore 2002).
Hagerman RJ, Hills J, Scharfenaker S, Lewis H: Fragile X syndrome and selective mutism. Am J med Genet 83:313–317 (1999).
Hagerman RJ, Hull CE, Safanda JF, Carpenter I, Staley LW, Oconnor RA, Seydel C, Mazzocco M, Snow K, Thibodeau SN, Kuhl D, Nelson DL, Caskey CT, Taylor AK: High functioning fragile X males: Demonstration of an unmethylated fully expanded FMR-1 mutation associated with protein expression. Am J med Genet 51:298–308 (1994).
Hansen RS, Gartler SM, Scott CR, Chen SH, Laird CD: Methylation analysis of CGG sites in the CpG island of the human FMR1 gene. Hum molec Genet 1:571–578 (1992).
Hinds HL, Ashley CT, Sutcliffe JS, Nelson DL, Warren ST, Housman DE, Schalling M: Tissue specific expression of FMR-1 provides evidence for a functional role in fragile X-syndrome Nature Genet 3:36–43 (1993).
Hinton VJ, Brown WT, Wisniewski K, Rudelli RD: Analysis of neocortex in three males with the fragile X syndrome. Am J med Genet 41:289–294 (1991).
Hirst M, Grewal P, Flannery A, Slatter R, Maher E, Barton D, Fryns JP, Davies K: Two new cases of FMR1 deletion associated with mental impairment. Am J hum Genet 56:67–74 (1995).
Hornstra IK, Nelson DL, Warren ST, Yang TP High resolution methylation analysis of the FMR1 gene trinucleotide repeat region in fragile X syndrome. Hum molec Genet 2:1659–1665 (1993).
Huber KM, Gallagher SM, Warren ST, Bear MF: Altered synaptic plasticity in a mouse model of fragile X mental retardation. Proc natl Acad Sci, USA 99:7746–7750 (2002).
Huber KM, Kayser MS, Bear MF: Role for rapid dendritic protein synthesis in hippocampal mGluR-dependent long-term depression. Science 288: 1254–1257 (2000).
Huber KM, Roder JC, Bear MF: Chemical induction of mGluR5- and protein synthesis-dependent long-term depression in hippocampal area CA1. J Neurophysiol 86:321–325 (2001).
Inoue S, Shimoda M, Nishinokubi I, Siomi M, Okamura M, Nakamura A, Kobayashi S, Ishida N, Siomi H: A role for the Drosophila fragile X-related gene in circadian output. Curr Biol 12:1331 (2002).
Irwin SA, Patel B, Idupulapati M, Harris JB, Crisostomo RA, Larsen BP, Kooy F, Willems PJ, Cras P, Kozlowski PB, Swain RA, Weiler IJ, Greenough WT: Abnormal dendritic spine characteristics in the temporal and visual cortices of patients with fragile-X syndrome: A quantitative examination. Am J med Genet 98:161–167 (2001).
Ivanco TL, Greenough WT: Altered mossy fiber distributions in adult *Fmr1* (FVB). knockout mice. Hippocampus 12:47–54 (2002).
Jakala P, Hanninen T, Ryynanen M, Laakso M, Partanen K, Mannermaa A, Soininen H: Fragile-X: Neuropsychological test performance, CGG triplet repeat lengths, and hippocampal volumes. J Clin Invest 100:331–338 (1997).
Jin P, Warren ST: Understanding the molecular basis of fragile X syndrome. Hum molec Genet 9:901–908 (2000).
Khandjian E, Corbin F, Woerly S, Rousseau F: The fragile X mental retardation protein is associated with ribosomes. Nature Genet 12:91–93 (1996).
Kirkpatrick LL, McIlwain KA, Nelson DL: Comparative genomic sequence analysis of the FXR gene family: FMR1, FXR1, and FXR2. Genomics 78: 169–177 (2001).
Kooy RF, Dhooge R, Reyniers E, Bakker CE, Nagels G, Deboulle K, Storm K, Clincke G, Dedeyn PP, Oostra BA, Willems PJ: Transgenic mouse model for the fragile X syndrome. Am J med Genet 64:241–245 (1996).
Kooy RF, Reyniers E, Verhoye M, Sijbers J, Bakker CE, Oostra BA, Willems PJ, Van Der Linden A: Neuroanatomy of the fragile X knockout mouse brain studied using in vivo high resolution magnetic resonance imaging. Eur J hum Genet 7:526–532 (1999).
Kremer EJ, Pritchard M, Lynch M, Yu S, Holman K, Baker E, Warren ST, Schlessinger D, Sutherland GR, Richards RI: Mapping of DNA instability at the fragile X to a trinucleotide repeat sequence $p(CCG)_n$. Science 252:1711–1714 (1991).
Kumar TR, Wang Y, Lu N, Matzuk MM: Follicle stimulating hormone is required for ovarian follicle maturation but not male fertility. Nature Genet 15:201–204 (1997).
Laggerbauer B, Ostareck D, Keidel EM, Ostareck-Lederer A, Fischer U: Evidence that fragile X mental retardation protein is a negative regulator of translation. Hum molec Genet 10:329–338 (2001).
Langenbeck U, Schmidtke J, Bartels I, Hansmann I, Knuppel H: Mean corpuscular hemoglobin is increased in Martin-Bell syndrome. Hum Genet 66:365–366 (1984).
Lavedan C, Grabczyk E, Usdin K, Nussbaum RL: Long uninterrupted CGG repeats within the first exon of the human FMR1 gene are not intrinsically unstable in transgenic mice. Genomics 50:229–240 (1998).
Lavedan CN, Garrett L, Nussbaum RL: Trinucleotide repeats $(CGG)_{22}TGG(CGG)_{43}TGG(CGG)_{21}$ from the fragile X gene remain stable in transgenic mice. Hum Genet 100:407–414 (1997).
Li J, Pelletier MR, Perez Velazquez JL, Carlen PL: Reduced cortical synaptic plasticity and GluR1 expression associated with fragile X mental retardation protein deficiency. Mol Cell Neurosci 19: 138–151 (2002).
Li Z, Zhang Y, Ku L, Wilkinson KD, Warren ST, Feng Y: The fragile X mental retardation protein inhibits its translation via interacting with mRNA. Nucl Acids Res 29:2276–2283 (2001).
Lu W, Man H, Ju W, Trimble WS, MacDonald JF, Wang YT: Activation of synaptic NMDA receptors induces membrane insertion of new AMPA receptors and LTP in cultured hippocampal neurons. Neuron 29:243–254 (2001).
Maes B, Fryns JP, Van Walleghem M, Van den Berghe H: Cognitive functioning and information processing of adult mentally retarded men with fragile-X syndrome. Am J med Genet 50:190–200 (1994).
Malter HE, Iber JC, Willemsen R, De Graaff E, Tarleton JC, Leisti J, Warren ST, Oostra BA: Characterization of the full fragile X syndrome mutation in fetal gametes. Nature Genet 15:165–169 (1997).
McConkie-Rosell A, Lachiewicz A, Spiridigliozzi GA, Tarleton J, Schoenwald S, Phelan MC, Goonewardena P, Ding X, Brown WT: Evidence that methylation of the FMR1 locus is responsible for variant phenotypic expression of the fragile X syndrome. Am J hum Genet 53:800–809 (1993).
McHugh TJ, Blum KI, Tsien JZ, Tonegawa S, Wilson MA: Impaired hippocampal representation of space in CA1-specific NMDAR1 knockout mice. Cell 87:1339–1349 (1996).
Meijer H, De Graaff E, Merckx DML, Jongbloed RJE, De Die-Smulders CEM, Engelen JJM, Fryns JP, Curfs PMG, Oostra BA: A deletion of 1.6 kb proximal to the CGG repeat of the FMR1 gene causes the clinical phenotype of the Fragile X syndrome. Hum molec Genet 3:615–620 (1994).
Merenstein SA, Sobesky WE, Taylor AK, Riddle JE, Tran HX, Hagerman RJ: Molecular-clinical correlations in males with an expanded FMR1 mutation. Am J med Genet 64:388–394 (1996).
Merlin LR, Bergold PJ, Wong RK: Requirement of protein synthesis for group I mGluR-mediated induction of epileptiform discharges. J Neurophysiol 80:989–993 (1998).
Mila M, Castellvi-Bel S, Sanchez A, Barcelo A, Badenas C, Mallolas J, Estivill X: Rare variants in the promoter of the fragile X syndrome gene (FMR1). Mol Cell Probes 14:115–119 (2000).
Miller LJ, McIntosh DN, McGrath J, Shyu V, Lampe M, Taylor AK, Tassone F, Neitzel K, Stackhouse T, Hagerman RJ: Electrodermal responses to sensory stimuli in individuals with fragile X syndrome: a preliminary report. Am J med Genet 83:268–279 (1999).

Mineur YS, Sluyter F, de WS, Oostra BA, Crusio WE: Behavioral and neuroanatomical characterization of the *Fmr1* knockout mouse. Hippocampus 12: 39–46 (2002).
Moore PS, Chudley AE, Winter JS: Pituitary-gonadal axis in prepubertal boys with the fragile X syndrome. Am J med Genet 39:374–375 (1991).
Morales J, Hiesinger PR, Schroeder AJ, Kume K, Verstreken PF, Jackson FR, Nelson DL, Hassan BA: Drosophila Fragile X protein, DFXR, regulates neuronal morphology and function in the brain. Neuron 34:961–972 (2002).
Morgan JI, Curran T: Proto-oncogene transcription factors and epilepsy. Trends Pharmacol Sci 12: 343–349 (1991).
Morris R Developments of a water-maze procedure for studying spatial learning in the rat. J Neurosci Meth 11:47–60 (1984).
Morris RG, Garrud P, Rawlins JN, O'Keefe J: Place navigation impaired in rats with hippocampal lesions. Nature 297:681–683 (1982).
Musleh W, Bi X, Tocco G, Yaghoubi S, Baudry M: Glycine-induced long-term potentiation is associated with structural and functional modifications of alpha-amino-3-hydroxyl-5-methyl-4-isoxazolepropionic acid receptors. Proc natl Acad Sci, USA 94:9451–9456 (1997).
Musumeci SA, Bosco P, Calabrese G, Bakker C, De Sarro GB, Elia M, Ferri R, Oostra BA: Audiogenic seizures susceptibility in transgenic mice with fragile X syndrome. Epilepsia 41:19–23 (2000).
Musumeci SA, Hagerman RJ, Ferri R, Bosco P, Dalla Bernardina B, Tassinari CA, De Sarro GB, Elia M: Epilepsy and EEG findings in males with fragile X syndrome. Epilepsia 40:1092–1099 (1999).
Nayak A, Zastrow DJ, Lickteig R, Zahniser NR, Browning MD: Maintenance of late-phase LTP is accompanied by PKA-dependent increase in AMPA receptor synthesis. Nature 394:680–683 (1998).
Nielsen DM, Derber WJ, McClellan DA, Crnic LS: Alterations in the auditory startle response in *Fmr1* targeted mutant mouse models of fragile X syndrome. Brain Res 927:8–17 (2002).
Nielsen KB, Tommerup N, Dyggve HV, Schou C: Macroorchidism and fragile X in mentally retarded males. Clinical, cytogenetic, and some hormonal investigations in mentally retarded males, including two with the fragile site at Xq28, fra(X)(q28). Hum Genet 61:113–117 (1982).
Nimchinsky EA, Oberlander AM, Svoboda K: Abnormal development of dendritic spines in *Fmr1* knock-out mice. J Neurosci 21:5139–5146 (2001).
Oberlé I, Rousseau F, Heitz D, Kretz C, Devys D, Hanauer A, Boue J, Bertheas MF, Mandel JL: Instability of a 550-base pair DNA segment and abnormal methylation in fragile X syndrome. Science 252:1097–1102 (1991).
Orth JM: Proliferation of Sertoli cells in fetal and postnatal rats: a quantitative autoradiographic study. Anat Rec 203:485–492 (1982).
Paradee W, Melikian HE, Rasmussen DL, Kenneson A, Conn PJ, Warren ST: Fragile X mouse: strain effects of knockout phenotype and evidence suggesting deficient amygdala function. Neuroscience 94:185–192 (1999).
Peier AM, McIlwain KL, Kenneson A, Warren ST, Paylor R, Nelson DL: (Over) correction of FMR1 deficiency with YAC transgenics: behavioral and physical features. Hum molec Genet 9:1145–1159 (2000).
Pelletier MR, Hablitz JJ: Tetraethylammonium-induced synaptic plasticity in rat neocortex. Cereb Cortex 6:771–780 (1996).
Pieretti M, Zhang FP, Fu YH, Warren ST, Oostra BA, Caskey CT, Nelson DL: Absence of expression of the FMR-1 gene in fragile X syndrome. Cell 66: 817–822 (1991).
Raymond CR, Thompson VL, Tate WP, Abraham WC: Metabotropic glutamate receptors trigger homosynaptic protein synthesis to prolong long-term potentiation. J Neurosci 20:969–976 (2000).
Reiss AL, Abrams MT, Greenlaw R, Freund L, Denckla MB: Neurodevelopmental effects of the FMR-1 full mutation in humans. Nature Med 1:159–167 (1995).
Reiss AL, Aylwarth E, Freund LS, Joshi PK, Bryan RN: Neuroanatomy of fragile X syndrome: the posterior fossa. Ann Neurol 29:26–32 (1991a).
Reiss AL, Freund L: Behavioral phenotype of fragile X syndrome: DSM-III-R autistic behavior in male children. Am J med Genet 43:35–46 (1992).
Reiss AL, Freund L, Tseng JE, Joshi PK: Neuroanatomy in fragile X females: the posterior fossa. Am J hum Genet 49:279–288 (1991b).
Reiss AL, Patel S, Kumar AJ, Freund L: Preliminary communication: neuronanatomical variations of the posterior fossa in men with the fragile X (Martin Bell) syndrome. Am J med Genet 31:407–414 (1988).
Reyniers E, Vanbockstaele DR, Deboulle K, Kooy RF, Bakker CE, Oostra BA, Willems PJ: Mean corpuscular hemoglobin is not increased in *Fmr1* knockout mice. Hum Genet 97:49–50 (1996).
Rousseau F, Heitz D, Biancalana V, Blumenfeld S, Kretz C, Boue J, Tommerup N, Van Der Hagen C, DeLozier-Blanchet C, Croquette MF, Gilgenkranz S, Jalbert P, Voelckel MA, Oberlé I, Mandel JL: Direct diagnosis by DNA analysis of the fragile X syndrome of mental retardation. New Engl J Med 325:1673–1681 (1991a).
Rousseau F, Heitz D, Oberlé I, Mandel JL: Selection in blood cells from female carriers of the fragile X syndrome: inverse correlation between age and proportion of active X chromosomes carrying the full mutation. J med Genet 28:830–836 (1991b).
Rudelli RD, Brown WT, Wisniewski K, Jenkins EC, Laure-Kamionowska M, Connell F, Wisniewski HM: Adult fragile X syndrome. Clinico-neuropathologic findings. Acta Neuropathol 67:289–295 (1985).
Sakimura K, Kutsuwada T, Ito I, Manabe T, Takayama C, Kushiya E, Yagi T, Aizawa S, Inoue Y, Sugiyama H, et al: Reduced hippocampal LTP and spatial learning in mice lacking NMDA receptor epsilon 1 subunit. Nature 373:151–155 (1995).
Schaefer GB, Thompson JN, Bodensteiner JB, McConnell JM, Kimberling WJ, Gay CT, Dutton WD, Hutchings DC, Gray SB: Hypoplasia of the cerebellar vermis in neurogenetic syndromes. Ann Neurol 39:382–385 (1996).
Schaeffer C, Bardoni B, Mandel JL, Ehresmann B, Ehresmann C, Moine H: The fragile X mental retardation protein binds specifically to its mRNA via a purine quartet motif. Embo J 20:4803–4813 (2001).
Schwegler H, Crusio WE: Correlations between radial-maze learning and structural variations of septum and hippocampus in rodents. Behav Brain Res 67:29–41 (1995).
Schwegler H, Crusio WE, Brust I: Hippocampal mossy fibers and radial-maze learning in the mouse: a correlation with spatial working memory but not with non-spatial reference memory. Neuroscience 34: 293–298 (1990).
Sharpe RM: Experimental evidence for Sertoli-germ cell and Sertoli-Leydig cell interactions, in Russel LGD, Griswold MD (eds): The Sertoli cell, pp 391–418 (Cache River Press, Clearwater 1993).
Shimizu M, Gellibolian R, Oostra BA, Wells RD: Cloning and characterization, and properties of plasmids containing CGG triplet repeats from the FMR-1 gene. J molec Biol 258:614–626 (1996).
Silva AJ, Paylor R, Wehner JM, Tonegawa S: Impaired spatial learning in alpha-calcium-calmodulin kinase II mutant mice. Science 257:206–211 (1992a).
Silva AJ, Stevens CF, Tonegawa S, Wang Y: Deficient hippocampal long-term potentiation in alpha-calcium-calmodulin kinase II mutant mice. Science 257:201–206 (1992b).
Siomi H, Choi M, Siomi MC, Nussbaum RL, Dreyfuss G: Essential role for KH domains in RNA binding: impaired RNA binding by a mutation in the KH domain of FMR1 that causes fragile X syndrome. Cell 77:33–39 (1994).
Siomi H, Siomi MC, Nussbaum RL, Dreyfuss G: The protein product of the fragile X gene, FMR1, has characteristics of an RNA-binding protein. Cell 74:291–298 (1993).
Sittler A, Devys D, Weber C, Mandel J-L: Alternative splicing of exon 14 determines nuclear or cytoplasmic localization of FMR1 protein isoforms. Hum molec Genet 5:95–102 (1996).
Slegtenhorst-Eegdeman KE, van de Kant HJG, Post M, Ruiz A, Uilenbroek JTJ, Bakker CE, Oostra BA, Grootegoed JA, de Rooij DG, Themmen APN: Macro-orchidism in FMR1 knockout mice is caused by increased Sertoli cell proliferation during testis development. Endocrinology 139:156–162 (1998).
Smeets H, Smits A, Verheij CE, Theelen J, Willemsen R, Losekoot M, Van de Burgt I, Hoogeveen AT, Oosterwijk J, Oostra BA: Normal phenotype in two brothers with a full FMR1 mutation. Hum molec Genet 4:2103–2108 (1995).
Smits A, Smeets D, Hamel B, Dreesen J, Dehaan A, Van Oost B: Prediction of mental status in carriers of the fragile X mutation using CGG repeat length. Am J med Genet 51:497–500 (1994).
Snyder EM, Philpot BD, Huber KM, Dong X, Fallon JR, Bear MF: Internalization of ionotropic glutamate receptors in response to mGluR activation. Nature Neurosci 4:1079–1085 (2001).
Steward O, Bakker CE, Willems PJ, Oostra BA: No evidence for disruption of normal patterns of mRNA localization in dendrites or dendritic transport of recently synthesized mRNA in FMR1 knockout mice, a model for human fragile-X mental retardation syndrome. Neuroreport 9:477–481 (1998).
Sung YJ, Conti J, Currie JR, Brown WT, Denman RB: RNAs That interact with the Fragile X syndrome RNA binding protein FMRP. Biochem biophys Res Commun 275:973–980 (2000).
Sutcliffe JS, Nelson DL, Zhang F, Pieretti M, Caskey CT, Saxe D, Warren ST: DNA methylation represses FMR-1 transcription in fragile X syndrome. Hum molec Genet 1:397–400 (1992).
Tamanini F, Meijer N, Verheij C, Willems PJ, Galjaard H, Oostra BA, Hoogeveen AT: FMRP is associated to the ribosomes via RNA. Hum molec Genet 5:809–813 (1996).
Tamanini F, Van Unen L, Bakker C, Sacchi N, Galjaard H, Oostra BA, Hoogeveen AT: Oligomerization properties of fragile-X mental-retardation protein (FMRP). and the fragile-X-related proteins FXR1P and FXR2P. Biochem J 343:517–523 (1999).
Tamanini F, Willemsen R, van Unen L, Bontekoe C, Galjaard H, Oostra BA, Hoogeveen AT: Differential expression of FMR1, FXR1 and FXR2 proteins in human brain and testis. Hum molec Genet 6:1315–1322 (1997).
Tassone F, Hagerman RJ, Taylor AK, Mills JB, Harris SW, Gane LW, Hagerman PJ: Clinical involvement and protein expression in individuals with the FMR1 premutation. Am J med Genet 91:144–152 (2000).
Taylor AK, Safanda JF, Fall MZ, Quince C, Lang KA, Hull CE, Carpenter I, Staley LW, Hagerman RJ: Molecular predictors of cognitive involvement in female carriers of fragile X syndrome. J Am med Assoc 271:507–514 (1994).

Taylor AK, Tassone F, Dyer PN, Hersch SM, Harris JB, Greenough WT, Hagerman RJ: Tissue heterogeneity of the FMR1 mutation in a high-functioning male with fragile X syndrome. Am J med Genet 84:233–239 (1999).

Tsien JZ: Linking Hebb's coincidence-detection to memory formation. Curr Opin Neurobiol 10:266–273 (2000).

Turner G, Webb T, Wake S, Robinson H: The prevalence of the fragile X syndrome. Am J med Genet 64:196–197 (1996).

Van Dam D, D'Hooge R, Hauben E, Reyniers E, Gantois I, Bakker CE, Oostra BA, Kooy RF, De Deyn PP: Spatial learning, contextual fear conditioning and conditioned emotional response in *Fmr1* knockout mice. Behav Brain Res 117:127–136 (2000).

van Heyningen V: Model organisms illuminate human genetics and disease. Mol Med 3:231–237 (1997).

Verheij C, Bakker CE, de Graaff E, Keulemans J, Willemsen R, Verkerk AJ, Galjaard H, Reuser AJ, Hoogeveen AT, Oostra BA: Characterization and localization of the FMR-1 gene product associated with fragile X syndrome. Nature 363:722–724 (1993).

Verkerk AJ, Pieretti M, Sutcliffe JS, Fu YH, Kuhl DP, Pizzuti A, Reiner O, Richards S, Victoria MF, Zhang FP, Eussen BE, Van Ommen GJB, Blonden LAJ, Riggins GJ, Chastain JL, Kunst CB, Galjaard H, Caskey CT, Nelson DL, Oostra BA, Warren ST: Identification of a gene (FMR-1). containing a CGG repeat coincident with a breakpoint cluster region exhibiting length variation in fragile X syndrome. Cell 65:905–914 (1991).

Wan L, Dockendorff TC, Jongens TA, Dreyfuss G: Characterization of dFMR1, a *Drosophila melanogaster* homolog of the Fragile X mental retardation protein. Mol Cell Biol 20:8536–8547 (2000).

Weiler IJ, Greenough WT: Synaptic synthesis of the Fragile X protein: possible involvement in synapse maturation and elimination. Am J med Genet 83:248–252 (1999).

Weiler IJ, Irwin SA, Klintsova AY, Spencer CM, Brazelton AD, Miyashiro K, Comery TA, Patel B, Eberwine J, Greenough WT: Fragile X mental retardation protein is translated near synapses in response to neurotransmitter activation. Proc natl Acad Sci, USA 94:5395–5400 (1997).

Wells RD: Molecular basis of genetic instability of triplet repeats. J biol Chem 271:2875–2878 (1996).

White EL, Weinfeld L, Lev DL: A survey of morphogenesis during the early postnatal period in PMBSF barrels of mouse SmI cortex with emphasis on barrel D4. Somatosens Mot Res 14:34–55 (1997).

White PJ, Borts RH, Hirst MC: Stability of the human fragile X $(CGG)_n$ triplet repeat array in *Saccharomyces cerevisiae* deficient in aspects of DNA metabolism. Mol Cell Biol 19:5675–5684 (1999).

Willemsen R, Bontekoe C, Tamanini F, Galjaard H, Hoogeveen AT, Oostra BA: Association of FMRP with ribosomal precursor particles in the nucleolus. Biochem biophys Res Comm 225:27–33 (1996).

Wöhrle D, Kotzot D, Hirst MC, Manca A, Korn B, Schmidt A, Barbi G, Rott HD, Poustka A, Davies KE, Steinbach P: A microdeletion of less than 250 kb, including the proximal part of the FMR-I gene and the fragile-X site, in a male with the clinical phenotype of fragile-X syndrome. Am J hum Genet 51:299–306 (1992).

Wöhrle D, Salat U, Glaser D, Mucke J, Meisel-Stosiek M, Schindler D, Vogel W, Steinbach P: Unusual mutations in high functioning fragile X males: apparent instability of expanded unmethylated CGG repeats. J med Genet 35:103–111 (1998).

Wohrle D, Schwemmle S, Steinbach P: DNA methylation and triplet repeat stability: New proposals addressing actual questions on the CGG repeat of fragile X syndrome – Letter to the editor. Am J med Genet 64:266–267 (1996).

Wong RK, Bianchi R, Taylor GW, Merlin LR: Role of metabotropic glutamate receptors in epilepsy. Adv Neurol 79:685–698 (1999).

Wu ZL, Thomas SA, Villacres EC, Xia Z, Simmons ML, Chavkin C, Palmiter RD, Storm DR: Altered behavior and long-term potentiation in type I adenylyl cyclase mutant mice. Proc natl Acad Sci, USA 92:220–224 (1995).

Zamanillo D, Sprengel R, Hvalby O, Jensen V, Burnashev N, Rozov A, Kaiser KM, Koster HJ, Borchardt T, Worley P, Lubke J, Frotscher M, Kelly PH, Sommer B, Andersen P, Seeburg PH, Sakmann B: Importance of AMPA receptors for hippocampal synaptic plasticity but not for spatial learning. Science 284:1805–1811 (1999).

Zhang YQ, Bailey AM, Matthies HJ, Renden RB, Smith MA, Speese SD, Rubin GM, Broadie K: Drosophila fragile X-related gene regulates the MAP1B homolog Futsch to Control synaptic structure and function. Cell 107:591–603 (2001).

Cytogenet Genome Res 100:124–128 (2003)
DOI: 10.1159/000072846

Expression of the FMR1 gene

F. Tassone and P.J. Hagerman

Department of Biological Chemistry, University of California, Davis, School of Medicine, Davis CA (USA)

Abstract. Fragile X syndrome is a trinucleotide repeat disorder in which a $(CGG)_n$ element located within the 5′ untranslated region of the Fragile X Mental Retardation 1 (FMR1) gene expands to more than 200 copies (full mutation) and becomes hypermethylated. Such expansions are accompanied by the failure to produce FMR1 protein (FMRP), resulting in the fragile X phenotype. For smaller (premutation) expansions (~55–200 repeats), FMR1 mRNA and FMRP levels had been assumed to be normal; however, our group and others have recently demonstrated that FMR1 mRNA levels are elevated in cells harboring premutation alleles. Moreover, mRNA levels remain elevated in fragile X males with partially methylated full mutation alleles. Finally, some fragile X males with hypermethylated, full mutation alleles continue to produce FMR1 mRNA, despite the expectation that those genes should be silent. These observations all point to a complex mechanism of expression of the FMR1 gene, one that provides a more consistent foundation for the spectrum of clinical involvement. An FMRP deficit is observed in all categories of fragile X individuals, including carriers of the premutation and partially methylated full mutation alleles. These results demonstrate that lowered FMRP levels, in the absence of methylation-coupled silencing of the FMR1 gene, are not caused by reduced transcriptional activity, but rather by a reduced efficiency of translation.

Recent studies have dramatically changed our understanding of the expression of the fragile X gene. According to the standard model, the FMR1 gene becomes silenced as a consequence of the expansion and the methylation of the CGG trinucleotide repeat element located in the 5′-UTR of the gene (Pieretti et al., 1991). The resulting absence (or deficit) of the fragile X mental retardation protein (FMRP) causes fragile X syndrome. Based on this model, carriers of premutation alleles were assumed to be clinically normal, since methylation-coupled silencing does not occur in the premutation range. However, contrary to expectation, a subgroup of male and female carriers does show some clinical involvement. These include characteristic physical features, learning disabilities, anxiety, developmental delay and emotional problems. In addition, approximately 20% of female carriers of premutation alleles have premature ovarian failure, which is not observed in females with hypermethylated full mutation (Allingham-Hawkins et al., 1999).

A new neurodegenerative syndrome, characterized by a progressive intention tremor, ataxia, cognitive decline, and generalized brain atrophy, has been described in a subgroup of older male carriers of premutation alleles (Hagerman et al., 2001, 2003; Jacquemont et al., 2003). These clinical manifestations of the fragile X-associated tremor/ataxia syndrome (FXTAS), which have not been observed in males with a full mutation, are strongly suggestive of a process specifically involving FMR1 premutation alleles. The most consistent biochemical abnormality within the premutation range is an elevation of FMR1 mRNA, which has been observed in male and female carriers of either premutation or unmethylated full mutation alleles (Tassone et al., 2000a, b, c). It has been suggested that the progressive neurodegenerative process may be caused, directly or indirectly, by the elevation of FMR1 message levels (Hagerman et al., 2001; Greco et al., 2002).

In full-mutation alleles, hypermethylation of CpG dinucleotides within the FMR1 promoter correlates with the lack of mRNA (Pieretti et al., 1991; Hornstra et al., 1993; Imbert et al.,

Supported by a UC Davis Health System Research Award (F.T.), and a grant from the National Institute of Child Health and Development (HD40661; P.J.H.). The authors also wish to thank the Boory and Cooper/Kraff/Fishman Family Funds for their support of this research. This research is also supported by the UC Davis M.I.N.D. Institute.

Received 11 November 2002; manuscript accepted 27 November 2002.

Request reprints from Flora Tassone PhD, Department of Biological Chemistry
University of California, Davis, School of Medicine
One Shields Avenue, Davis CA (USA); telephone: 530-754 7268
fax: 530-754-7269; e-mail: ftassone@ucdavis.edu

KARGER
Fax + 41 61 306 12 34
E-mail karger@karger.ch
www.karger.com

0301–0171/03/1004–0124$19.50/0

Accessible online at:
www.karger.com/cgr

1998). However, the assumed relationship between methylation (defined as resistance to cleavage by methyl-sensitive enzymes) and FMR1 gene silencing is not generally valid, as some fragile X males with methylated, fully expanded alleles continue to produce FMR1 mRNA (Tassone et al., 2001). The observation that hypermethylated, full-mutation alleles can escape silencing (i.e., remain transcriptionally active) is of direct relevance to our understanding of the molecular mechanism of transcriptional silencing and the development of fragile X syndrome. Such an understanding is also clinically important because it could lead to the identification of new therapeutic approaches designed to increase the level of FMR1 protein through translational activation.

FMR1 gene expression in male carriers of the fragile X premutation

Until recently, FMR1 gene activity and FMRP levels had been considered to be normal for alleles in the premutation range (Devys et al., 1993; Verheij et al., 1993; Feng et al., 1995a; Kaufmann et al., 1999; Tassone et al., 1999a). However, recent studies have demonstrated that FMRP expression levels are diminished, especially in the upper premutation range, and are negatively correlated with the CGG repeat number (Tassone et al., 2000a, d; Kenneson et al., 2001). Although the mechanism causing the spectrum of clinical involvement in premutation carriers is not completely understood, reduction of FMRP levels is consistent with the presence of mild fragile X features in some premutation carriers.

It has now become clear that the reduced FMRP expression in premutation carriers is not the direct consequence of decreased gene activity. Using a quantitative approach, we have recently demonstrated that FMR1 mRNA expression is elevated in premutation males (Tassone et al., 2000a), with message levels that are 5- to 10-fold higher than normal, despite the lowered FMRP levels. FMR1 mRNA levels are displayed for 57 premutation males in Fig. 1. Message elevations are most pronounced in the upper premutation range; however, the levels begin to increase only slightly above the upper end of the normal repeat range. The FMR1 mRNA levels measured by quantitative fluorescence RT-PCR are clearly correlated with the CGG repeat number. While a wide range of expression levels is observed in the low premutation range (<100 CGG repeats), from 8-fold higher to nearly normal (Fig. 1), overlap in FMR1 mRNA levels between premutation carriers and normal individuals is not seen in the upper premutation range. The observed elevation of FMR1 mRNA levels in males with a premutation provides direct evidence for altered regulation of the FMR1 gene, and supports the idea that premutation alleles may contribute to the disease phenotype among premutation carriers.

The reduction of FMRP in the premutation range is now known to be a result of reduced translational efficiency of the FMR1 mRNA (Primerano et al., 2002), presumably due to the effects of the expanded CGG repeat in the 5′ untranslated region of the FMR1 mRNA. Such an impediment had been shown to be responsible for the absence of FMRP in unmethylated full mutation alleles (Feng et al., 1995b). Within the premutation range, a strong correlation is observed between translation efficiency of FMR1 mRNA and the number of CGG repeats in premutation alleles (Primerano et al., 2002). The PMP (percent of message on polysomes) calculated for both FMR1 mRNA and for beta-actin mRNA (control), was much lower in premutation alleles than normal. In addition, FMR1 mRNA was observed to shift from heavy to lighter polysomes with increasing CGG repeat size, suggesting a defect in translation initiation. The strong correlation ($r^2 = 0.92$) between translational activity of FMR1 mRNA, expressed by the PMP value, and FMRP levels is demonstrated in Fig. 2.

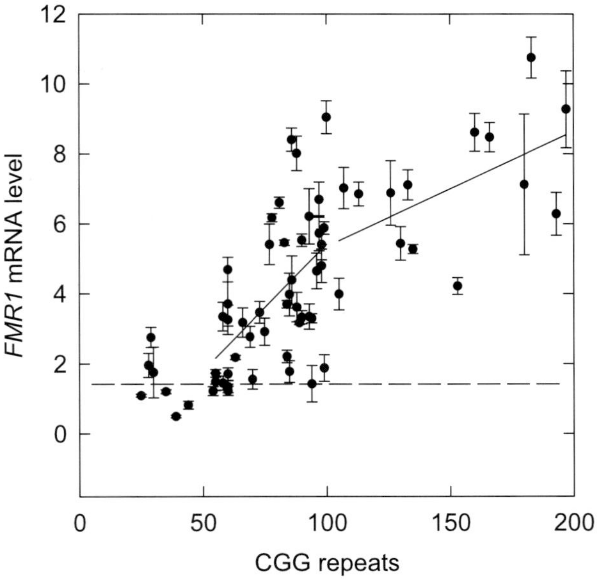

Fig. 1. Graph of FMR1 mRNA levels as a function of CGG repeat size for alleles within the premutation size range. Horizontal dashed line represents the mean mRNA level for normal controls. Solid lines represent separate linear regressions within the lower (55–100) and upper (100–200) repeat ranges. Message levels are reported relative to glucuronidase mRNA levels as control. Method details are described in Tassone et al. (2000a).

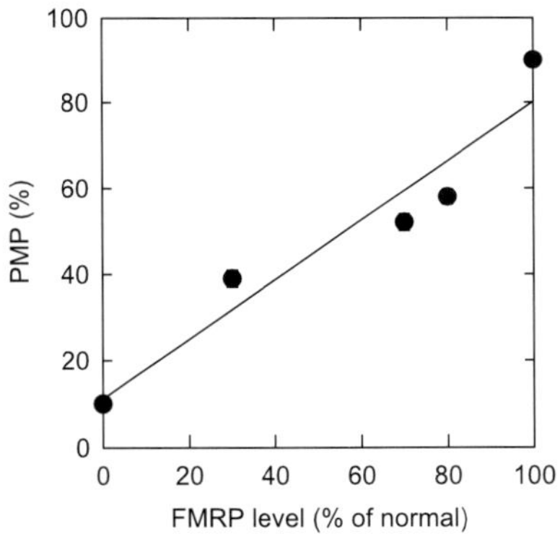

Fig. 2. Plot of the relationship between the percent of FMR1 mRNA on polysomes (PMP) and FMRP levels detected by Western blot (Primerano et al., 2002). ($r^2 = 0.92$).

Fig. 3. Hypothetical feedback mechanism whereby lowered FMRP levels lead to increased transcription of the FMR1 gene. The causal relationship between lowered FMRP levels and elevated mRNA levels has yet to be determined.

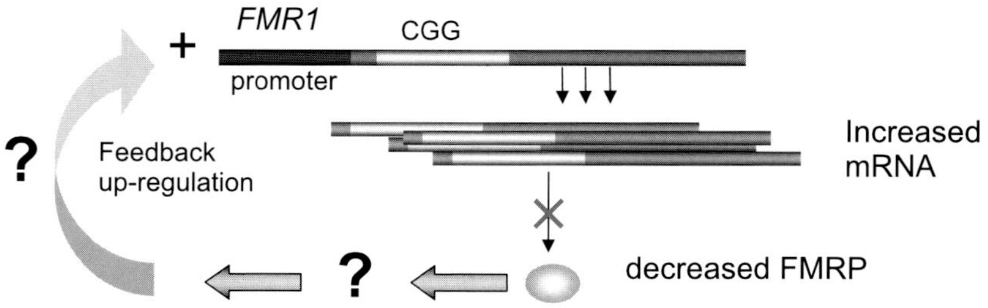

The combination of reduced FMRP levels, despite elevated mRNA, suggests that the elevated FMR1 levels may represent a feedback response to an impaired translation, which causes an FMRP deficit (Fig. 3). However, the hypothesized feedback mechanism still awaits direct confirmation. It is also possible that the expanded CGG repeat exerts its effect, in some fashion, at the DNA level. Studies to test these models are currently in progress.

Finally, while it has been hypothesized that elevated FMR1 mRNA levels may contribute to the neuropathology experienced by some older male carriers of premutation alleles via a gain-of-function toxicity of the mRNA (Hagerman et al., 2001, 2003; Greco et al., 2002), it is not known why only a subset (~ 20%) of male carriers present with the tremor/ataxia syndrome. It is likely that a synergism with other genetic loci may play a role.

Expression in males with mosaicism

Mosaicism of the FMR1 locus is common, and can take several different forms. In the context of the FMR1 gene, allele size mosaicism normally refers to the presence of both hypermethylated full mutation and premutation alleles, although for individuals with hypermethylated full mutation alleles, multiple alleles (mosaicism) are nearly always present within the full mutation size range. Males with methylation mosaicism possess methylated alleles in the full mutation range and unmethylated alleles in the premutation and/or full mutation range. For both allele-size and methylation mosaics, the fragile X phenotype is less severe, depending mainly on the percent of cells carrying alleles in the premutation range, or the percent of full mutation alleles that remain unmethylated (Hagerman et al., 1994; Merenstein et al., 1994; Rousseau et al., 1994; Smeets et al., 1995; de Vries et al., 1996; Lachiewicz et al., 1996; Wang et al., 1996; Wohrle et al., 1998; Burman et al., 1999; Tassone et al., 1999a; Taylor et al., 1999). The less severe phenotype observed in these individuals reflects either the production of FMRP from premutation alleles or the continued production of high levels of mRNA from unmethylated full mutation alleles, which implies that transcription of the FMR1 gene is still efficient in full mutation alleles.

Quantitative measures of FMR1 mRNA levels determined in both size and methylation mosaics have confirmed that the FMR1 gene is actively transcribing, with a broad range of mRNA levels, depending substantially on the fraction of alleles in the premutation size range. Up to six-fold elevations of message levels have also been observed for unmethylated alleles within the full mutation range, indicating that full mutation expansions do not block transcription of the FMR1 gene (Tassone et al., 2000b; unpublished results). Therefore, the lower FMRP levels detected in these individuals do not arise from gene silencing, but rather from a defect in the utilization (translation) of the FMR1 mRNA.

Expression in males with hypermethylated full mutation

Full mutation alleles are generally hypermethylated within the CGG repeat element and the upstream CpG island, resulting in transcriptional silencing of the FMR1 gene and absence of FMRP (Pieretti et al., 1991 Verheij et al., 1993). Because of the assumed association between methylation and silencing, methylation of the FMR1 gene was generally believed to *necessarily* lead to transcriptional silencing at this locus. However, we have recently reported that true transcriptional silencing occurs in fewer than half of full mutation individuals with hypermethylated, full mutation alleles (Tassone et al., 2001). From a group of 48 males with hypermethylated alleles (defined by complete resistance to cleavage by methyl-sensitive restriction enzymes), more than half showed some expression of the FMR1 gene, with a broad range of mRNA levels from near normal (six cases) to undetectable (Tassone et al., 2001).

Additional studies demonstrated that the observed FMR1 mRNA levels were not due to unmethylated alleles that had not been detected by Southern blot analysis (Tassone et al., 2001).

In our study more than 50% of the individuals with a hypermethylated full mutation showed the presence of FMR1 mRNA; nevertheless, the ones with an FMR1 mRNA level within the normal range were rare, and only in 15% of the cases were the levels >50% of normal. In all the remaining cases FMR1 message was detected only because of the highly sensitive method used in the analysis. This could also be the reason why previously reported data have not shown transcription in hypermethylated full mutation alleles. In addition, FMR1 mRNA levels were measured on peripheral blood leukocytes and not in cell lines, which have been the primary source in most previous studies.

The reason why many hypermethylated, full mutation alleles remain transcriptionally active is not understood, although preliminary data suggest that methylation of a specific subset of CpG elements located within the CpG island may be critical for silencing (Tassone, unpublished results).

Bisulfite-coupled sequencing of the proximal promoter elements is in progress, in an attempt to identify whether specific CpG methylation events may play a role in the silencing process. Alternatively, transcriptional activity may be a consequence of altered patterns of histone modification, including methylation and acetylation, that may be partially uncoupled from the methylation events in some individuals. It has been shown that histones H3 and H4 are hypoacetylated within the FMR1 promoter region in fragile X individuals, a common characteristic of transcriptionally silent loci (Coffee et al., 1999). In addition, high levels of histone H3 methylation at lysine 9, and decreased methylation of histone H3 at lysine 4, have also been observed in cells carrying fragile X hypermethylated alleles (Coffee et al., 2002). Thus, it is possible that the pathways involved with hyopacetylation and/or histone methylation may be altered in these individuals.

Although FMR1 mRNA was present in more than half of the samples with hypermethylated alleles (Tassone et al., 2001), none of the subjects analyzed showed significant FMRP production (<10% by immunocytochemistry). The absence of FMRP thus reflects a post-transcriptional effect, specifically, a defect in translational efficiency, as reported by Feng et al. (1995). The absence of FMRP explains why these individuals have the full fragile X phenotype despite the presence of varying levels of FMR1 transcript.

Conclusions

Recent studies have shown that FMR1 mRNA levels are elevated in white blood cells of carriers of fragile X premutation alleles. The highest levels are observed in premutation carriers with a CGG repeat number greater than 100, although smaller elevations are also present in the lower premutation range. No overlap in the FMR1 message levels is observed for alleles in the high premutation range or alleles in the normal range. Thus, dysregulation of FMR1 gene expression is present even for alleles approaching the normal repeat range. The elevation of FMR1 message is unlikely to be due to increased message stability (Tassone et al., 2000a), but rather by a direct increase in the transcription of the FMR1 gene (Tassone et al., unpublished results). The reduced translational efficiency of the FMR1 mRNA (Primerano et al., 2002), beginning in the low premutation range, suggests that the increased transcriptional activity of the FMR1 gene may represent a cellular response to decreased FMRP production. This mechanism may be at least partially compensatory within the premutation range, but would fail within the full mutation range as the translational impediment becomes more severe. The mechanism by which this response would be accomplished is not clear, but research efforts are focused on this issue.

High levels of FMR1 mRNA have also been detected in unmethylated, full mutation alleles, indicating that the size of the CGG repeat element does not represent a major impediment to transcription. The more surprising observation of transcriptionally active, hypermethylated full mutation alleles raises new questions as to the mechanisms of regulation of the FMR1 gene expression, and the nature and extent of silencing at the FMR1 locus. These findings have important clinical implications: understanding why the FMR1 gene remains active for some individuals with hypermethylated alleles may provide clues to guide possible therapeutic reactivation of the gene. Although the end result of both transcriptional silencing and translational suppression is essentially the same – absence of FMRP – separate consideration of the two processes is also important. Fragile X individuals who exhibit translational suppression of FMR1 mRNA, but are transcriptionally active, may be candidates for therapeutic intervention that is directly targeted to increase translational efficiency without the necessity of altering transcriptional activity.

References

Allingham-Hawkins DJ, Babul-Hirji R, Chitayat D, Holden JJ, Yang KT, Lee C, et al: Fragile X premutation is a significant risk factor for premature ovarian failure: The International Collaborative POF in Fragile X study-preliminary data. Am J med Genet 83:322–325 (1999).

Burman RW, Yates PA, Green LD, Jacky PB, Turker MS, Popovich BW: Hypomethylation of an expanded *FMR1* allele is not associated with a global DNA methylation defect. Am J hum Genet 65: 1375–1386 (1999).

Coffee B, Zhang F, Warren ST, Reines D: Acetylated histones are associated with *FMR1* in normal but not fragile X-syndrome cells. Nature Genet 22:98–101(1999).

Coffee B, Zhang F, Ceman S, Warren ST, Reines D: Histone modifications depict an aberrantly heterochromatinized *FMR1* gene in fragile X syndrome. Am J hum Genet 71:923–932 (2002).

De Vries BBA, Jansen CCAM, Duits AA, Verheij C, Willemsen R, van Hemel JO, van den Ouweland AMW, Niermeijer MF, Oostra BA, Halley DJJ: Variable *FMR1* gene methylation of large expansions leads to variable phenotype in three males from one fragile X family. J med Genet 33:1007–1010 (1996).

Devys D, Lutz Y, Rouyer N, Bellocq JP, Mandel JL: The *FMR-1* protein is cytoplasmic, most abundant in neurons and appears normal in carriers of a fragile X premutation. Nature Genet 4:335–340 (1993).

Feng Y, Lakkis L, Devys D, Warren ST: Quantitative comparison of *FMR1* gene expression in normal and premutation alleles. Am J hum Genet 56:106–113 (1995a).

Feng Y, Zhang F, Lokey LK, Chastain JL, Lakkis L, Eberhart D, Warren ST: Translational suppression by trinucleotide repeat expansion at *FMR1*. Science 268:731–734 (1995b).

Greco CM, Hagerman RJ, Tassone F, Chudley AE, Del Bigio MR, Jacquemont, Leehey M, Hagerman PJ: Neuronal intranuclear inclusions in a new cerebellar tremor/ataxia syndrome among fragile X carriers. Brain 125:1760–1771 (2002).

Hagerman RJ, Hull CE, Safanda JF, Carpenter I, Staley LW, O'Connor, RA, Seydel C, Mazzocco MM, Snow K, Thibodeau SN, Kuhl D, Nelson DL, Caskey TC, Taylor AK: High functioning fragile X males: demonstration of an unmethylated fully expanded *FMR-1* mutation associated with protein expression. Am J med Genet 51:298–308 (1994).

Hagerman RJ, Leehey M, Heinrichs W, Tassone F, Wilson R, Hills J, Grisby J, Gage B, Hagerman PJ: Intention tremor, parkinsonism, and generalized brain atrophy in male carriers of fragile X. Neurology 57:127–130 (2001).

Hagerman PJ, Greco CM, Hagerman RJ: A cerebellar tremor/ataxia syndrome among fragile X carriers. Cytogenet Genome Res 100:206–212 (2003).

Hornstra IK, Yang TP: Multiple *in vivo* footprints are specific to the active allele of the X-linked human hypoxanthine phosphoribosyltransferase gene 5′ region: implications for X chromosome inactivation. Mol Cell Biol 12:5345–5354 (1993).

Imbert G, Feng Y, Nelson D, Warren ST, Mandel JL: *FMR1* and mutations in Fragile X syndrome: molecular biology, biochemistry, and genetics, in Wells RD, Warren, ST (eds): Genetic Instabilities and Hereditary Neurological Disorders, pp27–53 (Academic Press, San Diego 1998).

Jacquemont S, Hagerman RJ, Leehy M, Grigsby J, Zhang L, Brunberg JA, Greco C, Des Portes V, Jardini T, Levine R, Berry-Kravis E, Brown WT, Schaeffer S, Kissel J, Tassone F, Hagerman PJ: Fragile X premutation tremor/ataxia syndrome: Molecular, clinical and neuroimaging correlates. Am J hum Genet 72:869–878 (2003).

Kaufmann WE, Abrams MT, Chen W, Reiss AL: Genotype, molecular phenotype, and cognitive phenotype: correlations in fragile X syndrome. Am J med Genet 83:286–295 (1999).
Kenneson A, Zhang F, Hagedorn CH, Warren ST: Reduced FMRP and increased *FMR1* transcription is proportionally associated with CGG repeat number in intermediate-length and premutation carriers. Hum molec Genet 14:1449–1454 (2001).
Lachiewicz AM, Spiridigliozzi GA, McConkie-Rosell A, Burgess D, Feng Y, Warren ST, Tarleton J: A fragile X male with a broad smear on Southern blot analysis representing 100–500 CGG repeats and no methylation at the *Eag*I site of the *FMR1* gene. Am J med Genet 64:278–282 (1996).
Merenstein SA, Shyu V, Sobesky WE, Staley L, Berry-Kravis E, Nelson DL, Lugenbeel KA, Taylor AK, Pennington BF, Hagerman RJ: Fragile X syndrome in a normal IQ male with learning and emotional problems. J Am Acad Child Adolesc Psychiatry 33:1316–1321 (1994).
Pieretti M, Zhang FP, Fu YH, Warren ST, Oostra BA, Caskey CT, Nelson DL: Absence of expression of the FMR-1 gene in fragile X syndrome. Cell 66:817–822 (1991).
Primerano B, Tassone F, Hagerman RJ, Hagerman PJ, Amaldi F, Bagni C: Reduced *FMR1* mRNA translation efficiency in fragile X patients with premutations. RNA 8:1482–1488 (2002).
Rousseau F, Robb LJ, Rouillard P, Der Kaloustian VM: No mental retardation in a man with 40% abnormal methylation at the *FMR1* locus and transmission of sperm cell mutations as premutations. Hum molec Genet 3:927–930 (1994).
Smeets HJM, Smits APT, Verheij CE, Theelen JPG, Willemsen R, van de Burgt I, Hoogeveen AT, Oosterwijk, Oostra BA: Normal phenotype in two brothers with a full *FMR1* mutation. Hum molec Genet 4:2103–2108 (1995).
Tassone F, Hagerman RJ, Ikle D, Dyer PN, Lampe M, Willemsen R, Oostra BA, Taylor AK: FMRP expression as a potential prognostic indicator in fragile X syndrome. Am J med Genet 84:250–261 (1999a).
Tassone F, Hagerman RJ, Taylor AK, Gane LW, Godfrey TE, Hagerman PJ: Elevated levels of *FMR1* messenger RNA in carriers males: a new mechanism of involvement in the fragile X syndrome. Am J hum Genet 66:6–15 (2000a).
Tassone F, Hagerman RJ, Loesch DZ, Lachiewicz A, Taylor AK, Hagerman PJ: Fragile X males with unmethylated, full mutation trinucleotides repeat expansions and elevated levels of *FMR1* messenger. Am J med Genet 94:232–236 (2000b).
Tassone F, Hagerman RJ, Chamberlain WD, Hagerman PJ: Transcription of the *FMR1* gene in individuals with fragile X syndrome. Am J med Genet 97:195–203 (2000c).
Tassone F, Hagerman RJ, Taylor AK, Mills JB, Harris SW, Gane LW, Hagerman PJ: Clinical involvement and protein expression in individuals with the *FMR1* premutation. Am J med Genet 91:144–152 (2000d).
Tassone F, Hagerman, RJ, Taylor, AK, Hagerman, PJ: A majority of fragile X males with methylated, full mutation alleles have significant levels of *FMR1* messenger RNA. J med Genet 38:453–456 (2001).
Taylor AK, Tassone F, Dyer PN, Hersch SM, Harris JB, Greenough WT, Hagerman RJ: Tissue heterogeneity of the *FMR1* mutation in a high-functioning male with fragile X syndrome. Am J med Genet 84:233–239 (1999).
Verheij C, Bakker CE, de Graaff E, Keulemans J, Willemsen R, Verkerk AJ, Galjaard H, et al: Characterization and localization of the FMR-1 gene product associated with fragile X syndrome. Nature 363:722–724 (1993).
Wang Z, Taylor AK, Bridge JA: *FMR1* fully expanded mutation with minimal methylation in a high functioning fragile X male. J med Genet 33:376–378 (1996).
Wohrle D, Salat U, Glaser D, Mucke J, Meisel-Stosiek M, Schindler D, Vogel W, Steinbach P: Unusual mutations in high functioning fragile X males: apparent instability of expanded unmethylated CGG repeats. J Med Genet 35:103–111 (1998).

Cytogenet Genome Res 100:129–139 (2003)
DOI: 10.1159/000072847

FMR2 function: insight from a mouse knockout model

Y. Gu and D.L. Nelson

Department of Molecular and Human Genetics, Baylor College of Medicine, Houston TX (USA)

Abstract. The FMR2 gene is dysregulated by the fragile X E triplet repeat expansion in patients with FRAXE mental retardation syndrome. A CCG triplet, located in the 5′ untranslated region of the FRAXE gene undergoes expansion and methylation in these patients, eliminating detectable gene transcription. FRAXE syndrome is distinct from fragile X syndrome, a more common genetic form of mental retardation caused by expansion and methylation of a similar repeat in the FMR1 gene located 600 kb proximal to FRAXE. FRAXE syndrome is rare, and patients' phenotypes are highly variable, leading to difficulties with predicting specific FMR2 functions based on the human disease. Recently, *Lilliputian (Lilli)*, a Drosophila FMR2 orthologue, was identified; this gene has been linked with several signal transduction pathways, including the transforming growth factor-β (TGF-β) pathway, the Raf/MEK/MAP kinase (MAPK) pathway, and the P13K/PKB pathway. Mutation of *Lilli* shows defects in germinal band extension, cytoskeletal structure, cell growth, and organ development. The *Lilli* gene suggests possible functions for FMR2 (and related genes) in humans and mice, but cannot predict specific functions. Modeling *Fmr2* mutation in the mouse will be useful to understand specific functions of this gene in vertebrates. This review presents what has been learned thus far from the *Fmr2* knockout mouse model and suggests future studies on this model in order to compare it with the human FRAXE mental retardation disorder, *Lilli* mutants in Drosophila and other mouse models of genes in this family.

FMR2 gene and family

FMR2 is located in Xq28, approximately 600 kilobases (kb) distal to the FMR1 (Fragile X mental retardation 1) gene in Xq27.3. The human genome sequence is mostly completed (Lander et al., 2001; Venter et al., 2001), however a gap remains between the FMR1 and FMR2 genes, so that the precise distance remains unknown. According to the most recently updated version of the National Center for Biotechnology Information (NCBI) database, human FMR2 spans 500,225 base pairs (bp) and consists of 21 or 22 exons (Accession number: NT_019686 for FMR2 contig, Accession number: NT_011537 for FMR1, http://www.ncbi.nlm.nih.gov). Several alternatively spliced forms of FMR2 exist according to our data and other group's work. The largest protein encoded by FMR2 is predicted to be 1311 amino acids, corresponding to a 141-kDa band in Western blot studies (Gecz et al., 1996; Gu et al., 1996; Miller et al., 2000). Most of the alternatively spliced forms contain the conserved domains found in the FMR2/AF4 gene family. However one form encodes about one third of the largest form of FMR2 and its message ends at exon 10. The function of this small form is of interest (Chakrabarti et al., 1996; Gecz et al., 1997), since it shows altered patterns of expression from the major forms.

FMR2 is conserved in mammals as evidenced by zoo blot and DNA sequence data. Human FMR2 and murine *Fmr2* share 77% identity at the nucleotide level and 86% homology at amino acid level (Chakrabarti et al., 1998). The genomic structure of murine *Fmr2* is very similar to that of human

Supported by NIH grant HD29256.

Received 17 February 2003; manuscript accepted 28 February 2003.

Request reprints from David L. Nelson, Ph.D.
Department of Molecular and Human Genetics
Room 904E, Baylor College of Medicine
Houston, TX 77030 (USA)
telephone: 713-798-4787; fax: 713-798-1116
e-mail: nelson@bcm.tmc.edu

KARGER Fax + 41 61 306 12 34
E-mail karger@karger.ch
www.karger.com

Accessible online at:
www.karger.com/cgr

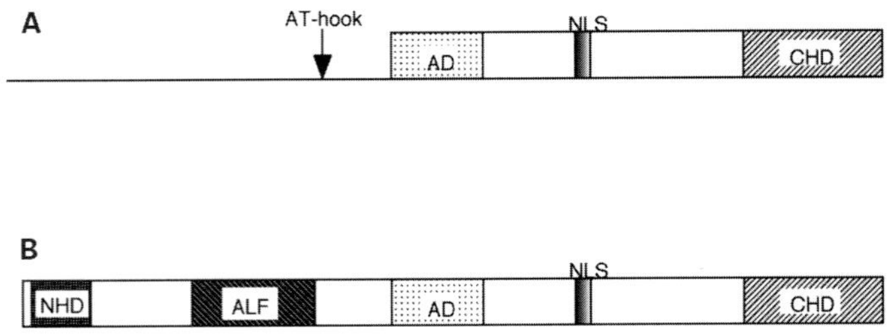

Fig. 1. Schematic comparison of Lilliputian protein and human FMR2/AF4 family. (**A**) Lilli protein. The square regions represent the homologous regions with human FMR2/AF4 family. The single line represents the region that shares no homology with the FMR2/AF4 family. (**B**) FMR2/AF4 gene family conserved domains: NHD: the N-terminal homologous domain, ALF: the AF4, LAF4, FMR2 homologous Domain, AD: The transactivation domain, NLS: the nuclear localization signal, CHD: the C-terminal domain.

FMR2 in several ways. First, the entire murine *Fmr2* spans about 510 kb of sequence, slightly larger than human FMR2, and is located 658 kb from murine *Fmr1*. Second, exons are conserved between the two species, with the sole exception of exon 5, which is not found in the murine genome. (All exons are named in this article according to human FMR2). Among the other 20 exons, 14 are conserved in size. These are exons 2–4, 6, 8–10, 12, and 15–20. The largest exon is exon 11 (1001 bp in human FMR2, 990 bp in murine *Fmr2*). Third, though the intron sizes between the two species show more variation, the two largest introns seem to have remained similarly large. In human FMR2, intron 1 is 150,855 bp. Due to the large size of this intron, the initial efforts in the identification of FMR2 were hampered. Spanning 147,110 bp, intron 3 is another large intron in human FMR2. In the murine gene, intron 1 spans about 162 kb and intron 3 about 149 kb. Similar to exon 5, intron 5 does not exist in murine *Fmr2*. These genomic data demonstrate significant conservation of both sequence and structure, and suggest that the function and regulation of FMR2 is similar between human and mouse.

FMR2 belongs to the AF4/FMR2 gene family. This family is composed of four genes: FMR2, MLLT2 (alias AF4) (Gu et al., 1992), LAF4 (Liao et al., 1996; Ma and Staudt 1996), and AF5q31 (Taki et al., 1999). In humans, these are located on human chromosomes Xq28 (FMR2), 4q21 (MLLT2/AF4), 2q11.2→q12 (LAF4), 5q31 (AF5q31). Recently, Estable et al. (2002) reported a gene named MECF of the AF4/FMR2 family, which appears to be AF5q31. Several lines of evidence suggest that MCEF and AF5q31 are actually the same gene. First, 99% identity is shared between the two nucleotide sequences. Second, the MCEF mRNA sequence contains several N nucleotides in the GenBank database (Accession #AF213987). Finally, the predicted protein in the database showed different C-terminal amino acids probably due to a frame shift caused by sequence errors, because the 3′ region sequence of MCEF still shares 99% identity with that of the AF5q31 gene.

The AF4/FMR2 family proteins contain several highly homologous domains ranging from 60–88%. These include a 90-amino acid N-terminal domain (NTD), a 175-amino acid AF-4, LAF-4, FMR2 homologous domain (ALF domain), a 64-amino acid serine-rich activation domain (AD), a 20-amino acid consensus bipartite nuclear localization sequence (NLS), as well as a 240-amino acid C-terminal homologous domain (CHD) (Fig. 1) (Nilson et al., 1997). The entire proteins share 20–35% homology with the highest between AF4 and AF5q31 (35%) among FMR2, AF4, LAF4 and MCEF/AF5q31. FMR2 shares 20% homology with LAF4, 24% with AF4, 30% with AF5q31 at the amino acid level. The ALF domain is the most highly conserved and the function of this domain is still not clear. The NTD domain possesses weak transcriptional activation in yeast but not in mammalian cells and the CTD domain can bind at least two protein partners in yeast two-hybrid system (Y. Gu and D.L. Nelson, unpublished data).

Functional studies of FMR2 and related genes

Several reports (including our unpublished data) have found that the products of MLLT2/AF4, LAF4 and FMR2 possess transcriptional activation potential in yeast and mammalian cells (Ma and Staudt, 1996; Hillman and Gecz, 2001). Despite these advances, their involvement and specific roles in the process of transcription remain unknown. Recent discoveries have shown that a new gene, EAF1 contains an activation domain homologous with the AF4/FMR2 family and interacts with RNA polymerase II transcriptional elongation factors: ELL and ELL2. Moreover, MCEF/AF5q31 has been found to interact with the elongation factor P-TEFb. These data provide some insight into how this family is involved in the transcription process (Simone et al., 2001; Estable et al., 2002). Currently, it is not clear which region of the protein is essential for interaction with ELL and ELL2 and whether a similar interaction applies to FMR2, AF4, and LAF4. There are several factors with elongation activity including TFIIS, P-TEFb, TFIIF, Elongin and FACT. It will be useful to determine whether FMR2 interacts with any of these elongation factors.

Recently, *Lilliputian (Lilli)*, a Drosophila FMR2/AF4 orthologue, was identified (Tang et al., 2001; Wittwer et al., 2001). Several lines of evidence show *Lilli* is very important during Drosophila embryogenesis, affecting germinal band extension, cytoskeletal structure, cell growth, and organ development. The Drosophila genome contains only a single gene with similarity to the AF4/FMR2 gene family, and loss of function of *Lilli* is lethal to flies, suggesting that there may be essential functions of the members of this family in vertebrates that can be compensated for by other members.

Although we have gained some knowledge about this gene family since AF4 and FMR2 were discovered in 1992 and 1996 respectively, the connection between mental retardation and FMR2 function is poorly understood. Here, we use a different approach to understand FMR2 function: a comparison between a mouse model of *Fmr2* deficiency, human FRAXE mental retardation disorder, the *Lilli* mutants in Drosophila and other mouse models carrying mutations in other genes of the AF4/FMR2 family.

FRAXE mental retardation syndrome in humans and *Fmr2* knockout model

FMR2 was isolated in 1996 due to efforts to identify the basis for FRAXE-linked mental retardation (Chakrabarti et al., 1996; Gecz et al., 1996; Gu et al., 1996). Mental retardation (MR) is a common disorder that affects about 1.5–3% of the human population worldwide. The most common form of inherited MR is fragile X syndrome, named due to the presence of a folate-sensitive fragile site at the FRAXA locus. Prior to the molecular identification of the FRAXA site and CGG repeat expansion, routine analysis of patients with possible X-linked patterns of mental retardation involved referral for cytogenetic analysis of the FRAXA site. Thus, large cohorts of patients with mental retardation and chromosomal fragile sites in distal Xq were collected in numerous centers. When the FMR1 gene was identified in 1991 and shown to be linked to the FRAXA site and CGG repeat expansion (Kremer et al., 1991; Oberle et al., 1991; Verkerk et al., 1991), it was noted that some individuals and families did not exhibit the expected CGG repeat expansion in the FMR1 gene. However, because the folate-sensitive fragile site in chromosome Xq27.3→q28 region was present, they had been characterized as being affected with fragile X syndrome. Further study made it clear that there were two additional folate-sensitive fragile sites in the Xq27.3→q28 region in addition to FRAXA. These two fragile sites were termed FRAXE and FRAXF, and both are located on chromosome Xq28 distal to FRAXA (Sutherland and Baker, 1992; Parrish et al., 1994).

In 1993, a DNA fragment containing FRAXE was isolated and the presence of an expanded GCC repeat was shown to be responsible for the fragile site. Numerous features were found to be similar to FRAXA (Knight et al., 1993), but some unique features were also found. Detailed description of the unstable CCG repeats at FRAXE is beyond the scope of this review. Interested readers can find more information in other reviews (Nelson, 1998; Gecz, 2000). The identification of FRAXE made it possible to examine in more detail the relationship between FRAXE and mental retardation. Most patients with FRAXE show mild to borderline mental retardation, with delays in language development a common problem. Some FRAXE patients also exhibit behavioral deficits, such as attention deficit, hyperactivity, autistic-like behavior, even schizophrenia and obsessive-compulsive disorder (OCD). These deficits are accompanied by inconsistent symptoms, including a long, narrow face, mild facial hypoplasia, a high-arched palate, irregular teeth, hair abnormality, angiomata, clinodactyly, thick lips and nasal abnormalities (Hamel et al., 1994; Knight et al., 1994; Mulley et al., 1995; Biancalana et al., 1996; Carbonell et al., 1996; Knight et al., 1996; Barnicoat et al., 1997; Russo et al., 1998).

In some families, the FRAXE fragile site was not clearly segregating with mental retardation and some cytogenetically positive FRAXE males carrying an expanded and fully methylated allele have been reported to not display intellectual disability (Sutherland and Baker, 1992; Murgia et al., 1996; Lo Nigro et al., 2000). This made it difficult to conclude whether there is a clear causal relationship between FRAXE and non-specific mental retardation. An alternative explanation lay in the possibility that the fragile site was coincident with MR in these patients, and that the use of cytogenetic analysis had introduced a bias in patient ascertainment.

In 1996, three independent groups using different approaches identified FMR2 as the gene associated with and affected by the FRAXE site and mutation. The Australian group further examined the relationship between FMR2 expression and mental retardation (Gecz et al., 1997b). Although absence of FMR2 expression is found in all FRAXE full mutation patients, mental retardation is not found in all patients lacking FMR2 expression, as shown by fibroblast analysis. These results still do not allow us to make conclusions about whether there is a causal relationship between FRAXE/absence of FMR2 expression and mental retardation.

Several hypotheses have been proposed to explain this phenomenon and these hypotheses are not mutually exclusive. One explanation is mosaicism of the FRAXE CCG repeat expansion and methylation. The status of expansion and methylation in the brain or CNS may be different from those in cells typically analyzed (lymphocytes and fibroblasts). In addition, FMR2 seems to play a role in early embryonic stages of the development of brain and other tissues. Currently, we do not know the time during development when methylation of expanded CCG repeats in the FRAXE fragile site occurs. It is possible that there is a time window in development when the expansion is unmethylated. Variation in the timing of methylation between individuals and/or within their tissues may cause different effects in cell differentiation and function, resulting in different phenotypes. Alternatively, Gecz (2000) proposed that the contribution of FMR2 gene expression to normal IQ may be minor and thus in its absence the drop in IQ points does not necessarily result in a drop below the mental retardation threshold (<70), such as in an originally high IQ (120) background family. A third explanation is gene redundancy. FMR2 is one member of a gene family with several members. Almost every gene of this family is expressed in the brain, especially in the fetal brain. It is therefore possible that the effects of deficiency of FMR2 may be partly compensated by other paralogues.

Several approaches can be used to test the above hypotheses. One is to screen for other mutations in the FMR2 region other than FRAXE. Non-FRAXE mutation can exclude the problem of CCG expansion and methylation mosaicism. At present, a few such cases have been reported. Four patients with extensive deletions in the region, often including FMR1, have been described, but these only complicate the picture due to possible effects from other genes (Albright et al., 1994; Wolff et al., 1997; Moore et al., 1999; Fengler et al., 2002). Only one case involving a discrete deletion in exon regions of FMR2 has been identified. This patient exhibited global developmental delay, including MR and speech delay (Gedeon et al., 1995). At least, three cases were found to carry a deletion in intron 3. These cases display X-linked MR and speech delay (Gedeon et al., 1995; Gecz and Mulley, 1999). It is not clear how deletion of intron 3 in FMR2 leads to MR or affects FMR2 function.

Comparison of human and murine FMR2 genomic sequences shows several regions with high homology besides the exons. The largest homologous region is located in intron 3

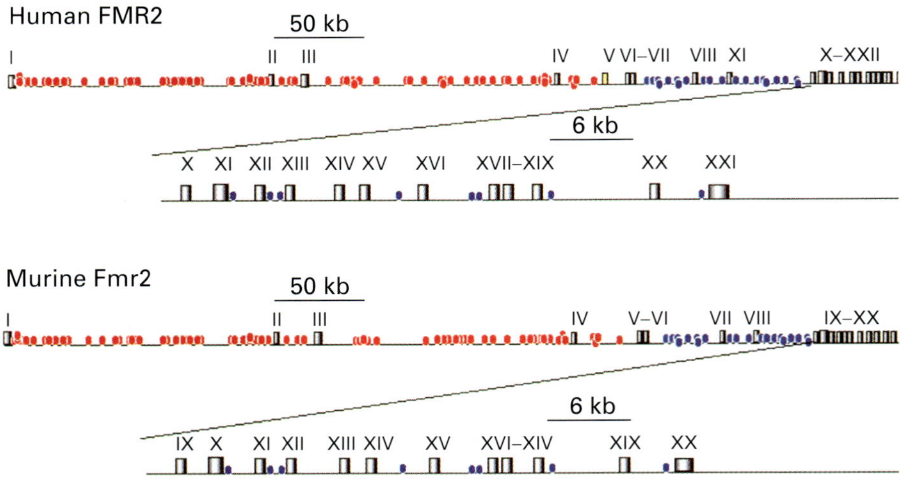

Fig. 2. Distribution of DNA homologous elements in the entire FMR2 gene of human and murine genomes. The red spots represent homologous elements in introns 1–4 in both human and murine DNA sequences. The blue spots represent homologous elements in introns 6–20 in human and introns 5–19 in mice because the murine genome lacks the human corresponding exon 5 and intron 5. The human intron 6 corresponds to murine intron 5, the human intron 7 to murine intron 6 and so on.

Table 1. List of homologous fragments in FMR2 introns

Region[a]	Total length (bp)[b]	Fragments	Repeat sequence fragment	Percent of homology	Length of intron (bp)
Intron 1	12,201	37	1 (449 bp, L1M3C–5)	81–93	150,855
Intron 2	844	3	None	82–86	9,776
Intron 3	8,620	25	None	81–96	147,110
Intron 4	2,063	4	None	81–88	27,893
Intron 5	N/A	N/A	N/A	N/A	5,234
Intron 6	0	0	N/A	N/A	329
Intron 7	2,091	8	None	83–89	42,449
Intron 8	1,411	3	None	86–95	18,233
Intron 9	3,258	9	None	83–95	49,331
Intron 10	0	0	N/A	N/A	1,863
Intron 11	282	1	None	85	1,723
Intron 12	653	2	None	86, 93	4,252
Intron 13	0	0	N/A	N/A	3,850
Intron 14	0	0	N/A	N/A	549
Intron 15	150	1	None	87	5,773
Intron 16	491	2	None	82, 92	4,320
Intron 17	0	0	N/A	N/A	357
Intron 18	0	0	N/A	N/A	2,282
Intron 19	228	1	None	86	6,598
Intron 20	218	1	None	84	3,653

[a] Region indicates the human FMR2 gene. In murine *Fmr2*: introns 1–4 correspond to the same intron of humann FMR2; introns 5–19 correspond to introns 6–20 of human FMR2 because murine *Fmr2* lacks the human exon 5 and intron 5.

[b] The total length means sum of all homologous fragments in that intron.

(1129 bp, 89% identity). Interestingly, it is larger and more homologous than the region in exon 3 (911 bp, 87% identity). In total, there are 25 homologous elements in intron 3, ranging in size from 129 bp to 1129 bp. The distribution of homologous regions between human and murine FMR2 genomic sequences is shown in Fig. 2. There is no evidence at present that these regions of homology encode other genes embedded into FMR2. These homologous elements may play an important role in the regulation of FMR2 expression and function (Table 1). Microdeletions within the FRAXE CCG repeat region have also been reported. In female patients, these microdeletions appear to be associated with premature ovarian failure (POF) (Murray et al., 1999). Boys, on the other hand, have language problems although the boundaries of these deletions have yet to be characterized (Mazzocco et al., 1998). Taken together, all these observations support the notion that FMR2 does play an important role in mental activity, especially in language development. However, the limited number of cases does not permit a final conclusion.

Mouse models of *Fmr2* deficiency

A second approach to understanding the contribution of FMR2 to MR is to use animal models. Recently, we created a mouse model carrying a mutation in the murine *Fmr2* gene by disrupting the mouse gene using homologous recombination. This model should help us clarify the function of *Fmr2* as all members of this family are conserved between human and mice (Gu et al., 2002). The *Fmr2* knockout mice showed several noteworthy phenotypes related to the nervous system. One phenotype is impaired performance in the conditioned fear test, a learning and memory test, which consists of two parts: the context test and the auditory conditioned stimulus (CS) test. During the training period, the mouse is placed in a test chamber, the bottom of which has a grid floor that is used to deliver a mild electric foot shock. The mouse is allowed to explore freely for 2 min. An 80 dB white noise, which serves as the CS, is then presented for 30 s, followed by a mild foot shock. The procedure is repeated 2 min later. After 24 h, the mouse is placed back in the same test chamber to record its reaction, and whether it recalls the presence of the electric foot shock during the training period. This serves as the context test. *Fmr2* mice showed significantly less freezing behavior (a measure of fear in rodents) during the context test, compared with the wildtype mice. The test was repeated three times with mice carrying *Fmr2* mutations on different genetic backgrounds. The results were always the same. This demonstrates that *Fmr2* plays an essential role in this part of the contextual fear test. In other words, *Fmr2* is indispensable for maintaining some form of mental activity or memory.

In the other part of the conditioned fear test: the auditory CS, the mouse is placed in a different chamber or in the same one with a different environment (changes in shape, floor, spatial cues, the color of light or even smell). Then the auditory CS is presented and freezing is recorded. Initially, the auditory CS test showed that *Fmr2* mutant mice reacted with less freezing than the wildtype controls. However, a third test of animals with a different genetic background showed no difference between *Fmr2* KO and controls. While it appears that *Fmr2* plays a role in the conditioned fear test, it is clearly affected by genetic background. This may also be a clue to the finding of variability among FRAXE patients.

The conditioned fear test also shows that *Fmr2* KO mice have normal learning process and can remember the context over a short delay interval. The fact that the levels of fear conditioning were similar between *Fmr2* KO and wildtype control mice when the test occurred 30 min after the initial training session supports this conclusion.

Other phenotypes of *Fmr2* KO mice affecting the nervous system are increased sensitivity to heat stimulus and enhanced long-term potentiation (LTP) in the hippocampus. Though no sensory changes have been reported in any FRAXE patients, in general, it is difficult for clinicians to find subtle changes in sensation except when patients have obvious pathological ones such as abnormal pain, numbness or sensory loss. The sensory changes in *Fmr2* KO mice may be caused by abnormal central nervous system function evidenced by enhanced LTP in the hippocampus. However, the possibility of peripheral nervous system dysfunction cannot be excluded.

LTP is assumed to be the basis of synaptic plasticity for memory formation and has been studied most extensively in the hippocampus. However the relationship between LTP in hippocampus and memory is hard to establish and remains inconclusive (Bliss and Collingridge, 1993; Holscher, 1999). Furthermore, the mechanism underlying LTP is not clear and very complicated; it may involve presynaptic modifications (regulation of transmitter release), postsynaptic modifications (alterations in the number and/or properties of the ion channels), extrasynaptic changes (changes in uptake of neurotransmitters by glial cells) as well as morphological modifications. Enhanced LTP in *Fmr2* knockout hippocampus is found in both NMDA receptor-dependent and NMDA receptor-independent stimuli as well as both at 25 °C and 32 °C. This demonstrates that *Fmr2* is required for normal long-term synaptic activity. The *Fmr2* model did not demonstrate a relationship between LTP in hippocampus and hippocampus-dependent memory, as there are no defects in spatial learning performance in the Morris water task. The *Fmr2* KO mice take longer to locate the platform during training, but are able to be trained similarly to wildtype mice. On the other hand, impaired fear conditioned learning may reflect an abnormality in LTP in the amygdala since there is a possible association between conditioned fear and LTP in the amygdala (McKernan and Shinnick-Gallangher, 1997; Rogan et al., 1997). It needs to be verified whether LTP in the amygdala changes similarly as in the hippocampus.

In all, the *Fmr2* KO mouse model confirms that as in humans, the primary function of *Fmr2* is in the nervous system and that *Fmr2* is involved with learning and memory. In addition, this model has provided new information about the function of *Fmr2*, data that cannot be derived from study of human patients, such as increased sensitivity to aversive heat stimulus and impaired memory over longer periods, but not over a short time interval. *Fmr2* is also necessary for normal long-term synaptic plasticity in the hippocampus. Whether neurotransmission abnormalities similar to the LTP findings in mice can be found in human FRAXE patients remains to be investigated.

Fmr2 knockout model and Drosophila *Lilliputian* mutants

The study of FMR2 paralogues and orthologues in a variety of organisms, especially in those that can be more easily manipulated, has facilitated our understanding of FMR2 gene function. In 2001, three groups independently identified the *Lililputian* gene in *Drosophila melanogaster. Lilliputian (Lilli)* is the only member of the FMR2/AF4 gene family found in the Drosophila genome, and thus it is simpler to study the function of this gene family in the fruit fly than in humans and mice (Su et al., 2001; Tang et al., 2001; Wittwer et al., 2001). *Lilli* contains 13 exons and spans approximately 68 kb. The full-length *Lilli* transcript is 8516 bp and encodes a 1673 amino acid protein. The Lilli protein has a C-terminal homology region (CHD) (Tang et al., 2001; Wittwer et al., 2001) that is 31–37% identical to the corresponding sequence of human AF4, FMR2, AF5q31 and LAF4. The CHD region extends from the end of exon 8 to the end of exon 13. Additionally, the position of intron/exon boundaries within this region is conserved between *Lilli* and the human gene family. Like other family members, Lilli is rich in proline and serine residues (9.0% and 12.7% of all amino acids, respectively) and serines are found in Lilli at the same relative positions in FMR2/AF4 family members. This indicates that the putative transactivation domain is conserved in Lilli although some domains in humans (NTD and ALF domains) are not present in Lilli (Fig. 1). Taken together, the function of *Lilli* should be similar to the function of the FMR2/AF4 family of genes and understanding its function in Drosophila should help us understand the function of this family in humans.

Lilli is essential for proper cellularization, gastrulation and segmentation during Drosophila embryogenesis (Tang et al., 2001). Most embryos deficient in *Lilli,* fail to hatch and subsequently die. A small percentage of embryos hatch and die as first or second instar larvae. Examination of these dead embryos shows two classes of phenotypes. One is involved with defects in segmentation and germband extension. The other defect is failure to secrete cuticle properly. These phenotypes are variable among individuals. Considering that genes known as pair-rule genes are required for segmental patterning and germband extension, Tang et al. (2001) examined the expression patterns of several of these genes in *Lilli* mutant embryos. These include the maternal coordinator gene: *bicoid*, the gap gene: *hunchback*, the pair rule genes: *fushi tarazu*, *even-skipped*, *hairy* and *runt*, the segmental polarity genes: *engrailed* and *wingless*, and the terminal gap genes: *tailless* and *huckebein*. Among these, two genes: *fushi tarazu* and *huckebein* display a change in expression patterns in *Lilli* mutant embryos. Failure to secrete cuticle properly is associated with specific defects in the maintenance of the actin network and defects in transport of organelles during cellularization. In the mutants, actin filaments showed uneven distribution in nuclei, ranging from abnormally large bundles to multinucleated cells lacking filaments. Expression of *Sry α*, a gene involved in actin network function, was reduced in *Lilli* mutant embryos.

Although *Fmr2* KO mice did not exhibit prenatal lethality, it is possible that FMR2 is involved in the regulation of the vertebrate homologues of *fushi tarazu, huckebein* and *Sry α.* The lack of phenotypes related to embryonic development such as segmentation and cellularization, may be due to compensation by other members of the AF4/FMR2 family and the increased complexity in function present in humans and mice. The distribution of actin in *Fmr2* KO mice has not been examined and is worthy of future study. Comparison of the Drosophila *huckebein* sequence with vertebrate cDNA by BLASTN showed that the Sp family in human and mouse are homologous to *huckebein.* Homologues of *fushi tarazu* correspond to Homeobox genes including Hox A3, Hox B5 and Hox C6 in human and mouse. There are six Sp genes in human and mouse, which are termed Sp1, Sp2, Sp3, Sp4, Sp5 and Sp6 (Kadonaga et al., 1987; Kingsley and Winoto, 1992; Ihn and Trojanowska, 1997; Harrison et al., 2000; Song et al., 2001).

In order to determine whether *Fmr2* controls Sp family members in mice, we have compared Sp1, Sp3, Sp5 and Sp6 mRNA levels in *Fmr2* KO adult mice with wildtype controls by Northern blot. Preliminary data show altered expression pattern of Sp1 and Sp6 in *Fmr2* KO mice. Expression of Sp1 is reduced in *Fmr2* KO liver and brain compared to control tissues. Expression of Sp6 is increased in *Fmr2* KO liver. Sp1 and Sp3 are expressed ubiquitously, whereas expression of Sp5 is associated with development of different tissues, including the neural tube, somites and their derivatives, and pharyngeal regions. Sp6 is a new member of the Sp1 family that is predicted from GenomeScan (Yeh et al., 2001) and EST matches suggest it is functional. Our data indicate that Sp6 generates a 4.4-kb message in liver, brain and spleen. The significance of changes in the Sp1 and Sp6 expression is still unknown and waits further study. Up to now, orthologues of *Sry α* have not been found in human and mouse databases (including genomic and EST databases) despite the near completion of the genomic sequences. As the human genome project approaches its final stage of complete sequencing, we may have a greater chance of finding them in the near future if the orthologues do exist in humans.

Due to the presence of defects in cell growth and differentiation, *Lilli* is predicted to be involved in several signal transduction pathways, including the transforming growth factor-β signal pathway (Su et al., 2001), the RAF/MEK/MAP kinase (MAPK) pathway, and the PI3K/PKB pathway (Tang et al., 2001; Wittwer et al., 2001). Mutations in RAF/MEK/MAPK pathway and the PI3K/PKB both cause the cells to grow slowly resulting in smaller cells in Drosophila eyes (Ommatidia) or wings. *Lilli* may be the convergent point of these pathways. Mutations in *Lilli* suppressed the rough eye phenotypes by causing constitutive activation of the Raf kinase. The effect of *Lilli* on PI3K/PKB is more complicated in Drosophila. PTEN is a negative regulator of the PI3K/PKB pathway. Mutations of *Pten* in Drosophila cause cell hyperplasia and hypertrophic growth, resulting in an increase in eye and head sizes. Wittwer et al. (2001) found that *Lilli* mutants show a considerable reduction in the size of the eyes when compared to the *Pten* mutant eyes, though loss of *Lilli* did not completely suppress the *Pten* phenotype. Tang et al. (2001) found that loss of *Lilli* did not prevent *Pten* induced cell enlargement and that photoreceptor cells were severely disorganized in the double mutants. These results indicate *Lilli* is indirectly involved in the PI3K/PKB pathway. As for the transforming growth factor β signal pathway, *Lilli* mutations enhanced the wing heldout phenotype in a *decapentaplegic (DPP)* mutant background. *DPP* is a member of the transforming growth factor family and the mechanism of DPP signal transduction is not understood.

In humans, several genes associated with non-specific X-linked mental retardation (MRX) have been identified and shown to be involved in different pathways, including the RAF/MEK/MAPK pathway, Calcium/calmodulin/calmodulin-dependent kinase (Camk) pathway and others. There are at least five genes associated with MRX from the RAF/MEK/MAPK pathway. The five genes are: GDI1, a Rab GDP-associated inhibitor (D'Adamo et al., 1998); oligophrenin 1, Rho-GTPase-activating protein (Billuart et al., 1998); PAK3, a p21 activated kinase (Allen et al., 1998); RPS6KA3, a MAPK activated CREB kinase (Merienne et al., 1999); and ARHGEF6, a guanine nucleotide exchange factor for Rho GTPases (Kutsche et al., 2000). In *Fmr2* KO mice, abnormal LTP was found in hippocampal slices. LTP is known to be associated with several signal pathways, including RAF/MEK/MAPK pathways (English and Sweatt, 1997), Calcium/calmodulin/calmodulin-dependent kinase (Camk) pathway (Mayford et al., 1996), and cAMP protein kinase (PKA)/CREB (Roberson et al., 1999). Although it is not yet known how mutations of FMR2 cause

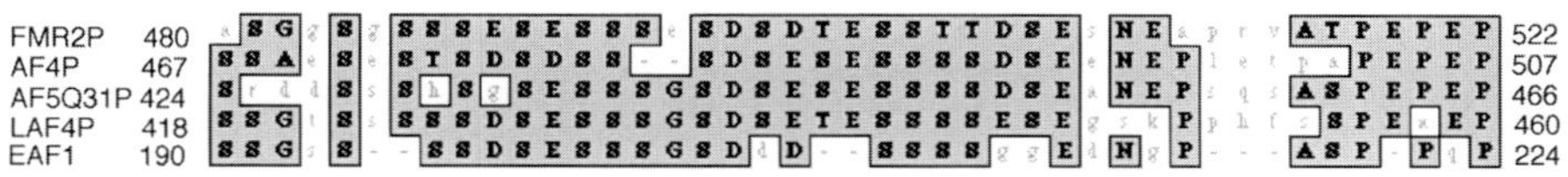

Fig. 3. Multiple sequence alignment of human EAF1 and AF4/FMR2 family that includes AF4, LAF4, FMR2 and AF5q31. This is the only domain conserved with the FMR2/AF4 family in EAF1. Except for this domain there is no significant homology between EAF1 and the FMR2/AF4 family.

mental retardation in humans and abnormal LTP in mice, we can reasonably suggest that FMR2 may be the convergent point of different pathways similar to what is observed with the Drosophila homolog, *Lilli*. The value of drawing parallels between the fly, mouse and human systems is clear, and provides additional clues to testing function in mouse models.

Fmr2 knockout model and other mouse knockout models of the FMR2/AF4 family

As described above, FMR2 is a member of a gene family and this family includes at least three other genes: AF4, AF5q31 and LAF4. The newly discovered EAF1, on the other hand, is much shorter than the others and several of the domains conserved in other members are not present (Fig. 3). The protein encoded by EAF1 is 286 amino acids while the other proteins are over 1000 amino acids long. It is therefore possible, that EAF1 is not highly conserved with respect to function. In the GenBank database, other names such as MECF and LAF1 have been linked to this family. However, these likely represent two additional names for the AF5q31 gene. Acute leukemia has been associated with the chromosome translocation associated fusion of the MLL gene on chromosome 11 with either AF4 or AF5q31 (Gu et al., 1992; Taki et al., 1999). The MLL gene possesses a highly conserved SET domain found in the Drosophila trithorax and polycomb group, and has been shown to regulate Hox gene expression and segmental development. Homozygous *Mll* mutant mice exhibit embryonic lethality and no expression of Hox a7 and Hox c9, while heterozygotes show growth delay and abnormal expression of Hox a7 and Hox c9 (Yu et al., 1995). More than 30 different cytogenetic aberrations that affect the MLL gene have been described in leukemias and, more than 20 MLL partner chromosomal breakpoints have been cloned. These MLL partner genes show no common features. In addition, a knock-in mouse model generating an MLL-*LacZ* fusion gene also causes leukemia (Dobson et al., 2000), indicating that AF4 and AF5q31 are unlikely to play specific roles in leukemia formation in their respective MLL fusion forms.

The expression patterns of AF4, AF5q31 and LAF4 are ubiquitous, particularly in fetal stages. Although it is expressed mainly in adult brain, FMR2 is also found in fetal heart, lung, liver and brain. The *Fmr2* KO model created by us contains the *LacZ* gene under the control of the *Fmr2* promoter. Ubiquitous *LacZ* expression was observed in embryonic stages, especially in embryonic day 12.5 (Fig. 4). These observations, combined with data from *Lilli* mutants in Drosophila suggest that this family may play an important role in cell or organ development. Human LAF4 was initially isolated from a lymphoma cell line by a subtractive cDNA cloning method (Ma and Staudt, 1996). Recently Britanova, et al. (2002) found that *Laf4* was one of several genes that were upregulated in the subventricular zone (SVZ) during development of the cerebral cortex. There are several forms of *Laf4* and the expression pattern of these forms is changed during brain development. In embryonic brain, one form of *Laf4* containing exon 6 was expressed at embryonic day (E) 13.5, while two forms were found in E15.5. The form without exon 6 is more abundant than that with exon 6 in later stages. In adult brain, there are at least four other forms detected. Some of these adult forms may be due to the variable size of exon 12 and the presence or absence of exon 11. The main difference between embryonic forms and adult forms is whether they contain exons 8–10. Exons 8–10 encode the strong transcriptional activation domain described in human LAF4. Embryonic forms of *Laf4* completely lack this domain. Expression of *Laf4* is mostly confined to the SVZ at E13 and SVZ and cortical plate in E15. This expression pattern is similar to that of *Fmr2*. It will be interesting to compare phenotypes of these two mutant mice, once the *Laf4* KO mouse model has been generated.

The phenotypes of the *Af5q31* and *Af4* KO mouse models have been reported by us and by others respectively (Isnard et al., 2000; Gu and Nelson, 2002). These two mouse models in addition to the *Fmr2* KO mouse model will help us further understand the function of this gene family in humans and mice. The *Af4* KO model was produced by deletion of a portion of exon 11 and 20% lethality was observed in homozygotes by postnatal day 10. This was accompanied by general growth retardation and a reduced number of cells in the thymus. Mutant mice with no growth retardation showed a normal lymphocyte compartment. Detailed analysis of the different subtypes of T-lymphocytes showed that *Af4* did not affect lymphocyte differentiation, rather it affected expansion/growth and selection of subtypes in early stages, but not in terminal stages. A similar phenomenon was found in B-lymphocyte cells of the bone marrow of *Af4* KO mice. Together with data from *Lilli*, which showed a redundant function in photoreceptor cell differentiation such as R7 and some R1-R6, essential for maintaining the cell size and number of photoreceptor cells, it would appear that each of the members of this gene family plays a role in expansion or growth of specific cell populations. Although it is still not clear which group of cells is affected in *Fmr2* KO mice, the expression pattern of *Fmr2* is related to neuronal

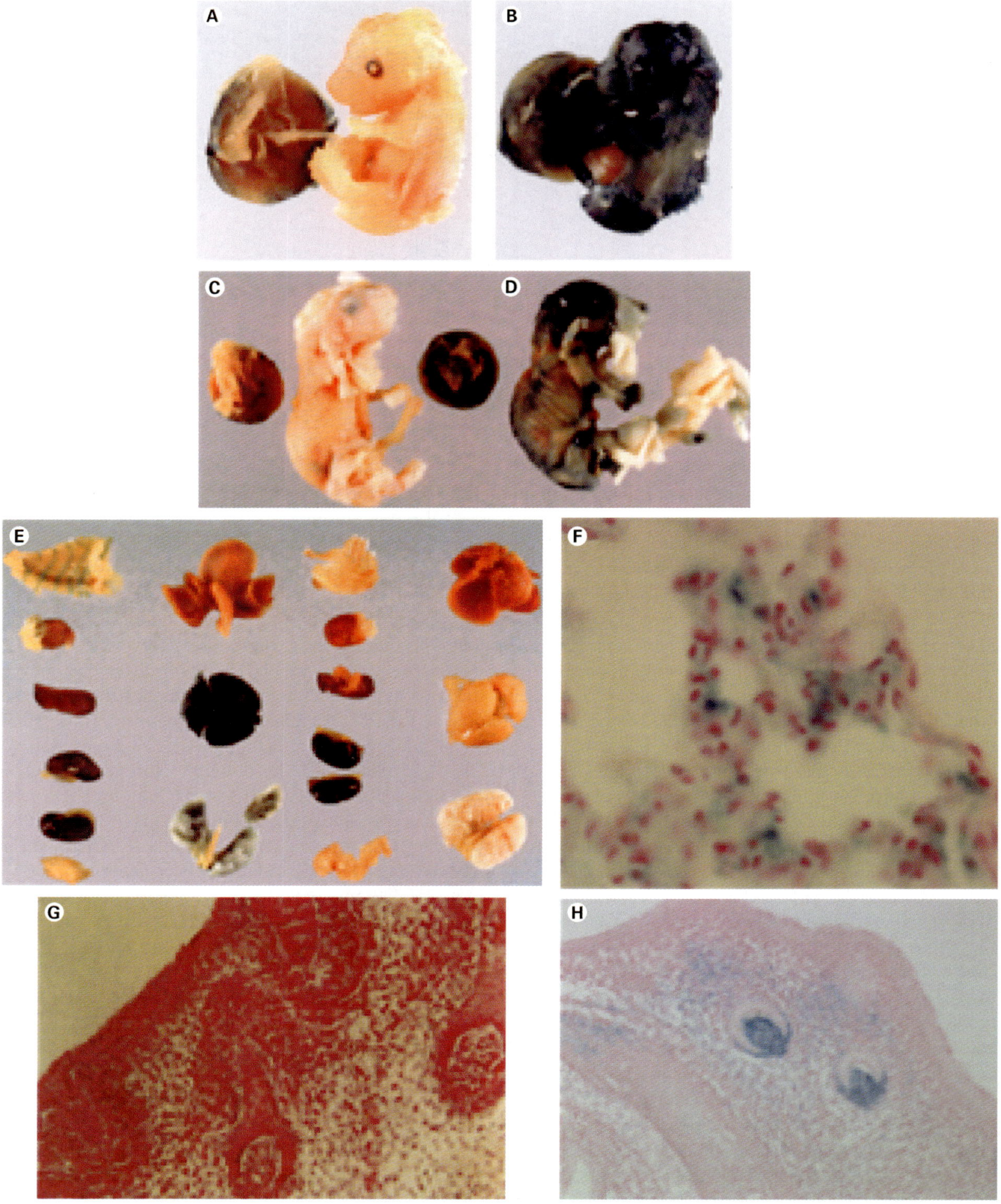

Fig. 4. *Fmr2* expression pattern from 5-bromo-4-chloro-3-indolyl-β-D-galactopyranoside (X-gal) staining of Lac Z protein under the control of the *Fmr2* promoter. (**A**) X-gal staining of a wild type control embryo 12.5 P.C., (**B**) X-gal staining of an *Fmr2* KO embryo 12.5 P.C., (**C**) a wild type control embryo 15.5 P.C., (**D**) an *Fmr2* KO embryo 15.5 P.C., (**E**) adult organs from the *Fmr2* KO and the wild type control. The organs are: rib and muscle, heart, spleen, two kidneys, and thymus from a KO mouse in order from top to bottom in the first column; liver, brain and lung from a KO mouse in the second column. In the third column, it is the same organ order from a wild-type control mouse as that in the first column and in the fourth column the same order from a wildtype control mouse as in the second column. (**F**) X-gal staining of the KO lung. Only some cells were stained. (**G**) X-gal staining of the control hair follicle. (**H**). X-gal staining of the KO hair follicle. The X-gal stained cells (*Fmr2* expressing cells) are located in mesenchymal cells of papilla derived from the dermis, but not in the cells of sheath, cortex and medulla, derived from the epidermis. The connective tissue derived from the dermis is not stained as well.

development. The gene starts to express when neuroepithelial cells in ventricular zones differentiate into neuroblasts at E10.5, and continues to express during neuroblast differentiation into neurons. We hypothesize that the expansion of some subtypes of neurons is affected and are working to define possible differences in neuronal architecture between normal and mutant mouse brains.

The *Af5q31* mouse model was created by deletion of exon 2, which contains the start-codon and encodes part of the conserved N-terminal domain of the FMR2/AF4 gene family. These animals show 50% embryonic lethality as homozygotes (Gu and Nelson, 2002). Some of the mice die in utero before E12.5. It is interesting that for each of the mouse models created thus far for this gene family, lethal phenotypes are observed at various stages of development (*Af5q31* KO mice as embryos, *Af4* KO mice before 10 days after birth, *Fmr2* KO mice at 3–9 months). However, no obvious segmentation or organ development defects such as the ones described in the Drosophila model have been found. It would appear that in mice (and possibly in humans) each member of this family functions at different stages through regulation of cell expansion and development. Failure of expansion of one or several subpopulations of cells leads to lethality, although the specific organ systems in each model remain to be determined. Some of *Af5q31* KO mice also exhibit much smaller spleens and stomachs when compared to normal littermates. It is possible that some cells in these organs have defects of expansion. Additional study of these mice should illuminate these questions.

Conclusion and future directions

Since the molecular cloning of the FRAXE site in 1993 (Knight et al., 1993) and the identification of FMR2 association with FRAXE in 1996 (Gecz et al., 1996; Gu et al., 1996), we have gained insight into the mechanism of how this X-linked nonsyndromic mental retardation occurs. The fact that the FMR2 gene contains nuclear localization signals (NLS) and conserved transactivation domains suggested that FMR2 is located in the nucleus. The intracellular location was confirmed by a GFP-FMR2 fusion protein experiment as well as FMR2 antibody staining (Gecz et al., 1997a; Miller et al., 2000). Data from human, mouse and fly studies have provided answers to several questions about the gene and mutations in it, but at the same time, many new questions and challenges have arisen and await solution. Based on the phenotypes of FRAXE patients and *Fmr2* KO mice, FMR2 is a gene important for mental activity and nervous system development. Why mutation of FMR2 primarily affects nervous system function remains unclear. The expression pattern of FMR2 is not limited to the brain, yet other tissues appear unaffected. In adult tissues, the brain is the main organ expressing *Fmr2* (Fig. 4). At the present time, several questions about the precise function of FMR2 need to be addressed. First, if the AF4/FMR2 family is a regulator of transcription, what is its mechanism of transcriptional regulation and how does each individual of this family regulate its targets? Considering that EAF1 and AF5q31/MECF have been found to associate with different transcription elongation factors, it is possible that FMR2 affects transcription through one of these transcription elongation factors. Identifying FMR2 partners will help resolve this problem and may elucidate phenotypes in humans and mice.

Another significant need is to identify the genes that are regulated by FMR2, and the direction and magnitude of changes in their expression in the absence of the FMR2 gene product. Preliminary results from *Fmr2* KO mice and *AF5q31* KO mice show that different genes are regulated in different tissues and possibly at different developmental stages. Additionally, some of these potential target genes have several alternatively spliced forms and the ratio of different forms seems to be regulated by members of AF4/FMR2 family. This latter finding offers significant challenges to identifying genes with expression alterations by typical cDNA subtractive hybridization or microarray methods. An improved and reliable method is required to approach this problem. Currently, we are examining candidate genes in *Fmr2* KO, *Af5q31* KO, *Fmr2* and *Af5q31* double KO, and control mice according to clues from the *Lilli* model. While this provides an entry point, a more general survey will also be required to ensure that crucial genes related to the nervous system are not missed.

It is also necessary to continue to search for other mutations in FMR2 in addition to the GCC expansion. These will be important to extend our understanding of the range of phenotypes resulting from FMR2 mutations in humans. The phenotype associated with the FRAXE mutation is variable and relatively subtle. In some FRAXE families, several other members in addition to the propositus showed the GCC expansion and methylation along with a normal IQ and no discernible phenotype. This might be explained by expression of FMR2 in these individual's brains during a developmental time window, since it remains unclear when methylation of expanded GCC and extinction of FMR2 expression occurs. There are two other types of mutations reported in the literature besides the GCC expansion: large deletions and microdeletions. Large deletions were detected by the probe pS8, which is located in intron 3. Among four patients with these deletions, only one carries an exon mutation, while the other three have deletions of only intron sequences. Each patient exhibits mental retardation, although other phenotypes of these patients were not described in detail. The mechanism by which deletion of intron 3 in these three patients leads to mental retardation is unknown. Sequence comparison between intron 3 in the human FMR2 and murine *Fmr2* genes suggests conserved elements are present. These may be important for FMR2 expression, regulating time, place or quantity of expression. Zoo blots showed pS8 is conserved in several species, including mouse, hamster, dog, chicken and others, further demonstrating likely importance (Gedeon et al., 1995). Definition of the conserved elements will provide additional ability to define the importance of expression patterns in the human FMR2 disease. Other patients have been found to carry microdeletions around the GCC repeat. Although some of them were inherited from parents, the key to understanding the phenotypes of these patients will be again to determine how these mutations affect FMR2 expression. Due to limited access to patient samples, these studies remain difficult.

Advances in the study of FMR2 and FRAXE syndrome have provided a first glimpse into the function of FMR2 with the assistance of studies in human, mouse and Drosophila. However, the data from animal models need to be interpreted with caution. Much more work is required to verify these hypotheses in the human disease, and such studies are limited by current methods and materials. With further advances of technology and additional efforts, the elucidation of the FMR2 function will become more complete, this will not only allow us to understand the basic cell biology processes, but also provide possible ways to treat FRAXE and FMR2 related mental retardation.

Acknowledgements

We are particularly indebted to Dr. Joannella Morales for useful suggestions and for critically reading the manuscript, and grateful to Allan Harris for assistance in sequence analysis.

References

Albright S, Lachiewicz AM, Tarleton JC, Rao KW, Schwartz CE, Richie R, Tennison MB, Aylsworth AS: Fragile X phenotype in a patient with a large de novo deletion in Xq27→q28. Am J med Genet 51:294–297 (1994).

Allen K, Gleeson JG, Bagrodia S, Partington MW, MacMillan JC, Cerione RA, Mulley JC, Walsh CA: PAK3 mutation in nonsyndromic X-linked mental retardation. Nature Genet 20:25–30 (1998).

Barnicoat A, Wang Q, Turk J, Green E, Mathew CG, Flynn G, Buckle V, Hirst M, Davies K, Bobrow M: Clinical, cytogenetic, and molecular analysis of three families with FRAXE. J med Genet 34:13–17 (1997).

Biancalana V, Taine L, Bouix JC, Finck S, Chauvin A, De Verneuil H, Knight SJ, Stoll C, Lacombe D, Mandel JL: Expansion and methylation status at FRAXE can be detected on *Eco*RI blots used for FRAXA diagnosis: analysis of four FRAXE families with mild mental retardation in males. Am J hum Genet 59:847–854 (1996).

Billuart P, Bienvenu T, Ronce N, des Portes V, Vinet MC, Zemni R, Roest Crollius H, Carrie A, Fauchereau F, Cherry M, Briault S, Hamel B, Fryns JP, Beldjord C, Kahn A, Moraine C, Chelly J: Oligophrenin-1 encodes a rhoGAP protein involved in X-linked mental retardation. Nature 392:923–926 (1998).

Bliss TVP, Collingridge GL: A synaptic model of memory long-term potentiation in the hippocamppus. Nature 361:31–39 (1993).

Britanova O, Lukyanov S, Gruss P, Tarabykin V: The Mouse *Laf4* gene: exon/intron organization, cDNA sequence, alternative splicing, and expression during central nervous system development. Genomics 80:31–37 (2002).

Carbonell P, Lopez I, Gabarron J, Bernabe MJ, Lucas JM, Guitart M, Gabau E, Glover G: FRAXE mutation analysis in three Spanish families. Am J med Genet 64:434–440 (1996).

Chakrabarti L, Knight SJ, Flannery AV, Davies KE: A candidate gene for mild mental handicap at the FRAXE fragile site. Hum molec Genet 5:275–282 (1996).

Chakrabarti L, Bristulf J, Foss GS, Davies KE: Expression of the murine homologue of FMR2 in mouse brain and during development. Hum molec Genet 7:441–448 (1998).

D'Adamo P, Menegon A, Lo Nigro C, Grasso M, Gulisano M, Tamanini F, Bienvenu T, Gedeon AK, Oostra B, Wu SK, Tandon A, Valtorta F, Balch WE, Chelly J, Toniolo D: Mutations in GDI1 are responsible for X-linked non-specific mental retardation. Nature Genet 19:134–139 (1998).

Dobson C, Warren AJ, Pannell R, Forster A, Rabbitts TH: Tumorigenesis in mice with a fusion of the leukaemia oncogene *Mll* and the bacterial *LacZ* gene. EMBO J 19:843–851 (2000).

English J, Sweatt JD: A requirement for the mitogen-activated protein kinase cascade in hippocampal long term potentiation. J biol Chem 272:19103–19106 (1997).

Estable M, Naghavi MH, Kato H, Xiao H, Qin J, Vahlne A, Roeder RG: MCEF, the newest member of the AF4 family of transcription factors involved in leukemia, is a positive transcription elongation factor-b-associated protein. J Biomed Sci 9:234–245 (2002).

Fengler S, Fuchs S, Konig R, Arnemann J: Mosaicism for FMR1 and FMR2 deletion: a new case. J med Genet 39:200–201 (2002).

Gecz J: The FMR2 gene, FRAXE and non-specific X-linked mental retardation: clinical and molecular aspects. Ann hum Genet 64:95–106 (2000).

Gecz J, Mulley JC: Characterisation and expression of a large, 13.7-kb FMR2 isoform. Eur J hum Genet 7:157–162 (1999).

Gecz J, Gedeon AK, Sutherland GR, Mulley JC: Identification of the gene FMR2, associated with FRAXE mental retardation. Nature Genet 13: 105–108 (1996).

Gecz J, Bielby S, Sutherland GR, Mulley JC: Gene structure and subcellular localization of FMR2, a member of a new family of putative transcription activators. Genomic 44:201–213 (1997a).

Gecz J, Oostra BA, Hockey A, Carbonell P, Turner G, Haan EA, Sutherland GR, Mulley JC: FMR2 expression in families with FRAXE mental retardation. Hum molec Genet 6:435–441 (1997b).

Gedeon A, Meinanen M, Ades LC, Kaariainen H, Gecz J, Baker E, Sutherland GR, Mulley JC: Overlapping submicroscopic deletions in Xq28 in two unrelated boys with developmental disorders: identification of a gene near FRAXE. Am J hum Genet 56:907–914 (1995).

Gu Y, McIlwain KL, Weeber EJ, Yamagata T, Xu B, Antalffy BA, Reyes C, Yuva-Paylor L, Armstrong D, Zoghbi H, Sweatt JD, Paylor R, Nelson DL: Impaired conditioned fear and enhanced long-term potentiation in *Fmr2* knock-out mice. J. Neurosci 22:2753–2763 (2002).

Gu Y, Nakamura T, Alder H, Prasad R, Canaani O, Cimino G, Croce CM, Canaani E: The t(4;11) chromosome translocation of human acute leukemias fuses the ALL-1 gene, related to Drosophila *trithorax*, to the AF-4 gene. Cell 71:701–708 (1992).

Gu Y, Nelson DL: AF5q31 knockout mouse model: exploring the functions of FMR2/AF4 gene family. 8th international fragile X conference Chicago: G9 (2002).

Gu Y, Shen Y, Gibbs RA, Nelson DL: Identification of FMR2, a novel gene associated with the FRAXE CCG repeat and CpG island. Nature Genet 13:109–113 (1996).

Hamel B, Smits AP, de Graaff E, Smeets DF, Schoute F, Eussen BH, Knight SJ, Davies KE, Assman-Hulsmans CF, Oostra BA: Segregation of FRAXE in a large family: clinical, psychometric, cytogenetic, and molecular data. Am J hum Genet 55:923–931 (1994).

Harrison S, Houzelstein D, Dunwoodie SL, Beddington RS: Sp5, a new member of the Sp1 family, is dynamically expressed during development and genetically interacts with Brachyury. Dev Biol 227:358–372 (2000).

Hillman M, Gecz J: Fragile XE-associated familial mental retardation protein 2 (FMR2) acts as a potent transcription activator. J hum Genet 46:251–259 (2001).

Holscher C: Synaptic plasticity and learning and memory: LTP and beyond. J Neurosci Res 58:62–75 (1999).

Ihn H, Trojanowska M: Sp3 is a transcriptional activator of the human alpha2(I) collagen gene. Nucl Acids Res 25:3712–3717 (1997).

Isnard P, Core N, Naquet P, Djabali M: Altered lymphoid development in mice deficient for the mAF4 proto-oncogene. Blood 96:705–710 (2000).

Kadonaga J., Carner KR, Masiarz FR, Tjian R: Isolation of cDNA encoding transcription factor Sp1 and functional analysis of the DNA binding domain. Cell 51:1079–1090 (1987).

Kingsley C, Winoto A: Cloning of GT box-binding proteins: a novel Sp1 multigene family regulating T-cell receptor gene expression. Mol cell Biol 12:4251–4261 (1992).

Knight S, Flannery AV, Hirst MC, Campbell L, Christodoulou Z, Phelps SR, Pointon J, Middleton-Price HR, Barnicoat A, Pembrey ME, et al: Trinucleotide repeat amplification and hypermethylation of a CpG island in FRAXE mental retardation. Cell 74:127–134 (1993).

Knight S, Ritchie RJ, Chakrabarti L, Cross G, Taylor GR, Mueller RF, Hurst J, Paterson J, Yates JR, Dow DJ, Davies KE: A study of FRAXE in mentally retarded individuals referred for fragile X syndrome (FRAXA) testing in the United Kingdom. Am J hum Genet 58:906–913 (1996).

Knight S, Voelckel MA, Hirst MC, Flannery AV, Moncla A, Davies KE: Triplet repeat expansion at the FRAXE locus and X-linked mild mental handicap. Am J hum Genet 55:81–86 (1994).

Kremer E, Pritchard M, Lynch M, Yu S, Holman K, Baker E, Warren ST, Schlessinger D, Sutherland GR, Richards RI: Mapping of DNA instability at the fragile X to a trinucleotide repeat sequence p(CCG)n. Science 252:1711–1714 (1991).

Kutsche K, Yntema H, Brandt A, Jantke I, Nothwang HG, Orth U, Boavida MG, David D, Chelly J, Fryns JP, Moraine C, Ropers HH, Hamel BC, van Bokhoven H, Gal A: Mutations in ARHGEF6, encoding a guanine nucleotide exchange factor for Rho GTPases, in patients with X-linked mental retardation. Nature Genet 26:247–250 (2000).

Lander E, Linton LM, Birren B, Nusbaum C, Zody MC, Baldwin J, Devon K, Dewar K, Doyle M, FitzHugh W, et al: International Human Genome Sequencing Consortium: Initial sequencing and analysis of the human genome. Nature 409:860–921 (2001).

Liao X, Ma C, Trask B, Massa H, Gilbert DJ, Staudt LM, Jenkins NA, Copeland NG: LAF4 maps to mouse chromosome 1 and human chromosome 2q11.2→q12. Mammal Genome 7:467–468 (1996).

Lo Nigro C, Faravelli F, Cavani S, Perroni L, Novello P, Vitali M, Bricarelli FD, Grasso M: FRAXE mutation in a mentally retarded subject and in his phenotypically normal twin brother. Eur J hum Genet 8:157–162 (2000).

Ma C, Staudt LM: LAF-4 encodes a lymphoid nuclear protein with transactivation potential that is homologous to AF-4, the gene fused to MLL in t(4;11) leukemias. Blood 87:734–745 (1996).

Mayford M, Bach ME, Huang YY, Wang L, Hawkins RD, Kandel ER: Control of memory formation through regulated expression of a CaMKII transgene. Science 274:1678–1683 (1996).

Mazzocco M, Myers GF, Hamner JL, Panoscha R, Shapiro BK, Reiss AL: The prevalence of the FMR1 and FMR2 mutations among preschool children with language delay. J Pediatr 132:795–801 (1998).

Mckernan M, Shinnick-Gallangher P: Fear conditioning induces a lasting potentiation of synaptic currents in vitro. Nature 390:607–611 (1997).

Merienne K, Jacquot S, Pannetier S, Zeniou M, Bankier A, Gecz J, Mandel JL, Mulley J, Sassone-Corsi P, Hanauer A: A missense mutation in RPS6KA3 (RSK2) responsible for non-specific mental retardation. Nature Genet 22:13–14 (1999).

Miller W, Skinner JA, Foss GS, Davies KE: Localization of the fragile X mental retardation 2 (FMR2) protein in mammalian brain. Eur J Neurosci 12:381–384 (2000).

Moore S, Strain L, Cole GF, Miedzybrodzka Z, Kelly KF, Dean JC: Fragile X syndrome with FMR1 and FMR2 deletion. J med Genet 36:565–566 (1999).

Mulley J, Yu S, Loesch DZ, Hay DA, Donnelly A, Gedeon AK, Carbonell P, Lopez I, Glover G, Gabarron I, et al: FRAXE and mental retardation. J med Genet 32:162–169 (1995).

Murgia A, Polli R, Vinanzi C, Salis M, Drigo P, Artifoni L, Zacchello F: Amplification of the Xq28 FRAXE repeats: extreme phenotype variability? Am J med Genet 64:441–444 (1996).

Murray A, Webb J, Dennis N, Conway G, Morton N: Microdeletions in FMR2 may be a significant cause of premature ovarian failure. J med Genet 36:767–770 (1999).

Nelson DL: FRAXE mental retardation and other folate-sensitive fragile sites. Genetic instabilities and hereditary neurological diseases, Wells RD, Warren ST (eds), pp 65–74 (Academic Press, San Diego 1998).

Nilson I, Reichel M, Ennas MG, Greim R, Knorr C, Siegler G, Greil J, Fey GH, Marschalek R: Exon/intron structure of the human AF-4 gene, a member of the AF-4/LAF-4/FMR-2 gene family coding for a nuclear protein with structural alterations in acute leukaemia. Br J Haematol 98:157–169 (1997).

Oberle I, Rousseau F, Heitz D, Kretz C, Devys D, Hanauer A, Boue J, Bertheas MF, Mandel JL: Instability of a 550-base pair DNA segment and abnormal methylation in fragile X syndrome. Science 252:1097–1102 (1991).

Parrish J, Oostra BA, Verkerk AJ, Richards CS, Reynolds J, Spikes AS, Shaffer LG, Nelson DL: Isolation of a GCC repeat showing expansion in FRAXF, a fragile site distal to FRAXA and FRAXE: Nature Genet 8:229–235 (1994).

Roberson E, English JD, Adams JP, Selcher JC, Kondratick C, Sweatt JD: The mitogen-activated protein kinase cascade couples PKA and PKC to cAMP response element binding protein phosphorylation in area CA1 of hippocampus. J Neurosci 19:4337–4348 (1999).

Rogan M, VS Ursula, JE LeDoux: Fear conditioned induces associative long-term potentiation in the amygdala. Nature 390:604–607 (1997).

Russo S, Selicorni A, Bedeschi MF, Natacci F, Viziello P, Fortuna R, Pagani G, Dalpra L, Larizza L: Molecular characterization of FRAXE-positive subjects with mental impairment in two unrelated Italian families. Am J med Genet 75:304–308 (1998).

Simone F, Polak PE, Kaberlein JJ, Luo RT, Levitan DA, Thirman MJ: EAF1, a novel ELL-associated factor that is delocalized by expression of the MLL-ELL fusion protein. Blood 98:201–209 (2001).

Song J, Mangold M, Suske G, Geltinger C, Kanazawa I, Sun K, Yokoyama KK: Characterization and promoter analysis of the mouse gene for transcription factor Sp4. Gene 264:19–27 (2001).

Su M, Wisotzkey RG, Newfeld SJ: A screen for modifiers of decapentaplegic mutant phenotypes identifies lilliputian, the only member of the Fragile-X/Burkitt's Lymphoma family of transcription factors in *Drosophila melanogaster*. Genetics 157:717–725 (2001).

Sutherland G, Baker E: Characterization of a new rare fragile site easily confused with the fragile X. Hum molec Genet 1:111–113 (1992).

Taki T, Kano H, Taniwaki M, Sako M, Yanagisawa M, Hayashi Y: AF5q31, a newly identified AF4-related gene, is fused to MLL in infant acute lymphoblastic leukemia with ins(5;11)(q31;q13q23). Proc natl Acad Sci, USA 96:14535–14540 (1999).

Tang A, Neufeld TP, Rubin GM, Muller HA: Transcriptional regulation of cytoskeletal functions and segmentation by a novel maternal pair-rule gene, lilliputian. Development 128:801–813 (2001).

Venter J, Adams MD, Myers EW, Li PW, Mural RJ, Sutton GG, Smith HO, Yandell M, Evans CA, Holt RA, et al: The sequence of the human genome. Science 291:1304–1351 (2001).

Verkerk A, Pieretti M, Sutcliffe JS, Fu YH, Kuhl DP, Pizzuti A, Reiner O, Richards S, Victoria MF, Zhang FP, et al: Identification of a gene (FMR-1) containing a CGG repeat coincident with a breakpoint cluster region exhibiting length variation in fragile X syndrome. Cell 65:905–914 (1991).

Wittwer F, van der Straten A, Keleman K, Dickson BJ, Hafen E: Lilliputian: an AF4/FMR2-related protein that controls cell identity and cell growth. Development 128:791–800 (2001).

Wolff D, Gustasshaw KM, Zurcher V, Ko L, White W, Weiss L, Van Dyke DL, Schwartz S, Willard HF: Deletions in Xq26.3→q27.3 including FMR1 result in a severe phenotype in a male and variable phenotypes in females depending upon the X inactivation pattern. Hum Genet 100:256–262 (1997).

Yeh R, Lim LP, Burge CB: Computational inference of homologous gene structures in the human genome. Genome Res 11:803–816 (2001).

Yu B, Hess JL, Horning SE, Brown GA, Korsmeyer SJ: Altered *Hox* expression and segmental identity in *Mll*-mutant mice. Nature 378:505–508 (1995).

Cytogenet Genome Res 100:140–146 (2003)
DOI: 10.1159/000072848

Instability of the fragile X syndrome repeat in mice: the effect of age, diet and mutations in genes that affect DNA replication, recombination and repair proficiency

K. Fleming, D.K. Riser, D. Kumari and K. Usdin

Section on Genomic Structure and Function, Laboratory of Molecular and Cellular Biology, National Institute of Diabetes, Digestive and Kidney Diseases, National Institutes of Health, Bethesda, MD (USA)

Abstract. Repeat expansion diseases such as fragile X syndrome (FXS) result from increases in the size of a specific tandem repeat array. In addition to large expansions, small changes in repeat number and deletions are frequently seen in FXS pedigrees. No mouse model accurately recapitulates all aspects of this instability, particularly the occurrence of large expansions. This may be due to differences between mice and humans in *cis* and/or *trans*-acting factors that affect repeat stability. The identification of such factors may help reveal the expansion mechanism and allow the development of suitable animal models for these disorders. We have examined the effect of age, dietary folate, and mutations in the Werner's syndrome helicase (*Wrn*) and *Trp53* genes on FXS repeat instability in mice. WRN facilitates replication of the FXS repeat and enhances Okazaki fragment processing, thereby reducing the incidence of processes that have been suggested to lead to expansion. p53 is a protein involved in DNA damage surveillance and repair. We find two types of repeat instability in these mice, small changes in repeat number that are seen at frequencies approaching 100%, and large deletions which occur at a frequency of about 10%. The frequency of these events was independent of WRN, p53, parental age, or folate levels. The large deletions occur at the same frequency in mice homozygous and heterozygous for the repeat suggesting that they are not the result of an interallelic recombination event. In addition, no evidence of large expansions was seen. Our data thus show that the absence of repeat expansions in mice is not due to a more efficient WRN protein or p53-mediated error correction mechanism, and suggest that these proteins, or the pathways in which they are active, may not be involved in expansion in humans either. Moreover, the fact that contractions occur in the absence of expansions suggests that these processes occur by different mechanisms.

The repeat expansion diseases are a group of genetic disorders that result from an expansion or increase in the number of repeats in a specific tandem array. Diseases in this group include fragile X syndrome, myotonic dystrophy, and Huntington's disease. The repeat expansion diseases fall into two classes with respect to expansion size. Diseases in the first class (Class I disorders) have expansions that involve the gain of only a small number of repeats, while the repeat tract in diseases in the second class (Class II disorders) show small increases in repeat number until a certain threshold is reached. Beyond this point much larger changes occur on intergenerational transmission of the allele resulting in the production of disease symptoms. The risk of expansion is directly related to the number of repeats in the parental allele, and in general, Class I disorders are paternally transmitted, while the larger expansions that cause the Class II disorders are maternally inherited. While large expansions are most frequently observed in these disorders, the repeat region is also a hotspot for deletions and small changes in repeat number reminiscent of microsatellite instability (de Graaff et al., 1995, 1996; Mila et al., 1996).

Received 2 October 2002; revision accepted 3 February 2003.

Request reprints from Karen Usdin, Building 8, Room 202
National Institutes of Health, 8 Center Drive MSC 0830
Bethesda, MD 20892-0830 (USA); telephone: 301-496-2189
fax: 301-402-0053; e-mail: ku@helix.nih.gov

K.F. and D.K.R. contributed equally to this work.

KARGER Fax + 41 61 306 12 34
E-mail karger@karger.ch
www.karger.com

0301–0171/03/1004–0140$19.50/0

Accessible online at:
www.karger.com/cgr

The molecular basis of these events is not known. The small changes in repeat number seen on transmission of the disease in the Class I disorders, as well as those seen in individuals with small premutation alleles in the case of the Class II disorders, could perhaps be explained by a simple strand-slippage event during replication or to aberrant repair in gametes (Kovtun and McMurray, 2001). However, the larger expansions and contractions that are characteristic of the transmission of the Class II disorders might involve quite different mechanisms.

Fragile X syndrome, the most common heritable form of mental retardation, belongs to the Class II subset of the repeat expansion disorders showing a biphasic pattern of instability. Small changes in repeat number occur in alleles with a small number of repeats. As the repeat number approaches 100, so the likelihood of large changes increases, with intergenerational expansions of ten fold or more being seen in some instances. We have previously generated transgenic mice containing the 5′ end of an FMR1 allele that is associated with an ~100% likelihood of expansion on maternal transmission in humans (Lavedan et al., 1998). This transgene contains a CGG·CCG tract with 97 uninterrupted CGG-repeats at its 3′ end. No large expansions that recapitulate the FXS expansion in humans were seen at birth in mice heterozygous for the repeat. Similar results have been reported for a knock-in mouse model of FXS as well as transgenic and knock-in mouse models of other repeat expansion diseases (Gourdon et al., 1997; Lavedan et al., 1997; Monckton et al., 1997; La Spada et al., 1998; Wheeler et al., 1999; Bontekoe et al., 2001; Miranda et al., 2002). In addition, not only is the rate of mouse minisatellite instability much lower than that of humans, but it differs too in being intra-allelic (Bois et al., 2002). These observations have led to the suggestion that the major mechanisms of instability in mice and humans might differ. This may be due to differences between mice and humans in important *cis-* or *trans-*acting factors that give rise to expansions in humans.

Identifying potential *trans-* or *cis-*acting factors that promote expansions in mammals is important since it may provide clues as to the molecular mechanism of expansion in FXS. It would also allow us to develop animal models that more accurately recapitulate the human disease. Elimination of potential factors from contention also provides useful information in that it points to other mechanisms and pathways that may ultimately be responsible. Many studies of FXS repeat stability have been carried out using different mutant bacterial and yeast strains. This has led to better understanding of which *cis-* and *trans-*acting factors do or do not affect instability in these microorganisms (see for examples: Jaworski et al., 1995; Freudenreich et al., 1998; Kokoska et al., 1998; Chen et al., 1999; Jakupciak and Wells, 1999; Kokoska et al., 1999; Morag et al., 1999; Schweitzer and Livingston, 1999; White et al., 1999; Ireland et al., 2000; Iyer et al., 2000). The stability of FXS repeats has been examined in the human cell lines HTC116 and LoVo which have mutations in the mismatch repair proteins, MLH1 and MSH2, respectively but contain FMR1 alleles with repeats in the normal range. They show no evidence of increased instability (Kramer et al., 1996). No similar studies with unstable FXS alleles have been reported in animals. Such studies are important particularly since FXS expansion occurs only on maternal transmission in humans, either as a germline or early post-fertilization event.

Most current models for the expansion event invoke defects in chromosomal replication, or other processes that involve DNA synthesis such as recombination or DNA repair (Richards and Sutherland, 1994; Usdin and Woodford, 1995; Gordenin et al., 1997; Kennedy and Shelbourne, 2000; Bowater and Wells, 2001; Kamath-Loeb et al., 2001; Kovtun and McMurray, 2001). We therefore chose to examine the effects of mutations in two genes involved in DNA replication and repair on the stability of the FXS repeat in transgenic mice.

The WRN helicase has been suggested to contribute to repeat stability since it reduces the incidence of polymerase stalling at the secondary structures that form in the FXS repeat region, thereby reducing the likelihood of events like strand-slippage or recombination that may lead to expansion (Kamath-Loeb et al., 2001). In addition WRN interacts with FEN-1, the human homolog of rad27, and stimulates its cleavage of the flap structures generated on Okazaki fragments during lagging strand DNA synthesis (Brosh et al., 2001). Secondary structures formed by the disease-causing repeats normally inhibit FEN-1 cleavage (Spiro et al., 1999). In addition, WRN has been implicated in both non-homologous and homologous recombination (reviewed in Brosh and Bohr, 2002). While there have been no reports of an increased likelihood of expansion in individuals with Werner's syndrome who have FXS repeat numbers that are in the normal range, it is possible that the effect of the WRN helicase is only seen when the repeat number exceeds the normal range, at which point the repeat tract can form longer and more stable secondary structures that provide more of an impediment to normal DNA processing events.

p53 is a multifunctional protein involved in maintaining genomic integrity. It mediates cell cycle arrest in response to genotoxic stress, and is a downstream activator of apoptosis in response to unrepaired DNA damage. In addition to its role in DNA damage surveillance, there is accumulating evidence that p53 may also be directly involved in replication, repair, and recombination (Lu and Lane, 1993; Bakalkin et al., 1994; Deng et al., 1995; Wang et al., 1995; Mummenbrauer et al., 1996; Sturzbecher et al., 1996; Kim et al., 1997; Lee et al., 1997). Loss of p53 function is associated with genome instability, including increased aneuploidy, microsatellite mutation (Ahrendt et al., 2000), gene amplification (Yin et al., 1992), and homologous recombination (Wiesmuller et al., 1996; Bertrand et al., 1997; Mekeel et al., 1997; Gebow et al., 2000). These events are observed when p53 is absent because not only are cells compromised in their ability to repair certain types of lesions, and perhaps forced to use other more mutation prone pathways, but since these cells are deficient in the G_1/S cell cycle checkpoint response and the G_2/M arrest is short-lived, they are also less likely to have an opportunity to complete any necessary DNA repairs before cell division occurs. In addition those cells that have failed to complete repair are also less likely to be lost from the population by apoptosis.

We also decided to examine the effect of folate deficiency on repeat instability. Folate is an important precursor for nucleotides needed for DNA synthesis. Folate deficiency, which is common in humans (Blount et al., 1997), not only affects the rate and fidelity of DNA synthesis but also causes DNA damage including nicks and breaks in chromosomes (Blount et al., 1997). Folate starvation also induces expression of the fragile site at Xq27.3 that gives fragile X syndrome its name, and such sites are associated with delayed replication (Hansen et al., 1997; Le Beau et al., 1998; Hellman et al., 2000), and elevated levels of chromosome breakage and recombination (Glover and Stein, 1988; Hirsch, 1991; Coquelle et al., 1997; Jones et al., 2000; Auer et al., 2001).

This article describes the germline and somatic instability of the FXS repeat in both the heterozygous and homozygous state, and the effect of parental age, dietary folate and *Wrn* and *Trp53* mutations on FXS repeat instability in mice. The consequences of our findings for mechanisms of repeat instability are discussed.

Materials and methods

Mice

Transgenic mice containing long uninterrupted CGG-tracts have been described elsewhere (Lavedan et al., 1998). They were bred for three generations to FVB/N mice lacking a transgene. After that point they were crossed to other mice that also contained the transgene, to WRN or p53-deficient mice (Donehower et al., 1992), or to CB6F1 mice (NIH Animal Facility, Bethesda, MD). The $Wrn^{-/-}$ mice were a kind gift of Dr Michel Lebel and Philip Leder (Lebel and Leder, 1998). $Trp53^{-/-}$ mice were purchased from Taconic (Germantown, NY). Folate stress was induced in some mice by maintaining them on either a folate-free diet (Clifford) or on a very low folate diet (Clifford Diet fortified with 0.45 mg folate/kg; Dyets, Inc, Bethlehem, PA) for at least three months before breeding. Mice were maintained on these diets for three generations.

DNA isolation

DNA was prepared either from tail snips taken at weaning or from different tissues taken immediately after euthanasia. DNA was isolated using the Qiagens DNeasy Tissue Kit (Qiagen, Valencia, CA) according to the manufacturer's instructions.

Genotyping

Genotyping was carried out on DNA isolated from tail snips taken at weaning prepared as described above. Primers were synthesized by Life Technologies, Inc (Gaithersburg, MD).

Trp53: The primers 10477 (5′-AAGAGGTGACTTTGGGGTGAAG-CTC-3′) and 10479 (5′-CTAGTTTACACACAGTCAGGATGG-3′) which amplifies exon 7 or primers 10680 (5′-CCCTACTCTACAACTAAAA-CTGAA-3′) and 10478 (5′-AAGCTG-GGGAAGAAACAGGCTAAC-3′) which amplifies exon 8 of the *Trp53* gene were used as internal controls in each PCR reaction. Primers 10480 (5′-ATGGGAGGCTGCCAGTCC-TAACCC-3′) and 10681 (5′-GTGTTTCATTAGTTCCCCACCTTGAC-3′) were used to detect the presence of the wild-type *Trp53* allele. Primers 10588 (5′-GTGGGAGGGACAAAAGTTCGAGGCC-3′) and 10930 (5′-TTTACGGAGCCCTGGCGCTCGATGT-3′) were used to detect the presence of the knockout allele. The PCR was carried out using DisplayTaq FL DNA polymerase (PGC, Gaithersburg, MD, USA), using the buffer provided by the supplier. Following an initial hot start at 95 °C for 5 min, 30 cycles of denaturation at 95 °C for 1.5 min, annealing at 65 °C for 1.5 min, and elongation at 72 °C for 2.5 min were performed using a PerkinElmer Cetus DNA Thermo Cycler. The internal PCR control produces a fragment of ~ 280 bp, the wild-type *Trp53* allele produces a fragment of ~ 320 bp and the mutant allele one of ~ 170 bp.

Wrn: Genotyping was carried out using NE-1 (5′-GTTTCCTCTATCT-GTAACAGG-3′), E-28 (5′-AGTGAGACATGTATGACTACC-3′) and PNT-1 (5′-GCGAAGGAGCAAAGCTGCTAT-3′). Briefly, the PCR was carried out using DisplayTaq FL DNA polymerase (PGC, Gaithersburg, MD, USA), using the buffer provided by the supplier. Following an initial hot start at 95 °C for 5 min, 30 cycles of 94 °C for 30 s, 55 °C for 30 s, and 72 °C for 50 s were performed using a PerkinElmer Cetus DNA Thermo Cycler. The wild-type allele produces a fragment of ~ 400bp while the mutant allele produces a fragment of ~ 450 bp.

CGG·CCG-containing transgene: Primers Eco3.3.12C (5′-GGAAGT-GAAACGGAAACG-3′) and FX5P (5′-ACCTCTGCAGAAATGGGCGTT-CTGGCCCT-3′) were used to screen for the presence of the transgene. These primers amplify a 300-bp region of the transgene downstream of the CGG-tract itself. This reaction is less prone to false negatives since it it is less likely to be affected by CGG-repeat-mediated deletions that remove flanking regions, or expansions that would be unamplifiable. RedTaq™ DNA polymerase (Sigma, St. Louis, MO, USA) was used for these reactions using the conditions recommended by the supplier. The reactions were subjected to 30 cycles of denaturation at 94 °C for 1 min, annealing at 60 °C for 1 min, and elongation at 72 °C for 30 s using a PerkinElmer Cetus DNA Thermo Cycler.

Analysis of CGG-repeat instability

The length of the CGG-repeat was monitored either by standard Southern blotting or by PCR analysis as previously described (Lavedan et al., 1998). FLZ2a (5′-dCATGTTCTCTTCAGCCCTGCTAGCGCCGGGAG-CCCG-3′) was used as the downstream primer in all instances. The experiment shown in Fig. 2 was carried out using frax-c (5′-dGCTCAGCT-CCGTTTCGGTTTCACTTCCGGT-3′) as the upstream primer. The experiment shown in Fig. 3 was carried out using frax-b (5′-dGGGGTTCGGCC-TCAGTCAGGCGCTCAGCTC-3′) as the upstream primer. Frax-b is located 20 bp 5′ of the frax-c primer, and unlike frax-c does not have 100 % homology to the mouse *Fmr1* gene. The PCR products were mixed with an equal volume of Stop buffer (99 % formamide, 5 mM EDTA, 0.01 % bromophenol blue, 0.01 % xylene cyanol). Samples were heated to 90 °C immediately before electrophoresis on a 6 % sequencing gel. The resultant gels were dried and autoradiographed according to standard procedures.

Results

The FXS-repeat containing mice used in this study were obtained from the progeny of a mouse having three to five copies of a transgene containing the sequence $(CGG)_9AGG(CGG)_9AGG(CGG)_{97}$ embedded in the immediate flanking sequence of a human FMR1 gene derived from a fragile X patient (Fig. 1; Lavedan et al., 1998). This repeat configuration corresponds to an allele with ~ 100 % likelihood of expansion on maternal transmission in humans.

PCR analysis of the first three generations of mice heterozygous for the CGG·CCG-containing transgene shows a pattern of stutter products that differed from mouse to mouse (Fig. 2). The stutter products reflect not only any in vivo instability but also problems that are typically encountered when amplifying long repeat tracts. A reproducible change in the distribution of these stutter products reflects the change in the average repeat size in that individual. The mutation rate observed in these mice thus approaches 100 %. However, in general these changes involve the gain or loss of fewer than three to five repeats. These small changes resemble the microsatellite instability seen in other tandem arrays of short repeated sequences including those seen in other mouse models for triplet repeat instability (Gourdon et al., 1997; Monckton et al., 1997). No evidence of larger expansions was seen by PCR or Southern blotting in more than 200 mice screened. This was true for both males and females, irrespective of whether the repeat was maternally or paternally transmitted or whether the parents were heterozygous or homozygous for the repeat. The effect of the age of the

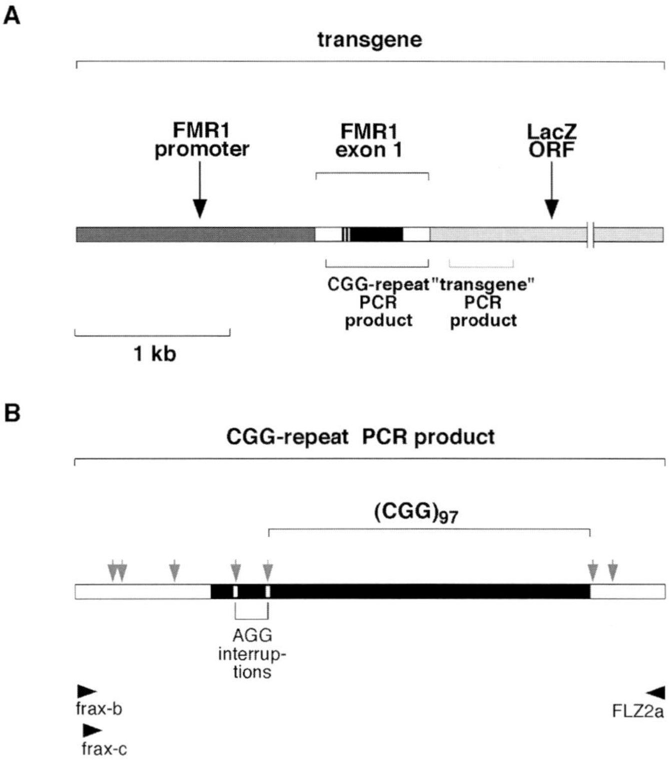

Fig. 1. The structure of the CGG-transgene. The white bar indicates the first exon of the FMR1 gene. The black bar indicates the CGG-repeat tract in the 5′ UTR of the transgene. The two AGG interruptions at repeats 10 and 23 are shown by the white lines. (**A**) Overall structure of the CGG-transgene. The dark grey bar indicates the human FMR1 promoter. The light grey bar delineates the lacZ reporter gene. The grey bracket indicates the location of the 300-bp PCR product used to identify mice containing the transgene. The black bracket indicates the so-called CGG-repeat PCR product used to examine the stability of the CGG-repeat tract. (**B**) Enlarged view of the CGG-repeat PCR product. The black arrows indicate the primers used to generate this product. Frax-c and frax-b are two alternative 5′ primers corresponding to sequences located 20 bp apart in the human FMR1 gene. The grey arrows indicate the locations of the *Mnl*I sites in this fragment.

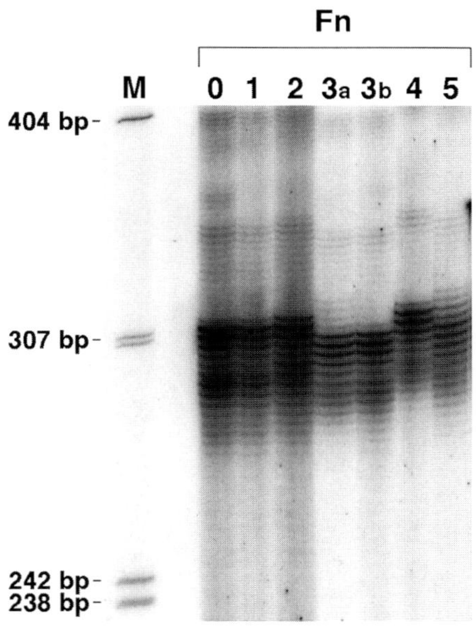

Fig. 2. Example of the microsatellite instability observed in the CGG-tract in five different generations of CGG-containing mice. Denaturing gel showing the *Mnl*I digestion products of the CGG-containing PCR fragment produced using the primer pair frax-c and FLZ2a. *Mnl*I cuts at the AGG interruptions at the promoter proximal end of the CGG-tract, and six bp 3′ of the promoter distal end. F0 indicates the founder mouse, and F1–F5, individuals from the first through fifth generation litters from this founder mouse.

parents at conception was also tested by examining the repeat size in the first litter produced at reproductive maturity to that of the last litter of mice produced before reproductive senescence. No age-dependent increase in instability was observed.

DNA purified from the brain, liver, gonads, kidneys, heart, spleen and lung of five mice of one to two years of age showed the same repeat profile seen in the initial tail snips taken at weaning. This was true even in the case of the tumor tissue from these mice (data not shown). This suggests that there is little, if any, postnatal somatic instability of the repeat. PCR of dilute solutions of a pure $(CGG)_9AGG(CGG)_9AGG(CGG)_{97}$ DNA fragment showed significant heterogeneity in the PCR products obtained (data not shown). This made it impossible to carry out small pool PCR to examine the somatic instability any further or to carry out single sperm analysis since we would not be able to distinguish heterogeneity due to instability from a PCR artifact.

While no large changes in repeat size were seen in the first three generations of transgenic mice, about 10% of the progeny of crosses of third generation littermates had large deletions. These deletions ranged in size from ~30 bp to more than 300 bp (see Fig. 3 for examples). Why the deletions only became visible in the fourth generation of mice is not clear at this time. It does not seem to be related to the methylation status of these alleles as they are all equally resistant to digestion with *Eag*I, a methylation-sensitive restriction enzyme that is often used to assess the methylation status of fragile X alleles (data not shown). There was also no correlation between the deletion frequency and whether or not the mice were homozygous for the transgene, nor was the deletion frequency affected by the sex or age of the transmitting parent (data not shown). Similar deletion frequencies were seen when these mice were crossed to CB6F1 mice, although the deletion frequency seen in first generation crosses with mice with a C57BL/6 strain background was higher (see below). Deletions occurring in mice heterozygous for the transgene suggest that this event is not the result of interchromosomal recombination. Moreover, the failure to see expansions when deletions are present suggests that the deletions were not the result of a reciprocal exchange of genetic information. Since the germline deletions were stably maintained and transmitted to subsequent generations at the normal Mendelian ratios, little, if any, homogenization of the transgene array occurs over the time scale of the experiment. Thus the absence of detectable expansions is probably not due to array homogenization either.

Figure 3 shows the pattern of deletions seen in a three-generation pedigree. The female progenitor of this lineage has no

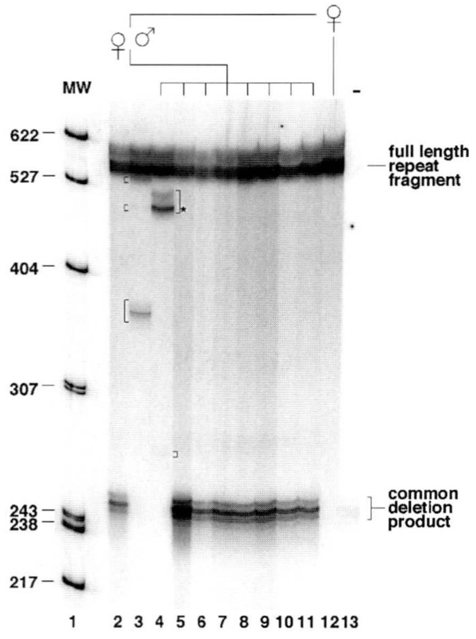

Fig. 3. CGG-repeat tract deletions in three generations of transgenic mice. The CGG-repeat PCR products from the tail DNA of three generations of mice are shown. The brackets indicate the deletion products. Lane –: Mouse lacking the transgene. The stutter products visible in Fig. 2 are difficult to discern in the larger PCR products shown here. They are more clearly seen when the gel is run longer, or when smaller fragments are generated by digestion of the PCR products with enzymes such as *Mnl*I used in the experiment shown in Fig. 2. As expected no stutter products are visible in the case of the very large deletions because the resultant PCR products contain only a small number of repeats.

large deletions (Fig. 3, lane 12). One of her female offspring has a large deletion indicated by the bracket in lane 2. This deletion is passed on to most of her offspring, indicating a germline mutation. Deletions of this size occurred in multiple independent lineages, and resulted in a PCR product very similar in size to those produced with primers for the murine *Fmr1* gene in wild-type mice (data not shown). This product may thus reflect a process that favors this size of deletion. The father of her offspring, one of her littermates, shows three sets of smaller deletions (brackets in lane 3). Two out of three deletions are not seen in any offspring (see lanes 4–11), demonstrating that these were somatic events. These two deletions presumably occurred shortly after fertilization since different organs of this individual either had no deletions or the two deletion products, with no further evidence of somatic heterogeneity (data not shown). A fragment with the same mobility as the third deletion in this individual is seen only in one of his offspring (indicated by the asterisk in Fig. 3). This deletion product may reflect parental germ line mosaicism.

The effect of WRN or p53 deficiency

Mice containing the FXS-repeat were crossed with mice containing a targeted disruption of the Werner's syndrome helicase gene (Lebel and Leder, 1998). The stability of the repeat was examined over four generations involving equivalent numbers of male and female transmissions of the repeat. A total of 66 *Wrn*$^{+/-}$ and 23 *Wrn*$^{-/-}$ mice were examined for evidence of repeat expansion. None was seen in tail snips taken at weaning, nor was there any evidence of expansion or increased microsatellite instability in the brains, liver, spleen, kidney or gonads of adult mice.

Mice containing the CGG·CCG-repeat were also crossed with mice containing a targeted disruption of the p53 gene (Donehower et al., 1992). The stability of the repeat was examined over five generations involving a total of 93 *Trp53*$^{+/-}$ and 30 *Trp53*$^{-/-}$ mice. As with the WRN-deficient mice, no evidence of expansion or increase in instability was seen in tail snips taken at weaning, or in the brains, liver, spleen, kidney or gonads of mice *Trp53*$^{+/-}$ or *Trp53*$^{-/-}$ mice. The repeat size in five spontaneously occurring tumors in these mice was also examined. They too showed no evidence of expansion or other changes in stability.

The effect of folate deprivation

Mice wildtype, heterozygous or nullizygous for *Trp53* were also maintained on either a zero folate or low folate diet for at least three months prior to the initial breeding and the offspring of this breeding were subsequently maintained on this diet from birth. Breeding pairs given a folate-free diet produced very few live births, and the litter size on the low folate diet was half that of mice on normal feed. The reduced litter sizes or failure to produce liveborn pups are characteristics of severe folate deprivation. Furthermore tail DNA isolated from mice on these diets was consistently more degraded than DNA isolated from comparable animals maintained on a normal folate diet, consistent with increased DNA damage (data not shown). Three generations of mice kept on these diets were analyzed, involving a total of 55 mice. No evidence was found for any expansions in any of the mice tested. The frequency of large deletions was also not changed relative to mice on diets with normal folate content. No significant effect was seen on the gain or loss of small numbers of repeats either.

Discussion

The CGG·CCG-tract in human fragile X pedigrees shows both small changes in repeat number, as wells as large expansions and contractions. We show here that an FMR1 allele with a repeat configuration $(CGG)_9AGG(CGG)_9AGG(CGG)_{97}$ that is extremely unstable in humans, does show a high level of repeat instability in mice. However, most of this instability involves the gain or loss of less than five repeat units. These small changes occur at a rate approaching 100%. No large, intergenerational expansions characteristic of the transmission of FXS in humans were seen. No significant postnatal somatic instability was seen either. This was true of the offspring of young as well as old mice, irrespective of whether the parents or offspring were heterozygous or homozygous for the repeat. The lack of a maternal age effect on the stability of the FXS repeat is significant since it has been suggested that expansion of the CTG·CAG-repeats in a model of Huntington's disease is a consequence of post-meiotic events occurring in haploid cells (Kovtun and McMurray, 2001). If small and large expansions of different repeats share a common mechanism, an increase with maternal age might be expected.

Deletions of between 30 and 300 bases were also seen but they occur at a frequency of 10%. In bacteria and yeast both expansions and contractions are seen. However, in wild-type *E. coli* and *S. cerevisiae* expansions are only seen within a narrow range of repeat numbers, and deletions predominate when the size of the repeat array exceeds a particular threshold, perhaps because of selection for rapidly replicating molecules. Therefore, one interpretation of our failure to see expansions in our transgenic mice could be that repeat tract in these mice is in excess of the threshold beyond which deletions predominate. We do not think that this interpretation is correct since we have previously shown that expansions are not seen in transgenic mice with an otherwise identical transgene having nine or 53 rather than 97 uninterrupted CGG·CCG repeats (Lavedan et al., 1998). The deletion frequency is the same in mice heterozygous and homozygous for the repeat suggesting that inter-chromosomal recombination is not a factor in this process. Deletions seen in the absence of expansions thus suggests that the deletions are not the result of a reciprocal recombination event, and that the mechanisms of deletion and expansion may be fundamentally different. The presence of little, if any, somatic mosiacism of the deletions suggests that they are germ-line or early post-fertilization events. In these respects, the deletions resemble what is seen in fragile X pedigrees. Therefore, while our CGG·CCG-transgenic mice do not show the large expansions characteristic of fragile X syndrome, they do seem to recapitulate the deletion events in certain respects.

No expansions, increases in deletion frequency, or evidence of somatic instability were seen in mice heterozygous or homozygous for a *Wrn* disruption. The same was true of mice heterozygous or nullizygous for *Trp53*. Thus the absence of expansions in wild-type mice is not due to more efficient WRN protein or the repair/elimination of the mutation mediated by a hyper-vigilant murine p53. Moreover since folate deprivation did not affect stability significantly even in p53 deficient mice, it suggests that the reduced dNTP pools and the mild DNA damage that results from folate deficiency do not increase repeat instability in mice either. Our data also raise the possibility that neither these proteins nor folate levels are significant modifiers of repeat instability in humans and that the molecular basis for the large expansions seen in fragile X syndrome may lie elsewhere.

References

Ahrendt SA, Decker PA, Doffek K, Wang B, Xu L, Demeure MJ, Jen J, Sidransky D: Microsatellite instability at selected tetranucleotide repeats is associated with p53 mutations in non-small cell lung cancer. Cancer Res 60:2488–2491 (2000).

Auer RL, Jones C, Mullenbach RA, Syndercombe-Court D, Milligan DW, Fegan CD, Cotter FE: Role for CCG-trinucleotide repeats in the pathogenesis of chronic lymphocytic leukemia. Blood 97:509–515 (2001).

Bakalkin G, Yakovleva T, Selivanova G, Magnusson KP, Szekely L, Kiseleva E, Klein G, Terenius L, Wiman KG: p53 binds single-stranded DNA ends and catalyzes DNA renaturation and strand transfer. Proc natl Acad Sci, USA 91:413–417 (1994).

Bertrand P, Rouillard D, Boulet A, Levalois C, Soussi T, Lopez BS: Increase of spontaneous intrachromosomal homologous recombination in mammalian cells expressing a mutant p53 protein. Oncogene 14:1117–1122 (1997).

Blount BC, Mack MM, Wehr CM, MacGregor JT, Hiatt RA, Wang G, Wickramasinghe SN, Everson RB, Ames BN: Folate deficiency causes uracil misincorporation into human DNA and chromosome breakage: implications for cancer and neuronal damage. Proc natl Acad Sci, USA 94:3290–3295 (1997).

Bois PR, Grant GR, Jeffreys AJ: Minisatellites show rare and simple intra-allelic instability in the mouse germ line. Genomics 80:2–4 (2002).

Bontekoe CJM, Bakker CE, Nieuwenhuizen IM, van der Linde H, Lans H, de Lange D, Hirst MC, Oostra BA: Instability of a $(CGG)_{98}$ repeat in the *Fmr1* promoter. Hum molec Genet 10:1693–1699 (2001).

Bowater RP, Wells RD: The intrinsically unstable life of DNA triplet repeats associated with human hereditary disorders. Prog Nucl Acid Res Mol Biol 66:159–202 (2001).

Brosh RM Jr, Bohr VA: Roles of the Werner syndrome protein in pathways required for maintenance of genome stability. Exp Gerontol 37:491–506 (2002).

Brosh RM Jr, von Kobbe C, Sommers JA, Karmakar P, Opresko PL, Piotrowski J, Dianova I, Dianov GL, Bohr VA: Werner syndrome protein interacts with human flap endonuclease 1 and stimulates its cleavage activity. EMBO J 20:5791–5801 (2001).

Chen C, Merrill BJ, Lau PJ, Holm C, Kolodner RD: *Saccharomyces cerevisiae pol30* (proliferating cell nuclear antigen) mutations impair replication fidelity and mismatch repair. Mol Cell Biol 19:7801–7815 (1999).

Coquelle A, Pipiras E, Toledo F, Buttin G, Debatisse M: Expression of fragile sites triggers intrachromosomal mammalian gene amplification and sets boundaries to early amplicons. Cell 89:215–225 (1997).

de Graaff E, Rouillard P, Willems PJ, Smits AP, Rousseau F, Oostra BA: Hotspot for deletions in the CGG repeat region of FMR1 in fragile X patients. Hum molec Genet 4:45–49 (1995).

de Graaff E, de Vries BB, Willemsen R, van Hemel JO, Mohkamsing S, Oostra BA, van den Ouweland AM: The fragile X phenotype in a mosaic male with a deletion showing expression of the FMR1 protein in 28% of the cells. Am J med Genet 64:302–308 (1996).

Deng C, Zhang P, Harper JW, Elledge SJ, Leder P: Mice lacking p21CIP1/WAF1 undergo normal development, but are defective in G1 checkpoint control. Cell 82:675–684 (1995).

Donehower LA, Harvey M, Slagle BL, McArthur MJ, Montgomery CA Jr, Butel JS, Bradley A: Mice deficient for p53 are developmentally normal but susceptible to spontaneous tumours. Nature 356:215–221 (1992).

Freudenreich CH, Kantrow SM, Zakian VA: Expansion and length-dependent fragility of CTG repeats in yeast. Science 279:853–856 (1998).

Gebow D, Miselis N, Liber HL: Homologous and nonhomologous recombination resulting in deletion: effects of p53 status, microhomology, and repetitive DNA length and orientation. Mol cell Biol 20:4028–4035 (2000).

Glover TW, Stein CK: Chromosome breakage and recombination at fragile sites. Am J hum Genet 43:265–273 (1988).

Gordenin DA, Kunkel TA, Resnick MA: Repeat expansion – all in a flap? Nature Genet 16:116–118 (1997).

Gourdon G, Radvanyi F, Lia AS, Duros C, Blanche M, Abitbol M, Junien C, Hofmann-Radvanyi H: Moderate intergenerational and somatic instability of a 55-CTG repeat in transgenic mice [see comments] Nature Genet 15:190–192 (1997).

Hansen RS, Canfield TK, Fjeld AD, Mumm S, Laird CD, Gartler SM: A variable domain of delayed replication in FRAXA fragile X chromosomes: X inactivation-like spread of late replication. Proc natl Acad Sci, USA 94:4587–4592 (1997).

Hellman A, Rahat A, Scherer SW, Darvasi A, Tsui LC, Kerem B: Replication delay along FRA7H, a common fragile site on human chromosome 7, leads to chromosomal instability. Mol cell Biol 20:4420–4427 (2000).

Hirsch B: Sister chromatid exchanges are preferentially induced at expressed and nonexpressed common fragile sites. Hum Genet 87:302–306 (1991).

Ireland MJ, Reinke SS, Livingston DM: The impact of lagging strand replication mutations on the stability of CAG repeat tracts in yeast. Genetics 155: 1657–1665 (2000).

Iyer RR, Pluciennik A, Rosche WA, Sinden RR, Wells RD: DNA polymerase III proofreading mutants enhance the expansion and deletion of triplet repeat sequences in *Escherichia coli*. J biol Chem 275:2174–2184 (2000).

Jakupciak JP, Wells RD: Genetic instabilities in (CTG.CAG) repeats occur by recombination. J biol Chem 274:23468–23479 (1999).

Jaworski A, Rosche WA, Gellibolian R, Kang S, Shimizu M, Bowater RP, Sinden RR, Wells RD: Mismatch repair in *Escherichia coli* enhances instability of $(CTG)_n$ triplet repeats from human hereditary diseases. Proc natl Acad Sci, USA 92:11019–11023 (1995).

Jones C, Mullenbach R, Grossfeld P, Auer R, Favier R, Chien K, James M, Tunnacliffe A, Cotter F: Co-localisation of CCG repeats and chromosome deletion breakpoints in Jacobsen syndrome: evidence for a common mechanism of chromosome breakage. Hum molec Genet 9:1201–1208 (2000).
Kamath-Loeb AS, Loeb LA, Johansson E, Burgers PM, Fry M: Interactions between the Werner syndrome helicase and DNA polymerase delta specifically facilitate copying of tetraplex and hairpin structures of the $d(CGG)_n$ trinucleotide repeat sequence. J biol Chem 276:16439–16446 (2001).
Kennedy L, Shelbourne PF: Dramatic mutation instability in HD mouse striatum: does polyglutamine load contribute to cell-specific vulnerability in Huntington's disease? Hum molec Genet 9:2539–2544 (2000).
Kim E, Albrechtsen N, Deppert W: DNA-conformation is an important determinant of sequence-specific DNA binding by tumor suppressor p53. Oncogene 15:857–869 (1997).
Kokoska RJ, Stefanovic L, Buermeyer AB, Liskay RM, Petes TD: A mutation of the yeast gene encoding PCNA destabilizes both microsatellite and minisatellite DNA sequences. Genetics 151:511–519 (1999).
Kokoska RJ, Stefanovic L, Tran HT, Resnick MA, Gordenin DA, Petes TD: Destabilization of yeast micro- and minisatellite DNA sequences by mutations affecting a nuclease involved in Okazaki fragment processing (rad27) and DNA polymerase delta (pol3-t). Mol cell Biol 18:2779–2788 (1998).
Kovtun IV, McMurray CT: Trinucleotide expansion in haploid germ cells by gap repair. Nature Genet 27:407–411 (2001).
Kramer PR, Pearson CE, Sinden RR: Stability of triplet repeats of myotonic dystrophy and fragile X loci in human mutator mismatch repair cell lines. Hum Genet 98:151–157 (1996).
La Spada AR, Peterson KR, Meadows SA, McClain ME, Jeng G, Chmelar RS, Haugen HA, Chen K, Singer MJ, Moore D, et al: Androgen receptor YAC transgenic mice carrying CAG 45 alleles show trinucleotide repeat instability. Hum molec Genet 7:959–967 (1998).
Lavedan CN, Garrett L, Nussbaum RL: Trinucleotide repeats $(CGG)_{22}TGG(CGG)_{43}TGG(CGG)_{21}$ from the fragile X gene remain stable in transgenic mice. Hum Genet 100:407–414 (1997).
Lavedan C, Grabczyk E, Usdin K, Nussbaum RL: Long uninterrupted CGG repeats within the first exon of the human FMR1 gene are not intrinsically unstable in transgenic mice. Genomics 50:229–240 (1998).
Le Beau MM, Rassool FV, Neilly ME, Espinosa R 3rd, Glover TW, Smith DI, McKeithan TW: Replication of a common fragile site, FRA3B, occurs late in S phase and is delayed further upon induction: implications for the mechanism of fragile site induction. Hum molec Genet 7:755–761 (1998).
Lebel M, Leder P: A deletion within the murine Werner syndrome helicase induces sensitivity to inhibitors of topoisomerase and loss of cellular proliferative capacity. Proc natl Acad Sci, USA 95:13097–13102 (1998).
Lee S, Cavallo L, Griffith J: Human p53 binds Holliday junctions strongly and facilitates their cleavage. J biol Chem 272:7532–7539 (1997).
Lu X, Lane DP: Differential induction of transcriptionally active p53 following UV or ionizing radiation: defects in chromosome instability syndromes? Cell 75:765–778 (1993).
Mekeel KL, Tang W, Kachnic LA, Luo CM, DeFrank JS, Powell SN: Inactivation of p53 results in high rates of homologous recombination Oncogene 14:1847–1857 (1997).
Mila M, Castellvi-Bel S, Sanchez A, Lazaro C, Villa M, Estivill X: Mosaicism for the fragile X syndrome full mutation and deletions within the CGG repeat of the FMR1 gene. J Med Genet 33:338–340 (1996).
Miranda CJ, Santos MM, Ohshima K, Smith J, Li L, Bunting M, Cossee M, Koenig M, Sequeiros J, Kaplan J, et al: Frataxin knockin mouse. FEBS Lett 512:291–297 (2002).
Monckton DG, Coolbaugh MI, Ashizawa KT, Siciliano MJ, Caskey CT: Hypermutable myotonic dystrophy CTG repeats in transgenic mice. Nature Genet 15:193–196 (1997).
Morag AS, Saveson CJ, Lovett ST: Expansion of DNA repeats in *Escherichia coli:* effects of recombination and replication functions. J molec Biol 289:21–27 (1999).
Mummenbrauer T, Janus F, Muller B, Wiesmuller L, Deppert W, Grosse F: p53 Protein exhibits 3′ to-5′ exonuclease activity. Cell 85:1089–1099 (1996).
Richards RI, Sutherland GR: Simple repeat DNA is not replicated simply. Nature Genet 6:114–116 (1994).
Schweitzer JK, Livingston DM: The effect of DNA replication mutations on CAG tract stability in yeast. Genetics 152:953–963 (1999).
Spiro C, Pelletier R, Rolfsmeier ML, Dixon MJ, Lahue RS, Gupta G, Park MS, Chen X, Mariappan SV, McMurray CT: Inhibition of FEN-1 processing by DNA secondary structure at trinucleotide repeats. Mol Cell 4:1079–1085 (1999).
Sturzbecher HW, Donzelmann B, Henning W, Knippschild U, Buchhop S: p53 is linked directly to homologous recombination processes via RAD51/RecA protein interaction. EMBO J 15:1992–2002 (1996).
Usdin K, Woodford KJ: CGG repeats associated with DNA instability and chromosome fragility form structures that block DNA synthesis in vitro. Nucl Acids Res 23:4202–4209 (1995).
Wang XW, Yeh H, Schaeffer L, Roy R, Moncollin V, Egly JM, Wang Z, Freidberg EC, Evans MK, Taffe BG, et al: p53 modulation of TFIIH-associated nucleotide excision repair activity. Nature Genet 10:188–195 (1995).
Wheeler VC, Auerbach W, White JK, Srinidhi J, Auerbach A, Ryan A, Duyao MP, Vrbanac V, Weaver M, Gusella JF, et al: Length-dependent gametic CAG repeat instability in the Huntington's disease knock-in mouse. Hum molec Genet 8:115–122 (1999).
White PJ, Borts RH, Hirst MC: Stability of the human fragile X (CGG)(n) triplet repeat array in *Saccharomyces cerevisiae* deficient in aspects of DNA metabolism. Mol cell Biol 19:5675–5684 (1999).
Wiesmuller L, Cammenga J, Deppert WW: In vivo assay of p53 function in homologous recombination between simian virus 40 chromosomes. J Virol 70:737–744 (1996).
Yin Y, Tainsky MA, Bischoff FZ, Strong LC, Wahl GM: Wild-type p53 restores cell cycle control and inhibits gene amplification in cells with mutant p53 alleles. Cell 70:937–948 (1992).

Cytogenet Genome Res 100:147–153 (2003)
DOI: 10.1159/000072849

Spinocerebellar ataxia type 6 and episodic ataxia type 2: differences and similarities between two allelic disorders

E. Mantuano,[a] L. Veneziano,[a] C. Jodice[b] and M. Frontali[a]

[a] Istituto di Neurobiologia e Medicina Molecolare del CNR;
[b] Dipartimento di Biologia Università Tor Vergata, Roma (Italia)

Abstract. Spinocerebellar ataxia type 6 (SCA6) is one of three allelic disorders caused by mutations of CACNA1A gene, coding for the pore-forming subunit of calcium channel type P/Q. SCA6 is associated with small expansions of a CAG repeat at the 3′ end of the gene, while point mutations are responsible for its two allelic disorders (Episodic Ataxia type 2 and Familial Hemiplegic Migraine). Genetic, clinical, pathological and pathophysiological data of SCA6 patients are reviewed and compared to those of other SCAs with expanded CAG repeats as well as to those of its allelic channelopathies, with particular reference to Episodic Ataxia type 2. Overall SCA6 appears to share features with both types of disorders, and the question as to whether it belongs to polyglutamine disorders or to channelopathies remains unanswered at present.

Spinocerebellar ataxia type 6 (SCA6) belongs to the group of autosomal dominant cerebellar ataxias (ADCAs), which have an estimated prevalence around 3/100,000 inhabitants (Skre, 1974; van de Warrenburg et al., 2002). SCA6 accounts for 1–11 % of ADCA families collected in European centres (Leggo et al., 1997; Riess et al., 1997; Stevanin et al., 1997; Takano et al., 1998; Pujana et al., 1999; Sinke et al., 2001; Silveira et al., 2002), 6–31 % in Japanese (Matsuyama et al., 1997; Takano et al., 1998; Watanabe et al., 1998; Maruyama et al., 2002) and 0–11 % in Chinese series (Tang et al., 2000; Soong et al., 2001a). As most ADCAs, SCA6 is due to expansions of a CAG repeat stretch, which is embedded in the 3′ coding region of a calcium channel gene, CACNA1A, on chromosome 19p13. The gene codes for the α_{1A} pore-forming subunit of voltage-gated calcium channels type P/Q, expressed predominantly in cerebellar Purkinje and granule cells. Point mutations at the CACNA1A gene are known to cause Episodic Ataxia type 2 (EA2) and Familial Hemiplegic Migraine (FHM) (Ophoff et al., 1996). SCA6 shares some features with these disorders, particularly EA2, and differs in many aspects from ADCAs with expansions of a CAG repeat.

Supported by Telethon grant E847, FISR "Neurobiotecnologie", Ministry of Health "Danno neuronale misfolding di proteine" to M.F.

Received 4 November 2002; revision accepted 30 January 2003.

Request reprints from Marina Frontali
Istituto di Neurobiologia e Medicina Molecolare del CNR
Via Fosso del Cavaliere, I–00044 Frascati Rome (Italy)
telephone: +39-06-49934222; fax: +39-06-49934257
e-mail: Marina.Frontali@ims.rm.cnr.it

Gene, mutation and inheritance

The CACNA1A gene maps on chromosome 19p13.2→p13.1 (Diriong et al., 1995) and covers about 300 kb with 48 exons (Ophoff et al., 1996; Trettel et al., 1998). Well conserved, the gene is expressed as a transcript of approximately 9.8 kb in the brain (Ophoff et al., 1996), more abundantly in the cerebellum than in other cerebral areas, and particularly in Purkinje cells.

The cDNA clones predict large peptides with molecular masses of 200 to 275 kDa with four homologous domains (I–IV), each containing six hydrophobic transmembrane segments, S1–S6 (Fig. 1A). The short N-terminal and the long C-terminal tails of the protein are located in the cytoplasm. In some isoforms a polymorphic polyglutamine sequence, coded by the CAG_n repeat expanded in SCA6, is present in the C-terminal tail.

KARGER
Fax + 41 61 306 12 34
E-mail karger@karger.ch
www.karger.com

0301–0171/03/1004–0147$19.50/0

Accessible online at:
www.karger.com/cgr

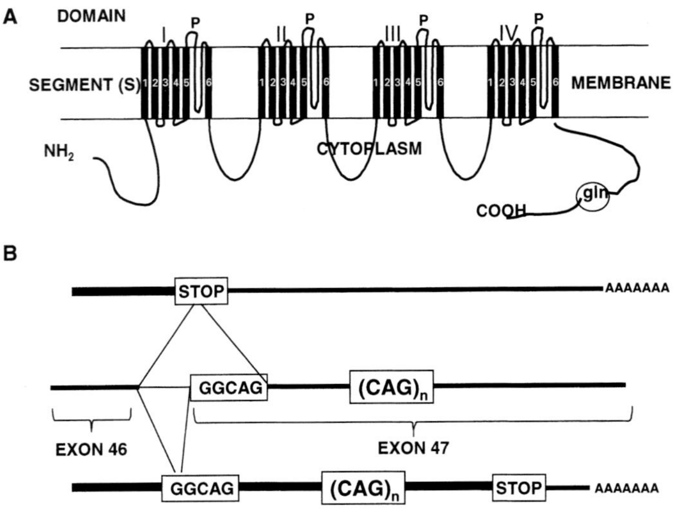

Fig. 1. (**A**) Schematic structure of the α_{1A} subunit of the voltage-gated calcium channel type P/Q. Each of the four domains (I–IV) has six α-helix transmembrane segments (S1–S6). The carboxyl terminus of the protein contains a stretch of polyglutamines (gln) in some of the isoforms (see text). (**B**) Alternative splicing between exons 46 and 47. If exon 46 is joined to exon 47 excluding the GGCAG segment, a TGA stop codon is encountered and the poly-CAG stretch is not expressed. Whereas, if exon 46 is joined to exon 47 including the GGCAG segment, the translation continues beyond the poly-CAG, until a TAA stop codon is encountered, thus producing a protein containing a polyglutamine stretch.

The transcript undergoes considerable alternative splicing, producing several isoforms (Mori et al., 1991; Zhuchenko et al., 1997). In particular, the presence or absence of a 5-nucleotide stretch, GGCAG, at the 3′ end of the gene (between exons 46 and 47) is critical for the expression of the CAG repeat stretch (Fig. 1B): when the 5 nucleotides are left in place, the CAG repeat is translated and expressed as a polyglutamine sequence at the protein level; when they are spliced out, a stop codon is encountered upstream the CAG_n stretch, and the polyglutamine repeat is not included in the resultant protein. Isoforms both with or without the GGCAG insertion, i.e. expressing or not expressing the polyglutamine tract, have been found in the cerebellar cortex (Ishikawa et al., 1999a). Whether these isoforms have distinct functional properties, what is their pattern of expression and how this is regulated in the human brain is largely unknown.

The normal allele size of the polymorphic CAG repeat ranges from 4–18 units (Zhuchenko et al., 1997; Shizuka et al., 1998a, b), while that of expanded alleles is from 20–30 repeats (Jodice et al., 1997; Matsuyama et al., 1997; Shizuka, et al., 1998a, b; Katayama et al., 2000). Data on the intermediate allele with 19 repeats are contradictory. Katayama et al. (2000) report an ataxic patient with a 19/7 genotype and lacunar lesions in the pons and basal ganglia. In other studies heterozygotes for this allele were normal even at old ages (Ishikawa et al., 1997; Mariotti et al., 2001).

When compared to other CAG_n expansion disorders the size of SCA6 expanded alleles appears much smaller, overlapping their wildtype allele distribution (Margolis, 2002). The size of SCA6 expanded alleles is more stable, as expected on the basis of their relatively low number of repeats. No variation is usually observed in families over successive generations, and no mosaicism is apparent in cells from different parts of the brain (Ishikawa et al., 1999b) or in sperm (Shizuka et al., 1998b). However, some degree of meiotic instability should be assumed since in three families an intergenerational jump of the expanded allele size has been reported (Jodice et al., 1997; Matsuyama et al., 1997; Mariotti et al., 2001).

SCA6 has a dominant pattern of inheritance. Homozygous patients for a repeat expansion, however, are slightly differing from heterozygous ones, showing an earlier onset and a more rapid course (Geschwind et al., 1997; Matsumura et al., 1997; Takiyama et al., 1998; Fukutake et al., 2002). With intermediate size (19 repeat) alleles, however a recessive rather than a dominant pattern has been suggested, based on the observation of an ataxic proband carrying a 19/19 genotype, and of a normal phenotype in its heterozygous relatives (Mariotti et al., 2001)

SCA6 expansions have been found in a number of sporadic ataxia cases (Ikeuchi et al., 1997; Matsumura et al., 1997; Riess et al., 1997; Zhuchenko et al., 1997; Shizuka et al., 1998b), but a new mutation has been documented in one patient only (Shizuka et al., 1998a). Should all the reported sporadic cases be new mutations, the mutability of normal alleles would be very high. If a mutation/selection equilibrium is assumed, a high frequency of SCA6 *de novo* mutations, would be in disagreement with the, presumably, small or absent selection against a disease, such as this one, with a late onset and a long life span. In addition, linkage disequilibrium among SCA6 families was also found (Dichgans et al., 1999; Soong et al., 2001a) which, again, seems in contrast with a high gene mutability. Incomplete penetrance or presence of neglected mild expression of the disorder in probands' relatives appear to be more likely explanations for the high number of sporadic cases.

Clinical features

The SCA6 clinical picture was initially described (Zhuchenko et al., 1997) as characterized by a late onset slowly progressive cerebellar ataxia with nystagmus and dysarthria, brainstem involvement (dysphagia), vibratory and proprioceptive sensory loss, often preceeded by episodes of "wooziness" and imbalance. This phenotype, similar to that reported by some later studies (Geschwind et al., 1997; Stevanin et al., 1997; Filla et al., 1999), matches that of ADCA type 1, according to the classification of Harding (1982), i.e. a cerebellar deficit with the involvement of other nervous functional systems. According to other reports, instead, SCA6 has the features typical of a "pure" cerebellar ataxia with no extracerebellar involvement (ADCA type III according to Harding, 1982) (Ikeuchi et al., 1997; Ishikawa et al., 1997; Matsumura et al., 1997; Stevanin et al., 1997; Nagai et al., 1998; Watanabe et al., 1998; Garcia-Planells et al., 1999). Still other studies, however, show a more complex clinical picture. Patients may initially experience episodes of ataxia and dysarthria, and/or of vertigo and nausea, accompanied by visual disturbance (such as diplopia or blurred vision), and tinnitus, lasting from minutes to days and triggered by head movements and physical or emotional stress (Calandriello et al., 1996; Geschwindt et al., 1997; Jodice et al., 1997; Koh et al., 2001; Sinke et al., 2001). Episodes have a variable frequency (from yearly to daily), and duration (from seconds to days). Interictally, patients may be neurologically normal or show mild cerebellar signs such as gaze-evoked nystagmus and/or saccadic pursuit, mild dysarthria and dysmetria, and mild gait imbalance (Calandriello et al., 1996; Geschwindt et al., 1997; Jodice et al., 1997; Koh et al., 2001; Sinke et al., 2001). This early phase can have a variable duration. It commonly lasts a few years before the onset of a permanent and progressive ataxia, usually of a "pure" cerebellar type. The neurological examination shows trunk and limb ataxia, dysmetria, dysarthria, hypotonia, gaze evoked nystagmus with or without a downbeat component, dysmetric saccades, saccadic pursuit and hyperactive vestibulo-ocular reflex (Gomez et al., 1997). The presence of extra-cerebellar signs, such as dysphagia, opthalmoplegia, hyperreflexia, sensory loss and/or dementia is sometimes reported (Geschwind et al., 1997; Zhuchenko et al., 1997) particularly in older subjects (Ikeuchi et al., 1997; Ishikawa et al., 1997; Stevanin et al., 1997) or in patients with other coexisting disorders (Takiyama et al., 1998). Vertigo episodes may continue during this second progressive phase (Gomez et al., 1997), and periodical exacerbations of the cerebellar signs can be present (Yabe et al., 1998). In some cases the disease may not progress towards a full-blown disorder and maintains its episodic character with mild non-progressive interictal cerebellar signs (Calandriello et al., 1996; Jodice et al., 1997; Takiyama et al., 1998; Koh et al., 2001). More frequently the disease very slowly progresses towards inability; autonomous walking is maintained even after 18–20 years from the onset (Geschwind et al., 1997). However a very rapid course has also been reported (Watanabe et al., 1998). The life span appears to be normal, although no statistical analysis is available.

The above picture of SCA6 phenotype strikingly overlaps that of the allelic disorder EA2, due to point mutations of the same gene. EA2 has been described as an early onset disorder characterised by episodes of vertigo/ataxia with dysartria, visual disturbances and variable interictal cerebellar signs, from nystagmus only to severe progressive cerebellar ataxia (Farmer and Mustian, 1967; Moon and Koller, 1991; Baloh et al., 1997). Acetazolamide, a carbonic anhydrase inhibitor, is a well-known treatment preventing EA2 episodes. Differences and similarities between the two disorders emerged by comparing the features of 315 SCA6 and 44 EA2 patients, reported in the literature, and carrying, respectively, a CAG expansion or a point mutation of CACNA1A gene (Frontali, 2001). Both can have an episodic and/or a progressive ataxia phase, but they differ in the proportion of patients with progressive cerebellar involvement, which is larger in SCA6, and of those with vertigo and/or ataxia episodes, which is larger in the EA2 group. As in EA2, SCA6 episodes respond to acetazolamide treatment (Calandriello et al., 1996; Jen et al., 1998; Koh et al., 2001), although no control for a placebo effect has ever been performed. Additional extracerebellar signs, such as decrease in vibration sense, hyperreflexia with or without Babinski sign, dysphagia and/or ophtalmoplegia, brainstem atrophy, appear to affect only a minority of SCA6 patients.

Overall, the main features distinguishing SCA6 from EA2 appear to be a higher frequency of progressive ataxia occasionally associated with extracerebellar signs. It should be noted, however, that the two groups have different ages at examination, SCA6 patients being older than EA2 patients (mean age 58.5 vs. 38.0 years). It is tempting to speculate that the phenotypic differences in SCA6 patients are the consequence either of other neurological disorders intervening with age, or to a longer disease duration, with a greater likelihood of progression to permanent ataxia. This would suggest the possibility that the two conditions might be expressions of the same highly variable disorder. A strong support for such a hypothesis comes from families carrying an SCA6 expansion mutation in which different members show either the typical features of EA2 with a prominent paroxystic ataxia or those of SCA6 with a progressive and permanent ataxia (Jodice et al., 1997; Koh et al., 2001).

Another possible discrepancy between SCA6 and EA2 concerns the age at onset. In most studies SCA6 age at onset refers to the beginning of the progressive ataxia phase and is reported around 50 years of age, with a range from 16–73 (Gomez et al., 1997; Matsumura et al., 1997; Watanabe et al., 1998; van de Warrenburg et al., 2002). In EA2, instead, the age at onset, referred to the beginning of vertigo/ataxia episodes, is usually in the first or second decade of life (Yue et al., 1997; Jen et al., 1998, 1999, 2001; Denier et al., 1999; Guida et al., 2001). It should be noted, however, that this discrepancy is probably in part, if not completely, due to an ascertainment bias: EA2 is more easily diagnosed in infancy or adolescence when prominent vertigo/ataxia episodes have fewer alternative causes, than in older ages, when such symptoms are more frequently (but inappropriately) ascribed to low blood pressure, cervical arthrosis, or other common disorders. On the other hand SCA6 evokes, by analogy with other SCAs, a permanent and progressive cerebellar ataxia and a possible early phase of ataxic/vertigo spells is overlooked both by patients and physicians, as a minor and non pertinent disturbance.

As in other CAG repeat expansion disorders, a significant inverse correlation between age at onset and size of expanded alleles has been reported in several studies (e.g. Ikeuchi et al., 1997; Ishikawa et al., 1997; Riess et al., 1997; Zhuchenko et al., 1997; Maruyama et al., 2002) as well as an anticipation of age at onset over successive generations (e.g. Ikeuci et al., 1997; Matsumura et al., 1997; Matsuyama et al., 1997; Watanabe et al., 1998; Sinke et al., 2001). The latter observation is rather surprising considering that no intergenerational variation of the expanded allele size was present. The phenomenon, hence, should be ascribed, again, to an ascertainment bias, being more likely to observe a parent-child pair with the offspring affected earlier than the parent, rather than vice versa (Penrose, 1948). Alternative explanations can rely on the exposure to different environmental factors affecting age at onset in different generations, or on the difficulty of assessing age at onset in older patients.

Neuropathology

Neuroimaging in SCA6 patients reveals a cerebellar vermis atrophy predominating in the anterior portion, which might later extend to cerebellar hemispheres, usually with preservation of brainstem (Calandriello et al., 1996; Gomez et al., 1997; Jodice et al., 1997; Nagai et al., 1998; Satoh et al., 1998; Shizuka et al., 1998a, b; Takiyama et al., 1998). An early involvement of the cerebellar vermis appears also to characterize EA2 (e.g. Denier et al., 1999, Guida et al., 2001). Occasionally, at MRI, a size reduction of pons as well as of red nucleus and middle cerebellar peduncle has been reported (Murata et al., 1998; Arpa et al., 1999; Nakagawa et al., 1999). A widespread reduction of glucose metabolism at PET scan, with particularly low values in the brainstem and cerebellar hemispheres, was observed in a group of SCA6 patients as compared to normal controls (Soong et al., 2001b). Unfortunately the controls were younger than patients, requiring a further confirmation of data.

Consistently with most neuroimaging studies, the brains of autopsied SCA6 patients showed a marked atrophy of the cerebellar vermis and, to a lesser extent, of the hemispheres. Histologically, the cerebellar cortex is characterised by a remarkable loss of Purkinje cells. Granule cells are also affected, although less severely (Subramony et al., 1996; Gomez et al., 1997; Sasaki et al., 1998; Ishikawa et al., 1999b; Tashiro et al., 1999). Loss of neurons in the dentate and inferior olivary nuclei has also been found in one study (Subramony et al., 1996). Atrophy of brainstem has been occasionally described (Zhuchenko et al., 1997).

By virtue of the analogy with other $(CAG)_n$ expansion disorders, ubiquitin and polyglutamine immunoreactive nuclear inclusions, have been looked for in SCA6 brains and in cultured cells transfected with the α_{1A} subunit cDNA coding for an expanded polyglutamine repeat. Non ubiquitinated cytoplasmic protein aggregates were detected by anti-α_{1A} antibodies (Ishikawa et al., 1999a, 2001) both in transfected cells and in patients' cerebella. In addition, the cytoplasm and nucleus of Purkinje cells showed small aggregates immunoreactive with 1C2, i.e. an antibody detecting polyglutamine sequences larger than 40 repeats. However, aggregates immunoreactive with the α_{1A} subunit were not reactive with 1C2. The interpretation of this finding is not straightforward. It should be noted, in fact, that protein aggregates in other polyglutamine disorders are most likely due to the tendency of polyglutamine stretches to form β-sheets by linking to one another through hydrogen bonds between their main chain and the side chain amides (Perutz, 1994). This process is strongly dependent on the number of repeat units: no aggregate formation is observed with polyglutamine stretches below 27 repeats (Scherzinger et al., 1999). The structural transition permitting aggregation begins, instead, in an expansion range between 32 and 37 glutamines, corresponding to the lower limit of the expansion range seen in patients affected with non-SCA6 polyglutamine expansion disorders. In light of these data the significance of aggregates in SCA6 brains remains obscure. Are the anti-polyglutamine reactive aggregates due to the SCA6 expansion inducing an aggregation of other polyglutamine proteins with longer numbers of units? And if so, what is the chemical basis for such an aggregation? Or, are aggregates aspecific by-products of Purkinje cell neurodegenerative process caused by a dysfunction of calcium channels? On the other hand the anti-α_{1A} reactive aggregates are hardly comparable with those found in other SCAs, being non-ubiquitinated and located in the cytosol only. It should be reminded that the α_{1A} subunit has many isoforms, only some of which contain the polyglutamine stretch (see below). Is the expression pattern of isoforms, or their likelihood of being included in the membrane or their turnover, different when the polyglutamine stretch is expanded compared to when it is not? If yes, then these aggregates could be due to the degradation of unused or obsolete isoforms. An increased expression of the α_{1A} subunit with a polyglutamine expansion, as compared to wildtype, has indeed been reported (Ishikawa et al., 1999, 2001; Piedras-Renteria et al., 2001). This would imply that SCA6 is completely different from other polyglutamine disorders for which a comparable expression level is reported for wildtype and mutated genes (see e.g. Persichetti et al., 1995; Servadio et al., 1995).

Pathogenesis

Several studies addressed the question whether the small polyglutamine expansion in SCA6 patients exerts its pathological effect by altering the calcium channel function or, as in other SCAs, by acquiring a new toxic function, independent from the channel kinetics. In other words the question to be answered is: does SCA6 belong to the group of channelopathies (Ptacek, 1997), thus influencing the level (decrease or increase) of calcium influx into the cell, or does it belong to the group of polyglutamine expansion disorders together with other SCAs?

So far experimental evidence (obtained by transfecting α_{1A} subunit cDNA with different number of CAG units in non neuronal cells, usually renal embryonic cells HEK293) is highly contradictory on this point. Two studies report that the α_{1A} protein with expanded polyglutamines induces a hyperpolarizing shift in the voltage dependence of channel inactivation (Matsuyama et al., 1999; Toru et al., 2000). The change is predicted to exert a considerable decrement on channel availability at resting potentials, approximately halving the Ca^{2+} influx, which may in turn lead, directly or indirectly, to neuronal cell death. This effect appears to be dependent on the type α_{1A} subunit isoform expressed and on the number of repeat units (Toru et al., 2000). It is notable that in these experiments the current density in transfected cells is not reduced, implying that the mutated protein is normally transported to the membrane and is not sequestered into aggregates (Matsuyama et al., 1999). Opposite results were reported by Piedras-Rentería et al. (2001). They observed a sharp increase of current density in cells expressing the mutated, as compared to wildtype, protein. Consistently, a higher expression of the mutated subunit was observed both in the membrane and in the cytoplasm. The effect, however, was not dependent on the number of polyglutamines. No significant alteration of channel function was observed either in the activation or in the inactivation kinetics. Although a confirmation of the functional analysis through sin-

Table 1. SCA6 as compared with EA2 and other SCAs with CAG expansions

Features	SCA 1, 2, 3, 7	SCA 6	EA2
Mutation	CAG expansion (large) Range 32–100	CAG expansion (small) Range 20–30	Point mutation
Meiotic/mitotic instability of mutation	High	Very low / absent	Absent
Frequency among sporadic ataxia cases	Low or absent	High	Not reported
Ataxia/vertigo episodes	Absent	Frequent	Very frequent
Response to acetazolamide	Not assessed	Positive	Positive
Progressive ataxia	Invariably present	Very frequent	Frequent
Extracerebellar deficits	Invariably present	Low frequency	Extremely rare
Neuropathology	Generalized cerebellar atrophy + extra-cerebellar involvement	Atrophy of cerebellar vermis + hemispheres later on + (occasionally) extra-cerebellar structures	Atrophy of cerebellar vermis + hemispheres later on + very rarely extracerebellar structures
Age at onset (range)	10–65 (progressive ataxia)	16–73 (progressive ataxia)	I–II decade (episodes)
Inverse correlation age at onset/allele size	yes	yes	–
Age at onset anticipation	Yes (intergenerational jump of allele size)	Yes (no intergenerational jump of allele size)	No
Nuclear and cytoplasmic aggregates	Yes (reactive to anti-ubiquitin, -polyglutamine, -protein)	Yes (non ubiquitinated, dissociated reaction to polyglutamine and protein antibodies)	Never assessed
Electrophysiological characteristics of mutated protein	–	Decrease of calcium influx, in some studies only	Decrease of calcium influx in all studies
Increased expression of mutated protein	No	Yes in some studies	No

gle channel recording would have been appropriate, these data would imply that SCA6 is induced by an abnormal expression or turnover of the mutated protein rather than by a channelopathy as previously suggested. Still distinct results were obtained by Restituito et al. (2000) in a different expression system (Xenopus oocytes). These authors observed an abnormal channel function, but in a direction opposite to that reported by Matsuyama et al. (1999) and Toru et al. (2000): an increased influx of Ca^{2+}, in fact, was predicted as a consequence of a hyperpolarizing shift in the voltage dependent channel activation and a slowed rate of inactivation. This effect was not obtained when the mutated protein had less than 30 glutamine units and an auxiliary β subunit different from β4 was coexpressed.

The above discrepancies could be ascribed to a number of variables in the experimental setting, namely to cDNAs used for transfecting cells (rabbit, human or chimeric rabbit/human cDNA), expression of different alternative exons (e.g. exon 37a or 37b), or coexpression of different auxiliary β units isoforms. Should these differences really account for the discrepant results obtained, this would indicate that the SCA6 mutations are interacting with the isoforms of the channel proteins. A more reliable picture of SCA6 pathogenesis, therefore, would require a detailed knowledge of functional characteristics and expression regulation of the different wildtype protein isoforms, in order to understand their interactions with the gene mutations.

Conclusions

SCA6 is one of three disorders due to mutations of the pore-forming subunit of calcium channels type P/Q and is due to small expansions of a polyglutamine sequence at the COOH terminal of the protein. Its clinical phenotype shows a wide overlap with its allelic disorder EA2, associated with point (missense or protein truncating) mutations. The two disorders share, among other things (see Table 1): a) a similar, highly variable phenotype ranging from vertigo/ataxia episodes with interictal nystagmus and, possibly, other mild cerebellar signs with no overt ataxia (Denier et al., 1999; Guida et al., 2001) to a severe progressive pure cerebellar ataxia preceded or not by episodes (Yue et al., 1997; Guida et al., 2001); b) a predominant atrophy of the cerebellar vermis (Denier et al., 1999; Guida et al., 2001); and c) a sensitivity of episodes to acetazolamide treatment (Calandriello et al., 1996; Jen et al., 1998; Koh et al., 2001). The issue of the phenotype similarities between SCA6 and EA2 has implications for the functional significance of the protein. In fact, when different mutations in the same gene are causing the same phenotype it is far more likely that they both lead to a loss rather than to a gain of protein function. Electrophysiological experiments have, indeed, assessed that CACNA1A point mutations leading to EA2 are associated with a loss of channel function (Guida et al., 2001; Jen et al., 2001; Wappl et al., 2002). Moreover, FHM point mutations show a loss of channel function when the phenotype includes a cerebellar deficit (Hans et al., 1999), while a gain of function can be present when the phenotype is purely migraineous. Overall the above evidence would suggest that a loss of channel function is underlying the cerebellar involvement in all the three CACNA1A allelic disorders. Electrophysiological evidence on SCA6 mutation, however, is far from clear on this point and no conclusions can be drawn.

With other SCAs (1, 2, 3, 7, 17) due to polyglutamine expansion, SCA6 shares the type of mutation and the protein aggregates (Table 1). However major differences are present: a) the range of repeat units in SCA6 expanded alleles falls within the distribution of other SCA normal alleles and, what is more, below the threshold for polyglutamine aggregation; b) the gene product is a membrane rather than a nuclear or cytoplasmic protein; c) the intracellular aggregates are not ubiquitinated and their relationship with the mutated protein are far from clear.

In this situation the pathogenic mechanism underlying SCA6 remains unanswered: i.e. whether a gain of a new toxic activity of the polyglutamine expansion per se should be envisaged or there exists an altered (possibly decreased) channel function, as in EA2. The presence of both mechanisms has been suggested in order to reconcile conflicting evidence (Gomez, 2002). However when no sufficient support is available for each separate hypothesis, the same holds true for their combination.

A solution for the above question will hopefully arrive when more information about the physiological properties of the channel protein isoforms and their interactions with different mutations will be attained through cell and animal models.

References

Arpa J, Cuesta A, Cruz-Martinez A, Santiago S, Sarrià J, Palau F: Clinical features and genetic analysis of a Spanish family with spinocerebellar ataxia 6. Acta Neurol Scand 99:43–47 (1999).

Baloh RW, Qing Yue MD, Furman JM, Nelson SF: Familial episodic ataxia: clinical heterogeneity in four families linked to chromosome 19p. Ann Neurol 41:8–16 (1997).

Calandriello L, Veneziano L, Francia A, Sabbadini G, Colonnese C, Mantuano E, Jodice C, Trettel F, Viviani P, Manfredi M, Frontali M: Acetazolamide-responsive episodic ataxia in an Italian family refines gene mapping on chromosome 19p13. Brain 120:805–812 (1996).

Denier C, Ducros A, Vahedi K, Joutel A, Thierry P, Rotz A, Castelnovo G, Deonna T, Gerard P, Devoize JL, Gayou A, Perrouty B, Soisson T, Autret A, Warter JM, Vighetto A, Van Bogaert P, Alamowitch S, Roullet E, Tournier-Lasserve E: High prevalence of CACNA1A truncations and broader clinical spectrum in episodic ataxia type 2. Neurology 52:1816–1821 (1999).

Dichgans M, Schols L, Herzog J, Stevanin G, Weirich-Schwaiger H, Rouleau G, Burk K, Klockgether T, Zuhlke C, Laccone F, Riess O, Gasser T: Spinocerebellar ataxia type 6 evidence for a strong founder effect among German families. Neurology 52:849–851 (1999).

Diriong S, Lory P, Williams ME, Ellis SB, Harpold MM, Taviaux S: Chromosomal localization of the human genes for α_{1A}, α_{1B}, and α_{1E} voltage-dependent Ca^{2+} channel subunits. Genomics 30:605–609 (1995).

Farmer TW, Mustian VM: Vestibulocerebellar ataxia. A newly defined hereditary syndrome with periodic manifestations. Arch Neurol 8:21–30 (1963).

Filla A, De Michele G, Santoro L, Calabrese O, Castaldo I, Giuffrida S, Restivo D, Serlenga L, Condorelli DF, Bonuccelli U, Scala R, Coppola G, Caruso G, Cocozza SL: Spinocerebellar ataxia type 2 in Southern Italy: a clinical and molecular study of 30 families. Neurology 246:467–471 (1999).

Frontali M: Spinocerebellar ataxia type 6: channelopathy or glutamine repeat disorder? Brain Res Bull 56:227–231 (2001).

Fukutake T, Kamitsukasa T, Arai K, Hattori T, Nakajima T: A patient homozygous for the SCA6 gene with retinitis pigmentosa. Clin Genet 61:375–379 (2002).

Garcia-Planells J, Cuesta A, Vilchez JJ, Martinez F, Prieto F, Palau F: Genetics of the SCA6 gene in a large family segregating an autosomal dominant "pure" cerebellar ataxia. J med Genet 36:148–151 (1999).

Geschwind DH, Perlman S, Figueroa BS, Karrim BS, Baloh RW, Pulst SM: Spinocerebellar ataxia type 6. Frequency of the mutation and genotype-phenotype correlations. Neurology 49:1247–1251 (1997).

Gomez CM: Polyglutamine aggregates in SCA6 Purkinje cells. A tail of two toxicities. Neurology 56:1618–1619 (2002).

Gomez CM, Thompson M, Gammack JT, Perlman SL, Dobyns WD, Truwit CL, Zee DS, Clark HB, Anderson JH: Spinocerebellar ataxia type 6: gaze-evoked and vertical nystagmus, Purkinje cell degeneration, and variable age of onset. Ann Neurol 42:933–950 (1997).

Guida S, Trettel F, Pagnutti S, Mantuano E, Tottene A, Veneziano L, Fellin T, Spadaro M, Stauderman KA, Williams ME, Volsen S, Ophoff RA, Frants RR, Jodice C, Frontali M, Pietrobon D: Complete loss of P/Q calcium channel activity caused by a CACNA1A missense mutation carried by episodic ataxia type 2 patients. Am J hum Genet 68:759–764 (2001).

Hans M, Luvisetto S, Williams ME, Spagnolo M, Urrutia A, Tottene A, Brust PF, Johnson EC, Harpold MM, Stauderman KA, Pietrobon D: Functional consequences of mutations in the human alpha1A calcium channel subunit linked to familial hemiplegic migraine. J Neurosci 19:1610–1619 (1999).

Harding AE: The clinical features and classification of the late onset autosomal dominant cerebellar ataxias: a study of 11 families, including descendants of the Drew family of Walworth. Brain 105:1–28 (1982).

Ikeuchi T, Takano H, Koide R, Horikawa Y, Honma Y, Onisji Y, Igarashi S, Tanaka H, Nakao N, Sahashi K, Tsukagoshi H, Inoue K, Takahashi H, Tsuji S: Spinocerebellar ataxia type 6: CAG repeat expansion in alpha1A voltage-dependent calcium channel gene and clinical variations in Japanese population. Ann Neurol 42:879–884 (1997).

Ishikawa K, Tanaka H, Saito M, Ohkoshi N, Fujita T, Yoshizawa K, Ikeuchi T, Watanabe M, Hayashi A, Takiyama Y, Nishizawa M, Nakano I, Matsubayashi K, Miwa M, Shoji S, Kanazawa I, Tsuji S, Mizusawa H: Japanese families with autosomal dominant pure cerebellar ataxia map to chromosome 19p13.1 → p13.2 and are strongly associated with mild CAG expansions in the spinocerebellar ataxia type 6 gene in chromosome 19p13.1. Am J hum Genet 61:336–346 (1997).

Ishikawa K, Fujigasaki H, Saegusa H, Ohwada K, Fujita T, Iwamoto H, Komatsuzaki Y, Toru S, Toriyama H, Watanabe M, Ohkoshi N, Shoji S, Kanazawa I, Tanabe T, Mizusawa H: Abundant expression and cytoplasmic aggregations of [alpha]1A voltage-dependent calcium channel protein associated with neurodegeneration in spinocerebellar ataxia type 6. Hum molec Genet 8:1185–1193 (1999a).

Ishikawa K, Watanabe M, Yoshizawa K, Fujita T, Iwamoto H, Yoshizawa T, Harada K, Nakamagoe K, Komatsuzaki Y, Satoh A, Doi M, Ogata T, Kanazawa I, Shoji S, Mizusawa H: Clinical, neuropathological, and molecular study in two families with spinocerebellar ataxia type 6 (SCA6). J Neurol Neurosurg Psychiatry 67:86–89 (1999b).

Ishikawa K, Ohwada K, Ishida K, Fujigasaki H, Shun Li M, Tsunemi T, Ohkoshi N, Toru S, Mizutani T, Hayashi M, Arai N, Hasegawa K, Kawanami T, Kato T, Makifuchi T, Shoji S, Tanabe T, Mizusawa H: Cytoplasmic and nuclear polyglutamine aggregates in SCA6 Purkinje cells. Neurology 56:1753–1756 (2001).

Jen JC, Yue Q, Karrim J, Nelson SF, Baloh RW: Spinocerebellar ataxia type 6 with positional vertigo and acetazolamide responsive episodic ataxia. J Neurol Neurosurg Psychiatry 65:565–568 (1998).

Jen JC, Yue Q, Nelson SF, Yu H, Litt M, Nutt J, Baloh RW: A novel nonsense mutation in CACNA1A causes episodic ataxia and hemiplegia. Neurology 53:34–37 (1999).

Jen JC, Wan J, Graves M, Yu H, Mock AF, Coulin CJ, Kim G, Yue Q, Papazian DM, Baloh RW: Loss of function EA2 mutations are associated with impaired neuromuscular transmission. Neurology 57:1843–1848 (2001).

Jodice C, Mantuano E, Veneziano L, Trettel F, Sabbadini G, Calandriello L, Francia A, Spadaro M, Pierelli F, Salvi F, Ophoff RA, Frants RR, Frontali M: Episodic ataxia type 2 (EA2) and spinocerebellar ataxia type 6 (SCA6) due to CAG repeat expansion in the CACNA1A gene on chromosome 19p. Hum molec Genet 6:1973–1978 (1997).

Katayama T, Ogura Y, Aizawa H, Kuroda H, Suzuki Y, Kuroda K, Kikuchi K: Nineteen CAG repeats of the SCA6 gene in a Japanese patient presenting with ataxia. J Neurol 415:711–712 (2000).

Koh SH, Kim HT, Kim SH, Lee GY, Kim J, Kim MH: Spinocerebellar ataxia type 6 and episodic ataxia type 2 in a Korean family. J Korean Med Sci 16:809–813 (2001).

Leggo J, Dalton A, Morrison PJ, Dodge A, Connarty M, Kotze MJ, Rubinsztein DC: Analysis of spinocerebellar ataxia type 1,2,3,6, dentate-rubro-pallidoluysian atrophy and Friedreich ataxia genes in spinocerebellar ataxia patients in the UK. J med Genet 34:982–985 (1997).

Margolis RL: Spinocerebellar Ataxias: order emerges from chaos. Curr Neurol Neurosi Rep 2:447–456 (2002).

Mariotti C, Gellera C, Grisoli M, Mineri R, Castucci A, Di Donato S: Pathogenic effect of an intermediate-size SCA-6 allele (CAG)(19) in a homozygous patient. Neurology 57:1502–1504 (2001).

Maruyama H, Izumi Y, Morino H, Oda M, Toji H, Nakamura S, Kawakami H: Difference in disease-free survival curve and regional distribution according to subtype of spinocerebellar ataxia: a study of 1,286 Japanese patients. Am J med Genet 114:578–583 (2002).

Matsumura R, Futamura N, Fujimoto Y, Yanagimoto S, Horikawa H, Suzumura A, Takayanagi T: Spinocerebellar ataxia type 6. Molecular and clinical features of 35 Japanese patients including one homozygous for the CAG repeat expansion. Neurology 49:1238–1243 (1997).

Matsuyama Z, Kawakami H, Maruyama H, Izumi Y, Komure O, Udaka F, Kameyama M, Nishio T, Kuroda Y, Nishimura M, Nakamura S: Molecular features of the CAG repeats of spinocerebellar ataxia 6 (SCA6). Hum molec Genet 6:1283–1287 (1997).

Matsuyama Z, Wakamori M, Mori Y, Kawakami H, Nakamura S, Imoto K: Direct alteration of the P/Q-type Ca^{2+} channel property by polyglutamine expansion in spinocerebellar ataxia 6. J Neurosci 19:RC14 (1999).

Moon SL, Koller WC: Hereditary periodic ataxias, in de Jong JM (ed): Handbook of Clinical Neurology, 16:433–443 (Elsevier, Amsterdam 1991).
Mori Y, Friedrich T, Kim MS, Mikami A, Nakai J, Ruth P, Bosse E, Hofmann F, Flockerzi V, Furuichi T, Mikishiba K, Imoto K, Tanabe T, Numa S: Primary structure of functional expression from complementary DNA of a brain calcium channel. Nature 350:398–402 (1991).
Murata Y, Kawakami H, Yamaguchi S, Nishimura M, Kohriyama T, Ishizaki F, Matsuyama Z, Mimori Y, Nakamura S: Characteristic magnetic resonance imaging findings in spinocerebellar ataxia 6. Arch Neurol 55:1348–1352 (1998).
Nagai Y, Azuma T, Unauchi M, Fujita M, Umi M, Hirano M, Matsubara T, Ueno S: Clinical and molecular genetic study in seven Japanese families with spinocerebellar ataxia type 6. J Neurol Sci 157:52–59 (1998).
Nakagawa N, Katayama T, Makita Y, Kuroda K, Aizawa H, Kikuchi K: A case of spinocerebellar ataxia type 6 mimicking olivopontocerebellar atrophy. Neuroradiology 41:501–503 (1999).
Ophoff RA, Terwindt GM, Vergouwe MN, van Eijk R, Oefner PJ, Hoffman SMG, Lamerdin JE, Mohrenweiser HW, Bulman DE, Ferrari M, Haan J, Lindhout D, van Ommen GJB, Hofker MH, Ferrari MD, Frants RR: Familial hemiplegic migraine and episodic ataxia type-2 are caused by mutations in the Ca^{2+} channel gene CACNL1A4. Cell 87:543–552 (1996).
Penrose LS: The problem of anticipation in pedigrees of Dystonia Myotonica. Ann Eugenics 14:125–132 (1948).
Persichetti F, Ambrose CM, Ge P, McNeil SM, Srinidhi J, Anderson MA, Jenkins B, Barnes GT, Duyao MP, Kanaley L, et al: Normal and expanded Huntington's disease gene alleles produce distinguishable proteins due to translation across the CAG repeat. Mol Med 1:374–383 (1995).
Perutz MF: Polar zippers: their role in human disease. Protein Sci 3:1629–37 (1994).
Piedras-Rentería ES, Watase K, Harata N, Zhuchenko O, Zoghbi H, Lee CC, Tsien RW: Increased expression of α_{1A} Ca^{2+} channel currents arising from expanded trinucleotide repeats in spinocerebellar ataxia type 6. J Neurosci 21:9185–9193 (2001).
Ptacek LJ: Channelopathies: ion channel disorders of muscle as a paradigm for paroxysmal disorders of the nervous system. Neuromusc Disord 7:250–255 (1997).
Pujana MA, Corral J, Gratacos M, Combarros O, Berciano J, Genis D, Banchs I, Estivill X, Volpini V: Spinocerebellar ataxia in Spanish patients: genetic analysis of familial and sporadic cases. The Ataxia Study Group. Hum Genet 104:516–522 (1999).
Restituito S, Thomson RM, Eliet J, Raike RS, Riedl M, Charnet P, Gomez C: The polyglutamine expansion in spinocerebellar ataxia type 6 causes a β subunit-specific enhanced activation of P/Q-type calcium channels in Xenopus oocytes. J Neurosci 20:6394–6430 (2000).
Riess O, Schols L, Bottger H, Nolte D, Vieira-Saecker AM, Schimming C, Kreuz F, Macek M Jr, Krebsovà A, Macek M Sen, Klockgether T, Zuhlke C, Laccone FA: SCA6 is caused by moderate CAG expansion in the alpha1A-voltage-dependent calcium channel gene. Hum molec Genet 6:1289–1293 (1997).
Sasaki H, Kojima H, Yabe I, Hamada T, Sawa H, Hiraga H, Nagashima K: Neuropathological and molecular studies of spinocerebellar ataxia type 6 (SCA6). Acta Neuropathol 95:199–204 (1998).
Satoh JI, Tokumoto H, Yukitake M, Matsui M, Matsuyama Z, Kawakami H, Nakamura S, Kuroda Y: Spinocerebellar ataxia type 6: MRI of three Japanese patients. Neuroradiology 40:222–227 (1998).
Scherzinger E, Sittler A, Schweiger K, Heiser V, Lurz R, Hasenbank R, Bates GP, Lehrach H, Wanker EE: Self assembly of polyglutamine containing huntingtin fragments into amiloid-like fibrils: implication for Huntington's disease pathology. Proc natl Acad Sci, USA 96:4604–4608 (1999).
Servadio A, Koshy B, Armstrong D, Antalffy B, Orr HT, Zoghbi HY: Expression analysis of the ataxin-1 protein in tissues from normal and spinocerebellar ataxia type 1 individuals. Nature Genet 10:94–98 (1995).
Shizuka M, Watanabe H, Ikeda Y, Mizushima K, Okamoto K, Shoji M: Molecular analysis of a de novo mutation for spinocerebellar ataxia type 6 and $(CAG)_n$ repeat units in normal elder controls. J Neurol Sci 161:85–87 (1998a).
Shizuka M, Watanabe H, Ikeda Y, Mizushima K, Kanai M, Tsuda T, Abe K, Okamoto K, Shoji M: Spinocerebellar ataxia type 6: CAG trinucleotide expansion, clinical characteristics and sperm analysis. Eur J Neurol 5:381–387 (1998b).
Silveira I, Miranda C, Guimaraes L, Moreira CC, Alonso I, Mendonca P, Ferro A, Pinto-Basto J, Coelho J, Ferreirinha F, Poirier J, Parreira E, Vale J, Januario C, Barbot C, Tuna, A, Barros J, Koide R, Tsuji S, Holmes SE, Margolis RL, Jardim L, Pandolfo M, Coutinho P, Sequeiros J: Trinucleotide repeats in 202 families with ataxia. A small expanded $(CAG)_n$ allele at the SCA17 locus. Arch Neurol 59:623–9 (2002).
Sinke RJ, Ippel EF, Diepstraten CM, Beemer FA, Wokke JHJ, Van Hilten BJ, Knoers NVAM, Ploos van Amstel HC, Kremer HPH: Clinical and molecular correlations in spinocerebellar ataxia 6. Arch Neurol 58:1839–1844 (2001).
Skre H: Spinocerebellar ataxia in western Norway. Clin Genet 6:265–288 (1974).
Soong BW, Lu YC, Choo KB, Lee HY: Frequency analysis of autosomal dominant cerebellar ataxias in Taiwanese patients and clinical molecular characterization of spinocerebellar ataxia type 6. Arch Neurol 58:1105–1109 (2001a).
Soong BW, Liu RS, Wu LC, Lu YC, Lee HY: Metabolic characterization of spinocerebellar ataxia type 6. Arch Neurol 58:300–304 (2001b).
Stevanin G, Durr A, David G, Didierjean O, Cancel G, Rivaud S, Tourbah A, Warter JM, Agid Y, Brice A: Clinical and molecular features of spinocerebellar ataxia type 6. Neurology 49:1243–1246 (1997).
Subramony SH, Fratkin JD, Manyam BV, Currier RD: Dominantly inherited cerebello-olivary atrophy is not due to a mutation at the spinocerebellar ataxia 1, Machado-Joseph disease, or dentato-rubro-pallido-luysian atrophy locus. Mov Disord 11:174–180 (1996).
Takano H, Cancel G, Ikeuchi T, Lorenzetti D, Mawad R, Stevanin G, Didierijean O, Durr A, Oyake M, Shimohata T, Sasaki R, Koide R, Igarashi S, Hayashi S, Takiyama Y, Nishizawa M, Tanaka H, Zoghbi H, Brice A, Tsuji S: Close associations between prevalences of dominant inherited ataxias with CAG-repeat expansions and frequencies of large normal CAG alleles in Japanese and Caucasian population. Am J hum Genet 63:1060–1066 (1998).
Takiyama Y, Sakoe K, Nemekawa M, Soutome M, Esumi E, Ogawa T, Ishikawa K, Mizusawa H, Nakano I, Nishizawa M: A Japanese family with spinocerebellar ataxia type 6 which includes three individuals homozygous for an expanded CAG repeat in the SCA6/CACNL1A4 gene. J Neurol Sci 158: 141–147 (1998).
Tang B, Liu C, Shen L, Dai H, Pan Q, Jing L, Ouyang S, Xia J: Frequency of SCA1, SCA2, SCA3/MJD, SCA6, SCA7, and DRPLA CAG trinucleotide repeat expansion in patients with hereditary spinocerebellar ataxia from Chinese kindreds. Arch Neurol 57:540–544 (2000).
Tashiro H, Suzuki SO, Hitotsumatsu T, Iwaki T: An autopsy case of spinocerebellar ataxia type 6 with mental symptoms of schizophrenia and dementia. Clin Neuropathol 18:198–204 (1999).
Toru S, Murakoshi T, Ishikawa K, Saegusa H, Fujigasaki H, Uchihara T, Nagayama S, Osanai M, Mizusawa H, Tanabe T: Spinocerebellar ataxia type 6 mutation alters P-type calcium channel function. J biol Chem 275:10893–10898 (2000).
Trettel F, Mantuano E, Calabresi V, Veneziano L, Olsen AS, Georgescu A, Gordon L, Sabbadini G, Frontali M, Jodice C: A fine physical map of the CACNA1A gene region on 19p13.2 → p13.1 chromosome. Gene 241:45–50 (2000).
Van de Warrenburg BPC, Sinke RJ, Verschuuren-Bemelmans CC, Scheffer H, Brunt ER, Ippel PF, Maat-Kievit JA, Dooijes D, Notermans NC, Lindhout D, Knoers NVAM, Kremer HPH: Spinocerebellar ataxias in the Netherlands. Neurology 58: 702–708 (2002).
Wappl E, Koschak A, Poteser M, Sinnegger MJ, Walter D, Eberhart A, Grosachner K, Glossman H, Kraus RL, Grabner M, Striessnig J: Functional consequences of P/Q-type Ca^{2+} channel Cav2.1 missense mutations associated with episodic ataxia type 2 and progressive ataxia. J biol Chem 277:6960–6966 (2002).
Watanabe H, Tanaka F, Matsumoto M, Doyu M, Ando T, Mitsuma T, Sobue G: Frequency analysis of autosomal dominant ataxias in Japanese patients and clinical characterization of spinocerebellar ataxia type 6. Clin Genet 53:13–19 (1998).
Yabe I, Sasaki H, Yamashita I, Takei A, Tashiro K: Clinical trial of acetazolamide in SCA6, with assessment using ataxia rating scale and body stabilometry. Acta Neurol Scand 104:44–47 (2001).
Yue Q, Jen CJ, Nelson SF, Baloh RW: Progressive ataxia due to a missense mutation in a calcium-channel gene. Am J hum Genet 61:1078–1087 (1997).
Zhuchenko O, Bailey J, Bonnen P, Ashizawa T, Stockton DW, Amos C, Dobyns WB, Subramony SH, Zoghbi HY, Chi LC: Autosomal dominant cerebellar ataxia (SCA6) associated with small polyglutamine expansion in the α_{1A}-voltage-dependent calcium channel. Nature Genet 15:62–69 (1997).

Cytogenet Genome Res 100:154–163 (2003)
DOI: 10.1159/000072850

Spinocerebellar ataxia 7 (SCA7)

A.-S. Lebre[a,b,c] and A. Brice[a,b,c,d]

[a] INSERM U289,
[b] Institut fédératif de recherche des Neurosciences,
[c] Département de Génétique, Cytogénétique et Embryologie,
[d] Fédération de Neurologie, Groupe hospitalier Pitié-Salpêtrière, Paris (France)

Abstract. Spinocerebellar ataxia 7 (SCA7) is a progressive autosomal dominant neurodegenerative disorder characterized clinically by cerebellar ataxia associated with progressive macular dystrophy. The disease affects primarily the cerebellum and the retina, but also many other CNS structures as the disease progresses. SCA7 is caused by expansion of an unstable trinucleotide CAG repeat encoding a polyglutamine tract in the corresponding protein, ataxin-7. Normal SCA7 alleles contain 4–35 CAG repeats, whereas pathological alleles contain from 36–306 CAG repeats. SCA7 has a number of features in common with other diseases with polyglutamine expansions: (i) the appearance of clinical symptoms above a threshold number of CAG repeats (>35); (ii) a correlation between the size of the expansion and the rate of progression of the disease: the larger the repeat, the faster the progression; (iii) instability of the repeat sequence (~ 12 CAG/transmission) that accounts for the marked anticipation of ~ 20 years/generation. The CAG repeat sequence is particularly unstable and de novo mutations can occur during paternal transmissions of intermediate size alleles (28–35 CAG repeats). This can explain the persistence of the disease in spite of the anticipation that should have resulted in its extinction.

Ataxin-7 is a protein of unknown function expressed in many tissues including the CNS, but its mutation leads to selective neuronal death only in the brain. The expanded polyglutamine tract renders the protein toxic and leads to its aggregation in a subset of vulnerable neurons in SCA7 brain. These aggregates, named neuronal intranuclear inclusions (NIIs), also stain positively for ubiquitin, chaperones and proteasome, as in other polyglutamine diseases, providing evidence that the mechanisms underlying these disorders have features in common. Cellular models and SCA7 transgenic mice have been generated which constitute valuable resources for studying the disease mechanism. They will also be useful for screening and evaluating possible therapeutic strategies.

Supported by the VERUM foundation, the Association Française contre les Myopathies (AFM) and the Institut National de la Santé et de la Recherche Médicale (INSERM) (Nr 4MR12F), the French Ministère de la Recherche and the Association pour le Développement de la Recherche sur les Maladies Génétiques Neurologiques et Psychiatriques (ADRMGNP).

Received 27 November 2002; manuscript accepted 14 December 2002.

Request reprints from Alexis Brice, Groupe hospitalier Pitié-Salpêtrière
47 bd de l'Hôpital, 75651 Paris Cedex 13 (France)
telephone: +33 1 42 16 21 82; fax: +33 1 44 24 36 58
e-mail: brice@ccr.jussieu.fr

Phenotype

SCA7 is an autosomal dominant cerebellar ataxia (ADCA), which has also been designated olivo-ponto-cerebellar-atrophy type III (Konigsmark and Weiner, 1970) or ADCA type II (Harding, 1982, 1993). This type of ADCA was initially described by Froment et al. (1937) and is characterized by very heterogeneous clinical presentations, disease severity and age at onset (Benomar et al., 1994; Enevoldson et al., 1994). Because of the retinopathy observed in most patients, this form of ADCA was considered to be a separate entity (OMIM database entry: 164500).

Visual impairment

SCA7, unlike other ADCAs, is characterised by a progressive macular degeneration that can be visualised in most patients as a pigmented central core in the macula that can

KARGER
Fax + 41 61 306 12 34
E-mail karger@karger.ch
www.karger.com

0301–0171/03/1004–0154$19.50/0

Accessible online at:
www.karger.com/cgr

extend, in latter stages, into the periphery. Visual failure is progressive, bilateral and symmetrical, and leads irreversibly to blindness (Enevoldson et al., 1994; Gouw et al., 1994). Aleman and colleagues performed a detailed study of the retinal phenotype of affected members of a family with the SCA7 mutation (45–47 CAG repeats) (Aleman et al., 2002). There was a spectrum of severity from mild to severe dysfunction. Early functional abnormalities were at both photoreceptor and postreceptoral levels. When cone and rod photoreceptor dysfunction was present, it was approximately equal. Regional retinal dysfunction was evident: there was more dysfunction centrally than peripherally with least effect in the midperiphery. In vivo cross-sectional retinal images with optical coherence tomography showed an early disease stage of altered foveal lamination (abnormal area of low reflectivity splitting the outer retina-choroidal complex) accompanied in the parafovea by reduced retinal thickness. Later disease stages showed foveal and parafoveal retinal thinning. The phenotype in this family with SCA7 is that of a cone-rod dystrophy.

Neurological signs

The neurological signs in ADCAs clearly overlap (Stevanin et al., 2000). Studies of large groups of patients have, however, revealed a constellation of signs that are frequently found in SCA7 patients (Table 1). Cerebellar ataxia is always associated with dysarthria, but patients present variably with pyramidal signs (increased reflexes and/or extensor plantar reflexes and/or lower limb spasticity), decreased vibration sense, dysphagia, sphincter disturbances and oculomotor abnormalities (supranuclear ophthalmoplegia and/or viscosity of eye movements). Extrapyramidal features (dystonia), myokymia, peripheral neuropathy and mental impairment are rare. The frequency of swallowing and sphincter disturbances significantly increases with disease duration (David et al., 1998a). The association of cerebellar ataxia and dysarthria with pyramidal signs, supranuclear ophthalmoplegia, slow saccades and decreased visual acuity is highly suggestive of SCA7.

Neuroimaging

Brain imaging shows marked atrophy in the cerebellum, particularly in the superior part of the vermis and the brainstem, which may be associated with moderate atrophy of the cerebral cortex (Fig. 1).

Neuropathology

Neuropathological examination reveals moderate to severe neuronal loss and gliosis in the cerebellum and associated structures (inferior olive, cerebellar cortex, dentate nucleus, pontine nuclei), the basal ganglia (globus pallidus, substantia nigra, subthalamic nucleus, red nucleus) and the spinal cord (Enevoldson et al., 1994; Gouw et al., 1994; Martin et al., 1994; Holmberg et al., 1998; Cancel et al., 2000). Spinocerebellar, olivocerebellar and efferent cerebellar tracts are severely affected. In the cerebellum, the vermis is more affected than the hemispheres

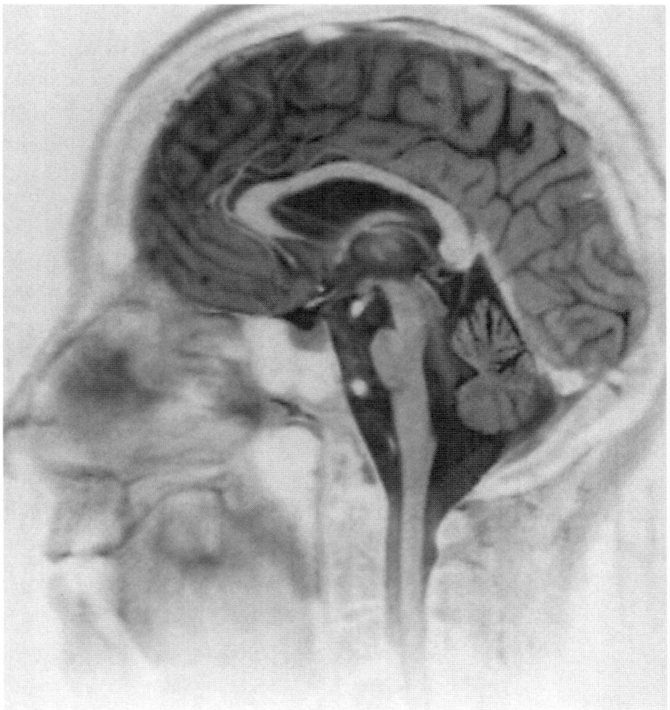

Fig. 1. Sagittal MRI of an SCA7 patient: T1 weighted sequence (TR = 450 ms, TE = 11 ms) showing obvious atrophy of the cerebellar vermis associated with mild atrophy of the pons (reproduced from David et al., 1998b).

Table 1. Major phenotypic characteristics of 69 SCA7 patients with a mean age at onset of 29 ± 16 years (1–70) and carrying 51 ± 13 CAG repeats (38–130) (reproduced from Stevanin et al., 2000b).

Clinical signs	%
Cerebellar ataxia	100
Dysarthria	98.5
Decreased visual acuity	81
Brisk reflexes	80
Diminished or abolished reflexes	3
Babinski sign	55
Ophthalmoplegia	54
Slow saccades	63
Deep sensory loss	60
Sphincter disturbances	55
Amyotrophy	25
Auditory impairment	24
Axonal neuropathy	18
Facial myokimia	13
Dementia	12
Extrapyramidal rigidity	14
Dystonia	9
Bulging eyes	6
Nystagmus	2

where Purkinje cells, and to a lesser extent granule cells, degenerate. Mild cell loss also occurs in the dentate nucleus which, as the result of Purkinje cell degeneration, has a reduced mantel. Extensive neuronal loss is observed in the inferior olive, with marked astrocytic gliosis. Mild cell loss is also observed in the substantia nigra and the basis pontis, whereas the thalamus and the striatum are spared.

The distinctive neuropathological features of SCA7 are degeneration of optic pathways and the retina. The pregeniculate visual pathways and the optic nerve are affected, probably as a consequence of retinal degeneration. In juvenile cases presenting with blindness, those systems may not be altered, probably due to the rapid course of the disease. Pathological examination of the retina shows early degeneration of photoreceptors and of bipolar and granular cells, particularly in the foveal and parafoveal regions. Later, patchy loss of epithelial pigment cells and their ectopic migration into the retinal layers are observed (Martin et al., 1994).

Identification of the SCA7 gene

Gene

Three independent groups mapped the responsible gene, designated SCA7 (spinocerebellar ataxia 7), to chromosome 3p21.1→p12 (Benomar et al., 1995; Gouw et al., 1995; Holmberg et al., 1995). As anticipation is particularly marked in SCA7, they postulated that a trinucleotide CAG repeat expansion might be involved in this disease. Lindblad et al. (1996) using the RED (repeat expansion detection) technique (Schalling et al., 1993), demonstrated that long stretches of CAG repeats co-segregated with the disease in several ADCA II pedigrees. These results were confirmed by Koob et al. (1998) by a strategy of cloning derived from the RED technique. A pure $(CAG)_{10}$ repeat, located between markers D3S1600 and D3S1287, was found in a sub-clone of YAC 882_d_9 (David et al., 1997; Krols et al., 1997). PCR analysis with primers flanking this CAG repeat showed the presence of an expansion in SCA7 patients (David et al., 1997; Del-Favero et al., 1998), permitting cloning of the SCA7 cDNA.

Diagnostic

Identification of SCA7 gene and its mutation enables direct detection of the mutation with routine laboratory tests in individuals who already present with symptoms of the disease. This molecular analysis is also useful to distinguish disorders which are clinically similar. Large series of controls and patients have now been analysed and have shown that the SCA7 CAG repeat is polymorphic, with sizes ranging from 4–35 units in control, and from 36–306 in SCA7 and at-risk carrier chromosomes (Benton et al., 1998; David et al., 1998a; Del-Favero et al., 1998; Gouw et al., 1998; Johansson et al., 1998; Koob et al., 1998; Lyoo et al., 1998; Stevanin et al., 1998; Giunti et al., 1999; Nardacchione et al., 1999). The largest expansions (between 114–306 CAG repeats) were found in juvenile or infantile cases (Benton et al., 1998; David et al., 1998a; Johansson et al., 1998; Giunti et al., 1999).

The strong anticipation in SCA7 and the rarity of contractions should have led to its extinction within a few generations (see below). However, this tendency is counterbalanced by the occurrence of de novo SCA7 expansions from intermediate alleles (IAs) with 28–35 repeats in individuals that were still unaffected at ages 50–84 (Stevanin et al., 1998). Should these cases become affected with time, the intermediate size alleles may cause a disease with incomplete penetrance, as is the case for small expansions carrying 36–41 repeats at the HD locus (Rubinsztein et al., 1996; Andrew et al., 1997). Indeed, individuals carrying 34 (Gouw et al., 1998) and 35 (Koob et al., 1998) CAG repeats have been reported, but their clinical status and ages at onset were not described.

Care should therefore be taken to detect both intermediate size and large alleles. The latter are difficult to amplify but should be suspected in juvenile or infantile forms of the disease (Benton et al., 1998; David et al., 1998a; Johansson et al., 1998; Giunti et al., 1999). Molecular diagnosis of isolated cases with a phenotype compatible with SCA7 is also crucial. Positive isolated cases rarely carry de novo mutation (Stevanin et al., 1998). More frequently, they reflect missing family histories if the transmitting parent died before the onset of symptoms or is still asymptomatic because of marked anticipation (Cancel et al., 1997; Schols et al., 1997).

Presymptomatic testing in SCA7 families is possible for adult at-risk individuals, but raises the same difficult ethical issues as other severe adult-onset disorders for which no treatment can be proposed and for which age at onset cannot be precisely predicted from the number of CAG repeats. The international guidelines established for Huntington's disease should also be followed for SCA7 (World Federation of Neurology Research Group on Huntington's Chorea, 1994).

Genotype/Phenotype correlations

First sign at onset

Cerebellar ataxia is usually the presenting symptom in adults with onset over thirty. In patients with earlier onset, however, decreased visual acuity, alone or associated with cerebellar ataxia, is the initial symptom (Benomar et al., 1994; Enevoldson et al., 1994). Some infantile cases may, however, result in early death without detectable retinal alterations (Martin et al., 1999). More than 45 years can elapse between the appearance of cerebellar symptoms and visual failure, while, in the reverse situation, the latency never exceeds nine years (David et al., 1998a; Giunti et al., 1999). In some patients with late onset, visual acuity may never decrease.

Age at onset, anticipation and disease duration

The clinical manifestations typically begin in the third or fourth decade, with mean age at onset close to 30, but a range of three months or less to over 70 years (Martin et al., 1994; Holmberg et al., 1995; Jöbsis et al., 1997; Benton et al., 1998; David et al., 1998a; Gouw et al., 1998; Giunti et al., 1999). Analysis of parent-child couples has revealed striking anticipation (~20 years/generation). Previous studies reported significantly greater anticipation in paternal than in maternal transmissions (Benomar et al., 1994, 1995; David et al., 1996). This was not confirmed in recent reports (Benton et al., 1998; David et al., 1998a; Del-Favero et al., 1998; Johansson et al., 1998; Giunti et al., 1999), although all juvenile cases are paternally transmitted. This correlates well with the marked tendency to increased expansion in the sperm of patients compared to their blood (David et al., 1998a). In a recent study using small pool PCR in sperm, very large expanded repeats were detected although they are under-represented in patients suggesting that a significant proportion of such alleles might be associated with embryonic lethality or dysfunctional sperm (Monckton et al., 1999).

There is a strong negative correlation between the size of the CAG repeat expansion and age at onset. The former accounts for ~75% of the variability of the latter, suggesting that other genetic and/or environmental factors play only a minor role in determining the SCA7 phenotype, unlike other disorders involving polyglutamine expansions (David et al., 1998a; Gouw et al., 1998; Giunti et al., 1999). The CAG length/age at onset correlation, together with the increase in expansion size in successive generations, is consistent with the marked anticipation observed in ADCA II families (David et al., 1998a).

In SCA7, anticipation is characterised by earlier onset, but also by more rapid disease progression and increased severity in successive generations. Disease duration until death is negatively correlated with the number of CAG repeats on the expanded allele and is limited to a few months or years in early onset patients (David et al., 1998a). The mean duration in subjects carrying less than 49 repeats is 15 years and differs significantly from the 11 year duration when the number of repeats is 49 or more (Giunti et al., 1999). Longer disease durations, up to 30 years or more, are observed only in late onset patients (Neetens et al., 1990). Anticipation is also associated with increasing severity of symptoms in successive generations. The frequency of decreased visual acuity, ophthalmoplegia, scoliosis and extensor plantar reflexes significantly increases with the size of the expansion (David et al., 1998a). The phenotype of a given SCA7 patient partly depends on both the size of the mutation and disease duration at the time of examination. In some infantile cases with very large repeat expansions, progression is extremely rapid and the heart can be affected (Benton et al., 1998; Neetens et al., 1990). It is surprising that the retina may also be affected in juvenile SCA2 patients (Babovic-Vuksanovic et al., 1998). The retina and the cardiac muscle may be sensitive only to large or very large expansions, respectively. The pathological threshold of the polyglutamine expansion may therefore be tissue-dependent.

Structure of ataxin-7 and localization in brain

Ataxin-7

A 7.5-kb transcript, detected by Northern-blot analysis, is expressed ubiquitously in adult and foetal tissues (David et al., 1997; Del-Favero et al., 1998). The 2727-bp open reading frame of the SCA7 cDNA encodes a protein of 892 amino acids designated ataxin-7 (David et al., 1997). Polymorphisms in the number of GCN and CCG repeats, upstream and downstream, respectively, of the CAG repeat have been observed (Stevanin et al., 1998), as well as an A/G3145 polymorphism resulting in a Met→Val substitution that does not affect function (Stevanin et al., 1999). Ataxin-7 contains several identifiable regions, such as polyalanine and polyglutamine repeats, a bipartite nuclear localization signal (NLS) and two serine-rich regions. Expression studies in several cellular models confirmed that the NLS directs normal and pathological ataxin-7 to the nucleus (Kaytor et al., 1999; Zander et al., 2001). Downstream of the polyglutamine tract, ataxin-7 contains four polyproline sequences predicted to be SH3-binding domains. Mushegian et al. identified a short motif in ataxin-7, homologous to a motif found in arrestins, that binds selectively to the phosphorylated activated forms of G protein-coupled receptors and quenches their signaling (Gurevich et al., 1995; Mushegian et al., 2000). The arrestin-like sequence in ataxin-7 binds phosphate in vitro (Mushegian et al., 2000). The murine SCA7 gene homolog was cloned: the mouse and human genes exhibit a high degree of identity at both DNA (88.2%) and protein (88.7%) level (Strom et al., 2002). The CAG repeat region is conserved in mouse but contained only five repeats in all mouse strains analyzed. The arrestin homology domain and the nuclear localization signal are also conserved in the murine homolog.

Localization of ataxin-7 in brain

The role of ataxin-7 remains unknown, but expression studies can provide information on its normal and pathological functions. Lindenberg and colleagues studied the regional and cellular expression patterns of ataxin-7 at the mRNA level by in situ hybridization in normal human brain (Lindenberg et al., 2000). At the mRNA level ataxin-7 was preferentially expressed in neurons; the regional distribution reflected neuronal packing density.

Several groups have used monoclonal and polyclonal antibodies raised against the normal ataxin-7 to establish the expression pattern of ataxin-7 in CNS and non-CNS tissue (Cancel et al., 2000; Lindenberg et al., 2000; Einum et al., 2001; Jonasson et al., 2002). Immunoblotting demonstrated that ataxin-7 is widely expressed but that expression levels vary among tissues (Einum et al., 2001). Immunohistochemical studies have confirmed that ataxin-7 is widely expressed, and have shown that in most neurons, ataxin-7 immunoreactivity (IR) is preferentially found in cell bodies and processes. Ataxin-7 is found in the cytoplasmatic compartment although some nuclear ataxin-7 IR has also been detected in most neurons. Double immunolabelling coupled with confocal microscopy showed that some ataxin-7 colocalizes with BiP/Grp78, a marker of the endoplasmic reticulum, but not with markers of mitochondria or the trans-Golgi network (Cancel et al., 2000). More intense and more prominently nuclear ataxin-7 IR was observed in neurons of the pons and the inferior olive, brain regions severely affected by the disease, suggesting that the subcellular localization and abundance of ataxin-7 is regulated in a region-specific manner. Lindenberg and colleagues suggested that since the neurons with more intense and more prominent nuclear ataxin-7 IR are those that are susceptible to SCA7, enrichment of normal ataxin-7 in the nuclear compartment may contribute to neurodegeneration (Lindenberg et al., 2000).

However, strong cytoplasmatic and nuclear ataxin-7 IR is not limited to sites of SCA7 pathology and is not always related to the severity of neuronal loss. For example, IR is low in some vulnerable populations of neurons, such as Purkinje cells (Cancel et al., 2000). Ataxin-7 IR has also been detected throughout the retina and is particularly intense within the cell bodies and photosensitive outer segments of cone photoreceptors (Einum et al., 2001).

Jonasson et al. (2002) compiled data on ataxin-7 expression from all SCA7 patients (n = 5) and controls (n = 7) published to date. Expression of ataxin-7 was found in neurons throughout the CNS and was highly abundant in Purkinje cells of the cerebellum, in regions of the hippocampus and in cerebral cortex. Ataxin-7 expression was not restricted to regions of pathology, and there were no apparent regional differences in ataxin-7 expression patterns between patients and controls. The subcellular distribution of ataxin-7 was primarily nuclear in all brain regions studied. In cerebellar Purkinje cells, however, differences in subcellular distribution of ataxin-7 were observed between patients and controls of different ages.

Expression of ataxin-7 was detected during mouse embryonic development and in all adult mouse tissues examined by Northern and Western blots (Strom et al., 2002). In brain, immunohistological staining revealed an ataxin-7 expression pattern similar to that in human, with ataxin-7 expression in cerebellum, several brainstem nuclei, cerebral cortex and hippocampus. These data showed high conservation of ataxin-7 both structurally and concerning its pattern of expression, suggesting a conserved role for the protein in mice and humans.

Abnormal gene function

The mutated protein was first reported to be enriched in the nucleus of lymphoblasts of SCA7 patients (Trottier et al., 1995). The 1C2 monoclonal antibody, which specifically recognizes long polyglutamine stretches on Western-blots, detected a specific 130-kDa protein in lymphoblasts (Trottier et al., 1995) and cerebral cortex (Stevanin et al., 1996) from SCA7 patients with predominantly early onset, that was not found in controls and patients with other ADCAs. The disease could result from a gain of function that occurs at the protein level and increases with repeat size after a threshold of approximately 36–40 glutamines. This is in agreement with the expression of both the mutated and the normal proteins as well as the dominant nature of the mutation. The expansion probably alters the conformation of the polyglutamine tract, as initially suggested by the specific detection of long repeats using the 1C2 antibody (Trottier et al., 1995; Stevanin et al., 1996). However, in Huntington's disease, recent results indicate that loss of the normal function of the protein might also contribute to the pathology (Cattaneo et al., 2001). This may also apply to SCA7, for which no knock-out models have yet been reported.

Neuronal nuclear inclusions (NIIs)

Ataxin-7-containing aggregates, called neuronal intranuclear inclusions (NIIs), first detected in neurons of several brain regions in a juvenile SCA7 patient (Holmberg et al., 1998), are pathological structures characteristic of polyglutamine disorders (Zoghbi and Orr, 2000). NIIs have been detected in the brain and retina of SCA7 patients with the 1C2 antibody and with anti-ataxin-7 antibodies (Holmberg et al., 1998; Mauger et al., 1999; Cancel et al., 2000; Einum et al., 2001; Jonasson et al., 2002). NIIs were most frequent in the inferior olivary complex, a site of severe neuronal loss in SCA7 (Holmberg et al., 1998). They were also observed in other brain regions, including the cerebral cortex, not considered to be affected in the disease. In addition, cytoplasmic staining by the 1C2 antibody was also observed, particularly in the supramarginal gyrus, the hippocampus, the thalamus, the lateral geniculate body and the pontine nuclei. These data confirm that, in SCA7, the NIIs are not restricted to the sites of severe neuronal loss. Conversely, no NIIs were detected in neurons of an SCA7 patient with 41 repeats after 40 years of disease duration (Einum et al., 2001). When observed by confocal microscopy, some inclusions were found to be ubiquitinated (Holmberg et al., 1998; Mauger et al., 1999), but to varying degrees, ranging from < 1 % in the cerebral cortex to 60% in the inferior olive (Holmberg et al., 1998). In electron microscopy studies using immunogold labeling, ataxin-7 immunoreactive NIIs appear as dense aggregates containing a mixture of granular and filamentary structures (Mauger et al., 1999). The inclusions may be only a hallmark of the diseases or reflect a cellular defense mechanism. Their implication in the degenerative process remains a matter of debate. Their presence in unaffected tissues in SCA7 patients does not seem sufficient to initiate the degenerative process (Holmberg et al., 1998).

NIIs contain not only the protein with the expanded polyglutamine stretch but also several chaperone proteins related to the ubiquitin proteasome pathway, transcription factors and nuclear matrix proteins. The composition of NIIs might reflect the process of NII formation and part of the pathogenesis of the disease. To investigate how these proteins relate to the pathogenesis of SCA7, Takahashi and colleagues performed immunohistochemical analyses of the composition of NIIs in two cases of SCA7 (Takahashi et al., 2002) (Fig. 2). They demonstrated that there are two types of NIIs in SCA7 that differ in size and immunoreactivity to promyelocytic leukaemia protein (PML), one of the essential components of nuclear bodies (NBs; also called PML oncogenic domains). Small and large NIIs contained ataxin-7, human DnaJ homologue 2 (HDJ-2) and proteasome subunit 19S. In contrast, PML was found only in small NIIs. CREB-binding protein (CBP), another component of NBs, was distributed like PML in NIIs. Authors suggest that NIIs are formed by the accumulation of ataxin-7 in NBs, which become enlarged as they recruit related proteins.

Cellular and animal models of disease

Cellular models

Immunocytochemical studies have been carried out on neuronal and non-neuronal cells expressing both wild type and mutated ataxin-7 (Kaytor et al., 1999; Zander et al., 2001). In both cases, intense labeling was observed almost exclusively in the nucleus. Nuclear transport was blocked by disruption of the nuclear localisation signal (Kaytor et al., 1999). In addition to diffuse distribution throughout the nucleus, ataxin-7 was asso-

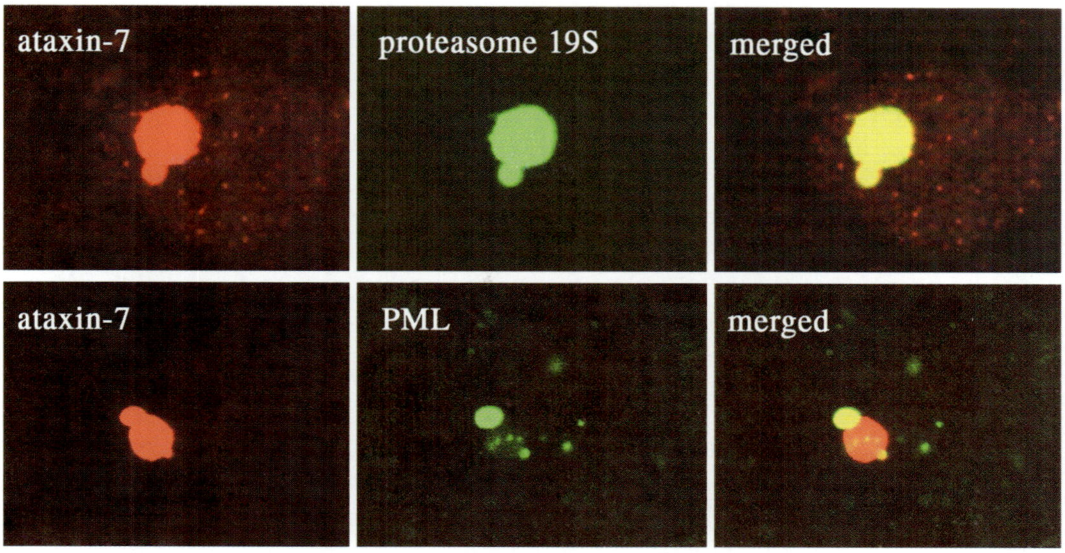

Fig. 2. Double labelling with antibodies against ataxin-7 and proteasome subunit 19S or PML in brain section of an SCA7 patient: The proteasome subunit 19S colocalized with ataxin-7, in both small and large NIIs. PML was localized consistently to the small NIIs, but was also detected in a small proportion of the large NIIs (reproduced from Takahashi et al., 2002).

ciated with the nuclear matrix, PML (promyelocytic leukemia protein) bodies and nucleoli (Kaytor et al., 1999). Thus, expanded ataxin-7 may exert its pathogenic effects in the nucleus by altering a matrix-associated nuclear structure or by disrupting nucleolar function.

Zander et al. (2001) have compared the inclusions in cells expressing mutant ataxin-7 and in human SCA7 brain tissue. The cells readily formed anti-ataxin-7 positive, fibrillar inclusions and small, nuclear electron dense structures. There were consistent signs of ongoing abnormal protein folding, including the recruitment of heat-shock proteins and proteasome subunits (Fig. 3). Occasionally, transcription factors were sequestered in the inclusions. Activated caspase-3 was recruited into the inclusions in both the cell models and human SCA7 brain, and its expression was upregulated in cortical neurons, suggesting that it may play a role in the disease process. Finally, on the ultrastructural level, there were signs of autophagy and nuclear indentations, indicative of a major stress response in cells expressing mutant ataxin-7.

Caspase cleavage of ataxin-7 might be implicated, since there are several potential cleavage sites in the ataxin-7 sequence (unpublished data). The presence of these sites in ataxin-7 is reminiscent of huntingtin, atrophin, ataxin-3 and the androgen receptor, which are suspected to be truncated by such proteases (Wellington et al., 1998) and, as a consequence, acquire a toxic property (Ikeda et al., 1996; Ellerby et al., 1999) or aggregate (Cooper et al., 1998; Martindale et al., 1998). Inhibition of caspases could, as in an HD model (Ona et al., 1999), prevent the progression of SCA7 in animal and cellular models.

To gain insight into ataxin-7 function and to decipher the molecular mechanisms of neurodegeneration in SCA7, two-hybrid assays were performed to identify ataxin-7 interacting proteins. As in other polyglutamine diseases, the selective pattern of degeneration contrasts with the ubiquitous expression of ataxin-7. Several explanations have been proposed to explain this, as such a difference in the level of ataxin-7 expression or the presence of specific ataxin-7 binding proteins, but none is as yet satisfactory. Lebre and colleagues screened a human retina cDNA library for ataxin-7 binding proteins (Lebre et al., 2001). A protein R85 was isolated and found to encode a splice variant of CAP (Cbl associated protein) (Ribon et al., 1998). R85 and CAP are generated by alternative splicing of the gene SORBS1 (alias SH3P12). The R85 cDNA contains a bipartite NLS, three SH3 domains and interacts with normal and mutated ataxin-7. SORBS1 gene products (SH3P12GP) are expressed in Purkinje cells in the cerebellum. Ataxin-7 colocalizes with an SH3P12GP in NIIs in brain from an SCA7 patient. This interaction may be part of a physiological pathway related to the function or turnover of ataxin-7 and may be also involved in the pathophysiological process of SCA7 disease. Matilla et al. (2001) showed interaction of ataxin-7 with the ATPase subunit S4 of the proteasomal 19S regulatory complex. The ataxin-7/S4 association is modulated by the length of the polyglutamine tract whereby S4 shows a stronger association with the wild-type allele of ataxin-7. Immunohistochemical analyses suggest alterations either of the distribution or the levels of S4 immunoreactivity in neurons that degenerate in SCA7 brains. Immunoblot analyses demonstrate reduced levels of S4 immunoreactivity in cerebellar protein extracts from two SCA7 patients. Authors suggest a role of S4 and proteasomal proteolysis in the molecular pathogenesis of SCA7.

Animal models

No SCA7 invertebrate model or knock-out mice are available. Yvert and colleagues described transgenic SCA7 mice that

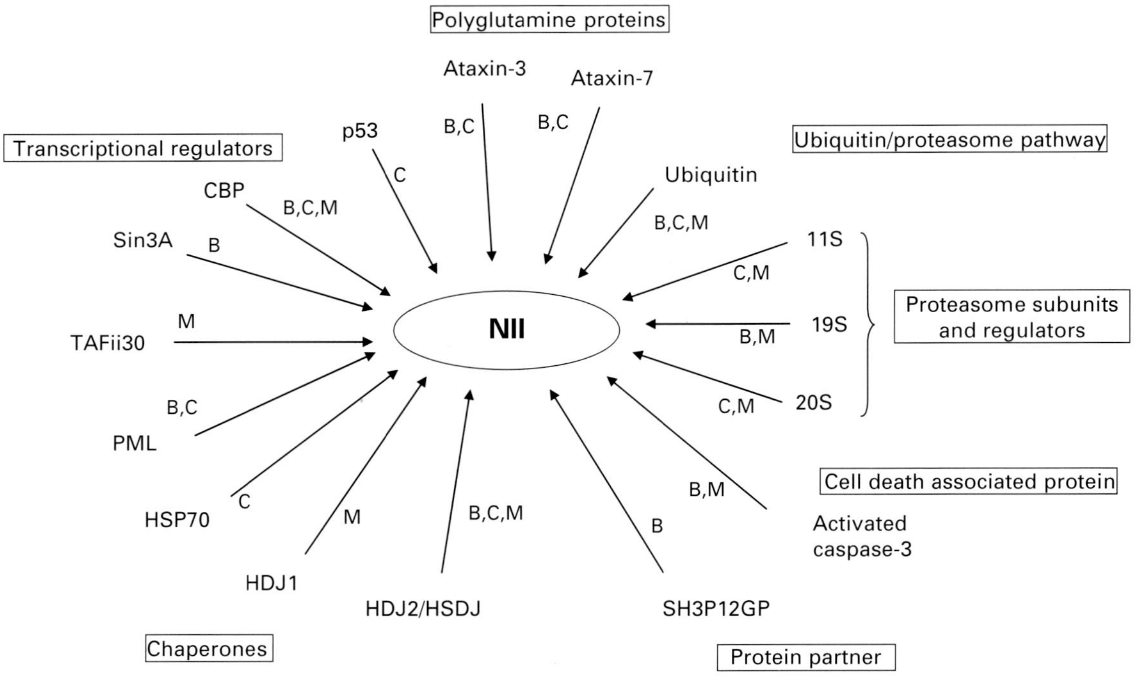

Fig. 3. Composition of intranuclear inclusions (NIIs) in human SCA7 brain (B), SCA7 cellular models (C) and SCA7 mouse models (M): CBP: CREB-binding protein, PML: promyelocytic leukaemia protein, HSP: Heat shock protein, HDJ1/2: proteins of the HSP40 family; SH3P12GP: SORBS1 gene products (Lebre et al., 2000; Yvert et al., 2000, 2001; Zander et al., 2001).

overexpress full-length mutant ataxin-7 Q90 in Purkinje cells (P7 mice), in rod photoreceptors (R7 mice) or with a more widespread expression of mutant ataxin-7 (B7 mice), including neuronal cell types that are unaffected in SCA7 patients (Yvert et al., 2000, 2001). P7 and R7 mice have deficiencies in motor coordination and vision, respectively, and display neurodegeneration with progressive ataxin-7 accumulation into ubiquitinated NIIs (Yvert et al., 2000). In R7 mice, severe degeneration of photoreceptors, but also secondary alterations of post-synaptic neurons was caused by overexpression of ataxin-7 Q90 in rods. Interestingly, ERG are abnormal in R7 mice before photoreceptor degeneration begins, thus providing a single means for assessing the protection of photoreceptors in future therapeutical trials (Yvert et al., 2000). Furthermore, these authors also suggest that proteolytic cleavage of mutant ataxin-7 and trans-neuronal responses are implicated in the pathogenesis of SCA7 (Yvert et al., 2000). An extensive screen for chaperones, proteasomal subunits and transcription factors sequestered in NIIs disclosed no pattern unique to neurons undergoing degeneration in SCA7 (Fig. 3). In particular, the mouse TAF(II)30 subunit of the TFIID initiation complex accumulated markedly in NIIs, even though this protein does not contain a polyglutamine stretch. A striking discrepancy between mRNA and ataxin-7 levels in B7 mice expressing wild-type but not mutant ataxin-7, indicates selective stabilization of mutant ataxin-7, both in this model and in P7 mice. B7 mice, therefore, provide in vivo evidence that the polyglutamine expansion mutation can increase protein half-life (Yvert et al., 2001).

La Spada and colleagues also produced transgenic SCA7 mice that express full length ataxin-7 in all neurons of the CNS, except for Purkinje cells (La Spada et al., 2001; Garden et al., 2002). Transgenic mice expressing mutant ataxin-7 Q92 developed a dramatic neurological phenotype presenting as a gait ataxia and culminating in premature death and a cone-rod dystrophy. NIIs were present, suggesting that the disease pathway involves the nucleus. Despite the absence of expression of polyglutamine-expanded ataxin-7 in Purkinje cells, they documented severe Purkinje cell degeneration in SCA7 transgenic mice (Garden et al., 2002). As in B7 mice, La Spada and colleagues detected an N-terminal truncation fragment of ataxin-7 in their transgenic mice and in SCA7 patient material with both anti-ataxin-7 and anti-polyglutamine specific antibodies. The appearance of truncated ataxin-7 in nuclear aggregates correlates with the onset of a disease phenotype in the SCA7 mice, suggesting that nuclear localization and proteolytic cleavage may be important features of SCA7 pathogenesis. The non-cell-autonomous nature of the Purkinje cell degeneration in their SCA7 mouse model indicates that polyglutamine-induced dysfunction in adjacent or connecting cell types contributes to the neurodegeneration.

Ataxin-7 interacted and colocalized with a cone-rod homeobox protein (CRX) and mutated ataxin-7 dramatically suppressed CRX transactivation (La Spada et al., 2001). In the La Spada transgenic SCA7 model, electrophoretic mobility shift assays indicated reduced CRX binding activity, and RT-PCR analysis showed reduced expression of CRX-regulated genes. Authors suggest that interference with CRX regulated transcription is responsible for retinal degeneration in this SCA7 model and may provide an explanation for how cell-type specificity is achieved in polyglutamine disease (La Spada et al., 2001).

Recent evidence indicates that transcriptional abnormalities may play an important role in the pathophysiology of polyglutamine diseases. Luthi-Carter et al. (2002) explored the extent to which polyglutamine-related changes in gene expression may be independent of protein context by comparing mouse models of dentatorubral-pallidoluysian atrophy (DRPLA) and Huntington's disease (HD) by microarray gene expression profiling. A large number of overlapping gene expression changes were observed in the cerebella of DRPLA mice (expressing full-length atrophin-1 with Q65) and HD mice (expressing partial huntingtin with Q82). They also conducted Northern analyses of cerebella from mouse models of SBMA and SCA7 (La Spada model). Four of the gene expression changes common to both DRPLA and HD mice were also observed in the cerebella of SBMA and SCA7 mouse models (enkephalin, insulin-like growth factor-binding protein 5 (IGFBP5), myristoylated, alanine-rich C-kinase substrate (MARCKS) and neuronal visinin-like protein 1 (VSNL1 alias NVP-1). In addition, changes in the expression of Snf1-related kinase (SNFK1) and neuronal acidic protein 22 (NAP22) were observed. The consistency of the cerebellar gene expression changes between five polyglutamine disease models clearly demonstrates that polyglutamine expansion leads to common gene expression events even in the context of distinct proteins. These results demonstrate that some mechanisms of transcriptional dysregulation can occur independently of protein context.

Conclusion

There is no specific drug therapy for this neurodegenerative disorder. Currently, therapy remains purely symptomatic. Therapy in gain of function diseases with adult onset is fraught with technical difficulties. Cellular models and transgenic SCA7 mice are now available. They constitute valuable resources for studying the physiopathology of the disease, and will be useful for screening and evaluating the potential therapeutic strategies that are currently being explored in other polyglutamine diseases.

Acknowledgments

The authors are grateful to Dr. Merle Ruberg for critical reading of the manuscript.

References

Aleman TS, Cideciyan AV, Volpe NJ, Stevanin G, Brice A, Jacobson SG: Spinocerebellar ataxia type 7 (SCA7) shows a cone-rod dystrophy phenotype. Exp Eye Res 74:737–745 (2002).

Andrew SE, Goldberg YP, Hayden MR: Rethinking genotype and phenotype correlations in polyglutamine expansion disorders. Hum molec Genet 6 2005–2010 (1997).

Babovic-Vuksanovic D, Snow K, Patterson MC, Michels VV: Spinocerebellar ataxia type 2 (SCA2) in an infant with extreme CAG repeat expansion. Am J med Genet 79:383–387 (1998).

Benomar A, Le Guern E, Dürr A, Ouhabi H, Stevanin G, Yahyaoui M, Chkili T, Agid Y, Brice A: Autosomal-dominant cerebellar ataxia with retinal degeneration (ADCA type II) is genetically different from ADCA type I. Ann Neurol 35:439–444 (1994).

Benomar A, Krols L, Stevanin G, Cancel G, Le Guern E, David G, Ouhabi H, Martin JJ, Dürr A, Zaim A, Ravise N, Busque C, Penet C, Van Regemorter N, Weissenbach J, Yahyaoui M, Chkili T, Agid Y, Van Broeckhoven C, Brice A: The gene for autosomal dominant cerebellar ataxia with pigmentary macular dystrophy maps to chromosome 3p12→p21.1. Nature Genet 10:84–88 (1995).

Benton CS, de Silva R, Rutledge SL, Bohlega S, Ashizawa T, Zoghbi HY: Molecular and clinical studies in SCA-7 define a broad clinical spectrum and the infantile phenotype. Neurology 51:1081–1086 (1998).

Cancel G, Dürr A, Didierjean O, Imbert G, Bürk K, Lezin A, Belal S, Benomar A, Abada-Bendib M, Vial C, Guimaraes J, Chneiweiss H, Stevanin G, Yvert G, Abbas N, Saudou F, Lebre A-S, Yahyaoui M, Hentati F, Vernant J-C, Klockgether T, Mandel J-L, Agid Y, Brice A: Molecular and clinical correlations in spinocerebellar ataxia 2: a study of 32 families. Hum molec Genet 6 709–715 (1997).

Cancel G, Duyckaerts C, Holmberg M, Zander C, Yvert G, Lebre AS, Ruberg M, Faucheux B, Agid Y, Hirsch E, Brice A: Distribution of ataxin-7 in normal human brain and retina. Brain 123:2519–2530 (2000).

Cattaneo E, Rigamonti D, Goffredo D, Zuccato C, Squitieri F, Sipione S: Loss of normal huntingtin function: new developments in Huntington's disease research. Trends Neurosci 24:182–188 (2001).

Cooper JK, Schilling G, Peters MF, Herring WJ, Sharp AH, Kaminsky Z, Masone J, Khan FA, Delanoy M, Borchelt DR, Dawson VL, Dawson TM, Ross CA: Truncated N-terminal fragments of huntingtin with expanded glutamine repeats form nuclear and cytoplasmic aggregates in cell culture. Hum molec Genet 7:783–790 (1998).

David G, Giunti P, Abbas N, Coullin P, Stevanin G, Horta W, Gemmill R, Weissenbach J, Wood N, Cunha S, Drabkin H, Harding AE, Agid Y, Brice A: The gene for autosomal dominant cerebellar ataxia type II is located in a 5-cM region in 3p12→p13: genetic and physical mapping of the SCA7 locus. Am J hum Genet 59:1328–1336 (1996).

David G, Abbas N, Stevanin G, Dürr A, Yvert G, Cancel G, Weber C, Imbert G, Saudou F, Antoniou E, Drabkin H, Gemmill R, Giunti P, Benomar A, Wood N, Ruberg M, Agid Y, Mandel J-L, Brice A: Cloning of the SCA7 gene reveals a highly unstable CAG repeat expansion. Nature Genet 17:65–70 (1997).

David G, Dürr A, Stevanin G, Cancel G, Abbas N, Benomar A, Belal S, Lebre A-S, Abada-Bendib M, Grid D, Holmberg M, Yahyaoui M, Hentati F, Chkili T, Agid Y, Brice A: Molecular and clinical correlations in autosomal dominant cerebellar ataxia with progressive macular dystrophy (SCA7). Hum molec Genet 7:165–170 (1998a).

David G, Abbas N, Durr A, Stevanin G, Cancel G, Yvert G, Imbert G, Saudou F, Mandel J-L, Agid Y, Brice A: Autosomal dominant cerebellar ataxia with macular dystrophy (SCA7) is caused by a highly unstable CAG repeat expansion, in Wells RD, Warren ST (eds): Genetic instabilities and hereditary neurological diseases, pp 273–282 (Academic Press, San Diego 1998b).

Del-Favero J, Krols L, Michalik A, Theuns J, Löfgren A, Goossens D, Wehnert A, van den Bossche D, van Zand K, Backhovens H, Van Regenmorter N, Martin JJ, Van Broeckhoven C: Molecular genetic analysis of autosomal dominant cerebellar ataxia with retinal degeneration (ADCA type II) caused by CAG triplet repeat expansion. Hum molec Genet 7:177–186 (1998).

Einum DD, Townsend JJ, Ptacek LJ, Fu YH: Ataxin-7 expression analysis in controls and spinocerebellar ataxia type 7 patients. Neurogenetics 3:83–90 (2001).

Ellerby LM, Andrusiak RL, Wellington CL, Hackam AS, Propp SS, Wood JD, Sharp AH, Margolis RL, Ross CA, Salvesen GS, Hayden MR, Bredesen DE: Cleavage of atrophin-1 at caspase site aspartic acid 109 modulates cytotoxicity. J biol Chem 274:8730–8736 (1999).

Enevoldson TP, Sanders MD, Harding AE: Autosomal dominant cerebellar ataxia with pigmentary macular dystrophy A clinical and genetic study of eight families. Brain 117:445–460 (1994).

Froment J, Bonnet P, Colrat A: Heredo-dégénérations rétinienne et spino-cérébelleuses: variantes ophtalmoscopiques et neurologiques présentées par trois générations successives. J Med Lyon 22:153–163 (1937).

Garden GA, Libby RT, Fu YH, Kinoshita Y, Huang J, Possin DE, Smith AC, Martinez RA, Fine GC, Grote SK, Ware CB, Einum DD, Morrison RS, Ptacek LJ, Sopher BL, La Spada AR: Polyglutamine-expanded ataxin-7 promotes non-cell-autonomous purkinje cell degeneration and displays proteolytic cleavage in ataxic transgenic mice. J Neurosci 22:4897–4905 (2002).

Giunti P, Stevanin G, Worth P, David G, Brice A, Wood NW: Molecular and clinical study of 18 families with ADCA type II: evidence for genetic heterogeneity and de novo mutation. Am J hum Genet 64:1594–1603 (1999).

Gouw LG, Digre KB, Harris CP, Haines JH, Ptacek LJ: Autosomal dominant cerebellar ataxia with retinal degeneration: clinical neuropathologic and genetic analysis of a large kindred. Neurology 44:1441–1447 (1994).

Gouw LG, Kaplan CD, Haines JH, Digre KB, Rutledge SL, Matilla A, Leppert M, Zoghbi HY, Ptacek LJ: Retinal degeneration characterizes a spinocerebellar ataxia mapping to chromosome 3p. Nature Genet 10:89–93 (1995).

Gouw LG,Castaneda MA, McKenna CK, Digre KB, Pulst SM, Perlman S, Lee MS, Gomez C, Fischbeck K, Gagnon D, Storey E, Bird T, Jeri FR, Ptacek LJ: Analysis of the dynamic mutation in the SCA7 gene shows marked parental effects on CAG repeat transmission Hum molec Genet 7:525–532 (1998).

Gurevich VV, Dion SB, Onorato JJ, Ptasienski J, Kim CM, Sterne-Marr R, Hosey MM, Benovic JL: Arrestin interactions with G protein-coupled receptors. Direct binding studies of wild type and mutant arrestins with rhodopsin beta 2-adrenergic and m2 muscarinic cholinergic receptors. J biol Chem 270:720–731 (1995).

Harding AE: The clinical features and classification of the late onset autosomal dominant cerebellar ataxias. A study of 11 families including descendants of the "the Drew family of Walworth". Brain 105:1–28 (1982).

Harding AE: Clinical features and classification of inherited ataxias. Adv Neurol 61:1–14 (1993).

Holmberg M, Johansson J, Forsgren L, Heijbel J, Sandgren O, Holmgren G. Localization of autosomal dominant cerebellar ataxia associated with retinal degeneration and anticipation to chromosome 3p12→p21.1. Hum molec Genet 4:1441–1445 (1995).

Holmberg M, Duyckaerts C, Durr A, Cancel G, Gourfinkel-An I, Damier P, Faucheux B, Trottier Y, Hirsch EC, Agid Y, Brice A: Spinocerebellar ataxia type 7 (SCA7): a neurodegenerative disorder with neuronal intranuclear inclusions. Hum molec Genet 7:913–918 (1998).

Ikeda H, Yamaguchi M, Sugai S, Aze Y, Narumiya S, Kakizuka A: Expanded polyglutamine in the Machado-Joseph disease protein induces cell death in vitro and in vivo. Nature Genet 13:196–202 (1996).

Jöbsis GJ, Weber JW, Barth PG, Keizers H, Baas F, van Schooneveld MJ, van Hilten JJ, Troost D, Geesink HH, Bolhuis PA: Autosomal dominant cerebellar ataxia with retinal degeneration (ADCA II): clinical and neuropathological findings in two pedigrees and genetic linkage to 3p12→p21.1. J Neurol Neurosurg Psychiatry 62:367–371 (1997).

Johansson J, Forsgren L, Sandgren O, Brice A, Holmgren G, Holmberg M: Expanded CAG repeat in Swedish Spinocerebellar ataxia type 7 (SCA7) patients: effect of CAG repeat length on the clinical manifestation. Hum molec Genet 7:171–176 (1998).

Jonasson J, Strom AL, Hart P, Brannstrom T, Forsgren L, Holmberg M: Expression of ataxin-7 in CNS and non-CNS tissue of normal and SCA7 individuals. Acta Neuropathol (Berl) 104:29–37 (2002).

Kaytor MD, Duvick LA, Skinner PJ, Koob MD, Ranum LP, Orr HT: Nuclear localization of the spinocerebellar ataxia type 7 protein ataxin-7. Hum molec Genet 8:1657–1664 (1999).

Konigsmark BW, Weiner LP: The olivopontocerebellar atrophies: a review. Medicine (Baltimore) 49:227–241 (1970).

Koob MD, Benzow KA, Bird TD, Day JW, Moseley ML, Ranum LPW: Rapid cloning of expanded trinucleotide repeat sequences from genomic DNA. Nature Genet 18:72–75 (1998).

Krols L, Martin JJ, David G, Van Regemorter N, Benomar A, Lofgren A, Stevanin G, Durr A, Brice A, Van Broeckhoven C: Refinement of the locus for autosomal dominant cerebellar ataxia type II to chromosome 3p21.1→p14.1. Hum Genet 99:225–232 (1997).

La Spada AR, Fu Y, Sopher BL, Libby RT, Wang X, Li LY, Einum DD, Huang J, Possin DE, Smith AC, Martinez RA, Koszdin KL, Treuting PM, Ware CB, Hurley JB, Ptacek LJ, Chen S: Polyglutamine-expanded ataxin-7 antagonizes crx function and induces cone-rod dystrophy in a mouse model of SCA7. Neuron 31:913–927 (2001).

Lebre AS, Jamot L, Takahashi J, Spassky N, Leprince C, Ravise N, Zander C, Fujigasaki H, Kussel-Andermann P, Duyckaerts C, Camonis JH, Brice A: Ataxin-7 interacts with a Cbl-associated protein that it recruits into neuronal intranuclear inclusions. Hum molec Genet 10:1201–1213 (2001).

Lindblad K, Savontaus ML, Stevanin G, Holmberg M, Digre K, Zander C, Ehrsson H, David G, Benomar A, Nikoskelainen E, Trottier Y, Holmgren G, Ptacek LJ, Anttinen A, Brice A, Schalling M: An expanded CAG repeat sequence in spinocerebellar ataxia type 7. Genome Res 6:965–971 (1996).

Lindenberg KS, Yvert G, Muller K, Landwehrmeyer GB: Expression analysis of ataxin-7 mRNA and protein in human brain: evidence for a widespread distribution and focal protein accumulation. Brain Pathol 10:385–394 (2000).

Luthi-Carter R,Strand AD, Hanson SA, Kooperberg C, Schilling G, La Spada AR, Merry DE, Young AB, Ross CA, Borchelt DR, Olson JM: Polyglutamine and transcription: gene expression changes shared by DRPLA and Huntington's disease mouse models reveal context-independent effects. Hum molec Genet 11:1927–1937 (2002).

Lyoo CH, Hun K, Choi YC, Lee SC, Stevanin G, David G, Brice A, Lee MS: CAG repeat expansion in the SCA7 gene in Korean families presenting with ADCA type II. J Korean Neurol Assoc 16:341–352 (1998).

Martin JJ, Van Regemorter N, Krols L, Brucher JM, de Barsy T, Szliwowski H, Evrard P, Ceuterick C, Tassignon MJ, Smet-Dieleman H, Hayez-Delatte F, Willems PJ, Van Broeckhoven C: On an autosomal dominant form of retinal-cerebellar degeneration: an autopsy study of five patients in one family. Acta Neuropathol (Berl) 88:277–286 (1994).

Martin J, Van Regemorter N, Del-Favero J, Lofgren A, Van Broeckhoven C: Spinocerebellar ataxia type 7 (SCA7) – correlations between phenotype and genotype in one large Belgian family. J Neurol Sci 168:37–46 (1999).

Martindale D, Hackam A, Wieczorek A, Ellerby L, Wellington C, McCutcheon K, Singaraja R, Kazemi-Esfarjani P, Devon R, Kim SU, Bredesen DE, Tufaro F, Hayden MR: Length of huntingtin and its polyglutamine tract influences localization and frequency of intracellular aggregates. Nature Genet 18:150–154 (1998).

Matilla A, Gorbea C, Einum DD, Townsend T, Michalik A, van Broeckhoven C, Jensen CC, Murphy KJ, Ptacek LJ, Fu Y-H: Association of ataxin-7 with the proteasome subunit S4 of the 19S regulatory complex. Hum mol Genet 10:2821–2831 (2001).

Mauger C, Del-Favero J, Ceuterick C, Lubke U, Van Broeckhoven C, Martin J: Identification and localization of ataxin-7 in brain and retina of a patient with cerebellar ataxia type II using anti-peptide antibody. Brain Res Mol Brain Res 74:35–43 (1999).

Monckton DG, Cayuela ML, Gould FK, Brock GJ, Silva R, Ashizawa T: Very large (CAG)(n) DNA repeat expansions in the sperm of two spinocerebellar ataxia type 7 males. Hum molec Genet 8:2473–2478 (1999).

Mushegian AR, Vishnivetskiy SA, Gurevich VV: Conserved phosphoprotein interaction motif is functionally interchangeable between ataxin-7 and arrestins. Biochemistry 39:6809–6813 (2000).

Nardacchione A, Orsi L, Brusco A, Franco A, Grosso E, Dragone E, Mortara P, Schiffer D, De Marchi M: Definition of the smallest pathological CAG expansion in SCA7. Clin Genet 56:232–234 (1999).

Neetens A, Martin JJ, Libert J, Van Den Ende P: Autosomal dominant cone-dystrophy-cerebellar atrophy (ADCoCA) (modified ADCA Harding II). Neuro-ophthalmology 10:261–275 (1990).

Ona VO, Li M, Vonsattel JP, Andrews LJ, Khan SQ, Chung WM, Frey AS, Menon AS, Li XJ, Stieg PE, Yuan J, Penney JB, Young AB, Cha JH, Friedlander RM: Inhibition of caspase-1 slows disease progression in a mouse model of Huntington's disease. Nature 399:263–267 (1999).

Ribon V, Printen JA, Hoffman NG, Kay BK, Saltiel AR: A novel multifuntional c-Cbl binding protein in insulin receptor signaling in 3T3-L1 adipocytes. Mol Cell Biol 18:872–879 (1998).

Rubinsztein DC, Leggo J, Coles R, Almqvist E, Biancalana V, Cassiman JJ, Chotai K, Connarty M, Craufurd D, Curtis A, Curtis D, Davidson MJ, Differ AM, Dode C, Dodge A, Frontali M, Ranen NG, Stine OC, Sherr M, Abbott MH, Franz ML, Graham CA, Harper PS, Hedreen JC, Jackson A, Kaplan JC, Losekoot M, MacMillan JC, Trottier Y, Novelleto A, Simpson SA, Theilman J, Whittaker JL, Folstein SE, Ross CA, Hayden MR: Phenotypic characterization of individuals with 30–40 CAG repeats in the Huntington disease (HD) gene reveals HD cases with 36 repeats and apparently normal elderly individuals with 36–39 repeats. Am J hum Genet 59:16–22 (1996).

Schalling M, Hudson TJ, Buetow KH, Housman DE: Direct detection of novel expanded trinucleotide repeats in the human genome. Nature Genet 4:135–139 (1993).

Schols L, Amoiridis G, Buttner T, Przuntek H, Epplen JT, Riess O: Autosomal dominant cerebellar ataxia: phenotypic differences in genetically defined subtypes? Ann Neurol 42:924–932 (1997).

Stevanin G, Trottier Y, Cancel G, Dürr A, David G, Didierjean O, Bürk K, Imbert G, Saudou F, Abada-Bendib M, Gourfinkel-An I, Benomar A, Abbas N, Klockgether T, Grid D, Agid Y, Mandel J-L, Brice A: Screening for proteins with polyglutamine expansions in autosomal dominant cerebellar ataxias. Hum molec Genet 5:1887–1892 (1996).

Stevanin G, Giunti P, Belal GDS, Durr A, Ruberg M, Wood N, Brice A: De novo expansion of intermediate alleles in spinocerebellar ataxia 7. Hum molec Genet 7:1809–1813 (1998).

Stevanin G, David G, Durr A, Giunti P, Benomar A, Abada-Bendib M, Lee MS, Agid Y, Brice A: Multiple origins of the spinocerebellar ataxia 7 (SCA7) mutation revealed by linkage disequilibrium studies with closely flanking markers including an intragenic polymorphism (G3145TG/A3145TG). Eur J hum Genet 7:889–896 (1999).

Stevanin G, Durr A, Brice A: Clinical and molecular advances in autosomal dominant cerebellar ataxias: from genotype to phenotype and physiopathology. Eur J hum Genet 8:4–18 (2000a).

Stevanin G, Durr A, Brice A: Spinocerebellar ataxia 7 (SCA7), in Klockgether T (ed): Handbook of ataxia disorders (Marcel Dekker Inc, New York 2000b).

Strom AL, Jonasson J, Hart P, Brannstrom T, Forsgren L, Holmberg M: Cloning and expression analysis of the murine homolog of the spinocerebellar ataxia type 7 (SCA7) gene. Gene 285:91–99 (2002).

Takahashi J, Fujigasaki H, Zander C, El Hachimi KH, Stevanin G, Durr A, Lebre AS, Yvert G, Trottier Y, The H, Hauw JJ, Duyckaerts C, Brice A: Two populations of neuronal intranuclear inclusions in SCA7 differ in size and promyelocytic leukaemia protein content. Brain 125:1534–1543 (2002).

Trottier Y, Lutz Y, Stevanin G, Imbert G, Devys D, Cancel G, Saudou F, Weber C, David G, Laszlo T, Agid Y, Brice A, Mandel J-L: Polyglutamine expansion as a pathological epitope in Huntington's disease and four dominant cerebellar ataxias. Nature 378:403–406 (1995).

Wellington CL, Ellerby LM, Hackam AS, Margolis RL, Trifiro MA, Singaraja R, McCutcheon K, Salvesen GS, Propp SS, Bromm M, Rowland KJ, Zhang T, Rasper D, Roy S, Thornberry N, Pinsky L, Kakizuka A, Ross CA, Nicholson DW, Bredesen DE, Hayden MR: Caspase cleavage of gene products associated with triplet expansion disorders generates truncated fragments containing the polyglutamine tract. J biol Chem 273:9158–9167 (1998).

World Federation of Neurology Research Group on Huntington's Chorea: International Huntington Association and the World Federation of Neurology Research Group on Huntington's Chorea Guidelines for the molecular genetics predictive test in Huntington's disease. J med Genet 31:555–559 (1994).

Yvert G, Lindenberg KS, Picaud S, Landwehrmeyer GB, Sahel JA, Mandel JL: Expanded polyglutamines induce neurodegeneration and trans-neuronal alterations in cerebellum and retina of SCA7 transgenic mice. Hum molec Genet 9:2491–2506 (2000).

Yvert G, Lindenberg KS, Devys D, Helmlinger D, Landwehrmeyer GB, Mandel JL: SCA7 mouse models show selective stabilization of mutant ataxin-7 and similar cellular responses in different neuronal cell types. Hum molec Genet 10:1679–1692 (2001).

Zander C, Takahashi J, El Hachimi KH, Fujigasaki H, Albanese V, Lebre AS, Stevanin G, Duyckaerts C, Brice A: Similarities between spinocerebellar ataxia type 7 (SCA7) cell models and human brain: proteins recruited in inclusions and activation of caspase-3. Hum molec Genet 10 2569–2579 (2001).

Zoghbi HY, Orr HT: Glutamine repeats and neurodegeneration. A Rev Neurosci 23:217–247 (2000).

Cytogenet Genome Res 100:164–174 (2003)
DOI: 10.1159/000072851

Insights into the molecular basis of polyglutamine neurodegeneration from studies of a spinocerebellar ataxia type 7 mouse model

S.K. Grote[a] and A.R. La Spada[a,b,c]

Departments of [a] Laboratory Medicine, [b] Medicine (Division of Medical Genetics), [c] Neurology (Division of Neurogenetics), University of Washington Medical Center, Seattle WA (USA)

Abstract. Spinocerebellar ataxia type 7 (SCA7) is one member of a growing list of neurodegenerative disorders that are all caused by CAG repeat expansions that produce disease by encoding elongated polyglutamine tracts in a variety of apparently unrelated proteins. In this review, we provide an overview of our efforts to determine the molecular basis of polyglutamine neurotoxicity in SCA7 by modeling this polyglutamine repeat disorder in mice. We discuss how our SCA7 mouse model develops a phenotype that is reminiscent of the retinal and cerebellar disease pathology seen in human patients. All of these findings are considered in the context of numerous other models of polyglutamine disease pathology in mice and other organisms, together with various other in vitro and biochemical studies. We present the competing hypotheses of polyglutamine disease pathogenesis, and explain how our studies of SCA7 brainstem and retinal degeneration using this mouse model have yielded insights into possible mechanisms and pathways of polyglutamine disease pathology. In addition to illustrating how our SCA7 mouse model has allowed us to develop and advance notions of disease pathogenesis, we propose a model of polyglutamine molecular pathology that attempts to integrate the key observations in the field. We close by describing why our SCA7 mouse model should be useful for the next phase of polyglutamine disease research – the development of therapies, and predict that this stage of experimentation will continue to rely heavily on the mouse.

SCA7 is a neurodegenerative CAG/polyglutamine repeat disease

Spinocerebellar ataxia type 7 (SCA7) is one of a number of diseases that have been classified as an autosomal dominant cerebellar ataxia (ADCA). The ADCAs are characterized by neuronal dysfunction in the cerebellum, and are clinically subdivided into three types based upon the presence or absence of extra-cerebellar findings (Harding, 1982). Ataxia, which comes from the Greek word "taxis" meaning order (so "a-taxis" is disorder), is defined as a disturbance in coordinated movement and is typically applied to gait, limb, speech, and ocular motor incoordination associated with cerebellar degeneration. In SCA7, the affected neuronal populations are located in the cerebellum (Purkinje cells and granule cells), the cerebellar outflow nuclei (the dentate nucleus), and the brainstem (the inferior olivary nucleus and pontine nucleus) (Enevoldson et al., 1994). Degeneration of the cerebellum and associated structures causes neurological deficits, first presenting as poor motor coordination and then progressing to loss of one's ability to maintain normal gait and posture. The disease eventually results in a complete loss of motor coordination, often culminating in death from diseases such as pneumonia secondary to physical disability and inanition. SCA7 can be distinguished from the other dominantly inherited cerebellar ataxias because of a cone-rod dystrophy type of retinal degeneration (To et al., 1993). Progressive loss of photoreceptors and bipolar cell func-

A.R.L. is supported by grant R01-EY14061 from the NIH for this work, and is the recipient of a Beeson Physician Scholar Award from the American Federation for Aging Research.

Received 22 November 2002; manuscript accepted 20 January 2003.

Request reprints from Albert R. La Spada, MD, PhD
Department of Laboratory Medicine
University of Washington Medical Center
Box 357110, Room NW 120, Seattle WA 98195-7110 (USA)
telephone: (206)-598-2138; fax: (206)-598-6189
e-mail: laspada@u.washington.edu

KARGER
Fax + 41 61 306 12 34
E-mail karger@karger.ch
www.karger.com

Accessible online at:
www.karger.com/cgr

tion throughout the retina is noted on full-field electroretinogram (ERG) testing, with cone cells that mediate color vision degenerating well before the rod cells that mediate black-and-white vision.

After ascertainment of numerous affected families and linkage of the SCA7 gene defect to the short arm of chromosome 3 (Benomar et al., 1995), the cause of SCA7 was found to be expansion of a CAG repeat in the ataxin-7 gene, leading to an abnormally long glutamine (Q) tract in the protein product (David et al., 1997). The expanded repeat is in the amino-terminus of ataxin-7, an 892-amino-acid protein of unknown function, which is ubiquitously expressed with highest levels in heart, skeletal muscle, and pancreas, and moderate levels of expression in brain, liver, and kidney. Within the central nervous system (CNS), ataxin-7 is expressed at highest levels in the cerebellum and the brainstem (Lindenberg et al., 2000). Normal SCA7 alleles contain 4–35 CAG repeats, and pathological lengths vary from 37–306 repeats, with the severity and age of onset being very sensitive to the repeat length (Benton et al., 1998; David et al., 1997). The length of the repeat affects not only the age of onset, but also the predominant symptoms. In one study, it was observed that SCA7 individuals with fewer than 59 repeats tend to display progressive ataxia as their major symptom, whereas individuals with more than 59 repeats usually presented with visual impairment due to the retinal degeneration (Johansson et al., 1998). The SCA7 CAG repeat is extremely unstable, especially in the male germ line, where it can rapidly expand in a single generation resulting in juvenile or infantile onset cases, a phenomenon known as anticipation (Monckton et al., 1999).

The discovery of CAG/polyglutamine repeat expansions in SCA7 patients places this disease in a growing family of neurodegenerative disorders. The pathological nature of CAG repeat expansion was first demonstrated in Kennedy's disease, which is also known as spinal and bulbar muscular atrophy (SBMA) (La Spada et al., 2001). In SBMA, neurodegeneration of brainstem motor nuclei and spinal cord motor neurons has been linked to CAG/polyglutamine repeat expansions in the androgen receptor (AR) gene. Within the last decade, it was quickly realized that CAG repeat expansions encoding elongated polyglutamine tracts are an important cause of neurodegenerative disease. The CAG/polyglutamine disease category now includes nine disorders, most prominent among them being Huntington's disease (HD). In addition to SBMA and HD, Dentatorubral Pallidoluysian Atrophy (DRPLA) together with six different spinocerebellar ataxias (SCA 1, 2, 3, 6, 7, and 17) rounds out this category of disorders (Nakamura et al., 2001; Zoghbi and Orr, 2000).

With the exception of SBMA (which is X linked), all the polyglutamine disorders show an autosomal dominant pattern of inheritance, suggestive of a toxic gain of function. In at least two of the disorders (HD and SBMA), loss of function mutations show phenotypes different from the polyglutamine expansion disease, reinforcing the theory that these mutations impart a toxic gain of function to the proteins within which they reside (Brown et al., 1988; Zeitlin et al., 1995). Although a toxic gain of function is the primary mechanism of pathology in these diseases, recent studies have suggested a potential role of loss of normal function as a contributing factor to the pathology, as well. Normal huntingtin protein may have neuroprotective properties, and was shown to regulate the production of a neurotrophic factor in cortical projection neurons that synapse with the susceptible medium spiny nerve cells in the striatum (Zuccato et al., 2001). Mice that completely lack the huntingtin gene (*Hdh*$^{-/-}$ "knock-outs") die in early gastrulation, while heterozygous *Hdh* knock-out mice show increased apoptosis (Zeitlin et al., 1995). Perhaps, one of the most powerful demonstrations that huntingtin's normal function may be important for preventing HD pathology was the finding that postnatal elimination of *Hdh* expression in the forebrain of tissue-specific knock-out mice yields a HD-like phenotype of basal ganglia degeneration (Dragatsis et al., 2000). Thus, the question of whether loss of function and gain of function conspire to produce the final phenotype in each of the polyglutamine disorders remains to be answered.

While the CAG/polyglutamine field is only twelve years old, it is clear that much has been learned about these neurodegenerative disorders by modeling them in mice. Although the mouse offers the advantage of ease of genetic manipulation and reasonable generation times, more importantly, mice provide us with an experimental system where it is possible to dissect the basis of a complex neurological dysfunction, such as ataxia, in an organism whose neuroanatomy is very similar to our own. Equally important is the fact that mice share most of their genes in common with us, and humans and mice have highly overlapping molecular pathways, transcriptomes, and proteomes. Our laboratory's interest in modeling SCA7 in the mouse stemmed from the fact that this CAG/polyglutamine repeat disease is remarkable in its unique involvement of the retina as a pathologically affected region of the CNS. In the CAG/polyglutamine repeat disorders, there is selective neuronal cell loss in different regions of the brain in each disease, although expression of the different disease gene products overlap (Ross, 1995). We hypothesized that recapitulation of the SCA7 disease phenotype in mice might allow us to understand the cell-type specific pattern of neurodegeneration in this disorder, and that such an understanding might have implications for a wider understanding of cell-type-specific degeneration in the rest of the CAG/polyglutamine repeat diseases.

Analysis of brainstem neurodegeneration in the SCA7 mouse model yields insight into the roles of aggregation and cell autonomy in polyglutamine neurotoxicity

In order to model SCA7 in mice, we decided to use a heterologous promoter-expression system since the regulatory elements controlling ataxin-7 gene expression are not characterized. To insure expression throughout the CNS and especially in the three neuronal layers of the retina, we obtained the murine prion promoter (MoPrP) system that had been developed by David Borchelt's group (Borchelt et al., 1996). We then generated two transgenic constructs, one carrying a control length CAG repeat (PrP-SCA7-c24Q), and one carrying a significantly expanded CAG repeat (PrP-SCA7-c92Q) (Fig. 1). Pronuclear microinjection into B6C3 F1 hybrid oocytes

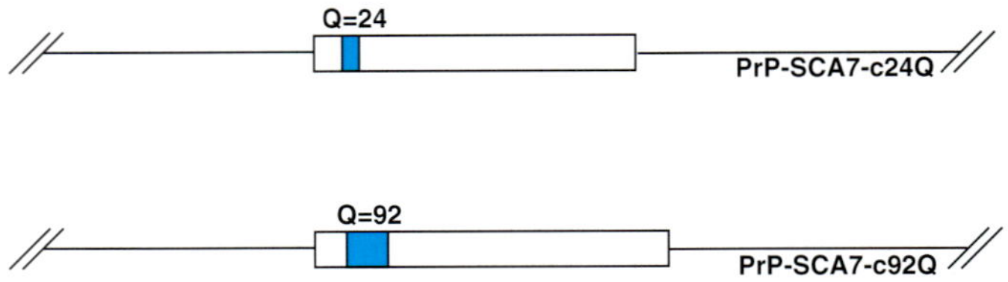

Fig. 1. Diagram of the SCA7 transgenic constructs. Ataxin-7 cDNAs containing either 24 CAGs (normal length) or 92 CAGs (disease length) were inserted into the murine prion protein (PrP) promoter-expression system to yield two SCA7 transgenic constructs, PrP-SCA7-c24Q and PrP-SCA7-c92Q.

yielded five transgene-positive founders, of which four (one PrP-SCA7-c24Q and three PrP-SCA7-c92Q) were subsequently maintained and expanded (La Spada et al., 2001). We devised an RT-PCR strategy to quantitate the expression level of the human transgene vs. endogenous murine ataxin-7, and were pleased to learn that expression levels ranged from 0.75 to just 2.5× that of endogenous, with the control line and one expansion line expressing nearly similar levels in the CNS. After confirming expression of full-length ataxin-7 protein in the CNS of the PrP-SCA7-c24Q line and the PrP-SCA7-c92Q lines, we turned our attention to the behavioral and neuropathology analysis of the different transgene-positive mice. The discovery of behavioral abnormalities did not take long, as the three highest-expressing PrP-SCA7-c92Q lines all manifested gait ataxia by 15 weeks of age (Garden et al., 2002). This gait ataxia progressed to a wide-based stance when walking, until the mice began to fall over. By this stage, mice were usually tremulous, but really never displayed overt seizures. The affected PrP-SCA7-c92Q mice also lost weight over time, appeared much smaller than their age-matched non-transgenic littermates, and exhibited early mortality.

Our current understanding of polyglutamine neurodegeneration is based upon a variety of biochemical, molecular, genetic, and organism studies. Although the CAG/polyglutamine repeat disorders share many features, the different disease-causing proteins appear to be unrelated except for the fact that they all contain an expanded polyglutamine tract. This observation raised the possibility that perhaps there is something inherently toxic about long polyglutamine stretches regardless of protein context. The pathological threshold for almost all the CAG/polyglutamine repeat diseases is nearly identical (between 35 and 40); and in all cases, the age of onset and severity of the disease are inversely related to the repeat length. In 1994, Max Perutz proposed the idea that once the number of glutamines in tandem crosses a threshold of around 35, then the polyglutamine tract acquires an altered conformation composed of beta-pleated sheets folding back on one another – the so-called "polar zipper" hypothesis (Perutz et al., 1994). Biochemical validation of altered conformation came one year later, when one group produced an antibody that could specifically detect expanded polyglutamine tracts and showed increasing avidity for proteins containing such polyglutamine tracts as they grew in length (Trottier et al., 1995). An interesting aside from this study was the observation that SCA7 patients expressed proteins recognized by this antibody (1C2), correctly predicting that a CAG repeat expansion would ultimately be found as the cause of SCA7. The production of abnormally folded polyglutamine-containing proteins as a key event in polyglutamine disease pathogenesis became clear when studies of a mouse model of HD and analysis of human brain material from patients with SCA3 detected protein aggregates in nuclei of neurons (Davies et al., 1997; Paulson et al., 1997). These so-called neuronal intranuclear inclusions (NIs) were subsequently found in most of the polyglutamine repeat diseases including SCA7 (Alves-Rodrigues et al., 1998; Becher et al., 1998; DiFiglia et al., 1997; Holmberg et al., 1998; Li et al., 1998), and were found to consist of the disease protein in combination with chaperone proteins and factors/subunits from the ubiquitin-proteasome protein turnover pathway (Cummings et al., 1998; Stenoien et al., 1999; Schmidt et al., 2002). Based upon this work, the theory has emerged that polyglutamine expansion tracts misfold, then become resistant to degradation, and accumulate as protein aggregates in nuclei and cytoplasm. The ability of Congo Red to prevent polyglutamine neurotoxicity by inhibiting oligomerization supports a model in which adoption of an altered conformation is an early event in polyglutamine disease pathogenesis (Sanchez et al., 2003).

In our mouse model of SCA7, we performed a variety of immunohistochemical studies to assess the role of NIs in disease pathogenesis. We began by noting that CNS sections from moderately to severely affected PrP-SCA7-c92Q mice revealed abundant NIs in almost all regions of the cortex, cerebellum, and brainstem (Garden et al., 2002). To assess the timing of aggregate formation relative to the onset of neurological dysfunction, we analyzed juvenile mice for behavioral abnormalities with a panel of tests, the most sensitive of which turned out to be the rotarod. For mice from our PrP-SCA7-c92Q line 6076 that express the mutant ataxin-7 protein at levels twice those of endogenous, we did not appreciate any visible gait ataxia but did nonetheless observe onset of the incoordination phenotype based upon rotarod testing at eight weeks of age. By 12 weeks of age, such mice become visibly ataxic and fail miserably on the rotarod. Immunostaining of cerebellar sections from line 6076 mice around this age revealed abundant NIs in neurons of the molecular layer and granule cell layer of the cerebellum. However, at eight weeks of age, such immunostaining studies only detected occasional NIs and also showed diffuse nuclear staining of granule cell layer neurons, suggesting accumulation of the polyglutamine-expanded ataxin-7 (Garden et al., 2002). Based on these results, we concluded that onset of neurological dysfunction in our SCA7 mouse model best correlates with the accumulation of misfolded ataxin-7 92Q protein and that formation of NIs occurs concomitantly. This finding suggests that NI formation is a marker for the presence of abnormally folded polyglutamine-expanded protein that has exceeded the limits of the chaperone–proteasome degradation pathway. That neurological dysfunction precedes onset of a significant aggregate burden in our SCA7 mice argues that the production of soluble misfolded polyglutamine-expanded ataxin-7 protein is the initial event in disease pathogenesis.

Another process that appears to be relevant to a number of CAG/polyglutamine repeat diseases is proteolysis. Cleavage of polyglutamine-expanded disease-causing proteins to yield amino-terminal peptide fragments has been documented in patients with HD, SBMA, and DRPLA (DiFiglia et al., 1997; Ross et al., 1997; Li et al., 1998; Gutekunst et al., 1999; Schilling et al., 1999b). In our mouse model of SCA7, we tested the ability of amino-terminal and carboxy-terminal directed ataxin-7 antibodies to detect ataxin-7-containing NIs, and found that only the amino-terminal directed antibody was capable of detecting NIs (Garden et al., 2002). Our results are consistent with studies performed on another mouse model of SCA7, where a carboxy-terminal directed antibody also was not immunoreactive with ataxin-7-containing NIs (Yvert et al., 2000). Of course, such immunostaining studies must be interpreted cautiously, as lack of immunoreactivity could be due to conformational changes that prevent accessibility to the C-terminal epitope. However, we went on to perform Western blot analysis on nuclear extracts of proteins from the CNS using our amino-terminal directed antibody (K), and observed a specific truncation fragment migrating at ~55 kDa in the PrP-SCA7-c92Q mice (Garden et al., 2002). Western blot analysis of a panel of tissues from transgenic mice and from humans, with the 1C2 antibody (Trottier et al., 1995), verified the presence of the amino-terminal truncation product as a polyglutamine-expanded peptide fragment. Interestingly, the abundance of the amino-terminal product was greatest in non-neuronal tissue, suggesting that the expression of the responsible protease may be high outside the CNS, but relatively low in the CNS, or that the cleavage product becomes rapidly insoluble in CNS tissues. While the exact role of proteolytic cleavage in SCA7 remains ill defined, much work has been performed upon the role of proteolysis in HD. Although the full-length huntingtin protein is quite large, consisting of 3,144 amino acids, the polyglutamine-expanded huntingtin protein appears to be subject to progressive proteolytic cleavage to ultimately yield a toxic polyglutamine-containing fragment with as few as 100–150 amino acids. Previous studies have established that proteolysis of huntingtin is mediated by caspases, cysteine-directed aspartyl-directed proteases well known for their role in executing the final phases of programmed cell death (Stennicke and Salvesen, 1998). In the case of HD, caspase 3 has been implicated as a likely protease, and is believed to be cleaving huntingtin at aspartic acid residues at positions 513 and 552 (Goldberg et al., 1996; Wellington et al., 1998, 2002). In addition to caspases, another family of proteases, known as the calpains, have been shown to cleave huntingtin (Kim et al., 2001; Gafni and Ellerby, 2002). Indeed, a recent study of huntingtin proteolytic cleavage showed that generation of the final amino-terminal 104–114 amino acid fragment involves a non-caspase aspartyl protease (Lunkes et al., 2002). In HD, proteolytic cleavage is viewed as an early event in disease pathogenesis, predisposing to the accumulation and nuclear localization of abnormally folded polyglutamine-containing huntingtin amino-terminal peptide fragments. Atrophin-1, the gene involved in DRPLA, is also cleaved near its amino-terminus by caspase-3 during apoptosis, although the cleavage event is not dependent on repeat length (Wellington et al., 1998). Like many other caspase targets, cleavage of atrophin-1 may allow it to induce the DNA degradation involved in apoptosis (Miyashita et al., 1997). In SBMA, proteolytic processing of expanded AR increases as a function of the repeat length, and is important for AR-induced cellular toxicity (Ellerby et al., 1999). Studies of ataxin-7 in vitro and of SCA7 patient material suggest that active caspase 3 may be recruited into NIs along with polyglutamine-expanded ataxin-7 (Zander et al., 2001). However, as the role of proteolysis in SCA7 pathogenesis remains unclear, further studies of our SCA7 mouse model and other available SCA7 mouse models are needed to address this question.

Another issue that remains to be fully clarified in the polyglutamine repeat diseases is whether or not neurodegeneration is cell autonomous ("suicide") or non-cell autonomous ("murder") or both. In amyotrophic lateral sclerosis (ALS), transgenic expression of disease-causing superoxide dismutase 1 (SOD1) proteins restricted to neurons is insufficient to cause neurodegeneration in mice (Pramatarova et al., 2001). In HD, recent studies have suggested the importance of trophic support from cortico-striatal projection neurons as a key factor in medium spiny neuron dysfunction (Zuccato et al., 2001). By utilizing the MoPrP expression system for our SCA7 mouse model, we created transgenic mice that express ataxin-7 24Q or 92Q throughout most tissues of the body and in all tissues of the CNS (neurons, glia, astrocytes), but with one notable exception. Due to a deletion of PrP flanking sequence in the creation of the final MoPrP vector, expression of the transgene is typically not observed in Purkinje cells of the cerebellum (Fischer et al., 1996). Thus, in our SCA7 mouse model, we were in a position to determine if lack of expression of polyglutamine-expanded ataxin-7 in Purkinje cells would protect them. Our immunohistochemical analysis of cerebellar sections from the PrP-SCA7-c92Q line 6076 mice indicated that, to the contrary, Purkinje cells are quite severely affected, with loss of calbindin staining becoming dramatically apparent by eight weeks of age – long before the onset of a visible phenotype (Garden et al., 2002). Indeed, by 13 weeks of age, when most of these mice are just becoming visibly ataxic, almost all of the calbindin staining is lost and marked degeneration of Purkinje cell dendritic arbors is apparent (Garden et al., 2002). The dramatic Purkinje cell degeneration is accompanied by extensive ataxin-7 antibody labeling of NIs in all neurons of the cerebellum, except for the Purkinje cells. Furthermore, ultrastructural analysis of the Purkinje cells reveals marked histopathology of the Purkinje cell soma with accompanying degenerative changes (Fig. 2). These findings led us to conclude that the Purkinje cell degeneration in our SCA7 mice is non-cell-autonomous, and thus that non-cell-autonomous processes could be operating in this and other polyglutamine repeat diseases. In another mouse model of SCA7 where expression of ataxin-7 90Q was restricted to photoreceptors, post-synaptic degenerative changes were observed in bipolar neurons, similarly supporting a role for non-cell-autonomous "'trans-neuronal" pathology in SCA7 (Yvert et al., 2000).

How then do we account for the non cell-autonomous Purkinje cell degeneration observed in our mouse model of SCA7? The Purkinje cells are the only neurons in the cerebellum not expressing the expanded ataxin-7 and yet they are one of the most severely affected populations. In SCA7 patients, the Purkinje cells are also severely affected, and yet NIs are not promi-

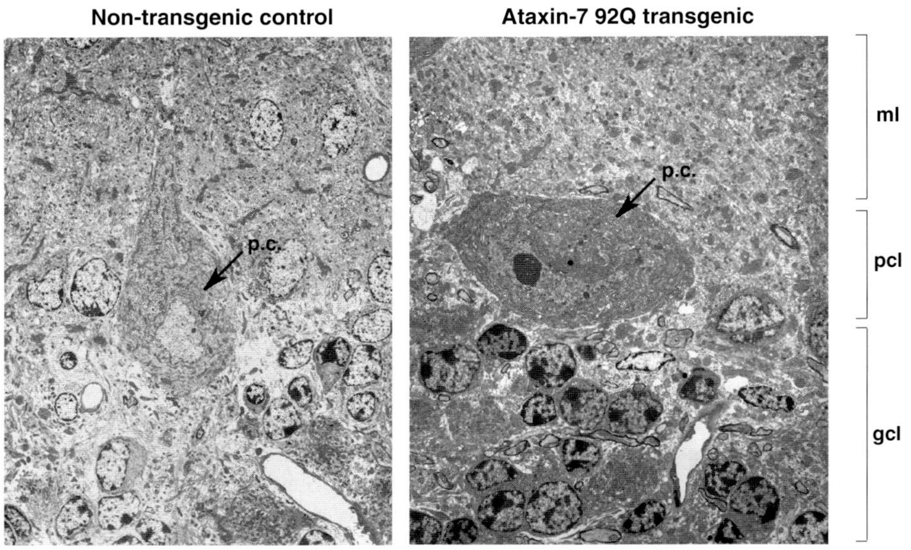

Fig. 2. Ultrastructural analysis of the SCA7 transgenic mice reveals evidence for Purkinje cell degeneration. To confirm the degenerative nature of the neuropathology in the PrP-SCA7-c92Q mice, transmission electron microscopy was performed on 20 week-old SCA7 line 6076 mice along with age-matched littermate controls. Although the Purkinje cells of non-transgenic mice exhibit normal cellular morphology, the Purkinje cells of the SCA7 transgenic mice display an altered appearance with soma aligned horizontally within the Purkinje cell layer, instead of radially. Purkinje cells from ataxin-7 92Q mice also show dilatation of the endoplasmic reticulum. p.c. = Purkinje cell; ml = molecular layer; pcl = Purkinje cell layer; gcl = granule cell layer.

nent in these cells (Holmberg et al., 1998). Indeed, examination of cerebellar sections from patients with SCA1, SCA2, SCA3, and DRPLA also indicated that NIs are uncommon in the Purkinje cell population (Koyano et al., 2002). The cause of the Purkinje cell degeneration in our SCA7 mouse model could be withdrawal of trophic factors as the surrounding cells are affected by the toxicity of the mutant protein. In the case of SCA7, candidate trophic factors such as brain derived neurotrophic factor (BDNF) or glia derived neurotrophic factor (GDNF) could be involved. Purkinje cells possess an enormous number of synaptic contacts, both from cells within the cerebellum and outside the cerebellum, suggesting that trophic support may be especially important for this neuron population. BDNF, which is known to be important for granule cell survival via the PI-3 kinase/Akt pathway (Dudek et al., 1997), may be normally supplied at high levels by these numerous contacts, and could diminish in the face of mutant ataxin-7 induced neurotoxicity, resulting in degeneration. GDNF, produced by the neighboring Bergmann glia, has actually been found to support the development and survival of cerebellar Purkinje cells (Mount et al., 1995). Complete ablation of Bergman glia in transgenic mice has further been shown to produce a cerebellar phenotype very similar to that seen in SCA7 patients and in our SCA7 mouse model (Cui et al., 2001). Polyglutamine-expanded ataxin-1, the protein involved in SCA1, can down-regulate the Purkinje cell-specific glutamate transporter EAAT4, suggesting that in our mouse model, ataxin-7 could be interfering with the ability of Bergmann glia to take up glutamate. Accumulation of glutamate in Purkinje cell synaptic clefts could produce glutamate excitotoxicity, a process that has long been implicated in HD (Ellerby, 2002). In HD, NMDA receptor subunit expression pattern has been suggested to predispose medium spiny neurons in the striatum to mutant huntingtin's toxic effects, as these neurons express the NR1A/NR2B type of NMDA receptors (Zeron et al., 2002). Mutant huntingtin was shown to increase apoptotic cell death when the neurons possessing such a receptor complement were exposed to NMDA. As this same complement of NMDA glutamate receptors is present in the medium spiny striatal neurons, selective vulnerability to glutamate excitotoxicity could explain why medium spiny neurons are specifically affected in HD. In a YAC transgenic mouse model of HD, medium spiny striatal neurons do appear more sensitive to excitotoxic insult than neurons in other regions of the brain expressing different NMDA-R subtypes, further supporting the role of NR2B in degeneration in HD (Zeron et al., 2002). Further studies should allow us to determine whether glutamate excitotoxicity due to impaired function of the Bergmann glia or loss of trophic factor support from neuronal afferents or efferents (or glia) accounts for the non-cell-autonomous Purkinje cell degeneration observed in our SCA7 transgenic mice.

Studies of retinal degeneration in the SCA7 mouse model implicate transcription dysregulation in polyglutamine disease pathogenesis

SCA7 is classified as an ADCA type II since affected patients display retinal involvement in addition to cerebellar neurodegeneration. Thus, we were very interested in evaluating our SCA7 mice for retinal histopathology and visual impairment. We began by performing immunohistochemistry on retinal sections, and demonstrated expression of polyglutamine-expanded ataxin-7 in all three nuclear layers of the retina: the outer nuclear (photoreceptor) layer (ONL); the inner nuclear (mostly bipolar cell) layer (INL); and the ganglion cell layer (GCL) of neurons that provide input to the occipital cortex via the optic nerve (La Spada et al., 2001). These results confirmed that ataxin-7 92Q localized to neuronal nuclei and did form nuclear aggregates in all three of these retinal layers. We next examined the retinas of transgenic mice for histological changes, and observed periodic thinning of the ONL beginning at around 11 weeks of age (La Spada et al., 2001). This periodic thinning produced areas within the ONL where the number of nuclei eventually decreased from a normal width of 12 nuclei to as few as 6 nuclei. To determine what cells were primarily being lost, we immunostained retinal sections with antibodies specific for cone pigments and showed that cone photoreceptors were dramatically decreased in number. We then performed ERGs, and found that cone cell responses were degrading prior to rod photoreceptor responses, although ultimately both types of responses disappeared, indicating that the mice were going blind. Importantly, SCA7 c92Q mice become blind long before an appreciable number of photoreceptors are lost, indicating that neuronal dysfunction accounts for the retinal phenotype. As the SCA7 mice were backcrossed onto the C57BL/6J background, we knew that the retinal degeneration could not be attributed to the rd6 mutation carried on most other inbred strains (Pittler and Baehr, 1991). We therefore concluded that the PrP-SCA7-c92Q mice accurately recapitulate the cone-rod dystrophy retinal degeneration seen in SCA7 patients.

As mentioned above, one key question in the polyglutamine disease field is why certain neurons degenerate in one disorder, while other neuron populations degenerate in other diseases, despite the fact that all the polyglutamine proteins appear to be expressed throughout the neuraxis. One possible explanation for selective vulnerability is that the protein context of each disease polypeptide accounts for these differences, due to disease-specific protein-protein interactions. Initial studies of amino-terminal huntingtin protein suggested that interaction with the huntingtin-associated protein 1 (HAP1) differed based on glutamine repeat length, and that this interaction might account for the pattern of neuronal loss in HD, based upon HAP1's restricted expression pattern (Li et al., 1995). As glutamine tracts often interact with one another when in transcription factors and co-activators, other workers have focused upon the likely interaction of polyglutamine-expanded disease proteins with transcription factors containing a glutamine tract, and have found that the CREB-binding protein (CBP) localizes to NIs in SBMA and HD (McCampbell et al., 2000; Steffan et al., 2000). To explain the selective vulnerability of the retina in SCA7 and to account for the rather unusual cone-rod dystrophy type of retinal degeneration seen in this disorder, we tested the hypothesis that the cone-rod homeobox protein (CRX) is involved in SCA7 retinal degeneration. CRX is a 299-amino-acid transcription factor with a homeodomain and glutamine rich region, and is expressed highest in retina and the pineal gland (Chen et al., 1997). Mutations in the CRX gene had been shown to cause an autosomal dominant form of cone-rod dystrophy (CORD 2) (Freund et al., 1997; Swain et al., 1997), suggesting that impairment in CRX function could be the pathway leading to cone-rod dystrophy in SCA7. We used the yeast two-hybrid assay and in vitro co-immunoprecipitation to demonstrate that ataxin-7 and CRX do indeed engage in a physical interaction (La Spada et al., 2001). We evaluated the functional significance of this interaction by determining if polyglutamine-expanded ataxin-7 could prevent CRX from activating transcription from its target consensus binding element in the rhodopsin promoter, and found that polyglutamine-expanded ataxin-7 (as well as normal ataxin-7 at high enough concentrations) could interfere with CRX's transactivation function. After verifying that CRX is not decreased or increased in the retinas of presymptomatic SCA7 transgenic mice, we used quantitative real-time RT-PCR to measure the level of expression of CRX targets and non-targets in the retinas of such presymptomatic mice (La Spada et al., 2001). We found that both cone-expressed genes and rod-expressed genes that are targets of CRX regulation were significantly decreased in retinas from presymptomatic SCA7 mice, noting that cone-expressed genes were much more decreased than the rod-expressed ones. These findings suggested to us that polyglutamine-expanded ataxin-7 interference of CRX function might explain the selective cell-type specific retinal degeneration seen in SCA7 patients. The inability of other transcription factors to interact with ataxin-7 and lack of interference of other polyglutamine-expanded disease proteins with CRX transactivation in our luciferase reporter assay led us to conclude that inhibition of CRX function by polyglutamine-expanded ataxin-7 is likely to be specific. Recently, another group produced SCA7 knock-in mice expressing ataxin-7 with 266 glutamines, and observed retinal degeneration in these mice without apparent involvement of CRX (Yoo et al., 2003). However, the question remains as to whether such an excessively long glutamine tract would alter a gene product's repertoire of protein-protein interactions, accounting for the discrepancy.

The interaction of ataxin-7 with CRX, and its resultant interference with CRX-mediated gene transcription supports one of the recent hypotheses concerning the mechanism of polyglutamine neurotoxicity. This hypothesis states that polyglutamine-expanded proteins interfere with transcription of genes important for neuronal survival by blocking the action of transcription factors, some of which are hypothesized to interact with polyglutamine disease proteins because the transcription factors themselves contain glutamine tracts or glutamine-rich regions. Thus, this "transcription dysregulation" hypothesis is predicated upon the expectation of a protein-protein interaction. To date, numerous such transcription interferences have been proposed. The transcription co-factor TAFII130 has been shown to be subject to interference by polyglutamine-expanded huntingtin and atrophin-1 (Shimohata et al., 2000). In SCA1, polyglutamine-binding protein 1 (PQBP1) has been shown to interact with polyglutamine-expanded ataxin-1 in unison with the carboxy-terminal domain of RNA polymerase II to yield a cessation of normal transcriptional function, suggesting that this might occur in the restricted population of cerebellar neurons where PQBP1 is believed to be expressed (Okazawa et al., 2002). As noted above, CBP has been implicated in HD and SBMA, and evidence for its transcriptional interference by polyglutamine-expanded versions of these two disease proteins has been reported (McCampbell et al., 2000; Nucifora et al., 2001). Recently, huntingtin was also shown to physically interact with the general transcription factor Sp1 and to inhibit its ability to transactivate physiologically important targets in striatal neurons (Dunah et al., 2002). Thus, our finding of polyglutamine-expanded ataxin-7 interference in the retinas of our SCA7 mouse model is consistent with a growing body of evidence that transcriptional alterations are fundamental to polyglutamine disease pathogenesis.

Although initially many investigators hypothesized that aberrant interaction of polyglutamine-expanded disease pro-

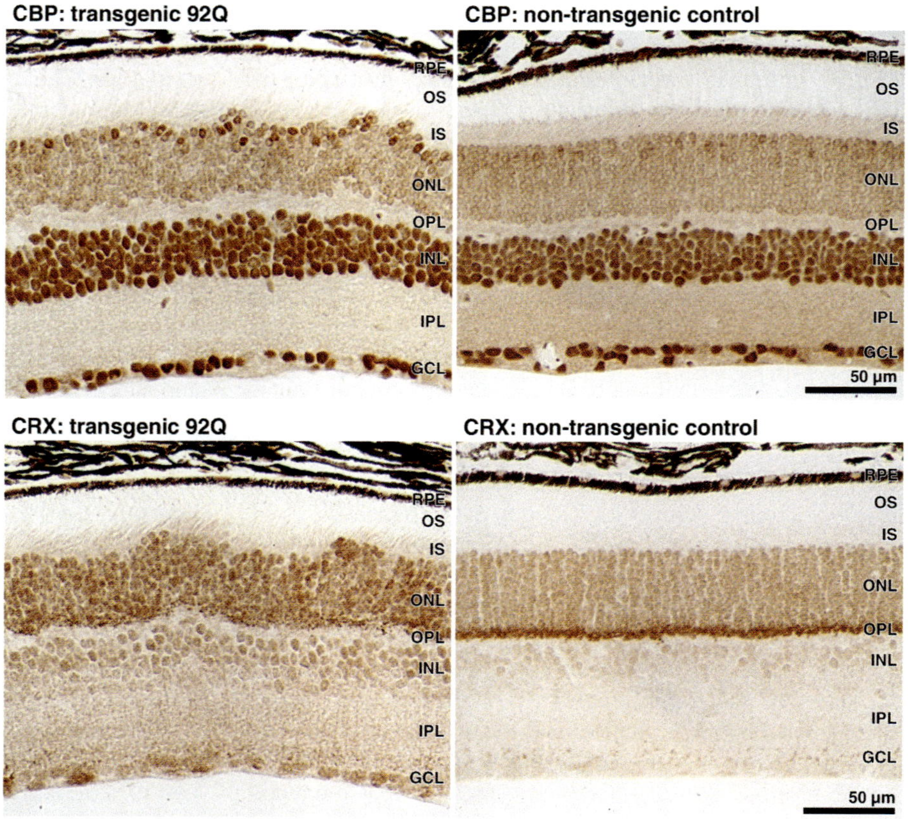

Fig. 3. Immunostaining of retinal sections from SCA7 transgenic and non-transgenic mice. Retinal sections from 19 week-old PrP-SCA7-c92Q line 6076 mice or age-matched non-transgenic littermate controls were mounted in OCT compound and frozen prior to cutting to a thickness of 12 µm on a cryostat. Immunohistochemistry was performed after blocking in 2% horse serum, 1% BSA, and 0.1% Triton X100 in PBS. Primary antibodies directed against CBP (Santa Cruz) or CRX (La Spada et al., 2001) were diluted to 1:200 in 0.3% Triton X100 in PBS, and then visualized with a secondary antibody conjugated with diaminobenzidine using the ABC Vector elite kit (Vector labs). Both the CBP immunostaining and the CRX immunostaining yield increased diffuse staining in the outer nuclear layer (ONL) of photoreceptors. Occasional punctate, inclusion-like bodies are seen upon CBP staining in the SCA7 c92Q transgenic mice. CRX immunostaining, however, reveals increased diffuse and particulate staining in the SCA7 c92Q transgenic mice, but does not label typical inclusions. This pattern of staining suggests limited sequestration of CRX into nuclear aggregates in the SCA7 transgenic mice, but instead supports the idea that an effect upon or interaction with these transcription factors is taking place in the soluble phase. RPE = retinal pigmented epithelium; OS = outer segments; IS = inner segments; ONL = outer nuclear layer; OPL = outer plexiform layer; INL = inner nuclear layer; IPL = inner plexiform layer; GCL = ganglion cell layer.

teins with transcription factors and other regulatory proteins may be causing these factors to be titrated away from the soluble phase by sequestration into insoluble aggregates, further work is now suggesting otherwise. Indeed, two recent studies of the interaction of Sp1 with huntingtin have both concluded that Sp1 is not being sequestered into NIs but rather is being interfered with in the soluble phase (Dunah et al., 2002; Li et al., 2002). Our studies of ataxin-7 interference of CRX transcription function in the SCA7 transgenic mice also support such a model. In the retinas of SCA7 transgenic mice, we have reported the presence of NIs in all three nuclear layers, and have noted the presence of various chaperone proteins and proteasome pathway members in the ataxin-7-containing NIs (La Spada et al., 2001). We have even detected CBP in these retinal NIs (Fig. 3). However, our immunostaining studies of SCA7 transgenic retinas with anti-CRX antibodies has not localized CRX to the NIs but rather to discrete regions of the nucleoplasm in a dense, but non-punctate distribution in the PrP-SCA7-c92Q mice (Fig. 3). If the interaction of polyglutamine-expanded proteins is occurring in the soluble phase, then what is the role of aggregates in the disease pathway? In our view of polyglutamine pathogenesis, the formation of aggregates may be simultaneously viewed as a protective "first line of defense" pathway taking toxic proteins out of the soluble phase AND as an overflow pathway when the chaperone/ubiquitin-proteasome pathway can no longer keep up with the accumulation of toxic protein in the soluble phase (Fig. 4). Evidence in support of this model comes from studies of the SCA1 mouse model in which ataxin-1 was mutated to prevent self-association. Despite losing their ability to form aggregates, these mice still developed the typical Purkinje cell degeneration, but without any sign of the hallmark NIs (Klement et al., 1998). Further evidence for a protective role for aggregation is the observation that crossing the SCA1 transgenic mice onto an E6-AP ubiquitin ligase null background actually reduced the amount of NIs, while worsening the severity of the neurodegeneration (Cummings et al., 1999). A cell culture model of HD has also demonstrated nuclear aggregates of mutant huntingtin protein with no apparent cellular toxicity (Saudou et al., 1998). In a recent SCA1 knock-in mouse model, it was shown that the amount of soluble polyglutamine-expanded ataxin-1 correlated best with cellular pathogenesis, and that sequestration into insoluble aggregates occurs in neuronal populations not subject to degeneration (Watase et al., 2002). All of this work is consistent with polyglutamine-expanded proteins causing pathogenesis when in the soluble phase (Fig. 4). At the same time, there is abundant evidence that formation of NIs is inextricably linked to disease pathogenesis in a variety of Drosophila, mouse, worm, and yeast models, and that aggregate burden closely correlates with disease severity (Ross, 1997; Schilling et al., 1999a; Warrick et al., 1999; Perutz and Windle, 2001). These observations point to a process whereby soluble toxic protein rapidly accumulates and overwhelms the ability of chaperones and the protein turnover pathway to process the soluble monomers, yielding nuclear and cytoplasmic aggregates. While these aggregates may not cause the primary pathology, it seems likely that their

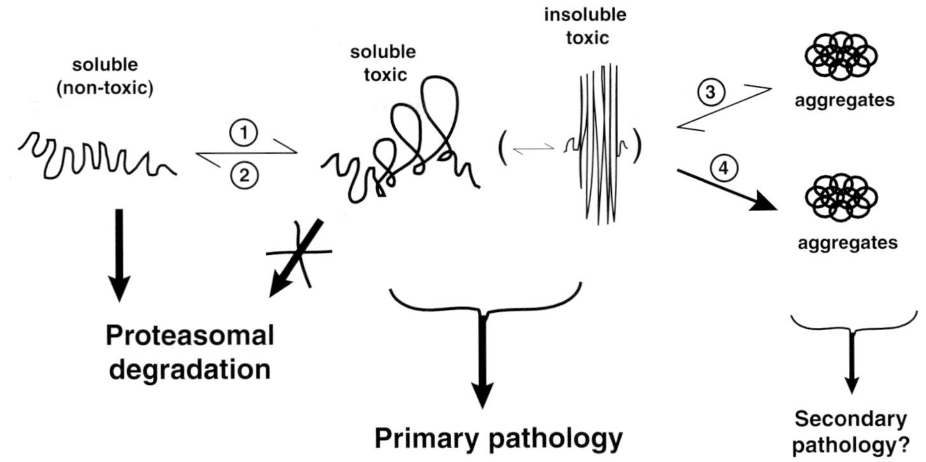

Fig. 4. An integrated model for polyglutamine neurotoxicity. Based on our studies and the work of many others, we propose that proteins or peptides containing expanded polyglutamine tracts exist in a variety of forms in both the soluble and insoluble phases. A soluble, non-toxic form of expanded polyglutamine-containing proteins exists in equilibrium with a soluble toxic form which may exist in equilibrium with an insoluble toxic form that can be converted into a stable insoluble conformer visualized as an aggregate or inclusion body. In this model, the soluble toxic form is subject to proteasomal degradation, but the soluble non-toxic form can not be degraded, due to an alteration in its tertiary structure, so it accumulates. In some cases, certain events (such as proteolytic cleavage; indicated as ①) catalyze the transition to the altered conformation, while other factors (such as refolding by chaperones; indicated as ②) favor the non-toxic, degradable state. Some studies suggest that an insoluble protofibrillar, perhaps amyloid-like, perhaps "microaggregated" state is transitioned to, but this remains to be proven. We propose that the soluble toxic form of the polyglutamine-expanded protein or peptide (in possible rapid equilibrium with a protofibrillar insoluble form) mediates the primary pathology through events, interactions, and pathways that vary from disease to disease (accounting for the cell-type specificity seen in each disorder). An insoluble stable sequestered form known as "aggregates" or "nuclear inclusions" also exists; however, our review of the data suggests two pathways for the production of the stable insoluble form: a reversible, "protective" pathway (indicated as ③), that is driven by the ubiquitination pathway and which accounts for findings suggesting that the amount of soluble polyglutamine-expanded protein is proportional to the degree of cell-type specific toxicity (Watase et al., 2002); and an "overflow" pathway, (indicated as ④), in which excessively accumulated polyglutamine-expanded protein simply overwhelms the ability of the cell to maintain it in its soluble state and thus is complexed into the aggregated state as a last resort. This finding is consistent with our observation of aggregates developing in the cerebellum of our SCA7 transgenic mice, just as the accumulation of polyglutamine-expanded ataxin-7 is accelerating. The aggregates are viewed as capable of producing toxicity in this model, but we believe that this toxicity is usually a secondary event, and that this type of toxicity employs a common pathway, accounting for the non-specific nature of polyglutamine neurotoxicity (Ordway et al., 1997). The non-specific nature of polyglutamine toxicity is suggested by the overlapping patterns of pathology seen in juvenile-onset versions of these diseases and mouse models thereof – where we believe that this secondary, common pathway overtakes the primary pathology pathway in importance.

continued production can be secondarily deleterious to nuclear and/or cytosolic function (Fig. 4).

Another aspect of our SCA7 mouse model that bears relevance to an understanding of polyglutamine pathology is the observation that polyglutamine-expanded ataxin-7 localizes to the nucleus early on in the disease process, and likely initiates the processes of retinal and cerebellar degeneration there. We have not yet tested the necessity of nuclear localization in driving SCA7 pathology, but, when using our anti-ataxin-7 antibodies, we do first note accumulation of ataxin-7 in the nucleus, just at the time when pathology is starting. In other mouse models of SCA7, translocation of polyglutamine-expanded ataxin-7 from the cytoplasm to the nucleus has similarly been shown to precede onset of the behavioral phenotype and histopathology (Yvert et al., 2001). Studies of SCA1 mice and HD neuron culture models have directly tested the importance of nuclear localization for toxicity, and have found that nuclear localization is absolutely required for neurodegeneration (Klement et al., 1998; Saudou et al., 1998). On the other hand, certain of the polyglutamine repeat diseases (namely, SCA2 and SCA6) feature cytoplasmic aggregates, suggesting that cellular pathogenesis may play out in the cytoplasm in some cases (Huynh et al., 2000; Ishikawa et al., 1999). Our mouse model, together with the detection of NIs in SCA7 patients and in other SCA7 mouse models, identifies the nucleus as the likely key site of disease pathogenesis in SCA7. The importance of the nucleus as a likely site of disease protein action for many of the polyglutamine disorders has been strengthened by the finding that expanded polyglutamine tracts induce a histone acetylation defect in yeast and fruitfly models. In yeast, expanded polyglutamines produce a defect that recapitulates the molecular alterations observed in mutants lacking components of the SAGA complex, a regulatory protein complex that activates transcription by acetylating histones (Hughes et al., 2001). In Drosophila, amino-terminal huntingtin fragments containing an expanded glutamine tract and the adjacent polyproline region were shown to directly interact with the acetyltransferase domain of two transcription factor co-activators, CBP and p/CAF (Steffan et al., 2001). Most interestingly, the retinal neurodegeneration induced by directing expression of these mutant huntingtin fragments to the fly eye could be prevented by feeding the flies histone deacetylase inhibitors (HDAC I's). All of these findings suggest that the polyglutamine expansion diseases that show nuclear localization and NIs may be causing disease by a general deleterious effect upon transcription regulation. In the case of SCA7 retinal degeneration, this transcriptional dysregulation is likely coupled to a specific defect of CRX-regulated transactivation based upon our mouse model findings.

Capitalizing upon the SCA7 mouse model: Towards polyglutamine disease therapies

With the creation of a representative mouse model of SCA7, development as well as testing of possible therapeutic interventions has become possible. Our SCA7 mouse model has three key features that make it an excellent choice for this work: i) the SCA7 mice recapitulate both brainstem degeneration and retinal degeneration; ii) the retinal degeneration in the SCA7 mice is an accurate representation of the human SCA7 phenotype, and iii) a retinal phenotype facilitates delivery of therapeutic agents as well as an assessment of their efficacy, since the retina is highly accessible and the disease process may be objectively followed by a non-invasive electrophysiological procedure (i.e. ERG) that can be repeated multiple times. Although therapeutic trials are laborious and require numerous mice, we believe that the SCA7 mouse model should serve as an excellent proving ground for evaluating therapies suggested by advances in our understanding of polyglutamine neurotoxicity. As adeno-associated virus (AAV) has been shown to be highly effective at transducing photoreceptor cells, recombinant AAV (rAAV) vectors containing gene products whose expression might be expected to prevent or reverse the disease pathology in the retinas of our mice could be delivered by subretinal injection. While the idea of reversing a degenerative process might sound unrealistic, one study has shown that turning off expression of mutant huntingtin protein in HD transgenic mice did permit symptomatic mice to regain apparently normal function and degrade aggregated polyglutamine-containing inclusions (Yamamoto et al., 2000). This finding holds out hope for recovery of function, even in instances of pronounced symptoms and argues for the importance of testing therapies for their ability to reverse as well as to prevent the polyglutamine disease process in mouse models, such as our SCA7 transgenic one. While it has been only a little more than a decade since the first polyglutamine repeat disease mutation was discovered, the progress towards understanding the disease mechanism has been remarkable. Mice have played a critically important role in moving the field forward during this time. It is our hope that over the course of the next decade we can make great strides towards the development of effective treatments for these disorders. It is our expectation that mouse models will play a central role in this process.

Acknowledgements

The authors wish to thank J. Huang and D. Possin for their assistance.

References

Alves-Rodrigues A, Gregori L, Figueiredo-Pereira ME: Ubiquitin, cellular inclusions and their role in neurodegeneration. Trends Neurosci 21:516–520 (1998).

Becher MW, Kotzuk JA, Sharp AH, Davies SW, Bates GP, Price DL, Ross CA: Intranuclear neuronal inclusions in Huntington's disease and dentatorubral and pallidoluysian atrophy: correlation between the density of inclusions and IT15 CAG triplet repeat length. Neurobiol Dis 4:387–397 (1998).

Benomar A, Krols L, Stevanin G, Cancel G, LeGuern E, David G, Ouhabi H, Martin JJ, Durr A, Zaim A, et al: The gene for autosomal dominant cerebellar ataxia with pigmentary macular dystrophy maps to chromosome 3p12→p21.1. Nature Genet 10:84–88 (1995).

Benton CS, de Silva R, Rutledge SL, Bohlega S, Ashizawa T, Zoghbi HY: Molecular and clinical studies in SCA-7 define a broad clinical spectrum and the infantile phenotype. Neurology 51:1081–1086 (1998).

Borchelt DR, Davis J, Fischer M, Lee MK, Slunt HH, Ratovitsky T, Regard J, Copeland NG, Jenkins NA, Sisodia SS, Price DL: A vector for expressing foreign genes in the brains and hearts of transgenic mice. Genet Anal 13:159–163 (1996).

Brown TR, Lubahn DB, Wilson EM, Joseph DR, French FS, Migeon CJ: Deletion of the steroid-binding domain of the human androgen receptor gene in one family with complete androgen insensitivity syndrome: evidence for further genetic heterogeneity in this syndrome. Proc natl Acad Sci, USA 85:8151–8155 (1988).

Chen S, Wang QL, Nie Z, Sun H, Lennon G, Copeland NG, Gilbert DJ, Jenkins NA, Zack DJ: Crx, a novel Otx-like paired-homeodomain protein, binds to and transactivates photoreceptor cell-specific genes. Neuron 19:1017–1030 (1997).

Cui W, Allen ND, Skynner M, Gusterson B, Clark AJ: Inducible ablation of astrocytes shows that these cells are required for neuronal survival in the adult brain. Glia 34:272–282 (2001).

Cummings CJ, Mancini MA, Antalffy B, DeFranco DB, Orr HT, Zoghbi HY: Chaperone suppression of aggregation and altered subcellular proteasome localization imply protein misfolding in SCA1. Nature Genet 19:148–154 (1998).

Cummings CJ, Reinstein E, Sun Y, Antalffy B, Jiang Y, Ciechanover A, Orr HT, Beaudet AL, Zoghbi HY: Mutation of the E6-AP ubiquitin ligase reduces nuclear inclusion frequency while accelerating polyglutamine-induced pathology in SCA1 mice. Neuron 24:879–892 (1999).

David G, Abbas N, Stevanin G, Durr A, Yvert G, Cancel G, Weber C, Imbert G, Saudou F, Antoniou E, Drabkin H, Gemmill R, Giunti P, Benomar A, Wood N, Ruberg M, Agid Y, Mandel JL, Brice A: Cloning of the SCA7 gene reveals a highly unstable CAG repeat expansion. Nature Genet 17:65–70 (1997).

Davies SW, Turmaine M, Cozens BA, DiFiglia M, Sharp AH, Ross CA, Scherzinger E, Wanker EE, Mangiarini L, Bates GP: Formation of neuronal intranuclear inclusions underlies the neurological dysfunction in mice transgenic for the HD mutation. Cell 90:537–548 (1997).

DiFiglia M, Sapp E, Chase KO, Davies SW, Bates GP, Vonsattel JP, Aronin N: Aggregation of huntingtin in neuronal intranuclear inclusions and dystrophic neurites in brain. Science 277:1990–1993 (1997).

Dragatsis I, Levine MS, Zeitlin S: Inactivation of Hdh in the brain and testis results in progressive neurodegeneration and sterility in mice. Nature Genet 26:300–306 (2000).

Dudek H, Datta SR, Franke TF, Birnbaum MJ, Yao R, Cooper GM, Segal RA, Kaplan DR, Greenberg ME: Regulation of neuronal survival by the serine-threonine protein kinase Akt. Science 275:661–665 (1997).

Dunah AW, Jeong H, Griffin A, Kim YM, Standaert DG, Hersch SM, Mouradian MM, Young AB, Tanese N, Krainc D: Sp1 and TAFII130 transcriptional activity disrupted in early Huntington's disease. Science 296:2238–2243 (2002).

Ellerby LM: Hunting for excitement: NMDA receptors in Huntington's disease. Neuron 33:841–842 (2002).

Ellerby LM, Hackam AS, Propp SS, Ellerby HM, Rabizadeh S, Cashman NR, Trifiro MA, Pinsky L, Wellington CL, Salvesen GS, Hayden MR, Bredesen DE: Kennedy's disease: caspase cleavage of the androgen receptor is a crucial event in cytotoxicity. J Neurochem 72:185–195 (1999).

Enevoldson TP, Sanders MD, Harding AE: Autosomal dominant cerebellar ataxia with pigmentary macular dystrophy. A clinical and genetic study of eight families. Brain 117:445–460 (1994).

Fischer M, Rulicke T, Raeber A, Sailer A, Moser M, Oesch B, Brandner S, Aguzzi A, Weissmann C: Prion protein (PrP) with amino-proximal deletions restoring susceptibility of PrP knockout mice to scrapie. EMBO J 15:1255–1264 (1996).

Freund CL, Gregory-Evans CY, Furukawa T, Papaioannou M, Looser J, Ploder L, Bellingham J, Ng D, Herbrick JA, Duncan A, Scherer SW, Tsui LC, Loutradis-Anagnostou A, Jacobson SG, Cepko CL, Bhattacharya SS, McInnes RR: Cone-rod dystrophy due to mutations in a novel photoreceptor-specific homeobox gene (CRX) essential for maintenance of the photoreceptor. Cell 91:543–553 (1997).

Gafni J, Ellerby LM: Calpain activation in Huntington's disease. J Neurosci 22:4842–4849 (2002).

Garden GA, Libby RT, Fu YH, Kinoshita Y, Huang J, Possin DE, Smith AC, Martinez RA, Fine GC, Grote SK, Ware CB, Einum DD, Morrison RS, Ptacek LJ, Sopher BL, La Spada AR: Polyglutamine-expanded ataxin-7 promotes non-cell-autonomous purkinje cell degeneration and displays proteolytic cleavage in ataxic transgenic mice. J Neurosci 22:4897–4905 (2002).

Goldberg YP, Nicholson DW, Rasper DM, Kalchman MA, Koide HB, Graham RK, Bromm M, Kazemi Esfarjani P, Thornberry NA, Vaillancourt JP, Hayden MR: Cleavage of huntingtin by apopain, a proapoptotic cysteine protease, is modulated by the polyglutamine tract. Nature Genet 13:442–449 (1996).

Gutekunst CA, Li SH, Yi H, Mulroy JS, Kuemmerle S, Jones R, Rye D, Ferrante RJ, Hersch SM, Li XJ: Nuclear and neuropil aggregates in Huntington's disease: relationship to neuropathology. J Neurosci 19:2522–2534 (1999).

Harding AE: The clinical features and classification of the late onset autosomal dominant cerebellar ataxias. A study of 11 families, including descendants of the "the Drew family of Walworth". Brain 105:1–28 (1982).

Holmberg M, Duyckaerts C, Durr A, Cancel G, Gourfinkel-An I, Damier P, Faucheux B, Trottier Y, Hirsch EC, Agid Y, Brice A: Spinocerebellar ataxia type 7 (SCA7): a neurodegenerative disorder with neuronal intranuclear inclusions. Hum molec Genet 7:913–918 (1998).

Hughes RE, Lo RS, Davis C, Strand AD, Neal CL, Olson JM, Fields S: Altered transcription in yeast expressing expanded polyglutamine. Proc natl Acad Sci, USA 98:13201–13206 (2001).

Huynh DP, Figueroa K, Hoang N, Pulst SM: Nuclear localization or inclusion body formation of ataxin-2 are not necessary for SCA2 pathogenesis in mouse or human. Nature Genet 26:44–50 (2000).

Ishikawa K, Fujigasaki H, Saegusa H, Ohwada K, Fujita T, Iwamoto H, Komatsuzaki Y, Toru S, Toriyama H, Watanabe M, Ohkoshi N, Shoji S, Kanazawa I, Tanabe T, Mizusawa H: Abundant expression and cytoplasmic aggregations of [alpha]1A voltage-dependent calcium channel protein associated with neurodegeneration in spinocerebellar ataxia type 6. Hum molec Genet 8:1185–1193 (1999).

Johansson J, Forsgren L, Sandgren O, Brice A, Holmgren G, Holmberg M: Expanded CAG repeats in Swedish spinocerebellar ataxia type 7 (SCA7) patients: effect of CAG repeat length on the clinical manifestation. Hum molec Genet 7:171–176 (1998).

Kim YJ, Yi Y, Sapp E, Wang Y, Cuiffo B, Kegel KB, Qin ZH, Aronin N, DiFiglia M: Caspase 3-cleaved N-terminal fragments of wild-type and mutant huntingtin are present in normal and Huntington's disease brains, associate with membranes, and undergo calpain-dependent proteolysis. Proc natl Acad Sci, USA 98:12784–12789 (2001).

Klement IA, Skinner PJ, Kaytor MD, Yi H, Hersch SM, Clark HB, Zoghbi HY, Orr HT: Ataxin-1 nuclear localization and aggregation: role in polyglutamine-induced disease in SCA1 transgenic mice. Cell 95:41–53 (1998).

Koyano S, Iwabuchi K, Yagishita S, Kuroiwa Y, Uchihara T: Paradoxical absence of nuclear inclusion in cerebellar Purkinje cells of hereditary ataxias linked to CAG expansion. J Neurol Neurosurg Psychiatry 73:450–452 (2002).

La Spada AR, Fu Y, Sopher BL, Libby RT, Wang X, Li LY, Einum DD, Huang J, Possin DE, Smith AC, Martinez RA, Koszdin KL, Treuting PM, Ware CB, Hurley JB, Ptacek LJ, Chen S: Polyglutamine-expanded ataxin-7 antagonizes CRX function and induces cone-rod dystrophy in a mouse model of SCA7. Neuron 31:913–927 (2001).

Li M, Miwa S, Kobayashi Y, Merry DE, Yamamoto M, Tanaka F, Doyu M, Hashizume Y, Fischbeck KH, Sobue G: Nuclear inclusions of the androgen receptor protein in spinal and bulbar muscular atrophy. Ann Neurol 44:249–254 (1998).

Li SH, Cheng AL, Zhou H, Lam S, Rao M, Li H, Li XJ: Interaction of Huntington disease protein with transcriptional activator Sp1. Mol cell Biol 22: 1277–1287 (2002).

Li XJ, Li SH, Sharp AH, Nucifora FC Jr, Schilling G, Lanahan A, Worley P, Snyder SH, Ross CA: A huntingtin-associated protein enriched in brain with implications for pathology. Nature 378:398–402 (1995).

Lindenberg KS, Yvert G, Muller K, Landwehrmeyer GB: Expression analysis of ataxin-7 mRNA and protein in human brain: evidence for a widespread distribution and focal protein accumulation. Brain Pathol 10:385–394 (2000).

Lunkes A, Lindenberg KS, Ben-Haiem L, Weber C, Devys D, Landwehrmeyer GB, Mandel JL, Trottier Y: Proteases acting on mutant huntingtin generate cleaved products that differentially build up cytoplasmic and nuclear inclusions. Mol Cell 10: 259–269 (2002).

McCampbell A, Taylor JP, Taye AA, Robitschek J, Li M, Walcott J, Merry D, Chai Y, Paulson H, Sobue G, Fischbeck KH: CREB-binding protein sequestration by expanded polyglutamine. Hum molec Genet 9:2197–2202 (2000).

Miyashita T, Okamura-Oho Y, Mito Y, Nagafuchi S, Yamada M: Dentatorubral pallidoluysian atrophy (DRPLA) protein is cleaved by caspase-3 during apoptosis. J biol Chem 272:29238–29242 (1997).

Monckton DG, Cayuela ML, Gould FK, Brock GJ, Silva R, Ashizawa T: Very large (CAG)(n) DNA repeat expansions in the sperm of two spinocerebellar ataxia type 7 males. Hum molec Genet 8: 2473–2478 (1999).

Mount HT, Dean DO, Alberch J, Dreyfus CF, Black IB: Glial cell line-derived neurotrophic factor promotes the survival and morphologic differentiation of Purkinje cells. Proc natl Acad Sci, USA 92: 9092–9096 (1995).

Nakamura K, Jeong SY, Uchihara T, Anno M, Nagashima K, Nagashima T, Ikeda S, Tsuji S, Kanazawa I: SCA17, a novel autosomal dominant cerebellar ataxia caused by an expanded polyglutamine in TATA-binding protein. Hum molec Genet 10: 1441–1448 (2001).

Nucifora FC Jr, Sasaki M, Peters MF, Huang H, Cooper JK, Yamada M, Takahashi H, Tsuji S, Troncoso J, Dawson VL, Dawson TM, Ross CA: Interference by huntingtin and atrophin-1 with cbp-mediated transcription leading to cellular toxicity. Science 291:2423–2428 (2001).

Okazawa H, Rich T, Chang A, Lin X, Waragai M, Kajikawa M, Enokido Y, Komuro A, Kato S, Shibata M, Hatanaka H, Mouradian MM, Sudol M, Kanazawa I: Interaction between mutant ataxin-1 and PQBP-1 affects transcription and cell death. Neuron 34:701–713 (2002).

Ordway JM, Tallaksen-Greene S, Gutekunst CA, Bernstein EM, Cearley JA, Wiener HW, Dure LSt, Lindsey R, Hersch SM, Jope RS, Albin RL, Detloff PJ: Ectopically expressed CAG repeats cause intranuclear inclusions and a progressive late onset neurological phenotype in the mouse. Cell 91:753–763 (1997).

Paulson HL, Perez MK, Trottier Y, Trojanowski JQ, Subramony SH, Das SS, Vig P, Mandel JL, Fischbeck KH, Pittman RN: Intranuclear inclusions of expanded polyglutamine protein in spinocerebellar ataxia type 3. Neuron 19:333–344 (1997).

Perutz MF, Windle AH: Cause of neural death in neurodegenerative diseases attributable to expansion of glutamine repeats. Nature 412:143–144 (2001).

Perutz MF, Johnson T, Suzuki M, Finch JT: Glutamine repeats as polar zippers: their possible role in inherited neurodegenerative diseases. Proc natl Acad Sci, USA 91:5355–5358 (1994).

Pittler SJ, Baehr W: Identification of a nonsense mutation in the rod photoreceptor cGMP phosphodiesterase beta-subunit gene of the rd mouse. Proc natl Acad Sci, USA 88:8322–8326 (1991).

Pramatarova A, Laganiere J, Roussel J, Brisebois K, Rouleau GA: Neuron-specific expression of mutant superoxide dismutase 1 in transgenic mice does not lead to motor impairment. J Neurosci 21:3369–3374 (2001).

Ross CA: When more is less: pathogenesis of glutamine repeat neurodegenerative diseases. Neuron 15: 493–496 (1995).

Ross CA: Intranuclear neuronal inclusions: a common pathogenic mechanism for glutamine-repeat neurodegenerative diseases? Neuron 19:1147–1150 (1997).

Ross CA, Becher MW, Colomer V, Engelender S, Wood JD, Sharp AH: Huntington's disease and dentatorubral-pallidoluysian atrophy: proteins, pathogenesis and pathology. Brain Pathol 7:1003–1016 (1997).

Sanchez I, Mahlke C, Yuan J: Pivotal role of oligomerization in expanded polyglutamine neurodegenerative disorders. Nature 421:373–379 (2003).

Saudou F, Finkbeiner S, Devys D, Greenberg ME: Huntingtin acts in the nucleus to induce apoptosis but death does not correlate with the formation of intranuclear inclusions. Cell 95:55–66 (1998).

Schilling G, Becher MW, Sharp AH, Jinnah HA, Duan K, Kotzuk JA, Slunt HH, Ratovitski T, Cooper JK, Jenkins NA, Copeland NG, Price DL, Ross CA, Borchelt DR: Intranuclear inclusions and neuritic aggregates in transgenic mice expressing a mutant N-terminal fragment of huntingtin. Hum molec Genet 8:397–407 (1999a).

Schilling G, Wood JD, Duan K, Slunt HH, Gonzales V, Yamada M, Cooper JK, Margolis RL, Jenkins NA, Copeland NG, Takahashi H, Tsuji S, Price DL, Borchelt DR, Ross CA: Nuclear accumulation of truncated atrophin-1 fragments in a transgenic mouse model of DRPLA. Neuron 24:275–286 (1999b).

Schmidt T, Lindenberg KS, Krebs A, Schols L, Laccone F, Herms J, Rechsteiner M, Riess O, Landwehrmeyer GB: Protein surveillance machinery in brains with spinocerebellar ataxia type 3: redistribution and differential recruitment of 26S proteasome subunits and chaperones to neuronal intranuclear inclusions. Ann Neurol 51:302–310 (2002).

Shimohata T, Nakajima T, Yamada M, Uchida C, Onodera O, Naruse S, Kimura T, Koide R, Nozaki K, Sano Y, Ishiguro H, Sakoe K, Ooshima T, Sato A, Ikeuchi T, Oyake M, Sato T, Aoyagi Y, Hozumi I, Nagatsu T, Takiyama Y, Nishizawa M, Goto J, Kanazawa I, Davidson I, Tanese N: Expanded polyglutamine stretches interact with TAFII130, interfering with CREB-dependent transcription. Nature Genet 26:29–36 (2000).

Steffan JS, Bodai L, Pallos J, Poelman M, McCampbell A, Apostol BL, Kazantsev A, Schmidt E, Zhu YZ, Greenwald M, Kurokawa R, Housman DE, Jackson GR, Marsh JL, Thompson LM: Histone deacetylase inhibitors arrest polyglutamine-dependent neurodegeneration in Drosophila. Nature 413: 739–743 (2001).

Steffan JS, Kazantsev A, Spasic-Boskovic O, Greenwald M, Zhu YZ, Gohler H, Wanker EE, Bates GP, Housman DE, Thompson LM: The Huntington's disease protein interacts with p53 and CREB-binding protein and represses transcription. Proc natl Acad Sci, USA 97:6763–6768 (2000).

Stennicke HR, Salvesen GS: Properties of the caspases. Biochim biophys Acta 1387:17–31 (1998).

Stenoien DL, Cummings CJ, Adams HP, Mancini MG, Patel K, DeMartino GN, Marcelli M, Weigel NL, Mancini MA: Polyglutamine-expanded androgen receptors form aggregates that sequester heat shock proteins, proteasome components and SRC-1, and are suppressed by the HDJ-2 chaperone. Hum molec Genet 8:731-741 (1999).

Swain PK, Chen S, Wang QL, Affatigato LM, Coats CL, Brady KD, Fishman GA, Jacobson SG, Swaroop A, Stone E, Sieving PA, Zack DJ: Mutations in the cone-rod homeobox gene are associated with the cone-rod dystrophy photoreceptor degeneration. Neuron 19:1329–1336 (1997).

To KW, Adamian M, Jakobiec FA, Berson EL: Olivopontocerebellar atrophy with retinal degeneration. An electroretinographic and histopathologic investigation. Ophthalmology 100:15–23 (1993).

Trottier Y, Lutz Y, Stevanin G, Imbert G, Devys D, Cancel G, Saudou F, Weber C, David G, Tora L, et al: Polyglutamine expansion as a pathological epitope in Huntington's disease and four dominant cerebellar ataxias. Nature 378:403–406 (1995).

Warrick JM, Chan HY, Gray-Board GL, Chai Y, Paulson HL, Bonini NM: Suppression of polyglutamine-mediated neurodegeneration in Drosophila by the molecular chaperone HSP70. Nature Genet 23:425–428 (1999).

Watase K, Weeber EJ, Xu B, Antalffy B, Yuva-Paylor L, Hashimoto K, Kano M, Atkinson R, Sun Y, Armstrong DL, Sweatt JD, Orr HT, Paylor R, Zoghbi HY: A long CAG repeat in the mouse *Sca1* locus replicates SCA1 features and reveals the impact of protein solubility on selective neurodegeneration. Neuron 34:905–919 (2002).

Wellington CL, Ellerby LM, Gutekunst CA, Rogers D, Warby S, Graham RK, Loubser O, van Raamsdonk J, Singaraja R, Yang YZ, Gafni J, Bredesen D, Hersch SM, Leavitt BR, Roy S, Nicholson DW, Hayden MR: Caspase cleavage of mutant huntingtin precedes neurodegeneration in Huntington's disease. J Neurosci 22:7862–7872 (2002).

Wellington CL, Ellerby LM, Hackam AS, Margolis RL, Trifiro MA, Singaraja R, McCutcheon K, Salvesen GS, Propp SS, Bromm M, Rowland KJ, Zhang T, Rasper D, Roy S, Thornberry N, Pinsky L, Kakizuka A, Ross CA, Nicholson DW, Bredesen DE, Hayden MR: Caspase cleavage of gene products associated with triplet expansion disorders generates truncated fragments containing the polyglutamine tract. J biol Chem 273:9158–9167 (1998).

Yamamoto A, Lucas JJ, Hen R: Reversal of neuropathology and motor dysfunction in a conditional model of Huntington's disease. Cell 101:57–66 (2000).

Yoo SY, Pennesi ME, Weeber EJ, Xu B, Atkinson R, Chen S, Armstrong DL, Wu S, M., Sweatt JD, Zoghbi HY: SCA7 Knockin mice model human SCA7 and reveal gradual accumulation of mutant ataxin-7 in neurons and abnormalities in short-term plasticity. Neuron 37:383–401 (2003).

Yvert G, Lindenberg KS, Devys D, Helmlinger D, Landwehrmeyer GB, Mandel JL: SCA7 mouse models show selective stabilization of mutant ataxin-7 and similar cellular responses in different neuronal cell types. Hum molec Genet 10:1679–1692 (2001).

Yvert G, Lindenberg KS, Picaud S, Landwehrmeyer GB, Sahel JA, Mandel JL: Expanded polyglutamines induce neurodegeneration and trans-neuronal alterations in cerebellum and retina of SCA7 transgenic mice. Hum molec Genet 9:2491–2506 (2000).

Zander C, Takahashi J, El Hachimi KH, Fujigasaki H, Albanese V, Lebre AS, Stevanin G, Duyckaerts C, Brice A: Similarities between spinocerebellar ataxia type 7 (SCA7) cell models and human brain: proteins recruited in inclusions and activation of caspase-3. Hum molec Genet 10:2569–2579 (2001).

Zeitlin S, Liu JP, Chapman DL, Papaioannou VE, Efstratiadis A: Increased apoptosis and early embryonic lethality in mice nullizygous for the Huntington's disease gene homologue. Nature Genet 11:155–163 (1995).

Zeron MM, Hansson O, Chen N, Wellington CL, Leavitt BR, Brundin P, Hayden MR, Raymond LA: Increased sensitivity to N-methyl-D-aspartate receptor-mediated excitotoxicity in a mouse model of Huntington's disease. Neuron 33:849–860 (2002).

Zoghbi HY, Orr HT: Glutamine repeats and neurodegeneration. A Rev Neurosci 23:217–247 (2000).

Zuccato C, Ciammola A, Rigamonti D, Leavitt BR, Goffredo D, Conti L, MacDonald ME, Friedlander RM, Silani V, Hayden MR, Timmusk T, Sipione S, Cattaneo E: Loss of huntingtin-mediated BDNF gene transcription in Huntington's disease. Science 293:493–498 (2001).

Cytogenet Genome Res 100:175–183 (2003)
DOI: 10.1159/000072852

Molecular genetics of spinocerebellar ataxia type 8 (SCA8)

A.K. Mosemiller,[a,c] J.C. Dalton,[a,c] J.W. Day[b,c] and L.P.W. Ranum[a,c]

Departments of [a] Genetics, Cell Biology, and Development, [b] Neurology, and the [c] Institute of Human Genetics, University of Minnesota, Minneapolis MN (USA)

Abstract. We previously reported that a transcribed but untranslated CTG expansion causes a novel form of ataxia, spinocerebellar ataxia type 8 (SCA8) (Koob et al., 1999). SCA8 was the first example of a dominant spinocerebellar ataxia that is not caused by the expansion of a CAG repeat translated into a polyglutamine tract. This slowly progressive form of ataxia is characterized by dramatic repeat instability and a high degree of reduced penetrance. The clinical and genetic features of the disease are discussed below.

Clinical features

Clinical and molecular analysis of over 200 individuals from 25 different SCA8 ataxia families from our collection and from the literature (Koob et al., 1999; Day et al., 2000; Ikeda et al., 2000a; Juvonen et al., 2000; Silveira et al., 2000; Brusco et al., 2002; Topisirovic et al., 2002) demonstrate that SCA8 consistently presents as a very slowly progressive ataxia that largely spares brainstem and cerebral function. Affected individuals have gait, limb, speech and oculomotor incoordination, spasticity, and sensory loss. In the largest SCA8 family reported to date (MN-A) the onset of gait abnormalities ranged from 13–60 years, and the age at which affected individuals needed a cane or a walker ranged from 35–50 years, with no one requiring a wheelchair prior to 45 years (Day et al., 2000). No affected MN-A family member with disease duration <20 years required mobility aids, while those requiring aids needed the support after 20–36 years of disease progression. Mild athetotic movements of extended fingers and intermittent low amplitude myoclonic jerks in the fingers and arms are sometimes present. Signs of oculomotor involvement are common in moderately and severely affected individuals (Day et al., 2000; Juvonen et al., 2000; Anderson et al., 2002). Speech is dysarthric with both ataxic and spastic components (Day et al., 2000; Juvonen et al., 2000). Hyperreflexia is common, with Babinski signs occasionally elicitable (Day et al., 2000; Juvonen et al., 2000). Mild sensory loss, primarily affecting vibratory perception was occasionally observed (Day et al., 2000).

MRI analysis of affected individuals reveals atrophy of the cerebellar hemispheres and vermis (Day et al., 2000; Ikeda et al., 2000a; Topisirovic et al., 2002). Figure 1 shows an individual with dramatic vermian atrophy in both superior and inferior aspects of the cerebellum. As is typical of other SCA8 patients, the cerebrum shows no definite abnormalities, with neither atrophy or white matter abnormalities, nor basal ganglia involvement. Brainstem atrophy is minimal. Consistent with slow disease progression, scans from the same patient nine years later show remarkably little change (Day et al., 2000).

RAPID cloning of an expanded CTG repeat from an ataxia patient

In contrast to most ataxia genes, the SCA8 expansion was isolated directly from a single DNA sample by the RAPID cloning method (Koob et al., 1998) without prior linkage data or clinical information. Unlike positional cloning approaches, RAPID cloning does not require large pedigrees or a highly penetrant inheritance pattern. Removal of these typical biases from the gene isolation strategy made it possible to find a mutation characterized by a high degree of reduced penetrance.

Received 23 December 2002; revision accepted 6 March 2003.

Request reprints from Laura Ranum, MMC 206 420 Delaware St. SE
University of Minnesota, Minneapolis, MN 55455 (USA)
telephone: (612) 624-0901; fax: (612) 626-2600
e-mail: ranum001@umn.edu

KARGER
Fax + 41 61 306 12 34
E-mail karger@karger.ch
www.karger.com

© 2003 S. Karger AG, Basel
0301–0171/03/1004–0175$19.50/0

Accessible online at:
www.karger.com/cgr

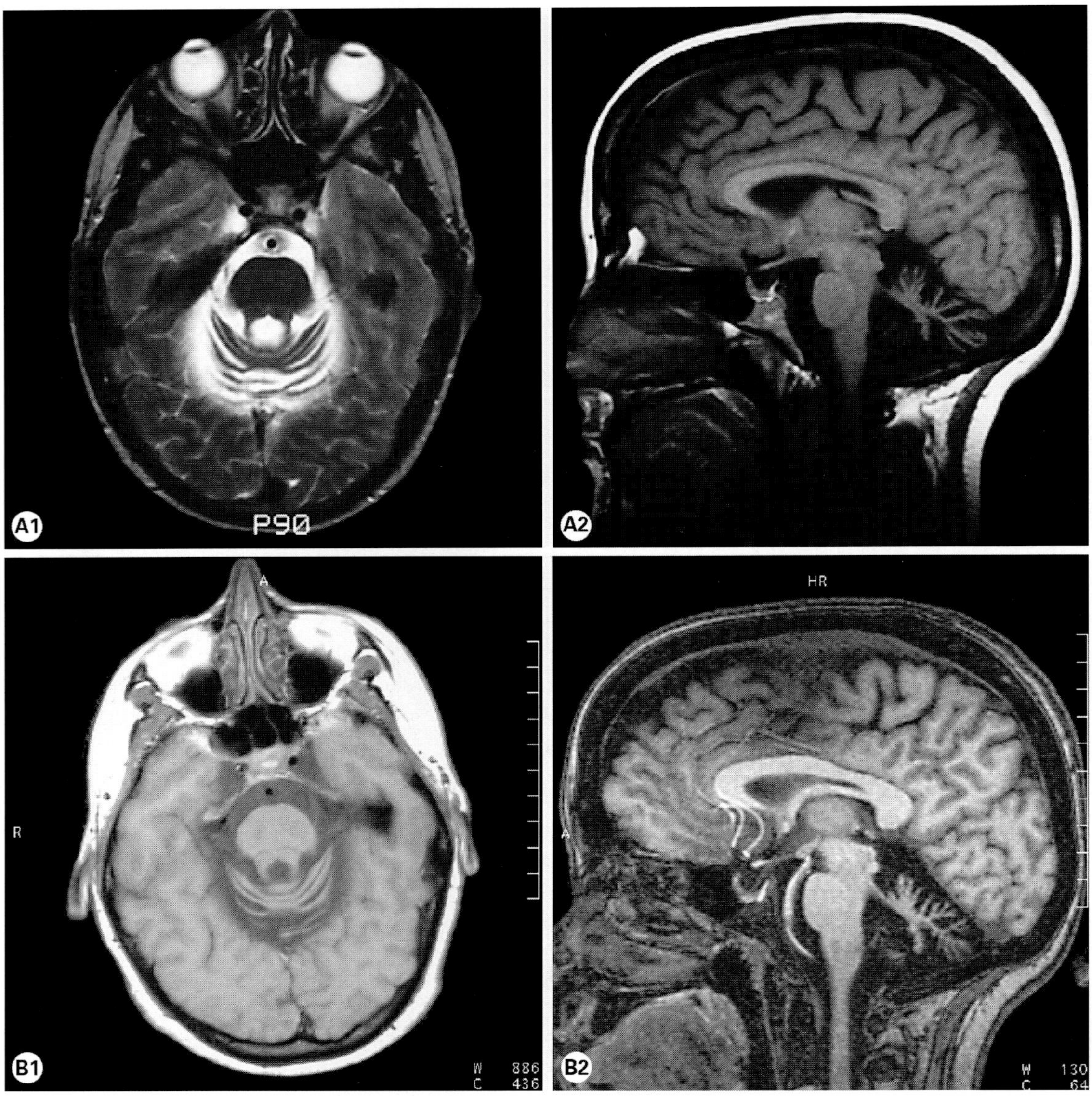

Fig. 1. Serial MRI scans of an affected individual. Horizontal (A1, B1) and sagittal (A2, B2) MRI scans from an affected individual at ages 26 (A) and 35 (B) years. The earlier image is 9 years after onset (17 years). There is marked cerebellar atrophy, minimal brainstem atrophy, and no evidence of cerebral involvement. There is very little change over the nine-year period between scans which is consistent with the slow progression of the disease. Figure reproduced with permission from Day et al. (2000).

As a first step in the RAPID cloning process, RED analysis was performed on DNA samples from an affected mother and an affected daughter with a genetically undefined form of ataxia. Both DNA samples generated RED products with 80 CTG repeats (RED80). RAPID cloning was then used to clone the expansion and unique sequence flanking the expanded repeat tract. Sequence analysis demonstrated that the expansion consisted of 80 uninterrupted CTG repeats preceded by a stretch of 11 CTA repeats. No significant open reading frames extend through this expansion. Most notably, in the reading frame that would have produced a polyglutamine expansion (CAG), 17 of the 30 codons immediately flanking the expansion are stop codons. In addition, detailed analysis of this sequence did not reveal any possible splice donor or acceptor signals that would allow an open reading frame to extend through the expansion in a spliced transcript of this genomic sequence. These data made it appear unlikely that the expansion could be translated into a polyglutamine tract (Koob et al., 1999).

PCR primers were designed from the genomic sequence to amplify across the repeat, and PCR analysis of a chromosome hybrid panel and the Centre d'Etude du Polymorphisme Humain (CEPH) yeast artificial chromosome (YAC) library physically mapped the expansion to chromosome 13q21 near the polymorphic markers D13S275 and D13S135. No ataxia genes had previously been mapped to this locus (Koob et al., 1999).

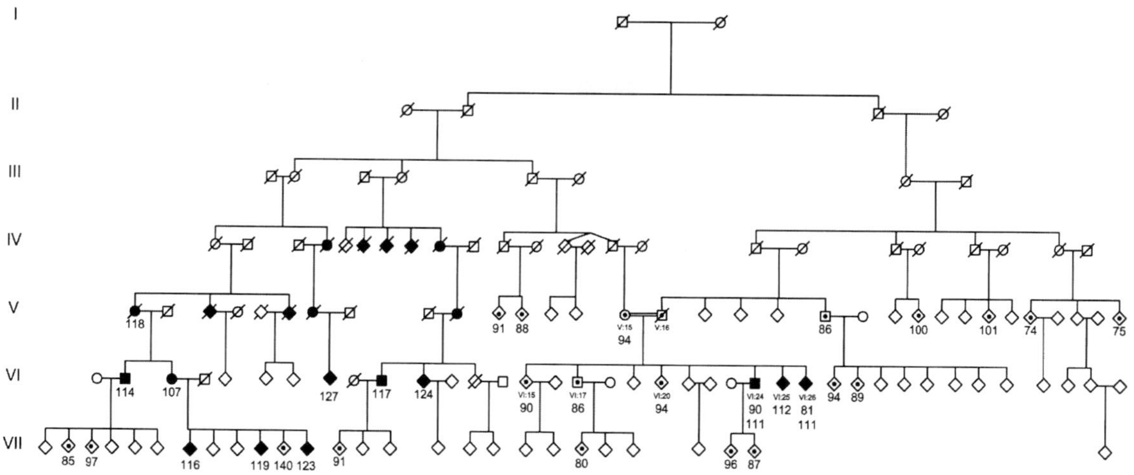

Fig. 2. The large SCA8 kindred (MN-A family). Filled symbols indicate individuals with ataxia, symbols with a dot indicate individuals that have inherited the CTG expansion but are not clinically affected by ataxia. The CTG repeat lengths of expanded alleles are indicated below the symbols. Haplotype analyses using five short tandem repeat markers confirm that both branches of the family inherited the expanded repeat from a common founder. Family members homozygous for the SCA8 expansion and their affected heterozygous sibling (individuals VI: 24–26) had similar clinical features, with comparable ages of onset and rates of disease progression. Figure reproduced with permission from Koob et al. (1999).

The expanded CTG repeat co-segregates with a novel dominant ataxia (SCA8)

PCR analysis of the CAG/CTG repeat was performed on genomic samples from the kindred from which it was originally isolated. Both of the affected individuals (each with 80 CTG repeats) and two at-risk individuals (87 and 113 CTG repeats, ages 38 and 35 years, respectively) were found to have an expansion on one of their alleles. Screening of DNA samples from our collection of 400 ataxia families revealed expansions in seven additional probands representing unrelated kindreds with dominant ataxia (Moseley et al., 1998). To investigate whether or not this triplet expansion was pathogenic, additional members of these kindred were examined and blood was collected. The largest of these is a seven-generation kindred (MN-A family in Fig. 2) from which 84 members were clinically evaluated and tested for expansions. PCR analyses showed that all of the affected individuals in the family had an expanded allele. Linkage analyses between ataxia and the expansion in this kindred gave a maximum lod score of 6.8 at $\theta = 0.00$, indicating that the expansion is tightly linked to a novel dominant SCA locus (SCA8) (Koob et al., 1999).

Affected MN-A family members have longer CTG expansions than unaffected carriers

Within the MN-A family there were 17 individuals who carried an expanded repeat but were not clinically affected at the time of evaluation. The ages of these asymptomatic carriers at time of evaluation ranged from 14 to 74 years, with a mean (43 ± 17 y) that was comparable to the mean age at exam of the affected family members. However, the repeat lengths among these asymptomatic carriers are significantly ($P < 10^{-8}$) shorter than those found among the affected individuals (means = 90 and 116 repeats, respectively). All but one of the individuals with a CTG repeat tract ≥ 107 repeats is clinically affected. The exception is a 42-year-old individual with 140 CTG repeats. Although this person's repeat length is larger than the apparent pathogenic threshold, a currently asymptomatic status is not surprising given that SCA8 is an adult onset disease with a documented age of onset as late as 65 years of age. These results demonstrate that, for the MN-A family, disease penetrance is affected by CTG repeat length. Because the penetrance of the disease depends on both age and repeat length, affected individuals within the MN-A family did not always have an obviously dominant family history of ataxia (Day et al., 2000).

Reduced penetrance in other families

The reduced penetrance seen in the MN-A family is also found in other families. However, the tight correlation between repeat size and pathogenesis present in the MN-A family cannot be extended to the broader panel of SCA8 families we have examined or that have been described in the literature (Koob et al., 1999; Day et al., 2000; Ikeda et al., 2000b; Cellini et al., 2001; Topisirovic et al., 2002). Several examples of reduced penetrance among families we have examined are described below. In two families, men (73 and 79 years) with 260 and 300 CTG repeats were unaffected, but transmitted shorter pathogenic alleles to four affected offspring. Three offspring in another ataxia family inherited expanded alleles of 400, 500 and 700 repeats (ages 24, 17 and 15, respectively) but are not clinically affected. Among the larger group of families we have studied, the repeat sizes most often associated with ataxia range from 80

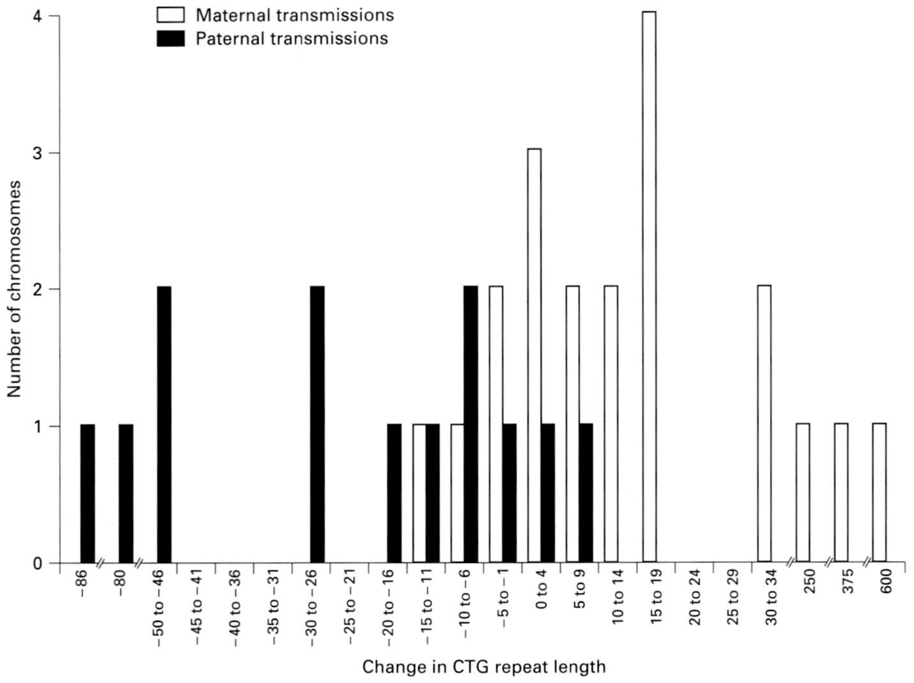

Fig. 3. Intergenerational variation in repeat number for maternal and paternal transmissions. Repeat variation is shown as a decrease or an increase of CTG repeat units. Maternal and paternal transmissions are represented by gray and black bars, respectively. Figure reproduced with permission from Koob et al. (1999).

to 250 combined CTA/CTG repeats; however, repeat sizes ranging from 71 to over 1,000 repeats have been found in some individuals with ataxia. Further analysis of other families showed that the pathogenic size range of the SCA8 expansion can vary between families and that a positive test for an SCA8 expansion cannot be used to predict whether or not an asymptomatic individual will develop ataxia, regardless of repeat size (Koob et al., 1999; Ranum et al., 1999; Day et al., 2000; Juvonen et al., 2000; Moseley et al., 2000b; Stevanin et al., 2000; Worth et al., 2000; Sobrido et al., 2001). Interestingly, MRI analysis of one clinically unaffected carrier (age 71) with 136 combined repeats revealed mild atrophy of the cerebellar hemispheres (Ikeda et al., 2000b). Surprisingly, SCA8 expansions >100 combined repeats have also been found among control chromosomes at a frequency of ~ 1/1,000, leading some to suggest the possibility that a separate mutation in linkage disequilibrium could exist among ataxia patients but not controls (Stevanin et al., 2000; Vincent et al., 2000; Worth et al., 2000).

Independently arising SCA8 expansions found among ataxia families

To test whether or not the CTG expansion could be a nonpathogenic polymorphism that happens to lie close enough to be in linkage disequilibrium with an ataxia locus, haplotype and linkage disequilibrium analyses were performed on individuals with SCA8 expansions from ataxia families as well as psychiatric patients and controls. Although initial studies with markers 100–200 kb from the CTG expansion did not detect a common haplotype, subsequent analyses with eight flanking STR markers detected a highly conserved haplotype among Caucasian ataxia and psychiatric patients as well as controls. TdT results for markers at 72 kb ($P = 0.04$), 56 kb ($P = 0.002$), 52 kb ($P = 0.03$), and 13 kb ($P = 0.001$) further support an ancestral origin of the Caucasian expansion. These data suggest that a separate mutation found only in the ataxia families, but not in the controls, is unlikely to account for the reduced penetrance of the disease (Moseley et al., 2000a; Ranum et al., 2002).

Instability and maternal penetrance bias

The intergenerational changes in CTG repeat number are typically larger for SCA8 than for the other dominant SCAs (Chung et al., 1993; Maciel et al., 1995; Maruyama et al., 1995; Cancel et al., 1997; David et al., 1997; Jodice et al., 1997; Zhuchenko et al., 1997) but are generally not as large as for DM1 (Tsilfidis et al., 1992). A histogram of the change in CTG repeat number in maternal and paternal transmissions of the SCA8 expansion is shown in Fig. 3. Among the ataxia families, paternal transmissions usually resulted in contractions in the CTG repeat (–86 to +7), while maternal transmissions typically resulted in expansions (–11 to +600). Three very large increases in repeat length (+250, +375, +600), similar in size to those seen in myotonic dystrophy, all resulted from maternal transmissions. The maternal expansion bias of the SCA8 repeat is similar to the non-coding expansion disorders FMR1 and DM1 but different from what had been reported for the other SCAs (Groenen and Wieringa, 1998; Koob et al., 1999; Jin and Warren, 2000).

Nineteen of the 21 documented transmissions of ataxia in the MN-A family resulted from maternal transmissions. The remaining two individuals inherited expanded alleles from both parents (Fig. 2). In contrast, of the 19 asymptomatic individuals with repeat expansions, only three were maternally transmitted while 16 were paternally transmitted. The maternal penetrance bias seen in the MN-A family is consistent with the higher frequency of expansions into the pathogenic range for female transmissions and contractions below the pathogenic threshold for paternal transmissions (Koob et al., 1999; Day et al., 2000); however, the maternal penetrance bias seen in the MN-A family is not seen in all families.

Massive deletions of large alleles in paternal transmissions and sperm samples

To further investigate the instability of very large alleles, Southern analysis was performed on sperm DNA from two unrelated men with repeats of 500 and 800 (Fig. 4A). In both cases, all or nearly all of the expanded alleles underwent massive deletions into size ranges (~90 and ~110 combined repeats) less frequently associated with ataxia. Southern blots of blood and sperm DNA from patients with smaller expansions reveal the same trend of contractions of the expanded allele in sperm to repeat sizes that are less often associated with ataxia (below ~100 combined repeats) (Fig. 4B). The equal intensities of the bands representing the normal and expanded alleles indicate that repeat contractions occurred in all or nearly all of the sperm with expanded alleles. PCR analysis of SCA8 contractions in two patients from an unusual family with paternal disease transmission also showed contracted repeats in sperm although the resulting alleles remain within the most penetrant size range (Fig. 4C). The preponderance of deletions in sperm may contribute to the reduced penetrance of SCA8 and the maternal penetrance bias seen in some families (Moseley et al., 2000c; Silveira et al., 2000).

Sequence variations at the SCA8 locus

In contrast to other triplet repeat diseases, pathogenic SCA8 expansions often have triplet interruptions within the repeat tracts. The SCA8 CTG repeat is preceded by a polymorphic but stable CTA tract, with the configuration $(CTA)_{3-29}(CTG)_n$ (Moseley et al., 2000c). The CTG portion of the repeat tract is elongated on pathogenic alleles and nearly always changes in size when transmitted from one generation to the next. Expanded alleles found in affected SCA8 individuals can have either a pure uninterrupted CTG repeat tract or an allele with one or more CCG, CTA, CTC, CCA or CTT interruptions. Sequence interruptions tend to be clustered at the 5′ end of the CTG portion of the repeat tract and the interruptions can duplicate when transmitted from one generation to the next resulting in a variety of alleles that vary both in overall CTG tract length and sequence configuration. Although both interrupted and pure repeat tracts are found among affected individuals in the various SCA8 families, the possible role that interruptions within the CTG repeat tract may play in disease pathogenesis is unclear (Moseley et al., 2000c).

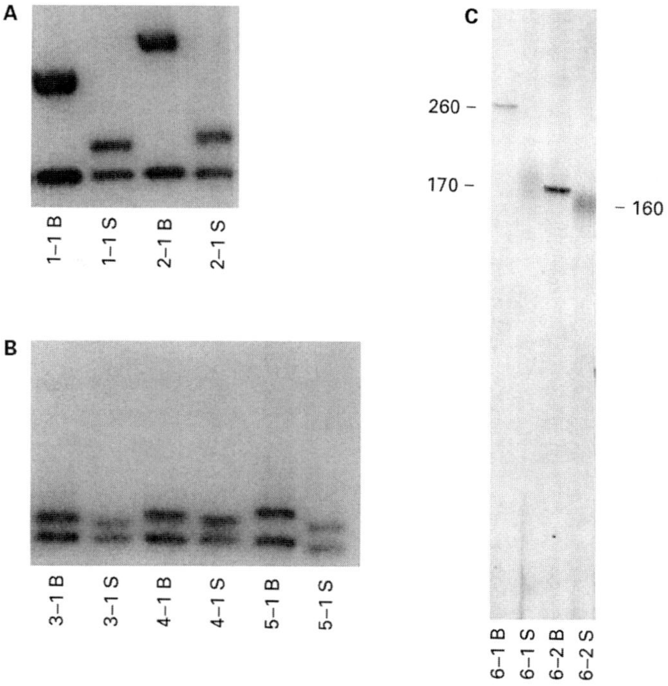

Fig. 4. En masse contraction of SCA8 alleles in sperm. (**A**) Dramatic repeat length changes in patients 1 and 2 detected by Southern blotting. The repeat length of patients 1 and 2 contract from 500–800 repeats in blood (B) to ~80–100 repeats in sperm (S), respectively. The probe used did not contain the CTG repeat. (**B**) Southern blots of blood and sperm DNA from patients with smaller expansions in their blood reveal the same trend of contractions of the expanded allele in sperm to repeat sizes that are less often associated with ataxia (below ~100 repeats). Again, the equal intensities of the bands representing the normal and expanded alleles indicate that repeat contractions occurred in all or nearly all of the sperm with expanded alleles. (**C**) PCR analysis of SCA8 contractions in two patients from a family with paternal disease transmission. Although contraction of repeats in sperm is again observed, the resulting alleles remain within a more penetrant size range (>100 CTGs). Figure reproduced with permission from Moseley et al. (2000c).

SCA8 gene organization and expression

The SCA8 expansion is transcribed in the CTG orientation as is seen for DM1, but not in the CAG orientation, as seen with the other dominant SCAs (Koob et al., 1999). As expected from analysis of the genomic sequence, random amplification of cDNA ends (RACE) experiments confirmed that the CTG repeat is present in the 3′ terminal exon that begins at the predicted splice-acceptor site. The longest transcripts identified are comprised of six exons with numerous alternatively spliced forms including variants that lack exon B (Benzow and Koob, 2002). These transcripts have no extended open reading frames and no significant homology to known genes. The genomic and cDNA organization of the SCA8 gene are summarized in Fig. 5.

To investigate the possibility of imprinting, cerebellar RNA isolated from two unaffected individuals was analyzed to determine whether or not both maternal and paternal SCA8 alleles are expressed in normal cerebellum. RT-PCR from exon C to exon A was performed on cerebellar RNA isolated from two

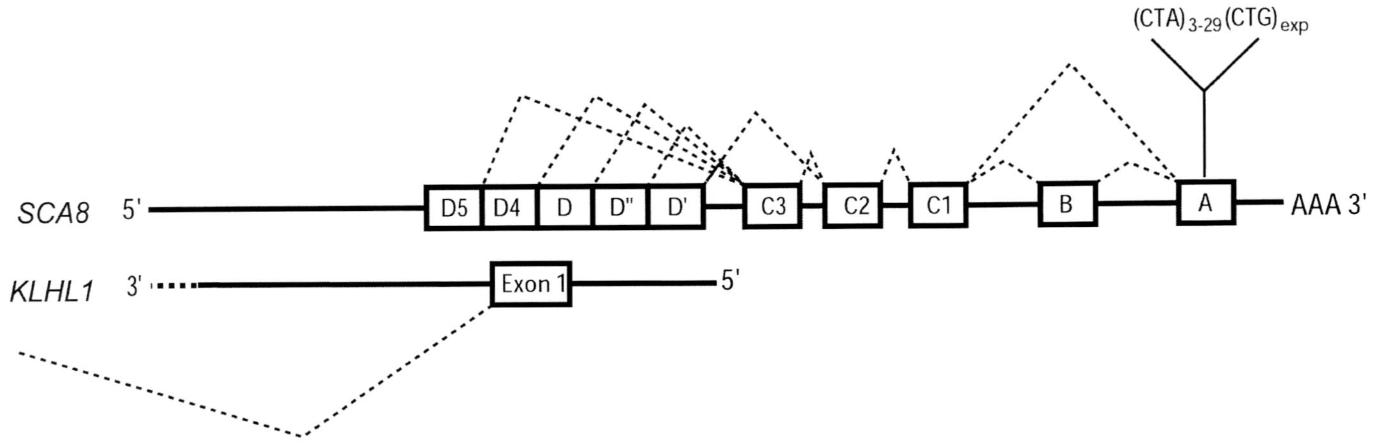

Fig. 5. SCA8 gene organization. Exons are shown as boxes and alternate splice forms are indicated by dashed lines. The SCA8 repeat tract is located in exon A and is transcribed in the CTG orientation. The region of the SCA8 gene that overlaps the 5′ end of KLHL1 is shown. Various alternative splice forms of the SCA8 transcript are indicated.

unaffected individuals with CTG repeats on each allele that differed in size. Transcripts from both of the SCA8 alleles were detected in each of the RNA samples, clearly indicating that normally both the maternal and paternal alleles are expressed in the cerebellum (Koob et al., 1999).

The steady state levels of the SCA8 transcript are very low and although detectable by RT-PCR are not present at high enough levels to be detected by Northern analysis. RT-PCR experiments indicate that the SCA8 transcript is expressed in brain tissue while being absent or present at much lower levels in non-CNS tissues examined (heart, placenta, liver, skeletal muscle, kidney and pancreas) (Janzen et al., 1999; Koob et al., 1999).

Surprisingly, RACE analysis using primers from exon D of the SCA8 transcript identified an overlapping transcript that is transcribed in the opposite orientation of the SCA8 gene (Koob et al., 1999). This overlapping gene is predicted to encode a protein that is 748 amino acids in length and highly homologous to the Drosophila kelch protein (Robinson and Cooley, 1997). This gene was named Kelch-like 1 (KLHL1) (Nemes et al., 2000).

Although a functional relationship between the SCA8 and KLHL1 transcripts has not been demonstrated, the genomic organization of these genes suggests the possibility that the normal function of the SCA8 transcripts may be as an antisense regulator of KLHL1 (Nemes et al., 2000; Benzow and Koob, 2002). Although it is possible that the SCA8 CUG expansion could affect KLHL1 expression through antisense interactions, it is not known whether both genes are co-expressed in any cells, whether levels of KLHL1 are actually altered in SCA8 patients nor whether altered levels of KLHL1 protein could plausibly contribute to cerebellar degeneration and ataxia.

Parallels between SCA8 and myotonic dystrophy suggest RNA mechanism

In contrast to the polyglutamine ataxias, the SCA8 repeat tract is transcribed in the CTG but not the CAG orientation. Although the clinical features of SCA8 are typical of the spinocerebellar ataxias, the only other triplet expansion disorder reported to involve a transcribed but untranslated CTG expansion is myotonic dystrophy type 1 (DM1) (Tapscott, 2000).

Although the DM1 mutation was identified as a CTG expansion located in the 3′ UTR of a kinase gene in 1992, the molecular mechanisms responsible for this dominantly inherited disorder have been a source of confusion and controversy because it was not understood how the CTG expansion mutation, which does not alter the protein-coding portion of a gene, could cause the broad constellation of clinical features associated with DM1 (Brook et al., 1992; Fu et al., 1992; Mahadevan et al., 1992). Suggested mechanisms that have been proposed for DM1 include: (i) haploinsufficiency of the DMPK protein (Fu et al., 1993); (ii) altered expression of neighboring genes, including SIX5 (Boucher et al., 1995; Otten and Tapscott, 1995; Klesert et al., 1997; Thornton et al., 1997; Alwazzan et al., 1999); and (iii) pathogenic effects of the CUG expansion in RNA which accumulates as nuclear foci (Taneja et al., 1995; Davis et al., 1997) and disrupts cellular function (Timchenko et al., 1996; Philips et al., 1998; Lu et al., 1999; Miller et al., 2000; Charlet et al., 2002; Mankodi et al., 2002). Several mouse models have developed different aspects of DM1 (Mankodi et al., 2000; Seznec et al., 2001), but none have accurately reflected the multisystemic nature of the disorder, suggesting that each theory may contribute to DM1 pathogenesis and that DM1 may be a regional gene disorder (Groenen and Wieringa, 1998; Tapscott, 2000).

Defining a second human mutation that causes the multisystemic effects of DM, and identifying what is common to these diseases at the molecular level has provided an independent means of determining the pathogenic pathway of DM. The myotonic dystrophy type 2 (DM2, also known as proximal myotonic myopathy, PROMM) locus was mapped to chromosome 3q21 (Ranum et al., 1998; Day et al., 1999), and the DM2 mutation was identified as a non-coding CCUG expansion in intron 1 of the zinc finger protein 9 (ZNF9) gene (Liquori et al., 2001). Although DMPK and ZNF9 and the genes in their respective regions bear no obvious relation to one another, DM1 and DM2 have strikingly similar clinical presentations including myotonia, myopathy, iridescent cataracts and similar cardiac features (Harper, 2001). In both disorders, the mutations involve expansions of similar repeat motifs (CTG or CCTG) that are transcribed into RNA but not translated in proteins. The CUG and CCUG expansions accumulate as nuclear RNA foci (Taneja et al., 1995; Liquori et al., 2001). These expanded repeat containing RNAs alter the binding or regulation of several proteins that normally recognize CUG sequences (Timchenko et al., 1996; Mankodi et al., 2002; Miller et al., 2000; Fardaei et al., 2002). Repeat induced alterations of one of these proteins, CUG-binding protein (CUG-BP) have been shown to cause altered RNA splicing of a number of genes including the insulin receptor (IR) and the chloride channel in both DM1 and DM2 (Philips et al., 1998; Ranum et al., 2001; Savkur et al., 2001; Charlet et al., 2002; Mankodi et al., 2002). The striking similarities between DM1 and DM2 at both the clinical and the molecular level provide strong support that CUG and CCUG expansions expressed at the RNA level can themselves disrupt cellular functions such as splicing and lead to the multisystemic features of DM (Ranum and Day, 2002a; Ranum and Day, 2002b; Day et al., 2003).

Molecular parallels between the SCA8, DM1 and DM2 mutations and the known toxic properties of transcripts with elongated CUGs suggest the possibility of a similar gain-of-function RNA mechanism at the cellular level. The difference in phenotype between these diseases may reflect the dramatically different expression patterns of their transcripts. Consistent with the multisystemic features of DM1 and DM2, the DMPK and ZNF9 transcripts are broadly expressed (Jansen et al., 1996; Liquori et al., 2001). In contrast, SCA8 transcripts are almost exclusively expressed in the brain, which is consistent with its central nervous system involvement (Janzen et al., 1999; Koob et al., 1999). Currently, animal models are being developed to examine what role the untranslated CTG expansion plays in the pathology of SCA8 (Moseley et al., 2002; Mutsuddi et al., 2002).

Conclusion

The linkage analysis (lod = 6.8, $\theta = 0.00$) and the strong correlation between CTG repeat size (107–127 CTGs) and affected status in the MN-A family, as well as the high prevalence of independently arising expanded alleles among ataxia patients demonstrate that the SCA8 CTG expansion underlies SCA8 pathogenesis. The reduced penetrance and unusual patterns of repeat instability seen for SCA8, however, complicate diagnostic testing. While large families with clearly dominant disease were used to localize and isolate the SCA1, SCA2, SCA3, SCA6 and SCA7 genes, we identified the SCA8 expansion from the DNA of a single affected individual. Given that the RAPID method used to isolate the SCA8 CTG expansion did not depend on the genetic bias needed for positional cloning, it is not surprising that the ataxia gene we found has so many distinctions from the other ataxia genes. RAPID cloning provides a useful tool to identify genes for diseases with reduced penetrance and complex patterns of inheritance that are caused by repeat expansions.

Although the reduced penetrance of SCA8 is not yet understood, the linkage data, the biological relationship between repeat length and disease status in the MN-A family, as well as the expression pattern of the SCA8 transcript support the hypothesis that the SCA8 CTG expansion is directly associated with ataxia. Several unique features of the SCA8 repeat tract could affect penetrance. These include the maternal expansion bias, the dramatic deletions of repeat size in male transmissions and in sperm, the length of the adjacent CTA repeat tract and the mutability of the CTG repeat tract itself (Koob et al., 1999; Moseley et al., 2000b; Moseley et al., 2000c).

The molecular similarities between SCA8 and myotonic dystrophy type 1 provide an opportunity to further define the pathogenic mechanisms mediated by the untranslated CTG expansions found in both diseases. Understanding how these untranslated CTG expansions can cause diseases as different as ataxia and myotonic dystrophy should further our understanding of the diversity of molecular mechanisms involved in neurodegenerative disease.

References

Alwazzan M, Newman E, Hamshere MG, Brook JD: Myotonic dystrophy is associated with a reduced level of RNA from the DMWD allele adjacent to the expanded repeat. Hum molec Genet 8:1491–1497 (1999).

Anderson JH, Yavuz MC, Kazar BM, Christova P, Gomez CM: The vestibular-ocular reflex and velocity storage in spinocerebellar ataxia 8. Arch Ital Biol 140:323–329 (2002).

Benzow KA, Koob MD: The KLHL1-antisense transcript (KLHL1AS) is evolutionarily conserved. Mammal Genome 13:134–141 (2002).

Boucher CA, King SK, Carey N, Krahe R, Winchester CL, Rahman S, Creavin T, Meghji P, Bailey MES, Chartier FL, Brown SD, Siciliano MJ, Johnson KJ: A novel homeodomain-encoding gene is associated with a large CpG island interrupted by the myotonic dystrophy unstable (CTG)n repeat. Hum molec Genet 4:1919–1925 (1995).

Brook JD, McCurrah ME, Harley HG, Buckler AJ, Church D, Aburatani H, Hunter K, Stanton VP, Thirion J-P, Hudson T, Sohn R, Zemelman B, Snell RG, Rundle SA, Crow S, Davies J, Shelbourne P, Buxton J, Jones C, Juvonen V, Johnson K, Harper PS, Shaw DJ, Housman DE: Molecular basis of myotonic dystrophy: Expansion of a trinucleotide (CTG) repeat at the 3′ end of a transcript encoding a protein kinase family member. Cell 68:799–808 (1992).

Brusco A, Cagnoli C, Franco A, Dragone E, Nardacchione A, Grosso E, Mortara P, Mutani R, Migone N, Orsi L: Analysis of SCA8 and SCA12 loci in 134 Italian ataxic patients negative for SCA1-3, 6, and 7 CAG expansions. J Neurol 249:923–929 (2002).

Cancel G, Durr A, Didierjean O, Imbert G, Burk K, Lezin A, Belal S, Benomar A, Abadabendib M, Vial C, Guimaraes J, Chneiweiss H, Stevanin G, Yvert G, Abbas N, Saudou F, Lebre A, Yahyaoui M, Hentati F, Vernant J, Klockgether T, Mandel J, Agrid Y, Brice A: Molecular and clinical correlations in spinocerebellar ataxia 2 – a study of 32 families. Hum molec Genet 6:709–715 (1997).

Cellini E, Nacmias B, Forleo P, Piacentini S, Guarnieri BM, Serio A, Calabro A, Renzi D, Sorbi S: Genetic and clinical analysis of spinocerebellar ataxia type 8 repeat expansion in Italy. Arch Neurol 58:1856–1859 (2001).

Charlet BN, Savkur RS, Singh G, Philips AV, Grice EA, Cooper TA: Loss of the muscle-specific chloride channel in type 1 myotonic dystrophy due to misregulated alternative splicing. Mol Cell 10:45–53 (2002).

Chung M-Y, Ranum LPW, Duvick LA, Servadio A, Zoghbi HY, Orr HT: Evidence for a mechanism predisposing to intergenerational CAG repeat instability in spinocerebellar ataxia type 1. Nature Genet 5:254–258 (1993).

David G, Abbas N, Stevanin G, Durr A, Yvert G, Cancel G, Weber C, Imbert G, Saudou F, Antoniou E, Drabkin H, Gemmill R, Giunti P, Benomar A, Wood NW, Ruberg M, Agid Y, Mandel J, Brice A: Cloning of the SCA7 gene reveals a highly unstable CAG repeat expansion. Nature Genet 17:65–70 (1997).

Davis BM, McCurrach ME, Taneja KL, Singer RH, Housman DE: Expansion of a CUG trinucleotide repeat in the 3′ untranslated region of myotonic dystrophy protein kinase transcripts results in nuclear retention of transcripts. Proc natl Acad Sci, USA 94:7388–7393 (1997).

Day JW, Roelofs R, Leroy B, Pech I, Benzow K, Ranum LPW: Clinical and genetic characteristics of a five-generation family with a novel form of myotonic dystrophy (DM2). Neuromuscul Disord 9:19–27 (1999).

Day JW, Schut LJ, Moseley ML, Durand AC, Ranum LPW: Spinocerebellar ataxia type 8: clinical features in a large family. Neurology 55:649–657 (2000).

Day JW, Ricker K, Jacobsen JF, Rasmussen LJ, Dick KA, Kress W, Schneider C, Koch MC, Beilman GJ, Harrison AR, Dalton JC, Ranum LPW: Myotonic dystrophy type 2: molecular, diagnostic and clinical spectrum. Neurology 60:657–664 (2003).

Fardaei M, Rogers MT, Thorpe HM, Larkin K, Hamshere MG, Harper PS, Brook JD: Three proteins, MBNL, MBLL and MBXL, co-localize in vivo with nuclear foci of expanded-repeat transcripts in DM1 and DM2 cells. Hum molec Genet 11:805–814 (2002).

Fu Y-H, Pizzuti A, Fenwick RGJ, King J, Rajnarayan S, Dunne PW, Dubel J, Nasser GA, Ashizawa T, De Jong P, Wieringa B, Korneluk R, Perryman MB, Epstein HF, Caskey CT: An unstable triplet repeat in a gene related to myotonic muscular dystrophy. Science 255:1256–1258 (1992).

Fu Y-H, Friedman DL, Richards S, Pearlman JA, Gibbs RA, Pizzuti A, Ashizawa T, Perryman MB, Scarlato G, Fenwick RG, Caskey CT: Decreased expression of myotonin-protein kinase messenger RNA and protein in adult form of myotonic dystrophy. Science 260:235–238 (1993).

Groenen P, Wieringa B: Expanding complexity in myotonic dystrophy. Bioessays 20:901–912 (1998).

Harper PS: Myotonic Dystrophy. 3rd ed (W.B. Saunders, London, 2001).

Ikeda Y, Shizuka M, Watanabe M, Okamoto K, Shoji M: Molecular and clinical analyses of spinocerebellar ataxia type 8 in Japan. Neurology 54:950–955 (2000a).

Ikeda Y, Shizuka-Ikeda M, Watanabe M, Schmitt M, Okamoto K, Shoji M: Asymptomatic CTG expansion at the SCA8 locus is associated with cerebellar atrophy on MRI. J Neurol Sci 182:76–79 (2000b).

Jansen G, Groenen PJTA, Bachner D, Jap PHK, Coerwinkel M, Oerlemans F, van den Broek W, Gohlsch B, Pette D, Plomp JJ, Molenaar PC, Nederhof MGJ, van Echted CJA, Dekker M, Berns A, Hameister H, Wieringa B: Abnormal myotonic dystrophy protein kinase levels produce only mild myopathy in mice. Nature Genet 13:316–324 (1996).

Janzen MA, Moseley ML, Benzow KA, Day JW, Koob MD, Ranum LPW: Limited expression of SCA8 is consistent with cerebellar pathogenesis and toxic gain of function RNA model. Am J hum Genet 65(Supplement):A267 (1999).

Jin P, Warren ST: Understanding the molecular basis of fragile X syndrome. Hum molec Genet 9:901–908 (2000).

Jodice C, Mantuano E, Veneziano L, Trettel F, Sabbadini G, Calandriello L, Francia A, Spadaro M, Pierelli F, Salvi F, Ophoff R, Frants R, Frontali M: Episodic ataxia type 2 (EA2) and spinocerebellar atxia type 6 (SCA6) due to CAG repeat expansion in the CACNA1A gene on chromosome 19p. Hum molec Genet 6:1973–1978 (1997).

Juvonen V, Hietala M, Paivarinta M, Rantamaki M, Hakamies L, Kaakkola S, Vierimaa O, Penttinen M, Savontaus ML: Clinical and genetic findings in Finnish ataxia patients with spinocerebellar ataxia 8 repeat expansion. Ann Neurol 48:354–361 (2000).

Klesert TR, Otten AD, Bird TD, Tapscott SJ: Trinucleotide repeat expansion at the myotonic dystrophy locus reduces expression of DMAHP. Nature Genet 16:402–406 (1997).

Koob MD, Benzow KA, Bird TD, Day JW, Moseley ML, Ranum LPW: Rapid cloning of expanded trinucleotide repeat sequences from genomic DNA. Nature Genet 18:72–75 (1998).

Koob MD, Moseley ML, Schut LJ, Benzow KA, Bird TD, Day JW, Ranum LPW: An untranslated CTG expansion causes a novel form of spinocerebellar ataxia (SCA8). Nature Genet 21:379–384 (1999).

Liquori C, Ricker K, Moseley ML, Jacobsen JF, Kress W, Naylor S, Day JW, Ranum LPW: Myotonic dystrophy type 2 caused by a CCTG expansion in intron 1 of ZNF9. Science 293:864–867 (2001).

Lu X, Timchenko NA, Timchenko LT: Cardiac elav-type RNA-binding protein (ETR-3) binds to RNA CUG repeats expanded in myotonic dystrophy. Hum molec Genet 8:53–60 (1999).

Maciel P, Gaspar C, DeStefano AL, Silveira I, Coutinho P, Radvany J, Dawson DM, Sudarsky L, Guimaraes J, Loureiro JEL, Nezarati MM, Corwin LI, Lopes-Cendes I, Rooke K, Rosenburg R, MacLeod D, Farrer LA, Sequeiros J, Roulean GA: Correlation between CAG repeat length and clinical features in Machado-Joseph disease. Am J hum Genet 57:54–61 (1995).

Mahadevan M, Tsilfidis C, Sabourin L, Shutler G, Amemiya C, Jansen G, Neville C, Narang M, Barcelo J, O'Hoy K, Leblond S, Earle-MacDonald J, De Jong PJ, Wieringa B, Koneluk RG: Myotonic dystrophy mutation: an unstable CTG repeat in the 3′ untranslated region of the gene. Science 255:1253–1255 (1992).

Mankodi A, Logigian E, Callahan L, McClain C, White R, Henderson D, Krym M, Thornton CA: Myotonic dystrophy in transgenic mice expressing an expanded CUG repeat. Science 289:1769–1773 (2000).

Mankodi A, Takahashi MP, Jiang H, Beck CL, Bowers WJ, Moxley RT, Cannon SC, Thornton CA: Expanded CUG repeats trigger aberrant splicing of ClC-1 chloride channel pre-mRNA and hyperexcitability of skeletal muscle in myotonic dystrophy. Mol Cell 10:35–44 (2002).

Maruyama H, Nakamura S, Matsuyama Z, Sakai T, Doyu M, Sobue G, Seto M, Tsujihata M, Oh-i T, Nishio T, Sunohara N, Takahashi R, Hayashi M, Nishino I, Ohtake T, Oda T, Nishimura M, Saida T, Matsumoto H, Baba M, Kawaguchi Y, Kakizuka A, Kawakimi H: Molecular features of the CAG repeats and clinical manifestation of Machado-Joseph disease. Hum molec Genet 4:807–812 (1995).

Miller JW, Urbinati CR, Teng-Umnuay P, Stenberg MG, Byrne BJ, Thornton CA, Swanson MS: Recruitment of human muscleblind proteins to (CUG)(n) expansions associated with myotonic dystrophy. EMBO J 19:4439–4448 (2000).

Moseley ML, Benzow KA, Schut LJ, Bird TD, Gomez CM, Barkhaus PE, Blindauer KA, Labuda M, Pandolfo M, Koob MD, Ranum LPW: Incidence of dominant spinocerebellar and Friedreich triplet repeats among 361 ataxia families. Neurology 51:1666–1671 (1998).

Moseley ML, Jacobsen JF, Liquori C, Durand A, Davis L, Ikeda Y, Bird TD, Spiegel R, Ashizawa T, Pandolfo M, Potter N, Schaefer A, Milunsky J, Kennedy JL, Seltzer WK, Atwood L, Vincent JB, Day JW, Ranum LPW: SCA8 CTG expansion: evidence for a common haplotype and highly mutable region on both ataxia and non-ataxia chromosomes. Am J hum Genet 67(Supplement):362 (2000a).

Moseley ML, Schut LJ, Bird TD, Day JW, Ranum LPW: Reply to Stevanin et al. and Worth et al. Nature Genet 24:215 (2000b).

Moseley ML, Schut LJ, Bird TD, Koob MD, Day JW, Ranum LPW: SCA8 CTG repeat: en masse contractions in sperm and intergenerational sequence changes may play a role in reduced penetrance. Hum molec Genet 9:2125–2130 (2000c).

Moseley ML, Weatherspoon M, Rasmussen L, Day JW, Ranum LPW: SCA8 BAC transgenic mice have a progressive and lethal neurological phenotype demonstrating pathogenicity of the CTG expansion. Am J hum Genet 71 (Supplement):178 (2002).

Mutsuddi M, Benzow K, Littleton JT, Koob MD, Rebay I: Using a Drosophila model to characterize spinocerebellar ataxia 8 (SCA8) induced neurodegeneration, Abstract Viewer/Itinerary Planner, Program No. 417.411. (Society for Neuroscience, Washington DC 2002).

Nemes JP, Benzow KA, Moseley ML, Ranum LPW, Koob MD: The SCA8 transcript is an antisense RNA to a brain-specific transcript encoding a novel actin-binding protein (KLHL1). Hum molec Genet 9:1543–1551 (Correction/Addition Hum molec Genet 9:2777) (2000).

Otten AD, Tapscott SJ: Triplet repeat expansion in myotonic dystrophy alters the adjacent chromatin structure. Proc natl Acad Sci, USA 92:5465–5469 (1995).

Philips AV, Timchenko LT, Cooper TA: Disruption of splicing regulated by a CUG-binding protein in myotonic dystrophy. Science 280:737–741 (1998).

Ranum LPW, Day JW: Dominantly inherited non-coding microsatellite expansion disorders. Curr Opin Genet Dev 12:266–271 (2002a).

Ranum LPW, Day JW: Myotonic dystrophy: Clinical and molecular parallels between DM1 and DM2. Curr Neurol Neurosci Rep 2:465–470 (2002b).

Ranum LPW, Rasmussen PF, Benzow KA, Koob MD, Day JW: Genetic mapping of a second myotonic dystrophy locus. Nature Genet 19:196–198 (1998).

Ranum LPW, Moseley ML, Leppert M, Guan den Eng MF, La Spada AR, Koob MD, Day JD: Massive CTG expansions and deletions reduce penetrance of spinocerebellar ataxia type 8. Am J hum Genet 65(Supplement):A466 (1999).

Ranum LPW, Liquori C, Moseley ML, Jacobsen JF, Philips AV, Savkur RS, Kress W, Naylor S, Cooper TA, Ricker K, Day JW: Myotonic dystrophy type 2 is caused by a CCTG expansion in intron 1 of ZNF9. Am J hum Genet 69(Supplement):A211 (2001).

Ranum LPW, Moseley ML, Jacobsen JF, Liquori C, Ikeda Y, Bird TD, Ashizawa T, Seltzer WK, Vincent JB, Day JW: The SCA8 CTG expansion causes ataxia: evidence from haplotype analysis. Neurology 58:A310–A311 (2002).

Robinson DN, Cooley L: Drosophila kelch is an oligomeric ring canal actin organizer. J Cell Biol 138: 799–810 (1997).
Savkur RS, Philips AV, Cooper TA: Aberrant regulation of insulin receptor alternative splicing is associated with insulin resistance in myotonic dystrophy. Nature Genet 29:40–47 (2001).
Seznec H, Agbulut O, Sergeant N, Savouret C, Ghestem A, Tabti N, Willer JC, Ourth L, Duros C, Brisson E, Fouquet C, Butler-Browne G, Delacourte A, Junien C, Gourdon G: Mice transgenic for the human myotonic dystrophy region with expanded CTG repeats display muscular and brain abnormalities. Hum molec Genet 10:2717–2726 (2001).
Silveira I, Alonso I, Guimaraes L, Mendonca P, Santos C, Maciel P, Figdalgo de Matos JM, Costa M, Barbot C, Tuna A, Barros J, Jardim LB, Coutinho P, Sequeiros J: High germinal instability of the (CTG)n at the SCA8 locus of both expanded and normal alleles. Am J hum Genet 66:830–840 (2000).
Sobrido M-J, Cholfin JA, Perlman S, Pulst SM, Geschwind DH: SCA8 repeats expansions in ataxia: a controversial association. Neurology 57:1310–1312 (2001).
Stevanin G, Herman A, Durr A, Jodice C, Frontali M, Agid Y, Brice A: Are (CTG)n expansions at the SCA8 locus rare polymorphisms? Nature Genet 24:213 (2000).
Taneja KL, McCurrach M, Schalling M, Housman D, Singer RH: Foci of trinucleotide repeat transcripts in nuclei of myotonic dystrophy cells and tissues. J Cell Biol 128:995–1002 (1995).
Tapscott SJ: Deconstructing myotonic dystrophy. Science 289:1701–1702 (2000).
Thornton CA, Wymer JP, Simmons Z, McClain C, Moxley RT: Expansion of the myotonic dystrophy CTG repeat reduces expression of the flanking DMAHP gene. Nature Genet 16:407–409 (1997).
Timchenko LT, Miller JW, Timchenko NA, DeVore DR, Datar KV, Lin L, Roberts R, Caskey CT, Swanson MS: Identification of a (CUG)n triplet repeat RNA-binding protein and its expression in myotonic dystrophy. Nucl Acids Res 24:4407–4414 (1996).
Topisirovic I, Dragasevic N, Savic D, Ristic A, Keckarevic M, Keckarevic D, Culjkovic B, Petrovic I, Romac S, Kostic VS: Genetic and clinical analysis of spinocerebellar ataxia type 8 repeat expansion in Yugoslavia. Clin Genet 62:321–324 (2002).
Tsilfidis C, MacKenzie AE, Mettler G, Barcelo J, Korneluk RG: Correlation between CTG trinucleotide repeat length and frequency of severe congenital myotonic dystrophy. Nature Genet 1:192–195 (1992).
Vincent JB, Yuan QP, Schalling M, Adolfsson R, Azevedo MH, Macedo A, Bauer A, DallaTorre C, Medeiros HM, Pato MT, Pato CN, Bowen T, Guy CA, Owen MJ, O'Donovan MC, Paterson AD, Petronis A, Kennedy JL: Long repeat tracts at SCA8 in major psychosis. Am J med Genet 96:873–876 (2000).
Worth PF, Houlden H, Giunti P, Davis MB, Wood NW: Large, expanded repeats in SCA8 are not confined to patients with cerebellar ataxia. Nature Genet 24:214–215 (2000).
Zhuchenko O, Bailey J, Bonnen P, Ashizawa T, Stockton DW, Amos C, Dobyns WB, Subramony SH, Zoghbi HY, Lee CC: Autosomal dominant cerebellar ataxia (SCA6) associated with small polyglutamine expansions in the alpha-1A-voltage-dependent calcium channel. Nature Genet 15:62–69 (1997).

Cytogenet Genome Res 100:184–188 (2003)
DOI: 10.1159/000072853

SCA10 and ATTCT repeat expansion: clinical features and molecular aspects

X. Lin and T. Ashizawa

Department of Neurology, University of Texas Medical Branch, Galveston, Texas (USA)

SCA10, a clinically distinct ADCA

In recent years, tremendous progress has been achieved in dissecting the genetic components of various inherited ataxias. Most of the success was made in understanding autosomal dominant cerebellar ataxias (ADCA), which are a heterogeneous group of neurodegenerative disorders. The ADCAs constitute a complex clinical picture: besides the typical signs and symptoms of cerebellar dysfunction, different ADCAs have various involvements in the retina, the brain stem and other parts of the brain. Based on these variable presentations, the ADCAs are classified into three subtypes, i.e. ADCA-I, ADCA-II and ADCA-III (Harding, 1993). Clinically, ADCA-I is cerebellar ataxia accompanied by complex phenotypes, such as dysarthria, ophthalmoplegia, peripheral neuropathy, pyramidal and extrapyramidal signs, whereas ADCA-II is specifically associated with retinal degeneration. ADCA-III is characterized by relatively pure cerebellar syndrome (Harding, 1993). Recent molecular studies of different ADCA subtypes have revealed further heterogeneity; and a number of spinocerebellar ataxias (SCA) have been recognized by unique genetic mutations within each clinically defined ADCA. Several SCAs are grouped under ADCA-I (SCA1–4, SCA8, SCA12–14, SCA17 and DRPLA), whereas ADCA-II has only one entity so far (SCA7). ADCA-III includes SCA5, SCA6, SCA10 and SCA11 (Evidente et al., 2000; Klockgether et al., 2000).

Like the other members of ADCA-III, SCA10 has predominant cerebellar phenotypes. Patients usually present with gait ataxia between the ages of 14 to 45 years (Grewal et al., 1998; Matsuura et al., 1999, 2000; Zu et al., 1999). The ataxia progressively deteriorates and eventually results in loss of the ability to walk. Cerebellar dysfunctions were detected as upper and lower limb dysmetria, dysdiadochokinesis and broad-based gait. Dysarthria occurs several years after the onset of gait ataxia, and has the typical scanning quality of ataxic speech. Other cerebellar phenotypes, such as dysphagia, fragmented ocular pursuit and ocular dysmetria, are also reported. But other neurological abnormalities observed in ADCA-I and II are rarely seen in SCA10 patients. What is unique about SCA10 is the recurrent generalized motor seizures and/or complex partial seizures that many patients develop. This unique combination of seizures and predominant cerebellar dysfunction makes the clinical diagnosis of SCA10. However, although seizures are quite common in SCA10, not all patients have it; and some SCA10 patients present with "pure" cerebellar signs and symptoms and are not easily differentiated from the other ADCA-IIIs. Therefore, SCA5, SCA6, as well as other SCAs, which present with only cerebellar ataxia at early stage, should be considered in the differential diagnosis and genetic analysis is needed for differentiation among these different SCAs.

Two recent reports on six Mexican families with SCA10 gave a more detailed and complete account of the clinical spectrum of SCA10 (Rasmussen et al., 2001; Grewal et al., 2002). In addition to pure cerebellar ataxia and seizures, patients often show mild pyramidal signs, ocular dyskinesia, cognitive impairment, and mood and personality changes. Brain magnetic resonance imaging (MRI) showed predominant atrophy of the cerebellum with little changes in the other brain structures. The electroencephalograms (EEG) are abnormal with diffuse cortical dysfunction, indicated by slow, fused and disorganized activities. Focal cortical irritability or slow activities are found in some patients. On neurological examinations, patients often show mild sensory loss in distal extremities, and nerve conduction studies reveal peripheral neuropathy in some cases. These new data suggest that many parts of the brain are affected in

Supported by a grant from NIH (NS41547).

Received 9 October 2002; revision accepted 17 February 2003.

Request reprints from Tetsuo Ashizawa, M.D.
Department of Neurology, University of Texas Medical Branch
301 University Boulevard, JSA9.128, Galveston, Texas 77555-0539 (USA)
telephone: 409-772-2466; fax: 409-772-2390; e-mail: teashiza@utmb.edu

KARGER
Fax + 41 61 306 12 34
E-mail karger@karger.ch
www.karger.com

Accessible online at:
www.karger.com/cgr

SCA10 pathogenesis, and the clinical classification of SCA10 as an ADCA-III is less clear-cut. Moreover, one family has been reported to have abnormalities outside the nervous system, including liver failure, anemia and thrombocytopenia; and some patients in this family have patent ductus arteriosus and cardiac murmurs in the pulmonic area. These observations raised the possibility that peripheral involvement is an inherent part of the complete spectrum of SCA10 pathology. More extensive clinical evaluation of SCA10 patients and families are needed to better understand the varied phenotypes pertinent to SCA10.

ATTCT repeat, a unique pentanucleotide expansion

By linkage analysis in two large families the SCA10 locus was mapped to chromosome 22q13→qter independently by Pulst's group at UCLA and Ashizawa's group at Baylor College of Medicine (Matsuura et al., 1999; Zu et al., 1999). The critical region was subsequently defined in a 3.8-cM interval between D22S1140 and D22S1160 by two recombination events. More polymorphic markers were then used to further refine the critical region down to 2.7 cM between D22S1140 and D22S1153 (Matsuura et al., 2000).

It is of note that, in these two large families, anticipation was observed: there are more severe phenotypes and earlier age of onset in successive generations of the affected families (Matsuura et al., 1999; Zu et al., 1999). Anticipation is well characterized in many SCAs. Genetic basis for this dynamic feature is thought to be the nucleotide-repeat expansion, which tends to be unstable. Trinucleotide repeat expansion has been well recognized as the genetic cause for many SCAs and several other neurodegenerative conditions, such as the polyglutamine-coding CAG repeat in SCA1, SCA2, SCA3/MJD, SCA6, SCA7, SCA17, DRPLA, Kennedy's disease and Huntington's disease; 5′ UTR CAG repeat in SCA12; 3′ UTR repeat in DM1 and SCA8; 5′ UTR CGG/CCG repeats in fragile X syndrome and FRAXE mental retardation; and an intron GAA repeat in Friedreich's ataxia (Cummings and Zoghbi 2000). Therefore, trinucleotide repeat expansion was first explored in SCA10. However, when all the possible trinucleotide repeats (14 different repeats longer than 4) identified from the available sequence of the critical region of SCA10 were examined, none of them showed any expansion in SCA10 patients. The possibility of CAG or CAA expansion present in the unsequenced gaps within the critical region was further excluded by repeat expansion analysis (RED) (Schalling et al., 1993) on genomic DNA of SCA10 patients. Additionally, Western blot analysis of proteins obtained from patients' lymphoblastoid cells, using a monoclonal antibody that recognizes expanded polyglutamine tracts (Trottier et al., 1995), failed to detect proteins containing expanded polyglutamine tracts. These negative findings prompted the attempt to search for repeat expansions other than trinucleotide repeats.

A novel ATTCT pentanucleotide repeat was thereof identified in the SCA10 critical region (Matsuura et al., 2000). PCR analysis using primers flanking the repeats indicated that the length of this ATTCT repeat is polymorphic in normal populations. Normal individuals carry 10–22 uninterrupted ATTCT repeats with the majority being heterozygous. Intriguingly, SCA10 patients seem to lose the heterozygosity as tested by PCR. Moreover, the single amplifiable allele seems always passed on by the unaffected parent; whereas the single allele of the affected parent is never transmitted. These interesting observations suggested that the mutant allele somehow could not be amplified by PCR. However, potential mutation in the priming sites of the PCR did not seem to be the cause, since multiple sets of primers consistently gave the same result. It is thus likely that the mutation is either an unusually large ATTCT repeat expansion or certain erroneous rearrangement at this locus. Indeed, Southern blot analysis using a probe adjacent to the ATTCT repeat revealed large alleles of variable sizes on top of the normal allele; the detection of the large allele was specific for SCA10 patients and was absent in the normal population. Although genomic rearrangement at this locus cannot be categorically excluded in theory, ATTCT repeat expansion seems to be a more likely mechanism for SCA10 mutation. This hypothesis is supported by several lines of evidence: First of all the ATTCT repeat is polymorphic, like the trinucleotide repeats whose expansion is responsible for many SCAs and several other neurodegenerative diseases. Secondly, anticipation, a common phenomenon for diseases caused by nucleotide repeat expansion, is also reported in SCA10 families (Grewal et al., 1998, 2002; Zu et al., 1999). Thirdly, the mutation in SCA10 is dynamic with variably expanded alleles for different individuals from the same family; and the expansion is diverse in size ranging from 4–20 kb, equivalent to 800–4,000 repeats of the ATTCT pentanucleotide. This makes ATTCT pentanucleotide repeat expansion a novel and one of the longest disease-causing microsatellite repeats.

Anticipation and the dynamic nature of the SCA10 mutation suggest that the ATTCT pentanucleotide repeat expansion is unstable. Instability has been well documented for trinucleotide repeats, which is presumably due to unfaithful DNA replication during mitosis and meiosis (Pearson and Sinden, 1998). A stereotypical DNA structure of repetitive nucleotides apparently plays an important role in repeat instability, which has been known to occur during DNA replication, repair and recombination. Several genetic and environmental factors are involved in controlling the extent of the instability. Repetitive DNA sequences are inherently prone to recombine (Pluciennik et al., 2002). When DNA repair machinery is at fault, stability of trinucleotide repeats is affected in an expansion-dependent manner. For example, MSH2, a core component of DNA mismatch repair machinery, is able to bind trinucleotide repeat DNA structures (Pearson et al., 1997); and deletion of *Msh2* stabilizes expanded CAG repeats in transgenic mice overexpressing mutant huntingtin (Kovtun and McMurray, 2001; Manley et al., 1999). MSH2 is involved in repairing different types of DNA mismatch by heterodimerizing with either MSH3 or MSH6 (Kolodner and Marsischky, 1999). Interestingly, MSH3 and MSH6 have opposite effects on the somatic instability of CTG expansion in a DM1 knock-in model: deletion of *Msh3* stabilizes expanded CTG repeats, whereas deletion of *Msh6* increases the expansion (van den Broek et al., 2002). Trinucleotide repeat instability can also be influenced by multiple *cis*-elements in the vicinity of the repeat; and one such element is the DNA replication origin, the location and orientation of which modulates the length of expanded CAG/CTG repeats (Gomes-Pereira et al., 2001; Cleary et al., 2002; Panigrahi et al., 2002). As a distinct type of nucleotide repeat, it will be interesting to see if similar mechanisms of instability exist for ATTCT pentanucleotide expansion.

SCA10, a novel protein with unknown functions

The SCA10-associated ATTCT pentanucleotide repeat resides within intron 9 of SCA10, a novel gene of unknown function. Spanning 172.8 kb, human SCA10 has 12 exons, which produce an mRNA of 2 kb. The open reading frame consists of 1428 bp encoding 475 amino acids. Analysis of the amino acid sequence of SCA10 protein does not reveal any conserved domains, except for the C-terminal armadillo repeats. The armadillo repeats are found in Drosophila armadillo protein and the β-catenins, and are responsible for membrane association (Hatzfeld 1999). However, the functional significance of these repeats in SCA10 has yet to be elucidated. It will be interesting to investigate if SCA10 is also associated with cellular membranous structure by the same mechanism, and if SCA10 has similar functions as β-catenin.

Human SCA10 is highly conserved to its homologues in mouse and rat over the entire 475 amino acids: it has 80% identity and 89% similarity to the mouse homologue and 81% identity and 88% similarity to the rat homologue. Potential homologues of SCA10 in other model organisms have also been identified by BLAST search against NCBI databases, which suggests that SCA10 and its homologues across different species are evolutionarily conserved. In Arabidopsis, a putative plant protein of unknown function is found to be 24% identical and 41% similar to human SCA10 over 409 amino acids. In Drosophila, a gene, designated as CG4975, is the potential SCA10 homologue. CG4975 is localized to 54F1 with a single exon encoding a protein of 352 amino acids. There is a significant conservation of 23% identity and 38% similarity comparing with human SCA10 over a region of 330 amino acids (136–445 aa for human vs. 13–331 aa for Drosophila). No specific mutants can be associated with CG4975 in FlyBase, although deficiency of a large region encompassing CG4975 is lethal. It is not known, however, if CG4975 plays a major role in the lethality. With the recent technological development of generating targeted mutation in Drosophila, it has become more feasible to knock-out a specific gene, such as CG4975 (Rong and Golic, 2000; Kalidas and Smith, 2002; Rong et al., 2002). Studying the biology of CG4975 in Drosophila will provide insight into the function of SCA10.

Potential mechanisms of pathogenesis in SCA10

The mutational mechanism of SCA10 has been established as an expansion of the ATTCT pentanucleotide repeat, which is present in the intron 9 of SCA10. But it is unclear why such a genetic change can lead to the typical phenotypes of SCA10. At this stage, all possible mechanisms need to be considered; SCA10 pathogenesis could involve alterations of genomic organization, a gain of function at RNA level and haploinsufficiency of SCA10 protein. In the following sections, we will discuss these hypotheses in more detail.

Genomic disruption

The disease-causing expansion of the ATTCT repeat in SCA10 is in the range of 800–4,000 pentanucleotides, which is equivalent to 4 kb to 20 kb DNA with an unusually high content of adenosine and thymidine. Intriguingly, ATTCT repeats are present in an extremely large intron of 66 kb, and no additional genes have been identified so far within the SCA10 locus (Matsuura et al., 2000). Therefore, direct disruption of intron-exon organization of SCA10 or another gene embedded in the locus seems less likely. But insertion of such a long stretch of AT-rich (80%) sequence could cause DNA structural changes, such as alterations in chromatin organization and aberrant recruitment of protein factors involved in transcription regulation. These changes could disturb the expression of adjacent genes, resulting in neuronal dysfunction and degeneration in SCA10 patients.

It is of interest to explore whether the chromatin structure around the SCA10 locus on chromosome 22 is altered, and if the expression of genes close to the locus is changed. Several genes identified in the immediate vicinity of SCA10 are over 200 kb away from the ATTCT repeat. Fibulin-1 (FBLN1) encodes an extracellular matrix protein carrying tandem EGF-like domains. FBLN1 is an essential gene presumably involved in cell adhesion and migration. In the brain fibulin-1 is confined to neurons with little expression in astrocytes and microglia, indicating that fibulin-1 might play important roles in neuronal function (Argraves et al., 1990; Ohsawa et al., 2001; Twal et al., 2001). It would be worthwhile to examine the expression of FBLN1 and the other genes close to the SCA10 locus in SCA10 patient tissues (preferably in affected tissues, such as the cerebellum). It is also likely that the ATTCT expansion might impose long-range chromosomal suppressing effects, which can affect genes further away. But with the limited availability of such tissues, cellular and animal models need to be established for these studies.

RNA gain-of-function

At least in lymphoblast cell lines of SCA10 patients, SCA10 transcription occurs from both alleles, and SCA10 mRNA expression does not seem to differ significantly from wild-type cells in RNA blot analysis on total RNA (Matsuura and Ashizawa, unpublished). Therefore, the expanded ATTCT repeat is possibly transcribed into a very long stretch of AUUCU repeat expansion in the SCA10 pre-mRNA. During maturation of SCA10 mRNA, the long stretch of AUUCU repeats is expected to be spliced as part of intron 9 and degraded in the nucleus. Thus, the AUUCU repeat expansion should be present in nuclear RNA within a temporal window after transcription, and this phase is presumably determined by the rate of transcription, the efficiency of splicing and the rate at which AUUCU repeats are degraded in the nucleus. Persistence of AUUCU repeat-bearing RNA species might result in a gain-of-function similar to the mutant DMPK transcript, which carries the expanded CUG repeats in DM1 (Wang et al., 1995; Reddy and Housman 1997; Philips and Cooper, 2000; Pluciennik et al., 2002). DM1 is an autosomal dominant degenerative disorder characterized by myotonia, muscle weakness/wasting, cardiac conduction defect, and cataracts, diabetes, mental retardation/dysfunction, male baldness and testicular atrophy. DM1 is caused by a large expansion of CTG repeats (ranging from 50 to several thousands) in the 3′ UTR of DMPK located in chromosome 19. The major pathogenic mechanism of DM1 has been firmly established as RNA gain-of-function of DMPK transcript containing expanded CUG repeats; and this notion is strongly supported by two lines of transgenic mouse models over-expressing transcripts carrying CUG expansion (Mankodi et al., 2000; Seznec et al., 2000, 2001). The CUG repeats at the 3′ UTR retain the DMPK mRNA in the nucleus forming distinct loci with reduced presence in the cytoplasm (Taneja et al., 1995). The mutant transcript also interact with a number of RNA binding proteins, such as CUG-BP1, ETR3, MBLL, MBNL and MBXL; and some of these proteins are known to be important in pre-mRNA processing, such as splicing and nuclear export (Timchenko et al., 1996; Philips et al., 1998; Miller et al., 2000). Accumulation of mutant DMPK transcript in the nucleus might disrupt the normal functions of these RNA-binding proteins, affecting the splicing efficiency and expression of other genes. For example, splicing of cardiac troponin T, insulin receptor and chloride channel are impaired in DM1 because of the direct binding of CUG-BP1 to the CUG repeats; and the expression of these genes are reduced, thus providing insights into the molecular mechanisms for the cardiac involvement, insulin insensitivity and myotonia in DM1 (Philips et al., 1998; Savkur et al., 2001; Charlet et al., 2002; Mankodi et al., 2002). The essential requirement for RNA gain-of-function hypothesis in SCA10 is the presence of AUUCU repeat expansion in the RNA isolated from SCA10 patient samples. By an RT-PCR method modified from a PCR genotyping protocol for SCA10 (Matsuura and Ashizawa 2002), AUUCU repeat expansion is indeed detectable in the lymphoblastoid cell lines established from SCA10 patients (Matsuura and Ashizawa, unpublished). It would be interesting to examine if AUUCU repeat-containing transcript is accumulated in the nucleus, and if it recruits any RNA binding proteins and affect mRNA processing.

SCA10 loss-of-function

SCA10 is evolutionarily conserved from Drosophila to mammals. In mammals, SCA10 mRNA is ubiquitously expressed with a much higher level in the brain. RNA in situ hybridization studies further suggest that SCA10 is predominantly expressed in the neuronal populations of brain (Matsuura et al., 2000). SCA10 is a novel protein, and neither the genetic nor the biochemical functions are known. To understand the biological functions of SCA10, we performed a yeast two hybrid screening, and identified several proteins as specific interactors of SCA10, such as heterotrimeric G-protein β subunits and the targeting subunit of myosin light chain phosphatase (MYPT) (Nagamitsu and Ashizawa, unpublished). Given the well-established physiological importance of these molecules in various neuronal functions, it is of interest to further characterize these interactions and to investigate the functional significance. By interacting with these proteins, SCA10 might serve critical functions in the neurophysiology of brain, and loss of SCA10 might be a critical step in SCA10 pathogenesis.

Due to the limit in getting access to the affected tissues in SCA10 patients, it is not clear yet whether the expression of SCA10 is reduced in the brain of SCA10 patients. But several mechanisms are likely to cause diminished expression of SCA10 in the presence of intron ATTCT repeat expansion. The large pentanucleotide repeat expansion could form a weak triplex, which might impose a structural constraint and suppress SCA10 transcription. This mechanism has been proposed for Friedreich's ataxia, where an intron GAA trinucleotide repeat interferes with transcription of the FRDA gene by DNA triplex formation (Ohshima et al., 1998; Bidichandani et al., 1999; Sakamoto et al., 1999). Alternatively, intron ATTCT expansion might compromise the processing efficiency of the SCA10 transcript. Like DM2, which is caused by a large CCTG tetranucleotide repeat expansion in intron 1 of the ZNF9 gene, reduced processing of pre-mRNA could result in accumulation of repeat-containing transcript in nuclear foci (Liquori et al., 2001). This scenario would imply the potential pathogenic effects by recruiting some RNA binding factors. Yet another possibility involves reduced accuracy of splicing, resulting in aberrant SCA10 translation and consequent loss of or altered function of SCA10. Resolving these different scenarios awaits either direct molecular analysis of the affected tissues from SCA10 patients, or establishing appropriate models.

Future direction

To date, six SCA10 families of Mexican origin have been reported (Grewal et al., 1998, 2002; Matsuura et al., 1999, 2000; Zu et al., 1999; Rasmussen et al., 2001). The Mexican origin argues for a founder effect of the ATTCT pentanucleotide repeat expansion (Fujigasaki et al., 2002). Apparently, screening for more SCA10 patients and identifying more families are important to test this hypothesis. However, we have recently identified additional SCA10 families in Brazil (unpublished). Additional patients from broader genetic backgrounds will help further refine the clinical picture and establish a more accurate genotype-phenotype relationship.

Regardless of the global prevalence of the ATTCT pentanucleotide repeat expansion, identification and subsequent investigation of this mutation will provide a unique opportunity for many fields of genetic and neurological research, such as repeat instability and chromatin disorganization. Many pathogenic mechanisms have been proposed, waiting to be proven or discarded. From this perspective, perhaps the most important studies in understanding the molecular mechanism of SCA10 pathogenesis are to establish cellular and animal models. Investigation of these models will surely yield important insights into the genetics and molecular neurobiology of SCA10 pathogenesis.

References

Argraves WS, Tran H, Burgess WH, Dickerson K: Fibulin is an extracellular matrix and plasma glycoprotein with repeated domain structure. J Cell Biol 111:3155–3164 (1990).

Bidichandani SI, Purandare SM, Taylor EE, Gumin G, Machkhas H, Harati Y, Gibbs RA, Ashizawa T, Patel PI: Somatic sequence variation at the Friedreich ataxia locus includes complete contraction of the expanded GAA triplet repeat, significant length variation in serially passaged lymphoblasts and enhanced mutagenesis in the flanking sequence. Hum molec Genet 8:2425–2436 (1999).

Charlet BN, Savkur RS, Singh G, Philips AV, Grice EA, Cooper TA: Loss of the muscle-specific chloride channel in type 1 myotonic dystrophy due to misregulated alternative splicing. Mol Cell 10:45–53 (2002).

Cleary JD, Nichol K, Wang YH, Pearson CE: Evidence of *cis*-acting factors in replication-mediated trinucleotide repeat instability in primate cells. Nature Genet 31:37–46 (2002).

Cummings CJ, Zoghbi HY: Trinucleotide repeats: mechanisms and pathophysiology. A Rev Genomics hum Genet 1:281–328 (2000).

Evidente VG, Gwinn-Hardy KA, Caviness JN, Gilman S: Hereditary ataxias. Mayo Clin Proc 75:475–490 (2000).

Fujigasaki H, Tardieu S, Camuzat A, Stevanin G, LeGuern E, Matsuura T, Ashizawa T, Durr A, Brice A: Spinocerebellar ataxia type 10 in the French population. Ann Neurol 51:408–409 (2002).

Gomes-Pereira M, Fortune MT, Monckton DG: Mouse tissue culture models of unstable triplet repeats: in vitro selection for larger alleles, mutational expansion bias and tissue specificity but no association with cell division rates. Hum molec Genet 10:845–854 (2001).

Grewal RP, Achari M, Matsuura T, Durazo A, Tayag E, Zu L, Pulst SM, Ashizawa T: Clinical features and ATTCT repeat expansion in spinocerebellar ataxia type 10. Arch Neurol 59:1285–1290 (2002).

Grewal RP, Tayag E, Figueroa KP, Zu L, Durazo A, Nunez C, Pulst SM: Clinical and genetic analysis of a distinct autosomal dominant spinocerebellar ataxia. Neurology 51:1423–1426 (1998).

Harding AE: Clinical features and classification of inherited ataxias. Adv Neurol 61:1–14 (1993).

Hatzfeld M: The armadillo family of structural proteins. Int Rev Cytol 186:179–224 (1999).

Kalidas S, Smith DP: Novel genomic cDNA hybrids produce effective RNA interference in adult Drosophila. Neuron 33:177–184 (2002).

Klockgether T, Wullner U, Spauschus A, Evert B: The molecular biology of the autosomal-dominant cerebellar ataxias. Mov Disord 15:604–612 (2000).

Kolodner RD, Marsischky GT: Eukaryotic DNA mismatch repair. Curr Opin Genet Dev 9:89–96 (1999).

Kovtun IV, McMurray CT: Trinucleotide expansion in haploid germ cells by gap repair. Nature Genet 27:407–411 (2001).

Liquori CL, Ricker K, Moseley ML, Jacobsen JF, Kress W, Naylor SL, Day JW, Ranum LP: Myotonic dystrophy type 2 caused by a CCTG expansion in intron 1 of ZNF9. Science 293:864–867 (2001).

Mankodi A, Logigian E, Callahan L, McClain C, White R, Henderson D, Krym M, Thornton CA: Myotonic dystrophy in transgenic mice expressing an expanded CUG repeat. Science 289:1769–1773 (2000).

Mankodi A, Takahashi MP, Jiang H, Beck CL, Bowers WJ, Moxley RT, Cannon SC, Thornton CA: Expanded CUG repeats trigger aberrant splicing of ClC-1 chloride channel pre-mRNA and hyperexcitability of skeletal muscle in myotonic dystrophy. Mol Cell 10:35–44 (2002).

Manley K, Shirley TL, Flaherty L, Messer A: *Msh2* deficiency prevents in vivo somatic instability of the CAG repeat in Huntington disease transgenic mice. Nature Genet 23:471–473 (1999).

Matsuura T, Ashizawa T: Polymerase chain reaction amplification of expanded ATTCT repeat in spinocerebellar ataxia type 10. Ann Neurol 51:271–272 (2002).

Matsuura T, Achari M, Khajavi M, Bachinski LL, Zoghbi HY, Ashizawa T: Mapping of the gene for a novel spinocerebellar ataxia with pure cerebellar signs and epilepsy. Ann Neurol 45:407–411 (1999).

Matsuura T, Yamagata T, Burgess DL, Rasmussen A, Grewal RP, Watase K, Khajavi M, McCall AE, Davis CF, Zu L, Achari M, Pulst SM, Alonso E, Noebels JL, Nelson DL, Zoghbi HY, Ashizawa T: Large expansion of the ATTCT pentanucleotide repeat in spinocerebellar ataxia type 10. Nature Genet 26:191–194 (2000).

Miller JW, Urbinati CR, Teng-Umnuay P, Stenberg MG, Byrne BJ, Thornton CA, Swanson MS: Recruitment of human muscleblind proteins to (CUG)(n) expansions associated with myotonic dystrophy. Embo J 19:4439–4448 (2000).

Ohsawa I, Takamura C, Kohsaka S: Fibulin-1 binds the amino-terminal head of beta-amyloid precursor protein and modulates its physiological function. J Neurochem 76:1411–1420 (2001).

Ohshima K, Montermini L, Wells RD, Pandolfo M: Inhibitory effects of expanded GAATTC triplet repeats from intron I of the Friedreich ataxia gene on transcription and replication in vivo. J biol Chem 273:14588–14595 (1998).

Panigrahi GB, Cleary JD, Pearson CE: In vitro (CTG)*(CAG) expansions and deletions by human cell extracts. J biol Chem 277:13926–13934 (2002).

Pearson CE, Sinden RR: Trinucleotide repeat DNA structures: dynamic mutations from dynamic DNA. Curr Opin Struct Biol 8:321–330 (1998).

Pearson CE, Ewel A, Acharya S, Fishel RA, Sinden RR: Human MSH2 binds to trinucleotide repeat DNA structures associated with neurodegenerative diseases. Hum mol Genet 6:1117–1123 (1997).

Philips AV, Cooper TA: RNA processing and human disease. Cell Mol Life Sci 57:235–249 (2000).

Philips AV, Timchenko LT, Cooper TA: Disruption of splicing regulated by a CUG-binding protein in myotonic dystrophy. Science 280:737–741 (1998).

Pluciennik A, Iyer RR, Napierala M, Larson JE, Filutowicz M, Wells RD: Long CTGCAG repeats from myotonic dystrophy are preferred sites for intermolecular recombination. J biol Chem 277:34074–34086 (2002).

Rasmussen A, Matsuura T, Ruano L, Yescas P, Ochoa A, Ashizawa T, Alonso E: Clinical and genetic analysis of four Mexican families with spinocerebellar ataxia type 10. Ann Neurol 50:234–239 (2001).

Reddy PS, Housman DE: The complex pathology of trinucleotide repeats. Curr Opin Cell Biol 9:364–372 (1997).

Rong YS, Golic KG: Gene targeting by homologous recombination in Drosophila. Science 288:2013–2018 (2000).

Rong YS, Titen SW, Xie HB, Golic MM, Bastiani M, Bandyopadhyay P, Olivera BM, Brodsky M, Rubin GM, Golic KG: Targeted mutagenesis by homologous recombination in *D. melanogaster.* Genes Dev 16:1568–1581 (2002).

Sakamoto N, Chastain PD, Parniewski P, Ohshima K, Pandolfo M, Griffith JD, Wells RD: Sticky DNA: self-association properties of long GAATTC repeats in RRY triplex structures from Friedreich's ataxia. Mol Cell 3:465–475 (1999).

Savkur RS, Philips AV, Cooper TA: Aberrant regulation of insulin receptor alternative splicing is associated with insulin resistance in myotonic dystrophy. Nature Genet 29:40–47 (2001).

Schalling M, Hudson TJ, Buetow KH, Housman DE: Direct detection of novel expanded trinucleotide repeats in the human genome. Nature Genet 4:135–139 (1993).

Seznec H, Agbulut O, Sergeant N, Savouret C, Ghestem A, Tabti N, Willer JC, Ourth L, Duros C, Brisson E, Fouquet C, Butler-Browne G, Delacourte A, Junien C, Gourdon G: Mice transgenic for the human myotonic dystrophy region with expanded CTG repeats display muscular and brain abnormalities. Hum molec Genet 10:2717–2726 (2001).

Seznec H, Lia-Baldini AS, Duros C, Fouquet C, Lacroix C, Hofmann-Radvanyi H, Junien C, Gourdon G: Transgenic mice carrying large human genomic sequences with expanded CTG repeat mimic closely the DM CTG repeat intergenerational and somatic instability. Hum molec Genet 9:1185–1194 (2000).

Taneja KL, McCurrach M, Schalling M, Housman D, Singer RH: Foci of trinucleotide repeat transcripts in nuclei of myotonic dystrophy cells and tissues. J Cell Biol 128:995–1002 (1995).

Timchenko LT, Miller JW, Timchenko NA, DeVore DR, Datar KV, Lin L, Roberts R, Caskey CT, Swanson MS: Identification of a $(CUG)_n$ triplet repeat RNA-binding protein and its expression in myotonic dystrophy. Nucl Acids Res 24:4407–4414 (1996).

Trottier Y, Lutz Y, Stevanin G, Imbert G, Devys D, Cancel G, Saudou F, Weber C, David G, Tora L, et al: Polyglutamine expansion as a pathological epitope in Huntington's disease and four dominant cerebellar ataxias. Nature 378:403–406 (1995).

Twal WO, Czirok A, Hegedus B, Knaak C, Chintalapudi MR, Okagawa H, Sugi Y, Argraves WS: Fibulin-1 suppression of fibronectin-regulated cell adhesion and motility. J Cell Sci 114:4587–4598 (2001).

van den Broek WJ, Nelen MR, Wansink DG, Coerwinkel MM, te Riele H, Groenen PJ, Wieringa B: Somatic expansion behaviour of the $(CTG)_n$ repeat in myotonic dystrophy knock-in mice is differentially affected by Msh3 and Msh6 mismatch-repair proteins. Hum molec Genet 11:191–198 (2002).

Wang J, Pegoraro E, Menegazzo E, Gennarelli M, Hoop RC, Angelini C, Hoffman EP: Myotonic dystrophy: evidence for a possible dominant-negative RNA mutation. Hum molec Genet 4:599–606 (1995).

Zu L, Figueroa KP, Grewal R, Pulst SM: Mapping of a new autosomal dominant spinocerebellar ataxia to chromosome 22. Am J hum Genet 64:594–599 (1999).

Cytogenet Genome Res 100:189–197 (2003)
DOI: 10.1159/000072854

Why is SCA12 different from other SCAs?

S.E. Holmes,[a] E. O'Hearn,[b,c] and R.L. Margolis[a,d]

Departments of [a] Psychiatry, [b] Neurology and [c] Neuroscience; [d] Program of Cellular and Molecular Medicine, Johns Hopkins University School of Medicine, Baltimore MD (USA)

Abstract. Spinocerebellar ataxia type 12 (SCA12), now described in European-American and Asian (Indian) pedigrees, is unique among the SCAs from clinical, pathological, and molecular perspectives. Clinically, the distinguishing feature is early and prominent action tremor with variability in other signs. Pathologically, brain MRIs also suggest variability, with prominent cortical as well as cerebellar atrophy. Genetically, SCA12 is caused by a CAG repeat expansion that does not encode polyglutamine; we speculate that the mutation may affect expression of the gene PPP2R2B, which encodes a brain-specific regulatory subunit of the protein phosphatase PP2A.

The hereditary spinocerebellar ataxias (SCAs) are a group of autosomal dominant adult onset progressive neurodegenerative disorders that have overlapping and variable phenotypes. Cerebellar dysfunction is a hallmark of all the SCAs, but most also involve other regions of the central or peripheral nervous system. It is often impossible to distinguish among the subtypes of SCA on the basis of clinical and even pathological findings, but 21 distinct forms to date (SCAs 1–8 and 10–22) have been delineated genetically, either through identification of the causative mutation or by linkage to a unique chromosomal locus (Margolis, 2002). These SCAs can be classified into three groups based on pathogenesis. In SCAs 1, 2, 3, 7, and 17, the expanded CAG repeat is translated into an abnormally long and presumably toxic polyglutamine tract (Zoghbi et al., 2000). The second group includes SCA6 and episodic ataxia types 1 and 2 (EA1 and EA2), in which the mutation appears to disrupt calcium or potassium channel function. The third group includes SCAs 8 (Koob et al., 1999), 10 (Matsuura et al., 2000), and 12 (Holmes et al., 1999), which are associated with repeat expansions in non-coding regions that may affect levels of gene expression. The remaining SCAs (4, 5, 9, 11, 13–16, 19, 21 and 22) are of unknown etiology, and may or may not fit into one of the three other groups (Margolis et al., 2002). These known subtypes comprise 60–80% of adult onset dominant cerebellar ataxias (ADCAs); 20–40% remain genetically unidentified (Grewal et al., 1998; Moseley et al., 1998). Here we review SCA12, an unusual form of SCA with distinctive features at the clinical, pathological, and molecular levels.

Clinical Aspects of SCA12

Epidemiology and Genetics

SCA12 is an autosomal dominant cerebellar ataxia caused by a CAG repeat expansion initially identified in a single large pedigree (R) of German-American descent (Figs. 1, 2) (Holmes et al., 1999; O'Hearn et al., 2001), and subsequently detected in 14 Indian pedigrees (S. Jain, unpublished observation; Fujigasaki et al., 2001; Srivastava et al., 2001; K.K. Sinha, personal communication).

Of 77 families with dominant cerebellar diseases systematically assessed in India, 6.5% had the SCA12 mutation, making it the third most common SCA in this patient population (SCA1 15.6%, SCA2 24.7%, SCA3/MJD 2.6%, SCA7 2.6%, and no mutations at the SCA6, SCA8, or DRPLA loci) (Srivastava et al., 2001). However, no cases of SCA12 have been found

Supported by grants from the Huntington's Disease Society of American, the National Ataxia Foundation, and NIH grants NS38054, NS16375, and MH01275.

Received 22 November 2002; manuscript accepted 19 December 2002.

Request reprints from Dr. Susan E. Holmes, Division of Neurobiology
Department of Psychiatry, Meyer 2-181, 600 N. Wolfe Street
Baltimore, MD 21287 (USA); telephone: 410-614-4262
fax: 410-955-8233; e-mail: seholmes@jhmi.edu

KARGER Fax + 41 61 306 12 34
E-mail karger@karger.ch
www.karger.com

Accessible online at:
www.karger.com/cgr

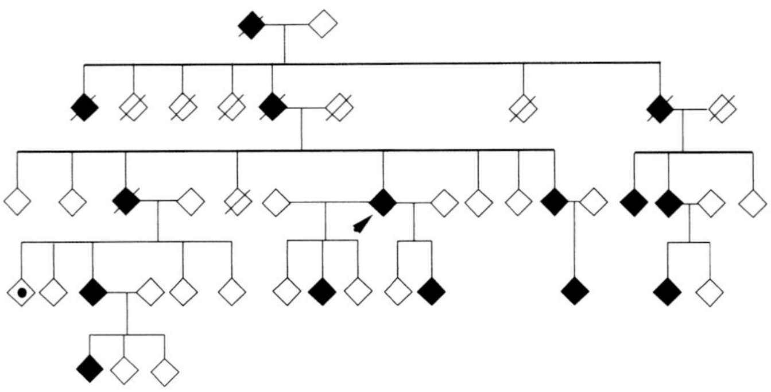

Fig. 1. Pedigree of the first described SCA12 family. The family is American of German descent. The proband is indicated by an arrow. One subject carries the mutation but is clinically unaffected (dot within diamond). The pedigree structure has been slightly altered to preserve confidentiality. Reproduced from Holmes et al. (2001), with permission from Elsevier Science.

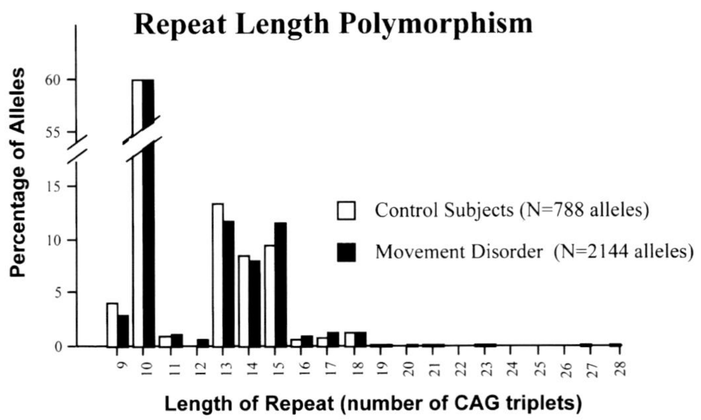

Fig. 3. Distribution of normal allele lengths in control subjects and movement disorder patients (including subjects with sporadic ataxia, familial ataxia, and Parkinson's disease) of European descent. Reproduced from Holmes et al. (2001), with permission from Elsevier Science.

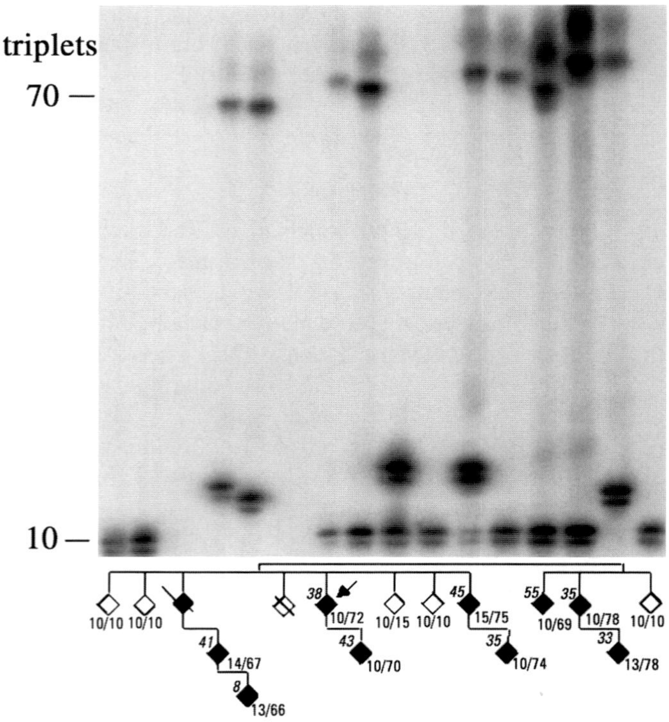

Fig. 2. Expansion of the SCA12 repeat in affected members of pedigree R. Allele sizes are noted to the bottom right of each individual and age of onset of affected members is at the top left of each filled diamond. Reproduced from Holmes et al. (1999) with permission from Nature Publishing Group.

in several sets of individuals with dominant familial ataxias from the U.S. and Europe (Holmes et al., 1999; Schols et al., 2000; Cholfin et al., 2001; Fujigasaki et al., 2001; Worth and Wood, 2001; Brusco et al., 2002; Maruyama et al., 2002; Silveira et al., 2002), suggesting that SCA12 is rare in these populations.

Normal repeat length has been assessed in control individuals of European, French, and Indian descent (Fig. 3) (Holmes et al., 1999; Fujigasaki et al., 2001; Srivastava et al., 2001). In all cases, normal repeat length is in the same range: 7–32 triplets, with a mode of 10. A greater proportion of alleles in India than in other populations contain more than 12 triplets (Fujigasaki et al., 2001), a finding now replicated in a second Indian study (Srivastava et al., 2001). As in the case of other SCAs, relative frequency of disease may parallel frequency of large normal alleles in a particular population (Fujigasaki et al., 2001).

The exact cutoff between repeats of normal and disease length is yet to be determined. Expanded alleles in clearly affected individuals range in length from 55–78 triplets. In a set of 933 European patients with familial and sporadic ataxia, alleles of 40 and 41 triplets were found in two patients with idiopathic late onset cerebellar ataxia (IDCA), and an allele of 49 triplets was found in an individual with SCA12-like phenotype and Creutzfeld-Jakob disease (C. Zuhlke, personal communication). Alleles of intermediate length have been described in two additional individuals of uncertain phenotype. An allele of 45 triplets was identified in a 28-year-old Indian individual with no family history of neurological or psychiatric disorder (Fujigasaki et al., 2001). This may represent the upper range of normal, or this individual may be at risk for the disease. A repeat expansion of 53 triplets has been identified in an Iranian woman with unipolar depression and her two monozygotic twin sons with schizophrenia (D. Wildenhauer, personal communication). No further information is currently available on this family, and it is not known whether they have any signs of neurodegenerative disease. While these cases suggest that the expansion may result in a psychiatric disorder without accompanying neurological symptoms, it is possible that this family has the SCA12 phenotype, or that the mutation is not penetrant for SCA12 in this family and their psychiatric disorders are coincidental. SCA12 expansions have not been detected in patients with bipolar disorder, schizophrenia, or Alzheimer's disease (Holmes, unpublished data).

There does not appear to be a correlation between repeat expansion size and age of onset in SCA12 (Holmes, unpublished), in contrast to the majority of SCAs caused by repeat expansion mutations, though the number of SCA12 individuals identified to date is still small, and age of onset can be difficult to define precisely in this disorder. Repeat length is modestly unstable during vertical transmission. In Pedigree R, four of five vertical transmissions showed slight contractions (the four contractions were seen in maternal transmissions, and allele size was stable in one paternal transmission) and variability was seen in each of the two sibships. In a large Indian pedigree, differences of three and five repeats were found in the two sibships with paternal transmission, and no changes were detected in maternally transmitted alleles (Fujigasaki et al., 2001). In five smaller Indian pedigrees, two slight contractions and one slight expansion were detected in three paternal transmissions (Srivastava et al., 2001).

Phenotype

SCA12 typically begins with action tremor of the upper extremities and progresses to include a wide and variable range of signs and symptoms, including mild cerebellar dysfunction, hyperreflexia, subtle Parkinsonian features, psychiatric symptoms, and dementia in some subjects with longstanding disease. The clinical manifestations observed in the initial pedigree (the R family) are listed in Table 1. Indian SCA12 patients have a similar phenotype except Parkinsonian features are not prominent. Of ten affected individuals examined in the R pedigree, eight have the combination of action tremor of head or arms, hyperreflexia, mild cerebellar dysfunction, subtle bradykinesia, and paucity of spontaneous movement. Age of onset ranges from 8 to 55, with most individuals noting disease onset in their 30's. Disease progression is quite slow.

SCA12 is the only SCA in which action tremor is the presenting and most common sign, as opposed to the initial symptom of gait ataxia seen in most of the SCAs. The tremor in SCA12 resembles essential tremor, and in fact several SCA12 patients were initially diagnosed with essential tremor. Therapy for essential tremor (beta-blockers or phenobarbital derivatives) has been helpful in some cases. In other patients, the tremor more closely resembles an intention tremor. As the disorder progresses, patients develop mild cerebellar dysfunction including gait ataxia and limb and eye movement dysmetria. They commonly display subtle Parkinsonian signs, including increased muscle tone, a paucity of spontaneous movement, and bradykinesia. In common with other SCAs, patients often display hyperreflexia and in older patients, extensor plantar responses. Signs seen in a minority of patients that may be manifestations of SCA12 include early onset nystagmus and dystonia, and late onset dementia. Several patients have psychiatric disorders, including anxiety and depression. A diagnosis of SCA12 should therefore be considered for a patient with action tremor of the upper extremities as the presenting sign, with later development of subtle cerebellar dysfunction, Parkinsonian features, hyperreflexia, or cognitive dysfunction.

Table 1. Clinical Features of SCA12

Clinical Sign	Number affected, of 10
Action tremor	10
Bradykinesia	9
Cerebellar signs, including ataxia	8
Hyperreflexia	8
Paucity of movement	8
Babinski present	5
Decreased tone	4
Psychiatric symptoms	4
Focal dystonia	2
Dementia	2
Incontinence	2

It should be noted that categorization of SCA12 as a spinocerebellar ataxia was based on clinical findings, but we recognize that the variability of the findings is such that the classification is somewhat arbitrary.

Case Report (reprinted from Holmes et al., 2003 with permission from Academic Press):

The 64-year-old proband of pedigree R (Fig. 1, arrow) developed postural and kinetic tremor of the arms at 38 years of age. Head tremor developed and increased in magnitude over the next decade. Her posture became less upright and her gait developed a staggering quality during her 50s. On examination at age 60, she had moderate upper extremity dysmetria, mild dysarthria, and abnormally brisk reflexes. A head CT (computed tomography) performed at age 62 (Fig. 4, A, B) revealed moderate, generalized atrophy of the cerebral hemispheres and cerebellum. Over the past two years, she has been unable to ambulate independently. She has been incontinent and cognitively impaired, and now requires assistance for most activities. She also has episodes of depression and irritability.

On neurologic exam at age 64, the patient displayed disorientation, memory loss, perseveration, and mild left sided neglect. She scored 8/28 on the Mini-mental State Exam (MMSE). Eye movements were normal. The most notable sign was a large amplitude, low frequency tremor of the head. She also had a similar tremor of outstretched arms, persisting during moderately dysmetric finger to nose testing. She was diffusely hyperreflexic with primitive reflexes, extensor plantar responses, and increased tone of arms and legs. Spontaneous movement was moderately reduced and she was mildly bradykinetic. Her speech was mildly dysarthric and hypophonic. When upright she was anteroflexed and had a tendency to retropulse. Her gait was ataxic and broad-based, and appeared to reflect a combination of cerebellar dysfunction, apraxia, and Parkinsonian features.

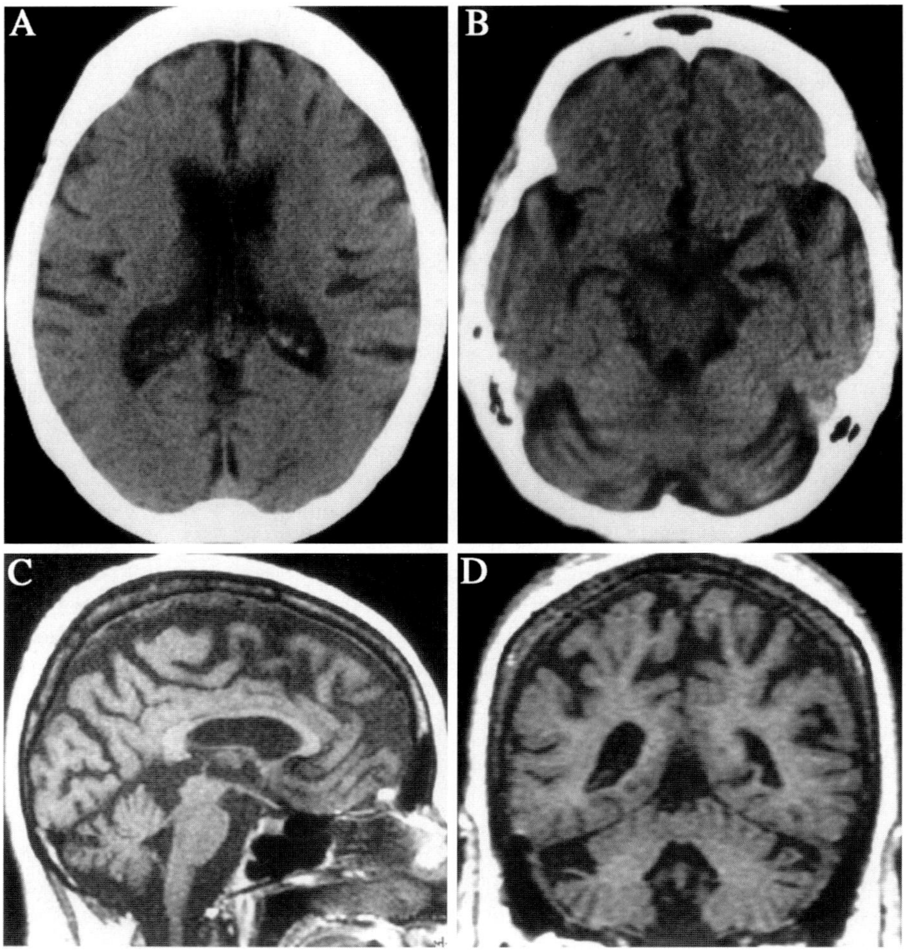

Fig. 4. Neuroradiologic images from two patients with SCA12. **A, B** Head CTs of the proband at age 62 years reveals cerebellar and diffuse cerebral cortical atrophy. **C** (sagittal), **D** (coronal): T-1 weighted MRI scans of a 59-year-old affected subject also shows cerebellar and cortical atrophy. Reproduced from Holmes et al. (2001), with permission from Elsevier Science.

Neuroimaging and neuropathology

Structural MRI findings in SCA12 typically reveal generalized atrophy of the brain, with the most prominent atrophy in the cerebral cortex. Atrophy of the cerebellum is modest, with the vermis somewhat more affected than the hemispheres. Other regions are variably and more mildly atrophic. This pattern of volume loss, in the presence of tremor and cerebellar signs and autosomal dominant inheritance, would justify inclusion of SCA12 in a differential diagnosis. In particular, the presence of greater atrophy of the cerebral cortex than the cerebellum may help distinguish SCA12 from other SCAs; a case can be made that SCA12 more nearly resembles multisystem atrophy than narrowly defined SCA.

As predicted from MRI findings, the preliminary analysis of the single SCA12 brain that has come to autopsy revealed diffuse atrophy, most marked in the cerebral and cerebellar cortex. There was definite loss of cerebellar Purkinje cells; the cause of volume loss in the cerebral cortex has yet to be determined, though findings typical of Alzheimer's disease were not observed (O'Hearn and Troncoso, unpublished observations).

The molecular basis of SCA12

A unique pathogenic mechanism

On the basis that each disorder presumably is a direct consequence of altered gene expression rather than a toxic protein (polyglutamine diseases) or a channel malfunction, SCA12 can be usefully grouped with SCA8 and SCA10. In each case, the causative repeat expansion is in a non-coding region. SCA8 is caused by a CTG expansion mutation in the 3′ untranslated region of the SCA8 gene, which is not translated. The SCA8 gene overlaps with KLHL-1, a gene on the opposite DNA strand. It is hypothesized that the SCA8 gene normally regulates expression of KLHL-1 through an RNA-RNA interaction, which is disrupted by a repeat expansion (Nemes et al., 2000). SCA10 is caused by very long expansions of an ATTCT pentanucleotide repeat found within intron 9 of a gene of unknown function termed SCA10 (Matsuura et al., 2000). It seems reasonable to suspect, based on the example of Friedreich's ataxia, that this massive mutation somehow disrupts SCA10 expression.

While the effect of the mutation in SCA12 may, like the SCA8 and SCA10 mutations, influence gene expression, SCA12 pathogenesis appears to be unique in several respects,

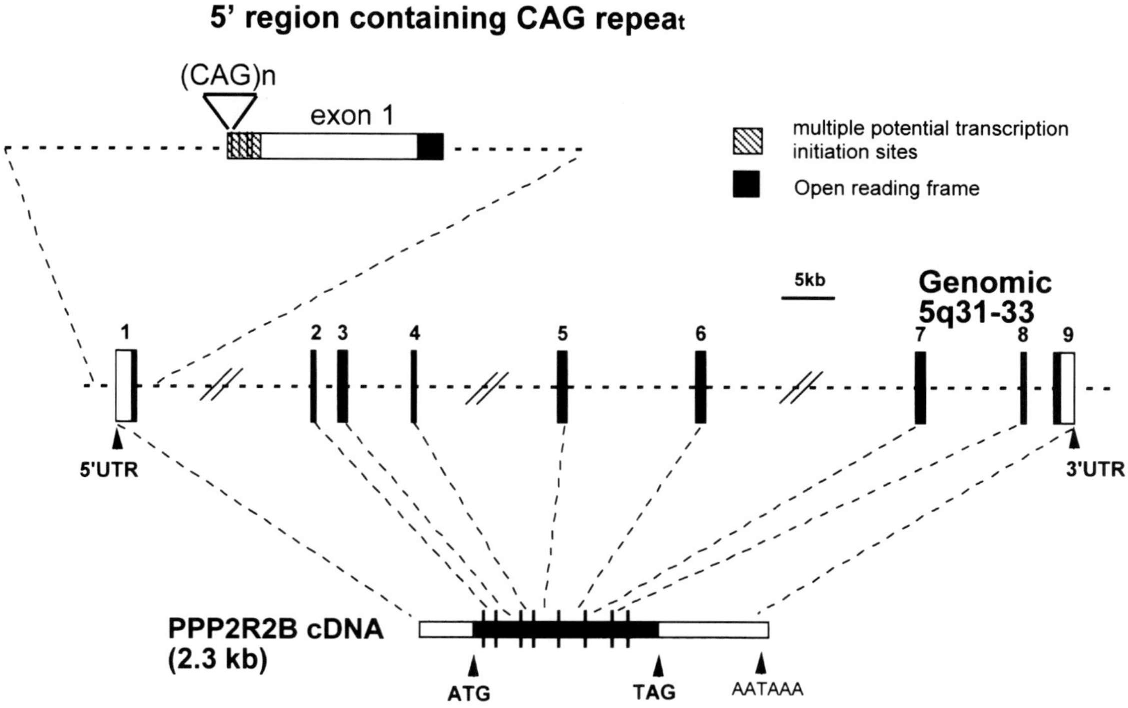

Fig. 5. Structure of PPP2R2B. Genomic sequence containing the nine known exons of PPP2R2B is shown aligned with the cloned cDNA. The gene stretches over approximately 290 kb (exons not to scale). Hatched box at the 5′ end of exon 1 indicates region containing multiple potential transcription start sites. The originally published PPP2R2B transcript is 3.4 kb long, but the majority of ESTs come from transcripts utilizing an upstream polyadenylation signal at approximately 2.3 kb. Modified from Holmes et al. (1999) with permission from Nature Publishing Group.

to be described in more detail below. Among other properties, it is the only known SCA caused by a triplet repeat expansion in a 5′ untranslated region. It is also the only SCA in which the mutation may increase gene expression; in this case, because the gene product is involved in regulation of protein phosphatase PP2A, abnormal phosphatase activity may be critical to SCA12 pathogenesis.

A CAG expansion mutation in PPP2R2B causes SCA12

The CAG repeat associated with SCA12 lies 133 nt upstream of the 5′ end of a cDNA of the gene PPP2R2B (GenBank accession no. M64930) (Mayer et al., 1991) and maps to 5q31→q33 (Fig. 5). However, the precise relationship between the repeat and the PPP2R2B transcript remains ambiguous. Several lines of evidence suggest that the actual start site of this gene is 5′ to that initially published, and that the expansion may in some instances lie within the 5′ UTR. First, ten clones derived by the 5′ rapid amplification of cDNA ends (RACE) technique from several different brain cDNA samples were sequenced. The products extended variable distances 5′ to the published cDNA, ending as close as nine nt 3′ to the CAG repeat (Fig. 5). Second, a mouse EST (AU051098) contains an interrupted CAG repeat $(CAG)_2GAG(CAG)_2CAC(CAG)_3$ that corresponds in position to the human repeat. The EST begins 22 nt 5′ to the human CAG repeat, extends 3′ into the protein-coding region, and has 91% homology to the human sequence. This suggests that the transcript, at least in some cases, includes the repeat. Third, two additional mouse ESTs (prepared using a Capfinder protocol) come to within 10 and 25 nts of the repeat. These results suggest the possibility of multiple alternative transcription start sites, indicated by hatching (Fig. 5), and are consistent with the presence of a MED-1 (multiple start-site element downstream-1) sequence (GCTCCC) (Ince and Scotto, 1995) 65 nt 3′ to the repeat. This promoter motif is typically found downstream of transcription initiation sites in genes that contain multiple initiation sites. Finally, RT-PCR and analysis of human genome sequence and additional ESTs indicates multiple upstream exons and additional splice variants, suggesting a complex picture at the 5′ end of the gene (Holmes et al., 2002). The repeat-containing exon is the first exon in some ESTs, but in other splice variants, this exon is skipped, and alternate upstream exons are used. Thus it appears likely that the repeat is variably included within the PPP2R2B transcript, and is located in various positions relative to the transcription start site depending on alternate patterns of splicing and alternate sites of transcription initiation.

The possibility that the CAG repeat might encode polyglutamine, either within an unidentified ORF of PPP2R2B, or within another unidentified adjacent or overlapping gene, has been examined in several ways, and no evidence has emerged to support the existence of such a transcript. Polyglutamine expansions were not detected on Western blots of protein

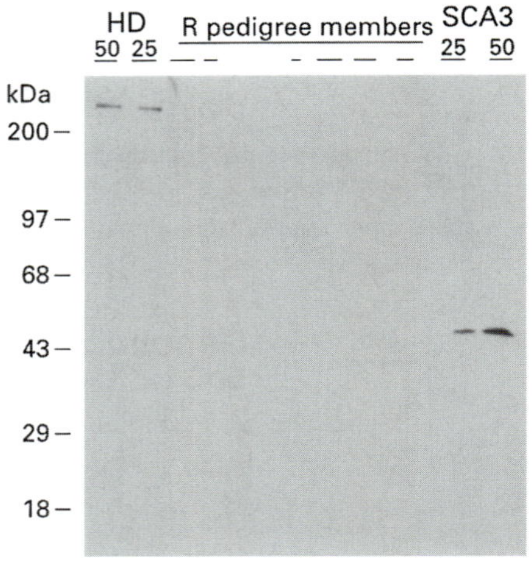

Fig. 6. Absence of polyglutamine expansion in protein from lymphoblastoid cells of SCA12 subjects. Protein from lymphoblastoid cell lines of the R pedigree was probed with the 1C2 antibody, which preferentially detects stretches of greater than 35–40 glutamines. Lanes 1 and 2 contain 50 and 25 µg respectively of protein extract from a lymphoblastoid cell line of an HD patient with a $(CAG)_{46}$ repeat, and lanes 10 and 11 contain 25 and 50 µg of protein from a cell line of an SCA3 subject with a $(CAG)_{76}$ repeat. Reproduced from Holmes et al. (2001), with permission from Elsevier Science.

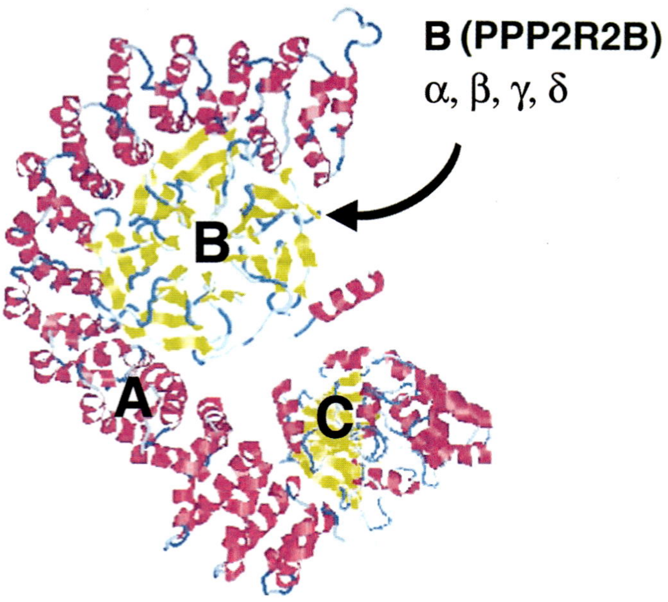

Fig. 7. Schematic diagram of the trimeric holoenzyme PP2A (protein phosphatase 2A): A and C form a core dimer with which one of multiple B regulatory subunits may combine. B subunits are thought to confer substrate and cell-type specificity on the ubiquitous enzyme. PPP2R2B encodes one of the brain-specific B subunits. PPP2R2B (Bβ) and related B subunits are thought to form a "beta-propeller" structure, composed of beta sheets and turns (yellow). (B subunits of other families (B′ and B′) also bind to A, but do not form beta-propeller structures.) The structural A subunit is composed primarily of alpha helixes (red), while the catalytic C subunit contains both beta-sheet and alpha-helix. See Strack et al. (2002)

derived from lymphoblastoid cell lines of affected family members and probed with an antibody (1C2) (Trottier et al., 1995) that specifically detects long stretches of polyglutamine (Fig. 6). This does not eliminate the possibility of a brain-specific unidentified ORF of PPP2R2B in which the repeat encodes polyglutamine, but provides support for the molecular evidence against the presence of such a transcript. As an additional approach to this issue, Northern blots containing brain mRNA and mRNA from neuroblastoma cell line LA-N-1, which expresses high levels of PPP2R2B, were probed using oligonucleotides immediately 5′ and 3′ to the CAG repeat expansion, and no signal was detected (unpublished results). Finally, GENSCAN (Burge and Karlin, 1997) does predict an exon in which the $(CAG)_{10}$ repeat encodes polyglutamine, but the prediction is of low probability, and no exon is predicted if the CAG repeat is reduced to a single triplet, suggesting that the prediction may be an artifact of a long triplet repeat.

Since the repeat apparently resides outside of an open reading frame, it does not appear that the pathogenesis of SCA12 involves a toxic effect of an expanded polyglutamine. A possible alternative mechanism for disease pathogenesis is an effect of the expansion on the level of gene expression, somewhat analogous to the 5′ UTR CGG expansion that causes the fragile X syndrome (Fu et al., 1991; Verkerk et al., 1991; Yu et al., 1991). As the repeat lies in the putative promoter region of PPP2R2B, it is plausible that expansion of the repeat alters expression of this gene, which could in turn affect activity of the trimeric enzyme PP2A. Preliminary results indicate that the repeat-containing region does function as a promoter for PPP2R2B, and that expansion of the repeat causes a several-fold increase in gene expression in some cell types (Holmes et al., 2000). Further experiments are in process to determine whether this effect is cell-specific and whether it can be attributed to altered spacing of promoter elements, as with the expansion of the dodecamer in the cystatin B promoter leading to EPM1 (Lalioti et al., 1999) or whether it depends on the specific CAG sequence.

PPP2R2B is a regulatory subunit of protein phosphatase PP2A

Protein phosphatase 2A (PP2A, also termed PP2) is a protein serine/threonine phosphatase that has been implicated in the regulation of many cellular processes involving protein phosphorylation, including cell growth and differentiation, DNA replication, cellular morphogenesis and cytokinesis, regulation of kinase cascades, ion channel function, neurotransmitter release, microtubule assembly, and apoptosis (Santoro et al., 1998; Price and Mumby, 1999; Virshup, 2000). It is an essential enzyme expressed in all eukaryotic cells, and its constitutive components are among the most highly conserved proteins in evolution (Shenolikar, 1994). The designation PP2A actually refers to a subfamily of enzymes, all of which are composed of three subunits: a catalytic subunit (C), a structural subunit (A), and a regulatory (B) subunit (Fig. 7). A and C form a complex to which one of multiple B regulatory subunits is recruited by subunit A. In mammals, two isoforms each of A and C exist, while there are at least 13 B regulatory subunits encoded by distinct genes. Theoretically, up to 40 different combinations of the three subunits could exist in mammalian cells. The A and C subunits of PP2A are expressed constitutively in all mammalian cells, whereas the B subunits are distributed differentially among different tissues and cell types, and are thought to modulate PP2A function through regulation of substrate specificity and intracellular targeting (e.g. nuclear, cytoplasmic, or mitochondrial) (Wera and Hemmings, 1995; McCright et al., 1996; Tehrani et al., 1996; Millward et al., 1999; Virshup, 2000).

One class of B subunits is referred to as PPP2R2, and includes PPP2R2A, PPP2R2C and PPP2R2B (which is also known as PR55β, B55β, and Bβ). Secondary structure prediction for this class of subunits indicates that the proteins form a seven-bladed β-propeller, with seven twisted, antiparallel β-sheets radially arranged around a common core (Strack et al., 2002). Deletion and site-directed mutagenesis was used to determine the PPP2R2C (also known as B-gamma) residues critical for holoenzyme formation (Strack et al., 2002). All of these critical residues are conserved between PPP2R2C and PPP2R2B.

PPP2R2B is brain specific (Mayer et al., 1991) and widely expressed in neurons throughout the brain, including constitutive expression in Purkinje cells of the cerebellar cortex (Strack et al., 1998), a pattern consistent with the clinical and neuroradiologic involvement of multiple CNS regions in SCA12. As expected, the PPP2R2 class of regulatory units, including PPP2R2B, regulates PP2A dephosphorylation activity for specific substrates, including vimentin (Turowski et al., 1999), histone-1 (Ferrigno et al., 1993), and tau (Sontag et al., 1996). PPP2R2 subunits presumably regulate PP2A activity against many other still unidentified substrates. It is clear from studies on multiple organisms that mutant forms of PPP2R2 subunits produce a wide variety of deleterious effects. Mutations in the yeast PPP2R2 homologue *CDC55* result in defective cell separation and septation (Healy et al., 1991). The Drosophila mutant *aar*^{1} (abnormal anaphase resolution), in which a P-element is integrated four base pairs downstream of the exon IV/intron 4 boundary of the Drosophila PPP2R2 homologue, has mitotic defects in anaphase (Mayer-Jaekel et al., 1993). The nearly identical Drosophila mutant *twins*^{P} displays abnormal specification of neuronal identity during development (Shiomi et al., 1994); *twins*^{P} has reduced dephosphorylation activity against some substrates (caldesmon and histone H1), but not others (phosphorylase a) (Mayer-Jaekel et al., 1994). Recent evidence suggests that down regulation of PP2A containing either PPP2R2A or PPP2R2B in cultured mammalian brain slices results in Alzheimer's-like accumulation of hyperphosphorylated tau. The relative contribution of PPP2R2A vs. PPP2R2B in vivo remains to be determined.

In addition to substrate specificity, subcellular localization of PP2A is regulated by B subunits. PPP2R2-containing PP2A enzyme complexes are usually cytoplasmic and localized in neuronal cell bodies and dendrites. However, association of the PPP2R2 subunit with HIV-encoded protein Vpr results in targeting of PP2A to the nucleus, where it increases dephosphorylation of nuclear cdc25 (Hrimech et al., 2000).

CAG repeat expansion
↓
increased PPP2R2B expression
↓
altered PP2A activity
↓
altered protein phosphorylation
↓
neuronal dysfunction
↓
neuronal death

Fig. 8. Hypothetical model of SCA12 pathogenesis. Reproduced from Holmes et al. (2003) with permission from Academic Press.

Taken together, these findings demonstrate that PPP2R2B may be an important regulator of phosphatase activity in the brain. It therefore follows that quantitative or qualitative change in PPP2R2B expression would be reflected in abnormal phosphatase activity.

Potential mechanisms of pathogenesis

Though preliminary evidence suggests that the SCA12 repeat expansion alters the level of expression of PPP2R2B, various configurations of the 5′ end of the transcript have been identified, and additional pathogenic mechanisms are possible. First, expansion could result in alteration of the splicing pattern of nearby exons, as the repeat could be part of a splice enhancer or inhibitor. Translation of identified splice variants indicates that there are at least six different potential N-termini of PPP2R2B, potentially affecting the subcellular localization of the PP2A enzyme (Holmes et al., 2002). Second, the mutation could exert a toxic effect at the RNA level, such as that proposed in the case of myotonic dystrophy (Mankodi et al., 2000; Liquori et al., 2001). This mechanism depends on the presence of the SCA12 repeat within the PPP2R2B transcript (5′ UTR). The DM repeat expansion is substantially larger than that of SCA12, and the repeat is always found within the transcript, so this mechanism seems less likely in the case of SCA12. Third, the mutation could directly inhibit transcription, as in Friedreich's ataxia (Patel et al., 2001). Fourth, PPP2R2B could associate with proteins in addition to the PP2A complex, such that altered expression of PPP2R2B could lead to toxicity through effects on pathways unrelated to PP2A. While this has not yet been demonstrated for PPP2R2B, the PP2A A subunit, and at least one type of B subunit, associate with PP5 (Lubert et al., 2001), suggesting that B subunits may have other roles beyond regulating PP2A.

Nonetheless, the most straightforward model of SCA12 pathogenesis remains an alteration of PP2A function as follows: 1) the CAG repeat expansion in PPP2R2B leads to abnormal levels of expression of PPP2R2B, 2) abnormal levels of PPP2R2B, a key regulatory subunit of the enzyme, result in abnormal PP2A function, 3) changes in PP2A function results in altered protein phosphorylation, and 4) altered phosphorylation of essential neuronal proteins leads to neurotoxicity (Fig. 8). Should this hypothesis of SCA12 pathogenesis prove correct, elucidation of the pathways that link PP2A to neuronal death should prove of considerable importance in understanding SCA12 and other neurodegenerative disorders.

Acknowledgements

We wish to thank the patients and their family members who have participated in our investigations of SCA12. We also thank S. Strack for providing Fig. 7; C. Callahan, H. Hwang, J. Kleiderlein, D. Gorelick-Feldman, and A. Sharp for their technical assistance; R. Seeger for LA-N-1 cells; and P.R. McHugh and J.R. DePaulo for support and encouragement. Portions of this paper were reprinted with minor modifications from Holmes et al. (2003) with permission from Academic Press.

References

Brusco A, Cagnoli C, Franco A, Dragone E, Nardacchione A, Grosso E, Mortara P, Mutani R, Migone N, Orsi L: Analysis of SCA8 and SCA12 loci in 134 Italian ataxic patients negative for SCA1-3 6 and 7 CAG expansions. J Neurol 249:923–929 (2002).

Burge C, Karlin S: Prediction of complete gene structures in human genomic DNA. J molec Biol 268:78–94 (1997).

Cholfin JA, Sobrido MJ, Perlman S, Pulst SM, Geschwind DH: The SCA12 mutation as a rare cause of spinocerebellar ataxia. Arch Neurol 58:1833–1835 (2001).

Ferrigno P, Langan TA, Cohen P: Protein phosphatase 2A1 is the major enzyme in vertebrate cell extracts that dephosphorylates several physiological substrates for cyclin-dependent protein kinases. Mol Biol Cell 4:669–677 (1993).

Fu Y-H, Kuhl DPA, Pizzuti A, Pieretti M, Sutcliffe JS, Richards S, Verkerk AJMH, Holden JJA, Fenmwick RG, Warren ST, Nelson DL, Caskey CT: Variation of the CGG repeat at the Fragile X site results in genetic instability: resolution of the Sherman paradox. Cell 67:1047–1058 (1991).

Fujigasaki H, Verma IC, Camuzat A, Margolis RL, Zander C, Lebre AS, Jamot L, Saxena R, Anand I, Holmes SE, Ross CA, Durr A, Brice A: SCA12 is a rare locus for autosomal dominant cerebellar ataxia: a study of an Indian family. Ann Neurol 49: 117–121 (2001).

Grewal RP, Tayag E, Figueroa KP, Zu L, Durazo A, Nunez C, Pulst SM: Clinical and genetic analysis of a distinct autosomal dominant spinocerebellar ataxia. Neurol 51:1423–1426 (1998).

Healy AM, Zolnierowicz S, Stapleton AE, Goebl M, DePaoli-Roach AA, Pringle JR: *CDC55* a *Saccharomyces cerevisiae* gene involved in cellular morphogenesis: identification characterization and homology to the B subunit of mammalian type 2A protein phosphatase. Mol Cell Biol 11:5767–5780 (1991).

Holmes SE, O'Hearn EE, McInnis MG, Gorelick-Feldman DA, Kleiderlein JJ, Callahan C, Kwak NG, Ingersoll-Ashworth RG, Sherr M, Sumner AJ, Sharp AH, Ananth U, Seltzer WK, Boss MA, Vieria-Saecker AM, Epplen JT, Riess O, Ross CA, Margolis RL: Expansion of a novel CAG trinucleotide repeat in the 5′ region of PPP2R2B is associated with SCA12. Nature Genet 23:391–392 (1999).

Holmes SE, Fujigasaki H, O'Hearn E, Antonarakis S, Cooper JK, Callahan C, Hwang J, Gorelick-Feldman D, Verma IC, Saxena R, Durr A, Brice A, Ross CA, Margolis RL: Spinocerebellar ataxia type 12 (SCA12): additional evidence for a causative role of the CAG repeat expansion in PPP2R2B. abstract Am J hum Genet 67:1074 (2000).

Holmes SE, O'Hearn E, Ross CA, Margolis RL: SCA12: an unusual mutation leads to an unusual spinocerebellar ataxia. Brain Res Bull 56:397–403 (2001).

Holmes SE, Hwang H, O'Hearn E, Antonarakis SE, Chip S, Strck S, Ross CA, Margolis RL: Genomic structure and splice variants of PPP2R2B the gene associated with spinocerebellar ataxia type 12 (SCA12). Abstract Am J hum Genet 71:973 (2002).

Holmes SE, O'Hearn E, Brahmachari SK, Choudhry S, Srivastava AK, Jain S, Ross CA, Margolis RL: SCA12, in Pulst SM (ed): Genetics of Movement Disorders, pp 121–132 (Academic Press, San Diego 2003).

Hrimech M, Yao XJ, Branton PE, Cohen EA: Human immunodeficiency virus type 1 Vpr-mediated G(2) cell cycle arrest: Vpr interferes with cell cycle signaling cascades by interacting with the B subunit of serine/threonine protein phosphatase 2A. EMBO J 19:3956–3967 (2000).

Ince TA, Scotto KW: A conserved downstream element defines a new class of RNA polymerase II promoters. J biol Chem 270:30249–30252 (1995).

Koob MD, Moseley ML, Schut LJ, Benzow KA, Bird TD, Day JW, Ranum LPW: An untranslated CTG expansion causes a novel form of spinocerebellar ataxia (SCA8). Nature Genet 21:379–384 (1999).

Lalioti MD, Scott HS, Antonarakis SE: Altered spacing of promoter elements due to the dodecamer repeat expansion contributes to reduced expression of the cystatin B gene in EPM1. Hum molec Genet 8:1791–1798 (1999).

Liquori CL, Ricker K, Moseley ML, Jacobsen JF, Kress W, Naylor SL, Day JW, Ranum LPW: Myotonic Dystrophy type 2 caused by a CCTG expansion in intron 1 of ZNF9. Science 293:864–867 (2001).

Lubert EJ, Hong YI, Sarge KD: Interaction between protein phosphatase 5 and the A subunit of protein phosphatase 2A: Evidence for a heterotrimeric form of protein phosphatase 5. J biol Chem 276:38582–38587 (2001).

Mankodi A, Logigian E, Callahan L, McClain C, White R, Henderson D, Krym M, Thornton CA: Myotonic dystrophy in transgenic mice expressing an expanded CUG repeat. Science 289:1701–1702 (2000).

Margolis RL: The spinocerebellar ataxias: Order emerges from chaos. Curr Neurol Neurosc Rep 2:447-456 (2002).

Maruyama H, Izumi Y, Morino H, Oda M, Toji H, Nakamura S, Kawakami H: Difference in disease-free survival curve and regional distribution according to subtype of spinocerebellar ataxia: A study of 1286 Japanese patients. Am J med Genet 114:578–583 (2002).

Matsuura T, Yamagata T, Burgess DL, Rasmussen A, Grewal RP, Watase K, Khajavi M, McCall AE, Davis CF, Zu L, Achari M, Pulst SM, Alonso E, Noebels JL, Nelson DL, Zoghbi HY, Ashizawa T: Large expansion of the ATTCT pentanucleotide repeat in spinocerebellar ataxia type 10. Nature Genet 26:191–194 (2000).

Mayer RE, Hendrix P, Cron P, Mattheis R, Stone SR, Goris J, Merlevede W, Hofsteenge J, Hemmings BA: Structure of the 55-kDa regulatory subunit of protein phosphatase 2A: evidence for a neuronal-specific isoforms. Biochem 30:3589–3597 (1991).

Mayer-Jaekel RE, Ohkura H, Gomes R, Sunkel CE, Baumgartner S, Hemmings BA, Glover DM: The 55 kDa regulatory subunit of Drosophila protein phosphatase 2A is required for anaphase. Cell 72:621–633 (1993).

Mayer-Jaekel RE, Ohkura H, Ferrigno P, Andjelkovic N, Shiomi K, Uemura T, Glover DM, Hemmings BA: Drosophila mutants in the 55 kDa regulatory subunit of protein phosphatase 2A show strongly reduced ability to dephosphorylate substrates of p34cdc2. J Cell Sci 107:2609–2616 (1994).

McCright B, Rivers AM, Audlin S, Virshup DM: The B56 family of protein phosphatase 2A (PP2A) regulatory subunits encodes differentiation-induced phosphoproteins that target PP2A to both nucleus and cytoplasm. J biol Chem 271:22081–22089 (1996).

Millward TA, Zolnierowicz S, Hemmings BA: Regulation of protein kinase cascades by protein phosphatase 2A. Trends Biochem Sci 24:186–191 (1999).

Moseley ML, Benzow KA, Schut LJ, Bird TD, Gomez CM, Barkhaus PE, Blindauer KA, Labuda M, Pandolfo M, Koob MD, Ranum LP: Incidence of dominant spinocerebellar and Friedreich triplet repeats among 361 ataxia families. Neurol 51:1666–1671 (1998).

Nemes JP, Benzow KA, Moseley ML, et al: The SCA8transcript is an antisense RNA to a brain-specific transcript encoding a novelactin-binding protein (KLHL1). Hum molec Genet 9:1543–1551 (2000).

O'Hearn E, Holmes SE Calvert PC, Ross CA, Margolis RL: SCA12: Tremor with cerebellar and cortical atrophy is associated with a CAG repeat expansion. Neurol 56:299–303 (2001).

Patel PI, Isaya G: Friedreich ataxia: from GAA triplet-repeat expansion to frataxin deficiency. Am J hum Genet 69:15–24 (2001).

Price NE, Mumby MC: Brain protein serine/threonine phosphatases. Curr Opin Neurobiol 9:336–342 (1999).

Santoro MF, Annand RR, Robertson MM, Peng Y-W, Brady MJ, Mankovich JA, Hackett MC, Ghayur T, Walter G, Wong WW, Giegel DA: Regulation of protein phosphatase 2A activity by caspase-3 during apoptosis. J biol Chem 273:13119–13128 (1998).

Schols L, Szymanski S, Peters S, Przuntek H, Epplen JT, Hardt C, Riess O: Genetic background of apparently idiopathic sporadic cerebellar ataxia. Hum Genet 107:132–137 (2000).

Shenolikar S: Protein serine/threonine phosphatases – new avenues for cell regulation. A Rev Cell Biol 10:55–86 (1994).

Shiomi K, Takeichi M, Nishida Y, Nishi Y, Uemura T: Alternative cell fate choice induced by low-level expression of a regulator of protein phosphatase 2A in the Drosophila peripheral nervous system. Devel 120:1591–1599 (1994).

Silveira I, Miranda C, Guimarães L, Moreira M-C, Alonso I, Mendonça P, Ferro A, Pinto-Basto J, Coelho J, Ferreirinha F, Poirier J, Parreira E, Vale J, Januário C, Barbot C, Tuna A, Barros J, Koide R, Tsuji S, Holmes SE, Margolis RL, Jardim L, Pandolfo M, Coutinho P, Sequeiros J: Trinucleotide repeats in 202 families with ataxia: a small expanded $(CAG)_n$ allele at the SCA17 locus. Arch Neur 59:623–629 (2002).

Sontag E, Nunbhakdi-Craig V, Lee G, Bloom GS, Mumby MC: Regulation of the phosphorylation state and microtubule-binding activity of Tau by protein phosphatase 2A. Neuron 17:1201–1207 (1996).

Sontag E, Nunbhakdi-Craig V, Lee G, Brandt R, Kamibayashi C, Kuret J, et al: Molecular interactions among protein phosphatase 2A, tau and microtubules. Implications for the regulation of tau phosphorylation and the development of tauopathies. J biol Chem 274:25490–25498 (1999).

Srivastava AK, Choudhry S, Gopinath MS, Roy S, Tripathi M, Brahmachari SK, Jain S: Molecular and clinical correlation in five Indian families with spinocerebellar ataxia 12. Ann Neur 50:796–800 (2001).

Strack S, Zaucha JA, Ebner FF, Colbran RJ, Wadzinski BE: Brain protein phosphatase 2A: developmental regulation and distinct cellular and subcellular localization by B subunits. J Comp Neurol 392:515–527 (1998).

Strack S, Ruediger R, Walter G, Dagda RK, Barwacz CA, Cribbs JT: Protein phosphatase 2A holoenzyme assembly: identification of contacts between B-family regulatory and scaffolding A subunits. J biol Chem 277:20750–20755 (2002).

Tehrani MA, Mumby MC, Kamibayashi C: Identification of a novel protein phosphatase 2A regulatory subunit highly expressed in muscle. J biol Chem 271:5164–5170 (1996).

Trottier Y, Lutz Y, Stevanin G, Imbert G, Devys D, Cancel G, Saudou F, Weber C, David G, Tora L, Agid Y, Brice A, Mandel J-L: Polyglutamine expansion as a pathological epitope in Huntington's disease and four dominant cerebellar ataxias. Nature 378:403–406 (1995).

Turowski P, Myles T, Hemmings BA, Fernandez A, Lamb NJ: Vimentin dephosphorylation by protein phosphatase 2A is modulated by the targeting subunit B55. Mol Biol Cell 10:1997–2015 (1999).

Verkerk AJMH, Pieretti M, Sutcliffe JS, Fu Y-H, Kuhl DPA, Pizzuti A, Reiner O, Richards S, Victoria MF, Zhang F, Eussen BE, van Ommen G-JB, Blonden LAJ, Riggins GJ, Chastain JL, Kunst CB, Galjaard H, Caskey CT, Nelson DL, Oostra BA, Warren ST: Identification of a gene (FMR-1) containing CGG repeat coincident with a breakpoint cluster region exhibiting length variation in Fragile X syndrome. Cell 65:905–914 (1991).

Virshup DM: Protein phosphatase 2A: a panoply of enzymes. Curr Opin Cell Biol 12:180–185 (2000).

Wera S, Hemmings BA: Serine/threonine protein phosphatases. Biochem J 311:17–29 (1995).

Worth PF, Wood NW: Spinocerebellar ataxia type 12 is rare in the United Kingdom. Neurol 56:419–420 (2001).

Yu S, Pritchard M, Kremer E, Lynch M, Nancarrow J, Baker E, Holman K, Mulley JC, Warren ST, Schlessinger D, Sutherland GR, Richards RI: Fragile X genotype characterized by an unstable region of DNA. Science 252:1179–1181 (1991).

Zoghbi HY, Orr HT: Glutamine repeats and neurodegeneration. Ann Rev Neurosci 23:217–247 (2000).

Cytogenet Genome Res 100:198–205 (2003)
DOI: 10.1159/000072855

The hereditary spinocerebellar ataxias in Japan

H. Sasaki, I. Yabe, and K. Tashiro

Department of Neurology, Hokkaido University Graduate School of Medicine, Sapporo (Japan)

Abstract. In Japan, multiple system atrophy (MSA) accounts for 40% of all spinocerebellar ataxias (SCAs) and hereditary disorders account for 30%. Among the latter, autosomal dominant disorders are common and recessive ataxias are rare. Although the frequency of SCA genotypes differs between geographic regions throughout Japan, SCA6, SCA3/MJD, and DRPLA are the three major disorders, while SCA7, SCA8, SCA10, SCA12, and SCA17 are infrequent or almost undetected. SCA1 predominantly occurs in the northern part of Japan. Overall, 20–40% of dominant SCAs are due to unknown mutations. From this cluster, pure cerebellar ataxias linked with the SCA4, SCA14, and SCA16 locus have been isolated. Among the recessive SCAs, patients with AVED and EAOH have been detected. However, FRDA associated with GAA repeat expansion in the frataxin gene has not been reported so far.

The spinocerebellar ataxias (SCAs) are a cluster of heterogeneous neurodegenerative disorders, in which progressive neuronal loss and gliosis predominantly affect the cerebellum and related structures in the CNS. In some of these disorders, the peripheral nerves, muscles, or autonomic nervous system are also affected. Several previous attempts have been made to classify the hereditary SCAs, but classification has on the basis of clinicopathological findings alone has proved difficult. Recent molecular genetic studies have determined the loci and mutations responsible for the majority of hereditary SCAs. So far, disease loci have been assigned for at least 18 dominant SCAs (SCA1–8; SCA10–14; SCA16; SCA17; SCA19; SCA21; and dentatorubropallidoluysian atrophy, DRPLA) and seven recessive SCAs (Friedreich ataxia, FRDA; early-onset ataxia with oculomotor apraxia and hypoalbuminemia, EAOH; ataxia-oculomotor apraxia 2, AOA2; Friedreich-like ataxia with selective vitamin E deficiency, AVED; autosomal recessive spastic ataxia of Charlevoix-Sagunay, ARSACS; infantile-onset spinocerebellar ataxia, IOSCA; and spinocerebellar ataxia with saccadic intrusions, SCASI, SCA24). In addition, another SCA locus has just been registered in the database of the Human Genome Organization (HUGO; http://www.hugo-international.org/hugo/). In this database, SCA9, SCA15, and SCA22 are currently reserved, while SCA18 and SCA23 are already mapped but their clinical and genetic data have not been published.

In ten of the dominant SCAs (SCA1–3, SCA6–8, SCA10, SCA12, SCA17, and DRPLA) and one recessive SCA (FRDA), the abnormal expansion of unstable tandem repeats has been demonstrated in each causative gene. Seven of these disorders (SCA1–3, SCA6, SCA7, SCA17, and DRPLA) are known to be caused by abnormal expansion of CAG repeats in the coding gene, and they are often classified as polyglutamine (Q) disorders since the translated protein of each gene contains abnormally elongated glutamine repeats. In these disorders, the size of the expanded tandem repeat is inversely correlated with the age of onset, rate of progression, and clinical severity. These disorders are also associated with the anticipation phenomenon. Although the typical clinical features of each disorder can contribute to making a diagnosis, phenotypic variation of these SCAs often means that it is difficult, especially in the polyglutamine disorders. The genetic and phenotypic profiles of the genetically defined dominant SCAs are summarized in Table 1.

In this review, we briefly introduce the prevalence of ataxic disorders in Japan, as well as the local frequency of each SCA

Received 28 October 2002; revision accepted 3 March 2003.

Request reprints from Hidenao Sasaki
Department of Neurology, Hokkaido University Graduate School of Medicine
N15 W7, Kita-ku, Sapporo, 060-8638 (Japan)
telephone: +81-11-706-6028; fax: +81-11-700-5356
e-mail: hsasaki@med.hokudai.ac.jp

KARGER Fax + 41 61 306 12 34
E-mail karger@karger.ch
www.karger.com

0301–0171/03/1004–0198$19.50/0

Accessible online at:
www.karger.com/cgr

Table 1. Genotype classification of dominant SCAs

Symbol	Locus	Gene name[a]	Mutation	Predominant phenotype
DRPLA	12p13.31	DRPLA protein	Expansion of $(CAG)_n$	Myoclonus epilepsy, ataxia, choreoathetosis, dementia
SCA1	6p23	Ataxin-1	Expansion of $(CAG)_n$	Ataxia, ophthalmoparesis, pyramidal and extrapyramidal findings
SCA2	12q24.1	Ataxin-2	Expansion of $(CAG)_n$	Ataxia, slow saccades, pyramidal and extrapyramidal findings (choreoathetosis, parkinsonism)
SCA3/MJD	14q32.1	Ataxin-3	Expansion of $(CAG)_n$	Ataxia, ophthalmoparesis, pyramidal and extrapyramidal findings (spasticity, dystonia, parkinsonism)
SCA4	16q22.1	N.D.	N.D.	Spinal ataxia with sensory axonal neuropathy (USA), pure ataxia (Japan)
SCA5	11p11 q11	N.D.	N.D.	Ataxia
SCA6	19p13.1	CACNA1A	Expansion of $(CAG)_n$	Ataxia
SCA7	3p12 p13	Ataxin-7	Expansion of $(CAG)_n$	Ataxia, macropathy, slow saccades, pyramidal and extrapyramidal findings
SCA8	13q21	N.D.	Expansion of $(CTG)_n$	Ataxia, pyramidal findings (spasticity)
SCA10	22q13	E46L	Expansion of $(ATTCT)_n$	Ataxia, seizures
SCA11	15q14 q21.3	N.D.	N.D.	Ataxia
SCA12	5q31 q33	PPP2R2B	Expansion of $(CAG)_n$	Action tremor, ataxia, pyramidal and extrapyramidal findings
SCA13	19q13.3 q13.4	N.D.	N.D.	Mild ataxia with very slow progression, mild intellectual impairment
SCA14	19q13.4 qter	N.D.	N.D.	Ataxia, intermittent axial myoclonus
SCA16	8q22.1 q24.1	N.D.	N.D.	Ataxia, head tremor
SCA17	6q27	TBP	Expansion of $(CAG)_n$	Ataxia, parkinsonism, pyramidal findings, dementia
SCA19	1p21 q21	N.D.	N.D.	Mild ataxia, cognitive impairment, myoclonus, tremor
SCA21	7p21.3 p15.1	N.D.	N.D.	Ataxia, hyporeflexia, parkinsonism

[a] Abbreviations: N.D. not determined; CACNA1A, P/Q type voltage-dependent calcium channel, subunit; PPP2R2B, brain-specific regulatory subunit of protein phosphatase 2A; TBP: TATA-binding protein.

genotype and the phenotypic variations that have been revealed through recent clinical, epidemiological, and molecular genetic studies.

Prevalence of SCA

In Japan, a nationwide survey was performed from 1988 to 1989 under the auspices of the Research Committee on Ataxia of the Ministry of Health and Welfare (Hirayama et al., 1994). According to this survey, more than 40% of registered cases of SCA were diagnosed as multiple system atrophy (MSA) and it was the most prevalent disorder in Japan. Approximately 30% of the patients were estimated to have various types of hereditary SCA. Most of this group had dominantly inherited disorders, and recessive SCA (diagnosed as FRDA) was rare, accounting for only 2.4%. At the time of the survey, the differential diagnosis of dominant SCAs was based on clinical findings alone. Moreover, the exact number of patients and the prevalence rates were unclear since the survey was restricted to hospitals with 200 beds or more. According to the current statistical data (from Research on Specific Disease by the Ministry of Health and Welfare of Japan), a total of 19,901 patients with SCA were registered at the end of fiscal year 2000, with the prevalence being 15.68/100,000. However, the frequency of each SCA subtype is still unknown at present.

Dominant SCA

Based on the mode of inheritance and the clinical phenotype, Harding (1993) classified late-onset autosomal dominant cerebellar ataxia (ADCA) into three categories: ADCA type I disorders cause ophthalmoplegia, optic atrophy, dementia, and extrapyramidal features; ADCA type II disorders cause macropathy with or without ophthalmoplegia and extrapyramidal features; and ADCA type III was late-onset pure cerebellar ataxia. According to this classification, the term, "Menzel-type olivopontocerebellar atrophy (OPCA)" and "Holmes-type hereditary ataxia", which were previously used in Japan, correspond to ADCA type I and type III, respectively. From this cluster of inherited SCAs, various genetically defined disorders have been distinguished through the application of molecular genetic methods.

SCA1

In 1974, Yakura et al. first reported a small two-generation family with a disease locus that was linked to HLA. Subsequently, this HLA-linked SCA was assigned to 6p23 as SCA1, and it is now known to be caused by abnormal CAG repeat expansion in the coding region of the ataxin-1 gene (Orr et al., 1993). The original "Yakura family" was characterized by (1) cerebellar ataxia predominating throughout the course; (2) pyramidal tract involvement, with an extensor plantar response, hyper-reflexia, or terminal hyporeflexia; (3) terminal amyotrophy; (4) slow saccades in the advanced stage; (5) atrophy of the facial and tongue muscles; (6) optic disc pallor; (7) terminal external ophthalmoparesis; (8) mydriasis and a sluggish light reflex; (9) mild peripheral neuropathy; (10) mild reduction of proprioceptive perception; (11) bulbar symptoms; and (12) emotional lability, irritation, or euphoria that becomes dominant terminally (Sasaki et al., 1990). Suzuki et al. (1995) studied CAG repeat expansion in the SCA1 gene among patients with dominant SCA and first reported the existence of SCA1 in Japan. In that study, they confirmed that the Yakura family also carried the SCA1 mutation. The SCA1 families in Hokkaido were shown to exhibit a common haplotype, with the majority of their ancestors coming from the northern part of

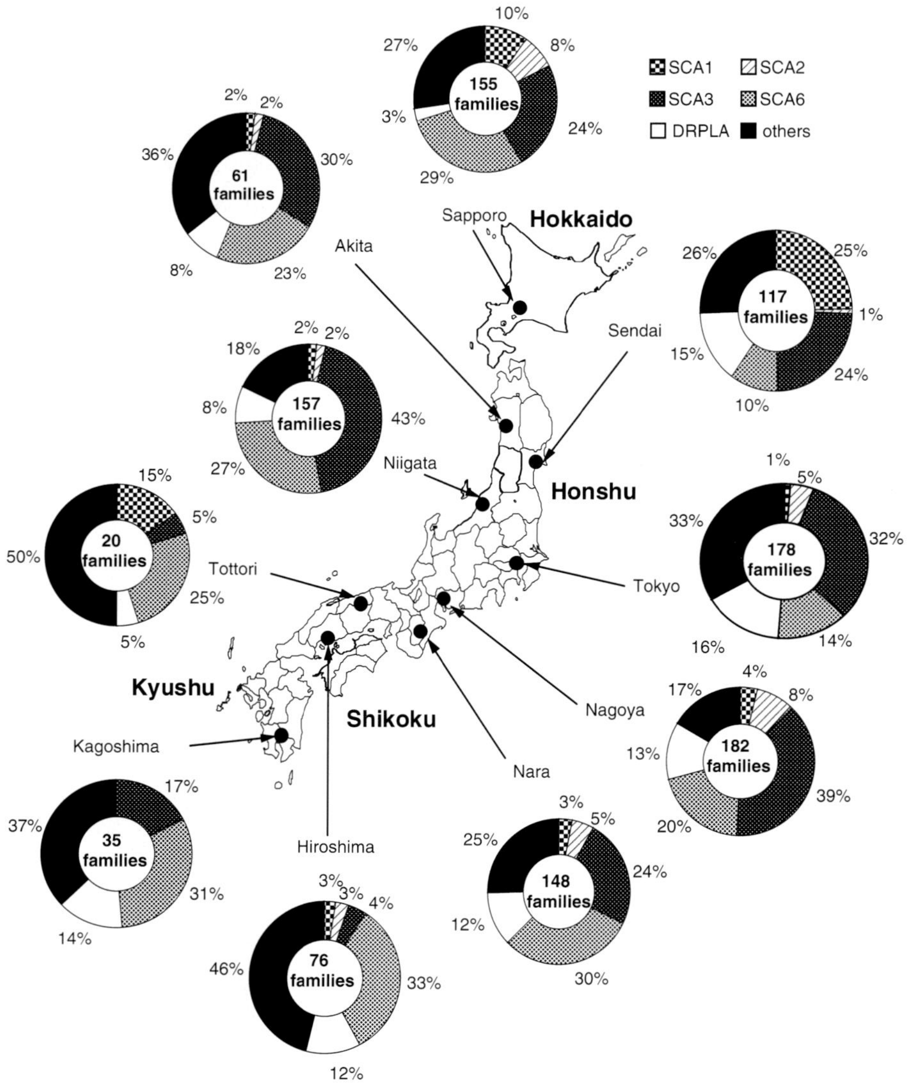

Fig. 1. Distribution of dominant SCA genotypes in Japan.

Honshu (Wakisaka et al., 1995). Later, a localized epidemic focus was confirmed by Onodera et al. (2000). SCA1 accounts for 10% of 155 SCA families in Hokkaido (Sasaki et al., 2000) and 25% of 117 SCA families in Sendai (Onodera et al., 2000), while it is a rare disorder that affects <5% in most of the rest of Japan (Fig. 1).

SCA2

SCA2 is caused by abnormal CAG repeat expansions in the coding sequence of the ataxin-2 gene on 12q24.1 (Imbert et al., 1996; Pulst et al., 1996; Sanpei et al., 1996). In Japan, SCA2 was first identified by a linkage study (Ihara et al., 1994). This family was then added to the cohort of subjects in which the SCA2 gene and unstable CAG repeat expansion were subsequently detected by Sanpei et al. (1996). Their data taken together with the findings from our family and the Cuban founders (Orozco Diaz et al., 1990) indicate that SCA2 features progressive ataxia, slow saccades, hyporeflexia, and an extensor plantar response. An exceptional case showed parkinsonism with minor cerebellar deficits, while juvenile-onset patients have severe ataxia that progress to choreoathetosis (Sasaki et al., 1996). Generally, early-onset patients tend to develop slow saccades and diminished limb reflexes even from an early stage, while choreiform movements occasionally occur in the advanced stage. When the onset is at an older age, these typical clinical features are less prominent. Thus, variations of the SCA2 phenotype depend on the age of onset and the duration of the disorder. SCA2 is an infrequent disorder that accounts for <5% of the patients in each area, and no epidemic focus has been found in Japan.

SCA3/MJD

Following the increase in reports on Machado-Joseph disease (MJD) in the English literature, various controversies have arisen regarding the classification of Japanese patients showing a similar phenotype. Sakai et al. (1983) first mentioned a Japanese family with a diagnosis of MJD in an English language report. The disease locus was first mapped to 14q by Takiyama et al. (1993), while abnormal CAG repeat expansion in the coding region of the MJD gene was first reported by Kawaguchi et al. (1994). Subsequent DNA analysis has shown that MJD is one of the common SCAs around the world. Linkage disequilibrium studies have indicated the presence of a founder effect in Japanese patients with MJD (Endo et al., 1996). MJD shows a variety of phenotypes, depending on the age of onset. Rosenberg (1992) classified the phenotypes of MJD into four categories, which were prominent pyramidal and extrapyramidal deficits (type I); pyramidal and extrapyramidal findings including definite cerebellar deficits (type II); cerebellar deficits with neuropathy (type III); and parkinsonism with neuropathy (type IV). Type II is the most common phenotype in Japan; it manifests with cerebellar ataxia, leg spasticity, and gaze nystagmus. Historically, dominant SCAs presenting with this phenotype, i.e., cerebellar ataxia with nystagmus and spasticity, used to be classified as "Marie's hereditary ataxia". In Japan, MJD accounts for 20–40% of SCA patients in central to northeastern Honshu and Hokkaido, but is less frequent in southwestern Honshu, Kyushu, and Shikoku.

SCA6

SCA6 is caused by moderate expansion of CAG repeats in the coding region of the α_{1A} subunit of the P/Q type voltage-dependent calcium channel (CACNA1A) on 19p13.1 (Zhunchenko et al., 1997). Besides rare atypical cases, SCA6 usually presents as late-onset pure cerebellar ataxia. Selective degeneration of cerebellar Purkinje cells is the hallmark of its pathology, and other CNS structures (including the deep cerebellar nuclei) are not involved by this disorder. Clinically, in addition to ataxia and obvious gaze-evoked nystagmus, positioning nystagmus occurs in association with attacks of vertigo (Geschwind et al., 1997; Yabe et al., 2003). A linkage study showed that ADCA III was linked with the SCA6 locus in some Japanese families, and these patients were subsequently demonstrated to carry the SCA6 mutation (Ishikawa et al., 1997). Haplotype analysis indicates that there are founder chromosomes in Japanese SCA6 patients (Yabe et al., 2001). SCA6 accounts for approximately half of the cases of ADCA III (Yabe et al., 1998) and for 10–33% of all dominant SCA patients in Japan.

DRPLA

DRPLA is an autosomal dominant disorder that is caused by abnormal CAG repeat expansion in the DRPLA gene on 12p13.31. Nagafuchi et al. (1994) and Koide et al. (1994) independently reported this specific mutation in the same year. Degeneration of the dentato-rubral and pallido-luysian pathways is the cardinal pathology. The clinicopathological profile of hereditary DRPLA has been established through a series of studies by Naito and Oyanagi (1982). Hereditary DRPLA shows marked phenotypic variation. Clinically, the juvenile type (onset at <20 years old) presents with myoclonus, epilepsy, and psychomotor retardation, whereas the late adult type (onset at >40 years old) is characterized by ataxia and choreoathetosis. The early adult type (onset at 20–40 years) shows a transitional phenotype between the other two types. In addition, there is a senile type (onset at >60 years old) that predominantly features ataxia associated with atrophy of the cerebellum and brain stem on neuroimaging studies. As a result, it is difficult to differentiate late-onset DRPLA from other types of SCA based on the clinical findings alone (Yabe et al., 2002). The prevalence of DRPLA ranges from 3% to 16%, with the frequency being higher along the Pacific coast than in other areas of Japan. DRPLA is exceedingly rare in other countries, although the same expansion of the DRPLA gene is found in Haw River syndrome (Burke et al., 1994). A single haplotype bearing the DRPLA expansion is found in Japanese patients and also in rare Caucasian patients, indicating that ancestral mutations arose from a common founder chromosome (Yanagisawa et al., 1996).

SCA7, SCA8, SCA10, SCA12, SCA17, and other disorders

SCA7 is a poly-Q disorder that is caused by abnormal expansion of CAG repeats in the ataxin-7 gene on 3p13→p12 (David et al., 1999). SCA7 is quite rare in Japan. No cases were found among 109 SCA patients from Tottori (Mori et al., 2001), 1,286 patients from Hiroshima (Maruyama et al., 2002), and the cohort of subjects studied so far in Hokkaido, but two cases from Sendai have been reported in the English literature (Abe et al., 2000).

SCA8 is caused by CTG repeat expansion in an unknown gene located on 13q21 and has the ADCA III phenotype (Koob et al., 1999). In Japan, this disorder was reported by Hokezu et al. (2000) and Ikeda et al. (2000), who found six cases among 22 SCA families and 41 sporadic ataxia patients. All of these patients had the ADCA III phenotype. SCA8 accounts for only 0.8% of the 117 families with dominant SCA in Sendai (Onodera et al., 2000), and for three (0.7%) out of 445 patients (including isolated cases) in the Kinki district (Futamura et al., 2000), while it has not been found in other areas.

SCA17 is caused by abnormal expansion of CAG repeats in the TATA-binding protein (TBP) on 6q27 (Nakamura et al., 2001). Prior to the report by Nakamura, Koide et al. (1999) reported a patient with de novo expansion of CAG repeats in this gene. This disorder presents in the 4th decade of life with a variable combination of ataxia, dementia, convulsions, and parkinsonism. In a cohort of 330 families with dominant SCA, SCA17 accounted for 0.3% of patients and there was no SCA7, indicating that these conditions are extremely rare in Japan (Maruyama et al., 2000).

Among the other dominant SCAs for which specific mutations have already been identified, neither SCA10 (Matsuura et al., 2000) nor SCA12 (Holmes et al., 1999) has been reported in Japan at present.

Dominant SCAs due to unknown mutations

Although the frequency shows regional variations, approximately 20–40% of dominant SCAs are still caused by unknown mutations. Among this group, Nagaoka et al. (2000) reported that Japanese families with the ADCA III phenotype showed linkage to the SCA4 locus. Patients in these families did not develop axonal sensory neuropathy, as described in the original Utah SCA4 kindred (Flanigan et al., 1996). Thus, the clinical difference between these two reported groups implies that they represent allelic variants of the same SCA4 locus. Only a single Japanese family was reported to show linkage with 19q13.4→qter (SCA14) and one family shows linkage with 8q22.1→

Table 2. Genotype frequencies of dominant SCA outside Japan

Country	Number of ADCA families	Frequency							Reference
		SCA1	SCA2	MJD	SCA6	DRPLA	SCA7	Others	
Italy	116	24%	47%	0%	2%	1%	2%	24%	Filla et al., 2000
Netherlands	145	6%	7%	28%	15%	NT	8%	36%	van del Warrenburg et al., 2001
Portugal	48	0%	4%	74%	1%	0%	NT	22%	Silveira et al., 1998
Spain	87	6%	15%	15%	1%	1%	3%	58%	Pujana et al., 1999
France	146	15%	10%	32%	1%	NT	NT	42%	Stevanin et al., 1997
USA	178	6%	15%	20%	15%	NT	4%	39%	Moseley et al., 1998
Australia	88	15%	6%	12%	17%	0%	2%	47%	Storey et al., 2000
India	39	8%	26%	5%	0%	0%	0%	62%	Saleem et al., 2000
China	85	5%	6%	48%	0%	0%	0%	41%	Tang et al., 2000
Taiwan	74	5%	11%	47%	11%	1%	3%	22%	Soong et al., 2001
Brazil	52	0%	0%	92%	0%	0%	2%	6%	Jardim et al., 2001

q24.1 (SCA16). The first family with SCA14 shows anticipation of the age of onset and intermittent axial tremor in the early-onset patients (Yamashita et al., 2000), while the family with SCA16 occasionally has patients who develop head tremor (Miyoshi et al., 2001). A second family with SCA14 from the USA shows pure cerebellar ataxia with a variable age of onset, but do not develop intermittent axial myoclonus as seen in the Japanese family (Brkanac et al., 2002). Since the mutations responsible for SCA4, SCA14, and SCA16 are unknown, the exact frequencies of these disorders are also unknown at present. No cases of SCA5 (Ranum et al., 1994), SCA11 (Worth et al., 1999), or SCA13 (Herman-Bert et al., 2000) have been reported in Japan so far.

Recessive SCA

As mentioned above, patients with recessive SCA account for only 2.4% of all cases of SCA in Japan and thus are rare in this country (Hirayama et al., 1994). From among this cluster with an FRDA-like phenotype, Gotoda et al. (1995) reported patients with the late-onset FRDA-like phenotype caused by a missense mutation of the αTTP gene. Retinitis pigmentosa is an additional characteristic of the disorder (Yokota et al., 1996). Recently, a novel gene known as aprataxin and its mutations were demonstrated in Japanese and Portuguese families. Mutations of this gene cause early-onset cerebellar ataxia and are associated with oculomotor apraxia, axonal neuropathy, hypoalbuminemia, mental retardation, and various involuntary movements (Date et al., 2001). Since another disorder that was reported as "ataxia with oculomotor apraxia (AOA)" is also caused by mutations of the same gene (Moreira et al., 2001), these disorders are designated by the term "early-onset ataxia with ocular motor apraxia and hypoalbuminemia" (EAOH) (Date et al., 2001). Generally, the age of onset is earlier and the phenotype is severer in patients with frame-shift mutations than in those with missense mutations. In contrast to FRDA, EAOH is not associated with cardiomyopathy and markedly affects the cerebellum. Current data on the available cases indicates that EAOH is the most frequent form of recessive SCA in Japan.

There are also other recessive SCAs with unknown mutations. Watanabe et al. (1998) reported four siblings with the onset of disease at 22–43 years in a single family, showing cerebellar ataxia, neuropathy, and elevated levels of creatine kinase, α-globulin, and γ-fetoprotein. The disease locus in this family was mapped to 9q33→q34 (Bomont et al., 2000), the same locus that is now assigned as AOA2 (Nemeth et al., 2000). However, the specific mutation that is involved in causing the symptoms is not known. At this time, no cases of ARSACS (Engert et al., 2000), IOSCA (Nikali et al., 1995), or SCA24 (Swart et al. 2002) have been reported in Japan. Furthermore, European-type FRDA with expansion of GAA repeats in the frataxin gene (Campuzano et al., 1996) has not been found, although it accounts for 11.4% of recessive and 5.2% of sporadic SCA in the USA (Mosely et al., 1998).

Molecular epidemiology of hereditary SCA

With the increasing number of hereditary SCAs in which the causative mutations have been determined, studies on molecular epidemiology have demonstrated the frequency and distribution of these disorders in Japan. At this time, nationwide survey data are not available. However, the regional frequency of SCA genotypes has been reported by neurology centers in Sapporo (Sasaki et al., 2000), Sendai (Onodera et al., 2000), Akita (Sugawara et al., 1999), Niigata (Shimohata et al., 2000), Tokyo (Ichikawa et al., 2000), Nagoya (Tanaka et al., 2000), Nara (Futamura et al., 2000), Tottori (Mori et al., 2001), Hiroshima (Maruyama et al., 2000), and Kagoshima (Ikeda et al., 2000). These centers have been engaging in clinical and genetic studies of SCA in each region. The data in each report were presented as family-based frequencies, that are plotted in Fig. 1. This shows that the distribution and frequency of SCA genotypes is not uniform throughout Japan, and that SCA3/MJD, SCA6, and DRPLA are the major disorders among the dominant SCAs.

Outside Japan, the family-based frequency of SCA genotypes in ADCA has been reported from Italy (Filla et al., 2000), the Netherlands (van del Warrenburg et al., 2001), Portugal (Silveira et al., 1998), Spain (Pujana et al., 1999), France (Stevanin et al., 1997), the USA (Moseley et al., 1998), Australia (Storey et al., 2000), India (Saleem et al., 2000), China (Tang et al., 2000), Taiwan (Soong et al., 2001), and Brazil (Jardim et al., 2001). The frequency in each country is summarized in Table 2. SCA1 accounts for 24% in Italy and for 15% in France and Australia, but is rare in Portugal and Brazil. SCA2 accounts for 47% in Italy and 26% in India, but is not found in Brazil. MJD is the most prevalent SCA, accounting for 74% in Portugal and for 92% in Brazil, but is almost unknown in Italy. SCA6 accounts for 11–17% in the Netherlands, USA, Australia, and Taiwan, but is extremely rare in other countries.

DRPLA is the most prevalent disorder in Japan, but these studies show that only a few cases of DRPLA are found outside Japan, even in nearby countries in Taiwan and China. In this regard, DRPLA was reported to occur in 3.4% of unrelated patients with SCA, but the family-based frequency is not known (Jin et al., 1999). In addition to DRPLA, SCA6 is also prevalent in Japan, but accounts for less than 20% in the Netherlands, USA, Australia, and Taiwan, and is rare in other countries. As for SCA8, 8% in the Netherlands is the highest frequency, while it is <4% in the other countries. These reports from different ethnic populations or regions indicate that, in addition to the low number of cases of European-type FRDA, a high prevalence of DRPLA and SCA6 is the outstanding feature of dominant SCAs in Japan. On a worldwide basis, the prevalence of hereditary SCA is reported to show a correlation with the frequency of intermediate-sized alleles for these SCAs in the general population (Takano et al., 1998). This might explain the high prevalence of SCA6 as well as DRPLA in Japan. On the other hand, the local prevalence of SCA might be dependent on independent ancestral mutations, including de novo mutations of founder chromosomes, immigration of ancestors, or the size of the founder population. This genetic and historical background underlies the uneven distribution of SCA genotypes not only in Japan but in different countries around the world.

References

Abe T, Tsuda T, Yoshida M, Wada Y, Kano T, Itoyama Y, Tamai M: Macular degeneration associated with aberrant expansion of trinucleotide repeat of the SCA7 gene in 2 Japanese families. Arch Ophthalmol 118:1415–1421 (2000).

Bomont P, Watanabe M, Gershoni-Barush R, Shizuka M, Tanaka M, Sugano J, Guiraud-Chaumeil C, Koenig M: Homozygosity mapping of spinocerebellar ataxia with cerebellar atrophy and peripheral neuropathy to 9q33→q34, and with hearing impairment and optic atrophy to 6p21→p23. Eur J hum Genet 8:986–990 (2000).

Brkanac Z, Bylenok L, Fernandez M, Matsushita M, Lipe H, Wolff J, Nochlin D, Raskind WH, Bird TD: A new dominant spinocerebellar ataxia linked to chromosome 19q13.4→qter. Arch Neurol 59:1291–1295 (2002).

Burke JR, Wingfield MS, Lewis KE, Roses AD, Lee JE, Hulette C, Pericak-Vance MA, Vance JM: The Haw River syndrome: dentatorubropallidoluysian atrophy (DRPLA) in an African-American family. Nature Genet 7:521–524 (1994).

Campuzano V, Montermini L, MoltÒ MD, Pianese L, Cossée M, Cavalcanti F, Monros E, Rodius F, Duclos F, Monticelli A, Zara F, Cañizares J, Koutnikova H, Bidichandani SI, Gellera C, Brice A, Trouillas P, De Michele G, Filla A, De Frutos R, Palau F, Patel PI, Di Donato S, Mandel J-L, Cocozza S, Koenig M, Pandolfo M: Friedreich's ataxia: autosomal recessive disease caused by an intronic GAA triplet repeat expansion. Science 271:1423–7 (1996).

Date H, Onodera O, Tanaka H, Iwabuchi K, Uekawa K, Igarashi S, Koike R, Hiroi T, Yuasa T, Awaya Y, Sakai T, Takahashi T, Nagatomo H, Sekijima Y, Kawachi I, Takiyama Y, Nishizawa M, Fukuhara N, Saito K, Sugano S, Tsuji S: Early-onset ataxia with ocular motor apraxia and hypoalbuminemia is caused by mutations in a new HIT superfamily gene. Nature Genet 29:184–188 (2001).

David G, Abbas N, Stevanin G, Durr A, Yvert G, Cancel G, Weber C, Imbert G, Saudou F, Antoniou E, Drabkin H, Gemmill R, Giunti P, Benomar A, Wood N, Ruberg M, Agid Y, Mandel JL, Brice A: Cloning of the SCA7 gene reveals a highly unstable CAG repeat expansion. Nature Genet 17:65–70 (1997).

Endo K, Sasaki H, Wakisaka A, Tanaka H, Saito M, Igarashi S, Takiyama Y, Sanpei K, Iwabuchi K, Suzuki Y, Onari K, Suzuki T, Weissenbach J, Weber JL, Nomura Y, Segawa M, Nishizawa M, Tsuji S: Strong linkage disequilibrium and haplotype analysis in Japanese pedigrees with Machado-Joseph disease. Am J med Genet 67:437–444 (1996).

Engert JC, Berube P, Mercier J, Dore C, Lepage P, Ge B, Bouchard JP, Mathieu J, Melancon SB, Schalling M, Lander ES, Morgan K, Hudson TJ, Richter A: ARSACS, a spastic ataxia common in northeastern Quebec, is caused by mutations in a new gene encoding an 11.5-kb ORF. Nature Genet 24:120–125 (2000).

Filla A, Mariotti C, Caruso G, Coppola G, Cocozza S, Castaldo I, Calabrese O, Salvatore E, De Michele G, Riggio MC, Pareyson D, Gellera C, Di Donato S. Relative frequencies of CAG expansions in spinocerebellar ataxia and dentatorubropallidoluysian atrophy in 116 Italian families. Eur Neurol 44:31–36 (2000).

Flanigan K, Gardner K, Alderson K, Galster B, Otterud B, Leppert MF, Kaplan C, Ptacek LJ: Autosomal dominant spinocerebellar ataxia with sensory axonal neuropathy (SCA4): clinical description and genetic localization to chromosome 16q22.1. Am J hum Genet 59:392–399 (1996).

Futamura N, Matsumura R, Ueno S: Regional feature of hereditary cerebellar degeneration in Kinki district of Japan. Sinkeinaika (Tokyo) 53:122–126 (2000).

Geschwind DH, Perlman S, Figueroa KP, Karrim J, Baloh RW, Pulst SM: Spinocerebellar ataxia type 6. Frequency of the mutation and genotype-phenotype correlations. Neurology 49:1247–1251 (1997).

Gotoda T, Arita M, Arai H, Inoue K, Yokota T, Fukuo Y, Yazaki Y, Yamada N: Adult-onset spinocerebellar dysfunction caused by a mutation in the gene for the alpha-tocopherol-transfer protein. New Eng J Med 333:1313–1318 (1995).

Harding AE: Clinical features and classification of inherited ataxias, in Harding AE, Deufel T (eds): Advances in Neurology, vol 61, Inherited ataxias, pp 1–14 (Raven Press, New York 1993).

Herman-Bert A, Stevanin G, Netter J-C, Rascol O, Brassat D, Calvas P, Camuzat A, Yuan Q, Schalling M, Dürr A, Brice A: Mapping of spinocerebellar ataxia 13 to chromosome 19q13.3→q13.4 in a family with autosomal dominant cerebellar ataxia and mental retardation. Am J hum Genet 67:229–235 (2000).

Hirayama K, Takayanagi T, Nakamura R, Yanagisawa N, Hattori T, Kita K, Yanagimoto S, Fujita M, Nagaoka M, Satomura Y, Sobue I, Iizuka R, Toyokura Y, Satoyoshi E: Spinocerebellar degenerations in Japan: a nationwide epidemiological and clinical study. Acta Neurol Scand Suppl 153:1–22 (1994).

Hokezu Y, Takiyama Y, Sakoe K, Nagamatsu K: A familial case of spinocerebellar ataxia type 8 (SCA8) – its clinical findings and an issue about the genetic basis. Rinsho Shinkeigaku (Tokyo) 40: 1116–1121 (2000).

Holmes SE, O'Hearn EE, McInnis MG, Gorelick-Feldman DA, Kleiderlein JJ, Callahan C, Kwak NG, Ingersoll-Ashworth RG, Sherr M, Sumner AJ, Sharp AH, Ananth U, Seltzer WK, Boss MA, Vieria-Saecker AM, Epplen JT, Riess O, Ross CA, Margolis RL: Expansion of a novel CAG trinucleotide repeat in the 5′ region of PPP2R2B is associated with SCA12. Nature Genet 23:391–392 (1999).

Ichikawa Y, Goto J: Regional feature of hereditary spinocerebellar degeneration in Kanto district of Japan. Sinkeinaika (Tokyo) 53:104–110 (2000).

Ihara T, Sasaki H, Wakisaka A, Takada A, Yoshiki T, Matsuura T, Hamada T, Suzuki Y, Tashiro K: Genetic heterogeneity of dominantly inherited olivoponto-cerebellar atrophy (OPCA) in the Japanese: Linkage study of two pedigrees and evidence for the disease locus on chromosome 12q (SCA2). Jpn J hum Genet 39:305–313 (1994).

Ikeda K, Nakagawa M, Osame M: Regional feature of hereditary spinocerebellar degeneration in South-Kyushu and Okinawa island of Japan. Sinkeinaika (Tokyo) 53:130–133 (2000).

Ikeda Y, Shizuka M, Watanabe M, Okamoto K, Shoji M: Molecular and clinical analyses of spinocerebellar ataxia type 8 in Japan. Neurology 54:950–955 (2000).

Imbert G, Saudou F, Yvert G, Devys D, Trottier Y, Garnier JM, Weber C, Mandel JL, Cancel G, Abbas N, Dürr A, Didierjean O, Stevanin G, Agid Y, Brice A: Cloning of the gene for spinocerebellar ataxia 2 reveals a locus with high sensitivity to expanded CAG/glutamine repeats. Nature Genet 14:285–291 (1996).

Ishikawa K, Tanaka H, Saito M, Ohkoshi N, Fujita T, Yoshizawa K, Ikeuchi T, Watanabe M, Hayashi A, Takiyama Y, Nishizawa M, Nakano I, Matsubayashi K, Miwa M, Shoji S, Kanazawa I, Tsuji S, Mizusawa H: Japanese families with autosomal dominant pure cerebellar ataxia map to chromosome 19p13.1→p13.2 and are strongly associated with mild CAG expansions in the spinocerebellar ataxia type 6 gene in chromosome 19p13.1. Am J hum Genet 61:336–346 (1997).

Jardim LB, Silveira I, Pereira ML, Ferro A, Alonso I, do Ceu Moreira M, Mendonca P, Ferreirinha F, Sequeiros J, Giugliani R: A survey of spinocerebellar ataxia in South Brazil – 66 new cases with Machado-Joseph disease, SCA7, SCA8, or unidentified disease-causing mutations. J Neurol 248: 870–876 (2001).

Jin DK, Oh MR, Song SM, Koh SW, Lee M, Kim GM, Lee WY, Chung CS, Lee KH, Im JH, Lee MJ, Kim JW, Lee MS: Frequency of spinocerebellar ataxia types 1, 2, 3, 6, 7 and dentatorubral pallidoluysian atrophy mutations in Korean patients with spinocerebellar ataxia. J Neurol 246:207–10 (1999).

Kawaguchi Y, Okamoto T, Taniwaki M, Aizawa M, Inoue M, Katayama S, Kawakami H, Nakamura S, Nishimura M, Akiguchi I, Kimura J, Narumiya S, Kakizuka A: CAG expansions in a novel gene for Machado-Joseph disease at chromosome 14q32.1. Nature Genet 8:221–228 (1994).

Koide R, Ikeuchi T, Onodera O, Tanaka H, Igarashi S, Endo K, Takahashi H, Kondo R, Ishikawa A, Hayashi T, Saito M, Tomoda A, Miike T, Naito H, Ikuta F, Tsuji S: Unstable expansion of CAG repeat in hereditary dentatorubral-pallidoluysian atrophy (DRPLA). Nature Genet 6:9–13 (1994).

Koide R, Kobayashi S, Shimohata T, Ikeuchi T, Maruyama M, Saito M, Yamada M, Takahashi H, Tsuji S: A neurological disease caused by an expanded CAG trinucleotide repeat in the TATA-binding protein gene: a new polyglutamine disease? Hum molec Genet 8:2047–2053 (1999).

Koob MD, Moseley ML, Schut LJ, Benzow KA, Bird TD, Day JW, Ranum LP: An untranslated CTG expansion causes a novel form of spinocerebellar ataxia (SCA8). Nature Genet 21:379–84 (1999).

Maruyama H, Kawakami H, Nakamura S: Regional feature of hereditary spinocerebellar degeneration in Chugoku – Shikoku district of Japan. Sinkeinaika (Tokyo) 53:127–129 (2000).

Maruyama H, Izumi Y, Morino H, Oda M, Toji H, Nakamura S, Kawakami H: Difference in disease free survival curve and regional distribution according to subtype of spinocerebellar ataxia: A study of 1,286 Japanese patients. Am J med Genet 114:578–583 (2002).

Matsuura T, Yamagata T, Burgess DL, Rasmussen A, Grewal RP, Watase K, Khajavi M, McCall AE, Davis CF, Zu L, Achari M, Pulst SM, Alonso E, Noebels JL, Nelson DL, Zoghbi HY, Ashizawa T: Large expansion of the ATTCT pentanucleotide repeat in spinocerebellar ataxia type 10. Nature Genet 26:191–194 (2000).

Miyoshi Y, Yamada T, Tanimura M, Taniwaki T, Arakawa K, Ohyagi Y, Furuya H, Yamamoto K, Sakai K, Sasazuki T, Kira J: A novel autosomal dominant spinocerebellar ataxia (SCA16) linked to chromosome 8q22.1→q24.1. Neurology 57:96–100 (2001).

Moreira MC, Barbot C, Tachi N, Kozuka N, Uchida E, Gibson T, Mendonca P, Costa M, Barros J, Yanagisawa T, Watanabe M, Ikeda Y, Aoki M, Nagata T, Coutinho P, Sequeiros J, Koenig M: The gene mutated in ataxia-ocular apraxia 1 encodes the new HIT/Zn-finger protein aprataxin. Nature Genet 29:189–193 (2001).

Mori M, Adachi Y, Kusumi M, Nakashima K: A genetic epidemiological study of spinocerebellar ataxias in Tottori prefecture, Japan. Neuroepidemiol 20:144–149 (2001).

Moseley ML, Benzow KA, Schut LJ, Bird TD, Gomez CM, Barkhaus PE, Blindauer KA, Labuda M, Pandolfo M, Koob MD, Ranum LP. Incidence of dominant spinocerebellar and Friedreich triplet repeats among 361 ataxia families. Neurology 51:1666–1671 (1998).

Nagafuchi S, Yanagisawa H, Sato K, Shirayama T, Ohsaki E, Bundo M, Takeda T, Tadokoro K, Kondo I, Murayama N, Tanaka Y, Kikushima H, Umino K, Kurosawa H, Furukawa T, Nihei K, Inoue T, Sano A, Komure O, Takahashi M, Yoshizawa T, Kanazawa I, Yamada M: Dentatorubral and pallidoluysian atrophy expansion of an unstable CAG trinucleotide on chromosome 12p. Nature Genet 6:14–18 (1994).

Nagaoka U, Takashima M, Ishikawa K, Yoshizawa K, Yoshizawa T, Ishikawa M, Yamawaki T, Shoji S, Mizusawa H: A gene on SCA4 locus causes dominantly inherited pure cerebellar ataxia. Neurology 54:1971–1975 (2000).

Naito H, Oyanagi S: Familial myoclonus epilepsy and choreoathetosis: hereditary dentatorubral-pallidoluysian atrophy. Neurology 32:798–807 (1982).

Nakamura K, Jeong SY, Uchihara T, Anno M, Nagashima K, Nagashima T, Ikeda S, Tsuji S, Kanazawa I: SCA17, a novel autosomal dominant cerebellar ataxia caused by an expanded polyglutamine in TATA-binding protein. Hum molec Genet 10: 1441–1448 (2001).

Nemeth AH, Bochukova E, Dunne E, Huson SM, Elston J, Hannan MA, Jackson M, Chapman CJ, Taylor AM: Autosomal recessive cerebellar ataxia with oculomotor apraxia (ataxia-telangiectasia-like syndrome) is linked to chromosome 9q34. Am J hum Genet 67:1320–1326 (2000).

Nikali K, Suomalainen A, Terwilliger J, Koskinen T, Weissenbach J, Peltonen L: Random search for shared chromosomal regions in four affected individuals: the assignment of a new hereditary ataxia locus. Am J hum Genet 56:1088–1095 (1995).

Onodera Y, Aoki M, Tsuda T, Kato H, Nagata T, Kameya T, Abe K, Itoyama Y: High prevalence of spinocerebellar ataxia type 1 (SCA1) in an isolated region of Japan. J Neurol Sci 178:153–158 (2000).

Orozco Diaz G, Nodarse Fleites A, Cordovés Sagaz R, Auburger G: Autosomal dominant cerebellar ataxia: clinical analysis of 263 patients from a homogeneous population in Holguín, Cuba. Neurology 40:1369–1375 (1990).

Orr HT, Chung MY, Banfi S, Kwiatkowski TJ Jr, Servadio A, Beaudet AL, McCall AE, Duvick LA, Ranum LP, Zoghbi HY: Expansion of an unstable trinucleotide CAG repeat in spinocerebellar ataxia type 1. Nature Genet 4:221–226 (1993).

Pujana MA, Corral J, Gratacos M, Combarros O, Berciano J, Genis D, Banchs I, Estivill X, Volpini V. Spinocerebellar ataxias in Spanish patients: genetic analysis of familial and sporadic cases. The Ataxia Study Group. Hum Genet 104:516–522 (1999).

Pulst SM, Nechiporuk A, Nechiporuk T, Gispert S, Chen XN, Lopes-Cendes I, Pearlman S, Starkman S, Orozco-Diaz G, Lunkes A, DeJong P, Rouleau GA, Auburger G, Korenberg JR, Figueroa C, Sahba S: Moderate expansion of a normally biallelic trinucleotide repeat in spinocerebellar ataxia type 2. Nature Genet 14:269–276 (1996).

Ranum LPW, Schut LJ, Lundgren JK, Orr HT, Livingston DM: Spinocerebellar ataxia type 5 in a family descended from the grandparents of President Lincoln maps to chromosome 11. Nature Genet 8:280–284 (1994).

Rosenberg RN: Machado-Joseph disease: an autosomal dominant motor system degeneration. Mov Disord 7:193–203 (1992).

Sakai T, Ohta M, Ishino H: Joseph disease in a non-Portuguese family. Neurology 33:74–80 (1983).

Saleem Q, Choudhry S, Mukerji M, Bashyam L, Padma MV, Chakravarthy A, Maheshwari MC, Jain S, Brahmachari SK: Molecular analysis of autosomal dominant hereditary ataxias in the Indian population: high frequency of SCA2 and evidence for a common founder mutation. Hum Genet 106:179–187 (2000).

Sanpei K, Takano H, Igarashi S, Sato T, Oyake M, Sasaki H, Wakisaka A, Tashiro K, Ishida Y, Ikeuchi T, Koide R, Saito M, Sato A, Tanaka T, Hanyu S, Takiyama Y, Nishizawa M, Shimizu N, Nomura Y, Segawa M, Iwabuchi K, Eguchi I, Tanaka H, Takahashi H, Tsuji S: Identification of the gene for spinocerebellar ataxia type 2 (SCA2) using a direct identification of repeat expansion and cloning technique (DIRECT). Nature Genet 14:277–284 (1996).

Sasaki H, Hamada T, Wakisaka A, Tashiro K: A clinical study of a family affected with HLA-linked hereditary spinocerebellar ataxia. No To Shinkei (Tokyo) 42:1103–1111 (1990).

Sasaki H, Fukazawa T, Wakisaka A, Hamada K, Hamada T, Koyama T, Tsuji K, Tashiro K: Central phenotype and related varieties of spinocerebellar ataxia 2 (SCA2) – a clinical and genetic study with a pedigree in the Japanese. J Neurol Sci 144:176–181 (1996).

Sasaki H, Yabe I, Yamashita I, Tashiro K: Prevalence of triplet repeat expansion in ataxia patients in Hokkaido, the northernmost island of Japan. J Neurol Sci 175:45–51 (2000).

Shimohata M, Onodera O, Tsuji S: Regional feature of hereditary spinocerebellar degeneration in Shinetsu – Hokuriku district of Japan. Sinkeinaika (Tokyo) 53:111–115 (2000).

Silveira I, Coutinho P, Maciel P, Gaspar C, Hayes S, Dias A, Guimaraes J, Loureiro L, Sequeiros J, Rouleau GA: Analysis of SCA1, DRPLA, MJD, SCA2, and SCA6 CAG repeats in 48 Portuguese ataxia families. Am J med Genet 81:134–138 (1998).

Soong BW, Lu YC, Choo KB, Lee HY: Frequency analysis of autosomal dominant cerebellar ataxias in Taiwanese patients and clinical and molecular characterization of spinocerebellar ataxia type 6. Arch Neurol 58:1105–1109 (2001).

Stevanin G, Dürr A, David G, Didierjean O, Cancel G, Rivaud S, Tourbah A, Warter JM, Agid Y, Brice A: Clinical and molecular features of spinocerebellar ataxia type 6. Neurology 49:1243–1246 (1997).

Storey E, du Sart D, Shaw JH, Lorentzos P, Kelly L, McKinley Gardner RJ, Forrest SM, Biros I, Nicholson GA: Frequency of spinocerebellar ataxia types 1, 2, 3, 6, and 7 in Australian patients with spinocerebellar ataxia. Am J med Genet 95:351–357 (2000).

Sugawara M, Toyoshima I, Kato K, Wada C, Imota T, Hirota K, Ishiguro H, Kagaya H, Hirata A, Ogasawara M, Masamune I: Hereditary ataxias in Akita prefecture. Rinsho Shinkeigaku (Tokyo) 39:763–766 (1999).

Suzuki Y, Sasaki H, Wakisaka A, Takada A, Yoshiki T, Iwabuchi K, Tashiro K, Fukazawa T, Hamada T: Spinocerebellar ataxia 1 (SCA1) in the Japanese: Analysis of CAG trinucleotide repeat expansion and instability of the repeat for paternal transmission. Jpn J hum Genet 40:131–143 (1995).

Swartz BE, Burmeister M, Somers JT, Rottach KG, Bespalova IN, Leigh RJ: A form of inherited cerebellar ataxia with saccadic intrusions, increased saccadic speed, sensory neuropathy, and myoclonus. Ann NY Acad Sci 956:441–444 (2002).

Takano H, Cancel G, Ikeuchi T, Lorenzetti D, Mawad R, Stevanin G, Didierjean O, Durr A, Oyake M, Shimohata T, Sasaki R, Koide R, Igarashi S, Hayashi S, Takiyama Y, Nishizawa M, Tanaka H, Zoghbi H, Brice A, Tsuji S: Close associations between prevalences of dominantly inherited spinocerebellar ataxias with CAG-repeat expansions and frequencies of large normal CAG alleles in Japanese and Caucasian populations. Am J hum Genet 63:1060–1066 (1998).

Takiyama Y, Nishizawa M, Tanaka H, Kawashima S, Sakamoto H, Karube Y, Shimazaki H, Soutome M, Endo K, Ohta S, Kagawa Y, Kanazawa I, Mizuno Y, Yoshida M, Yuasa T, Horikawa Y, Oyanagi K, Nagai H, Kondo T, Inuzuka T, Onodera O, Tsuji S: The gene for Machado-Joseph disease maps to human chromosome 14q. Nature Genet 4:300–304 (1993).

Tanaka F, Watanabe H, Sobue G: Regional feature of hereditary spinocerebellar degeneration in Tokai district of Japan. Sinkeinaika (Tokyo) 53:116–121 (2000).

Tang B, Liu C, Shen L, Dai H, Pan Q, Jing L, Ouyang S, Xia J. Frequency of SCA1, SCA2, SCA3/MJD, SCA6, SCA7, and DRPLA CAG trinucleotide repeat expansion in patients with hereditary spinocerebellar ataxia from Chinese kindreds. Arch Neurol 57:540–544 (2000).

Verbeek DS, Schelhaas JH, Ippel EF, Beemer FA, Pearson PL, Sinke RJ: Identification of a novel SCA locus (SCA19) in a Dutch autosomal dominant cerebellar ataxia family on chromosome region 1p21→q21. Hum Genet 111:388–393 (2002).

Vuillaume I, Devos D, Schraen-Maschke S, Dina C, Lemainque A, Vasseur F, Bocquillon G, Devos P, Kocinski C, Marzys C, Destee A, Sablonniere B: A new locus for spinocerebellar ataxia (SCA21) maps to chromosome 7p21.3→p15.1. Ann Neurol 52:666–670 (2002).

Wakisaka A, Sasaki H, Takada A, Fukazawa T, Suzuki Y, Hamada T, Iwabuchi K, Tashiro K, Yoshiki T: Spinocerebellar ataxia 1 (SCA1) in the Japanese in Hokkaido may derive from a single common ancestry. J med Genet 32:590–592 (1995).

van de Warrenburg BP, Sinke RJ, Verschuuren-Bemelmans CC, Scheffer H, Brunt ER, Ippel PF, Maat-Kievit JA, Dooijes D, Notermans NC, Lindhout D, Knoers NV, Kremer HP: Spinocerebellar ataxias in the Netherlands: prevalence and age at onset variance analysis. Neurology 58:702–708 (2002).

Watanabe M, Sugai Y, Concannon P, Koenig M, Schmitt M, Sato M, Shizuka M, Mizushima K, Ikeda Y, Tomidokoro Y, Okamoto K, Shoji M: Familial spinocerebellar ataxia with cerebellar atrophy, peripheral neuropathy, and elevated level of serum creatine kinase, γ-globulin, and a-fetoprotein. Ann Neurol 44:265–269 (1998).

Worth PF, Giunti P, Gardner-Thorpe C, Dixon PH, Davis MB, Wood NW: Autosomal dominant cerebellar ataxia type III: linkage in a large British family to a 7.6-cM region on chromosome 15q14→21.3. Am J hum Genet 65:420–426 (1999).

Yabe I, Sasaki H, Matsuura T, Takada A, Wakisaka A, Suzuki Y, Fukazawa T, Hamada T, Oda T, Ohnishi A, Tashiro K: SCA6 mutation analysis in a large cohort of the Japanese patients with late-onset pure cerebellar ataxia. J Neurol Sci 156:89–95 (1998).

Yabe I, Sasaki H, Yamashita, Takei A, Suzuki Y, Kida H, Takiyama Y, Nishizawa M, Hokezu Y, Nagamatsu K, Oda T, Ohnishi A, Inoue I, Hata A, Tashiro K: Predisposing chromosome for spinocerebellar ataxia type 6 (SCA6) in Japanese. J med Genet 38:328–333 (2001).

Yabe I, Sasaki H, Kikuchi S, Nonaka M, Moriwaka F, Tashiro K: Late onset ataxia phenotype in dentatorubro-pallidoluysian atrophy (DRPLA). J Neurol 249:432–436 (2002).

Yabe I, Sasaki H, Takeichi N, Takei A, Hamada T, Fukushima K, Tashiro K: Positional vertigo and macroscopic downbeat positioning nystagmus in spinocerebellar ataxia type 6 (SCA6). J Neurol 250:440–443 (2003).

Yakura H, Wakisaka A, Fujimoto S, Itakura K: Hereditary ataxia and HL-A. New Eng J Med 291:154–155 (1974).

Yamashita I, Sasaki H, Yabe I, Fukazawa T, Nogoshi S, Komeichi K, Takada A, Shiraishi K, Takiyama Y, Nishizawa M, Kaneko J, Tanaka H, Tsuji S, Tashiro K: A novel locus for dominant cerebellar ataxia (SCA14) maps to a 10.2-cM interval flanked by D19S206 and D19S605 on chromosome 19q13.4→qter. Ann Neurol 48:156–163 (2000).

Yanagisawa H, Fujii K, Nagafuchi S, Nakahori Y, Nakagome Y, Akane A, Nakamura M, Sano A, Komure O, Kondo I, Jin DK, Sorensen SA, Potter NT, Young SR, Nakamura K, Nukina N, Nagao Y, Tadokoro K, Okuyama T, Miyashita T, Inoue T, Kanazawa I, Yamada M: A unique origin and multistep process for the generation of expanded DRPLA triplet repeats. Hum molec Genet 5:373–379 (1996).

Yokota T, Shiojiri T, Gotoda T, Arai H: Retinitis pigmentosa and ataxia caused by a mutation in the gene for the α-tocopherol-transfer protein. New Eng J Med 335:1770–1771 (1996).

Zhuchenko O, Bailey J, Bonnen P, Ashizawa T, Stockton DW, Amos C, Dobyns WB, Subramony SH, Zoghbi HY, Lee CC: Autosomal dominant cerebellar ataxia (SCA6) associated with small polyglutamine expansions in the α_{1A}-voltage-dependent calcium channel. Nature Genet 15:62–69 (1997).

Cytogenet Genome Res 100:206–212 (2003)
DOI: 10.1159/000072856

A cerebellar tremor/ataxia syndrome among fragile X premutation carriers

P.J. Hagerman,[a] C.M. Greco,[b] and R.J. Hagerman[c]

[a] Department of Biological Chemistry,
[b] Department of Pathology, and
[c] M.I.N.D. Institute and Department of Pediatrics, University of California School of Medicine, Davis, CA (USA)

Abstract. Fragile X syndrome is a neurodevelopmental disorder that is not known to have any progressive neurological sequelae in adulthood. However, a neurological condition involving intention tremor, ataxia, and cognitive decline has recently been identified among older male carriers of premutation alleles of the FMR1 gene. This condition is clinically distinct from fragile X syndrome and arises through a different molecular mechanism involving the same gene (FMR1). Characteristic findings on magnetic resonance imaging include cerebral and cerebellar volume loss and altered signal intensities of the middle cerebellar peduncles. A striking feature of this fragile X-associated tremor/ataxia syndrome is the presence of ubiquitin-positive neuronal and astroglial intranuclear inclusions. Unlike the CAG repeat expansion diseases, which lead to altered protein products, there is no known protein abnormality among FMR1 premutation carriers. Thus, inclusion formation may reflect a gain-of-function effect of the FMR1 mRNA or the CGG repeat itself. Finally, since this syndrome may represent one of the more common single-gene causes of tremor, ataxia, and dementia among older males, FMR1 DNA testing should be considered when evaluating adult patients with tremor/ataxia.

Fragile X syndrome (MIM 309550) is a trinucleotide (CGG) repeat expansion disorder and the most common heritable form of mental retardation, with carrier frequencies of 1 in 260 females and 1 in 813 males (Rousseau et al., 1995; Dombrowski et al., 2002; reviewed by Bardoni and Mandel, 2002; Hagerman, 2002); other screening studies place the carrier frequency in females as high as 1 in 113 (Toledano-Alhadef et al., 2001). Within the normal population, FMR1 alleles generally possess between 5 and 50 CGG repeats, whereas individuals with fragile X syndrome have alleles with more than 200 repeats. In the latter instance, the CGG repeat and upstream promoter elements are generally methylated, although transcription is often not completely silenced (Tassone et al., 2001). Most individuals with fragile X syndrome have little or no FMR1 protein (FMRP) (Kaufmann et al., 1999; Tassone et al., 1999).

Clinical involvement in fragile X syndrome consists of developmental delays from birth onward, with associated clinical features that include hypotonia, motor incoordination, prominent ears, long face, macroorchidism, and flat feet (Hagerman, 2002). Findings on magnetic resonance imaging (MRI) include macrocephaly, with particular enlargement of the hippocampus, caudate, and thalamus, and reductions in the volume of the cerebellar vermis (Reiss et al., 1994, 1995). IQ in both males and females with fragile X syndrome is correlated with the level of FMRP in peripheral blood leukocytes (Tassone et al., 1999).

Premutation carriers (55 to 200 repeats) are typically normal intellectually, although a limited number of children have

Supported by grants from the National Institute of Child Health and Development (HD40661 to P.J.H. and HD36071 to R.J.H.) and the National Institute of Neurological Diseases and Stroke (NS43532 to P.J.H.), by the Boory Family Fund, and by the UC Davis M.I.N.D. Institute.

Received 11 November 2002; manuscript accepted 17 February 2003.

Request reprints from Dr. Paul J Hagerman
Department of Biological Chemistry, Tupper Hall
University of California, Davis
School of Medicine, One Shields Avenue, Davis, CA 95616 (USA)
telephone: 530-754-7266; fax: 530-754-7269
e-mail: pjhagerman@ucdavis.edu

KARGER
Fax +41 61 306 12 34
E-mail karger@karger.ch
www.karger.com

Accessible online at:
www.karger.com/cgr

been reported to have either learning disabilities or more significant cognitive deficits (Hagerman et al., 1996; Tassone et al., 2000c). This affected subgroup typically has a mild deficiency of FMRP, which is thought to cause their clinical involvement.

Clinical phenotype of the fragile X-associated tremor/ataxia syndrome

The clinical phenotype of the fragile X-associated tremor/ataxia syndrome (FXTAS) is quite distinct from fragile X syndrome. Thus far, 31 cases of FXTAS have been reported, with essentially all cases involving males over 50 years of age. All cases have been ascertained through known families with fragile X syndrome, where those affected with FXTAS are grandfathers of fragile X-affected children (Hagerman et al., 2001; Greco et al., 2002; Jacquemont et al., 2002; Leehey et al., 2002). These men are typically very bright, with over 60% having a college education, and a higher proportion with M.D.s or Ph.D.s than the general population. Patients normally present with either intention tremor or ataxia; complaints include tremor with activities such as handwriting or eating, and ataxia manifested by unsteadiness with walking and frequent falling. Approximately 20% may not display any tremor, although when tremor is present, it may also include postural or even resting components.

The tremor typically starts in the dominant hand and progresses, over 1–3 years, to involve the other hand. The tremor worsens over several years, with the gradual development of gait problems. Initial symptoms usually appear within the sixth to eighth decades, with an average age of onset of 63 ± 6 years in the 20 patients reported by Jacquemont et al. (2002). Patients became bedridden as early as 55 years of age and as late as 80 years. Thus far, of nine patients who have died, the average age of death was 74 ± 6 years, with congestive heart failure the cause of death in eight cases and pneumonia the cause of death in the remaining case. Problems with the coordination of swallowing late in the disease have lead to choking and subsequent pneumonia in several cases.

The clinical features associated with FXTAS include parkinsonism, which is seen in slightly more than half of the cases. The parkinsonism is generally mild and includes stiffness of the extremities, masked facies, and resting tremor. Some FXTAS patients have received a diagnosis of Parkinson's disease, although the cases are usually noted to be "atypical" (e.g., Parkinson's disease plus cerebellar findings). Patients may respond well, at least initially, to medications such as levodopa for Parkinson's disease. Additional patient complaints include muscle pain (35%) and lower extremity weakness (55%) and numbness (35%) in the cases reported by Jacquemont et al. (2002). On neurological testing, peripheral neuropathy is observed (61%) in the lower extremity, with decreased sensation to pinprick, lack of deep tendon reflexes, poor position sense, and decreased vibration sense (Jacquemont et al., 2002). Nerve conduction studies, performed on four patients with peripheral neuropathy, demonstrated mild reductions of nerve conduction in the lower extremities in all four cases. Finally, most FXTAS patients have associated autonomic dysfunction, with impotence (80%) and incontinence (bowel, 30%; bladder, 55%) being the most frequent complaints (Jacquemont et al., 2002).

Patients typically present initially with tremor alone, often leading to a preliminary diagnosis of essential tremor (Leehey et al., 2003). As the symptoms of FXTAS gradually progress, patients may become aware of cognitive deficits, first in memory and word retrieval and then in executive function tasks. Cognitive testing demonstrates an early decline in performance IQ, likely due to the interference of the tremor with visuomotor tasks. In addition, problems with working memory, information processing, inhibition, and distractibility are common and cause an overall decline in executive function abilities (Grigsby et al., 2002). These latter declines are significantly more rapid than those experienced by age-, sex-, and education-matched normal controls (Jacquemont et al., 2002). Eventually, many patients develop a dementia (Hagerman et al., 2001; Jacquemont et al., 2002). In the terminal phases of the disorder, the clinical features of FXTAS are similar to those of multiple system atrophy (MSA), which includes a combination of extrapyramidal, cerebellar, and dysautonomic features (Gilman et al., 1999). The remarkable clinical similarities of FXTAS and the cerebellar MSA subtype (MSA-C), coupled with the similar radiologic findings (see below), suggest that some MSA-C patients may in fact be premutation carriers. It is therefore important to screen the MSA population for carrier status. One important distinction between MSA and FXTAS appears to be in the distribution and composition of the brain inclusions (see below). Whereas the inclusions associated with FXTAS appear to be exclusively intranuclear, the inclusions in MSA are predominantly in the cytoplasm of astroglial cells. However, given the small number of brains examined in each disorder, there may yet turn out to be overlaps between the two disorders.

Molecular correlates

In the 31 patients published thus far (Hagerman et al., 2001; Greco et al., 2002; Jacquemont et al., 2002; Leehey et al., 2002, 2003), the CGG expansion ranges from 69 to 135 repeats. Although this range does not include any of the largest premutation alleles, it is likely that a selection bias occurs in those who present clinically. All of these reported patients are grandfathers; that is, their cognitive functioning and level of social interaction allows for marriage and childrearing, whereas males at the high end of the premutation range may have more significant cognitive deficits and may not father children. Remarkably, clinical involvement has been observed for alleles under 70 repeats, which underscores the mechanistic distinction between FXTAS and fragile X syndrome. All patients with FXTAS reported so far have either normal or only mildly lowered FMRP levels in peripheral blood leukocytes, consistent with the earlier findings among the male premutation population as a whole (Tassone et al., 2000b; Tassone and Hagerman, 2003).

The most consistent biochemical abnormality associated with premutation alleles is elevated FMR1 mRNA (Tassone et al., 2000b; Tassone and Hagerman, 2003). This elevation has

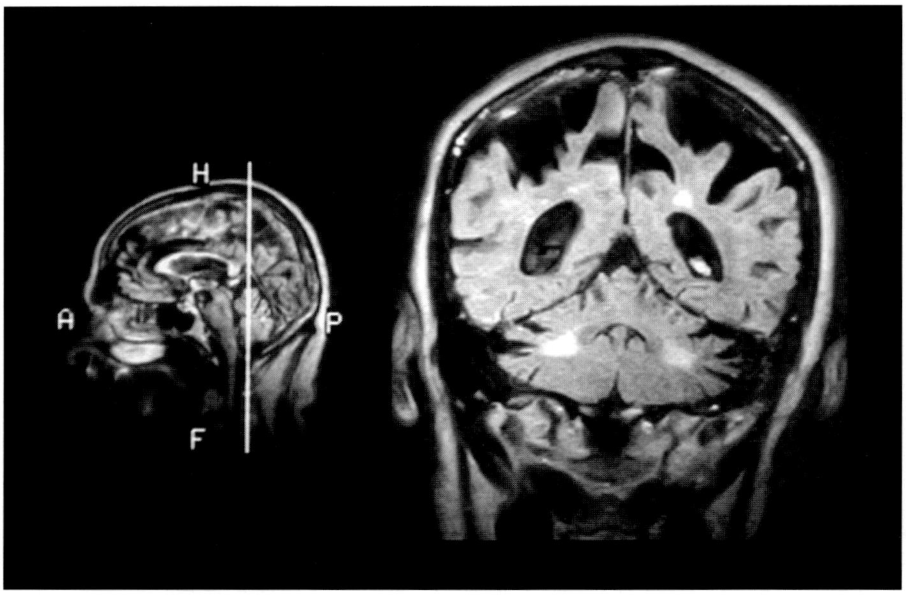

Fig. 1. T2-weighted magnetic resonance image of a 76-year-old male with FXTAS, demonstrating T2 signal hyperintensities of the cerebellar white matter and a marked reduction in cerebral volume within the parietal cortex.

been proposed to be a response to reduced efficiency of translation in the premutation range (Tassone et al., 2000b; Bardoni and Mandel, 2002), although the model has yet to be confirmed. Moreover, whether there exists a causal relationship between the elevated mRNA levels per se and FXTAS has not been determined, although this possibility is raised by the observation that the CUG repeat expansion of myotonic dystrophy can induce myotonia as an RNA element in transgenic mice (Mankodi et al., 2000) (see below).

Neuroimaging

Magnetic resonance neuroimaging reveals significant cortical atrophy in all cases and cerebellar atrophy in most cases. Severe atrophy is seen in approximately 20% of cases (Fig. 1). Brunberg et al. (2002) compared MRI findings in 17 patients with FXTAS to14 age- and sex-matched controls. They found significant reductions in the size of the middle cerebellar peduncles and pons and several indices of global brain atrophy, including ventricular enlargement. In addition, mild to severe thinning of the corpus callosum, not present in controls, was seen in 14 patients with FXTAS. Brunberg et al. (2002) demonstrated that the brain changes observed in FXTAS are not simply related to the aging process but, instead, are unique findings to a subgroup of males with the premutation. White matter disease is almost always present in patients with FXTAS and includes decreased T1 and increased T2 signal intensity in deep white matter of the cerebellar hemispheres and the middle cerebellar peduncles (MCP), while sparing the dentate nuclei of the cerebellum (Fig. 1). In addition, approximately three quarters of the cases have cortical white matter changes that include T2 hyperintensities of the frontal and parietal deep white matter and subependymal areas that are disproportionate for age (Brunberg et al., 2002).

The uniqueness of the white matter changes and, in particular, the MCP finding have lead to the inclusion of these changes in the diagnostic criteria proposed by Jacquemont et al. (2002). These proposed criteria include three major criteria (the MCP sign, intention tremor, and gait ataxia) and five minor criteria (cerebral white matter lesions, moderate to severe brain atrophy, parkinsonism, moderate to severe short-term memory deficits, and executive function deficits disproportionate for age). In those with the premutation, a definite diagnosis requires one radiological major criterion plus one clinical major criterion, whereas a probable diagnosis requires one radiological major criterion plus one clinical minor criterion or two clinical major criteria. Occasionally, a patient will have all of the clinical signs, but not the radiological (MCP) sign (Brunberg et al., 2002). It is not yet known why the white matter changes are not as apparent in these cases or if they will have the same neuropathology reported herein and by Greco et al. (2002).

Neuropathology of FXTAS

Thus far, a total of seven brains from male premutation carriers with FXTAS have been examined; all seven exhibited characteristic brain pathology, which includes substantial numbers of intranuclear inclusions in both neurons and astrocytes throughout the brain. Only small samples of brain tissue were available in three cases, whereas a single hemisphere or whole brain was analyzed in the remaining four cases. Four of these cases have been previously reported (Greco et al., 2002). In addition, the brains from three individuals unaffected by FXTAS were examined for characteristic neuropathology, including two female premutation carriers (17 and 75 years old), and a 25-year-old full mutation male. None of these individuals demonstrated any of the neuropathological changes associated with FXTAS. In particular, no intranuclear inclusions were noted in any brain tissues, nor was there any evidence of spongiform changes in the cerebrum or cerebellum or signs of cerebellar atrophy.

For the affected male carriers, external examination of fixed brains reveals mild to moderate, nonspecific cortical gyral atrophy involving the frontal and parietal lobes and, to a lesser extent, the temporal lobes. In two cases where the brainstem and cerebellum were received intact, prominent atrophy of the ventral pons and the middle cerebellar peduncles was apparent. Substantia nigra pallor was seen only in one individual who was concomitantly affected by Parkinson's disease (Greco et al., 2002). Cortical atrophy with hydrocephalus ex vacuo was identified in coronal sections of fixed hemispheres. Three cases from individuals with longstanding disease showed gray discoloration in non-periventricular white matter in subcortical locations and in the centrum semiovale (Greco et al., in preparation).

By conventional light microscopy, several major findings in brains of clinically affected individuals are (Fig. 2): (1) neuronal and astrocytic intranuclear inclusions, (2) white matter spongiosis in middle cerebellar peduncles and deep cerebellar white matter, and (3) evidence of Purkinje cell degeneration and drop-out (Greco et al., 2002). The neuronal and astrocytic intranuclear inclusions, which appear similar in both cell types, are discrete, round, refractile bodies ranging in size from 2 to 5 µm. The inclusions are eosinophilic on hematoxylin and eosin staining and ubiquitin positive in all cases. They stain negatively on Bielschowsky (silver stain for neuronal cytoskeletal components) and Congo red (amyloid) stains and are immunocytochemically negative for neurofilament, tau, synuclein, and polyglutamine proteins in preliminary studies on initial cases. The composition of the inclusions is currently unknown and is under active investigation. However, the presence of ubiquitin in the inclusions suggests that inclusion formation may reflect the abnormal accumulation of protein products that have been targeted for degradation. Ultrastructural examination shows compact, non–membrane-bound collections of granulofilamentous material (Greco et al., 2002).

Determination of the fraction of inclusion-positive nuclei in two cases reveals that the pyramidal neurons of the hippocampus possess the greatest numbers of inclusions (~ 40%), with various regions of the cerebrum displaying greater numbers (~ 5–10%) than various regions of the brain stem and cerebellum (0–3%) in both cases (Greco et al., 2002). Whereas the numbers of neuronal inclusions are generally concordant between the two cases, counts of astrocytic inclusions are quite distinct, with ~40% inclusion-positive astrocytic nuclei throughout the brain in one case, and ~ 10% inclusion-positive nuclei in the second case. Thus, there are two puzzling features of the astrocyte counts: first, the frequencies are relatively constant in different brain structures within the same case, in contrast to the striking variation within the neuronal populations; second, the numbers of astrocytic inclusions are quite different between the two cases, despite close agreement for the neuronal inclusion counts between the two cases.

The spongiform changes in the middle cerebellar peduncles (Greco et al., 2002) are entirely consistent with the hyperintense signals observed by MRI (Fig. 1) (Brunberg et al., 2002). Specific stains show depletion of both axons (Bielschowsky stain) and myelin (PAS) in cerebellar white matter (Greco et al., 2002). In addition to small numbers of intranuclear inclusions in the dentate nucleus and throughout the astrocytic population in the cerebellum, and spongiosis in the distribution of the middle cerebellar peduncle in the cerebellar white matter, there is also a dropout of Purkinje cells with torpedo formation in the cerebellar cortex (Fig. 2) (Greco et al., 2002). Spinal-cord sections from available cases show no inclusions within neuronal nuclei, although astroglial intranuclear inclusions are diffusely present.

Similar appearing intranuclear inclusions have been reported in other pathological conditions, including several of the CAG repeat disorders. The CAG repeat disorders represent a class of hereditary neurodegenerative disorders caused by expansion of the CAG triplet repeat in exonic sequence within the affected genes (Cummings and Zoghbi, 2000), which leads to an expanded polyglutamine tract in the disease protein. This group of disorders, currently classified on the basis of their genetic determinants at loci specific to each disorder, includes the spinocerebellar ataxias (SCAs), Huntington's disease (HD), dentatorubral-pallidoluysian atrophy (DRPLA), and spinal and bulbar muscular atrophy (SBMA). Though these disorders are genetically distinct and unique in their neurological and pathological characteristics, they all show (except for SCA 6) the presence of similar-appearing intranuclear inclusions in neuronal and astrocytic populations throughout the brain. By comparison, inclusions in both CAG and CGG repeat disorders are ubiquitin positive, but the regional increase of neuronal inclusions seen in the FXTAS hippocampus has not been reported in the CAG repeat disorders, nor have the white matter abnormalities. Another (rare) disorder showing diffuse intranuclear inclusions is neuronal intranuclear inclusion disease (NIID). Eosinophilic, ubiquinated, and polyglutamine-positive intranuclear inclusions are seen in neurons throughout the peripheral, central, and autonomic nervous system (Michaud and Gilbert, 1981; Kimber et al., 1998; Zannolli et al., 2002). The symptoms of multiple system degeneration in NIID appear in both sporadic and familial forms, with an onset primarily in childhood to young adulthood.

The inclusions seen in multiple system atrophy are entirely different from those seen in the trinucleotide repeat disorders. Cytoplasmic inclusions are present in oligodendroglia (GCIs) in specific locations throughout the brain and spinal cord, and are ubiquitin-, τ-, and α-synuclein positive (Gilman et al., 1999).

Perspectives

FXTAS and fragile X syndrome

The fragile X-associated tremor ataxia syndrome appears to be both clinically and mechanistically distinct from fragile X syndrome, although both conditions arise as a consequence of a CGG repeat expansion within the FMR1 gene. Clinically, FXTAS appears to be largely limited to male carriers of premutation expansions, with no cases having been identified, thus far, among full mutation individuals. Whereas the distribution of premutation alleles toward smaller expansions undoubtedly

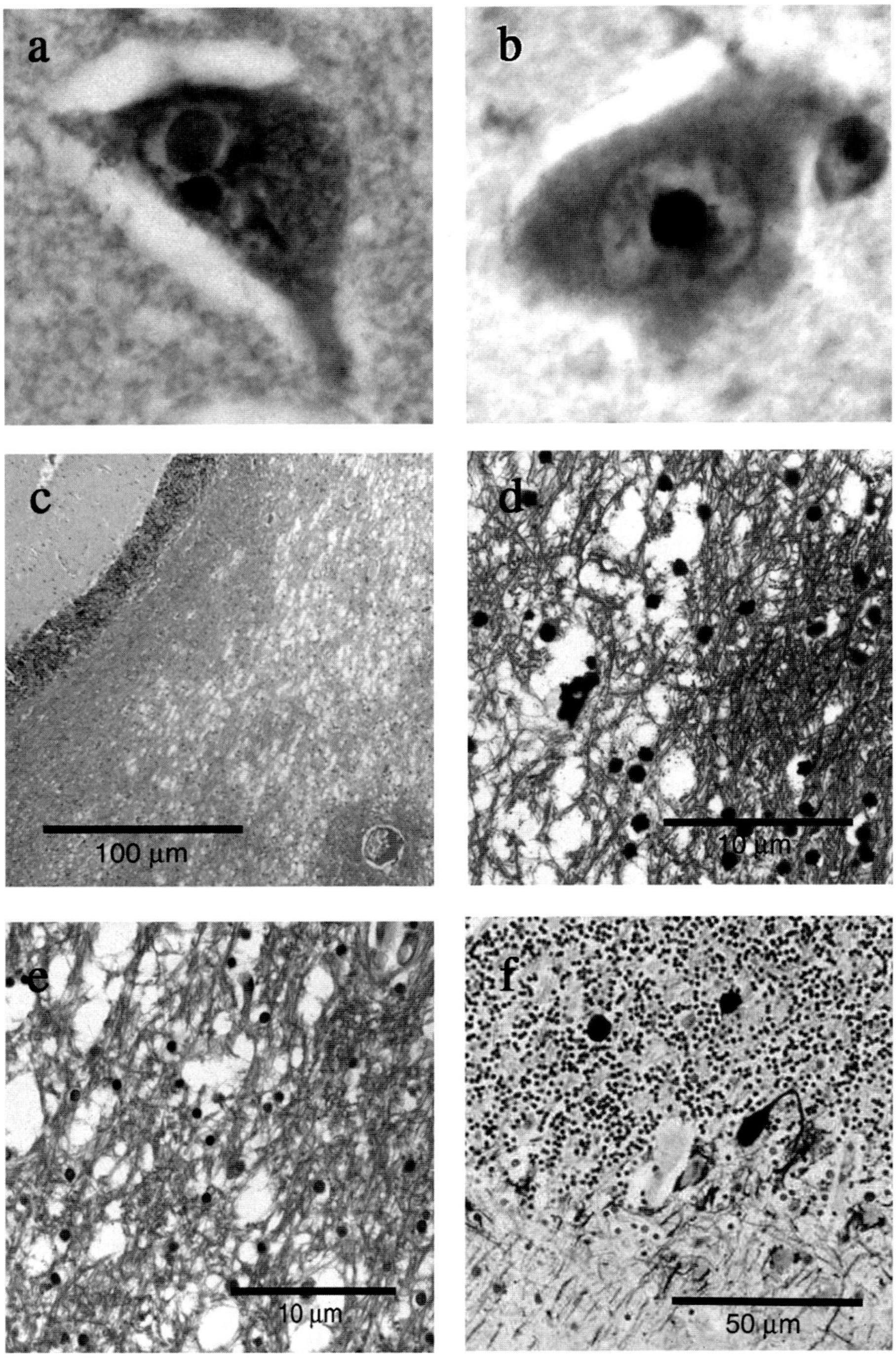

Fig. 2. Histopathology of FXTAS. (**a**) Intranuclear inclusion in a cortical neuron (hematoxylin and eosin). (**b**) Intranuclear inclusions in a hippocampal neuron (left) and astrocyte (right) (anti-ubiquitin antibody). (**c**) Spongiform changes in the white matter of the cerebellum (hematoxylin and eosin). (**d**) Silver staining (Bielschowsky) of the cerebellar white matter, demonstrating axonal loss. (**e**) Myelin staining (LFB-PAS) of the same region, showing myelin loss. (**f**) Purkinje cell dropout and the presence of axonal torpedoes in the cerebellum (anti-neurofilament antibody) (Greco et al., 2002).

reflects various ascertainment biases, including identification through affected grandchildren, the fact that the neurological phenotype has been observed for alleles as small as 69 CGG repeats is important from a mechanistic standpoint, since protein (FMRP) levels are normal or near normal in the lower premutation range, and mRNA levels are elevated only about two fold (Tassone et al., 2000b; Kenneson et al., 2001; Tassone and Hagerman, 2003). Furthermore, the absence of known cases of FXTAS in the full mutation range would appear to rule out a deficiency of FMRP as the basis for the neurological disorder.

There are at least two unexplained aspects of the clinical association of FXTAS and CGG repeat expansions: first, only a subset (~ 30%) of males appears to develop the tremor/ataxia syndrome; second, the syndrome may only rarely manifest in female premutation carriers. It is possible that the partial penetrance in males is a consequence of either genetic background effects and/or environmental factors. Although our preliminary evidence suggests that the penetrance may increase with age, there is no indication at this point that even a majority of male carriers will develop FXTAS. For this reason, it is essential that much larger patient populations be screened to better define the prevalence, and precipitating factors, of FXTAS. The relative paucity (or absence) of FXTAS in female carriers may reflect the requirement for both the genetic and/or possible environmental factors that determine clinical involvement in males and the requirement for highly skewed X-inactivation in the susceptible brain regions. Again, an examination of a much larger group of carrier females will be required to understand the nature and extent of the sex bias.

Association and prevalence

Although much of the data pertaining to the association and prevalence of FXTAS remain unpublished, early reports (Jacquemont et al., 2002; Smits et al., 2002; Berry-Kravis et al., 2003) have consistently noted a highly significant association between premutation status (in males) and both cerebellar tremor and ataxia compared to controls. Other features of the syndrome (e.g., dementia) are currently under study, but are clearly correlated with tremor/ataxia and with expanded CGG repeats. Early estimates of prevalence (Jacquemont et al., 2002; Smits et al., 2002; Berry-Kravis et al., 2003), based on cohorts of older male carriers ranging from 30 to 50 individuals, are in the range of 20–30%, but these numbers will need to be refined using much larger sample sizes. Nevertheless, there is now clear evidence for a clinical association between the tremor/ataxia syndrome and premutation status.

The more compelling case for association is based on the MRI findings (Brunberg et al., 2002) and the presence of intranuclear inclusions in post-mortem brain tissue (Greco et al., 2002). Eighty-eight percent (15/17) of male patients with FXTAS showed T2 signal hyperintensities in the cerebellar white matter, usually accompanied by extension of the signal hyperintensities into the middle cerebellar peduncles; none of these findings was seen in age- and sex-matched controls (Brunberg et al., 2002). While these findings are not unique to FXTAS, other occurrences are relatively rare. Thus, it is highly unlikely that the symmetric hyperintensities are a coincidental finding within the affected premutation carriers or that they are coincidental to the neurological condition. However, additional studies are required to define the progression of the cerebellar pathology, as well as the extent to which such imaging abnormalities occur among premutation males in the absence of overt disease. A similar case for association can be made on the basis of the unusual neuropathology associated with FXTAS. As of this writing, all (7/7) of the post-mortem brains from FXTAS patients possessed broadly distributed neuronal and astroglial intranuclear inclusions. Such inclusions are quite unusual, having been described mainly in the rare instances of NIID and several of the CAG-repeat expansion disorders (reviewed by Greco et al., 2002). Furthermore, the inclusions found in the brains of the premutation carriers appear to be polyglutamine-, τ-, and α-synuclein-negative, which sets them apart from the other intranuclear inclusion disorders. Thus, inclusion formation is highly associated with the male premutation carrier status, as well as disease formation. The nature of the association between clinical involvement (FXTAS) and inclusion formation is unknown at present; that is, in the absence of observations on brains from older premutation males who are clinically normal, it is not known whether the presence of inclusions always heralds broader neuropathology and clinical involvement.

Mechanism of disease formation

One of the most intriguing aspects of the neuropathology of FXTAS is the presence of intranuclear inclusions in the absence of any (known) structurally abnormal or dysregulated protein products. The premutation-associated neuropathology is thus unlike the situation with the CAG-repeat expansion disorders, where expanded polyglutamine stretches within the mutated proteins are believed to confer a dominant, toxic gain-of-function in susceptible cell populations. The only known biochemical abnormality associated with premutation alleles is an increased level of FMR1 mRNA in peripheral blood leukocytes (Tassone et al., 2000a, b; Kenneson et al., 2001). Thus, as we have proposed previously (Hagerman et al., 2001; Greco et al., 2002), it is possible that the elevated mRNA, per se, is responsible for inclusion formation and the broader neuropathology. Such a gain-of-function effect has already been demonstrated for myotonic dystrophy, where a CTG repeat was placed within the non-coding region of a heterologous gene in a transgenic mouse; expression of an RNA possessing the expanded CUG repeat induced myotonia (Mankodi et al., 2000). The CUG repeat was found to be associated with small intranuclear particles. Similar experiments should be performed using CGG repeats in the premutation range. Since both inclusion formation and FXTAS have been observed for alleles as small as 69 repeats, such an experiment is feasible.

Efforts to identify the contents of the inclusions are in progress; however, the intense ubiquitin staining may signal a general stress response to both neuronal and glial dysfunction, a response that may have features in common with a wide range of other inclusion disorders, where formation had been thought to be a direct consequence of abnormal protein accumulation. Thus, elucidation of the mechanism of inclusion formation in FXTAS may lead to a broader understanding of the neuropathology in a number of other inclusion disorders. One other puzzling aspect of the distribution of inclusions in FXTAS is their presence in the nuclei of astroglia. Previous reports, which were based on immunocytochemical staining in situ for FMRP, suggested that there was little or no FMR1 expression in glial cells (Devys et al., 1993; Hinds et al., 1993). However, primary astroglial cells in culture do express FMR1 mRNA (Tassone and Hagerman, unpublished results). Thus, while the dividing cells in culture cannot be directly compared with the post-replicative glial cells in brain, low levels of astroglial FMR1 mRNA expression (and hence low FMRP levels) in brain may have been missed.

Clinicopathological correlates

The multisystem character of the clinical presentation in FXTAS is clearly reflected in the global brain pathology: spongiform changes within the cerebellar white matter and middle cerebellar peduncles, Purkinje cell dropout, and cerebellar atrophy are all consistent with a predominant (initial) cerebellar component of the syndrome; the high inclusion density within the hippocampus is consistent with mood changes, including irritability and anxiety, and a moderate to severe decline in memory; and the executive function deficits and dementia are consistent with moderate to severe loss of cerebral volume. However, despite the concordance between the pathological and clinical features, little is known about the natural history of the disorder, what defines the subgroup of premutation carriers at risk for FXTAS and, in particular, the relationship between the onset of the neuropathic changes and clinical involvement. This last issue is of particular importance to the practice of both clinical neurology and psychiatry, since FXTAS may represent one of the more common monogenic causes of tremor, ataxia, and dementia. Determination of the lifetime risk of developing FXTAS will most likely come about through studies of larger populations of premutation carriers within known families with fragile X syndrome and the identification of carriers within the neurological populations with tremor and/or ataxia.

References

Bardoni B, Mandel JL: Advances in understanding of fragile X pathogenesis and FMRP function, and in identification of X linked mental retardation genes. Curr Opin Genet Dev 12:284–293 (2002).

Berry-Kravis E, Lewin F, Wuu J, Leehey M, Hagerman RJ, Hagerman P, Goetz C: Tremor and ataxia in adult fragile X premutation carriers: blinded videotape evaluation. Annls Neurol 53:616–623 (2003).

Brunberg JA, Jacquemont S, Hagerman RJ, Berry-Kravis EM, Grigsby J, Leehey MA, Tassone F, Brown WT, Greco CM, Hagerman PJ: Fragile X premutation carriers: characteristic MR imaging findings of adult male patients with progressive cerebellar and cognitive dysfunction. Am J Neuroradiol 23:1757–1766 (2002).

Cummings CJ, Zoghbi HY: Trinucleotide repeats: mechanisms and pathophysiology. A Rev Genomics Hum Genet 1:281–328 (2000).

Devys D, Lutz Y, Rouyer N, Bellocq JP, Mandel JL: The FMR-1 protein is cytoplasmic, most abundant in neurons and appears normal in carriers of a fragile X premutation. Nature Genet 4:335–340 (1993).

Dombrowski C, Levesque S, Morel ML, Rouillard P, Morgan K, Rousseau F: Premutation and intermediate-size FMR1 alleles in 10,572 males from the general population: loss of an AGG interruption is a late event in the generation of fragile X syndrome alleles. Hum molec Genet 11:371–378. (2002).

Gilman S, Low PA, Quinn N, Albanese A, Ben-Shlomo Y, Fowler CJ, Kaufmann H, Klockgether T, Lang AE, Lantos PL, Litvan I, Mathias CJ, Oliver E, Robertson D, Schatz I, Wenning GK: Consensus statement on the diagnosis of multiple system atrophy. J neurol Sci 163:94–98 (1999).

Greco CM, Hagerman RJ, Tassone F, Chudley AE, Del Bigio MR, Jacquemont S, Leehey M, Hagerman PJ: Neuronal intranuclear inclusions in a new cerebellar tremor/ataxia syndrome among fragile X carriers. Brain 125:1760–1771 (2002).

Grigsby J, Leehey M, Hagerman RJ, Epstein J, Wilson R, Jacquemont S, Greco C, Brunberg J, Tassone F, Hagerman PJ: Dementia and a tremor-ataxia disorder among older male carriers of the fragile X premutation. Proceedings of the 2002 Annual Meeting of the American Geriatrics Society, Washington, DC (2002).

Hagerman RJ: Physical and behavioral phenotype, in Hagerman RJ, Hagerman PJ (eds): Fragile X Syndrome: Diagnosis, Treatment and Research, 3rd Ed, pp 3–109 (The Johns Hopkins University Press, Baltimore 2002).

Hagerman RJ, Leehey M, Heinrichs W, Tassone F, Wilson R, Hills J, Grigsby J, Gage B, Hagerman PJ: Intention tremor, parkinsonism, and generalized brain atrophy in male carriers of fragile X. Neurology 57:127–130 (2001).

Hagerman RJ, Staley LW, O'Connor R, Lugenbeel K, Nelson D, McLean SD, Taylor A: Learning-disabled males with a fragile X CGG expansion in the upper premutation size range. Pediatrics 97:122–126 (1996).

Hinds HL, Ashley CT, Sutcliffe JS, Nelson DL, Warren ST, Housman DE, Schalling M: Tissue specific expression of FMR-1 provides evidence for a functional role in fragile X syndrome. Nature Genet 3:36–43 (1993).

Jacquemont S, Hagerman RJ, Leehey M, Greco C, Brunberg J, Tassone F, Gane LW, Jardini T, Harris SW, Zhang L, Grigsby J, Des Portes V, Berry-Kravis E, Brown WT, Hagerman PJ: Characterization of a progressive neurological condition in older adult male carriers of the fragile X premutation. Am J hum Genet 71:185 (2002).

Kaufmann WE, Abrams MT, Chen W, Reiss AL: Genotype, molecular phenotype, and cognitive phenotype: correlations in fragile X syndrome. Am J med Genet 83:286–295 (1999).

Kenneson A, Zhang F, Hagedorn CH, Warren ST: Reduced FMRP and increased FMR1 transcription is proportionally associated with CGG repeat number in intermediate-length and premutation carriers. Hum molec Genet 10:1449–1454 (2001).

Kimber TE, Blumbergs PC, Rice JP, Hallpike JF, Edis R, Thompson PD, Suthers G: Familial neuronal intranuclear inclusion disease with ubiquitin positive inclusions. J neurol Sci 160:33–40 (1998).

Leehey MA, Hagerman RJ, Landau WM, Grigsby J, Tassone F, Hagerman PJ: Tremor/ataxia syndrome in fragile X carrier males. Mov Disord 17:744–745 (2002).

Leehey MA, Munholz RP, Lang AE, Brunberg JA, Grigsby J, Greco CM, Jacquemont S, Tassone F, Lozano AM, Hagerman PJ, Hagerman RJ: The fragile X premutation presenting as essential tremor. Archs Neurol 60:117–121 (2003).

Mankodi A, Logigian E, Callahan L, McClain C, White R, Henderson D, Krym M, Thornton CA: Myotonic dystrophy in transgenic mice expressing an expanded CUG repeat. Science 289:1769–1772 (2000).

Michaud J, Gilbert JJ: Multiple system atrophy with neuronal intranuclear hyaline inclusions: report of a new case with light and electron microscopic studies. Acta Neuropathol 54:113–119 (1981).

Reiss AL, Abrams MT, Greenlaw R, Freund L, Denckla MB: Neurodevelopmental effects of the FMR-1 full mutation in humans. Nature Med 1:159–167 (1995).

Reiss AL, Lee J, Freund L: Neuroanatomy of fragile X syndrome: the temporal lobe. Neurology 44:1317–1324 (1994).

Rousseau F, Rouillard P, Morel ML, Khandjian EW, Morgan K: Prevalence of carriers of premutation-size alleles of the FMRI gene—and implications for the population genetics of the fragile X syndrome. Am J hum Genet 57:1006–1018 (1995).

Smits A, van der Bruggen W, Sistermans E, Kiemeney B, Renier W, Kremer B: Striking neurological symptoms in normal transmitting males of the fragile X syndrome. Proceedings of the 8th International Fragile X Conference, Chicago, IL (2002).

Tassone F, Hagerman P: Expression of the FMR1 gene. Cytogenet Genome Res (2003, in press).

Tassone F, Hagerman RJ, Chamberlain WD, Hagerman PJ: Transcription of the FMR1 gene in individuals with fragile X syndrome. Am J hum Genet 97:195–203 (2000a).

Tassone F, Hagerman RJ, Iklé DN, Dyer PN, Lampe M, Willemsen R, Oostra BA, Taylor AK: FMRP expression as a potential prognostic indicator in fragile X syndrome. Am J hum Genet 84:250–261 (1999).

Tassone F, Hagerman RJ, Taylor AK, Gane LW, Godfrey TE, Hagerman PJ: Elevated levels of *FMR1* mRNA in carrier males: a new mechanism of involvement in fragile X syndrome. Am J hum Genet 66:6–15 (2000b).

Tassone F, Hagerman RJ, Taylor AK, Hagerman PJ: A majority of fragile X males with methylated, full mutation alleles have significant levels of FMR1 messenger RNA. J med Genet 38:453–456 (2001).

Tassone F, Hagerman RJ, Taylor AK, Mills JB, Harris SW, Gane LW, Hagerman PJ: Clinical involvement and protein expression in individuals with the FMR1 premutation. Am J hum Genet 91:144–152 (2000c).

Toledano-Alhadef H, Basel-Vanagaite L, Magal N, Davidov B, Ehrlich S, Drasinover V, Taub E, Halpern GJ, Ginott N, Shohat M: Fragile-X carrier screening and the prevalence of premutation and full-mutation carriers in Israel. Am J hum Genet 69:351–360 (2001).

Zannolli R, Gilman S, Rossi S, Volpi N, Bernini A, Galluzzi P, Galimberti D, Pucci L, D'Ambrosio A, Morgese G, Giannini F: Hereditary neuronal intranuclear inclusion disease with autonomic failure and cerebellar degeneration. Archs Neurol 59: 1319–1326 (2002).

Cytogenet Genome Res 100:213–223 (2003)
DOI: 10.1159/000072857

The epilepsy, the protease inhibitor and the dodecamer: progressive myoclonus epilepsy, cystatin b and a 12-mer repeat expansion

M.D. Lalioti,[a] S.E. Antonarakis[b] and H.S. Scott[c]

[a] Department of Genetics, Yale University School of Medicine, New Haven CT (USA); [b] Division of Medical Genetics University of Geneva Medical School, and University Hospitals of Geneva, Geneva (Switzerland); [c] Genetics and Bioinformatics Division, The Walter and Eliza Hall Institute of Medical Research, Parkville (Australia)

Abstract. Progressive myoclonus epilepsy 1 (EPM1) or Unverricht-Lundborg disease is a human autosomal recessive neurodegenerative disorder caused by mutations in cystatin B (CSTB). The CSTB gene maps to human chromosome 21 and encodes an inhibitor of lysosomal cysteine proteases. Five point mutations have been found, two of which are seen in numerous unrelated patients. However, the main CSTB mutation in EPM1, even among patients of different ethnic origins, is an expansion of a dodecamer repeat (CCCCGCCCCGCG) in the 5′ flanking area of CSTB. Most normal alleles contain either two or three repeats, while rarer normal alleles that are highly unstable contain between 12 and 17 repeats. Mutant expanded alleles have been reported to contain between 30 and 80 copies and are also highly unstable, particularly via parental transmission. There is no apparent correlation between mutant repeat length and disease phenotype. While the repeat expansion is outside the CSTB transcriptional unit, it results in a marked decrease in CSTB expression, at least in certain cell types in vitro. *Cstb* homozygous knockout mice show some parallels to the phenotype of human EPM1 including myoclonic seizures, development of ataxia and neuropathological changes associated with cell loss via apoptosis. Loss of CSTB function due to mutations is consistent with the observed neurodegenerative pathology and phenotype, but the functional link to the epileptic phenotype of EPM1 remains largely unknown.

The progressive myoclonus epilepsies (PMEs) are a group of rare genetically heterogeneous hereditary disorders characterized by generalized epileptic seizures, myoclonus (brief contraction of a muscle or a group of muscles), and progressive neurological deterioration including ataxia and dementia (Berkovic et al., 1986, 1991, 1993; Delgado-Escueta et al., 2001).

The Unverricht-Lundborg type of progressive myoclonus epilepsy (EPM1; OMIM 254800) was first described by Unverricht in 1891 in Estonia (Unverricht, 1891) and Lundborg in Sweden in 1903 (Lundborg, 1903). Also known as Baltic myoclonus and Mediterranean myoclonus, EPM1 is an autosomal recessive disorder with an incidence of approximately one in 20,000 in Finland (Norio and Koskiniemi, 1979). Also found in the western Mediterranean (Southern France, North Africa) (Norio and Koskiniemi, 1979; Genton et al., 1990; Labauge et al., 1997) and eastern Mediterranean (Mazarib et al., 2001), there is frequent consanguinity in the parents of affected individuals.

Clinically, EPM1 is characterized by severe stimulus-sensitive myoclonus, generalized tonic-clonic seizures and distinctive electroencephalogram findings, with marked sensitivity to photic stimulation. The age of onset is between 6 and 18 years of age, with most diagnoses occurring at around 11 years. The disorder has been described as showing a progressive course

The EPM1 project was supported by the Swiss National Science Foundation.

Received 31 January 2003; manuscript accepted 1 March 2003.

Request reprints from Dr Hamish S. Scott
Genetics and Bioinformatics Division
The Walter and Eliza Hall Institute of Medical Research
1G Royal Pde, Parkville, 3050 (Australia)
telephone: 61-3-9345 2485/2555; fax: 61-3-9347-0852
e-mail: hscott@wehi.edu.au

KARGER
Fax + 41 61 306 12 34
E-mail karger@karger.ch
www.karger.com

0301–0171/03/1004–0213$19.50/0

Accessible online at:
www.karger.com/cgr

with the severity of the symptoms and the rate of progression variable both between and within families (Koskiniemi et al., 1974a, 1974b; Norio and Koskiniemi, 1979; Lehesjoki, 2002). Mental deterioration, dementia, an intention tremor, and dysarthria may develop in addition to mild ataxia late in the course of the disease, which is usually 10 to 20 years in duration. Intelligence is only slightly affected and psychotic symptoms are not found, but emotional lability is a typical feature and suicide has often been reported. More recently, prognosis has improved significantly with modern anticonvulsive therapy controlling many symptoms fairly well and limiting disease progression. The myoclonus is resistant to treatment, whereas the generalized seizures respond to antiepileptic therapy (valproate and clonazepam).

The most consistent pathological findings in EPM1 are a marked loss of Purkinje cells in the cerebellum, neuronal loss in medial thalamus and spinal cord, and proliferation of Bergmann glia. There may also be a loss of cells in the granular cell layer (Haltia et al., 1969; Koskiniemi et al., 1974a, 1974b; Meldium and Bruton, 1992; Mascalchi et al., 2002; Takuma et al., 2003). Membrane-bound vacuoles with clear contents in eccrine cells have been described in EPM1 (Cochius et al., 1994).

The molecular genetic findings described below have now made possible a genetic diagnosis of EPM1.

Linkage mapping of EPM1

Using DNA from 12 Finnish EPM1 families, linkage was first detected with a highly informative VNTR (variable number of tandem repeats) marker mapped to chromosome 21q (D21S112) (Lehesjoki et al., 1991). Additional analyses suggested that the EPM1 locus could be refined to an 11.8-cM interval on 21q22.3 around markers D21S112, D21S49, and D21S154 in these Finnish families. Subsequent studies showed locus homogeneity for EPM1, first between the Baltic and Mediterranean myoclonus epilepsies (Malafosse et al., 1992) and then with EPM1 families from around the world (Lehesjoki et al., 1992, 1993a, 1993b; Cochius et al., 1993; Lalioti et al., 1995).

The advanced knowledge of the genomic infrastructure of chromosome 21q in the mid 1990's, plus the sole use of Finnish EPM1 patients, facilitated the fine mapping of the EPM1 locus by linkage disequilibrium to only a 175-kb interval between markers D21S2040 and D21S1259 (Lehesjoki et al., 1993b; Virtaneva et al., 1996). The implications of the haplotype analyses and linkage disequilibrium studies in the Finnish EPM1 were that there was a single major Finnish EPM1 or Baltic myoclonus mutation derived from a common ancestral origin. It was estimated that the original Finnish EPM1 mutation was at least 2000 years old (or about 60–70 generations) (de la Chapelle, 1993; Virtaneva et al., 1996). Comparable studies of Mediterranean EPM1 predicted a similar founder effect in Mediterranean myoclonus and a common Mediterranean EPM1 mutation(s). It was not clear whether this was the same as the Baltic myoclonus mutation (Lehesjoki et al., 1994; Labauge et al., 1997; Parmeggiani et al., 1997).

Positional cloning of the EPM1 gene

Since there was no biochemical clue as to the pathophysiology of the disease, positional cloning strategies were employed by several laboratories to clone the gene for EPM1. As was standard at the time, physical mapping (Lafreniere et al., 1995; Stone et al., 1996) was followed by gene and transcript identification using a number of techniques including exon trapping and cDNA selection (Cheng et al., 1994; Chen et al., 1996). Following the exclusion of several positional candidate genes, mutations were found in the cystatin B gene (CSTB or CST6; also called stefin B or STFB) in a few patients with EPM1. The CSTB cDNA had been cloned in 1990 but had not been mapped to any human chromosome until this study, which provided the first convincing evidence that CSTB is responsible for EPM1 (Pennacchio et al., 1996), though the mutations detected only accounted for a small number of the EPM1 alleles studied.

Cystatin B

Cystatins are reversible inhibitors of cysteine proteases of the papain superfamily, and are found both intra and extracellularly. The name cystatin was introduced by Barrett (1981) for a protein from chicken egg white that inhibits various cysteine proteases. Cystatins are believed to control the activities of endogenous proteases and to protect organisms from protein degradation by exogenous cysteine proteases. The cystatins are classified into three families based on their size and structural characteristics such as the presence of internal disulphide bonds. Type 1 cystatins (or stefins), the family to which CSTB belongs, are small (~ 100 amino acids), non-glycosylated proteins without internal disulphide bonds. They are intracellular inhibitors where type 2 cystatins and kininogens (the third family) are extracellular (reviewed in Turk and Bode, 1991; Turk et al., 1997, 2001, 2002b).

The human CSTB protein is 98 amino acids long and highly conserved between species e.g. rat, mouse, and cow (Ritonja et al., 1985; Sato et al., 1990; Turk et al., 1992; Pennacchio and Myers, 1997). CSTB binds to and inhibits lysosomal cysteine proteases such as the cathepsins B, H, L and S. Lysosomal cathepsins are involved in many processes that require intracellular proteolysis such as extracellular matrix remodelling. Although they are primarily intracellular proteins, cystatins have also been detected in extracellular fluids (Abrahamson et al., 1986). Northern blot analysis detected CSTB mRNA in all tissue examined (Pennacchio et al., 1996).

The crystal structures of chicken CSTB and the CSTB-papain complex have been determined (Bode et al., 1988; Stubbs et al., 1990). Chicken CSTB consists of a five stranded antiparallel β-sheet (A to E) wrapped around a five-turn α-helix. The most amino-terminal part of CSTB (Met1 to Pro6 equivalent in human) and the loops connecting the β-sheets B to C and D to E were shown to interact with the papain active site (PPB entry 1STF) (Stubbs et al., 1990). Truncation experiments with chicken and human CSTB proteins showed that N-terminal deletions result in a 5000-fold decrease in binding affinity to

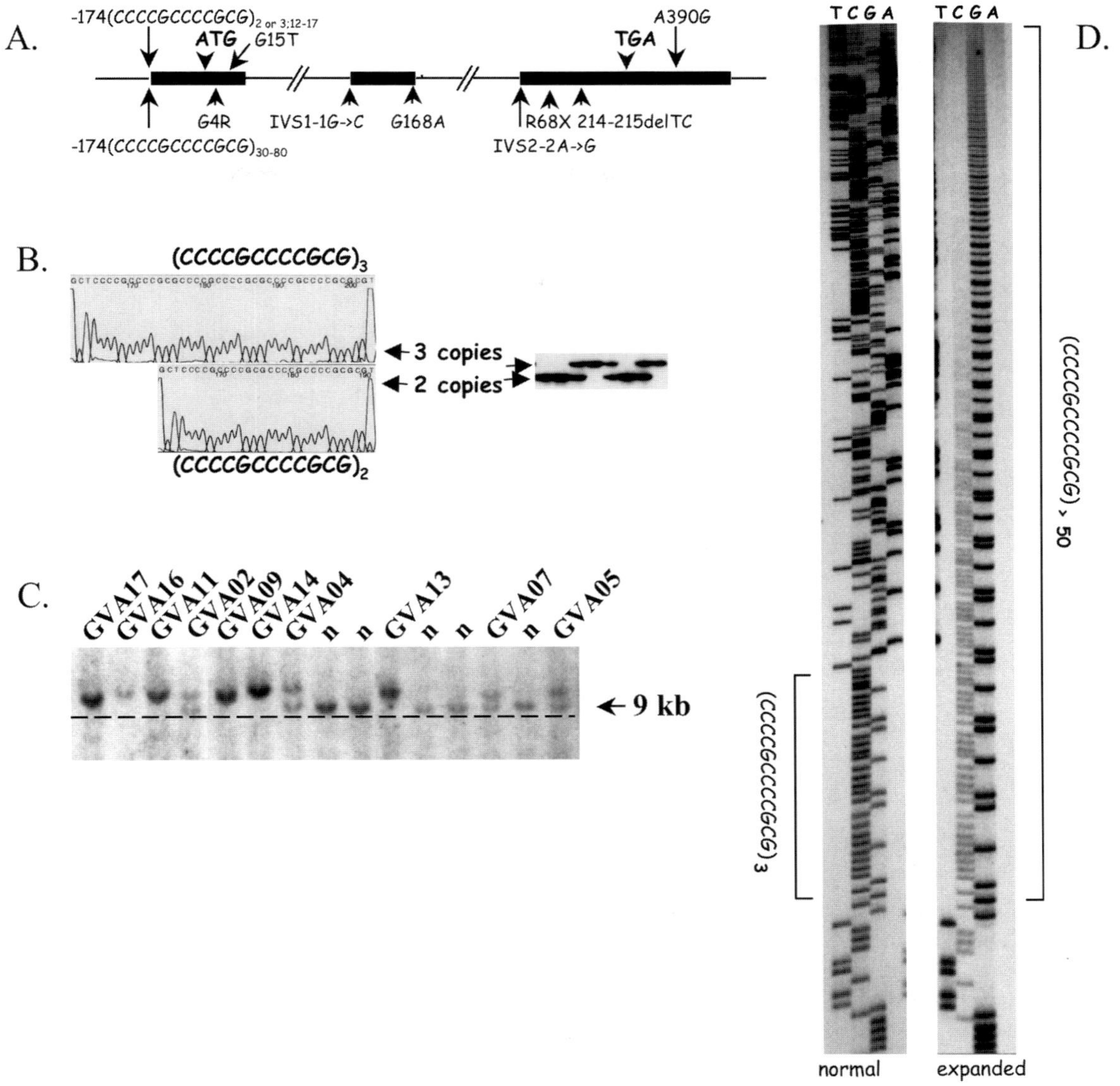

Fig. 1. A summary of the discovery of the dodecamer repeat expansion in EPM1. (**A**) Schematic representation of the CSTB gene. Below the gene, point mutations that cause EPM1 are shown. The missense and nonsense mutations are annotated by G4R and R68X respectively. Above the gene, polymorphic variants in normal individuals are shown. G168A is discussed in a footnote to Table 1. (**B**) Nucleotide sequence of alleles with two or three copies of the 12-mer (CCCCGCCCCGCG) repeat in the 5′ flanking region of CSTB. To the right is an autoradiogram of the PCR-amplified region that contains the polymorphic repeat in different individuals. (**C**) Southern blot analysis of EPM1 patients (GVA) and normals (n). DNA was digested with *Eco*RI and probed with the CSTB cDNA. The normal fragment is 9 kb. Some patients show only an abnormal (larger) fragment. The larger fragment size varies between patients. Patients showing two different fragments have defined point mutations in the normal-sized CSTB allele. (**D**) Nucleotide sequence analysis of a normal and a mutant clone that contain the region of the dodecamer repeat in the 5′ flanking region of the CSTB gene. The mutant clone contains more than 50 identical, uninterrupted copies of the 12-mer (CCCCGCCCCGCC) repeat.

papain (Abrahamson et al., 1987; Machleidt et al., 1989; Thiele et al., 1990). The only CSTB missense mutation in EPM1 described to date, G4R (Fig. 1A, Table 1 and below), is predicted by molecular modeling to strongly affect the papain-binding capacity of CSTB (Lalioti et al., 1997a).

Point mutations and DNA polymorphisms in CSTB

The CSTB gene is small, containing only three exons (Fig. 1A). Five point mutations, some recurrent, that are likely to cause abnormal gene expression and protein function have been identified by analysis of many EPM1 patients (Bespalova et al., 1997a, 1997b; Lafreniere et al., 1997; Lalioti et al., 1997a; Pennacchio et al., 1998). These mutations are shown in Fig. 1A and are listed in Table 1. Surprisingly, these CSTB mutations only accounted for approximately 10% (23/236) of the EPM1 alleles examined (Pennacchio et al., 1996; Lafreniere et al., 1997; Lalioti et al., 1997a, 1997b; Virtaneva et al., 1997). Additionally, no point mutations could be found associated with the most common Baltic/Finnish and Mediterranean myoclonus haplotypes. Absent or greatly reduced CSTB expression had been shown in a number of EPM1 cell lines, including those with no known point mutations (Pennacchio et al., 1996;

Table 1. Polymorphisms and mutations in CSTB

Position	Mutation originally named[a]	Nucleotide position[b]/ amino acid position	Remark	References
Promoter	$207(CCCCGCCCCGCG)_{2\ or\ 3;\ 12\text{-}17}$	c-174/ -	Polymorphic dodecamer repeat	Lalioti et al., 1997a Lalioti et al., 1997b
Promoter	$207(CCCCGCCCCGCG)_{30\ to\ 80}$	c-174/ -	Pathogenic dodecamer repeat expansion	Lalioti et al., 1997b
Exon 1	426G→C	c10G→C/ G4R	pathogenic	Lalioti et al., 1997a
Exon 1	431T→G	c15G→T/ -	polymorphic	Lalioti et al., 1997a
Intron 1	1925G→C	IVS1-1G→C	pathogenic	Pennacchio et al., 1996 Bespalova et al., 1997b Lafreniere et al., 1997 Lalioti et al., 1997a
Exon 2	-	c168G→C	pathogenic?[c]	Kagitani-Shimono et al., 2002
Intron 2	2352AG/→GG/ or 2352A→G	IVS2-2A→G	pathogenic	Lalioti et al., 1997a
Exon 3	2388C→T	c202C→T/ R68X	pathogenic	Pennacchio et al., 1996 Lafreniere et al., 1997
Exon 3	del2399tc	c214-215delTC/ truncation	pathogenic	Bespalova et al., 1997a Lafreniere et al., 1997 Lalioti et al., 1997a
3'-UTR	2575A→G	c390A→G/ -	polymorphic	Lalioti et al., 1997a

[a] Nucleotide positions were according to the position in genomic sequence U46692. Some original positions are one nucleotide incorrect.
[b] Mutation nomenclature is according to the recommendations of Dunnen and Antonarakis (2000). If nucleotide positions are annotated with c, the A of the ATG is position 1 and the position is respective of the cDNA (cDNA and protein sequence for CSTB; GenBank RefSeq: NM_000100, SwissProt P04080).
[c] This is the last nucleotide of exon 2 and does not change an amino acid. The patient has one allele with an expansion and this is the only other nucleotide change reported. While not addressed in the paper, it may affect splicing and reduce the quantity of CSTB RNA and thus be pathogenic or it may be a polymorphism and the other pathogenic CSTB allele has not been detected in this patient.

Lafreniere et al., 1997). The unidentified mutations were presumed to lie outside the coding regions, in unknown regulatory elements, so disrupting the normal expression of the gene.

Several single nucleotide polymorphisms were identified as shown in Fig. 1A (Lalioti et al., 1997a). The most significant polymorphism, however, is the presence of two or three copies of the dodecamer sequence (CCCCGCCCCGCG) in the 5′ flanking region of the gene shown in Figs. 1B and C (Lalioti et al., 1997a).

A dodecamer repeat expansion in CSTB

Southern blot analysis of DNAs from EPM1 patients with unknown CSTB mutations revealed that restriction fragments containing the entire CSTB gene were larger than normal. The larger fragments varied in size between patients, segregated with the EPM1 phenotype in families, and were not present in normal alleles. Moreover, all the EPM1 patients with a known point mutation in one allele were heterozygous for one normal and one larger restriction fragment and study of their families showed that the point mutation was present in the normal-sized fragment (Lafreniere et al., 1997; Lalioti et al., 1997b; Virtaneva et al., 1997).

Further analysis localized the region of the abnormal sequence, as detected by Southern blot analysis, to an area resistant to PCR amplification between nucleotides –400 and –80 upstream of the ATG translation initiation codon of CSTB. The abnormal sequence was difficult to clone, amplify or sequence and mapped precisely to the 12-bp polymorphic tandem repeat located in the 5′ flanking region of the CSTB gene (Lafreniere et al., 1997; Lalioti et al., 1997b; Virtaneva et al., 1997). Genomic clones in bacterial plasmids containing the abnormal region were unstable and showed a high frequency of insert loss under normal culture conditions. Sequencing of these cloned DNA fragments did not reveal genomic rearrangement, translocation, or new sequences which were among the possible explanations. Instead, under special conditions, the nucleotide sequence from the genomic clones derived from French and Swiss EPM1 patients clearly showed a long stretch of uninterrupted perfect repeats of the dodecamer $(CCCCGCCCCGCG)_n$ sequence (Lalioti et al., 1997b) (Fig. 1D). In the initial sequencing experiments more than 50 repeats were observed in the clones studied. The Southern blot analysis of DNAs from all available patients indicated that the expansion of the dodecamer repeat $(CCCCGCCCCGCG)_n$, which is located 174 nucleotides upstream of the initiation ATG codon of CSTB, is the most common mutation in all populations examined and accounted for all the previously unknown mutations (213 of 236 alleles) (Lafreniere et al., 1997; Lalioti et al., 1997b; Virtaneva et al., 1997; Kagitani-Shimono et al., 2002).

At the time, the expansion of the dodecamer $(CCCCGCCCGCG)_n$ repeat in EPM1 represented the first case of expansion of a repeat unit other than trinucleotides in association with a human disorder. While the dodecamer contains a shorter repeat of CCCCG within it, the expansion involves the entire dodecamer and not the shorter pentamer.

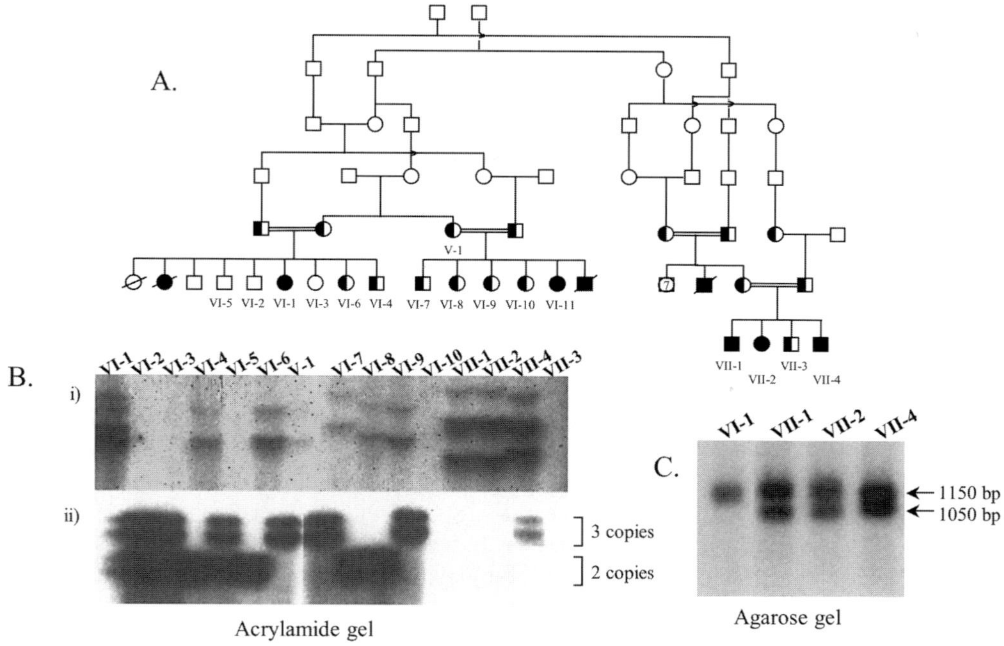

Fig. 2. Examples of the EPM1 dodecamer repeat expansion instability. (**A**) Pedigree of Swiss family with EPM1, showing repeat instability. Affected and heterozygous individuals in the last three generations are indicated. (**B**) PCR amplification, and incorporation of label, run on a denaturing acrylamide gel. Panel i, showing the expanded alleles, was exposed for several days. Panel ii, showing normal alleles, represents a shorter exposure of a lower region of the same gel. Because of the denaturation, each allele shows two bands. The large allele of individual V-1 (mother) has further expanded, by one repeat unit, on transmission to her carrier daughter, VI-8. Individual VII-4 has expanded allele sizes that are different from those of his affected siblings, VII-1 and VII-2. (**C**) Detection of expanded EPM1 alleles in selected affected individuals, by the alternative methods of PCR amplification and hybridization.

Instability of the dodecamer repeat expansion

At the level of resolution of the Southern blots used, no instances of intergenerational – parent to child – changes in the expansion size were detected (Lafreniere et al., 1997; Lalioti et al., 1997b; Virtaneva et al., 1997). However, several observations indicated at least a moderate level of germline instability. They were:

I. The dodecamer expansion was observed in EPM1 patients of different geographical and ethnic origins with different haplotypes near the CSTB gene (Lafreniere et al., 1997; Virtaneva et al., 1997). This suggested that a number of independent dodecamer repeat expansions had occurred.

II. The number of the dodecamer repeat expansions varies between distant affected relatives in which the haplotypes of affected alleles are the same (Fig. 2) (Lalioti et al., 1998).

III. In the Finnish population, a putative founder haplotype was detected in 75 of 87 chromosomes with the expansion (Virtaneva et al., 1997). This striking linkage disequilibrium suggests a single origin of the expansion mutation in this population. However, the number of the expanded repeats varies in these patients from the same ethnic group with the same haplotype (Lafreniere et al., 1997; Virtaneva et al., 1997). This strongly indicates that the expanded dodecamer repeat is unstable and changes of repeat number occur frequently.

Two studies with substantial data on the instability of the dodecamer repeat expansions are summarized in Table 2 (Lalioti et al., 1997b; Larson et al., 1999). Other studies that show unique events not documented in the table are also discussed below (Lalioti et al., 1998; Mazarib et al., 2001).

Besides the normal alleles with two or three copies of the dodecamer repeat and the EPM1 alleles with many copies of the repeat, intermediate alleles of 12–17 copies of the repeat were observed in two CEPH collection pedigrees (102 and 104) from Venezuela (Lalioti et al., 1997b). Genotyping in these families revealed that the alleles with 12 or 13 copies were often transmitted unstably from parents to offspring. From 21 transmissions of these alleles, six repeat expansions (29%) to alleles with 13, 14, 15 or 17 repeats were observed, with the largest observed expansion in a given meiosis being four repeats (from 13 to 17). The instability of these intermediate (perhaps premutation) alleles of the dodecamer repeat in this family is 0.29, whereas the mutation rate of some so-called hypermutable polymorphisms is between 0.067 and 0.4 (Vergnaud and Denoeud, 2000). Haplotype analysis revealed that all six expansions were paternally transmitted (Lalioti et al., 1997b). Haplotype data in the CEPH families showed that there were no contractions of the premutation alleles. Individuals homozygous for the 12-repeat allele or compound heterozygous for 12 and 17 repeats had no clinical phenotype of EPM1 and therefore these alleles are considered noncausative of EPM1. The clinical significance of intermediate alleles remains unknown and large population studies are needed to establish the probability of expansion to the disease range.

Table 2. Instability of the dodecamer repeat expansion

Reference	Transmissions[a]			Unstable			Expansions			Contractions			Comment
	Total	M	P	Total	M	P	Total	M	P	Total	M	P	
Lalioti et al., 1997b	21	?[b]	?	6	0	6	6	0	6	0	0	0	13-17 repeat "normal" alleles
Larson et al., 1999	74[c]	33	35	35	7	22	11	0	11	18	7	11	EPM1 alleles

[a] Transmissions of dodecamer repeat alleles of size greater that 2 or 3. M = maternal and P = paternal.
[b] The number of parental origins of all transmissions was not recorded.
[c] The parental origin of all the EPM1 allele transmissions could not be determined.

By Southern blot analysis, restriction fragments containing the dodecamer repeat expansions were between 0.5 and 1.5-kb larger than normal alleles (Lafreniere et al., 1997; Lalioti et al., 1997b; Virtaneva et al., 1997). This corresponds to estimated expansions of between 40 and 125 dodecamers. Special PCR methods were subsequently developed in two studies to allow accurate measurement of the size of the repeat expansions in EPM1 patients. The number of dodecamer repeats found in affected individuals ranged from 30 to 80 (Lalioti et al., 1998; Larson et al., 1999).

In the first study using PCR of EPM1 expanded alleles, different numbers of repeats were observed among affected siblings, between parent and child, and between more distant relatives, which directly confirmed the instability of the disease-related expansion (Fig. 2). All affected individuals in the consanguineous family, shown in Fig. 2, share the same haplotype around CSTB (Lalioti et al., 1998). However, there are six different sizes of expanded alleles (containing 51, 52, 55, 56, 58, and 59 repeats), all of which must have originated from a single, ancestral expanded allele. It is impossible from these data to determine how many expansions and contractions have occurred in this family although an expansion has occurred in heterozygous individual VI-8 with the expanded allele transmitted from her mother, V-I (Fig. 2). This expansion of the dodecamer repeat was the first to be associated with maternal transmission (Lalioti et al., 1998) although others have now been observed (Mazarib et al., 2001).

The second study examined dodecamer stability in 17 EPM1 family pedigrees (Larson et al., 1999). They followed 74 transmissions of expanded alleles from 32 parental carriers. Fifty-two expanded alleles were transmitted to affected individuals while 22 alleles were transmitted to carriers. There were 35 unstable transmissions of expanded alleles, including 20 contractions and 15 expansions, giving a mutation frequency for expanded EPM1 alleles of 0.47, even higher than that of the 12–17 repeat containing permutation alleles. While the maternal or paternal origin of some expansions and contractions could not be determined, seven of 33 maternal and 22 of 35 paternal transmissions of expanded alleles resulted in changes in allele size. All the unstable maternal transmissions in this study were contractions. As has been found for CAG repeat expansion disorders, there appears to be greater instability associated with male transmissions. There were six expansions or contractions of greater than one repeat, ranging from three repeats gained to two repeats lost, with all the remaining expansions or contractions being by a single dodecamer repeat. No mutations of wild-type alleles (two or three repeats) were observed in 30 transmissions and the instability of the dodecamer was locus-specific, as another GC-rich minisatellite was stable in the same families when typed in the same way.

Larson et al. (1999) also studied mitotic instability by genotyping lymphoblastoid cell lines derived from six EPM1 patients heterozygous for expanded alleles. Three cell lines showed genotypes that differed from the three corresponding patients blood. In one patient, a compound heterozygote (69/78), cell line DNA yielded a (69/76) genotype. In cell lines from two patients, three or more alleles were detected. These data strongly suggest the mitotic instability of expanded EPM1 alleles. However, mitotic instability has never been observed directly in patient's DNA.

Expansion size and age of onset of EPM1 phenotype

The accurate measurement of the size of the dodecamer repeat expansion permitted the study of the potential correlation between the age of onset and the size of the repeat. Data from 28 patients were collected and analyzed. Unlike many other repeat expansion disorders, it appears that there is no correlation between the size of the expanded EPM1 repeat alleles and the age of onset of the disease (Lalioti et al., 1998). This is consistent with the observation that siblings that share the same repeat expansion may have different ages of onset. It is possible that once the dodecamer repeat expansion extends beyond a critical threshold, it reduces CSTB expression in certain cells, with pathological consequences. In this case, the age of onset of EPM1 would no longer depend on the size of the alleles after this threshold was passed by both of the patient's alleles. Modifier genes and environmental factors may influence the age of onset, severity and evolution of the phenotypes.

Compound heterozygous patients with one point mutation and one expanded CSTB allele did not show any particular phenotypic abnormality and their history of seizures was not different from that of the general EPM1 population. Normal carriers of expanded EPM1 alleles also have no apparent phenotype. In addition, no anticipation on the age of onset of the disease was observed, mainly because EPM1 is a recessive phenotype and all patients are usually in one generation.

The effect of expansion on CSTB expression

A single ubiquitously expressed 0.8 kb CSTB mRNA was detected by Northern blot analysis in all tissues studied (Pennacchio et al., 1996). Many recent microarray experiments have shown that CSTB is expressed at a relatively high level and, as much as any other gene, can be considered a housekeeping gene (Hsiao et al., 2001). Absence of, or greatly reduced CSTB expression had been shown by Northern blot analysis in a number of EPM1 lymphoblastoid cell lines homozygous for expansions of the dodecamer repeat (Pennacchio et al., 1996; Bespalova et al., 1997a; Lafreniere et al., 1997).

Using RNase protection assays, a marked reduction of CSTB mRNA levels was observed in blood leucocytes from patients with dodecamer expansion compared to normals. In contrast, in fibroblasts and lymphoblastoid cell lines, the amount of CSTB mRNA was either normal or only slightly reduced (Fig. 3). The low level of mRNA in blood leucocytes compared with the normal level in lymphoblastoid cell lines suggested that the repeat expansion modulates the expression of the CSTB gene only in some cell types (Lalioti et al., 1997b).

A subsequent study showed a marked reduction of CSTB mRNA in lymphoblastoid cell lines from EPM1 patients compared to normals and also in lymphoblastoid cell lines from individuals from the CEPH families with the 12–17 copy dodecamer repeat "premutation" alleles (Alakurtti et al., 2000). Biochemical studies in lymphoblastoid lines from EPM1 patients appear to support these findings; papain inhibitory activity (cystatin) was significantly decreased or absent and a corresponding increase in papain-like (cathepsin) activity, specifically cathepsins B, L and S was observed (Kinne et al., 2002). Cathepsin S is also transcriptionally upregulated at the transcriptional level in brains from *Cstb* deficient (–/–) mice (Lieuallen et al., 2001).

The apparent differences in these studies, that is detecting or not detecting reduced CSTB expression in EPM1 lymphoblastoid cell lines, is curious. It is possible that there are some EPM1 cell lines with the expansion that still express CSTB, or perhaps more likely, the variable expression is related to the tissue culture conditions or other experimental variables. This would raise interesting possibilities for therapy of EPM1. In general, however, all studies agree that the dodecamer repeat expansion results in a reduction of CSTB expression.

The dodecamer repeat expansion is in the CSTB promoter

The repeat expansions in all other repeat expansion disorders lie within the transcriptional units of the genes affected (Richards, 2001). The dodecamer repeat expansion in EPM1 is 174-nt 5′ of the CSTB initiating methionine codon, lying either in the 5′-UTR or promoter of CSTB. Using RNase protection, two transcription start sites of CSTB, were observed, 97 and 108-nt upstream of the ATG or 66 and 77-nts downstream of the dodecamer repeat (Lalioti et al., 1997b). Thus, unlike in all the other disorders due to repeat expansions, the dodecamer repeat is not within the primary transcript of the CSTB gene.

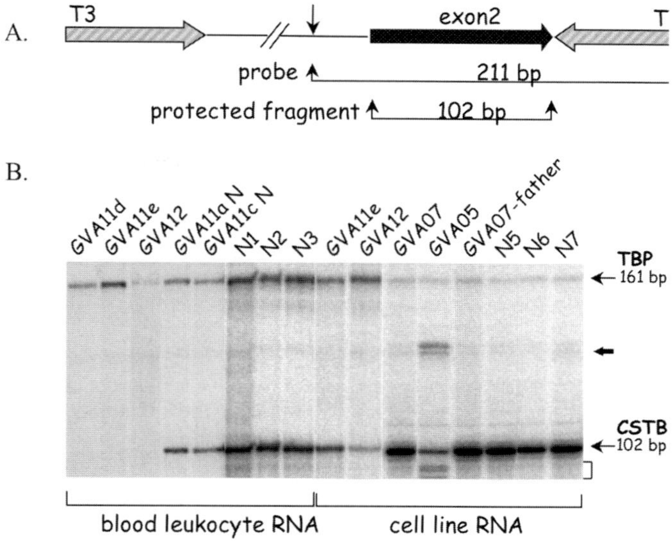

Fig. 3. Quantitation of CSTB RNA in EPM1 and normal blood samples and cell lines. (**A**) Schematic representation of a plasmid containing exons 1 and 2 of CSTB used for RNase protection assays of exon 2 of CSTB. The expected protected fragment is 102 bp. The expected control TBP protected fragment is 161 bp. (**B**) Autoradiogram of RNase protection showing the expression of CSTB relative to TBP in blood leucocytes and lymphoblastoid cell lines. Different lanes from the same gel have been combined for convenience of presentation. The arrow and bracket on the right depict the alternatively spliced protected fragments from patient GVA05 who is a compound heterozygote for an expansion and IVS1-1G→C. The sizes of the protected fragments are indicated. GVA11d, e, a, and c are four siblings from family GVA11; the first two are homozygous for the repeat expansion, while the last two are unaffected. In blood leucocytes, there is a marked reduction of CSTB RNA in patients; in contrast, in most of the cell lines from patients its expression is not different from normal. Note that in the blood sample of GVA11e there is almost undetectable CSTB RNA, while in his cell line the CSTB RNA is within the normal limits. Patient GVA12 shows more CSTB RNA in the LBL than in blood leukocytes, however the expression in LBL is lower than normal.

The mechanism(s) by which the dodecamer repeat expansion results in reduced CSTB transcription is still largely unknown. Among possible explanations are hypermethylation, altered chromatin structure, recruitment of transcription repressors on the dodecamer repeat and altered spacing of promoter elements. It is also possible that more than one of these mechanisms contributes to the reduction of gene expression.

In terms of methylation, while the repeat expansion itself has not been studied, DNA methylation differences were not observed between normal and EPM1 samples for a limited number of *Hpa*II/*Msp*I sites both within the CSTB gene and 5′ of the dodecamer repeat (Lalioti et al., 1997b). The dodecamer repeat forms more secondary structures than any of the disease associated triplet repeats, including hairpins stabilized primarily by G·G base pairs, G-tetraplexes and i-tetraplexes, and both Y·R·Y and R·R·Y triplexes (Jithesh et al., 2001; Pataskar et al., 2001a 2001b; Saha and Usdin, 2001). Both intrastrand tetraplexes, which can be thought of as folded hairpins or hairpin dimers, and G-tetraplexes have extraordinary stability relative to hairpins. Some of these structures are similar to those formed by the GC-rich fragile X syndrome triplet repeat (Chen

A.

B.

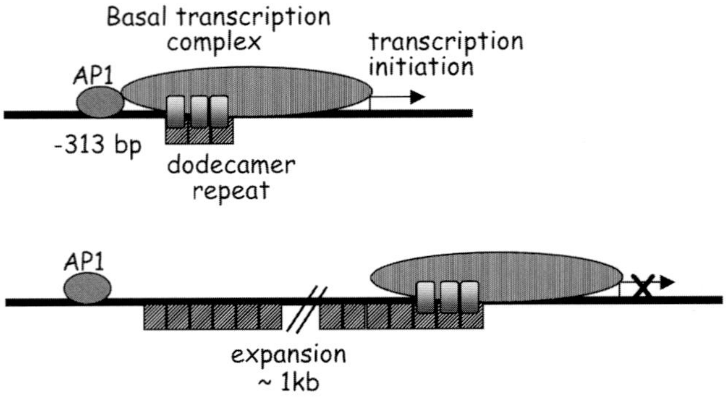

Fig. 4. A model for the transcriptional repression of CSTB gene in EPM1 due to the dodecamer repeat expansion. Sp1-like and AP1 transcription factors are indicated by grey shaded rectangles and left hatched ovals respectively. For simplicity, all other regulators that contribute to transcription are referred to as the "basal transcription complex" and are represented by the large vertically striped oval. The dodecamers are represented by right-hatched boxes. The transcription activator site of AP1 that is identified by both studies at position –312 is indicated (Lalioti et al., 1999; Alakurtti et al., 2000). The position is calculated using the A of the ATG as position +1 to the C of the core AP1 binding site (TGAC). (**A**) When two to three repeats of the dodecamer are present, the transcription activator AP1 can interact with the basal transcription complex including Pol II. (**B**) The second panel illustrates the inability of the transcription activator to interact with the complex, due to the expansion, which shifted its binding site to a non-permissive distance.

et al., 1995) and contain C·C mismatches that may be hotspots for methylation (Laayoun and Smith, 1995; Kho et al., 1998).

Altered chromatin structure has long been suspected in repeat expansion disorders. The GC-rich fragile X syndrome repeat has been shown to exclude nucleosomes while methylation increases their exclusion, both of which could clearly alter chromatin structure (Wang et al., 1996; Wang and Griffith, 1996). Alteration in chromatin structure due to the dodecamer repeat expansion in EPM1 would be expected to result in altered expression of neighboring genes in EPM1 cells lines and patients.

The dodecamer repeat sequence contains a GC box or Sp1 transcription factor-binding site. In common with most transcription factors, Sp1 can act as a repressor depending on the promoter and on other transcription factors present. In fact, several other Sp1-like transcription factors were originally described as transcription repressors (Cook et al., 1999). The dodecamer repeat expansion may recruit Sp1 transcription repressors in a cell specific manner. However, a severe depletion of Sp1 transcription factors may also be expected to alter the expression of additional genes.

These last two hypotheses have yet to be tested but, as well as affecting the expression of other genes, both might partially explain the variation in EPM1 such as age of onset. Against these hypotheses are the apparent lack of correlation of age of onset with expansion size, the lack of apparent phenotypic difference between EPM1 patients with point mutations and expansions, and the fact that *Cstb*$^{-/-}$ mice also develop an EPM1-like disease. However, the phenotypic description of EPM1 phenotypes and disease progression in both humans and *Cstb*$^{-/-}$ mice may not be sufficiently detailed to support either of these hypotheses.

There is considerable evidence from two studies to support the fact that the reduction in CSTB mRNA in EPM1 is due, in part, to a simple increase in the distance between DNA binding sites of cell-specific transcription factors and transcription initiation sites. Both studies looked at the effect of the dodecamer repeat expansion on CSTB promoter activity by comparing the activity of reporter constructs containing wildtype and expanded alleles. In the first study, a 3.8-kb fragment, containing the putative CSTB promoter with a 600-bp repeat expansion, showed a 2- to 4-fold reduction of luciferase activity in a reporter plasmid compared to an identical fragment with a normal repeat. This reduction was observed only in certain cell types. Introduction of heterologous DNA fragments of 730 and 1000 bp into the normal promoter in place of the repeat expansion resulted in a similar reduction in activity (Lalioti et al., 1999). In the second study, an increase in the number of dodecamer units from three to 19 repeats reduced transcription 10-fold in the only cell type tested (COS-1) (Alakurtti et al., 2000). Both studies implicated AP1 and Sp1 binding sites via different techniques. The first study looked for transcription factor binding sites (TFBS) conserved between the human, mouse and rat CSTB promoters as defined by TFBS matrices and TFBS position relative to the ATG. This identified Sp1 sites in the dodecamer repeat and an AP1 site at position –312 (Lalioti et al., 1999). The second study identified the same TFBSs and extra AP1, Sp1 and other TFBS using EMSAs with purified recombinant Sp1 and c-Jun proteins (Alakurtti et al., 2000). Terminal deletions of the promoter in both studies implicated the AP-1 binding site at position –312, upstream of the repeat, in CSTB transcription activation. These findings are consistent with the conclusion that an alteration in the spacing of TFBSs, and/or the spacing of the transcription initiation site and TFBSs, due to repeat expansion, are among the causes of reduction of CSTB expression and thus, EPM1 (Fig. 4) (Lalioti et al., 1999; Alakurtti et al., 2000).

Mechanism of repeat expansion and instability

It is not the purpose of this review to cover all proposed models of repeat expansion, but rather to discuss some aspects specifically applicable to the dodecamer repeat and EPM1.

Repeat expansion in human disease is affected by a number of variables that have been divided into *cis*-acting elements and *trans*-acting factors. *Cis*-acting elements are implicated as the stability of the same repeat may be different at different loci. Also, specific haplotypes around disease expansion loci are predisposed to expansions in many disorders implying a role of specific flanking sequences. Other *cis*-acting elements include the composition and number of repeats. EPM1 is similar to many other repeat expansion disorders in that the mutant alleles have common haplotypes in several different populations, presumably due to a founder effect (Lehesjoki et al., 1993b, 1994; Richards, 2001; Moulard et al., 2002). Recently it was shown in mice that changing the flanking genomic sequences of the SCA7 CAG repeat dramatically altered repeat stability (Libby et al., 2003). Polymorphisms in flanking sequences affecting the mutability of microsatellites have also been noted (Vergnaud and Denoeud, 2000). Both results suggest that *cis*-information present on the genomic fragment and/or the genomic context is responsible for repeat instability and the common haplotypes of EPM1 may be an unstable genomic context. This raises the possibility that the repeat expansion has occurred multiple independent times on the same susceptible haplotype as opposed to having occurred once on a haplotype that has become the common disease haplotype.

There is only indirect evidence, such as parental gender bias in repeat instability, for the existence of *trans*-acting factors. However, obvious candidates include the proteins involved in DNA replication and recombination whose interaction with a repeat may contribute to its instability. Secondary structures of the repeats may play an important role in interacting with trans-acting factors (Usdin and Grabczyk, 2000). As described above, the dodecamer repeat forms many secondary structures and these may constitute a barrier to DNA or RNA polymerases and DNA methyltransferases. Single-stranded regions of these secondary structures may also be prone to strand cleavage.

How does a loss of CSTB cause EPM1?

The fact that there are five different point mutations distributed throughout the three exons, of CSTB in EPM1 patients, and the autosomal recessive inheritance of EPM1, clearly implicate a reduction in, or absence of, the level of functional CSTB protein. This was confirmed by the observation that *Cstb*$^{-/-}$ mice develop progressive ataxia and myoclonic seizures at a young age. *Cstb*-deficient mice, a mammalian model for the human disease, display extensive apoptotic cell death in cerebellar granule cells, the likely cause of the rapid progression of ataxia in these mice (Pennacchio et al., 1998). However *Cstb*$^{-/-}$ mice also show an ~ one-third decrease in body weight compared to wildtype, a phenotype not described in human EPM1 (Shannon et al., 2002).

Transcriptional profiling of brains from *Cstb*$^{-/-}$ mice showed increased expression of genes involved in proteolysis, apoptosis and glial activation (Lieuallen et al., 2001). In agreement with this, more detailed histopathological examination of the *Cstb*$^{-/-}$ mice shows extensive neuronal loss by apoptosis and gliosis in many parts of the brain, with granule cells being particularly sensitive to Cstb deficiency. Hippocampal and cortical neuronal apoptosis was observed in young mutant mice and older mutant mice exhibited hippocampal and cortical gliosis (Shannon et al., 2002).

Cstb$^{-/-}$ mice, however, only show myoclonic seizures on an isogenic background, while developing ataxia and abnormal brain histopathology on both a mixed and isogenic background (Pennacchio et al., 1998; Shannon et al., 2002). These findings indicate that while neuronal cell death is an important consequence of CSTB deficiency, it is independent of the epileptic seizure events (Shannon et al., 2002).

While the described studies strongly suggest a physiological role for CSTB in the maintenance of normal neuronal structure, it is still unclear how the loss of CSTB activity in neurons is linked to cell-specific apoptosis and the EPM1 phenotype. Apoptosis or programmed cell death is the major mechanism used by multicellular organisms to remove cells. Cysteine proteases from the caspase family play a crucial role in the process. However, there is increasing evidence that lysosomal proteases (cathepsins), known to be inhibited by cystatins, are also involved in apoptosis (Turk et al., 2002a).

Is there a function of CSTB other than as protease inhibitor? Recent studies identified five recombinant proteins interacting with CSTB in vitro and none was a protease. Three of these proteins (RACK-1, beta-spectrin and NF-L), co-immunoprecipitated with CSTB in rat cerebellum. Immunofluorescent confocal microscopy showed that the same proteins are present in the granule cells of developing cerebellum, as well as in Purkinje cells of adult rat cerebellum. This raises the possibility that a CSTB multiprotein complex may have a specific cerebellar function and that the loss of this function might contribute to disease in EPM1 patients (Di Giaimo et al., 2002).

Similar to other PMEs, EPM1 is one of the epilepsies associated with reduced neuronal survival. These are in contrast to the idiopathic Mendelian epilepsies that have emerged as a new category of channelopathies. How defects in CSTB lead to decreased neuronal survival, and whether this subsequently leads to the hyperexcitable neurons in EPM1, remains largely a mystery.

Acknowledgements

We thank previous collaborators, and members of the Antonarakis laboratory of the mid and late 1990s for their contributions to the discovery of the 12-mer repeat expansions. We also thank the patients and their families for their collaboration and donation of their precious tissues. A special thanks to Pauline Crewther for her expert help in editing and revising the manuscript.

References

Abrahamson M, Barrett AJ, Salvesen G, Grubb A: Isolation of six cysteine proteinase inhibitors from human urine. Their physicochemical and enzyme kinetic properties and concentrations in biological fluids. J biol Chem 261:11282–11289 (1986).

Abrahamson M, Ritonja A, Brown MA, Grubb A, Machleidt W, Barrett AJ: Identification of the probable inhibitory reactive sites of the cysteine proteinase inhibitors human cystatin C and chicken cystatin. J biol Chem 262:9688–9694 (1987).

Alakurtti K, Virtaneva K, Joensuu T, Palvimo JJ, Lehesjoki AE: Characterization of the cystatin B gene promoter harboring the dodecamer repeat expanded in progressive myoclonus epilepsy EPM1. Gene 242:65–73 (2000).

Barrett AJ: Cystatin the egg white inhibitor of cystein proteinases. Meth Enzym 80:771–778 (1981).

Berkovic SF, Andermann F, Carpenter S, Wolfe LS: Progressive myoclonus epilepsies: specific causes and diagnosis. New Engl J Med 315:296–305 (1986).

Berkovic SF, So NK, Andermann F: Progressive myoclonus epilepsies: clinical and neurophysiological diagnosis. J Clin Neurophys 8:261–274 (1991).

Berkovic SF, Cochius J, Andermann E, Andermann F: Progressive myoclonus epilepsies: clinical and genetic aspects. Epilepsia 34 Suppl 3:S19–30 (1993).

Bespalova IN, Adkins S, Pranzatelli M, Burmeister M: Novel cystatin B mutation and diagnostic PCR assay in an Unverricht-Lundborg progressive myoclonus epilepsy patient. Am J med Genet 74:467–471 (1997a).

Bespalova IN, Pranzatelli M, Burmeister M: G to C transversion at a splice acceptor site causes exon skipping in the cystatin B gene. Mutat Res 382:67–74 (1997b).

Bode W, Engh R, Musil D, Thiele U, Huber R, Karshikov A, Brzin J, Kos J, Turk V: The 2.0 Å X-ray crystal structure of chicken egg white cystatin and its possible mode of interaction with cysteine proteinases. EMBO J 7:2593–2599 (1988).

Chen H, Chrast R, Rossier C, Morris MA, Lalioti MD, Antonarakis SE: Cloning of 559 potential exons of genes of human chromosome 21 by exon trapping. Genome Res 6:747–760 (1996).

Chen X, Mariappan SV, Catasti P, Ratliff R, Moyzis RK, Laayoun A, Smith SS, Bradbury EM, Gupta G: Hairpins are formed by the single DNA strands of the fragile X triplet repeats: structure and biological implications. Proc natl Acad Sci, USA 92:5199–5203 (1995).

Cheng JF, Boyartchuk V, Zhu Y: Isolation and mapping of human chromosome 21 cDNA: progress in constructing a chromosome 21 expression map. Genomics 23:75–84 (1994).

Cochius J, Carpenter S, Andermann E, Rouleau G, Nousiainen U, Kalviainen R, Farrell K, Andermann F: Sweat gland vacuoles in Unverricht-Lundborg disease: a clue to diagnosis? Neurology 44:2372–2375 (1994).

Cochius JI, Figlewicz DA, Kalviainen R, Nousiainen U, Farrell K, Patry G, Soderfeldt B, Frydman M, Lerman P, Andermann F: Unverricht-Lundborg disease: absence of nonallelic genetic heterogeneity. Ann Neurol 34:739–741 (1993).

Cook T, Gebelein B, Urrutia R: Sp1 and its likes: biochemical and functional predictions for a growing family of zinc finger transcription factors. Ann NY Acad Sci 880:94–102 (1999).

de la Chapelle A: Disease gene mapping in isolated human populations: the example of Finland. J med Genet 30:857–865 (1993).

Delgado-Escueta AV, Ganesh S, Yamakawa K: Advances in the genetics of progressive myoclonus epilepsy. Am J med Genet 106:129–138 (2001).

Di Giaimo R, Riccio M, Santi S, Galeotti C, Ambrosetti DC, Melli M: New insights into the molecular basis of progressive myoclonus epilepsy: a multiprotein complex with cystatin B. Hum molec Genet 11:2941–2950 (2002).

Dunnen JT, Antonarakis SE: Mutation nomenclature extensions and suggestions to describe complex mutations: a discussion. Hum Mutat 15:7–12 (2000).

Genton P, Michelucci R, Tassinari CA, Roger J: The Ramsay Hunt syndrome revisited: Mediterranean myoclonus versus mitochondrial encephalomyopathy with ragged-red fibers and Baltic myoclonus. Acta Neurol Scan 81:8–15 (1990).

Haltia M, Kristensson K, Sourander P: Neuropathological studies in three Scandinavian cases of progressive myoclonus epilepsy. Acta Neurol Scan 45:63–77 (1969).

Hsiao LL, Dangond F, Yoshida T, Hong R, Jensen RV, Misra J, Dillon W, Lee KF, Clark KE, Haverty P, Weng Z, Mutter GL, Frosch MP, Macdonald ME, Milford EL, Crum CP, Bueno R, Pratt RE Mahadevappa M, Warrington JA, Stephanopoulos G, Gullans SR: A compendium of gene expression in normal human tissues. Physiol Genomics 7:97–104 (2001).

Jithesh PV, Singh P, Joshi R: Molecular dynamics studies of trinucleotide repeat DNA involved in neurodegenerative disorders. J biomol Struct Dyn 19:479–495 (2001).

Kagitani-Shimono K, Imai K, Okamoto N, Ono J, Okada S: Unverricht-Lundborg disease with cystatin B gene abnormalities. Pediatr Neurol 26:55–60 (2002).

Kho MR, Baker DJ, Laayoun A, Smith SS: Stalling of human DNA (cytosine-5) methyltransferase at single-strand conformers from a site of dynamic mutation. J molec Biol 275:67–79 (1998).

Kinne R, Saukko P, Jarvinen M, Lehesjoki AE: Reduced cystatin B activity correlates with enhanced cathepsin activity in progressive myoclonus epilepsy. Ann Med 34:380–385 (2002).

Koskiniemi M, Donner M, Majuri H, Haltia M, Norio R: Progressive myoclonus epilepsy. A clinical and histopathological study. Acta Neurol Scan 50:307–332 (1974a).

Koskiniemi M, Toivakka E, Donner M: Progressive myoclonus epilepsy. Electroencephalographical findings. Acta Neurol Scan 50:333–359 (1974b).

Laayoun A, Smith SS: Methylation of slipped duplexes snapbacks and cruciforms by human DNA(cytosine-5)methyltransferase. Nucl Acids Res 23:1584–1589 (1995).

Labauge P, Ouazzani R, M'Rabet A, Grid D, Genton P, Dravet C, Chkili T, Beck C, Buresi C, Baldy-Moulinier M, Malafosse A: Allelic heterogeneity of Mediterranean myoclonus and the cystatin B gene. Ann Neurol 41:686–689 (1997).

Lafreniere RG, de Jong PJ, Rouleau GA: A 405-kb cosmid contig and *Hin*dIII restriction map of the progressive myoclonus epilepsy type 1 (EPM1) candidate region in 21q22.3. Genomics 29:288–290 (1995).

Lafreniere RG, Rochefort DL, Chretien N, Rommens JM, Cochius JI, Kalviainen R, Nousiainen U, Patry G, Farrell K, Soderfeldt B, Federico A, Hale BR, Cossio OH, Sorensen T, Pouliot MA, Kmiec T, Uldall P, Janszky J, Pranzatelli MR, Andermann F, Andermann E, Rouleau GA: Unstable insertion in the 5′ flanking region of the cystatin B gene is the most common mutation in progressive myoclonus epilepsy type 1 EPM1. Nature Genet 15:298–302 (1997).

Lalioti MD, Bottani A, Morris MA, Antonarakis SE: The Swiss type of progressive myoclonous epilepsy (EPM1) maps tp 21q22.3. Med Genet 2:249 (1995).

Lalioti MD, Mirotsou M, Buresi C, Peitsch MC, Rossier C, Ouazzani R, Baldy-Moulinier M, Bottani A, Malafosse A, Antonarakis SE: Identification of mutations in cystatin B the gene responsible for the Unverricht-Lundborg type of progressive myoclonus epilepsy (EPM1). Am J hum Genet 60:342–351 (1997a).

Lalioti MD, Scott HS, Buresi C, Rossier C, Bottani A, Morris MA, Malafosse A, Antonarakis SE: Dodecamer repeat expansion in cystatin B gene in progressive myoclonus epilepsy. Nature 386:847–851 (1997b).

Lalioti MD, Scott HS, Genton P, Grid D, Ouazzani R, M'Rabet A, Ibrahim S, Gouider R, Dravet C, Chkili T, Bottani A, Buresi C, Malafosse A, Antonarakis SE: A PCR amplification method reveals instability of the dodecamer repeat in progressive myoclonus epilepsy (EPM1) and no correlation between the size of the repeat and age at onset. Am J hum Genet 62:842–847 (1998).

Lalioti MD, Scott HS, Antonarakis SE: Altered spacing of promoter elements due to the dodecamer repeat expansion contributes to reduced expression of the cystatin B gene in EPM1. Hum molec Genet 8:1791–1798 (1999).

Larson GP, Ding S, Lafreniere RG, Rouleau GA, Krontiris TG: Instability of the EPM1 minisatellite. Hum molec Genet 8:1985–1988 (1999).

Lehesjoki AE: Clinical features and genetics of Unverricht-Lundborg disease. Adv Neurol 89:193–197 (2002).

Lehesjoki AE, Koskiniemi M, Sistonen P, Miao J, Hastbacka J, Norio R, de la Chapelle A: Localization of a gene for progressive myoclonus epilepsy to chromosome 21q22. Proc natl Acad Sci, USA 88:3696–3699 (1991).

Lehesjoki AE, Koskiniemi M, Pandolfo M, Antonelli A, Kyllerman M, Wahlstrom J, Nergardh A, Burmeister M, Sistonen P, Norio R: Linkage studies in progressive myoclonus epilepsy: Unverricht-Lundborg and Lafora's diseases. Neurology 42:1545–1550 (1992).

Lehesjoki AE, Eldridge R, Eldridge J, Wilder BJ, de la Chapelle A: Progressive myoclonus epilepsy of Unverricht-Lundborg type: a clinical and molecular genetic study of a family from the United States with four affected sibs. Neurology 43:2384–2386 (1993a).

Lehesjoki AE, Koskiniemi M, Norio R, Tirrito S, Sistonen P, Lander E, de la Chapelle A: Localization of the EPM1 gene for progressive myoclonus epilepsy on chromosome 21: linkage disequilibrium allows high resolution mapping. Hum molec Genet 2:1229–1234 (1993b).

Lehesjoki AE, Tassinari CA, Avanzini G, Michelucci R, Franceschetti S, Antonelli A, Rubboli G, de la Chapelle A: PME of Unverricht-Lundborg type in the Mediterranean region: linkage and linkage disequilibrium confirm the assignment to the EPM1 locus. Hum Genet 93:668–674 (1994).

Libby RT, Monckton DG, Fu YH, Martinez RA, McAbney JP, Lau R, Einum DD, Nichol K, Ware CB, Ptacek LJ, Pearson CE, La Spada AR: Genomic context drives SCA7 CAG repeat instability while expressed SCA7 cDNAs are intergenerationally and somatically stable in transgenic mice. Hum molec Genet 12:41–50 (2003).

Lieuallen K, Pennacchio LA, Park M, Myers RM, Lennon GG: Cystatin B-deficient mice have increased expression of apoptosis and glial activation genes. Hum molec Genet 10:1867–1871 (2001).

Lundborg H: Die progressive Myoclonous-Epilepsie (Unverricht's Myoclonie). (Almquist and Wiksell, Uppsala 1903).

Machleidt W, Thiele U, Laber B, Assfalg-Machleidt I, Esterl A, Wiegand G, Kcs J, Turk V, Bode W: Mechanism of inhibition of papain by chicken egg white cystatin. Inhibition constants of N-terminally truncated forms and cyanogen bromide fragments of the inhibitor. FEBS Letters 243:234–238 (1989).
Malafosse A, Lehesjoki AE, Genton P, Labauge P, Durand G, Tassinari CA, Dravet C, Michelucci R, de la Chapelle A: Identical genetic locus for Baltic and Mediterranean myoclonus. Lancet 339:1080–1081 (1992).
Mascalchi M, Michelucci R, Cosottini M, Tessa C, Lolli F, Riguzzi P, Lehesjoki AE, Tosetti M, Villari N, Tassinari CA: Brainstem involvement in Unverricht-Lundborg disease (EPM1): An MRI and (1)H MRS study. Neurology 58:1686–1689 (2002).
Mazarib A, Xiong L, Neufeld MY, Birnbaum M, Korczyn AD, Pandolfo M, Berkovic SF: Unverricht-Lundborg disease in a five-generation Arab family: instability of dodecamer repeats. Neurology 57:1050–1054 (2001).
Meldium BS, Bruton CJ: Epilepsy, in Duchen LW (ed): Greenfield's Neuropathology, pp 1246–1283 (E. Arnold, London 1992).
Moulard B, Genton P, Grid D, Jeanpierre M, Ouazzani R, Mrabet A, Morris M, LeGuern E, Dravet C, Mauguiere F, Utermann B, Baldy-Moulinier M, Belaidi H, Bertran F, Biraben A, Ali Cherif A, Chkili T, Crespel A, Darcel F, Dulac O, Geny C, Humbert-Claude V, Kassiotis P, Buresi C, Malafosse A: Haplotype study of West European and North African Unverricht-Lundborg chromosomes: evidence for a few founder mutations. Hum Genet 111:255–262 (2002).
Norio R, Koskiniemi M: Progressive myoclonus epilepsy: genetic and nosological aspects with special reference to 107 Finnish patients. Clin Genet 15:382–398 (1979).
Parmeggiani A, Lehesjoki AE, Carelli V, Posar A, Santi A, Santucci M, Gobbi G, Pini A, Rossi PG: Familial Unverricht-Lundborg disease: a clinical neurophysiologic and genetic study. Epilepsia 38:637–641 (1997).
Pataskar SS, Dash D, Brahmachari SK: Intramolecular i-motif structure at acidic pH for progressive myoclonus epilepsy (EPM1) repeat d(CCCCGCCCCGCG)n. J biomol Struct Dyn 19:307–313 (2001a).
Pataskar SS, Dash D, Brahmachari SK: Progressive myoclonus epilepsy [EPM1] repeat d(CCCCGCCCCGCG)n forms folded hairpin structures at physiological pH. J biomol Struct Dyn 19:293–305 (2001b).
Pennacchio LA, Bouley DM, Higgins KM, Scott MP, Noebels JL, Myers RM: Progressive ataxia myoclonic epilepsy and cerebellar apoptosis in cystatin B-deficient mice. Nature Genet 20:251–258 (1998).
Pennacchio LA, Lehesjoki AE, Stone NE, Willour VL, Virtaneva K, Miao J, D'Amato E, Ramirez L, Faham M, Koskiniemi M, Warrington JA, Norio R, de la Chapelle A, Cox DR, Myers RM: Mutations in the gene encoding cystatin B in progressive myoclonus epilepsy (EPM1). Science 271:1731–1734 (1996).
Pennacchio LA, Myers RM: Isolation and characterization of the mouse cystatin B gene. Genome Res 6:1103–1109 (1997).
Richards RI: Dynamic mutations: a decade of unstable expanded repeats in human genetic disease. Hum molec Genet 10:2187–2194 (2001).
Ritonja A, Machleidt W, Barrett AJ: Amino acid sequence of the intracellular cysteine proteinase inhibitor cystatin B from human liver. Biochem biophys Res Commun 131:1187–1192 (1985).
Saha T, Usdin K: Tetraplex formation by the progressive myoclonus epilepsy type-1 repeat: implications for instability in the repeat expansion diseases. FEBS Lett 491:184–187 (2001).
Sato N, Ishidoh K, Uchiyama Y, Kominami E: Molecular cloning and sequencing of cDNA for rat cystatin beta. Nucl Acids Res 18:6698 (1990).
Shannon P, Pennacchio LA, Houseweart MK, Minassian BA, Myers RM: Neuropathological changes in a mouse model of progressive myoclonus epilepsy: cystatin B deficiency and Unverricht-Lundborg disease. J Neuropath Exp Neurol 61:1085–1091 (2002).
Stone NE, Fan JB, Willour V, Pennacchio LA, Warrington JA, Hu A, de la Chapelle A, Lehesjoki AE, Cox DR, Myers RM: Construction of a 750-kb bacterial clone contig and restriction map in the region of human chromosome 21 containing the progressive myoclonus epilepsy gene. Genome Res 6:218–225 (1996).
Stubbs MT, Laber B, Bode W, Huber R, Jerala R, Lenarcic B, Turk V: The refined 2.4 Å X-ray crystal structure of recombinant human stefin B in complex with the cysteine proteinase papain: a novel type of proteinase inhibitor interaction. EMBO J 9:1939–1947 (1990).
Takuma K, Kiriu M, Mori K, Lee E, Enomoto R, Baba A, Matsuda T: Roles of cathepsins in reperfusion-induced apoptosis in cultured astrocytes. Neurochem Int 42:153–159 (2003).
Thiele U, Assfalg-Machleidt I, Machleidt W, Auerswald EA: N-terminal variants of recombinant stefin B: effect on affinity for papain and cathepsin B. Biol Chem Hoppe-Seyler 371 Suppl:125–136 (1990).
Turk B, Krizaj I, Turk V: Isolation and characterization of bovine stefin B. Biol Chem Hoppe-Seyler 373:441–446 (1992).
Turk B, Turk V, Turk D: Structural and functional aspects of papain-like cysteine proteinases and their protein inhibitors. Biol Chem 378:141–150 (1997).
Turk B, Stoka V, Rozman-Pungercar J, Cirman T, Droga-Mazovec G, Oreic K, Turk V: Apoptotic pathways: involvement of lysosomal proteases. Biol Chem 383:1035–1044 (2002a).
Turk B, Turk D, Salvesen GS: Regulating cysteine protease activity: essential role of protease inhibitors as guardians and regulators. Curr Pharm Des 8:1623–1637 (2002b).
Turk V, Bode W: The cystatins: protein inhibitors of cysteine proteinases. FEBS Lett 285:213–219 (1991).
Turk V, Turk B, Turk D: Lysosomal cysteine proteases: facts and opportunities. EMBO J 20:4629–4633 (2001).
Unverricht H: Die Myoclonie. (Franz Deuticke, Leipzig 1891).
Usdin K, Grabczyk E: DNA repeat expansions and human disease. Cell Mol Life Sci 57:914–931 (2000).
Vergnaud G, Denoeud F: Minisatellites: mutability and genome architecture. Genome Res 10:899–907 (2000).
Virtaneva K, D'Amato E, Miao J, Koskiniemi M, Norio R, Avanzini G, Franceschetti S, Michelucci R, Tassinari CA, Omer S, Pennacchio LA, Myers RM, Dieguez-Lucena JL, Krahe R, de la Chapelle A, Lehesjoki AE: Unstable minisatellite expansion causing recessively inherited myoclonus epilepsy EPM1. Nature Genet 15:393–396 (1997).
Virtaneva K, Miao J, Traskelin AL, Stone N, Warrington JA, Weissenbach J, Myers RM, Cox DR, Sistonen P, de la Chapelle A: Progressive myoclonus epilepsy EPM1 locus maps to a 175-kb interval in distal 21q. Am J hum Genet 58:1247–1253 (1996).
Wang YH, Griffith J: Methylation of expanded CCG triplet repeat DNA from fragile X syndrome patients enhances nucleosome exclusion. J biol Chem 271:22937–22940 (1996).
Wang YH, Gellibolian R, Shimizu M, Wells RD, Griffith J: Long CCG triplet repeat blocks exclude nucleosomes: a possible mechanism for the nature of fragile sites in chromosomes. J molec Biol 263:511–516 (1996).

Cytogenet Genome Res 100:224–229 (2003)
DOI: 10.1159/000072858

Spontaneous chromosome loss and colcemid resistance in lymphocytes from patients with myotonic dystrophy type 1

M. Casella,[a] M. Lucarelli,[a] M. Simili,[a] P. Beffy,[a] R. Del Carratore,[a] F. Minichilli,[b] C. Chisari[c] and S. Simi[a]

[a] Institute of Clinical Physiology, CNR, Cellular Biology and Cytogenetics Unit, Pisa;
[b] Institute of Clinical Physiology, CNR, Epidemiology Unit, Pisa;
[c] Department of Neurosciences, University of Pisa, Pisa (Italy)

Abstract. Myotonic Dystrophy type 1 (DM1) is one of the many inherited human diseases whose molecular defect is the expansion of a trinucleotide DNA sequence. DM1 shares with fragile X syndrome (FMR1), another "unstable triplet syndrome", several molecular features not present in the remaining triplet diseases. As FMR1 is also characterised by chromosome instability at the site of the expanded triplet, lymphocytes from DM1 patients and healthy donors were cultured for micronucleus (MN) analysis, in order to verify if DM1 is also prone to chromosome instability. A FISH analysis was also carried out to detect the presence of centromeric sequences in the observed MN. The data indicate that DM1 patients present a percentage of centromere-positive MN significantly higher than controls, suggesting that chromosome loss is the main mechanism underlying the origin of the increased spontaneous instability. To further assess the proneness to instability of cells of DM1 patients, cultures from patients and controls were treated in vitro with growing concentrations of two different mutagens: colcemid, a "pure" aneugen compound whose target is tubulin, and mytomicin C, a strong clastogen. The results show that the patient group is significantly less sensitive to colcemid. These data, together with FISH analysis, suggest the presence, in DM1 patients, of an already damaged tubulin, which becomes no more sensitive to the effect of colcemid and which could be the main defect underlying the aneugenic effects in DM1.

DM1 is a multisystemic disease associated with trinucleotide expansion in the 3′ untranslated region of a DM gene which codes for a protein kinase (DMPK) and has been mapped on the long arm of chromosome 19, band 13.3 (19q13.3). In DM1 patients, the trinucleotide CTG is present in up to 2,000 copies while normal individuals have 10–30 copies. Fragile X syndrome (FMR1) is another triplet expansion disease which is also associated with high chromosome instability at the site of the expanded repeats (Xq27.3), although the gene locus and the trinucleotide are different from DM1.

It has been demonstrated that the DM locus at 19q13.3 is a site specific for adeno-associated virus, suggesting a specific proneness to genomic instability at this locus (Giraud et al., 1994), while more recently it has been shown that expanded CTG repeats are sites of chromosome breaks in yeast (Freudenreich et al., 1998), highlighting other parallels between fragile sites and myotonic dystrophy.

Although two previous studies failed to find cytological evidence either of folate-sensitive fragile sites (FS) (Jalal et al., 1993) or other rare and common FS at the DM locus (Wenger et al., 1994), the analysis of cytogenetic endpoints other than chromosome aberrations at 19q13.3 may reveal different features of genomic instability, either at the DMPK locus or in the rest of the genome.

For these reasons, lymphocytes from DM1 patients and from healthy, age, sex and smoking habit matched donors were

This work was partially supported by Telethon grant no. 1138.

Received 31 July 2002; revision accepted 2 January 2003.

Request reprints from Silvana Simi, Istituto di Fisiologia Clinica, CNR
Area della Ricerca, Via Moruzzi, 156124 Pisa (Italy)
telephone: +39 050 315 2779; fax: 39 315 3327; e-mail: s.simi@ifc.cnr.it

M.C. and M.L. contributed equally to the work.

KARGER
Fax + 41 61 306 12 34
E-mail karger@karger.ch
www.karger.com

0301–0171/03/1004–0224$19.50/0

Accessible online at:
www.karger.com/cgr

cultured in vitro, and the frequency of micronuclei (MN), which allows the analysis of a larger number of cells, was investigated. As DM1 patients showed a significant increase of spontaneous MN over the control population, a FISH analysis with an all human pancentromeric probe was carried out in order to discriminate between MN bearing positive centromeric signals (indicating the presence of one entire chromosome, due to malsegregation events) and MN without signals (indicating the presence of acentric chromosome fragments, due to clastogenic events). The sensitivity of DM1 patients to both an aneugenic (colcemid, COLC) and a clastogenic (mytomicin C, MCC) compound, with respect to control individuals, was also tested.

Materials and methods

Subjects

Twenty-one patients were identified (11 females and 10 males). Individuals 7 and 20 were son and father, 1 and 8 daughter and mother, and 5 and 19 son and mother. All the patients were drug-free for at least two months before the cytogenetic analysis. The diagnosis was based both on clinic and molecular data. Patients were classified as belonging to classes of expansion 1–4 on the basis of the triplet amplification number: E_1, 200–500 triplets; E_2, 500–1,000; E_3, 1,000–1,500; E_4, >1,500 (Tsifildis et al., 1992). In some patients lymphocyte heterogeneity was also present: in this case the class of expansion was reported in brackets.

A control group of 30 healthy volunteers, matched with respect to age, sex and smoking habit (in this study, all individuals were non-smoking), was also tested. Written informed consent was obtained from patients before their inclusion in this study.

Cell culture and cytogenetic tests

MN test was carried out according to the cytokinesis-block micronucleus technique. Lymphocytes from whole blood were stimulated by phytohaemagglutinin and cultured in duplicate at 37 °C in RPMI-1640 (Gibco, Life Technologies) supplemented with 15 % fetal calf serum for 72 h. To block the cytokinesis, cytochalasin B (3 μg/ml) was added after 44 h of culture. Treatments with colcemid and mytomicin C were carried out adding the drugs after 24 h of culture till harvesting (72 h). Coded slides were prepared according to standard techniques. MN were scored in 1,000 binucleated cells, following standard criteria (Fenech, 1993) and their frequency was expressed as the ratio between binucleated cells presenting one or more MN and the total number of cells scored.

For the SCE analysis, 6 μg/ml of 5-bromodeoxyuridine (BrdU, Sigma-Aldrich, Steinheim, Germany) were added after 24 h of culture and the FPG method was used to differentially stain sister chromatids (Wolff and Perry, 1974). Colcemid (Sigma-Aldrich, Steinheim, Germany, 0.05 μg/ml) was added during the last 2 h and a minimum of 30 well-stained second division metaphases was scored per individual.

As an indicator of cell cycle progression, the proliferation index (PI) was calculated from BrdU-labeled metaphases according to the formula where M_1, M_2 and M_3 represent the number of first, second and third metaphases respectively:

$$PI = \frac{M_1 + 2\,M_2 + 3\,M_3}{M_1 + M_2 + M_3}$$

To distinguish between MN derived from clastogenic (centromere negative) or from aneugenic (centromere positive) events, FISH analysis with all-human pancentromeric probe (Appligene-Oncor, CP5095-DG.5, Illkirch, France) was carried out according to instructions. At least 50 MN per individual and per drug concentration were scored for the presence (centromere-positive micronucleus, MN/C+) or the absence (centromere-negative micronucleus, MN/C–) of fluorescent signals. A preliminary FISH analysis with a chromosome 1/5/19 centromeric probe (Appligene-Oncor, CP5008-DG.5, Illkirch, France) was also performed, in order to verify the presence of chromosome 19 in the MN.

Statistics

Decoded MN data were statistically elaborated by using multifactorial analysis of variance (ANOVA) to assess differences in MN frequency between DM1 patients and controls after checking for confounding factors (gender as categorical factor and age as quantitative factor).

Since the basal MN frequency was significantly different between DM1 patients and controls, the difference of induced MN frequency between every two following doses (Δf) was calculated for both the groups. Δf differences between DM1 patients and controls and the presence of a growing trend of induced MN frequency at growing doses within each group were evaluated by multifactorial analysis of variance (ANOVA), using disease as categorical factor and dose as quantitative factor, checking for confounding factors (gender as categorical factor and age as quantitative factor).

The spontaneous frequency of molecularly characterised MN was analysed by t test, evaluating the two mean values. The drug-induced frequency of molecularly characterised MN was analysed by Poisson regression analysis using disease and drug concentrations as categorical factors.

Results

Table 1 reports some demographic and clinical features together with the spontaneous frequency of MN in 21 DM1 patients and 30 control individuals. The analysis shows a highly significant correlation ($P < 0.001$) between the presence of the disease and MN increase. The effects of age ($P < 0.001$) and gender ($P < 0.001$) also significantly correlated to MN increase.

Although it was not possible to carry out a molecular quantification of the triplet number in all the patients, it could be interesting to note that six patients belonging to the intermediate-high (E_2–E_3) class of expansion, supposed to correlate with the disease severity, presented a MN frequency higher than 22 % (never found in control individuals).

To verify if a higher presence of MN might affect the cell cycle progression, an SCE analysis was performed and the correlated proliferation index (PI) calculated in a subset of eight patients, selected only on the basis of their arrival to the hospital, and eight controls. No significant differences either in the level of SCE or in the PI were found in the two populations, indicating that neither disturbance in the DNA replication nor a slowing down of cell cycle time occur in DM1 patients (Table 2).

In order to analyse whether the increased frequency of MN was the result of chromosome loss or clastogenic events, a molecular approach was used to discriminate between MN bearing positive centromeric signals and MN without signals. The results of FISH experiments with a pancentromeric probe in the two populations are shown in Fig. 1. The data indicate that the observed increase of MN frequency in DM1 patients is due to a significant higher percentage of centromere-positive MN (MN/C+, $P < 0.001$), suggesting an increased frequency of chromosome loss in this disease.

To further assess the proneness to instability of DM1 cells or the presence of a more sensitive subset of T cells in the peripheral blood of these patients, lymphocytes from 11 DM1 patients, selected once again only on the basis of their arrival to the hospital, and 11 controls were treated in vitro with colcemid (COLC, an aneugen) and mytomicin C (MCC, a clastogen), and the concentration-effect curves were evaluated.

Figure 2 shows the mean MN frequencies in the two populations after induction with four concentrations of MCC (0.1, 0.3,

Table 1. Demographic, clinical and cytogenetic features of the two populations

A. DM1 patients

Individuals	Age	Sex	MN[a]	E[b]
1	14	F	16	
2	27	M	12	E_2
3	28	F	6	$E_1(E_2)$
4	29	F	14	E_1
5	30	M	23	
6	33	M	11	E_2
7	34	M	26	E_2
8	35	F	14	
9	36	F	22	$E_2(E_4)$
10	37	M	5	
11	41	F	30	$E_1(E_2)$
12	50	M	11	
13	51	M	16	$E_2(E_3)$
14	54	F	18	E_1
15	54	F	22	E_1
16	57	F	13	
17	57	F	29	$E_2(E_3)$
18	60	M	15	$E_1(E_2)$
19	62	F	40	E_2
20	63	M	26	
21	64	M	11	
Mean ± s.d.	43.62 ± 15.6		18.1 ± 8.7[c]	

[a] Number of micronucleated binucleated cells.
[b] Range of triplet amplification: E_1, 200-500; E_2, 500-1000; E_3, 1000-1500; E_4, >1500.
[c] MN frequency of DM1 patients is significantly different from controls (*P*<0.001, ANOVA test).

B. Controls

Individuals	Age	Sex	MN[a]
22	14	F	8
23	25	F	5
24	27	M	6
25	27	M	5
26	27	F	6
27	28	F	8
28	28	M	13
29	29	F	6
30	30	M	6
31	31	M	5
32	31	M	7
33	31	M	4
34	36	M	6
35	40	M	8
36	44	F	18
37	44	M	10
38	52	F	16
39	53	F	12
40	53	F	14
41	55	F	18
42	60	M	11
43	60	M	13
44	61	M	10
45	64	M	15
46	25	F	5
47	27	F	6
48	28	M	13
49	31	M	4
50	44	M	10
51	60	M	13
Mean ± s.d.	38.83 ± 14.6		9.37 ± 4.2

[a] Number of micronucleated binucleated cells.

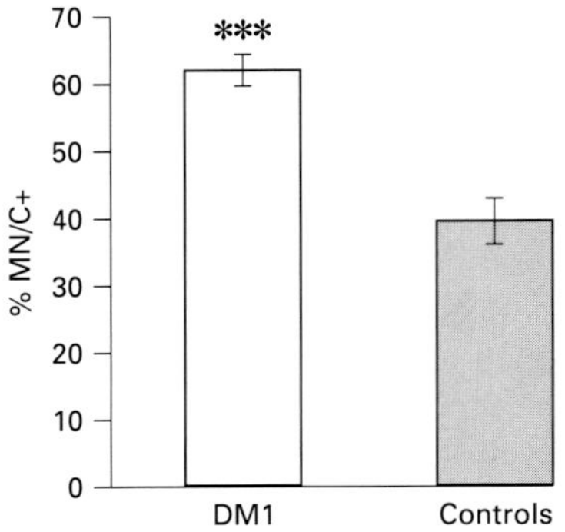

Fig. 1. Frequency of micronuclei bearing positive centromeric signals (MN/C+) in the two populations. *** This value is significantly different from control value ($P < 0.001$, t test).

Table 2. Frequencies of sister chromatid exchanges (SCE) in the two populations

A. DM1 patients	Age	Sex	SCE[a]	PI[b]
2	26	M	4.8 2.1	1.7
4	29	F	5.1 2.6	1.8
5	30	M	4.3 2.3	2.3
3	27	F	4.9 1.5	2.2
6	34	M	5.2 2.0	1.8
10	37	M	6.4 2.0	1.8
14	54	F	4.6 1.9	2.2
Mean ± s.d.	33.8 ± 9.7		5.3 ± 2.2	2.0 ± 0.3
B. Controls	Age	Sex	SCE[a]	PI[b]
25	27	M	5.6 2.0	1.6
26	27	F	5.5 2.9	2.5
27	28	F	5.7 3.1	1.8
28	28	M	7.0 2.6	1.7
31	31	M	6.5 2.3	1.8
33	31	M	5.9 1.8	1.7
40	53	F	6.1 2.3	1.8
Mean ± s.d.	32.1 ± 9.3		6.0 ± 2.4	1.8 ± 0.3

[a] Mean of the number of sister chromatid exchanges ± standard deviation.
[b] Proliferation index.

0.5, 0.7 µg/ml). MCC, as expected, is able to induce an increase of MN frequency in both populations. The statistical analysis of these data shows that the observed increases are highly significant ($P < 0.001$) and concentration dependent ($P < 0.001$) in both populations, which are therefore equally sensitive to this substance ($P = 0.11$).

Figure 3 shows the mean frequencies of MN in the two populations after induction with four concentrations of COLC (0.01, 0.02, 0.025, 0.03 µg/ml). While a concentration-dependent increase of MN frequency is highly significant ($P < 0.001$) in the control population, no increase is present in the patient population.

A molecular characterisation of MCC- and COLC-induced MN was performed with a pancentrometic probe at an inter-

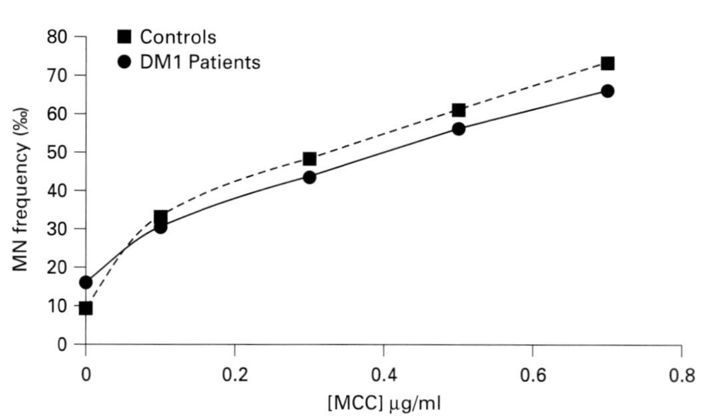

Fig. 2. Mean frequencies of micronuclei (MN) induced by mytomicin C (MCC, 0.1, 0.3, 0.5, 0.7 μg/ml) in cell cultures from 11 DM1 patients and in 11 controls. The increase of MN frequencies is significant in both the populations ($P < 0.001$, ANOVA).

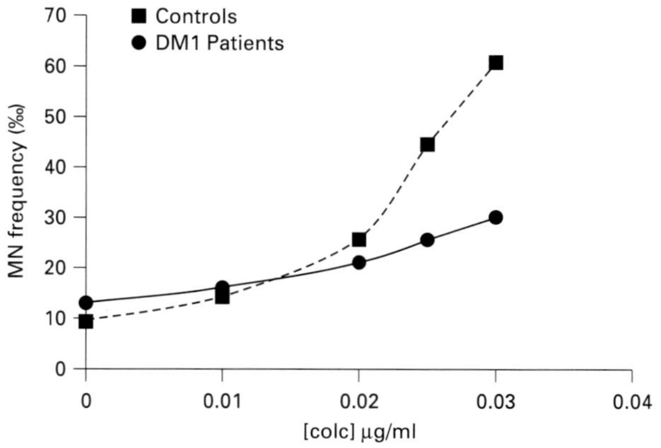

Fig. 3. Mean frequencies of micronuclei (MN) induced by colcemid (COLC, 0.01, 0.02, 0.025, 0.03 μg/ml) in cell cultures from 11 DM1 patients and in 11 controls. The increase of MN frequencies is significant only in control populations ($P < 0.001$, ANOVA).

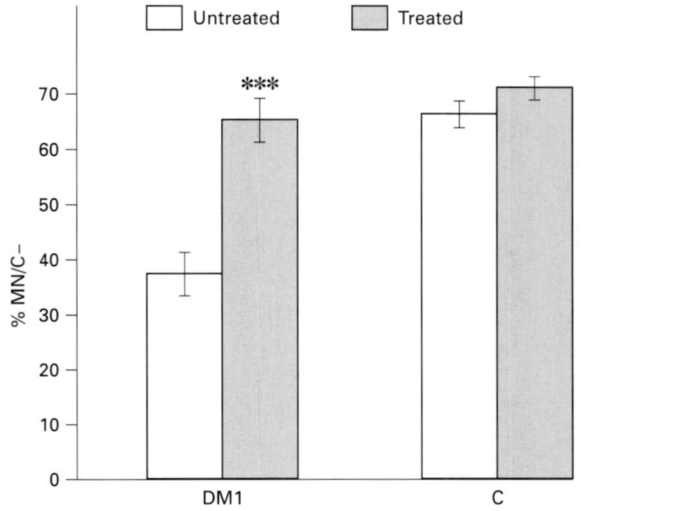

Fig. 4. Mean frequencies of micronuclei bearing negative centromeric signals (MN/C–) induced by mytomicin (MCC, 0.3 μg/ml) in cell cultures from 11 DM1 patients and in 11 controls. *** This value is significantly different from the spontaneous value ($P < 0.001$, Poisson regression analysis).

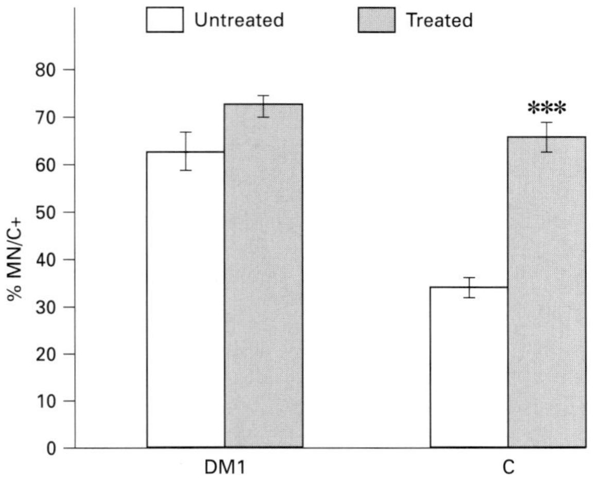

Fig. 5. Mean frequencies of micronuclei bearing positive centromeric signals (MN/C+) induced by colcemid (COLC, 0.025 μg/ml) in cell cultures from 11 DM1 patients and in 11 controls. *** This value is significantly different from the spontaneous value ($P < 0.001$, Poisson regression analysis).

mediate concentration of the drugs (MCC 0.3 μg/ml; COLC 0.025 μg) in the same 11 DM1 patients and controls. The data in Fig. 4 show that MCC was able to induce MN/C– ($P < 0.001$) in patients. The control population, whose spontaneous MN are essentially MN/C–, showed a weak, although not significant, increase of MN/C–, reaching anyway the same final value as the patients. On the contrary, COLC was able to induce a significant increase of MN/C+ in the control population ($P < 0.001$), while it was unable to increase the basal frequency of MN/C+ in the patient population (Fig. 5).

Discussion

Recently, an increasing number of reports have suggested the presence of chromosome instability in degenerative diseases other than cancer and chromosome breakage syndromes: systemic sclerosis (Jabs et al., 1993), Nijmegen syndrome (Shiloh, 1997) Alzheimer disease (Migliore et al., 1999), immunodeficiency syndrome (Xu et al., 1999), Parkinson's disease (Petrozzi et al., 2001) and coronary artery disease (Botto et al., 2001).

One of the most important outcomes of the Human Genome Project has been the discovery of the instability of simple DNA triplet repeat sequences (Sutherland and Richards, 1995), which underlies a new type of mutation known as dynamic mutation. Nearly 20 diseases caused by trinucleotide expansion, including Huntington disease, two forms of FRAX, and DM1, have been identified so far.

DM1 and fragile X mental retardation syndrome (FMR1) share some peculiarities not present in the other triplet-repeat diseases: at both DM and FMR1 loci the expansion occurs in the transcribed non-translated regions embedding CpG islands, and it is associated with altered chromatin, aberrant methylation, and suppressed expression of DMPK and FMR1 genes (Tapscott et al., 1998). Furthermore, FMR1 is cytologically characterised by chromosome instability at the site of expanded trinucleotide repeats, which is the localisation of the fragile site FRAXA.

In this work, we found that DM1 also shares chromosome instability with FMR1, which in DM1 is characterised by a consistent and significant increase of chromosome loss, as revealed by the MN test and FISH analysis. The apparent discrepancy between our data and previous work failing to find cytological evidence of instability (Jalal et al., 1993, Wenger et al., 1994) could be explained by the fact that the MN test allows the analysis of a much larger number of cells with respect to aberration, and at the same time it also avoids possible confounding effects of artefactual chromosome loss in metaphase preparations (Heddle et al., 1991).

Although it is not easy to explain improper chromosome segregation in terms of triplet expansion, some hypotheses could be drawn. It can be hypothesised that the dynamic DM mutation could indirectly affect the machinery of chromosome segregation or cell cycle checkpoints, leading to a more general and random chromosome instability. The fact that the fission yeast kinase Orb6, closely related to the human myotonic dystrophy protein kinase (DMPK), has been shown to have a role during the cell cycle, and Orb6 mutants have disorganised microtubule arrays and delocalized actin dots (Verde et al., 1998) appears to sustain this hypothesis and suggests a possible role of DMPK in coordinating cell cycle and chromosome segregation. Furthermore, the observation of DM1 patients presenting genetic syndromes associated with chromosome numerical abnormalities, such as Klinefelter's syndrome (Rolland et al., 1999), seems to confirm the proneness to malsegregation of DM1 disease.

Alternatively, it could be suggested that the expansion, localised distally in chromosome 19, may alter some normal functions, such as the telomere function, leading to great mitotic instability of this chromosome. Studies in yeast showed that mutations in genes involved in telomere stabilisation such as *est 1* and *tel 1* are related to an increase of chromosome loss (Cheng and Loeb, 1993; Greenwell et al., 1995). As in DM1 patients, alterations of cholesterol metabolism and insulin resistance occur, it is intriguing to note that on chromosome 19 the low density lipoprotein (LDL) receptor gene (Francke et al., 1984), whose mutations are the primary defect in familial hypercholesterolemia, the gene for apolipoprotein E, a ligand for LDL receptor (Scott et al., 1985), and the insulin receptor gene (Yang-Feng et al., 1985) are also located.

In an attempt to clarify if chromosome 19 is preferentially lost, we performed FISH experiments using a chromosome 19 specific centromeric probe. This probe is specific also for chromosomes 1 and 5, and, as only binucleated cells carrying the total number of fluorescent signals have to be analysed, this meant that we had to take into account only binucleated cells carrying 12 fluorescent signals. In our hand, the fraction of cells presenting such a high number of fluorescent signals was quite low, making this analysis extremely difficult. Anyway, preliminary results seem to indicate the absence of a specific loss of chromosomes 1, 5 and 19 (data not shown), making random chromosome loss the most likely explanation.

When DM1 and control lymphocytes were treated in vitro with drugs, DM1 cells were shown to be as sensitive as control cells to MCC. On the contrary, DM1 cells appear more resistant to COLC and the expected increase in the MN frequency was not found. As the target of colcemid is β-tubulin, the apparent colcemid-resistance could be due to the presence, in DM1 patients, of an already damaged tubulin, which is no longer sensitive to the effect of colcemid. The observation that COLC is unable not only to increase the total number of MN, but also the basal frequency of MN/C+ in DM1 cells further stresses the possibility that aberrant tubulin is the main cause of the colcemid resistance and underlies the chromosome loss observed in DM1 cells.

Aberrant regulation of microtubule associated and tau protein has been found in DM1 patients (Sergeant et al., 2001), and tau protein alterations are present in Alzheimer disease, which is also characterized by increased chromosome loss (Migliore et al., 1999).

Recently, a second myotonic dystrophy locus has been mapped on the long arm of chromosome 3 (3q21). A CCTG expansion located in intron 1 of the zinc finger protein 9 is the cause of myotonic dystrophy type 2 (DM2) (Liquori et al., 2001). The chromosomal analysis of these patients would be helpful to verify some of our hypothesis. If chromosome instability were also present in DM2, it could be argued that the repeat expansion is the most likely etiological factor of chromosome loss; the opposite result would instead support a role of DMPK in chromosome segregation.

Acknowledgements

Prof. R. Barale, Prof. L Migliore and Prof. B. Rossi (Pisa University) are gratefully acknowledged for their critical reading of the manuscript.

References

Botto N, Rizza A, Colombo MG, et al: Evidence for DNA damage in patients with coronary artery disease. Mutat Res 493:23–30 (2001).

Cheng KC, Loeb LA: Genomic instability and tumour progression: mechanistic considerations. Review. Adv Cancer Res 60:121–156 (1993).

Fenech M: A detailed description of the method and its applications to genotoxicity studies in human population. Mutat Res 285:35–44 (1993).

Francke U, Brown MS, Goldstein JL: Assignment of the human gene for the low density lipoprotein receptor to chromosome 19: synteny of a receptor, a ligand and a genetic disease. Proc natl Acad Sci, USA 81:2826–2830 (1984).

Freudenreich CH, Kantrow SM, Zakian VA: Expansion and length-dependent fragility of CTG repeats in yeast. Science 279:853–856 (1998).

Giraud C, Winocour E, Berns KI: Site-specific integration by adeno-associated virus is directed by a cellular DNA sequence. Proc natl Acad Sci, USA 91:10039–10043 (1994).

Greenwell PW, Kronmal SL, Porter SE, Gassenhuber J, Obermaier B, Petes TD: TEL1, a gene involved in controlling telomere length in *S. cerevisiae*, is homologous to the human ataxia telangiectasia gene. Cell 82:823–829 (1995).

Heddle JA, Cimino MC, Hayashi M, et al: Micronuclei as an index of cytogenetic damage: past, present, and future. Environ Mol Mutagen 18:277–291 (1991).

Jabs EW, Tuck-Muller CM, Anhalt GJ, Earnshaw W, Wise RA, Wigley F: Cytogenetic survey in systemic sclerosis: correlation of aneuploidy with the presence of anticentromere antibodies. Cytogenet Cell Genet 63:169–175 (1993).

Jalal SM, Lindor NM, Michels VV, et al: Absence of chromosome fragility at 19q13.3 in patients with myotonic dystrophy. Am J med Genet 46:441–443 (1993).

Liquori CL, Ricker K, Moseley ML, Jacobsen JF, et al: Myotonic dystrophy type 2 caused by a CCTG expansion in intron 1 of ZNF9. Science 293:864–867 (2001).

Migliore L, Botto N, Scarpato R, Petrozzi L, Cipriani G, Bonuccelli U: Preferential occurrence of chromosome 21 malsegregation in peripheral blood lymphocytes of Alzheimer disease patients. Cytogenet Cell Genet 87:41–46 (1999).

Petrozzi L, Lucetti C, Gambaccini G, et al: Cytogenetic analysis oxidative damage in lymphocytes of Parkinson's disease patients. Neurol Sci 22:83–84 (2001).

Rolland Y, Mazieres B, Laroche M: Steinert disease associated with Klinefelter's syndrome. Rev Neurol 155:229–230 (1999).

Scott J, Knott TJ, Shaw DJ, Brook JD: Localization of genes encoding apolipoproteins CI, CII, and E to the p13→cen region of human chromosome 19. Hum Genet 71:144–146 (1985).

Sergeant N, Sablonniere B, Schraen-Maschke S, et al: Dysregulation of human brain microtubule-associated tau mRNA maturation in myotonic dystrophy type 1. Hum molec Genet 10:2143–2155 (2001).

Shiloh Y: Ataxia telangiectasia and the Nijmegen breakage syndrome: related disorders but gene apart. A. Rev Genet 31:635–662 (1997).

Sutherland GR, Richards RI: Simple tandem DNA repeats and human genetic diseases. Proc natl Acad Sci, USA 92:3636–3641 (1995).

Tapscott SJ, Klesert TR, Widrow RJ, Stoger R, Laird CD: Fragile X syndrome and myotonic dystrophy: parallels and paradoxes. Curr Opin Genet Dev 8:245–253 (1998).

Tsilfidis C, MacKenzie AE, Mettler G, Barcelo J Korneluk RG: Correlation between CTG trinucleotide repeat length and frequency of severe congenital myotonic dystrophy. Nature Genet 1:192–195 (1992).

Verde F, Wiley DJ, Nurse P: Fission yeast orb6, a ser/thr protein kinase related to mammalian rho kinase and myotonic dystrophy kinase, is required for maintenance of cell polarity and coordinate cell morphogenesis with cell cycle. Genetics 13:7526–7531 (1998).

Wenger SL, Giangreco CA, Tarleton J, Wessel HB: Inability to induce fragile sites at CTG repeats in congenital myotonic dystrophy. Am J med Genet 66:60–66 (1994).

Wolff S, Perry P: Differential Giemsa staining of sister chromatids and the study of chromatid exchanges without autoradiography. Chromosoma 8:341–53 (1974).

Xu GL, Bestor TH, Hsieh CL, et al: Chromosome instability and immunodeficiency syndrome caused by mutations in a DNA methyltransferase gene. Nature 402:187–191 (1999).

Yang-Feng T, Francke U, Ullrich A: Gene for human insulin receptor: localization to site on chromosome 19 involved in pre-B-cell leukemia. Science 228:728–731 (1985).

Cytogenet Genome Res 100:230–242 (2003)
DOI: 10.1159/000072859

Transgenic mouse models for myotonic dystrophy type 1 (DM1)

D.G. Wansink and B. Wieringa

Department of Cell Biology, NCMLS, University Medical Center, Nijmegen (The Netherlands)

Abstract. The study of animal models for myotonic dystrophy type 1 (DM1) has helped us to 'de- and reconstruct' our ideas on how the highly variable multisystemic constellation of disease features can be caused by only one type of event, i.e., the expansion of a perfect $(CTG)_n$ repeat in the DM1 locus on 19q. Evidence is now accumulating that cell type, cell state and species dependent activities of the DNA replication/repair/recombination machinery contribute to the intergenerational and somatic behavior of the $(CTG)_n$ repeat at the DNA level. At the RNA level, a gain-of-function mechanism, with dominant toxic effects of $(CUG)_n$ repeat containing transcripts, probably has a central role in DM1 pathology. Parallel study of DM2, a closely related form of myotonic dystrophy, has revealed a similar mechanism, but also made clear that part of the attention should remain focused on a possible role for candidate loss-of-function genes from the DM1 locus itself (like DMWD, DMPK and SIX5) or elsewhere in the genome, to find explanations for clinical aspects that are unique to DM1. This review will focus on new insight regarding structure-function features of candidate genes involved in DM1 pathobiology, and on the mechanisms of expansion and disease pathology that have now partly been disclosed with the help of transgenic animal models.

Myotonic dystrophy (DM), also known as dystrophia myotonica or Steinert disease, and first described about one hundred years ago (Batten and Gibb, 1909; Steinert, 1909), is the most common form of muscular dystrophy among adults, with an estimated incidence of 1:8000 (Harper, 2001). Manifestation of disease is highly variable and typically consists of a combination of muscular, neural and endocrine features, each of which may vary in severity with much variation in the age of onset. Congenital DM constitutes a very severe disorder, characterized by a high neonatal mortality and symptoms like hypotonia, mental retardation and respiratory distress. About one decade ago, the genetic mutation underlying myotonic dystrophy (now called DM1, MIM 160900) was identified as an expanding $(CTG)_n$ repeat on chromosome 19q13.3 (Aslanidis et al., 1992; Brook et al., 1992; Fu et al., 1992; Mahadevan et al., 1993). Above a critical threshold, the $(CTG)_n$ repeat has a greater than 95% chance to expand from one generation to the next, causing "anticipation", i.e., the progressive earlier onset of the disease and worsening of the symptoms in subsequent generations (Groenen and Wieringa, 1998; Harper, 2001).

Here it is important to note that shortly after the development of $(CTG)_n$ repeat diagnostics, a category of patients with a myotonic dystrophy phenotype similar to DM1, but without an expanded $(CTG)_n$ repeat, were identified (reviewed in IDMC, 2000; Meola, 2000; Ranum and Day, 2002). The mutation in patients with this form of DM, designated DM2 or PROMM (MIM 602668), was recently characterized as an expanding $(CCTG)_n$ repeat (Liquori et al., 2001), and mapped to chromosome 3q21.3. How $(CTG)_n$ or $(CCTG)_n$ repeat expansions cause the overtly similar, but clinically distinct, constellations of clinical features in DM1 and DM2 is still an open question.

To "deconvolute" the molecular mechanisms that could possibly be associated with the complex disease manifestation of DM, a number of different transgenic DM1 mouse models have now been generated. Broadly speaking, these models can be classified into two main categories, namely (i) transgenics

B.W. is supported by the Dutch Prinses Beatrix Fonds (PBF), the American Muscular Dystrophy Association (MDA), l'Association Française contre les Myopathies (AFM), the Netherlands Organization of Scientific Research (NWO), and the Dutch Cancer Society (KWF).

Received 28 October 2002; manuscript accepted 18 December 2002.

Request reprints from Bé Wieringa, PhD
Department of Cell Biology (code 163)
NCMLS, University Medical Center
PO Box 9101, 6500 HB Nijmegen (The Netherlands)
telephone: +31-24-3614329; fax: +31-24-3615317
e-mail: b.wieringa@ncmls.kun.nl

KARGER
Fax + 41 61 306 12 34
E-mail karger@karger.ch
www.karger.com

0301–0171/03/1004–0230$19.50/0

Accessible online at:
www.karger.com/cgr

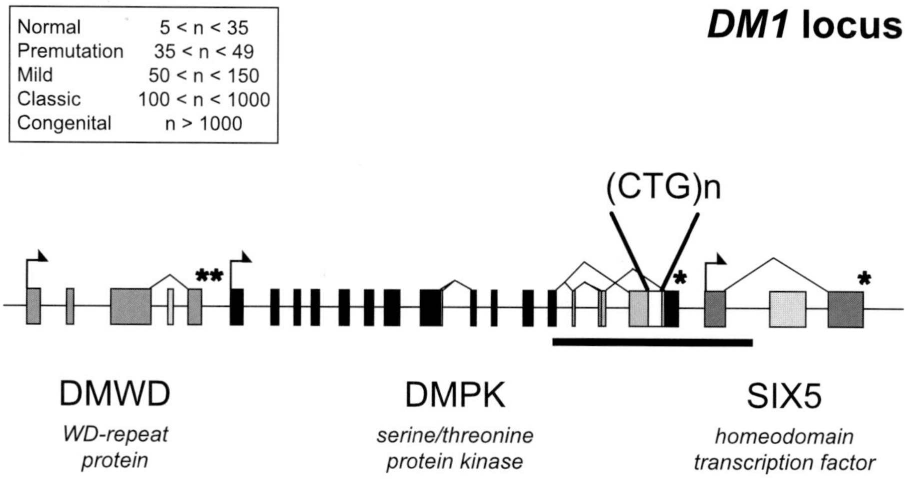

Fig. 1. Schematic representation of the DM1 locus: The DM1 locus is strongly conserved between mouse and man and contains three genes, DMWD, DMPK and SIX5, located closely together in a gene-dense region. DMWD encodes a WD-repeat protein, DMPK a serine/threonine protein kinase, and SIX5 a homeodomain transcription factor. Transcription initiation sites are indicated with arrows; exons with boxes, and introns and intergenic sequences with a straight line; alternative splice modes with connecting lines between exons, and polyadenylation sites with asterisks (two for DMWD). In the human situation, and to a lesser extent also in the mouse locus, a CpG island containing a DNaseI hypersensitive site overlaps the 3′-end of DMPK and the 5′-end of SIX5 (black bar). The unstable $(CTG)_n$ repeat is located in exon 15 of the human DMPK gene, in contrast, the mouse locus contains a $(CTG)_2(CAG)_2(CTG)$-sequence. The length of the repeat ranges from five to several thousand CTG-triplets, and is strongly correlated with age of onset and severity of DM1 disease symptoms (see box and text).

developed to test hypotheses regarding DM1 disease pathology, and (ii) transgenics developed to study the mechanisms underlying DNA instability, i.e., the $(CTG)_n$ repeat expansion characteristic for DM1. In a few cases, transgenic mouse lines have proved suitable to study both aspects simultaneously. Mouse models for DM2 have not been reported yet, presumably because not enough time has elapsed since the discovery of the $(CCTG)_n$ repeat. In this review, we shall provide an overview of the current DM1 mouse models, and do so against a background of relevant knowledge on clinical, genetic and pathobiological aspects in human patients. For details on the many intricate aspects of genotype-phenotype relationships in DM1 one is referred to any of the recent comprehensive reviews that have been written on this subject (Brewster et al., 1998; Groenen and Wieringa, 1998; Korade-Mirnics et al., 1998; Ueda et al., 2000; Larkin and Fardaei, 2001; Harper, 2001). We will begin by giving a (non-comprehensive) survey of recent literature on the characterization of gene loci and gene products that are of direct relevance for understanding DM1 pathobiology.

Myotonic dystrophy type 1 (DM1)

Correlations between $(CTG)_n$ repeat expansion and disease pathology

One of the most challenging aspects of DM1 research is to explain how $(CTG)_n$ repeat length at the DNA level, or $(CUG)_n$ repeat dosage at the RNA expression level, are determined and correlated to the severity and variability in manifestation of clinical features like myotonia, muscle fiber degeneration, cataracts, testicular atrophy, endocrine abnormalities, cardiac conduction defects, hypersomnia, neuropsychiatric disturbances, smooth muscle dysfunction and insulin resistance (reviewed by Brewster et al., 1998; Harper, 2001; IDMC, 2000; Meola, 2000).

What we know now is mostly based on data obtained soon after the discovery of the DM1 mutation: Unaffected individuals have fewer than 37 repeats, while repeat sizes in affected individuals range from ~100 into the thousands (see box, Fig. 1). Global disease severity and age of onset correlate reasonably well with repeat length in blood: alleles in the 35–49 repeats range are considered premutation alleles, mild or very late onset of disease is seen in individuals with $(CTG)_{50-150}$ repeats, "classic" disease manifestation is associated with $(CTG)_{100-1000}$ and in congenital DM1 patients the repeat contains several thousands of CTG triplets (Groenen and Wieringa, 1998; Harper, 2001).

What we also know is that patients with moderate or long expansions demonstrate age-dependent, expansion-biased mosaicism in repeat length, presumably mediated via multiple small length change events (Thornton et al., 1994; Monckton et al., 1995; Wong et al., 1995; Martorell et al., 1997). Besides being highly unstable in germ line tissue, repeat length also varies considerably within and between somatic tissues. For example, repeats in muscle cells (skeletal muscle and heart) are generally greater than in blood lymphocytes, but currently no clear picture exists for the relative stability in other tissues, simply because this has not been studied in a large enough cohort of patients. This may become a serious problem, because the question how the distribution of (average) repeat lengths throughout individual tissues and organs correlates to the disease spectrum and severity in patients must soon be answered. If we fail, we may be generating mouse models without knowing precisely which ones come closest in mimicking the complex and variable DM1 condition in human patients.

DM1 locus, transcripts and proteins

There are many obvious reasons for taking mouse as the animal of choice for generating models of DM1. Firstly, mouse is a mammal with proven suitability for the study of muscle-related physiological phenomena. Secondly, genetic manipulation is relatively easy in mouse, and the chromosomal segment in which the unstable $(CTG)_n$ repeat is located is relatively well conserved between the two species, in a large piece of syntenic DNA (Kwitek et al., 2001; Gregory et al., 2002). The pertinent segment represents a gene-dense chromosome area, on human 19q13.3 and mouse proximal 7, with exactly similar gene order and orientation (reviewed in Brewster et al., 1998; Groenen and Wieringa, 1998; Larkin and Fardaei, 2001). The $(CTG)_n$ repeat itself is contained within the last 3′-untranslated exon of the human myotonic dystrophy protein kinase (DMPK) gene, but also located just 5′ of the homeodomain protein-encoding gene SIX5 (Fig. 1). Furthermore, immediately upstream of the DMPK gene in both species lies the DMWD gene. It has been reported that, possibly due to the presence of weak polyadenylation sites, transcription of the DMWD gene can extend into the DMPK gene (Jansen et al., 1992; Mahadevan et al., 1993; Shaw et al., 1993), thus creating one large transcript (Fig. 1, see also, Jansen et al., 1995).

In the human situation, the $(CTG)_n$ repeat is located within a relatively short segment of DNA with distinct features, characterized by the presence of a CpG island and by a DNaseI hypersensitive site (Boucher et al., 1995; Brock et al., 1999). Presumably, this segment, which is flanked by two binding sites of the zinc-finger protein CTCF, serves as an insulator element by blocking enhancer promoter communication necessary for independent transcription regulation of SIX5 and DMPK (Filippova et al., 2001). In congenital DM1, in contrast to classic DM1 with an adult onset, CpGs in this insulator are hypermethylated (Steinbach et al., 1998), and the insulator function is abolished (Filippova et al., 2001). Unlike the human DM1 locus, the cognate mouse DNA region does not contain a $(CTG)_n$ repeat, but instead a $(CTG)_2(CAG)_2(CTG)$ sequence is present, and the CpG-rich island is much smaller (Heath et al., 1997). In spite of these differences, it has been shown that expression of each of the three genes *Dmwd*, *Dmpk*, and *Six5* is independently and similarly regulated in mice and humans (Eriksson et al., 2000).

(i) DMPK: Central in the DM1 area is the DMPK gene (MIM 605377), specifying a serine/threonine protein kinase, which is expressed mainly in skeletal muscle, heart, stomach, testis and brain (reviewed by Groenen and Wieringa, 1998; Ueda et al., 2000). DMPK transcripts are subject to extensive cell-type dependent alternative splicing. This process is conserved between mouse and man and results in six major splice isoforms (Fig. 1, Groenen et al., 2000). These isoforms all contain a $(CUG)_n$ repeat in the 3′-UTR of the mRNA. A minor DMPK splice isoform, recently only identified in human, is characterized by lack of a $(CUG)_n$ repeat, due to use of an alternative splice acceptor site in exon 15 (Fig. 1, Tiscornia and Mahadevan, 2000). Alternative splicing defines presence or absence of the five amino acid motif val–ser–gly–gly–gly (VSGGG), and the nature of the C-terminus, for which three different variants exist among the major isoforms (Groenen and Wieringa, 1998; Groenen et al., 2000). The VSGGG-motif and different C-termini confer distinct properties to the six DMPK isoforms (Wansink et al., unpublished). Next to the two alternatively spliced domains, all DMPK isoforms share an N-terminal leucine-rich domain, a ser/thr protein kinase domain and a coiled coil domain.

According to sequence homology in the kinase domain, DMPK is a member of the AGC group of ser/thr kinases (Hanks and Hunter, 1995) and is most homologous to the p21-activated kinases myotonic dystrophy kinase-related Cdc42-binding kinase (MRCK) (Leung et al., 1998) and ROCK/Rho-kinase/ROK (Leung et al., 1996; Matsui et al., 1996; Nakagawa et al., 1996). By analogy, it is suggested that DMPK may have a role in actin cytoskeleton remodeling, but evidence to sustain this assumption is lacking. Various in vitro substrates were reported for DMPK: the dihydropyridine receptor (Timchenko et al., 1995), DMAP (Fu, 1996), MKBP (Suzuki et al., 1998), phospholemman (Mounsey et al., 2000), CUG-BP (Roberts et al., 1997), and myosin phosphatase targeting subunit (Muranyi et al., 2001, Wansink et al., unpublished). Whether members of this rather diverse set of proteins are also phosphorylated by DMPK in vivo needs to be investigated.

An important stimulus for understanding DM1 molecular pathogenesis was the finding that DMPK transcripts bearing long $(CUG)_n$ repeats in their 3′-UTR were retained in nuclear foci of cells from myotonic dystrophy patients (Davis et al., 1997; Mankodi et al., 2001; Taneja et al., 1995). As a consequence, cytoplasmic DMPK transcripts with long $(CUG)_n$ repeats drop to almost undetectable levels (discussed in first paragraph of DMPK knock-out mice section), which has been shown to be essentially a post-transcriptional process (Krahe et al., 1995; Wang et al., 1995). In addition, next to the *cis*-effect of the $(CUG)_n$ repeat on DMPK mRNA production, dramatic *trans*-effects on production and use of other mRNAs with profound influence on alternative splicing and translation of mRNAs have been reported (Philips et al., 1998; Savkur et al., 2001; Sergeant et al., 2001; Buj-Bello et al., 2002; Charlet-B. et al., 2002; Mankodi et al., 2002;). This RNA gain-of-function mechanism, termed “RNA dominance”, is currently the favored hypothesis to explain the major features of DM1 disease manifestation. The recent characterization of a $(CCTG)_n$ repeat as the genetic defect in DM2 (Liquori et al., 2001), the DM symptoms in the HSA^{LR} mouse (see below, Mankodi et al., 2000) and the finding of effects of long $(CUG)_n$ repeats in differentiating myoblast cultures with severe consequences for late myotube differentiation (Amack et al., 1999, 2002; Amack and Mahadevan, 2001), further strengthen the RNA dominance theory.

RNA dominance is best explained by a mechanism whereby, at the level of RNA, the expanded $(CUG)_n$ repeat interferes with processing or translation of many (primary) non-DMPK transcripts. In this mechanism, interaction with RNA-binding proteins, like CUG-BP, ETR-3, or muscleblind proteins (Timchenko et al., 1996; Lu et al., 1999; Michalowski et al., 1999; Miller et al., 2000; Fardaei et al., 2002), may play a determining role, leading to overall perturbation of nuclear RNA metabolism (reviewed in Korade-Mirnics et al., 1998; Galvao et al., 2001; Larkin and Fardaei, 2001; Tapscott and Thornton, 2001; Ranum and Day, 2002). Circumstantial evidence suggests that $(CUG)_n$ repeat directed misregulation of CUG-BP may have the most severe consequences, with effects on late skeletal muscle differentiation, chloride-channel evoked myotonia, and insulin resistance via transcription and splice factor regulation (Philips et al., 1998; Savkur et al., 2001; Charlet-B. et al., 2002). In another scenario, a $(CUG)_n$ length-dependent activation of dsRNA-activated protein kinase (PKR), with toxic effects on myocyte differentiation, or survival, has been proposed (Tian et al., 2000). Since also the neurodegenerative disease spinocerebellar ataxia type 8 (SCA8) is caused by an untranslated $(CTG)_n$ repeat (Koob et al., 1999), it may be that SCA8, DM1 and DM2 (where the $(CCUG)_n$ repeat is expressed in the intron of primary ZNF9 transcripts) share a common disease pathway. However, in the case of DM1, not only DMPK transcripts with long $(CUG)_n$ repeats, but also the specific protein products of the DM1 locus proper (DMPK, DMWD and SIX5) are strong candidates for playing an important role in pathogenesis.

(ii) DMWD: The DMWD gene (formerly *Dmr-N9* or gene 59 in mouse and human, respectively), is located immediately 5′ to the DMPK gene (Fig. 1), and is mainly expressed in testis and brain, with low levels of expression in many other tissues, starting already early during embryogenesis (Jansen et al., 1995; Eriksson et al., 2000). The Dmwd protein was, based on computer analysis, predicted to be a protein with WD-repeat elements, 44–60 amino acid residue sequences that usually contain a GH dipeptide at position 11–24 from its N-terminus and a C-terminal WD dipeptide at the C-terminal end (Neer et al., 1994; Smith et al., 1999). Dmwd contains five segments with WD-homology, of which only the first four genuine repeats are thought to form a so-called β-propeller structure with four blades, in a closed ring-like arrangement (Westerlaken and Wieringa, unpublished). Each of these blades consists of a four-stranded twisted β sheet, formed by adjacent repeats. It is thought that all WD-proteins form these kinds of structures, which places them into one large structural family. Propeller β forming WD-repeat segments provide to proteins an interface for association to more than one protein partner.

A database search for full-length sequences has revealed that mouse and human DMWD are evolutionary strongly conserved genes, with homologues in *C. elegans* (C08B6.7), *S. pombe* (YDE3), *A. thaliana* (T2N18.8), *D. melanogaster* (CG6420), and *A. nidulans* (CreC). Recently, it was discovered that the CreC protein interacts with another protein, designated CreB, with a role in metabolic regulation in *A. nidulans*. CreB is a deubiquinating enzyme and has a human homologue called ubiquitin hydrolysing enzyme 1 (USP12). CreC most likely stabilizes the CreB enzyme by masking several internal PEST sequences, signals for proteolytic breakdown, in the protein. CreB in its turn stabilizes yet another protein, CreA, by removing covalently bound ubiquitin moieties. CreA itself is a DNA-binding regulatory protein involved in carbon catabolite repression (Lockington and Kelly, 2002). Extrapolating this complex interaction scheme to the situation in mouse and human, suggests a role for the Dmwd protein in proteolytic breakdown pathways, with a role in (gene) regulation.

In DM1 patients with large repeats, it has been demonstrated that DMWD expression was lowered, with $(CTG)_n$ expansion size being correlated inversely with DMWD expression levels (Hamshere et al., 1997; Alwazzan et al., 1999; Eriksson et al., 1999; but see also Frisch et al., 2001). It is now of utmost importance to define the repeat expansion thresholds that could be involved, and to develop animal models to study the pathogenic effects that a decreased Dmwd level could have.

(iii) SIX5: The SIX5 gene (formerly known as DMAHP), first identified by Johnson and coworkers (Boucher et al., 1995), is located 3′ of the DMPK gene (Fig. 1) and is a vertebrate homologue of *sine oculis*, a gene essential for development of the visual system of Drosophila. SIX5 encodes a homeodomain transcription factor and is a member of the Six (sine oculis homeobox) family (Kawakami et al., 2000). Several different SIX5 transcription initiation sites were identified, conserved between man and mouse, which were all downstream of the $(CTG)_n$ repeat, excluding inclusion of the $(CUG)_n$ repeat in SIX5 mRNA (Murakami et al., 1998). Expression of *Six5* was already detectable during early mouse stage E7 and continued to E17 (Murakami et al., 1998). In adult tissues, *Six5* is mainly expressed in skeletal muscle, heart, eye, kidney, liver, lung and small intestine (Kawakami et al., 1996; Heath et al., 1997; Murakami et al., 1998; Ohto et al., 1998; Winchester et al., 1999; Eriksson et al., 2000). Remarkably, expression patterns of *Six5* and *Dmpk* are essentially overlapping, albeit that *Six5* levels are generally 5–50 fold lower than those of *Dmpk* (Heath et al., 1997; Eriksson et al., 2000).

A screen for downstream targets of *Six5* revealed several candidate genes expressed in somites, skeletal muscle, brain and meninges, one of which was *Igfbp5*, encoding a component of IGF signaling (Sato et al., 2002). In an independent study, it was shown that recombinant proteins of *Six5* domains fused to GST bind to elements in the Na^+/K^+-ATPase α1 subunit gene, but not to putative *Six5* binding sites in the *Dmpk* promoter (Harris et al., 2000). In Drosophila, the *Six5* homologue D *Six4* was found to play a role in gonad and muscle development (Kirby et al., 2001). In conclusion, the presumed function of Six5 in tissues involved in DM1 pathology (e.g., skeletal muscle, eyes, testis), and the finding that expansion of the $(CTG)_n$ repeat has a negative effect on SIX5 expression levels (discussed below) has opened the possibility that DM1 must be considered a multigenic (locus) disease. Evidence to support this hypothesis has been sought in $Six5^{-/-}$ transgenic mouse models (see section SIX5 knock-out mice).

Knock-out and overexpressor models

To investigate involvement of gene products of DMPK and SIX5 in DM1 molecular pathogenesis, several "first-generation" mouse models for DM1 were developed. Thus far, no mouse model has been reported to investigate the putative involvement of DMWD.

Table 1. Transgenic DM1 mouse models[a]

Transgenic line[b]	Targeting strategy	Pathology phenotype	Number of $(CTG)_n$ repeats in founder(s)[c]	Intergenerational instability (change in repeats)	Somatic instability (change in repeats)
Knock-out and overexpressor mice					
Dmpk KO[1]	replacement first 7 exons of *Dmpk* gene with hygromycin cassette	mild myopathy, altered Ca^{2+} homeostasis	n.a.	n.a.	n.a.
Dmpk KO[2]	replacement first 7 exons of *Dmpk* gene with neomycin cassette	mild myopathy, reduced skeletal muscle force, cardiac conduction defects, abnormal Na^+ channel gating	n.a.	n.a.	n.a.
DMPK Tg overexpressor[3]	insertion 14 kb human DM1 locus fragment, containing DMPK gene (~20 copies)	hypertrophic cardiomyopathy, enhanced neonatal mortality	11	n.d.	n.d.
Six5 KO[4]	replacement complete *Six5* gene with neomycine cassette	ocular cataract	n.a.	n.a.	n.a.
Six5 KO[5]	replacement exon 1 of *Six5* gene with beta-gal reporter	ocular cataract	n.a.	n.a.	n.a.
Expanded $(CTG)_n$ repeat mice					
Dmt-A-E[6]	random insertion human genomic DMPK $(CTG)_{162}$ fragment	n.d.	143–163	contractions and expansions (–7 to +7), parent-of-origin effects	in various tissues; highest in kidney (–20 to +540), age dependent
DM20, DM55, DM300[7]	random insertion 45 kb genomic fragment carrying DMWD, DMPK and SIX5, with either $(CTG)_{20}$, $(CTG)_{55}$ or $(CTG)_{\sim 360}$	DM300: myopathy, myotonia, altered tau splice forms, growth retardation (mouse line dependent)	55 or >300 (20 in control line)	mainly expansions, DM55 (–1 to +6), DM300 (–30 to +60), parent-of-origin effects	in various tissues; highest in kidney, pancreas, liver (–10 to + >100), age and mouse line dependent
$(CTG)_{11}$, $(CTG)_{84}$[8]	replacement mouse 3'-end of *Dmpk* gene by human DMPK sequence containing either $(CTG)_{11}$ or $(CTG)_{84}$	no overt abnormalities	84 (11 in control line)	contractions and expansions (–16 to +8), mouse strain effects	in various tissues; highest in stomach, liver, kidney, (up to +45), age and cell type dependent
HSA^{LR} HSA^{SR}[9]	random insertion genomic fragment containing human skeletal actin gene, bearing $(CTG)_5$ or $(CTG)_{250}$ in its 3'-UTR	HSA^{LR}: myotonia, myopathy, altered chloride channel splice forms, increased early mortality	~250 (5 in control line)	n.d.	n.d.

[a] Listing of all available DM1 mouse models, including Dmpk and Six5 knock-outs, and DMPK Tg overexpressor in the upper panel, and several mouse lines carrying expanded (CTG)n repeats in the lower panel.
[b] 1. (Benders et al., 1997; Jansen et al., 1996); 2. (Berul et al., 1999; Mounsey et al., 2000; Reddy et al., 1996; Reddy et al., 2002); 3. (Jansen et al., 1996); 4. (Mistry et al., 2001; Sarkar et al., 2000); 5. (Klesert et al., 2000); 6. (Fortune et al., 2000; Monckton et al., 1997; Zhang et al., 2002); 7. (Gourdon et al., 1997; Lia et al., 1998; Seznec et al., 2000; Seznec et al., 2001); 8. (van den Broek et al., 2002); 9. (Mankodi et al., 2000; Mankodi et al., 2002).
[c] n.a. = not applicable, n.d. = not determined.

DMPK knock-out mice

After the discovery of the $(CTG)_n$ repeat in the early nineties, the mere observation that DMPK was central in the DM1 locus and $(CTG)_n$ repeat expansion resulted in an altered – according to most researchers, decreased – cytoplasmic DMPK level in patients triggered the speculation that the DM1 phenotype was caused by DMPK (haplo)insufficiency (Carango et al., 1993; Fu et al., 1993; Hofmann-Radvanyi et al., 1993; Novelli et al., 1993; Krahe et al., 1995; Wang et al., 1995; Hamshere et al., 1997; Eriksson et al., 1999; Eriksson et al., 2001; Frisch et al., 2001; but see also Sabouri et al., 1993; Inukai et al., 2000; Narang et al., 2000). As was discussed above (section (i) DMPK), any decrease in cytoplasmic DMPK level in DM1 patients is presumably due to nuclear retention of DMPK transcripts bearing long $(CUG)_n$ repeats.

To test the effect of lowered dosage of *Dmpk* gene products on murine development and DM1 disease manifestation, independently two different *Dmpk* knock-out mouse lines were created (Table 1; Jansen et al., 1996; Reddy et al., 1996). Both knock-out mouse lines were generated by targeted replacement of the first seven exons of the *Dmpk* gene by a resistance cassette (Table 1), resulting in complete absence of Dmpk protein. $Dmpk^{-/-}$ mouse lines thus established seemed healthy and overtly normal. On closer examination, $Dmpk^{-/-}$ mice created by the Reddy lab displayed a late onset progressive skeletal myopathy, characterized by decreased force generation and increased fiber degeneration, suggesting that *Dmpk* is involved in maintenance of muscle fiber structural integrity (Reddy et al., 1996). In *Dmpk* null mice from our own lab, only a mild myopathy in head and neck muscles at advanced age was observed (Jansen et al., 1996). In myoblasts isolated from these mice and in in vitro differentiated $Dmpk^{-/-}$ myotubes, we found an altered Ca^{2+} response when compared to wild type cells, providing evidence for a role for *Dmpk* in modulation of Ca^{2+} homeostasis (Benders et al., 1997). Reddy and coworkers found cardiac conduction abnormalities which include first, second, and third degree atrioventricular block in their $Dmpk^{-/-}$ mice (Berul et al., 1999). Remarkably, also in $Dmpk^{+/-}$ mice a first degree heart block was observed, which is surprisingly similar to what is observed in DM1 patients. Furthermore, in skeletal muscle cells isolated from *Dmpk*$^{-/-}$ and *Dmpk*$^{+/-}$ mice an abnormal Na^+ channel opening was observed (Mounsey et al., 2000; Mistry et al., 2001; Reddy et al., 2002). This feature was reminiscent of abnormal Na^+ channel gating found in muscle biopsies from DM1 patients (Franke et al., 1990) and may contribute to muscle hyperexcitability. A link between Na^+ channels and DMPK activity has already been shown earlier in vitro (Mounsey et al., 1995; Chahine and George, 1997) and also by Benders et al. (1997) in *Dmpk*$^{-/-}$ skeletal muscle cells. Although the evidence is compelling, a direct interaction between DMPK and Na^+ channels still needs to be demonstrated.

Finally, it should be noted that insertion of the hygromycin cassette into the DM1 locus resulted in formation of an altered *Dmwd* mRNA, albeit that Dmwd protein levels remained unchanged (Jansen et al., 1996). This clearly demonstrates that targeting a gene-dense locus, such as the DM1 region, may have unexpected and unwanted side effects which can make straightforward interpretation of phenotypic effects difficult (see also section SIX5 knock-out mice for effects of *Six5* targeting on *Dmpk* levels). Despite subtle differences between the two knock-out lines, which may be attributed to background differences, one must conclude that *Dmpk*$^{-/-}$ mice display a relatively mild picture of a number of symptoms observed in DM1 patients, in particular with regard to the cardiac conduction defects and abnormal skeletal muscle physiology. Other characteristic DM1 features such as myotonia, cataract and male infertility seemed to be fully absent.

DMPK overexpressor mouse

To examine cell biological and physiological consequences of DMPK overexpression, a transgenic DMPK overexpressor mouse model was created, using a genomic fragment, encoding all human DMPK exons (including $(CTG)_{11}$ repeats) and the last exon of DMWD, in transgenesis (Table 1; Jansen et al., 1996). Several founders were obtained with 2– ~25 tandemly integrated copies of the transgene. Highest expression was demonstrated in lines with >20 copies, showing approximately 5–10 fold expression of hDMPK versus endogenous m*Dmpk*. The hDMPK tissue distribution exactly matched the endogenous expression pattern, indicating that all major *cis*-acting regulatory elements were present in the genomic construct. Also, the various cell-type dependent modes of hDMPK and m*Dmpk* alternative splicing were conserved (Groenen et al., 2000). During breeding, an increased prenatal and/or neonatal mortality and a high incidence of illness in pregnant and lactating females was observed in lines with a high transgene copy number, indicating a detrimental effect of high DMPK levels during pregnancy and development. No histological and ultrastructural abnormalities were observed in various tissues examined and also detailed investigation of skeletal muscle physiology revealed no abnormal features. In heart muscle, however, a clear but mild myopathy was detected, reminiscent of human cardiac hypertrophy (Jansen et al., 1996). Thus, mice expressing multiple copies of the human DMPK gene, carrying $(CTG)_{11}$ repeats, develop hypertrophic cardiomyopathy, but do not show any truly characteristic feature of DM1.

SIX5 knock-out mice

Apart from affecting the DMPK transcript level, $(CTG)_n$ repeat expansion also lowers expression of the adjacent gene SIX5 (Klesert et al., 1997; Thornton et al., 1997; Gennarelli et al., 1999; Korade-Mirnics et al., 1999; Inukai et al., 2000; Frisch et al., 2001; but see also Hamshere et al., 1997; Eriksson et al., 1999). Since the $(CTG)_n$ repeat is located outside the SIX5 gene, yet is positioned very close to its enhancer/promoter region and the DNaseI-hypersensitive site, the underlying mechanism for loss of SIX5 transcripts is probably not a *trans*-effect (posttranscriptional), like for DMPK mRNA, but a repression effect in *cis*. Evidence has been provided that expansion of the $(CTG)_n$ repeat has a strong positive effect on nucleosome assembly and chromatin organization, thus resulting in heterochromatin formation and SIX5 transcriptional silencing (Wang et al., 1994; Otten and Tapscott, 1995; Wang and Griffith, 1995; Godde and Wolffe, 1996).

To determine whether SIX5 dosage effects could possibly contribute to the DM1 phenotype, *Six5* knock-out mice models were created independently by two labs (Table 1; Klesert et al., 2000; Sarkar et al., 2000). Both *Six5*$^{-/-}$ lines developed ocular cataract at a higher rate than wild type animals; in fact, in one of the two lines cataract formation was already significantly enhanced in *Six5*$^{+/-}$ mice (Sarkar et al., 2000). The latter finding is particularly interesting in view of the fact that loss of SIX5 mRNA expression generally will not exceed the 50% level (Klesert et al., 1997; Thornton et al., 1997). Another interesting observation was that the penetrance of the cataract phenotype was dependent on the mouse strain (Sarkar et al., 2000), suggesting that genetic modifiers play a role in its manifestation, as is probably also the case in DM1 patients. A point of caution here is the observation that the cataract in DM1 patients and *Six5* knockouts is clinically not entirely similar (Ranum and Day, 2002). Although *Six5* is expressed in (developing) skeletal muscle and is involved in muscle development in Drosophila (see above), no abnormal skeletal muscle function in *Six5*$^{-/-}$ mice was observed. In addition, *Six5*$^{+/-}$ mice did not show the Na^+ channel gating abnormality of *Dmpk*$^{+/-}$ mice (Mistry et al., 2001). *Six5* being a transcription factor, both knock-out mouse lines were screened for down-stream effects of *Six5* loss, in particular on the transcript levels of *Atp1a1*, encoding the Na^+/K^+ ATPase α1 subunit. Remarkably, conflicting results were obtained: Sarkar et al. (2000) report an increase in *Atp1a1* expression, whereas Klesert et al. (2000) find no change in *Atp1a1* levels. The reason for this discrepancy is unknown. Despite all this work, the actual mechanism of how *Six5* deficiency results in cataract formation in *Six5*$^{-/-}$ mice remains unclear. One possibility is that loss of *Six5* leads to abnormal expression of one or more target genes, resulting in an altered ion homeostasis within the lens, which then may lead to cataract formation. Further studies are necessary to clarify this point.

In the context of DM1 being multigenic, it should be noted that both of the abovementioned *Six5*$^{-/-}$ lines showed a 2–3 fold down-regulation of *Dmpk*. This may be a *cis*-effect of the targeting strategy in the gene-dense DM1 locus, or an effect of *Six5* loss on *Dmpk* transcription. Whether this decrease in *Dmpk* levels has contributed in any way to the phenotype in *Six5*$^{-/-}$ mice remains unknown, but seems unlikely because no cataract was revealed in animals with complete *Dmpk* deficiency. Furthermore, given the fact that ocular cataract is also found in DM2 (Moxley, 1996; Moxley et al., 1998), and DM2 is a disorder, which has been linked to a locus which is completely different from the DMPK-SIX5 region, it is likely that other mechanisms than SIX5 (or DMPK) deficiency proper contribute to cataract formation in DM1 (and DM2) patients.

Mouse models with expanded $(CTG)_n$ repeats

Four groups have attempted to create mouse models that reproduce the very high germ line mutation rate, the considerable intergenerational length changes, and the massive expansion bias seen for the CTG triplet repeat in DM1 families. All transgenic mouse lines display a varying number of repeat instability characteristics also observed in DM1 patients. Only the two models expressing long $(CUG)_n$ repeats, designated DM300 (n > 300) and HSA^{LR} (n ~ 250), show pathological symptoms similar to those found in DM1 patients.

DMT-A-E

Monckton's group was the first to generate a series of five different transgenic founders, named Dmt-A-E, with randomly integrated one to three copy DNA inserts, consisting of a relatively short 3′-terminal area (exon 15) of the human DMPK gene spanning a $(CTG)_{162}$ repeat (Table 1; Monckton et al., 1997, 1998). Analysis of DNA from offspring of these founders, and monitoring lineages over several generations, clearly demonstrated repeat segregation distortion in favor of the transgene, hypermutability, sex of parent-of-origin effects, and a general bias to expansion (Monckton et al., 1997; Fortune et al., 2000). Application of small pool PCR analysis (originally developed by Monckton and co-workers, Monckton et al., 1995) also demonstrated somatic instability, manifested by mosaic length distribution profiles for most, but not all tissues. Interestingly, although the gross instability seen in DM1 patients was not replicated, some transgenes were significantly more unstable than others, suggesting influence of *cis*-effects imposed by DNA sequences flanking the transgene at the site of integration. In one transgenic lineage, Dmt-D, a rather dramatic, cell type dependent increase in repeat length was observed in kidney in adult mice during ageing. Repeat instability measured in in vitro cell cultures derived from the Dmt-D mice confirmed tissue-specific effects and progressive accumulation of larger alleles (Gomes-Pereira et al., 2001). Remarkably, no correlation between repeat instability and in vitro cell proliferation rate was found, indicating that *cis*- and/or *trans*-specific factors other than just DNA replication or cell division are important. Transcription, for example, may play a role, but it is not clear whether the Dmt162 transgene is transcribed (Fortune et al., 2000). Just like kidney, the male germ line tissue of Dmt-D mice showed an age dependence in the accumulation of expansions of $(CTG)_n$ repeats (Zhang et al., 2002). This can be most easily explained by (mitotic) mutation mechanisms contributing to the accumulation of mutations with time in spermatogenic stem cells (Leeflang et al., 1999). Finally, to date, no data on molecular pathogenesis are available for the Dmt mouse lines.

It may be relevant to mention here, that Monckton's lab also generated another transgenic model, designated Te162, in which a $(CTG)_{162}$ repeat containing minigene was coupled to the human EF1α promoter, to drive widespread expression of $(CUG)_n$ repeat containing RNA. Here, phenotypic problems, including female and male sterility, but no other features of DM1 were encountered and no follow-up studies on these animals have been published until now (Monckton et al., 1998).

DM20, DM55, DM300

Gourdon and co-workers generated a series of different mouse lines carrying large (45 kb) human genomic DNA inserts spanning the entire DMWD–DMPK–SIX5 area and carrying repeats consisting of either 20, 55 or >300 CTG triplets (named DM20, DM55, and DM300, respectively, Table 1; Gourdon et al., 1997; Seznec et al., 2000). No intergenerational length changes were detected in the DM20 lines. In contrast, moderate germ line instability was observed for the DM55 lineages, whereas a dramatically increased instability was seen for the DM300 lines, suggesting repeat-size related probability of mutation. In the DM300 transgenics, repeat length was expanded (with 1–60 CTG triplets) in 86–95% and contracted in 2–5% of descendants. Again, like for the Dmt models, for these mice sex of parents and age effects were reported, with greater variations upon paternal transmission. Unlike in congenital DM1, no saltatory expansions of several hundreds to thousands of repeats were observed (Seznec et al., 2000).

Somatic instability was observed in both the DM55 and DM300 lineages, but by far the highest instability was observed in tissues of the DM300 mice, suggesting expansion bias (Lia et al., 1998; Seznec et al., 2000). Instability was clearly age related and becoming truly prominent 6–9 months after birth. Length mosaicism was most obvious in liver, pancreas and kidney but also seen in other tissues of both DM55 and DM300, including sperm. Between animals there was clear lineage-dependence in the tissue distribution profiles, whereas within animals belonging to the same lineage always the same pattern of mosaic length distribution was observed (Seznec et al., 2000). Importantly, the large transgenic insert in these mice, although randomly integrated, presumably contains all control elements necessary for correct tissue-type regulation of transcription of the human DMWD, DMPK and SIX5 genes. Interestingly, however, no relationship between $(CTG)_n$ transgene somatic mutation rate and tissue differences in transcription levels of the three genes were seen (Lia et al., 1998).

The DM300 mouse lines displayed, in addition to the above-mentioned repeat expansion characteristics, also a great number of typical DM1 disease features (Seznec et al., 2001). Histological muscle abnormalities included centered nuclei, muscle de- and regeneration (e.g., neonatal MyHC expression), and altered mitochondrial morphology. EMG measurements of mouse legs indicated myotonia preferentially in the fore legs. In situ hybridization with a $(CAG)_{30}$ oligoprobe on isolated primary myoblasts revealed nuclear aggregates of $(CUG)_n$ repeat containing transcripts, as have been observed in DM1 patient material (Taneja et al., 1995). The DM300 lineage is currently the only DM1 transgenic line with obvious effects of the $(CTG)_{>300}$ repeat in brain. Similar to what was reported in brains from DM1 patients (Sergeant et al., 2001; Vermersch et al., 1996), DM300 mice display an altered tau isoform production, suggesting a role for transcripts containing long $(CUG)_n$ repeats in tau transcript splicing and processing.

Most of the above studies concerning DM1 pathology were done on homozygous DM300 mice, although many symptoms were also observed in hemizygous DM300 mice (Seznec et al., 2001). DM1 symptoms were found in many, but not in all mice investigated, and, in addition, also not in all different DM300 lines. This demonstrates that DM1 pathogenesis depends on $(CUG)_n$ repeat expression dosage and on $(CTG)_n$ repeat length (DM20 mice as control, did not show any symptoms).

HSA^{SR}, HSA^{LR}

Unfortunately, no data on triplet repeat instability are available for yet another series of transgenic mice with $(CTG)_n$ repeats, the mice with $(CTG)_5$ or $(CTG)_{250}$ repeats inserted into the middle of the last exon of the human skeletal muscle actin gene. These transgenes were designated HSA^{SR} and HSA^{LR}, respectively, and served to study the effect of expression of an untranslated expanded $(CUG)_n$ repeat in a DM1-unrelated transcript (Mankodi et al., 2000). Several lines, with random integrations of genomic fragments containing the HSA^{SR} or HSA^{LR} assemblies were generated. Some HSA^{LR} lines showed silencing of the transgene; lineage LR32a was analyzed most extensively because the transgene remained highly expressed. Expression was only observed in skeletal muscle, and transcripts were fully spliced and polyadenylated. It is important to note that actin levels were unchanged and that human and mouse actin share the same amino acid sequence, thus excluding effects on actin biology proper. A mortality in HSA^{LR} mice of 41% by 44 weeks was observed (< 5% in HSA^{SR}), the cause of which was unknown. EMG recordings revealed myotonia as early as four weeks of age, when normal skeletal muscle histology was (still) observed. Myopathy – i.e., central nuclei, ring fibers, fiber size variability – developed during ageing, the severity of which was strongly correlated with transgene mRNA levels. As in the DM300 mice of the Gourdon group, discrete nuclear foci were determined with fluorescence in situ hybridization, indicating retention of $(CUG)_n$ containing transcripts in the nucleus. No DM-like symptoms were detected in the HSA^{SR} mice.

In summary, HSA^{LR} mice display many of the characteristic skeletal muscle symptoms of DM1 patients, such as myotonia and muscle wasting, demonstrating a central, toxic role for RNA containing long $(CUG)_n$ repeats in DM1 pathology. The fact that severity of the symptoms strongly correlated with transgene expression levels, and that there was a general lack of typical (non-muscle) features (i.e., cardiac arrythmia, cataract) in tissues where the repeat was not expressed, strongly suggests that cell-type distribution and expression dosage, but not sequence context, of the expanded $(CUG)_n$ repeat is most crucial for development of aspects of the DM1 phenotype. Thus, the results obtained for the HSA^{LR} mice form a strong argument in favour of the RNA dominance mechanism in DM1 pathology. As a point of caution, we should mention, however, that muscle-directed interference with essential cell biological pathways (RNA processing pathways in this case), almost by default will affect muscle functions, producing all kinds of secondary effects. It is therefore of utmost importance to establish whether phenotypic effects in transgenic models truly reflect DM1 pathogenesis in all possible aspects.

Transcription of the $(CTG)_n$ repeats in the Gourdon and Thornton models is driven by human promoters present on the transgenic insert, i.e., from the DMPK and HSA promoters, respectively. It is good to realize that these promoters may be subject to silencing over time due to histone modification and/or DNA methylation, as already demonstrated for some of their lineages by the Thornton group (Mankodi et al., 2000). Epigenetically-driven silencing usually depends on breeding scheme, transgene size and genomic insertion site. Following their expression behavior over many generations in many individual lineages may thus yield highly valuable information on the possible relationship between transcriptional yield and instability, as well as between nuclear and cytosolic $(CUG)_n$ repeat dosage and severity of manifestation. Almost surely, therefore, these mouse models will remain a valuable source of information during many years to come.

$(CTG)_{11}$, $(CTG)_{84}$

Our own group has created a "second generation" series of transgenic knock-in mouse models with humanized DMPK alleles in which the entire 3′-gene region was replaced by the exact cognate human segment, with either a normal $(CTG)_{11}$ repeat or an unstable $(CTG)_{84}$ repeat (van den Broek et al., 2002). Use of a knock-out knock-in gene replacement strategy for introducing the $(CTG)_n$ repeat element has the obvious advantage that there is exact control of the chromosomal position and transcriptional fate of the repeat. In our models, we expected $(CTG)_n$ transcription and replication/repair accessibility to be ruled by the normal activity of the regulatory elements in and around the endogenous *Dmpk* gene and, for transcription, this proved indeed to be the case (Wansink et al., unpublished results). Still, the intergenerational and somatic instability behavior of the $(CTG)_n$ repeats in our models was remarkably similar to that of repeats in the other models. Whereas the $(CTG)_{11}$ repeat was completely stable as anticipated, the $(CTG)_{84}$ repeat slowly expanded during intergenerational segregation (at present $(CTG)_{\sim 110}$ in the F10 generation) and showed profound somatic instability, which was progressive upon aging, in kidney, stomach, small intestine and liver (van den Broek et al., 2002). Within these tissues, there was clear cell-type dependency of instability. In addition, both intergenerational and somatic instability showed a strong mouse strain background effect, indicative of trans-acting modifiers of repeat instability.

Involvement of proteins from the DNA repair and recombination machinery is now a salient feature of many studies that have pointed to the importance of postmitotic events in trinucleotide instability (reviewed in Bowater and Wells, 2001; Kovtun et al., 2001). In our model, introducing the $(CTG)_{84}$ repeat into an *$Msh3^{-/-}$* background, completely blocked somatic repeat instability, whereas *Msh6* deficiency had a stimulating effect on somatic mutation frequency (van den Broek et al., 2002). The first demonstration in animal models came from the group of Manley et al., who showed that absence of the mismatch repair

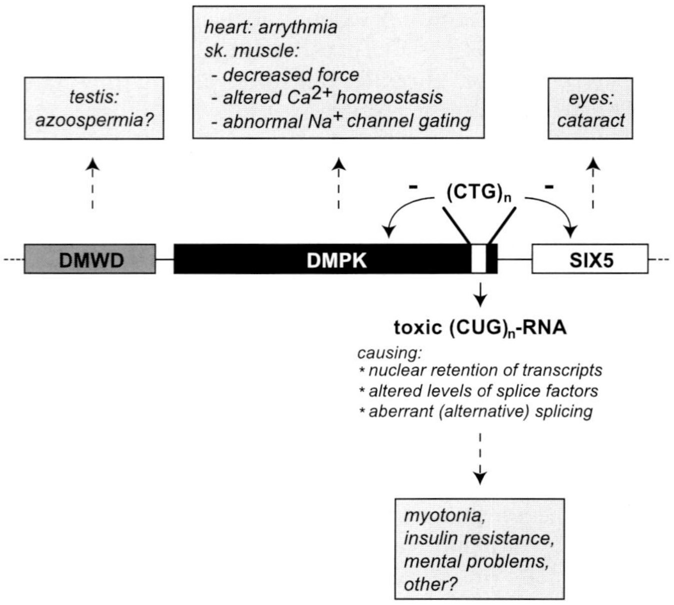

Fig. 2. DM1: consequences of an unstable $(CTG)_n$ repeat: DM1 molecular pathology is caused by an expanded $(CTG)_n$ repeat. This schematic representation summarizes presumed effects of the $(CTG)_n$ repeat in *cis*, via aberrant expression of individual gene products from the DM1 locus (above locus); or in *trans*, via toxic effects of $(CUG)_n$ containing DMPK transcripts (below locus).

core protein Msh2 had a stabilizing effect on the Huntington $(CAG)_n$ repeat in their transgenics (Manley et al., 1999b). Others have demonstrated a role for MMR protein Mlh3 in trinucleotide repeat instability in cell lines (Lipkin et al., 2000). Based on these and other findings, there has been ample speculation about the involvement of multiprotein assemblies that serve in (i) the initiation of origin-of-replication or leading/lagging strand replication during DNA synthesis, or in (ii) DNA repair and DNA recombination mediated restart of stalled or paused replication forks (Hoeijmakers, 2001; Klein and Kreuzer, 2002). Especially events where formation of anomalous DNA structures impair replication fork progression may be relevant to the problem of DNA instability (reviewed in Bowater and Wells, 2001; Galvao et al., 2001; Kovtun et al., 2001). In order to understand their role "in *trans*" in $(CTG)_n$ repeat hypermutation, it thus becomes imperative to obtain a better understanding of the cell-type specificity and stoichiometry of expression of these DNA repair/recombination proteins. As noticed earlier (Table 1; Lia et al., 1998; Fortune et al., 2000; Seznec et al., 2000; van den Broek et al., 2002), it is quite remarkable that kidney, liver and intestinal layers show the highest instability, and it may be not entirely coincidental that these tissues are predominantly of endodermal origin and contain cells that may be more than average exposed to genotoxic influences from the external milieu. Also here, cell and animal models will prove indispensable for studying the putative "*trans*" relationship between repeat instability and the cell-type dependent composition of DNA repair and recombination machinery.

Table 2. Repeat instability in transgenic mouse models. Mouse models with triplet repeat lengths known to be unstable in humans are listed for their (in)stability (either germ line or somatic)

Disease[a]	Triplet repeat	Repeat instability	References
DM1	$(CTG)_{84}$	yes	van den Broek et al., 2002
DM1	$(CTG)_{>300}$	yes	Seznec et al., 2000
DM1	$(CTG)_{55}$	yes	Gourdon et al., 1997; Lia et al., 1998; Seznec et al., 2000
DM1	$(CTG)_{162}$	yes	Fortune et al., 2000; Monckton et al., 1997; Zhang et al., 2002
Fragile X	$(CGG)_{98}$	yes	Bontekoe et al., 2001
Fragile X	$(CGG)_{76 \text{ or } 120}$	no	Lavedan et al., 1998
Fragile X	$(CGG)_{88}$	no	Lavedan et al., 1997
Fragile X	$(CGG)_{81}$	no	Bontekoe et al., 1997
Friedreich ataxia	$(GAA)_{230}$	no	Miranda et al., 2002
Huntington	$(CAG)_{72 \text{ or } 80}$	yes	Kennedy and Shelbourne, 2000; Shelbourne et al., 1999
Huntington	$(CAG)_{110-165}$	yes	Kovtun et al., 2000; Mangiarini et al., 1997; Manley et al., 1999a
Huntington	$(CAG)_{48, 90 \text{ or } 109}$	yes	Wheeler et al., 1999
Huntington	$(CAG)_{77}$	yes	Ishiguro et al., 2001
Huntington	$(CAG)_{44}$	no	Goldberg et al., 1996
SCA1	$(CAG)_{78}$	yes	Lorenzetti et al., 2000
SCA1	$(CAG)_{82}$	yes	Kaytor et al., 1997
SCA3	$(CAG)_{79}$	no	Ikeda et al., 1996
DRPLA	$(CAG)_{76-78}$	yes	Sato et al., 1999
SBMA	$(CAG)_{239}$	yes	Adachi et al., 2001
SBMA	$(CAG)_{97}$	no	Katsuno et al., 2002
SBMA	$(CAG)_{45}$	yes	La Spada et al., 1998
SBMA	$(CAG)_{45}$	no	Bingham et al., 1995

[a] DM1, myotonic dystrophy type 1; SCA, spinocerebellar ataxia; DRPLA, dentatorubropallidoluysian atrophy; SBMA, spinobulbar muscular atrophy. (Adapted from a table supplied by Ben Oostra, with permission.)

Concluding remarks and future outlook

Transgenic mouse models have been very instrumental in unraveling the many possible mechanisms involved in the multisystemic disease pathology of DM1, but clear differences with the situation in human patients remain for many aspects.

The $(CTG)_n$ repeat instability in the mouse animal and cell models now available never attained the level of instability encountered for repeats in somatic and germ line tissues in humans. This may partly be due to the fact that relatively short repeats (55 to 300 CTG triplets) were used, but comparison with many other mouse models for triplet repeat expansion disorders also suggests other possible explanations (Table 2). Steps in DNA repair or recombination involved in repeat expansion (see the reviews by Pearson and Freudenreich elsewhere in this issue) may simply occur at a higher rate in cells of human origin. Transgenic overproduction of DNA handling enzymes combined with knockout approaches will therefore be very informative in attempts to mimic the same level of somatic and meiotic instability as seen in humans. Already now, there is firm belief that the mechanistic aspects of repeat instability in mice and human patients, although different in magnitude, have much in common.

Also with respect to disease manifestation, ideal DM1 models do not yet exist. Evidence is accumulating that the RNA gain-of-function mechanism, exerted via toxic effects on RNA binding proteins may have a central role in DM1 pathology. $(CUG)_n$ repeat-based RNA toxicity may drive inappropriate expression of the insulin receptor, cardiac troponin T, tau isoforms and a muscle-specific chloride channel in muscle tissue, thus providing tantalizing explanations for the insulin resistance, mental problems and myotonia in DM1 patients. Important independent support for the RNA dominance theory comes from the discovery that in DM2 an abnormally expanded $(CCUG)_n$ repeat is expressed in the primary ZNF9 transcript, which accumulates in the nucleus, just as was demonstrated for $(CUG)_n$ repeat bearing DMPK mRNAs. We should not forget, however, that features like mental problems and myotonia could also be caused by distortion of many other cellular functions (including those mediated by other ion-channels, transcription factors). It is therefore imperative, that studies of single candidate genes in animal, cellular or molecular models should not lead to oversimplification, and negligence to study additional candidate gene products. Results obtained with knockout models for candidate loss-of-function genes *Dmpk* and *Six5* have already revealed that these genes are presumably involved in some disease features of DM1. A schematic outline summarizing molecular mechanisms in DM1 disease manifestation is given in Fig. 2. Based on new insight, it now becomes important to develop and study animal models for clinical aspects that are unique to DM1, like the congenital manifestation.

We also need further refinement of studies that draw direct parallels between tissue manifestation of disease in animal models and in DM1 patients. Clearly, one important hurdle that should be overcome soon is that only limited data on the cell-type, developmental profiles, or isoform distributions of RNA and protein products of transgenes and candidate genes are available. Because we have only restricted knowledge on the expression distribution of the $(CTG)_n$ containing DMPK gene and candidate genes from the DM1 locus and elsewhere in the genome, it is intrinsically difficult to draw a direct comparison between $(CUG)_n$ dosage, length distribution profiles and graded severity of disease manifestation in animal models and human patients. Ultimately, a series of mice that produce RNA repeats and protein products in different dosages in a range of cell-types, with expression driven from own regulatory elements or well-defined transgenic promoters, will be needed to unravel the still bewildering complexity of DM1 pathobiology.

Acknowledgements

We thank René van Herpen for critical reading of the manuscript, and Ben Oostra (Erasmus University, Rotterdam, The Netherlands) for help with Table 2.

References

Adachi H, Kume A, Li M, Nakagomi Y, Niwa H, Do J, et al: Transgenic mice with an expanded CAG repeat controlled by the human AR promoter show polyglutamine nuclear inclusions and neuronal dysfunction without neuronal cell death. Hum molec Genet 10:1039–1048 (2001).

Alwazzan M, Newman E, Hamshere MG, Brook JD: Myotonic dystrophy is associated with a reduced level of RNA from the DMWD allele adjacent to the expanded repeat. Hum molec Genet 8:1491–1497 (1999).

Amack JD, Paguio AP, Mahadevan MS: *Cis* and *trans* effects of the myotonic dystrophy (DM) mutation in a cell culture model. Hum molec Genet 8:1975–1984 (1999).

Amack JD, Mahadevan MS: The myotonic dystrophy expanded CUG repeat tract is necessary but not sufficient to disrupt C2C12 myoblast differentiation. Hum Mol Genet 10:1879–1887 (2001).

Amack JD, Reagan SR, Mahadevan MS: Mutant DMPK 3′-UTR transcripts disrupt C2C12 myogenic differentiation by compromising MyoD. J Cell Biol 159:419–429 (2002).

Aslanidis C, Jansen G, Amemiya C, Shutler G, Mahadevan M, Tsilfidis C, et al: Cloning of the essential myotonic dystrophy region and mapping of the putative defect. Nature 355:548–551 (1992).

Batten FE, Gibb HP: Myotonica atrophica. Brain 32:187–205 (1909).

Benders AGM, Groenen PJTA, Oerlemans FTJJ, Veerkamp JH, Wieringa B: Myotonic dystrophy protein kinase is involved in the modulation of the Ca^{2+} homeostasis in skeletal muscle cells. J Clin Invest 100:1440–1447 (1997).

Berul CI, Maguire CT, Aronovitz MJ, Greenwood J, Miller C, Gehrmann J, et al: DMPK dosage alterations result in atrioventricular conduction abnormalities in a mouse myotonic dystrophy model. J Clin Invest 103:R1–R7 (1999).

Bingham PM, Scott MO, Wang S, McPhaul MJ, Wilson EM, Garbern JY, et al: Stability of an expanded trinucleotide repeat in the androgen receptor gene in transgenic mice. Nature Genet 9:191–196 (1995).

Bontekoe CJ, de Graaff E, Nieuwenhuizen IM, Willemsen R, Oostra BA: FMR1 premutation allele $(CGG)_{81}$ is stable in mice. Eur J hum Genet 5:293–298 (1997).

Bontekoe CJ, Bakker CE, Nieuwenhuizen IM, van der Linde H, Lans H, de Lange D, et al: Instability of a $(CGG)_{98}$ repeat in the Fmr1 promoter. Hum molec Genet 10:1693–1699 (2001).

Boucher CA, King SK, Carey N, Krahe R, Winchester CL, Rahman S, et al: A novel homeodomain-encoding gene is associated with a large CpG island interrupted by the myotonic dystrophy unstable $(CTG)_n$ repeat. Hum molec Genet 4:1919–1925 (1995).

Bowater RP, Wells RD: The intrinsically unstable life of DNA triplet repeats associated with human hereditary disorders. Prog Nucl Acid Res Mol Biol 66:159–202 (2001).

Brewster B, Groenen P, Wieringa B: Myotonic dystrophy: clinical and molecular aspects, in Emery AEH (ed.): Neuromuscular Disorders: Clinical and Molecular Genetics, pp 323–364 (John Wiley & Sons New York 1998).

Brock GJ, Anderson NH, Monckton DG: *Cis*-acting modifiers of expanded CAG/CTG triplet repeat expandability: associations with flanking GC content and proximity to CpG islands. Hum molec Genet 8:1061–1067 (1999).

Brook JD, McCurrach ME, Harley HG, Buckler AJ, Church D, Aburatani H, et al: Molecular basis of myotonic dystrophy: expansion of a trinucleotide (CTG) repeat at the 3′ end of a transcript encoding a protein kinase family member. Cell 68:799–808 (1992).

Buj-Bello A, Furling D, Tronchere H, Laporte J, Lerouge T, Butler-Browne GS, et al: Muscle-specific alternative splicing of myotubularin-related 1 gene is impaired in DM1 muscle cells. Hum molec Genet 11:2297–2307 (2002).

Carango P, Noble JE, Marks HG, Funanage VL: Absence of myotonic dystrophy protein kinase (DMPK) mRNA as a result of a triplet repeat expansion in myotonic dystrophy. Genomics 18:340–348 (1993).

Chahine M, George AL Jr: Myotonic dystrophy kinase modulates skeletal muscle but not cardiac voltage-gated sodium channels. FEBS Lett 412:621–624 (1997).

Charlet-B N, Savkur RS, Singh G, Philips AV, Grice EA, Cooper TA: Loss of the muscle-specific chloride channel in type 1 myotonic dystrophy due to misregulated alternative splicing. Mol Cell 10:45–53 (2002).

Davis BM, McCurrach ME, Taneja KL, Singer RH, Housman DE: Expansion of a CUG trinucleotide repeat in the 3′ untranslated region of myotonic dystrophy protein kinase transcripts results in nuclear retention of transcripts. Proc natl Acad Sci, USA 94:7388–7393 (1997).

Eriksson M, Ansved T, Edstrom L, Anvret M, Carey N: Simultaneous analysis of expression of the three myotonic dystrophy locus genes in adult skeletal muscle samples: the CTG expansion correlates inversely with DMPK and 59 expression levels, but not DMAHP levels. Hum molec Genet 8:1053–1060 (1999).

Eriksson M, Ansved T, Edstrom L, Wells DJ, Watt DJ, Anvret M, et al: Independent regulation of the myotonic dystrophy 1 locus genes postnatally and during adult skeletal muscle regeneration. J biol Chem 275:19964–19969 (2000).

Eriksson M, Hedberg B, Carey N, Ansved T: Decreased DMPK transcript levels in myotonic dystrophy 1 type IIA muscle fibers. Biochem biophys Res Commun 286:1177–1182 (2001).

Fardaei M, Rogers MT, Thorpe HM, Larkin K, Hamshere MG, Harper PS, et al: Three proteins, MBNL, MBLL and MBXL, co-localize in vivo with nuclear foci of expanded-repeat transcripts in DM1 and DM2 cells. Hum molec Genet 11:805–814 (2002).

Filippova GN, Thienes CP, Penn BH, Cho DH, Hu YJ, Moore JM, et al: CTCF-binding sites flank CTG/CAG repeats and form a methylation-sensitive insulator at the DM1 locus. Nature Genet 28:335–343 (2001).

Fortune MT, Vassilopoulos C, Coolbaugh MI, Siciliano MJ, Monckton DG: Dramatic, expansion-biased, age-dependent, tissue-specific somatic mosaicism in a transgenic mouse model of triplet repeat instability. Hum molec Genet 9:439–445 (2000).

Franke C, Hatt H, Iaizzo PA, Lehmann-Horn F: Characteristics of Na^+ channels and Cl conductance in resealed muscle fibre segments from patients with myotonic dystrophy. J Physiol 425:391–405 (1990).

Frisch R, Singleton KR, Moses PA, Gonzalez IL, Carango P, Marks HG, et al: Effect of triplet repeat expansion on chromatin structure and expression of DMPK and neighboring genes, SIX5 and DMWD, in myotonic dystrophy. Mol Genet Metab 74:281–291 (2001).

Fu Y-H: Identification of a novel protein, DMAP, which interacts with the myotonic dystrophy protein kinase and shows strong homology to D1 snRNP. Genetica 97:117–125 (1996).

Fu Y-H, Pizzuti A, Fenwick Jr RG, King J, Rajnarayan S, Dunne PW, et al: An unstable triplet repeat in a gene related to myotonic muscular dystrophy. Science 255:1256–1258 (1992).

Fu YH, Friedman DL, Richards S, Pearlman JA, Gibbs RA, Pizzuti A, et al: Decreased expression of myotonin-protein kinase messenger RNA and protein in adult form of myotonic dystrophy. Science 260:235–238 (1993).

Galvao R, Mendes-Soares L, Camara J, Jaco I, Carmo-Fonseca M: Triplet repeats, RNA secondary structure and toxic gain-of-function models for pathogenesis. Brain Res Bull 56:191–201 (2001).

Gennarelli M, Pavoni M, Amicucci P, Angelini C, Menegazzo E, Zelano G, et al: Reduction of the DM-associated homeo domain protein (DMAHP) mRNA in different brain areas of myotonic dystrophy patients. Neuromuscul Disord 9:215–219 (1999).

Godde JS, Wolffe AP: Nucleosome assembly on CTG triplet repeats. J biol Chem 271:15222–15229 (1996).

Goldberg YP, Kalchman MA, Metzler M, Nasir J, Zeisler J, Graham R, et al: Absence of disease phenotype and intergenerational stability of the CAG repeat in transgenic mice expressing the human Huntington disease transcript. Hum molec Genet 5:177–185 (1996).

Gomes-Pereira M, Fortune MT, Monckton DG: Mouse tissue culture models of unstable triplet repeats: in vitro selection for larger alleles, mutational expansion bias and tissue specificity, but no association with cell division rates. Hum molec Genet 10:845–854 (2001).

Gourdon G, Radvanyi F, Lia AS, Duros C, Blanche M, Abitbol M, et al: Moderate intergenerational and somatic instability of a 55-CTG repeat in transgenic mice. Nature Genet 15:190–192 (1997).

Gregory SG, Sekhon M, Schein J, Zhao S, Osoegawa K, Scott CE, et al: A physical map of the mouse genome. Nature 418:743–750 (2002).

Groenen P, Wieringa B: Expanding complexity in myotonic dystrophy. BioEssays 20:901–912 (1998).

Groenen PJTA, Wansink DG, Coerwinkel M, van den Broek W, Jansen G, Wieringa B: Constitutive and regulated modes of splicing produce six major myotonic dystrophy protein kinase (DMPK) isoforms with distinct properties. Hum molec Genet 9:605–616 (2000).

Hamshere MG, Newman EE, Alwazzan M, Balwinder SA, Brook JD: Transcriptional abnormality in myotonic dystrophy affects DMPK but not neighboring genes. Proc natl Acad Sci, USA 94:7394–7399 (1997).

Hanks SK, Hunter T: The eukaryotic protein kinase superfamily: kinase (catalytic) domain structure and classification. FASEB J 9:576–596 (1995).

Harper PS: Myotonic dystrophy (WB Saunders, London 2001).

Harris SE, Winchester CL, Johnson KJ: Functional analysis of the homeodomain protein SIX5. Nucl Acids Res 28:1871–1878 (2000).

Heath SK, Carne S, Hoyle C, Johnson KJ, Wells DJ: Characterisation of expression of mDMAHP, a homeodomain-encoding gene at the murine DM locus. Hum molec Genet 6:651–657 (1997).

Hoeijmakers JH: Genome maintenance mechanisms for preventing cancer. Nature 411:366–374 (2001).

Hofmann-Radvanyi H, Lavedan C, Rabes JP, Savoy D, Duros C, Johnson K, et al: Myotonic dystrophy: absence of CTG enlarged transcript in congenital forms, and low expression of the normal allele. Hum molec Genet 2:1263–1266 (1993).

IDMC: New nomenclature and DNA testing guidelines for myotonic dystrophy type 1 (DM1). Neurology 54:1218–1221 (2000).

Ikeda H, Yamaguchi M, Sugai S, Aze Y, Narumiya S, Kakizuka A: Expanded polyglutamine in the Machado-Joseph disease protein induces cell death in vitro and in vivo. Nature Genet 13:196–202 (1996).

Inukai A, Doyu M, Kato T, Liang Y, Kuru S, Yamamoto M, et al: Reduced expression of DMAHP/SIX5 gene in myotonic dystrophy muscle. Muscle Nerve 23:1421–1426 (2000).

Ishiguro H, Yamada K, Sawada H, Nishii K, Ichino N, Sawada M, et al: Age-dependent and tissue-specific CAG repeat instability occurs in mouse knock-in for a mutant Huntington's disease gene. J Neurosci Res 65:289–297 (2001).

Jansen G, Mahadevan M, Amemiya C, Wormskamp N, Segers B, Hendriks W, et al: Characterization of the myotonic dystrophy region predicts multiple protein isoform-encoding mRNAs. Nature Genet 1:261–266 (1992).

Jansen G, Bachner D, Coerwinkel M, Wormskamp N, Hameister H, Wieringa B: Structural organization and developmental expression pattern of the mouse WD-repeat gene DMR-N9 immediately upstream of the myotonic dystrophy locus. Hum molec Genet 4:843–852 (1995).

Jansen G, Groenen PJ, Bachner D, Jap PH, Coerwinkel M, Oerlemans F, et al: Abnormal myotonic dystrophy protein kinase levels produce only mild myopathy in mice. Nature Genet 13:316–324 (1996).

Katsuno M, Adachi H, Kume A, Li M, Nakagomi Y, Niwa H, et al: Testosterone reduction prevents phenotypic expression in a transgenic mouse model of spinal and bulbar muscular atrophy. Neuron 35:843–854 (2002).

Kawakami K, Ohto H, Takizawa T, Saito T: Identification and expression of six family genes in mouse retina. FEBS Lett 393:259–263 (1996).

Kawakami K, Sato S, Ozaki H, Ikeda K: Six family genes – structure and function as transcription factors and their roles in development. BioEssays 22:616–626 (2000).

Kaytor MD, Burright EN, Duvick LA, Zoghbi HY, Orr HT: Increased trinucleotide repeat instability with advanced maternal age. Hum molec Genet 6:2135–2139 (1997).

Kennedy L, Shelbourne PF: Dramatic mutation instability in HD mouse striatum: does polyglutamine load contribute to cell-specific vulnerability in Huntington's disease? Hum molec Genet 9:2539–2544 (2000).

Kirby RJ, Hamilton GM, Finnegan DJ, Johnson KJ, Jarman AP: Drosophila homolog of the myotonic dystrophy-associated gene, SIX5, is required for muscle and gonad development. Curr Biol 11:1044–1049 (2001).

Klein HL, Kreuzer KN: Replication, recombination, and repair: going for the gold. Mol Cell 9:471–480 (2002).

Klesert TR, Otten AD, Bird TD, Tapscott SJ: Trinucleotide repeat expansion at the myotonic dystrophy locus reduces expression of DMAHP. Nature Genet 16:402–406 (1997).

Klesert TR, Cho DH, Clark JI, Maylie J, Adelman J, Snider L, et al: Mice deficient in *Six5* develop cataracts: implications for myotonic dystrophy. Nature Genet 25:105–109 (2000).

Koob MD, Moseley ML, Schut LJ, Benzow KA, Bird TD, Day JW, et al: An untranslated CTG expansion causes a novel form of spinocerebellar ataxia (SCA8). Nature Genet 21:379–384 (1999).

Korade-Mirnics Z, Babitzke P, Hoffman E: Myotonic dystrophy: molecular windows on a complex etiology. Nucl Acids Res 26:1363–1368 (1998).

Korade-Mirnics Z, Tarleton J, Servidei S, Casey RR, Gennarelli M, Pegoraro E, et al: Myotonic dystrophy: tissue-specific effect of somatic CTG expansions on allele-specific DMAHP/SIX5 expression. Hum molec Genet 8:1017–1023 (1999).

Kovtun IV, Therneau TM, McMurray CT: Gender of the embryo contributes to CAG instability in transgenic mice containing a Huntington's disease gene. Hum molec Genet 9:2767–2775 (2000).

Kovtun IV, Goellner G, McMurray CT: Structural features of trinucleotide repeats associated with DNA expansion. Biochem Cell Biol 79:325–336 (2001).

Krahe R, Ashizawa T, Abbruzzese C, Roeder E, Carango P, Giacanelli M, et al: Effect of myotonic dystrophy trinucleotide repeat expansion on DMPK transcription and processing. Genomics 28:1–14 (1995).

Kwitek AE, Tonellato PJ, Chen D, Gullings-Handley J, Cheng YS, Twigger S, et al: Automated construction of high-density comparative maps between rat, human, and mouse. Genome Res 11:1935–1943 (2001).

La Spada AR, Peterson KR, Meadows SA, McClain ME, Jeng G, Chmelar RS, et al: Androgen receptor YAC transgenic mice carrying CAG 45 alleles show trinucleotide repeat instability. Hum molec Genet 7:959–967 (1998).

Larkin K, Fardaei M: Myotonic Dystrophy – A multigene disorder. Brain Res Bull 56:389–395 (2001).

Lavedan CN, Garrett L, Nussbaum RL: Trinucleotide repeats $(CGG)_{22}TGG(CGG)_{43}TGG(CGG)_{21}$ from the fragile X gene remain stable in transgenic mice. Hum Genet 100:407–414 (1997).

Lavedan C, Grabczyk E, Usdin K, Nussbaum RL: Long uninterrupted CGG repeats within the first exon of the human FMR1 gene are not intrinsically unstable in transgenic mice. Genomics 50:229–240 (1998).

Leeflang EP, Tavare S, Marjoram P, Neal CO, Srinidhi J, MacFarlane H, et al: Analysis of germline mutation spectra at the Huntington's disease locus supports a mitotic mutation mechanism. Hum molec Genet 8:173–183 (1999).

Leung T, Chen X-Q, Manser E, Lim L: The p160 RhoA-binding kinase ROKa is a member of a kinase family and is involved in the reorganization of the cytoskeleton. Mol Cell Biol 16:5313–5327 (1996).

Leung T, Chen X-Q, Tan I, Manser E, Lim L: Myotonic dystrophy kinase-related Cdc42-binding kinase acts as a Cdc42 effector in promoting cytoskeletal reorganization. Mol Cell Biol 18:130–140 (1998).

Lia AS, Seznec H, Hofmann-Radvanyi H, Radvanyi F, Duros C, Saquet C, et al: Somatic instability of the CTG repeat in mice transgenic for the myotonic dystrophy region is age dependent but not correlated to the relative intertissue transcription levels and proliferative capacities. Hum molec Genet 7:1285–1291 (1998).

Lipkin SM, Wang V, Jacoby R, Banerjee-Basu S, Baxevanis AD, Lynch HT, et al: MLH3: a DNA mismatch repair gene associated with mammalian microsatellite instability. Nature Genet 24:27–35 (2000).

Liquori CL, Ricker K, Moseley ML, Jacobsen JF, Kress W, Naylor SL, et al: Myotonic dystrophy type 2 caused by a CCTG expansion in intron 1 of ZNF9. Science 293:864–867 (2001).

Lockington RA, Kelly JM: The WD40-repeat protein CreC interacts with and stabilizes the deubiquitinating enzyme CreB in vivo in *Aspergillus nidulans*. Mol Microbiol 43:1173–1182 (2002).

Lorenzetti D, Watase K, Xu B, Matzuk MM, Orr HT, Zoghbi HY: Repeat instability and motor incoordination in mice with a targeted expanded CAG repeat in the Sca1 locus. Hum molec Genet 9:779–785 (2000).

Lu X, Timchenko NA, Timchenko LT: Cardiac elav-type RNA-binding protein (ETR-3) binds to RNA CUG repeats expanded in myotonic dystrophy. Hum molec Genet 8:53–60 (1999).

Mahadevan M, Amemiya C, Jansen G, Sabourin L, Baird S, Neville C, et al: Structure and genomic sequence of the myotonic dystrophy (DM kinase) gene. Hum molec Genet 2:299–304 (1993).

Mangiarini L, Sathasivam K, Mahal A, Mott R, Seller M, Bates GP: Instability of highly expanded CAG repeats in mice transgenic for the Huntington's disease mutation. Nature Genet 15:197–200 (1997).

Mankodi A, Logigian E, Callahan L, McClain C, White R, Henderson D, et al: Myotonic dystrophy in transgenic mice expressing an expanded CUG repeat. Science 289:1769–1772 (2000).

Mankodi A, Urbinati CR, Yuan QP, Moxley RT, Sansone V, Krym M, et al: Muscleblind localizes to nuclear foci of aberrant RNA in myotonic dystrophy types 1 and 2. Hum molec Genet 10:2165–2170 (2001).

Mankodi A, Takahashi MP, Jiang H, Beck CL, Bowers WJ, Moxley RT, et al: Expanded CUG repeats trigger aberrant splicing of ClC-1 chloride channel pre-mRNA and hyperexcitability of skeletal muscle in myotonic dystrophy. Mol Cell 10:35–44 (2002).

Manley K, Pugh J. Messer A: Instability of the CAG repeat in immortalized fibroblast cell cultures from Huntington's disease transgenic mice. Brain Res 835:74–79 (1999a).

Manley K, Shirley TL, Flaherty L, Messer A: *Msh2* deficiency prevents in vivo somatic instability of the CAG repeat in Huntington disease transgenic mice. Nature Genet 23:471–473 (1999b).

Martorell L, Johnson K, Boucher CA, Baiget M: Somatic instability of the myotonic dystrophy $(CTG)_n$ repeat during human fetal development. Hum molec Genet 6:877–880 (1997).

Matsui T, Amano M, Yamamoto T, Chihara K, Nakafuku M, Ito M, et al: Rho-associated kinase, a novel serine/threonine kinase, as a putative target for the small GTP binding protein Rho. EMBO J 15:2208–2216 (1996).

Meola G: Myotonic dystrophies. Curr Opin Neurol 13:519–525 (2000).

Michalowski S, Miller JW, Urbinati CR, Paliouras M, Swanson MS, Griffith J: Visualization of double-stranded RNAs from the myotonic dystrophy protein kinase gene and interactions with CUG-binding protein. Nucl Acids Res 27:3534–3542 (1999).

Miller JW, Urbinati CR, Teng-Umnuay P, Stenberg MG, Byrne BJ, Thornton CA, et al: Recruitment of human muscleblind proteins to (CUG)(n) expansions associated with myotonic dystrophy. EMBO J 19:4439–4448 (2000).

Miranda CJ, Santos MM, Ohshima K, Smith J, Li L, Bunting M, et al: Frataxin knockin mouse. FEBS Lett 512:291–297 (2002).

Mistry DJ, Moorman JR, Reddy S, Mounsey JP: Skeletal muscle Na currents in mice heterozygous for *Six5* deficiency. Physiol Genomics 6:153–158 (2001).

Monckton DG, Wong LJ, Ashizawa T, Caskey CT: Somatic mosaicism, germline expansions, germline reversions and intergenerational reductions in myotonic dystrophy males: small pool PCR analyses. Hum molec Genet 4:1–8 (1995).

Monckton DG, Coolbaugh MI, Ashizawa KT, Siciliano MJ, Caskey CT: Hypermutable myotonic dystrophy CTG repeats in transgenic mice. Nature Genet 15:193–196 (1997).

Monckton DG, Ashizawa T, Siciliano MJ: Murine models for myotonic dystrophy, in Wells RD, Warren ST (eds): Genetic Instabilities and Hereditary Neurological Diseases, pp 181–193 (Academic Press San Diego 1998).

Mounsey JP, Xu P, John JE 3rd, Horne LT, Gilbert J, Roses AD, et al: Modulation of skeletal muscle sodium channels by human myotonin protein kinase. J Clin Invest 95:2379–2384 (1995).

Mounsey JP, Mistry DJ, Ai CW, Reddy S, Moorman JR: Skeletal muscle sodium channel gating in mice deficient in myotonic dystrophy protein kinase. Hum molec Genet 9:2313–2320 (2000).

Moxley RT: Proximal myotonic myopathy: mini-review of a recently delineated clinical disorder. Neuromuscul Disord 6:87–93 (1996).

Moxley RT, Udd B, Ricker K: 54th ENMC International Workshop: PROMM (proximal myotonic myopathies) and other proximal myotonic syndromes. 10–12th October 1997, Naarden, The Netherlands. Neuromuscul Disord 8:508–518 (1998).

Murakami Y, Ohto H, Ikeda U, Shimada K, Momoi T, Kawakami K: Promoter of mDMAHP/*Six5*: differential utilization of multiple transcription initiation sites and positive/negative regulatory elements. Hum Mol Genet 7:2103–2112 (1998).

Muranyi A, Zhang R, Liu F, Hirano K, Ito M, Epstein HF, et al: Myotonic dystrophy protein kinase phosphorylates the myosin phosphatase targeting subunit and inhibits myosin phosphatase activity. FEBS Lett 493:80–84 (2001).

Nakagawa O, Fujisawa K, Ishizaki T, Saito Y, Nakao K, Narumiya S: ROCK-I and ROCK-II, two isoforms of Rho-associated coiled-coil forming protein serine/threonine kinase in mice. FEBS Lett 392:189–193 (1996).

Narang MA, Waring JD, Sabourin LA, Korneluk RG: Myotonic dystrophy (DM) protein kinase levels in congenital and adult DM patients. Eur J hum Genet 8:507–512 (2000).

Neer EJ, Schmidt CJ, Nambudripad R, Smith TF: The ancient regulatory-protein family of WD-repeat proteins. Nature 371:297–300 (1994).

Novelli G, Gennarelli M, Zelano G, Pizzuti A, Fattorini C, Caskey CT, et al: Failure in detecting mRNA transcripts from the mutated allele in myotonic dystrophy muscle. Biochem molec Biol Int 29:291–297 (1993).

Ohto H, Takizawa T, Saito T, Kobayashi M, Ikeda K, Kawakami K: Tissue and developmental distribution of *Six* family gene products. Int J Dev Biol 42:141–148 (1998).

Otten AD, Tapscott SJ: Triplet repeat expansion in myotonic dystrophy alters the adjacent chromatin structure. Proc natl Acad Sci, USA 92:5465–5469 (1995).

Philips AV, Timchenko LT, Cooper TA: Disruption of splicing regulated by a CUG-binding protein in myotonic dystrophy. Science 280:737–740 (1998).

Ranum LPW, Day JW: Dominantly inherited, non-coding microsatellite expansion disorders. Curr Opin Genet Dev 12:266–271 (2002).

Reddy S, Smith DBJ, Rich MM, Leferovich JM, Reilly P, Davis BM, et al: Mice lacking the myotonic dystrophy protein kinase develop a late onset progressive myopathy. Nature Genet 13:325–335 (1996).

Reddy S, Mistry DJ, Wang QC, Geddis LM, Kutchai HC, Moorman JR, et al: Effects of age and gene dose on skeletal muscle sodium channel gating in mice deficient in myotonic dystrophy protein kinase. Muscle Nerve 25:850–857 (2002).

Roberts R, Timchenko NA, Miller JW, Reddy S, Caskey CT, Swanson MS, et al: Altered phosphorylation and intracellular distribution of a $(CUG)_n$ triplet repeat RNA-binding protein in patients with myotonic dystrophy and in myotonin protein kinase knockout mice. Proc natl Acad Sci, USA 94:13221–13226 (1997).

Sabouri LA, Mahadevan MS, Narang M, Lee DS, Surh LC, Korneluk RG: Effect of the myotonic dystrophy (DM) mutation on mRNA levels of the DM gene. Nature Genet 4:233–238 (1993).

Sarkar PS, Appukuttan B, Han J, Ito Y, Ai C, Tsai W, et al: Heterozygous loss of *Six5* in mice is sufficient to cause ocular cataracts. Nature Genet 25:110–114 (2000).

Sato S, Nakamura M, Cho DH, Tapscott SJ, Ozaki H, Kawakami K: Identification of transcriptional targets for *Six5*: implication for the pathogenesis of myotonic dystrophy type 1. Hum molec Genet 11:1045–1058 (2002).

Sato T, Oyake M, Nakamura K, Nakao K, Fukusima Y, Onodera O, et al: Transgenic mice harboring a full-length human mutant DRPLA gene exhibit age-dependent intergenerational and somatic instabilities of CAG repeats comparable with those in DRPLA patients. Hum molec Genet 8:99–106 (1999).

Savkur RS, Philips A, Cooper TA: Aberrant regulation of insulin receptor alternative splicing is associated with insulin resistance in myotonic dystrophy. Nature Genet 29:40–47 (2001).

Sergeant N, Sablonniere B, Schraen-Maschke S, Ghestem A, Maurage C, Wattez A, et al: Dysregulation of human brain microtubule-associated tau mRNA maturation in myotonic dystrophy type 1. Hum molec Genet 10:2143–2155 (2001).

Seznec H, Lia-Baldini AS, Duros C, Fouquet C, Lacroix C, Hofmann-Radvanyi H, et al: Transgenic mice carrying large human genomic sequences with expanded CTG repeat mimic closely the DM CTG repeat intergenerational and somatic instability. Hum molec Genet 9:1185–1194 (2000).

Seznec H, Agbulut O, Sergeant N, Savouret C, Ghestem A, Tabti N, et al: Mice transgenic for the human myotonic dystrophy region with expanded CTG repeats display muscular and brain abnormalities. Hum molec Genet 10:2717–2726 (2001).

Shaw DJ, McCurrach M, Rundle SA, Harley HG, Crow SR, Sohn R, et al: Genomic organization and transcriptional units at the myotonic dystrophy locus. Genomics 18:673–679 (1993).

Shelbourne PF, Killeen N, Hevner RF, Johnston HM, Tecott L, Lewandoski M, et al: A Huntington's disease CAG expansion at the murine *Hdh* locus is unstable and associated with behavioural abnormalities in mice. Hum molec Genet 8:763–774 (1999).

Smith TF, Gaitatzes C, Saxena K, Neer EJ: The WD repeat: a common architecture for diverse functions. Trends Biochem Sci 24:181–185 (1999).

Steinbach P, Glaser D, Vogel W, Wolf M, Schwemmle S: The DMPK gene of severely affected myotonic dystrophy patients is hypermethylated proximal to the largely expanded CTG repeat. Am J hum Genet 62:278–285 (1998).

Steinert H: Myopathologische Beiträge 1. Über das klinische und anatomische Bild des Muskelschwunds der Myotoniker. Dtsch Z Nervenheilkd 37:58–104 (1909).

Suzuki A, Sugiyama Y, Hayashi Y, Nyu-i N, Yoshida M, Nonaka I, et al: MKBP, a novel member of the small heat shock protein family, binds and activates the myotonic dystrophy protein kinase. J Cell Biol 140:1113–1124 (1998).

Taneja KL, McCurrach M, Schalling M, Housman D, Singer RH: Foci of trinucleotide repeat transcripts in nuclei of myotonic dystrophy cells and tissues. J Cell Biol 128:995–1002 (1995).

Tapscott SJ, Thornton CA: Reconstructing myotonic dystrophy. Science 293:816–817 (2001).

Thornton CA, Johnson K, Moxley RT: Myotonic dystrophy patients have larger CTG expansions in skeletal muscle than in leukocytes. Ann Neurol 35:104–107 (1994).

Thornton CA, Wymer JP, Simmons Z, McClain C, Moxley RT: Expansion of the myotonic dystrophy CTG repeat reduces expression of the flanking DMAHP gene. Nature Genet 16:407–409 (1997).

Tian B, White RJ, Xia T, Welle S, Turner DH, Mathews MB, et al: Expanded CUG repeat RNAs form hairpins that activate the double-stranded RNA-dependent protein kinase PKR. RNA 6:79–87 (2000).

Timchenko L, Nastainczyk N, Schneider T, Patel B, Hofmann F, Caskey CT: Full-length myotonin protein kinase (72 kDa) displays serine kinase activity. Proc natl Acad Sci, USA 92:5366–5370 (1995).

Timchenko LT, Miller JW, Timchenko NA, DeVore DR, Datar KV, Lin L, et al: Identification of a $(CUG)_n$ triplet repeat RNA-binding protein and its expression in myotonic dystrophy. Nucl Acids Res 24:4407–4414 (1996).

Tiscornia G, Mahadevan MS: Myotonic dystrophy: the role of the CUG triplet repeats in splicing of a novel DMPK exon and altered cytoplasmic DMPK mRNA isoform ratios. Mol Cell 5:959–967 (2000).

Ueda H, Ohno S, Kobayashi T: Myotonic dystrophy and myotonic dystrophy protein kinase. Prog Histochem Cytochem 35:187–251 (2000).

van den Broek WJ, Nelen MR, Wansink DG, Coerwinkel MM, te Riele H, Groenen PJ, et al: Somatic expansion behaviour of the $(CTG)_n$ repeat in myotonic dystrophy knock-in mice is differentially affected by Msh3 and Msh6 mismatch-repair proteins. Hum molec Genet 11:191–198 (2002).

Vermersch P, Sergeant N, Ruchoux MM, Hofmann-Radvanyi H, Wattez A, Petit H, et al: Specific tau variants in the brains of patients with myotonic dystrophy. Neurology 47:711–717 (1996).

Wang J, Pegoraro E, Menegazzo E, Gennarelli M, Hoop RC, Angelini C, et al: Myotonic dystrophy: evidence for a possible dominant-negative RNA mutation. Hum molec Genet 4:599–606 (1995).

Wang YH, Amirhaeri S, Kang S, Wells RD, Griffith JD: Preferential nucleosome assembly at DNA triplet repeats from the myotonic dystrophy gene. Science 265:669–671 (1994).

Wang YH, Griffith J: Expanded CTG triplet blocks from the myotonic dystrophy gene create the strongest known natural nucleosome positioning elements. Genomics 25:570–573 (1995).

Wheeler VC, Auerbach W, White JK, Srinidhi J, Auerbach A, Ryan A, et al: Length-dependent gametic CAG repeat instability in the Huntington's disease knock-in mouse. Hum molec Genet 8:115–122 (1999).

Winchester CL, Ferrier RK, Sermoni A, Clark BJ, Johnson KJ: Characterization of the expression of DMPK and SIX5 in the human eye and implications for pathogenesis in myotonic dystrophy. Hum molec Genet 8:481–492 (1999).

Wong LJ, Ashizawa T, Monckton DG, Caskey CT, Richards CS: Somatic heterogeneity of the CTG repeat in myotonic dystrophy is age and size dependent. Am J hum Genet 56:114–122 (1995).

Zhang Y, Monckton DG, Siciliano MJ, Connor TH, Meistrich ML: Age and insertion site dependence of repeat number instability of a human DM1 transgene in individual mouse sperm. Hum molec Genet 11:791–798 (2002).

Cytogenet Genome Res 100:243–251 (2003)
DOI: 10.1159/000072860

Transgenic mouse models of spinal and bulbar muscular atrophy (SBMA)

M. Katsuno, H. Adachi, A. Inukai and G. Sobue

Department of Neurology, Nagoya University Graduate School of Medicine, Nagoya (Japan)

Abstract. Spinal and bulbar muscular atrophy (SBMA) is a late-onset motor neuron disease characterized by proximal muscle atrophy, weakness, contraction fasciculations, and bulbar involvement. Only males develop symptoms, while female carriers usually are asymptomatic. A specific treatment for SBMA has not been established. The molecular basis of SBMA is the expansion of a trinucleotide CAG repeat, which encodes the polyglutamine (polyQ) tract, in the first exon of the androgen receptor (AR) gene. The pathologic hallmark is nuclear inclusions (NIs) containing the mutant and truncated AR with expanded polyQ in the residual motor neurons in the brainstem and spinal cord as well as in some other visceral organs. Several transgenic (Tg) mouse models have been created for studying the pathogenesis of SBMA. The Tg mouse model carrying pure 239 CAGs under human AR promoter and another model carrying truncated AR with expanded CAGs show motor impairment and nuclear NIs in spinal motor neurons. Interestingly, Tg mice carrying full-length human AR with expanded polyQ demonstrate progressive motor impairment and neurogenic pathology as well as sexual difference of phenotypes. These models recapitulate the phenotypic expression observed in SBMA. The ligand-dependent nuclear localization of the mutant AR is found to be involved in the disease mechanism, and hormonal therapy is suggested to be a therapeutic approach applicable to SBMA.

Numerous animal models of neurodegenerative diseases have been created to elucidate the molecular pathogenesis and develop therapeutic approaches. Some transgenic mouse models of SBMA, the first polyglutamine (polyQ) disease to be discovered, recapitulate the symptoms and pathologic features of the disease, and provide insight into its pathogenesis. Moreover, with our latest model, we developed an effective treatment for SBMA. Here we review transgenic mouse models of SBMA, and discuss the therapeutic strategies against SBMA as well as other polyQ diseases.

Supported by a Center-of-Excellence (COE) grant from the Ministry of Education, Culture, Sports, Science and Technology, Japan, and grants from the Ministry of Health, Labor and Welfare, Japan.

Received 30 September 2002; accepted 30 January 2003.

Request reprints from Gen Sobue, MD, PhD, Department of Neurology Nagoya University Graduate School of Medicine, 65 Tsurumai-cho Showa-ku, Nagoya, 466-8550 (Japan); telephone: +81-52-744-2385 fax: +81-52-744-2384; e-mail: sobueg@med.nagoya-u.ac.jp

Spinal and bulbar muscular atrophy (SBMA)

SBMA, also known as Kennedy's disease, is a late-onset motor neuron disease characterized by proximal muscle atrophy, weakness, contraction fasciculations, and bulbar involvement (Kennedy et al., 1968; Sobue et al., 1989). Lower motor neurons are markedly depleted through all spinal segments and in brainstem motor nuclei except for the third, fourth and sixth cranial nerves (Sobue et al., 1989). Sensory neurons in the dorsal root ganglia are less severely affected, and there is also a distally accentuated sensory axonopathy in the peripheral nervous system. This disease affects males, and male patients often show signs of androgen insensitivity such as gynecomastia, testicular atrophy, and decreased fertility (Fig. 1). Female carriers are usually asymptomatic, although some express subclinical phenotypes, including high amplitude motor unit potentials on electromyography (Sobue et al., 1993; Mariotti et al., 2000; Schmidt et al., 2002). No specific treatment for SBMA has been established. Testosterone may improve motor function in some patients, although it has no effect on the progression of SBMA (Danek et al., 1994; Goldenberg et al., 1996; Neuschmid-Kaspar et al., 1996).

KARGER Fax + 41 61 306 12 34
E-mail karger@karger.ch
www.karger.com

0301–0171/03/1004–0243$19.50/0

Accessible online at:
www.karger.com/cgr

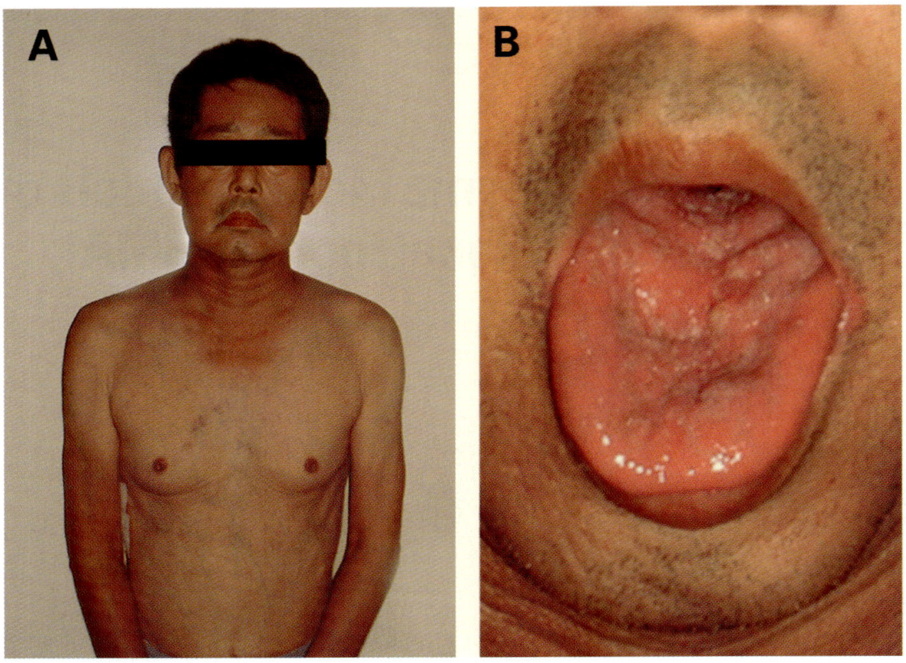

Fig. 1. SBMA patient. SBMA patients show muscle atrophy of the upper arms and gynecomastia (**A**) as well as tongue atrophy (**B**).

The molecular basis of SBMA is the expansion of a trinucleotide CAG repeat, which encodes the polyQ tract, in the first exon of the androgen receptor (AR) gene (La Spada et al., 1991). The CAG repeat within AR ranges in size from 5 to 33 repeats in normal subjects, but from 40 to 62 in SBMA patients (Tanaka et al., 1996, Merry, 2001). Expanded polyQ tracts have been found to cause several neurodegenerative diseases, including SBMA, Huntington's disease (HD), several forms of spinocerebellar ataxia, and dentatorubral and pallidoluysian atrophy (DRPLA) (Zoghbi and Orr, 2000). These disorders, known as polyQ diseases, share salient clinical features such as anticipation, somatic mosaicism (Tanaka et al., 1999), and selective neuronal and non-neuronal involvement despite widespread expression of the mutant gene (Doyu et al., 1994). There is also an inverse correlation between the CAG repeat size and the age at onset, or the disease severity adjusted by the age at examination in SBMA (Doyu et al., 1992; La Spada et al., 1992) as well as other polyQ diseases (Zoghbi and Orr, 2000).

Previously we reported nuclear inclusions (NIs) containing the mutant and truncated AR with expanded polyQ in the residual motor neurons in the brain stem and spinal cord (Li et al., 1998a) as well as in the skin, testis and some other visceral organs of SBMA patients (Li et al., 1998b). These inclusions had similar epitope features detectable by antibodies that recognize a small portion of the N-terminus of the AR protein only, and they were ubiquitinated. The presence of NIs is a pathologic hallmark of most other polyQ diseases, and is likely to be related to pathogenesis (Zoghbi and Orr, 2000). Although considerable controversy surrounds the importance of NIs in the pathogenesis of the polyQ diseases (Tobin and Signer, 2000), several studies have implied that nuclear localization of mutant protein is essential for inducing neuronal cell dysfunction and degeneration in the majority of polyQ diseases (Ross 2002). Although molecular mechanisms still remain to be elucidated, several studies have revealed that nuclear accumulation of the mutant protein resulted in sequestration of transcriptional regulatory proteins in polyQ diseases (Steffan et al., 2000; Nucifora et al., 2001).

SBMA is unique among polyQ diseases in that the mutant protein, AR, has a specific ligand, testosterone, which alters the subcellular localization of the protein by favoring its nuclear uptake. No ligands of causative protein have been found in other polyQ diseases. AR is normally confined to a multi-heteromeric inactive complex in the cell cytoplasm, and translocates into the nucleus in a ligand-dependent manner (Zhou et al., 1994; Zhou et al., 1995). This intracellular trafficking of AR may play an important role in the pathogenesis of SBMA.

Early mouse models of SBMA failed to show phenotypes

All of the available animal models of SBMA are transgenic mice with truncated or full-length human AR (Table 1). Transgenic mice were first created using the full-length AR containing 45 CAGs, which is equivalent to the repeat length observed in SBMA patients, driven by the interferon-inducible Mx promoter or the neuron-specific enolase (NSE) promoter (Bingham et al., 1995). Expression of mutant AR was found in mice with the inducible Mx promoter, but at a lower level than normal endogenous expression. The mice demonstrated neither phenotype nor repeat length instability.

Another transgenic mouse model was created with yeast artificial chromosomes (YACs) carrying the AR gene in the context of flanking non-coding sequences (La Spada et al., 1998). Studies on independent lines of AR YAC transgenic mice carrying 45 CAGs revealed intergenerational instability, which was greater with maternal transmission and with age of the transmitting mother. However, this model failed to show

Table 1. Comparison between transgenic mouse lines and SBMA patients

Reference	Transgene construct	CAG repeat instability	Motor impairment		Neuropathology		Muscle pathology
			Symptoms	Gender effect	Nuclear inclusions	Cell loss	
Bingham et al., 1995	Mx or NSE promoter, full-length human AR with 45 CAGs	(–)	(–)	(–)	(–)	(–)	(–)
LaSpada et al., 1998	YAC transgenic full-length human AR, with 45 CAGs	mild	(–)	(–)	(–)	(–)	(–)
Merry et al., 1996	NSE or NFL promoter, full-length human AR with 66 CAGs	(–)	(–)	(–)	(–)	(–)	(–)
Adachi et al., 2001	human AR promoter, sole 239 CAGs	mild	weakness, amyotrophy, incoordination	(–)	spinal cord, cerebrum cerebellum	(–)	(–)
Abel et al., 2001	NFL promoter, truncated human AR with 112 CAGs	(–)	weakness, foot clasping	(–)	spinal cord, cerebrum brainstem	(–)	(–)
Abel et al., 2001	PrP promoter, truncated human AR with 112 CAGs	(–)	hypoactivity, foot clasping, tremor, seizure	(–)	all neurons	(–)	(–)
Katsuno et al., 2002	chicken ß-actin promoter, full-length human AR with 97 CAGs	(–)	weakness, amyotrophy	significant	spinal cord, cerebrum brainstem	(–)	grouped atrophy
McManamny et al., 2002	cytomegalovirus promoter, full-length human AR with 120 CAGs	(–)	weakness, amyotrophy, foot clasping	mild	(–)	(+)	grouped atrophy fiber-type grouping hypertrophic fiber
Kennedy et al., 1968 Sobue et al., 1989	SBMA patients	mild	weakness, amyotrophy, fasciculation	significant	spinal cord, brainstem	(+)	grouped atrophy fiber-type grouping hypertrophic fiber

the expression of mutant AR in RT-PCR or Western blot analysis.

In order to enhance the toxicity of mutant AR, a transgenic mouse model was created with human AR containing 66 CAGs, which was longer than the longest repeat observed in SBMA patients, driven by the NSE promoter or the neurofilament light chain (NFL) promoter (Merry et al., 1996). Although expression levels of mutant AR were 2–5 times the endogenous AR levels, these mice showed no neurologic symptoms, presumably because the CAG repeat was not long enough.

Transgenic mouse models with truncated AR recapitulate neurologic symptoms

Since mouse models of SBMA could not be created using full-length AR, truncated AR with an expanded CAG repeat was used in later models. This strategy was based on the results reported on transgenic mice of Huntington disease and Machado-Joseph disease, where truncated protein had a particularly pronounced effect (Ikeda et al., 1996; Mangiarini et al., 1996). In SBMA and other polyQ diseases, NIs are detected by antibodies against an N-terminal epitope, but not by antibodies against a C-terminal epitope (Li et al., 1998a, b; DiFiglia et al., 1997). These findings suggest that truncated polyQ-containing proteins confer the toxicity in polyQ diseases. It should be noted that truncated mutant AR is more toxic than the full-length AR in an SBMA cell model (Merry et al., 1998; Kobayashi et al., 2000). Additionally, in vitro translated full-length AR protein with an expanded polyQ tract is cleaved by caspase-3, liberating a polyQ-containing fragment, and the susceptibility to cleavage is polyQ repeat length-dependent (Kobayashi et al., 1998). Thus, cleavage of polyQ-containing proteins is likely to contribute the toxicity of polyQ tracts.

Not like previous models, a transgenic mouse model carrying 239 CAGs driven by the human AR promoter demonstrated motor impairment and revealed that the polyQ tract is sufficient to induce the pathogenic process of SBMA (Adachi et al., 2001). Symptomatic lines exhibited small body size, weakness, truncal and limb incoordination, reduced activity and short lifespan. These phenotypes apparently developed within 4–8 weeks of birth in both lines, and gradually became severe at 8–16 weeks. Impairment of the rotarod task appeared within 8 weeks of birth, and the subsequent progression was monophasic. The most striking pathologic observation was widespread occurrence of NIs, which were distributed in the neuronal cell nuclei in the cerebrum, cerebellum, brainstem and spinal cord, but to a lesser extent or not at all in the basal ganglia. This pathologic distribution was limited by the promoter used, and was more widely spread than that of human SBMA. In the regions with their marked occurrence, the NIs were also prominently observed in the nuclei of the glial cells. The electron microscopic structure of NIs was very similar to that observed in SBMA patients; granular materials densely aggregate without a limiting membrane or filamentous materials. These NIs were positive for ubiquitin and colocalized with proteasome components (20s, PA28α, and PA28γ). Despite abundant NIs, there was no evidence of active neuronal degeneration or reactive astrogliosis. Thus neuronal dysfunction, rather than neuronal degeneration, is likely to be the pathogenesis of this mouse model. Expression of the transgene assessed by RT-PCR was revealed in the cerebrum, cerebellum, spinal cord, pituitary, lung, eye and skin; its distribution was compatible to that of NIs as well as mouse AR distribution. The mice showed subtle meiotic instability of the CAG repeat, in agreement with SBMA.

Simultaneously, another SBMA mouse model was created with truncated human AR containing 112 CAGs, driven by the

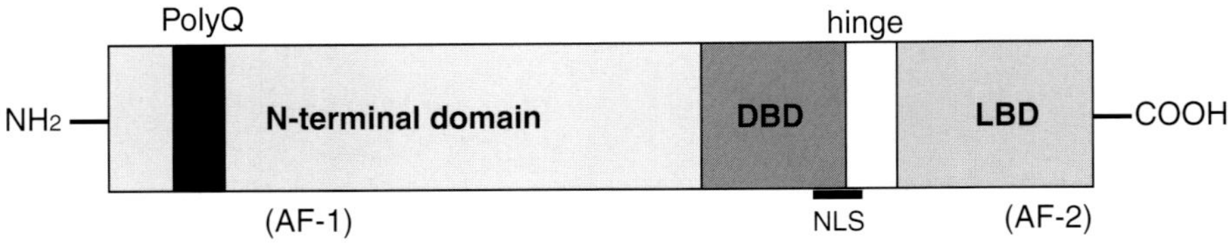

Fig. 2. Structure of AR protein. The AR protein consists of three major domains: N-terminal transactivating, DNA-binding, and ligand-binding domains. The polyglutamine tract is located in an N-terminal domain, which possesses major transactivating function (AF-1). The ligand binding domain (LBD) in the C-terminus also contains weak transactivating function (AF-2). The nuclear localization signal (NLS) in the DNA-binding domain (DBD) and hinge lesion is unmasked upon ligand-binding.

NFL or prion protein promoter (Abel et al., 2001). The mice with the prion protein promoter showed non-specific features such as tremor, seizure and loss of body weight, whereas those with the NFL promoter demonstrated motor impairment similar to SBMA patients, accompanied by upper motor deficits. Widely expressed mutant AR may account for these neurological phenotypes, because the distribution of histopathological involvement, which depended on the expression level of the transgene, was more extensive than that of SBMA. Immunohistochemical analysis of AR expression revealed transgenic AR-positive nuclear inclusions in isolated neurons in several restricted regions of the central nervous system including the brainstem and the cortex and at lower frequency in spinal cord motor neurons. NIs were ubiquitinated and contained several molecular chaperones, including HDJ-2 and Hsc70. About half of the NIs were positive for CREB-binding protein, a transcriptional coactivator (McCampbell et al., 2000). In spite of neurologic symptoms and NIs, neither neuronal loss nor neurogenic muscle atrophy was demonstrated in this model.

The fact that neuronal loss is not evident despite abundant NIs has been demonstrated in various mouse models of other polyQ diseases (Rubinsztein, 2002). Formation of NIs preceded cerebellar ataxia in a transgenic mouse model of SCA1 (Burright et al., 1995). Loss of cerebellar Purkinje cells, however, was not revealed until the symptoms progressed in this model. The life span of these model mice may not be long enough to demonstrate neuronal cell loss, although some other models have shown neuronal degeneration (Reddy et al., 1998; Hodgson et al., 1999; McManamny et al., 2002). These findings suggest that the pathogenesis of human polyQ disease could be neuronal dysfunction rather than neuronal loss in early stages.

Symptomatic full-length model provides therapeutic approach to SBMA

Unlike the profound gender difference of phenotypes in SBMA patients, neither a Tg mouse model of SBMA expressing expanded pure 239 CAGs (Adachi et al., 2001) nor another model carrying truncated AR with 112 CAGs (Abel et al., 2001) showed any remarkable phenotypic difference with gender, because the transgenes of these Tg mice did not contain the ligand-binding domain located in the C-terminus of AR (Fig. 2).

Recently, we generated Tg mice expressing the full-length human AR containing 24 or 97 CAGs under the control of the cytomegalovirus enhancer and the chicken β-actin promoter (Katsuno et al., 2002). This model recapitulated not only the neurologic disorder, but also the phenotypic difference with gender which is a specific feature of SBMA. Three out of five lines with 97 CAGs (AR-97Q) exhibited progressive motor impairment, although no lines with 24 CAGs showed any manifested phenotypes. All symptomatic lines showed small body size, muscle atrophy, weakness, reduced activity and short life-span; all of which were markedly pronounced and accelerated in the male AR-97Q mice, but either not observed or far less severe in the female AR-97Q mice regardless of the line. Early mortality of male mice, which is not common in SBMA, appears to be caused by cachexia due to progressive amyotrophy. Western blot analysis revealed the transgenic protein retained in the stacking gel in all symptomatic lines. We detected these proteins in the spinal cord, cerebrum, heart, muscle and pancreas. Although the male AR-97Q mice had more protein within the stacking gel than their female counterparts, the female AR-97Q mice had more monomeric AR protein. The nuclear fraction contained most of the mutant AR within the stacking gel. Despite the profound sexual difference of the mutant AR protein expression, there was no significant difference in the expression of the transgene mRNA between the male and female AR-97Q mice. These observations indicate that the testosterone level plays important roles in the sexual difference of phenotypes, especially in the post-transcriptional stage of the mutant AR.

In the AR-97Q mice, we detected diffuse nuclear staining and less frequent NIs with 1C2, an antibody specifically recognizing the expanded polyQ, in the neurons of the spinal cord, cerebrum, cerebellum, brainstem and dorsal root ganglia as well as non-neuronal tissue such as the heart, muscle and pancreas. The regions with diffuse nuclear staining and NIs also showed immunoreactivity to an antibody to AR (N-20). Neither 1C2 nor N-20 revealed immunoreactivity in the cytoplasm. The male AR-97Q mice showed markedly more abundant diffuse nuclear staining and NIs than females, in agreement with the symptomatic and Western blot profile differences with gender. Muscle histology revealed significant grouped atrophy and small angulated fibers in the male AR-97Q mice as well as mild myopathic change. Although neuronal cell loss was not evident in the spinal cord, morphologic studies

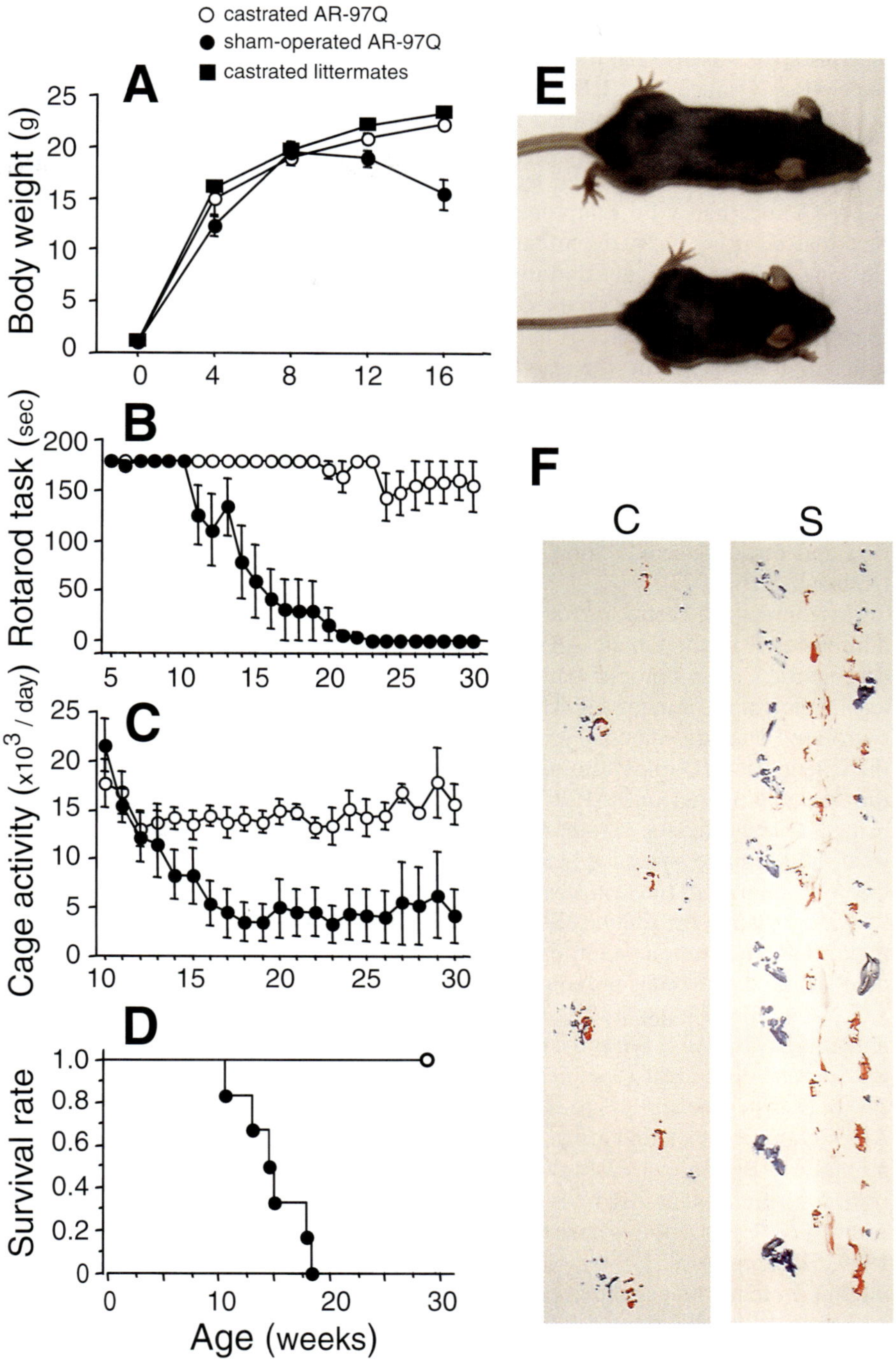

Fig. 3. Effects of castration on symptomatic phenotypes of male AR-97Q mice. (**A**, **B**, **C** and **D**). Body weight (**A**), rotarod task (**B**), cage activity (**C**), and survival rate (**D**) of castrated (í, $n = 6$) and sham-operated (λ, $n = 6$) male AR-97Q mice. All parameters are significantly different between sham-operated male AR-97Q mice and castrated male AR-97Q mice or castrated male littermates (■, $n = 2$) ($P = 0.0001$, $P < 0.0001$, $P = 0.006$, and $P = 0.0006$, respectively). (**E**) A castrated AR-97Q mouse (top) shows no muscular atrophy, which is striking in a sham-operated male AR-97Q mouse (bottom) (#2-6, 12-week-old). (**F**) Footprints of 12-week-old castrated (C) and sham-operated (S) male AR-97Q mice. Front paws are in red, and hind paws in blue.

revealed axonal atrophy of the ventral nerve root as well as shrinkage of the spinal motor neurons. The female AR-97Q mice showed no neurogenic changes. The neuronal cell populations in the cerebrum, cerebellum and dorsal root ganglia were also fairly well preserved despite the abundant diffuse nuclear staining and NIs. These results indicate that dysfunction of the lower motor neurons contributes to the motor impairment in this model, and this dysfunction is far more severe in the male Tg mice than in the females.

Because of such dramatic sexual difference of phenotype, we examined the effects of hormonal interventions in this mouse model. Castrated male AR-97Q mice, which had a decreased testosterone level, showed significant improvement of symptoms, pathologic findings, and nuclear localization of the mutant AR compared with the sham-operated male AR-97Q mice. The castrated male AR-97Q mice weighed the same as their castrated male littermates, whereas the sham-operated male AR-97Q mice showed progressive emaciation (Fig. 3A). Motor impairment assessed by rotarod and cage activity was significantly less or virtually absent in the castrated male AR-97Q mice as compared with the sham-operated male AR-97Q mice (Fig. 3B and C). The castrated male AR-97Q mice showed motor impairment similar to that of the female AR-97Q mice. The life span was also significantly prolonged in the castrated male AR-97Q mice (Fig. 3D). Castration ameliorated muscle atrophy and body size reduction (Fig. 3E). In a foot print analysis, the sham-operated male AR-97Q mice exhibited motor weakness with dragging of their legs, while the castrated male AR-97Q mice had mild symptoms (Fig. 3F). In the Western blot analysis using N-20, the mutant AR appearing within the stacking gel was markedly diminished in the castrated male AR-97Q mice compared with the sham-operated male AR-97Q mice (Fig. 4A). The mutant AR in the nuclear fraction also significantly decreased in the castrated male AR-97Q mice (Fig. 4B). Castration markedly diminished diffuse nuclear staining and NIs (Fig. 4C). These observations suggested that castration markedly prevented nuclear localization of the mutant AR protein. Leuprorelin, an LHRH agonist that reduces testosterone release from the testis, also rescued motor dysfunction and nuclear accumulation of mutant AR in this model (Katsuno et al., 2003).

In contrast to castration of the male mice, testosterone administered to the female AR-97Q mice caused significant exacerbation of symptoms, pathologic features, and nuclear localization of the mutant AR. The motor impairment assessed by rotarod and cage activity was significantly exacerbated in the female AR-97Q mice administered testosterone, similar to the case of untreated male AR-97Q mice. Western blot analysis using N-20 revealed the mutant AR in the stacking gel in whole tissue homogenates as well as in the nuclear fraction, which was larger in amount in the testosterone-administered female AR-97Q mice than in the sesame oil-administered female AR-97Q mice. The testosterone-administered female AR-97Q mice demonstrated markedly pronounced diffuse nuclear staining and NIs with 1C2 compared with the sesame oil-administered female AR-97Q mice. Since the nuclear translocation of AR is testosterone-dependent (Stenoien et al., 1999; Simeoni et al., 2000), testosterone appears to show toxic effects in the female AR-97Q mice by accelerating nuclear translocation of the mutant AR. By contrast, castration prevented the nuclear localization of the mutant AR by reducing the testosterone level. Although a few exceptions have been reported (Hodgson et al., 1999; Hackam et al., 1999), the nuclear localization of the mutant protein with expanded polyQ is important in inducing neuronal cell dysfunction and degeneration in the majority of polyQ diseases. Addition of a nuclear export signal to the mutant huntingtin eliminated aggregate formation and cell death in cell models of HD (Saudou et al., 1998; Peters et al., 1999), whereas a nuclear localization signal had the opposite effect (Peters et al., 1999). In Tg mice of SCA1 having a mutated nuclear localization signal, ataxin-1 was distributed in the cytoplasm, and the mice did not show any neurologic disorders (Klement et al., 1998). These findings suggest that reduction of testosterone ameliolates phenotypic expression by preventing nuclear localization of the mutant AR, while castration may enhance the protective effects of heat shock proteins (HSPs), which are normally associated with AR and dissociate upon ligand binding.

The castrated AR-97Q mice showed phenotypes similar to those of the female AR-97Q mice, implying that motor impairment in SBMA patients can be reduced to the level in female carriers. SBMA has been considered to be an X-linked disease, whereas other polyQ diseases show autosomal dominant inheritance. In fact, SBMA female carriers hardly manifest clinical phenotypes, although they possess similar numbers of a CAG repeat in the disease allele as their siblings with SBMA (Sobue et al., 1993; Mariotti et al., 2000). Indeed the lower level of mutant AR expression in female carriers due to X inactivation may cause the escape from the manifestation, but our present study also strongly suggests that the low level of testosterone prevents the nuclear localization of the expressed mutant AR, resulting in a lack of phenotypic manifestations in the female carriers. This hypothesis is supported by the finding that manifestation of symptoms is minimal even in homozygous females of SBMA (Schmidt et al., 2002). The testosterone-dependent neurodegenaration was also recently revealed in a fly model of SBMA (Takeyama et al., 2002). Thus, hormonal intervention to diminish testosterone level appears to be applicable to human therapy.

Another Tg mouse model carrying the full-length AR with 120 CAGs driven by the cytomegalovirus promoter showed slowly progressive motor impairment and neurogenic muscle atrophy (McManamny et al., 2002). The affected mice also displayed a progressive reduction in sperm production consistent with androgen insensitivities in SBMA patients. Although this model showed no neuronal inclusions throughout the nervous system, loss of motor neurons was demonstrated in the spinal cord. Notably, as expected, mild but evident sexual difference of phenotypes was observed. This finding supports the hypothesis that testosterone level is implicated in the phenotypic expression of SBMA. The above study and ours indicate that the C-terminus of mutant AR is necessary to recapitulate the testosterone-dependent pathogenesis of SBMA in mouse models.

"Loss of function" model of SBMA

As in other autosomal dominant polyQ diseases, a toxic gain of the mutant AR function has been considered the mainstream of SBMA pathogenesis. Although the expansion of polyQ tract in AR deteriorates the transcriptional activities of AR (Mhatre et al., 1993; Brooks et al., 1997), motor impairment has not been observed in severe testicular feminization (Tfm) patients lacking AR function (Gottlieb et al., 1999). Thus, the neurologic impairment in SBMA cannot be attributed to the loss of AR function (Maclean et al., 1995), a reason why testosterone shows insufficient and transient effects when used as a therapeutic agent for SBMA (Danek et al., 1994; Goldenberg et al., 1996; Neuschmid-Kaspar et al., 1996). Animal models of Tfm demonstrate a decreased number of motor neurons in the spinal nucleus of the bulbocavernosus (Sengelaub et al., 1989) without any other motor neuron involvement.

However, re-examination of earlier studies and recent data has revealed that loss of normal protein function also plays a role in the neuronal involvement in polyQ diseases. In a huntingtin knockout mouse model, the heterozygotes demonstrated motor impairment and basal ganglia degeneration (Na-

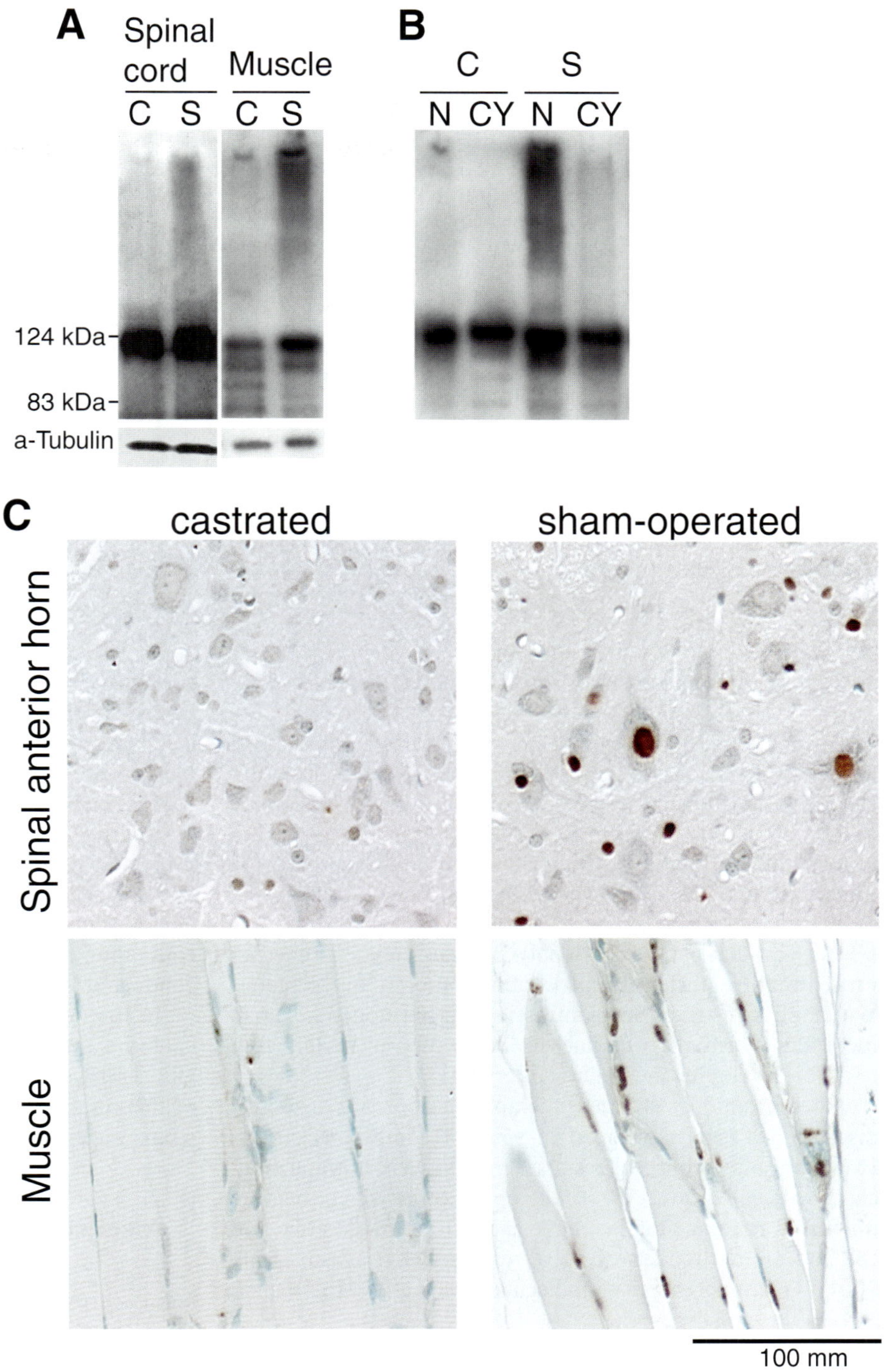

Fig. 4. Effects of castration on transgene expression and neuropathology of male AR-97Q mice. (**A**) Western blot analysis of total homogenates from the spinal cord and muscle of castrated (C) and sham-operated (S) male AR-97Q mice, that were immunolabeled by N-20. (**B**) Western blot analysis of nuclear (N) and cytoplasmic (CY) fraction from muscle of castrated (C) and sham-operated (S) male AR-97Q mice, immunolabeled by N-20. (**C**) Immunohistochemical study using 1C2 showed marked differences of diffuse nuclear staining and nuclear inclusions between castrated and sham-operated AR-97Q mice in the spinal anterior horn and muscle.

sir et al., 1995). More recently, conditional inactivation of wild-type huntingtin selectively in forebrain and testis resulted in a progressive degenerative neuronal phenotype and sterility (Dragatsis et al., 2000). Additionally, an anti-apoptotic effect of wild-type huntingtin was revealed in a cell model (Rigamonti et al., 2000). These results imply that loss of normal huntingtin function may contribute to the neurodegeneration in HD.

Androgens have been found to have neuroprotective effects. Administration of testosterone immediately after nerve injury impacts positively on functional recovery through actions mediated by the androgen receptor (Jones at al., 2001). In a cell culture model, AR with 24 CAGs showed trophic effects, whereas AR with 65 CAGs did not demonstrate any neuroprotection (Lieberman et al., 2002). The role of normal AR function in the pathogenesis of SBMA should be further studied in cell and animal models.

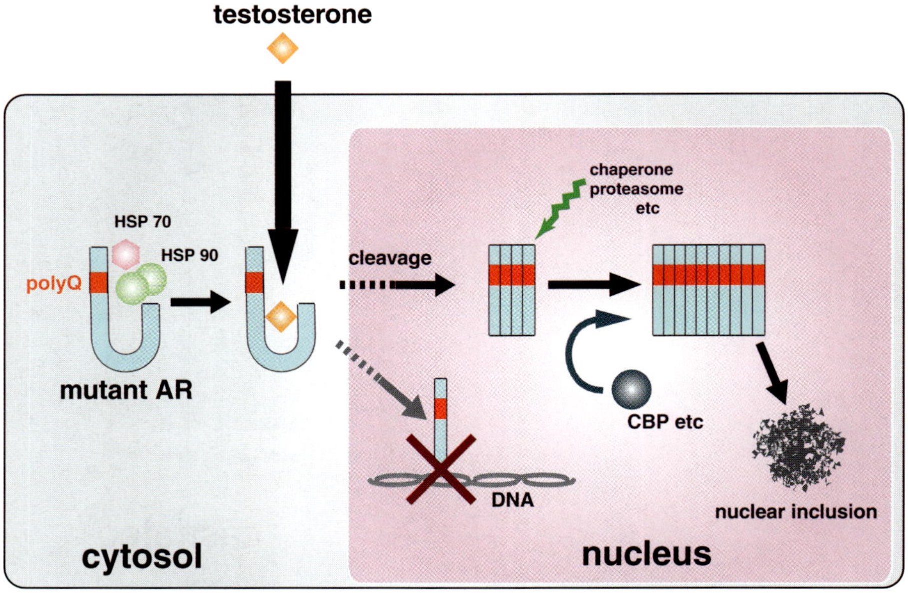

Fig. 5. Hypothetical dynamics of mutant AR in SBMA. In the absence of ligand, mutant AR is confined to a multi-heteromeric inactive complex with heat shock proteins (HSPs) and immunophilins in the cell cytoplasm. Upon testosterone-binding, the conformational change of mutant AR facilitates its dissociation from the complex and translocation into the nucleus. Mutant AR is cleaved and aggregates in the nucleus, whereas cellular mechanisms such as molecular chaperone and ubiquitin-proteasome system attempt to mitigate its toxicity. Aggregation sequestrates critical cellular proteins such as CREB-binding protein (CBP) resulting in aberrant transcription, and finally forms nuclear inclusions. On the other hand, decreased transactivating function of mutant AR may contribute to the neurodegeneration and androgen insensitivity in SBMA.

Toward therapy for SBMA and other polyQ diseases

As mentioned above, our recent study indicated that testosterone reduction exerts therapeutic effects by preventing nuclear translocation of mutant AR in the SBMA transgenic mouse model (Fig. 5). This approach can easily be applied to human SBMA therapy. Although no specific ligand of the mutant protein has been revealed in other polyQ diseases, the striking therapeutic effects of castration in our SBMA mice further suggest that patients with polyQ disease can be rescued by preventing the nuclear translocation of the mutant proteins. We emphasize the need of investigating hormone-like small molecules which alter the nuclear localization of mutant proteins for developing therapeutic intervention.

No substantially effective therapeutic approach to polyQ diseases has been developed in spite of continuous efforts. However, some promising results using transgenic animal models have been reported. Molecular chaperones, which renature misfolded mutant proteins, have exerted beneficial effects in cell and animal models of polyQ diseases (Kobayashi et al., 2001). Over-expression of molecular chaperone HSP70 had preventive effects in cell and Tg mouse models of SBMA (Kobayashi et al., 2000; Adachi et al., 2003) as well as in an SCA1 Tg mouse model (Cummings et al., 2001). It may also be possible to treat SBMA by increasing the expression level or enhancing the function of molecular chaperones.

Alternatively, histone deacetylase inhibitors could be a promising candidate therapy for polyQ diseases (Steffan et al., 2001; Hockly et al., 2003). These drugs potentially correct altered transcription due to toxic effects of the mutant protein containing expanded polyQ, although their toxicity should be conquered before clinical application.

In the near future, an ideal treatment for polyQ diseases could be a combination of these and other therapeutic strategies. Transgenic mice will be useful in testing the effectiveness of therapies, and further contribute to the development of a strategy against polyQ diseases including SBMA.

Acknowledgements

Figures 3 and 4 are reprinted from Katsuno et al. (2002) with permission from Elsevier Science.

References

Abel A, et al: Expression of expanded repeat androgen receptor produces neurologic disease in transgenic mice. Hum molec Genet 10:107–116 (2001).

Adachi H, et al: Transgenic mice with an expanded CAG repeat controlled by the human AR promoter show polyglutamine nuclear inclusions and neuronal dysfunction without neuronal cell death. Hum molec Genet 10:1039–1048 (2001).

Adachi H, et al: HSP70 chaperone over-expression ameliorates phenotypes of the SBMA transgenic mouse model by reducing nuclear-localized mutant AR protein. J Neurosci 23:2203–2211 (2003).

Bingham PM, et al: Stability of an expanded trinucleotide repeat in the androgen receptor gene in transgenic mice. Nature Genet 9:191–196 (1995).

Brooks BP, et al: Characterization of an expanded glutamine repeat androgen receptor in a neuronal cell culture system. Neurobiol Dis 3:313–323 (1997).

Burright EN, et al: SCA1 transgenic mice: a model for neurodegeneration caused by an expanded CAG trinucleotide repeat. Cell 82:937–948 (1995).

Cummings CJ, et al: Over-expression of inducible HSP70 chaperone suppresses neuropathology and improves motor function in SCA1 mice. Hum molec Genet 10:1511–1518 (2001).

Danek A, et al: Decrease in androgen binding and effect of androgen treatment in a case of X-linked bulbospinal neuronopathy. Clin Investig 72:892–897 (1994).

DiFiglia M, et al: Aggregation of huntingtin in neuronal intranuclear inclusions and dystrophic neurites in brain. Science 277:1990–1993 (1997).

Doyu M, et al: Severity of X-linked recessive bulbospinal neuronopathy correlates with size of the tandem CAG repeat in androgen receptor gene. Ann Neurol 32:707–10 (1992).

Doyu M, et al: Androgen receptor mRNA with increased size of tandem CAG repeat is widely expressed in the neural and nonneural tissues of X-linked recessive bulbospinal neuronopathy. J Neurol Sci 127:43–47 (1994).

Dragatsis I, et al: Inactivation of Hdh in the brain and testis results in progressive neurodegeneration and sterility in mice. Nature Genet 26:300–306 (2000).

Goldenberg JN, et al: Testosterone therapy and the pathogenesis of Kennedy's disease (X-linked bulbospinal muscular atrophy). J Neurol Sci 135:158–161 (1996).

Gottlieb B, et al: Androgen insensitivity. Am J med Genet 89:210–217 (1999).

Hackam AS, Singaraja R, Zhang T, Gan L, Hayden MR: In vitro evidence for both the nucleus and cytoplasm as subcellular sites of pathogenesis in Huntington's disease. Hum molec Genet. 8:25–33 (1999).

Hockly E, et al: Suberoylanilide hydroxamic acid, a histone deacylase inhibitor, ameliorates motor deficits in a mouse model of Huntington's disease. Proc natl Acad Sci, USA 100:2041–2046 (2003).

Hodgson JG, et al: A YAC mouse model for Huntington's disease with full-length mutant huntingtin, cytoplasmic toxicity, and selective striatal neurodegeneration. Neuron 23:181–192 (1999).

Ikeda H, et al: Expanded polyglutamine in the Machado-Joseph disease protein induces cell death in vitro and in vivo. Nature Genet 13:196–202 (1996).

Jones KJ, et al: Neuroprotective effects of gonadal steroids on regenerating peripheral motoneurons. Brain Res Brain Res Rev 37:372–382 (2001).

Katsuno M, et al: Testosterone reduction prevents phenotypic expression in a transgenic mouse model of spinal and bulbar muscular atrophy. Neuron 35:843–854 (2002).

Katsuno M, et al: Leuprorelin rescues polyglutamine-dependent phenotypes in a transgenic mouse model of spinal and bulbar muscular atrophy. Nature Med 9:768–773 (2003).

Kennedy WR, et al: Progressive proximal spinal and bulbar muscular atrophy of late onset. A sex-linked recessive trait. Neurology 18:671–680 (1968).

Klement IA, et al: Ataxin-1 nuclear localization and aggregation: role in polyglutamine-induced disease in SCA1 transgenic mice. Cell 95:41–53 (1998).

Kobayashi Y, et al: Caspase-3 cleaves the expanded androgen receptor protein of spinal and bulbar muscular atrophy in a polyglutamine repeat length-dependent manner. Biochem biophys Res Commun 252:145–150 (1998).

Kobayashi Y, et al: Chaperones Hsp70 and Hsp40 suppress aggregate formation and apoptosis in cultured neuronal cells expressing truncated androgen receptor protein with expanded polyglutamine tract. J biol Chem 275:8772–8778 (2000).

Kobayashi Y, Sobue G: Protective effect of chaperones on polyglutamine diseases. Brain Res Bull 56:165–168 (2001).

La Spada AR, et al: Androgen receptor gene mutations in X-linked spinal and bulbar muscular atrophy. Nature 352:77–79 (1991).

La Spada AR, et al: Meiotic stability and genotype-phenotype correlation of the trinucleotide repeat in X-linked spinal and bulbar muscular atrophy. Nature Genet 2:301–304 (1992).

La Spada AR, et al: Androgen receptor YAC transgenic mice carrying CAG 45 alleles show trinucleotide repeat instability. Hum molec Genet 7:959–967 (1998).

Li M, et al: Nuclear inclusions of the androgen receptor protein in spinal and bulbar muscular atrophy. Ann Neurol 44:249–254 (1998).

Li M, et al: Nonneural nuclear inclusions of androgen receptor protein in spinal and bulbar muscular atrophy. Am J Pathol 153:695–701 (1998).

Lieberman AP, et al: Altered transcriptional regulation in cells expressing the expanded polyglutamine androgen receptor. Hum molec Genet 11:1967–1976 (2002).

MacLean HE, et al: Defects of androgen receptor function: from sex reversal to motor neuron disease. Mol Cell Endocr 112:133–141 (1995).

Mangiarini L, et al: Exon 1 of the HD gene with an expanded CAG repeat is sufficient to cause a progressive neurological phenotype in transgenic mice. Cell 87:493–506 (1996).

Mariotti C, et al: Phenotypic manifestations associated with CAG-repeat expansion in the androgen receptor gene in male patients and heterozygous females: a clinical and molecular study of 30 families. Neuromuscul Disord 10:391–397 (2000).

McCampbell A, et al: CREB-binding protein sequestration by expanded polyglutamine. Hum molec Genet 9:2197–2202 (2000).

McManamny P, et al: A mouse model of spinal and bulbar muscular atrophy. Hum molec Genet 11:2103–2111 (2002).

Merry DE, et al: Toward a mouse model for spinal and bulbar muscular atrophy: effect of neuronal expression of androgen receptor in transgenic mice. (abstract) Am J hum Genet 59 (suppl):A271 (1996).

Merry DE, et al: Cleavage, aggregation and toxicity of the expanded androgen receptor in spinal and bulbar muscular atrophy. Hum molec Genet 7:693–701 (1998).

Merry DE: Molecular pathogenesis of spinal and bulbar muscular atrophy. Brain Res Bull 56:203–207 (2001).

Mhatre AN, et al: Reduced transcriptional regulatory competence of the androgen receptor in X-linked spinal and bulbar muscular atrophy. Nature Genet 5:184–188 (1993).

Nasir J, et al: Targeted disruption of the Huntington's disease gene results in embryonic lethality and behavioral and morphological changes in heterozygotes. Cell 81:811–823 (1995).

Neuschmid-Kaspar F, et al: CAG-repeat expansion in androgen receptor in Kennedy's disease is not a loss of function mutation. Mol Cell Endocrinol 117:149–156 (1996).

Nucifora FC Jr, et al: Interference by huntingtin and atrophin-1 with CBP-mediated transcription leading to cellular toxicity. Science 291:2423–2428 (2001).

Peters MF, et al: Nuclear targeting of mutant Huntingtin increases toxicity. Mol Cell Neurosci 14:121–128 (1999).

Reddy PH, et al: Behavioural abnormalities and selective neuronal loss in HD transgenic mice expressing mutated full-length HD cDNA. Nature Genet. 20:198–202 (1998).

Rigamonti D, et al: Wild-type huntingtin protects from apoptosis upstream of caspase-3. J Neurosci 20:3705–3713 (2000).

Ross CA: Polyglutamine pathogenesis: emergence of unifying mechanisms for Huntington's disease and related disorders. Neuron 35:819–822 (2002).

Rubinsztein DC: Lessons from animal models of Huntington's disease. Trends Genet 18:202–209 (2002).

Saudou F, et al: Huntingtin acts in the nucleus to induce apoptosis but death does not correlate with the formation of intranuclear inclusions. Cell 95:55–66 (1998).

Schmidt BJ, et al: Expression of X-linked bulbospinal muscular atrophy (Kennedy disease) in two homozygous women. Neurology 59:770–772 (2002).

Sengelaub DR, et al: Hormonal control of neuron number in sexually dimorphic spinal nuclei of the rat: II. Development of the spinal nucleus of the bulbocavernosus in androgen-insensitive (Tfm) rats. J Comp Neurol 280:630–636 (1989).

Simeoni S, et al: Motoneuronal cell death is not correlated with aggregate formation of androgen receptors containing an elongated polyglutamine tract. Hum molec Genet 9:133–144 (2000).

Sobue G, et al: X-linked recessive bulbospinal neuronopathy. A clinicopathological study. Brain 112:209–232 (1989).

Sobue G, et al: Subclinical phenotypic expressions in heterozygous females of X-linked recessive bulbospinal neuronopathy. J Neurol Sci 117:74–78 (1993).

Steffan JS, et al: The Huntington's disease protein interacts with p53 and CREB-binding protein and represses transcription. Proc natl Acad Sci, USA 97:6763–6768 (2000).

Steffan JS, et al: Histone deacetylase inhibitors arrest polyglutamine-dependent neurodegeneration in Drosophila. Nature 413:739–743 (2001).

Stenoien DL, et al: Polyglutamine-expanded androgen receptors form aggregates that sequester heat shock proteins, proteasome components and SRC-1, and are suppressed by the HDJ-2 chaperone. Hum molec Genet 8:731–741 (1999).

Takeyama K, et al: Androgen-dependent neurodegeneration by polyglutamine-expanded human androgen receptor in *Drosophila*. Neuron 35:855–864 (2002).

Tanaka F, et al: Founder effect in spinal and bulbar muscular atrophy (SBMA). Hum molec Genet 5:1253–1257 (1996).

Tanaka F, et al: Tissue-specific somatic mosaicism in spinal and bulbar muscular atrophy is dependent on CAG-repeat length and androgen receptor-gene expression level. Am J hum Genet 65:966–973 (1999).

Tobin AJ, Signer ER: Huntington's disease: the challenge for cell biologists. Trends Cell Biol. 10:531–536 (2000).

Zhou ZX, et al: The androgen receptor: an overview. Recent Prog Horm Res 49:249–274 (1994).

Zhou ZX, et al: Specificity of ligand- dependent androgen receptor stabilization: receptor domain interactions influence ligand dissociation and receptor stability. Mol Endocrinol 9:208–218 (1995).

Zoghbi HY, Orr HT: Glutamine repeats and neurodegeneration. A Rev Neurosci 23:217–247 (2000).

Cytogenet Genome Res 100:252–260 (2003)
DOI: 10.1159/000072861

Oculopharyngeal muscular dystrophy: a late-onset polyalanine disease

B. Brais

Faculté de Médecine, Université de Montréal, Centre de recherche du CHUM, Hôpital Notre-Dame-CHUM, Montréal, Québec (Canada)

Abstract. Oculopharyngeal muscular dystrophy (OPMD) is a muscle disease of late onset associated with progressive ptosis of the eyelids, dysphagia, and unique tubulofilamentous intranuclear inclusions (INIs). OPMD is usually transmitted as an autosomal dominant trait (OMIM 164300). A rarer allelic autosomal recessive form has also been observed (OMIM 257950). Both forms are caused by short $(GCG)_{8-13}$ expansions in the polyadenylate-binding protein nuclear 1 gene (PABPN1) located on chromosome 14q11.1. The mutations cause the lengthening of an *N*-terminal polyalanine domain. Both slippage and unequal recombination have been proposed as the mutation mechanisms. The size of the mutation has not yet been conclusively shown to inversely correlate with the severity of the phenotype. Mutated PABPN1 proteins have been shown to be constituents of the INIs. The INIs also contain ubiquitin, proteasome subunits, HSP 40, HSP 70, SKIP, and abundant poly(A)-mRNA. The exact mechanism responsible for polyalanine toxicity in OPMD is unknown. Various intranuclear inclusion dependent and independent mechanisms have been proposed based on the major known function of PABPN1 in polyadenylation of mRNA and its shuttling from the nucleus to the cytoplasm. OPMD is one of the few triplet-repeat diseases for which the function of the mutated gene is known. Because of the increasing number of diseases caused by polyalanine expansions and the pathological overlap with CAG/polyglutamine diseases, what pathological insight is gained by the study of OPMD could lead to a better understanding of a much larger group of developmental and degenerative diseases.

Clinical features

Oculopharyngeal muscular dystrophy (OPMD) was first clearly described by Taylor in 1915 in four members of a French Canadian family (Taylor, 1915). They had late-onset ptosis associated with progressive swallowing difficulties (dysphagia), which led to starvation and death. This paper and a few other case reports (Amyot, 1948a, b; Saucier, 1954) were largely overlooked until Victor, Hayes and Adams' 1962 classical publication (Victor et al., 1962). They were the first to refer to this myopathy as oculopharyngeal muscular dystrophy. These authors studied 10 cases (one sporadic and nine familial) in three generations of a North American Jewish family of Eastern European origin. André Barbeau found the largest cluster of OPMD in the Canadian Province of Quebec, with an estimated prevalence of 1:1,000 (Barbeau, 1966, 1969; Brais et al., 1995). Other large clusters have been observed in Bukhara Jews (1:600) (Blumen et al., 1997), Uruguayans (Medici et al., 1997), and Spanish Americans living in New Mexico (Becher et al., 2001) and California (Grewal et al., 1999). Cases of OPMD have now been reported in more than 30 countries.

Oculopharyngeal muscular dystrophy usually manifests itself in the fifth or sixth decade by two cardinal symptoms: ptosis of the eyelid and dysphagia, both of which have a slowly progressive course. Later, all extraocular and other voluntary muscles may become affected. In the majority of cases, the ptosis is the first symptom (Fig. 1). Initially, OPMD is often restricted to the levator palpebrae and pharyngeal muscles. As the disease evolves, there may be impairment of eye movements and, occasionally, diplopia; nevertheless, complete external ophthalmoplegia is infrequent. With time, the dysphagia will result in undernutrition and may cause death due to aspiration pneumonia. Other symptoms and signs are associated with myopathic involvement of other muscles. Most patients will

Received 11 November 2002; revision accepted 17 February 2003.

Request reprints from Dr. Bernard Brais, Faculté de Médecine
Université de Montréal, Centre de recherche du CHUM
Hôpital Notre-Dame-CHUM, M-4211-L3, 1560, rue Sherbrooke Est
Montréal, Québec, H2L 4M1 (Canada)
telephone: 1-514-890-8000 ext 25560; fax: 1-514-412-7525
e-mail: Bernard.Brais@umontreal.ca

KARGER
Fax + 41 61 306 12 34
E-mail karger@karger.ch
www.karger.com

0301–0171/03/1004–0252$19.50/0

Accessible online at:
www.karger.com/cgr

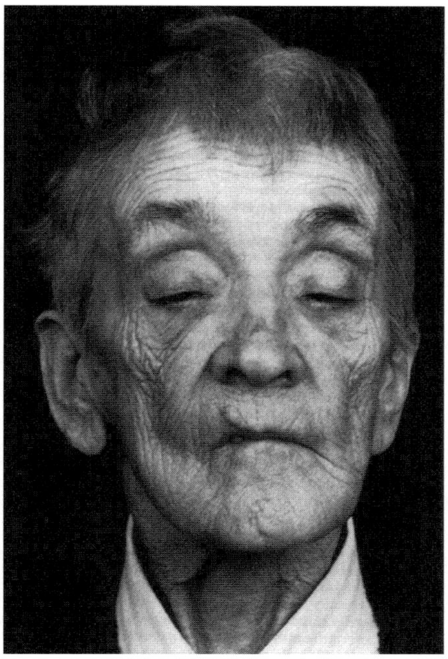

Fig. 1. Characteristic facies of OPMD: bilateral ptosis, raised eyebrows, and wrinkled forehead. This patient also exhibits wasting of the masseter muscles (photograph courtesy of Dr. J.-P. Bouchard).

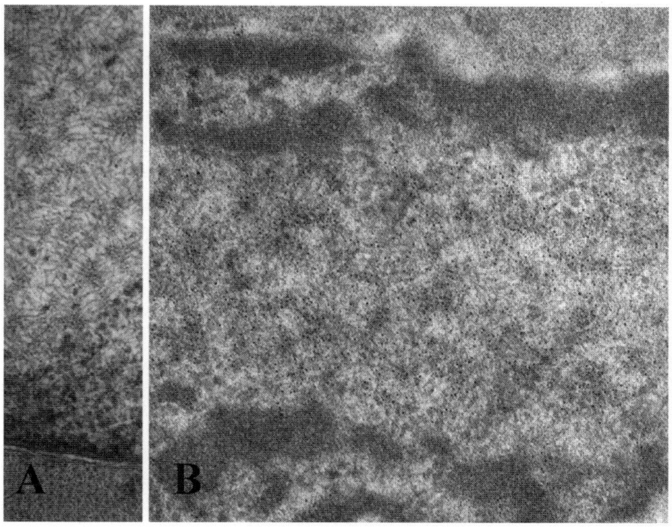

Fig. 2. Immuno-electron microscopy of PABPN1 in OPMD nuclei. Biopsy from deltoid muscle from a 60-year-old OPMD patient. Subsarcolemmal nuclei contain inclusions consisting of unique filaments. These filaments converge to form tangles or palisades, distinctly seen on sections taken from muscle specimens prepared according to standard electron microscopic technique (**A**). Using anti-PABPN1 antibodies, the filaments are decorated by immuno-gold particles. The labeling is particularly intense at the sites of filament convergence (**B**). Electronmicrographs graciously provided by F.M.S. Tomé and M. Carmo-Fonseca.

develop a nasal voice due to palatal weakness. Weakness and atrophy of the tongue is usually observed in the vast majority of patients (Bouchard et al., 1997). Mild facial, temporal, and masseter muscle involvement is often observed when the ptosis is obvious. Weakness also often occurs in pelvic girdle muscles and to a lesser extent in the shoulder girdle. The disease has a slowly progressive course. Life expectancy is not diminished by the condition (Becher et al., 2001). Until recently, death occurred at an advanced age as a result of starvation or aspiration pneumonia. With progress in the treatment of pharyngeal dysfunction and better nutrition, the life span of patients is longer, and the prognosis and quality of life are much improved.

Pathology

The changes seen in the muscle fibers of extraocular and other voluntary muscles vary according to the stage of the disease and the muscle examined. Probably all skeletal muscles are affected, but extraocular, lingual, pharyngeal, and diaphragmatic muscles are more severely involved at autopsy (Rebeiz et al., 1969; Schmitt and Krause, 1981; Little and Perl, 1982). Muscles studied using classical histological methods show changes that are common to many muscular dystrophies. These include loss of muscle fibers, abnormal variation in fiber size, increase in the number of nuclei, internal nuclei, and increased interstitial fibrous and fatty connective tissues. Fibers undergoing necrosis and phagocytosis are rare. Histochemical studies of frozen sections reveal two particular changes: (1) small angulated fibers that often react strongly for oxidative enzymes and are more frequently type 1 than type 2, as seen with ATPase reactions (these fibers are scattered throughout the biopsy), and (2) rimmed vacuoles within the muscle fibers. Although the presence of small angulated fibers suggests an underlying denervation process, their occurrence may be attributed simply to ageing. The rimmed vacuoles consist of irregularly round or polygonal clear spaces lined by a ring of material that is basophilic (hematoxylin and eosin staining) and stains red with Gomori's trichrome stain. The rimmed vacuoles have acid phosphatase activity. Electron microscopy also suggests that they may be autophagic in character. They vary in size and shape and contain membranous and myeloid structures, a few glycogen granules, and debris arising from the breakdown of unidentified muscle components. Rimmed vacuoles have been observed in several other disorders, particularly inclusion body myositis (Lotz et al., 1989). However, their presence in a biopsy supports a clinical diagnosis of OPMD.

The most significant ultrastructural change is the presence of intranuclear tubular filaments with an 8.5-nm outer diameter and a 3-nm inner diameter (Fig. 2). The filaments are unbranched, often course rectilinearly, are sometimes striated with 7–7.5 nm of periodicity, and are up to 0.25 µm in length. They are disposed in various directions and frequently form tangles or palisades. Large collections of filaments appear as clear zones surrounded by chromatin in the affected nuclei. These nuclei can often be identified in semithin epoxy sections by phase-contrast microscopy, where they appear as clear zones. Studies of serial semithin sections suggest that in some specimens the filamentous inclusions occur in all muscle fiber nuclei (Tomé et al., 1997). However, the spaces occupied by the inclusions vary greatly from nucleus to nucleus. In ultrathin sections, the frequency of nuclei with filamentous intranuclear inclusions (INIs) varied from 2% to 5% (mean, 4%) in 29 cases (Tomé et al., 1997). The highest percentage of muscle nuclei containing INI were observed in seven cases of homozygotes for $(GCG)_9$ mutations in the polyadenylate-binding protein nuclear 1 gene (PABPN1, formerly PABP2) (Blumen et al., 1999). In these more severe OPMD cases, 9.4% of nuclei contained INIs, compared to 4.9% in heterozygotes for the same mutation. One study reported the presence of inclusions in about 4% of the muscle nuclei in 12 limb muscle specimens and in 8% of the muscle nuclei in a specimen of pharyngeal muscle (Coquet et al., 1983). The inclusions occurred only in muscle fibers and not in the nuclei of any other cells (including satellite cells) in muscle. They were never observed in fibroblasts, smooth muscle cells, endothelial cells, pericytes, adipocytes, Schwann cells, or perineurial epithelial cells. The filaments persist in nuclear fractions of muscle homogenates and are resistant to high salt concentrations (Tomé and Fardeau, 1986; Calado et al., 2000).

Other mostly cytoplasmic filaments can be observed in OPMD muscle. Inclusion body myositis (IBM) tubular filaments have been observed in OPMD (Serratrice and Pellissier, 1987; Coquet et al., 1990). They are not of the same size as the OPMD filaments. They are no longer considered to be specific for IBM. In one study of OPMD, IBM filaments were found in 8 of 10 muscle specimens (Coquet et al., 1990). The IBM-type filaments are much less frequent in OPMD than in IBM and are mainly cytoplasmic, appearing close to or within the rimmed vacuoles. Occasionally, however, the IBM-type filaments are intranuclear, as reported in two cases (Smith and Chad, 1984; Coquet et al., 1990)

Morphological studies of biopsies of peripheral nerves from two elderly patients showed changes suggestive of an axonal neuropathy (Porschke et al., 1997; Boukriche et al., 2002). Autopsy studies have revealed no changes in either the central or peripheral nervous systems, including the brainstem motor nuclei, that could account for the changes in muscle (Bouchard et al., 1997; Tomé et al., 1997). In one autopsy, some loss of spinal cord anterior horn cells was observed in a man who died at age 81 of a pulmonary embolism (Schmitt and Krause, 1981). It is still unclear whether the changes might be fortuitous or if there is true axonal involvement in OPMD (Probst et al., 1982).

Molecular genetics

Autosomal dominant and recessive forms of OPMD have been described (OMIM 164300 and 257950, respectively). The OPMD locus was first mapped to chromosome 14q11.1 by completing a linkage analysis using large French Canadian families (Brais et al., 1995). Linkage studies suggested that dominant OPMD was a genetically homogeneous condition (Porschke et al., 1997; Stajich et al., 1997; Teh et al., 1997; Grewal et al., 1998). A positional cloning strategy led to the identification of short $(GCG)_{8-13}$ expansions in the PABPN1 gene in all dominant OPMD cases (Brais et al., 1998b). They consist of mitotically and meiotically stable expansions of a $(GCG)_n$ repeat in the first exon of the gene. Dominant mutations consist of the addition of two to seven (GCG) repeats to the usual $(GCG)_6$ stretch (Fig. 3A). Until recently, six different mutation sizes had been observed. The percentages of 81 families sharing the different sizes of dominant mutations were 5% $(GCG)_8$, 40% $(GCG)_9$, 26% $(GCG)_{10}$, 21% $(GCG)_{11}$, 7% $(GCG)_{12}$, and 1% $(GCG)_{13}$ (Brais et al., 1998b). The relative percentage will vary from one population to another (Table 1). In populations with founder effects, such as in French Canadians or Bukhara Jews, most families will share the same historical mutation, which will only rarely have changed size (Brais et al., 1998b; Blumen et al., 2000). Two studies have raised the possibility that some of the OPMD mutations are not due to expansions of the GCG stretch but are secondary to insertion of nonidentical polyalanine coding triplets (Scacheri et al., 1999; Nakamoto et al., 2002). It was suggested that both types of OPMD mutations are caused by unequal crossing-over (Nakamoto et al., 2002). This has been considered to be the major cause of polyalanine expansion mutations in other polyalanine diseases (Warren, 1997).

Gene dosage has a clear influence on the age of onset and severity of the OPMD phenotype (Brais et al., 1998a). The most severe OPMD phenotype is reported for individuals homozygous for a dominant OPMD mutation (Blumen et al., 1996; Brais et al., 1998b; Blumen et al., 1999). The study of four French Canadian and three Bukhara Jewish OPMD homozygotes documented that, on average, the onset was 18 years earlier than in $(GCG)_9$ heterozygotes (Blumen et al., 1999). Severity of the dominant OPMD phenotype is also variable (Bouchard et al., 1997). Severe cases have an earlier onset, before the age of 45, of ptosis and dysphagia and an incapacitating proximal leg weakness, which starts before age 60. In a large French Canadian cohort of $(GCG)_9$ mutation carriers, 5–10% of cases have a severe phenotype (Brais et al., 1998a). Approximately 20% of these severe cases are compound heterozygotes for the dominant mutation and have a $(GCG)_7$ polymorphism in their other copy of the PABPN1 gene (Brais et al., 1998b). This polymorphism has a 1–2% prevalence in North America, Europe, and Japan. Patients with a more severe compound heterozygote OPMD phenotype have also been observed in Italy and the UK (Mirabella et al., 2000; Hill et al., 2001). The cause of the increased severity in the other 80% is unknown. Strikingly, severe cases cluster in families, suggesting that other genetic factors are modulators of severity (Brais et al., 1998a). The more severe phenotypes observed in homozygotes and compound heterozygotes for a dominant and a recessive mutation suggest a clear gene-dosage effect (Brais et al., 1998a, b, 1999; Blumen et al., 1999). The recessive OPMD clinical phenotype is similar, but has an older age of onset. The only genetically proven recessive OPMD case was a homozygote carrier of two $(GCG)_7$ PABPN1 mutations (Brais et al., 1998b)

Following the discovery of a mutated gene in OPMD, much work centered on the identification of the other molecules present in the INIs (Table 2). A mutated PABPN1 was shown to be an integral part of the muscle OPMD inclusions (Becher et al., 2000; Calado et al., 2000; Uyama et al., 2000). The

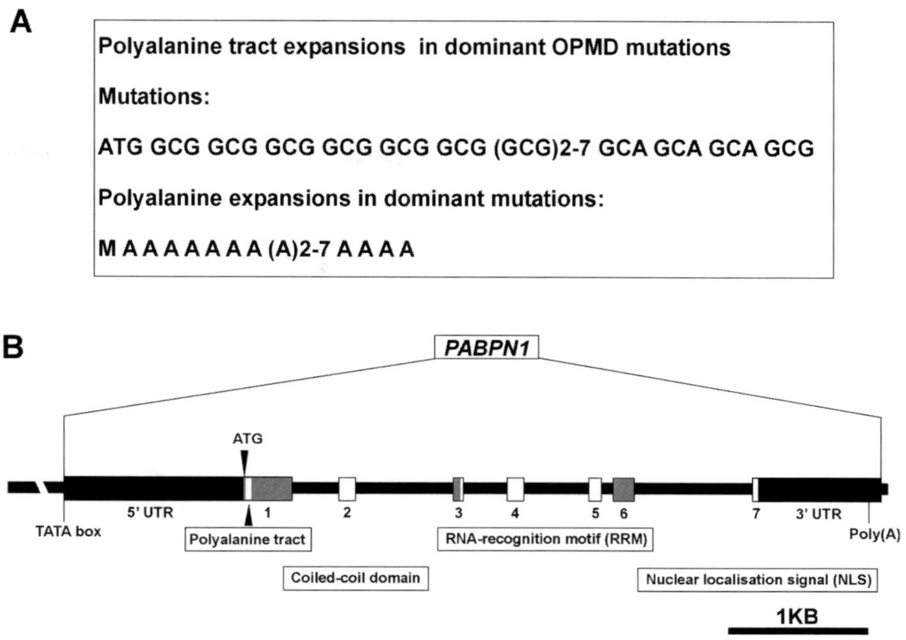

Fig. 3. (**A**) Polyalanine tract expansions caused by OPMD mutations in the PABPN1 gene. (**B**) Structure of the PABPN1 gene with its four-region coding for functional domains: polyalanine tract, coiled-coil, RNA-recognition motif (RRM), and nuclear localization signal (NLS).

Table 1. Percentage of families and cases sharing OPMD $(GCG)_n$ mutations in different populations

Mutation	Families		Cases			
	World[a] (144 families)	UK[b] (31 cases)	Canada[c] (402 cases)	USA[c] (24 cases)	Germany[d] (16 cases)	Italy[e] (11 cases)
$(GCG)_8$	3		<1	0	6	37
$(GCG)_9$	69	40	95	83	37	45
$(GCG)_{10}$	13	36	1	4	37	18
$(GCG)_{11}$	11		2	13	19	0
$(GCG)_{12}$	3		1	0	0	0
$(GCG)_{13}$	<1		0	0	0	0

[a] Brais et al. (1998) (n = 1).
[b] Hill et al. (2001) (n = 88).
[c] Brais et al. (unpublished data).
[d] Müller et al. (2001) (n = 87).
[e] Mirabella et al. (1999) (n = 131).

n = Total number of families or cases examined in each study.

Table 2. Characteristics of intranuclear inclusions in OPMD

Characteristic	Description
Cellular location	Nucleus only
Size	8.5 nm in external diameter, up to 0.25 m in length
Spatial organization	In tangles or palisades
Appearance on semithin sections	Clear zones
Disease association	OPMD only
PABPN1 content	Present
Poly(A) mRNA	Present
Ubiquitin	Present
Proteasome	Present
HSP 40 and 70	Present
SKIP	Present

OPMD-specific nuclear inclusions are decorated with anti-PABPN1 antibodies under both immuno-electron microscopy and fluorescence confocal microscopy (Calado et al., 2000). The inclusions are also decorated with antibodies directed against ubiquitin and proteasomal subunits. Previous work had detected ubiquitin by immunostaining in vacuolated muscle in rare fibers in OPMD and localized the ubiquitin immunoreactivity to cytoplasmic tubulofilamentous inclusions of the IBM type (Askanas et al., 1991; Leclerc et al., 1993). These observations suggest that the polyalanine expansions in PABPN1 induce misfolding and aggregation of ubiquitin-tagged proteins which can not be degraded by the proteasome machinery. Interestingly, the nuclear inclusions were also found to sequester poly(A)-RNA and not to contain DNA (Calado et al., 2000)

Pathogenesis

Our present knowledge of the function of PABPN1 leads to different pathogenic hypotheses. At the protein level, the OPMD mutations cause the lengthening of a predicted *N*-terminus polyalanine domain (Fig. 3B). PABPN1 is an abundant, mostly nuclear protein involved in the polyadenylation of all mRNAs (Wahle, 1991; Bienroth et al., 1993; Wahle et al., 1993; Krause et al., 1994; Nemeth et al., 1995). PABPN1 has at least the following four distinct domains: polyalanine, RNA binding, dimerization, and nuclear localization signal (NLS) (Fig. 3B). Although the PABPN1 RNA recognition motif (pfam00076) is highly conserved in many species, a variable-length polyalanine domain has been found only in vertebrates: *Homo sapiens*

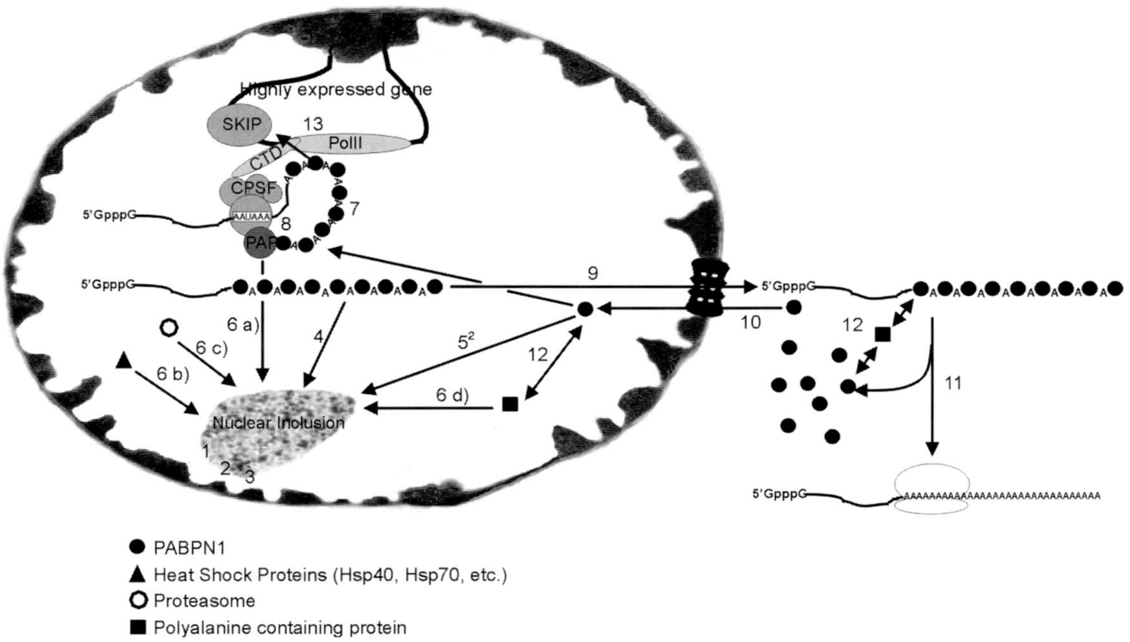

Fig. 4. Cellular trafficking of PABPN1 and possible sites of interference of the mutated PABPN1 gene with normal cellular processes. PABPN1 is involved in the polyadenylation of all messenger RNAs (7 and 8). PABPN1 travels with the mRNA to the cytoplasm through nuclear pores (9). It is released from the mRNA on the initiation of translation (11). It is actively transported back to the nucleus to partake in the polyadenylation of other mRNA molecules (10). Based upon our understanding of the major role and cellular trafficking and breakdown of PABPN1, mutated forms could interfere with different cellular processes (1–12). Some of the interference could be caused by intranuclear inclusion-dependent mechanisms: (1) physical rupture of the nuclear membrane; (2) disruption of transcriptional domains; (3) disruption of chromosomal domains; (4) sequestering of mRNAs coding for proteins vital for cell survival; (5) sequestering of PABPN1 in sufficient quantity to interfere with normal mRNA processing; and/or (6) sequestering of other proteins: (a) proteins interacting normally with PABPN1, (b) proteins involved in protein folding (e.g., chaperones such as Hsp70 and Hsp40), (c) proteins of the ubiquitin/proteasome pathway, or (d) other proteins with polyalanine domains. In addition, intranuclear inclusion-independent mechanisms could interfere with: (7) normal function of PABPN1 in mRNA polyadenylation; (8) the mRNA processing machinery; (9) exit of PABPN1 and mRNA from the nucleus; (10) import of PABPN1 into the nucleus; (11) initiation of translation; and/or (12) other soluble polyalanine-containing proteins that interact with PABPN1 in the nucleus or the cytoplasm. Abbreviations: *Pol*II = RNA polymerase II; CTD = *C*-terminal domains of *Pol*II; CPSF = cleavage and polyadenylation specificity factor; PAP = poly(A) polymerase.

(10Ala), *Mus musculus* (10Ala), *Bos taurus* (10Ala) and *Xenopus laevis* (2Ala). PABPN1 is an abundant nuclear protein of ~ 49 kDa that binds with high affinity to nascent poly(A) tails at the 3′ end of mRNAs (Wahle et al., 1993). The poly(A) tail is posttranscriptionally added to the mRNA by a number of *trans*-acting protein factors, including PABPN1, the cleavage and polyadenylation specificity factor (CPSF), the cleavage stimulation factor (CstF), the cleavage factors I_m and II_m (CFI_m and $CFII_m$), and the poly(A) polymerase (PAP) (Fig. 4) (Colgan and Manley, 1997; Minvielle-Sebastia and Keller, 1999; Wahle and Ruegsegger, 1999; Zhao et al., 1999). Polyadenylation is a two-step nuclear process. Endonucleolytic cleavage precedes the poly(A) tail synthesis. Following the cleavage, CPSF binds the poly(A) site AAUAAA and stimulates PAP to elongate poly(A) tails of mRNAs carrying this sequence (Keller et al., 1991). When 10–11 nucleotides have been added, PABPN1 oligomers bind the nascent poly(A) tail and further stimulate its elongation (Wahle, 1991). When the poly(A) tail reaches ~ 250 nucleotides, the elongation becomes slow. It has been proposed that the stoichiometric binding of PABPN1 to poly(A) tails could be involved in a counting mechanism that determines the length of the tail (Wahle, 1995). It was demonstrated that PABPN1 shuttles between nucleus and cytoplasm (Chen et al., 1999; Calado et al., 2000). Nuclear export of PABPN1 is temperature sensitive and depends on RNA binding and ongoing transcription. Nuclear import of PABPN1 is an active transportin-mediated process (Calado et al., 2000)

We and others have proposed different models of polyalanine toxicity gain-of-function pathogenesis (Brais et al., 1998b, 1999; Galvao et al., 2001). In these models, PABPN1 is suggested to carry a pathogenic expanded polyalanine domain with physical characteristics that cause both its accumulation and interference with normal cellular processes. This is reminiscent of carrier models proposed in CAG repeat diseases (Sisodia, 1998). Different lines of evidence suggest that polyalanine oligomers form resistant macromolecules in vivo and in vitro. Polyalanine oligomers are known to be very resistant to protease digestion or chemical degradation (Forood et al., 1995). They form a β-sheet structure in vitro (Forood et al., 1995). Polyalanine oligomers containing more than eight alanines in a row form fibrils spontaneously (Billingsley et al., 1994). Furthermore, PABPN1 molecules in OPMD muscle INIs are more resistant to salt extraction than the protein dispersed in the nucleoplasm (Calado et al., 2000). Various fusion proteins with long polyalanine domains accumulate as INIs (Gaspar et al., 2000; Rankin et al., 2000). In one transfection experiment, a long, 37 Ala-GFP fusion protein caused nuclear inclusion formation and cell death (Rankin et al., 2000)

One possible pathogenetic hypothesis is that the expanded PABPN1 polyalanine domains polymerize to form stable β-sheets that are resistant to nuclear proteasomal degradation. Polymerization of PABPN1 is clearly important for the formation of INIs and its toxicity, as demonstrated by the decreased toxicity and absence of INI formation when the PABPN1 oligomerization domain was removed in transfection experiments (Fan et al., 2001). The polyalanine macromolecules grow with time to form the OPMD PABPN1-containing intranuclear filaments that are seen on electron microscopy (Tomé and Fardeau, 1980; Tomé et al., 1997; Calado et al., 2000). When a significant portion of a nucleus is occupied by the inclusions, they sufficiently alter normal cellular processes to lead to fatal nuclear dysfunction. The altered function could be both nuclear and cytoplasmic. For instance, the observed rimmed vacuoles in OPMD may correspond to cytoplasmic PABPN1 toxicity (Tomé and Fardeau, 1980; Askanas et al., 1991; Tomé et al., 1997). Rimmed vacuoles appear as cells try to degrade abnormal proteins through autophagy, as has been suggested by the study of polyglutamine and polyalanine toxicity cell models (Ravikumar et al., 2002)

How the PABPN1 gain-of-function ultimately causes cellular demise still has to be uncovered. The possible pathophysiological mechanisms can be broken down into INI dependent and INI independent dysfunctions based on our present understanding of the protein's normal function and its nuclear/cytoplasmic shuttling. Figure 4 illustrates the subcellular localization of these two types of possible dysfunctions. Transfection-based cellular models of OPMD and the study of pathological muscles have already shed some light on the possible pathways involved in the polyalanine toxicity in OPMD. The observations that INIs in patient muscles and different cellular muscles contain ubiquitin, proteasome subunits, heat-shock protein 40, and heat-shock protein 70 support the hypothesis that PABPN1 misfolding may play a pathogenic role. The reduction of cell death and aggregation in a cellular OPMD model by overexpression of heat-shock proteins or the addition of the chemical chaperone Me_2SO further support the hypothesis that abnormal protein conformation plays an important role in polyalanine toxicity in OPMD (Bao et al., 2002). The presence of a large quantity of poly(A)-RNA has raised the possibility that the INI interfere with cellular functions by trapping the mRNA (Calado et al., 2000). Furthermore, the expanded repeat region of the PABPN1 mRNA may interfere with its normal processing, leading to an RNA toxic gain-of-function (Galvao et al., 2001). Work by a Japanese group has shown that PABPN1 interacts with SKIP, a transcription factor, to stimulate muscle-specific gene expression through MyoD and myogenin increased expression (Kim, Noguchi et al., 2001). The three proteins were shown to associate in a single complex. These results suggest that the mutated PABPN1, though ubiquitously expressed, may be more toxic in muscle because of its significant role in regulating the expression of muscle-specific genes. Using yeast two-hybrid screening, these workers also identified that PABPN1 interacts with heat-shock protein 40, the human SWI-SNF component of SMARCA4 (alias BRG1), and an unknown protein with zinc-finger motifs. It is likely that more than one of the proposed altered mechanisms is at play to time the segmental cell death—a cell death that could be non-apoptotic, as has been described in some models of CAG triplet disease (Evert et al., 1999; Turmaine et al., 2000). Autophagia could also play a role in the ultimate cellular demise (Ravikumar et al., 2002)

The degree of cellular differentiation also clearly plays a role in INI formation, as was demonstrated by the appearance of unusual INIs in cultured OPMD muscle cells innervated by fetal rat spinal-cord neurons (Tomé et al., 1989). These inclusions were tubulofilamentous and had the same diameter as the typical INIs observed in OPMD muscles, but they were less distinct and lacked the spatial arrangement characteristic of the OPMD inclusions. The nuclear inclusions were found in innervated OPMD muscle cultures only and occurred only in cells with well-developed myofibrils (myotubes or muscle fibers). The inclusions were not found in innervated muscle cell cultures obtained from normal subjects or disease controls. Inclusions similar to those found in culture were also found in the OPMD muscles. Thus, the inclusions detected in culture could represent an early stage of the typical intranuclear OPMD inclusions.

Diagnosis and treatment

Until the identification of PABPN1 mutations in OPMD, diagnosis relied on clinical evaluation and the presence of INIs on electron microscopy (Tomé and Fardeau, 1994). This approach has now been replaced by DNA testing (Brais et al., 1998b). Since autosomal dominant and recessive OPMD is allelic, the molecular diagnosis of both conditions is straightforward. A single polymerase chain reaction (PCR) is required to establish the carrier status of an individual (Brais et al., 1998b). The test has a sensitivity and specificity close to 100%. The test is available through many diagnostic laboratories worldwide. The major indications for DNA testing of symptomatic individuals are: (1) confirmation of the diagnosis in a previously untested family; (2) the clinical picture presents a diagnostic dilemma; (3) evaluation of the size of the mutation as a possible indicator of severity; (4) to exclude a compound heterozygote status in a patient with a severe early onset form of the disease; and (5) the patient may suffer from recessive OPMD. Much care should be taken before requesting the predictive testing of an asymptomatic individual; because no preventive treatment is available, it is unclear if these individuals will benefit from the test. Presymptomatic testing should be performed in the context in which genetic counseling and psychological support can be offered. Considering the late onset and relatively mild physical limitations caused by this disease, to our knowledge prenatal testing for OPMD has never been requested.

There is no medical treatment available for OPMD. A high-protein diet is recommended. This is particularly important because as the dysphagia becomes severe, patients shy away from sources of animal proteins, such as meat. Special attention should be paid to prevent the frequent social withdrawal of patients as their dysphagia progresses. With aspiration pneumonia being a frequent cause of death, patients should be advised to seek medical attention early if they have a productive cough accompanied by fever. Exercises that maintain good cardiovascular functionality should be encouraged, but strenuous exercises should not be promoted. Late in the course of the disease, a large percentage of patients will use either a cane or a walker. Rarely, patients with more severe disease will require a wheelchair, mostly to cover long distances.

At this time, only surgical treatments to correct some of the symptoms of OPMD are available. Two types of operations are used to correct the ptosis with overall good results: resection of the levator palpebral aponeurosis and frontal suspension of the eyelids (Codère, 1993; Codère et al., 2001) Surgery is recommended when the ptosis interferes with vision or cervical pain appears secondary to the constant dorsiflexion of the neck. Surgical evaluation of symptomatic dysphagia should be prompted by severe dysphagia, marked weight loss, near-fatal choking, or recurrent pneumonia. Cricopharyngeal myotomy will alleviate symptoms in most cases. Unfortunately, dysphagia will slowly reappear over years. Severe dysphonia and lower esophageal sphincter incompetence are contraindications to surgery (Duranceau, 1997)

Discussion

Oculopharyngeal muscular dystrophy is one of the few triplet repeat diseases in which it is not known if there is an inverse correlation between the severity of the phenotype and the size of the mutation (Zoghbi and Orr, 2000). It is clear, however, that the severity of the phenotype varies even with carriers of the same size of $(GCG)_n$ PABPN1 mutation (Bouchard et al., 1997; Brais et al., 1998a, b). Preliminary studies have not demonstrated whether the size of the mutation influences the severity of the phenotype, at least when assessed by age of onset (Mirabella et al., 2000; Hill et al., 2001; Muller et al., 2001). Only carriers of the smallest $(GCG)_8$ PABPN1 mutation clearly appear to have a milder phenotype, with a later age of onset in the seventh decade characterized by ptosis and only mild dysphagia (Brais et al., 1998a, b; Hill et al., 2001; Muller et al., 2001). Compound heterozygotes for dominant and recessive mutations have also been shown to have more severe phenotypes (Brais et al., 1998b; Hill et al., 2001). Comparison of large cohorts of carriers of different sizes of mutations should establish if there is or is not an inverse correlation between the size of the PABPN1 mutation and OPMD severity. This question is particularly interesting, considering that although there appears to be a size effect in other polyalanine diseases (Muragaki et al., 1996), OPMD is a far more common disease, which should determine statistically significant differences, if indeed they exist.

The mechanism responsible for the OPMD mutations is still unknown. Although slippage during replication was proposed in the first paper describing the PABPN1 mutations in OPMD, two reports suggest that unequal recombinations may be responsible for some, if not all, OPMD mutations (Brais et al., 1998b; Nakamoto et al., 2002). This would correspond to what is believed to be the mutation mechanism in the other polyalanine diseases (Warren, 1997). If unequal recombination is ultimately demonstrated to be the underlining mechanism, this would render inappropriate the suggestion of calling OPMD a short triplet repeat disease (Brais et al., 1998b). In this case, it would be preferable to refer to OPMD as a polyalanine domain expansion disease or simply a polyalanine disease.

The differential involvement of certain muscles in OPMD, as in most muscular dystrophies, has yet to be understood. This issue is particularly interesting in OPMD, because the mutated gene is ubiquitously expressed in all tissues. Furthermore, individuals that are homozygotes for two dominant mutations have a normal development but clearly have a more diffuse muscular involvement (Blumen et al., 1999). The gene dosage observed in homozygotes and compound heterozygotes, as discussed previously, are in favor of a gene-dosage effect. These observations raise the possibility that skeletal muscles may have a greater requirement of PABPN1, which could make them more vulnerable to its toxicity. The postmitotic nature of skeletal muscles may also play a role. However, how can one then explain why there is no clear neuronal involvement? Perhaps some of the reported abnormal electromyography, biopsy, and autopsy findings are indeed showing central and peripheral nervous system involvement in OPMD. The availability of muscle tissue in OPMD and the apparently simple production of INIs in various OPMD cellular models will undoubtedly allow the unraveling of this fascinating enigma.

Lastly, despite our extensive knowledge of the structure and function of PABPN1, we still do not know how expansions of its short polyalanine domain cause muscle demise. As discussed above, even the pathogenic role of the INIs still needs to be elucidated. What is clear, however, is that because of the increasing number of diseases caused by polyalanine expansions and the pathological overlap with CAG/polyglutamine diseases, whatever pathological insights are gained by the study of OPMD could lead to a better understanding of a much larger group of developmental and degenerative diseases (Gaspar et al., 2000)

Acknowledgements

We would like to thank Michel Fardeau, Jean-Pierre Bouchard, Fernando M.S. Tomé, and Guy A. Rouleau for their precious collaboration.

References

Amyot R: Hereditary, familial and acquired ptosis of late onset. Can med Assoc J 59:434–438 (1948).

Amyot R: Ptosis héréditaire familial et tardif des paupières supérieures, in Pharyngoplégie également héréditaire et familiales concomittante. l'Union Médical du Canada 77:1287 (1948).

Askanas V, Serdaroglu P, Engel WK, Alvarez RB: Immunolocalization of ubiquitin in muscle biopsies of patients with inclusion body myositis and oculopharyngeal muscular dystrophy. Neurosci Lett 130:73–76 (1991).

Bao Y, Cook LJ, O'Donovan D, Uyama E, Rubinsztein DC: Mammalian, yeast, bacterial and chemical chaperones reduce aggregate formation and death in a cell model of oculopharyngeal muscular dystrophy. J biol Chem 277:12263–12269 (2002).

Barbeau A: The syndrome of hereditary late onset ptosis and dysphagia in French Canada, in Kuhn E (ed): Symposium über progressive Muskeldystrophie, pp 102–109 (Springer-Verlag, Berlin 1966).

Barbeau A: Oculopharyngeal muscular dystrophy in French Canada, in Brunette J-R, Barbeau A (eds): Progress in Neuro-Ophthalmology, Vol 2, p 3 (Excerpta Medica, Amsterdam 1969).

Becher MW, Kotzuk JA, Davis LE, Bear DG: Intranuclear inclusions in oculopharyngeal muscular dystrophy contain poly(A) binding protein 2. Annls Neurol 48:812–815 (2000).

Becher MW, Morrison L, Davis LE, Maki WC, King MK, Bicknell JM, Reinert BL, Bartolo C, Bear DG: Oculopharyngeal muscular dystrophy in Hispanic New Mexicans. JAMA 286:2437–2440 (2001).

Bienroth S, Keller W, Wahle E: Assembly of a processive messenger RNA polyadenylation complex. EMBO J 12:585–594 (1993).

Billingsley GD, Cox DW, Duncan AM, Goodfellow PJ, Grzeschik KH: Regional localization of loci on chromosome 14 using somatic cell hybrids. Cytogenet Cell Genet 66:33–38 (1994).

Blumen SC, Brais B, Korczyn AD, Medinsky S, Chapman J, Asherov A, Nisipeanu P, Codere F, Bouchard JP, Fardeau M, Tome FM, Rouleau GA: Homozygotes for oculopharyngeal muscular dystrophy have a severe form of the disease. Annls Neurol 46:115–118 (1999).

Blumen SC, Korczyn AD, Lavoie H, Medynski S, Chapman J, Asherov A, Nisipeanu P, Inzelberg R, Carasso RL, Bouchard JP, Tomé FM, Rouleau GA, Brais B: Oculopharyngeal MD among Bukhara Jews is due to a founder $(GCG)_9$ mutation in the PABP2 gene. Neurology 55:1267–1270 (2000).

Blumen SC, Nisipeanu P, Sadeh M, Asherov A, Blumen N, Wirguin Y, Khilkevich O, Carasso RL, Korczyn AD: Epidemiology and inheritance of oculopharyngeal muscular dystrophy in Israel. Neuromuscul Disord 7:S38–40 (1997).

Blumen SC, Sadeh M, Korczyn AD, Rouche A, Nisipeanu P, Asherov A, Tomé FM: Intranuclear inclusions in oculopharyngeal muscular dystrophy among Bukhara Jews. Neurology 46:1324–1328 (1996).

Bouchard J-P, Brais B, Brunet D, Gould PV, Rouleau GA: Recent studies on oculopharyngeal muscular dystrophy in Quebec. Neuromuscul Disord 7:S22–29 (1997).

Boukriche Y, Maisonobe T, Masson C: Neurogenic involvement in a case of oculopharyngeal muscular dystrophy. Muscle Nerve 25:98–101 (2002).

Brais B, Bouchard JP, Tomé FMS, Fardeau M, Codère F, Duranceau A, Rouleau GA: Genetic evidence for the involvement of other genes in modulating the severity of oculopharyngeal muscular dystrophy (OPMD). Annls Neurol 44:455–456 (1998a).

Brais B, Bouchard JP, Xie YG, Rochefort DL, Chretien N, Tome FM, Lafreniere RG, Rommens JM, Uyama E, Nohira O, Blumen S, Korczyn AD, Heutink P, Mathieu J, Duranceau A, Codere F, Fardeau M, Rouleau GA, Korcyn AD: Short GCG expansions in the PABP2 gene cause oculopharyngeal muscular dystrophy. Nature Genet 18:164–167 (1998b).

Brais B, Rouleau GA, Bouchard JP, Fardeau M, Tomé FM: Oculopharyngeal muscular dystrophy. Sem Neurol 19:59–66 (1999).

Brais B, Xie YG, Sanson M, Morgan K, Weissenbach J, Korczyn AD, Blumen SC, Fardeau M, Tomé FM, Bouchard JP, Rouleau GA: The oculopharyngeal muscular dystrophy locus maps to the region of the cardiac α and β myosin heavy chain genes on chromosome 14q11.2→q13. Hum molec Genet 4:429–434 (1995).

Calado A, Kutay U, Kuhn U, Wahle E, Carmo-Fonseca M: Deciphering the cellular pathway for transport of poly(A)-binding protein II. RNA 6:245–256 (2000).

Calado A, Tomé FM, Brais B, Rouleau GA, Kuhn U, Wahle E, Carmo-Fonseca M: Nuclear inclusions in oculopharyngeal muscular dystrophy consist of poly(A) binding protein 2 aggregates which sequester poly(A) RNA. Hum molec Genet 9:2321–2328 (2000).

Chen Z, Li Y, Krug RM: Influenza A virus NS1 protein targets poly(A)-binding protein II of the cellular 3ʹ-end processing machinery. EMBO J 18:2273–2283 (1999).

Codère F: Oculopharyngeal muscular dystrophy. Can J Ophthal 28:1–2 (1993).

Codère F, Brais B, Rouleau G, Lafontaine E: Oculopharyngeal muscular dystrophy: what's new? Orbit 20:259–266 (2001).

Colgan DF, Manley JL: Mechanism and regulation of mRNA polyadenylation. Genes Dev 11:2755–2766 (1997).

Coquet M, Vallat JM, Vital C, Fournier M, Barat M, Orgogozo JM, Julien J, Loiseau P: Nuclear inclusions in oculopharyngeal dystrophy: an ultrastructural study of 6 cases. J neurol Sci 60:151–156 (1983).

Coquet M, Vital C, Julien J: Presence of inclusion body myositis-like filaments in oculopharyngeal muscular dystrophy: ultrastructural study of 10 cases. Neuropath appl Neurobiol 16:393 (1990).

Duranceau A: Cricopharyngeal myotomy in the management of neurogenic and muscular dysphagia. Neuromuscul Disord 7:S85–S89 (1997).

Evert BO, Wullner U, Schulz JB, Weller M, Groscurth P, Trottier Y, Brice A, Klockgether T: High level expression of expanded full-length ataxin-3 in vitro causes cell death and formation of intranuclear inclusions in neuronal cells. Hum molec Genet 8:1169–1176 (1999).

Fan X, Dion P, Lafrenière J, Brais B, Rouleau GA: Oligomerization of polyalanine expanded PABPN1 facilitates nuclear protein aggregation that is associated with cell death. Hum molec Genet 10:2341–2351 (2001).

Forood B, E Pérez-Payá, Houghten RA, Blondelle SE: Formation of an extremely stable polyalanine beta-sheet macromolecule. Biochem biophys Res Commun 211:7–13 (1995).

Galvao R, Mendes-Soares L, Camara J, Jaco I, Carmo-Fonseca M: Triplet repeats, RNA secondary structure and toxic gain-of-function models for pathogenesis. Brain Res Bull 56:191–201 (2001).

Gaspar C, Jannatipour M, Dion P, Laganiere J, Sequeiros J, Brais B, Rouleau GA: CAG tract of MJD-1 may be prone to frameshifts causing polyalanine accumulation. Hum molec Genet 9:1957–1966 (2000).

Grewal RP, Cantor R, Turner G, Grewal RK, Detera-Wadleigh SD: Genetic mapping and haplotype analysis of oculopharyngeal muscular dystrophy. Neuroreport 9:961–965 (1998).

Grewal RP, Karkera JD, Grewal RK, Detera-Wadleigh SD: Mutation analysis of oculopharyngeal muscular dystrophy in Hispanic American families. Arch Neurol 56:1378–1381 (1999).

Hill ME, Creed GA, McMullan TF, Tyers AG, Hilton-Jones D, Robinson DO, Hammans SR: Oculopharyngeal muscular dystrophy: phenotypic and genotypic studies in a UK population. Brain 124(Pt 3): 522–526 (2001).

Keller W, Bienroth S, Lang KM, Christofori G: Cleavage and polyadenylation factor CPF specifically interacts with the pre-mRNA 3ʹ processing signal AAUAAA. EMBO J 10:4241–4249 (1991).

Kim YJ, Noguchi S, Hayashi YK, Tsukahara T, Shimizu T, Arahata K: The product of an oculopharyngeal muscular dystrophy gene, poly(A)-binding protein 2, interacts with SKIP and stimulates muscle-specific gene expression. Hum molec Genet 10:1129–1139 (2001).

Krause S, Fakan S, Weis K, Wahle E: Immunodetection of poly(A) binding protein II in cell nucleus. Expl Cell Res 214:75–82 (1994).

Leclerc A, Tomé FMS, Fardeau M: Ubiquitin and beta-amyloid-protein in inclusion body myositis (IBM), familial IBM-like disorder and oculopharyngeal muscular dystrophy: an immunocytochemical study. Neuromuscul Disord 3:283–291 (1993).

Little BW, Perl DP: Oculopharyngeal muscular dystrophy. An autopsied case from the French-Canadian kindred. J neurol Sci 53:145–158 (1982).

Lotz BP, Engel AG, Nishino H, Stevens JC, Litchy WJ: Inclusion body myositis: observation in 40 patients. Brain 112:727 (1989).

Medici M, Pizzarossa C, Skuk D, Yorio D, Emmanuelli G, Mesa R: Oculopharyngeal muscular dystrophy in Uruguay. Neuromuscul Disord 7:S50–S52 (1997).

Minvielle-Sebastia L, Keller W: mRNA polyadenylation and its coupling to other RNA processing reactions and to transcription. Curr Opin Cell Biol 11:352–357 (1999).

Mirabella M, Silvestri G, de Rosa G, Di Giovanni S, Di Muzio A, Uncini A, Tonali P, Servidei S: GCG genetic expansions in Italian patients with oculopharyngeal muscular dystrophy. Neurology 54: 608–614 (2000).

Muller T, Schroder R, Zierz S: GCG repeats and phenotype in oculopharyngeal muscular dystrophy. Muscle Nerve 24:120–122 (2001).

Muragaki Y, Mundlos S, Upton J, Olsen BR: Altered growth and branching patterns in synpolydactyly caused by mutations in HOXD13. Science 272:548–551 (1996).

Nakamoto M, Nakano S, Kawashima S, Ihara M, Nishimura Y, Shinde A, Kakizuka A: Unequal crossing-over in unique PABP2 mutations in Japanese patients: a possible cause of oculopharyngeal muscular dystrophy. Arch Neurol 59:474–477 (2002).

Nemeth A, Krause S, Blank D, Jenny A, Jeno P, Lustig A, Wahle E: Isolation of genomic and cDNA clones encoding bovine poly(A) binding protein II. Nucl Acids Res 23:4034–4041 (1995).

Porschke H, Kress W, Reichmann H, Goebel HH, Grimm T: Oculopharyngeal muscular dystrophy and carnitine deficiency in a Northern German family. Neuromuscul Disord 7:S57–S62 (1997).

Probst A, Tackmann W, Stoeckli HR, Jerusalem F, Ulrich J: Evidence for a chronic axonal atrophy in oculopharyngeal "muscular dystrophy." Acta neuropath 57:209–216 (1982).

Rankin J, Wyttenbach A, Rubinsztein DC: Intracellular green fluorescent protein-polyalanine aggregates are associated with cell death. Biochem J 348:15–19 (2000).
Ravikumar B, Duden R, Rubinsztein DC: Aggregate-prone proteins with polyglutamine and polyalanine expansions are degraded by autophagy. Hum molec Genet 11:1107–1117 (2002).
Rebeiz JJ, Caulfield JB, Adams RD: Oculopharyngeal dystrophy—a presenescent myopathy: a clinicopathological study, in Brunette J-R, Barbeau A (eds): Progress in Neuro-Ophthalmology, Vol 2, p 12–31 (Excerpta Medica, Amsterdam 1969).
Saucier J: The clinical significance of ptosis with special reference to ptosis of late onset. J nerv ment Dis 119:148–158 (1954).
Scacheri PC, Garcia C, Hebert R, Hoffman EP: Unique *PABP2* mutation in "Cajuns" suggest multiple founders of oculopharyngeal muscular dystrophy in populations with French ancestry. Am J med Genet 86:477–481 (1999).
Schmitt HP, Krause KH: An autopsy study of a familial oculopharyngeal muscular dystrophy (OPMD) with distal spread and neurogenic involvement. Muscle Nerve 4:296–305 (1981).
Serratrice G, Pellissier JF: Myopathies oculaires: étude nosographique de 49 cas. Presses méd 16:1969–1974 (1987).
Sisodia SS: Nuclear inclusions in glutamine repeat disorders: are they pernicious, coincidental, or beneficial? Cell 95:1–4 (1998).
Smith TW, Chad D: Intranuclear inclusions in oculopharyngeal dystrophy. Muscle Nerve 7:339 (1984).
Stajich JM, Gilchrist JM, Lennon F, Lee A, Yamaoka L, Helms B, Gaskell PC, Donald L, Roses AD, Vance JM, Pericak-Vance MA: Confirmation of linkage of oculopharyngeal muscular dystrophy to chromosome 14q11.2-q13. Annls Neurol 40:801–804 (1997).
Taylor EW: Progressive vagus-glossopharyngeal paralysis with ptosis: a contribution to the group of family diseases. J nerv ment Dis 42:129–139 (1915).
Teh BT, Sullivan AA, Farnebo F, Zander C, Li FY, Strachan N, Schalling M, Larsson C, Sandstrom P: Oculopharyngeal muscular dystrophy; report and genetic studies of an Australian kindred. Clin Genet 51:52–55 (1997).
Tomé FMS, Fardeau M: Nuclear inclusions in oculopharyngeal muscular dystrophy. Acta neuropath 49:85–87 (1980).
Tomé FMS, Fardeau M: Nuclear changes in muscle disorders. Meth Achiev exp Pathol 12:261 (1986).
Tomé FMS, Fardeau M: Oculopharyngeal muscular dystrophy, in Engel AG, Franzini-Armstrong C (eds): Myology, Vol 2, pp 1233–1245 (McGraw-Hill, New York 1994).
Tomé FMS, Askanas V, Engel WK, Alvarez RB, Lee CS: Nuclear inclusions in innervated cultured muscle fibers from patients with oculopharyngeal muscular dystrophy. Neurology 39:926–932 (1989).
Tomé FMS, Chateau D, Helbling-Leclerc A, Fardeau M: Morphological changes in muscle fibers in oculopharyngeal muscular dystrophy. Neuromuscul Disord 7:S63–69 (1997).
Turmaine M, Raza A, Mahal A, Mangiarini L, Bates GP, Davies SW: Nonapoptotic neurodegeneration in a transgenic mouse model of Huntington's disease. Proc natl Acad Sci, USA 97:8093–8097 (2000).
Uyama E, Tsukahara T, Goto K, Kurano Y, Ogawa M, Kim YJ, Uchino M, Arahata K: Nuclear accumulation of expanded PABP2 gene product in oculopharyngeal muscular dystrophy. Muscle Nerve 23:1549–1554 (2000).
Victor M, Hayes R, Adams RD: Oculopharyngeal muscular dystrophy: a familial disease of late life characterized by dysphagia and progressive ptosis of the eyelids. New Engl J Med 267:1267–1272 (1962).
Wahle E: A novel poly(A)-binding protein acts as a specificity factor in the second phase of messenger RNA polyadenylation. Cell 66:759–768 (1991).
Wahle E: 3′-End cleavage and polyadenylation of mRNA precursors. Biochim biophys Acta 1261: 183–194 (1995).
Wahle E, Ruegsegger U: 3′-End processing of pre-mRNA in eukaryotes. FEMS Microbiol Rev 23: 277–295 (1999).
Wahle E, Lustig A, Jeno P, Maurer P: Mammalian poly(A)-binding protein II: physical properties and binding to polynucleotides. J biol Chem 268:2937–2945 (1993).
Warren ST: Polyalanine expansion in synpolydactyly might result from unequal crossing-over of HOXD13. Science 275:408–409 (1997).
Zhao J, Hyman L, Moore C: Formation of mRNA 3′ ends in eukaryotes: mechanism, regulation, and interrelationships with other steps in mRNA synthesis. Microbiol molec Biol Rev 63:405–445 (1999).
Zoghbi HY, Orr HT: Glutamine repeats and neurodegeneration. A Rev Neurosci 23:217–247 (2000).

Cytogenet Genome Res 100:261–275 (2003)
DOI: 10.1159/000072862

Molecular analyses of Machado-Joseph disease

T. Kobayashi and A. Kakizuka

Graduate School of Biostudies, Kyoto University, Kyoto, and CREST, Japan Science and Technology Corporation (Japan)

Introduction

Machado-Joseph disease (MJD) is an inherited neurodegenerative disorder with an autosomal dominant inheritance. The brain regions affected by MJD include the Clarke column and spinocerebellar tracts, pons, dentate nuclei, substantia nigra, anterior horn cells, red nuclei, and peripheral nerves, but MJD is distinguished from other inherited cerebellar ataxias because it spares the olives and cerebellar cortex. The established phenotypes of MJD include gait and limb ataxia, ophthalmoplegia, dystonia, amyotrophy, dysarthria, rigidity, pyramidal signs, facial and lingual fasciculations, and bulging eyes (the most characteristic feature).

In 1977 Romanul proposed the existence of a new disease entity based on apparent common clinical symptoms in four family pedigrees, including the Machado family and Joseph family (Nakano et al., 1972; Rosenberg et al., 1976). Since these pedigrees were all of Portuguese Azorean origin, the new disease was referred to as "Azorean disease" at the time (Romanul et al., 1997; Nakamoto et al., 1998). Later, the existence of several families with similar clinicopathological phenotypes of non-Azorean origin was reported, and therefore this disease has come to be more preferably known as Machado-Joseph disease (Healton et al., 1980; Lima and Coutinho, 1980; Sakai et al., 1983; Suite et al., 1986; Yuasa et al., 1986; Takiyama et al., 1989).

In 1993, a linkage study of large Japanese pedigrees identified the MJD locus on a distal region of the long arm of chromosome 14 (Takiyama et al., 1993). Almost simultaneously, a locus covering a similar region on chromosome 14 was reported as the thirdly identified locus of spinocerebellar ataxia (Stevanin et al., 1994). In this review, we introduce our long-term studies of MJD, covering the identification of MJD1 (the gene responsible for MJD) to the recent molecular analysis of MJD, and from the data obtained from MJD analyses, we discuss the potential general molecular mechanisms that underlie a broad spectrum of neurodegenerative disorders, if not all of them.

Identification of the gene responsible for MJD

Neurodegenerative disorders have a considerable range of signs and symptoms: dementia, ataxia, movement disorders, and so forth. However, from a more general point of view, they share several common features, e.g., the inheritances that result in these disorders are usually autosomal dominant, in many cases disease symptoms appear after middle age, and these symptoms progress after appearing. The most prominent pathology is neuronal cell loss and degeneration, although each disorder has its own regions of the central nervous system that are susceptible. Furthermore, for several inherited diseases, such as Huntington's disease (HD) and dentatorubral-pallidoluysian atrophy (DRPLA), clinical symptoms worsen and the age of onset becomes earlier in each succeeding generation. This phenomenon is called "anticipation." These features suggest that common mechanisms underlie these inherited neurodegenerative disorders, but the real molecular bases of these mechanisms have long remained unknown.

In retrospect, the first breakthrough was made in 1991 by La Spada et al. (1991). They reported the androgen receptor (AR) gene as the gene responsible for spinobulbar muscular atrophy (SBMA), an X-linked recessive neurodegenerative disorder. The gene responsible for HD, the most typical autosomal dominant neurodegenerative disease, was identified in 1993 (The

Received 20 January 2003; revision accepted 18 February 2003.

Request reprints from Dr. Akira Kakizuka, Laboratory of Functional Biology Graduate School of Biostudies, Kyoto University, Yosidakonoe, Sakyo-ku Kyoto 606-8501 (Japan), telephone: +81-75-753-7675; fax: +81-75-753-7676 e-mail: kakizuka@lif.kyoto-u.ac.jp

KARGER
Fax + 41 61 306 12 34
E-mail karger@karger.ch
www.karger.com

0301–0171/03/1004–0261$19.50/0

Accessible online at:
www.karger.com/cgr

A

```
a.a.                                                                        n.t.
    ATG GAG TCC ATC TTC CAC GAG AAA CAA GAA GGC TCA CTT TGT GCT CAA CAT TGC 54
  1 M   E   S   I   F   H   E   K   Q   E   G   S   L   C   A   Q   H   C
    CTG AAT AAC TTA TTG CAA GGA GAA TAT TTT AGC CCT GTG GAA TTA TCC TCA ATT 108
 19 L   N   N   L   L   Q   G   E   Y   F   S   P   V   E   L   S   S   I
    GCA CAT CAG CTG GAT GAG GAG GAG AGG ATG AGA ATG GCA GAA GGA GGA GTT ACT 162
 37 A   H   Q   L   D   E   E   E   R   M   R   M   A   E   G   G   V   T
    AGT GAA GAT TAT CGC ACG TTT TTA CAG CAG CCT TCT GGA AAT ATG GAT GAC AGT 216
 55 S   E   D   Y   R   T   F   L   Q   Q   P   S   G   N   M   D   D   S
    GGT TTT TTC TCT ATT CAG|GTT ATA AGC AAT GCC TTG AAA GTT TGG GGT TTA GAA 270
 73 G   F   F   S   I   Q   V   I   S   N   A   L   K   V   W   G   L   E
    CTA ATC CTG TTC AAC AGT CCA GAG TAT CAG AGG CTC AGG ATC GAT CCT AT|AAAT 324
 91 L   I   L   F   N   S   P   E   Y   Q   R   L   R   I   D   P   I   N
    GAA AGA TCA TTT ATA TGC AAT TAT AAG GAA CAC TGG TTT ACA GTT AGA AAA TTA 378
109 E   R   S   F   I   C   N   Y   K   E   H   W   F   T   V   R   K   L
    GGA AAA CAG TGG TTT AAC TTG AAT TCT CTC TTG ACG GGT CCA GAA TTA ATA TCA 432
127 G   K   Q   W   F   N   L   N   S   L   L   T   G   P   E   L   I   S
    GAT ACA TAT CTT GCA CTT TTC TTG GCT CAA TTA CAA CAG GAA GGT TAT TCT ATA 486
145 D   T   Y   L   A   L   F   L   A   Q   L   Q   Q   E   G   Y   S   I
    TTT GTT GTT AAG GGT GAT CTG CCA GAT TGC GAA GCT GAC CAA CTC CTG CAG ATG 540
163 F   V   V   K   G   D   L   P   D   C   E   A   D   Q   L   L   Q   M
    ATT AGG GTC CAA CAG ATG CAT CGA CCA AAA CTT ATT GGA GAA GAA TTA GCA CAA 594
181 I   R   V   Q   Q   M   H   R   P   K   L   I   G   E   E   L   A   Q
    CTA AAA GAG CAA AGA GTC CAT AAA ACA GAC CTG GAA CGA ATG TTA GAA GCA AAT 648
199 L   K   E   Q   R   V   H   K   T   D   L   E   R   M   L   E   A   N
    GAT GGC TCA GGA ATG TTA GAC GAA GAT GAG GAG GAT TTG CAG AGG GCT CTG GCA 702
217 D   G   S   G   M   L   D   E   D   E   E   D   L   Q   R   A   L   A
    CTA AGT CGC CAA GAA ATT GAC ATG GAA GAT GAG GAA GCA GAT CTC CGC AGG GCT 756
235 L   S   R   Q   E   I   D   M   E   D   E   E   A   D   L   R   R   A
    ATT CAG CTA AGT ATG CAA GGT AGT TCC AGA AAC ATA TCT CAA GAT ATG ACA CAG 810
253 I   Q   L   S   M   Q   G   S   S   R   N   I   S   Q   D   M   T   Q
    ACA TCA GGT ACA AAT CTT ACT TCA GAA GAG CTT CGG AAG AGA CGA GAA GCC TAC 864
271 T   S   G   T   N   L   T   S   E   E   L   R   K   R   R   E   A   Y
    TTT GAA AA|ACAG CAG CAA AAG CAG CAA CAG CAG CAG CAG CAG CAG CAG CAG CAG 918
289 F   E   K   Q   Q   Q   K   Q   Q   Q   Q   Q   Q   Q   Q   Q   Q   Q
    CAG CAG CAG CAG CAG CAG CAG CAG CAG CAG CAG CGG GAC CTA TCA GGA CAG AGT 972
307 Q   Q   Q   Q   Q   Q   Q   Q   Q   Q   Q   R   D   L   S   G   Q   S
    TCA CAT CCA TGT GAA AGG CCA GCC ACC AGT TCA GGA GCA CTT GGG AGT GAT CTA 1026
325 S   H   P   C   E   R   P   A   T   S   S   G   A   L   G   S   D   L
    G|GT AAG GCC TGC TCA CCA TTC ATC ATG TTC GCT ACC TTC ACA CTT TAT CTG ACA 1080
343 G   K   A   C   S   P   F   I   M   F   A   T   F   T   L   Y   L   T
    TAA
    *
```

B

```
    G|GTAAG GCC TGC TCA CCA TTC ATC ATG TTC GCT ACC TTC ACA CTT TAT CTG ACA 1080
343 G   K   A   C   S   P   F   I   M   F   A   T   F   T   L   Y   L   T
    G|GTGAT GCT ATG AGT GAA GAA GAC ATG CTT CAG GCA GCT GTG ACC ATG TCT TTA
    G   D   A   M   S   E   E   D   M   L   Q   A   A   V   T   M   S   L

    TAC GAG CTC CAT GTG ATT TTT GCT TTA CAT TAT TCT TCA TTC CCT CTT TAA TCA 1134
361 Y   E   L   H   V   I   F   A   L   H   Y   S   S   F   P   L   *
    GAA ACT GTC AGA AAT GAT TTG AAA ACA GAA GGA AAA AAA TAA TAC CTT TAA TAC
    E   T   V   R   N   D   L   K   T   E   G   K   K   *
```

C

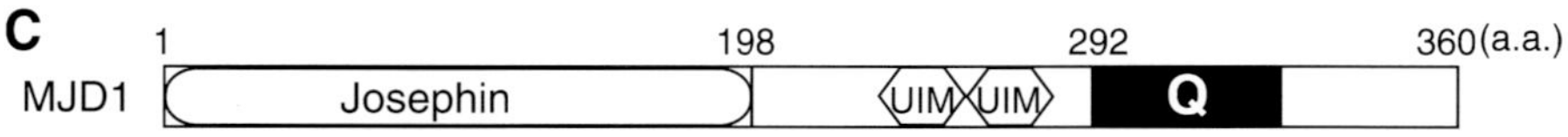

Fig. 1. Identified MJD1a cDNA and MJD1 protein. (**A**) Nucleotide and amino acid sequence of the coding region of MJD1a cDNA isolated from a human brain cDNA library. The CAG repeat is shown in bold type. Variant triplets, CAA and AAG, are boxed. Bars show the identified exon-intron boundaries. Based on Ikeda et al. (1996), with modifications. (**B**) Two additional, different *C*-terminal encoding portions were created by a polymorphic nucleotide change (from A to C) or alternative splicing. The A-to-C nucleotide change converts a stop codon to a tyrosine-coding codon and therefore adds 16 amino acids in the *C*-terminal to the MJD1a protein. (**C**) Schematic structure of the MJD1 protein. MJD1 contains a Josephin motif and two ubiquitin-interacting motifs. A closed box indicates the polyQ region.

Huntington's Disease Collaborative Research Group, 1993). Both disorders have totally different symptoms, inheritances, and chromosomal loci (on chromosomes X and 4 in SBMA and HD, respectively), and they both are caused by genes encoding nonhomologous proteins. However, patient-related mutations were expansions (more than 40 repeats) of polyglutamine (polyQ)-coding CAG repeats. The polyglutamines were located near the *N*-terminal portion in the resulting proteins. Surprisingly, in both diseases, patients with longer expansions appeared to have more severe clinical symptoms and to manifest an earlier disease onset. Even more surprisingly, the highly expanded CAG repeats in HD appeared to be unstably transmitted to succeeding generations, with a tendency toward further elongation, typically in paternal transmission (Duyao et al., 1993). This observation provided a simple but convincing molecular basis of anticipation. Thus, suddenly the possibility was raised that expansions of CAG repeats or other triplet repeats are also responsible for other inherited neurodegenerative disorders, especially those with anticipation.

On the assumption that several other inherited neurodegenerative disorders are also caused by CAG triplet expansions in yet unknown genes, we first searched for novel CAG-containing genes that are expressed in the human brain. Using an oligonucleotide with a $(CTG)_{13}$ repeat as a probe, which we expected to hybridize strongly to CAG repeats and weakly to CGG repeats, we screened a human brain cDNA library. We succeeded in isolating about 30 clones containing CAG repeats, but not CGG repeats (Kawaguchi et al., 1994). One of them consisted of 1,776 bp with a long open reading frame encoding a soluble cytoplasmic protein with an amino acid length of 360 (Kawaguchi et al., 1994) (Fig. 1A). The CAG repeat was predicted to be translated into a glutamine tract, as observed in SBMA and HD, but, instead, was located near the *C*-terminal portion. Northern blot analysis revealed that transcripts were faintly detectable in all of the tissues examined, with the exception of testis, in which strong expression was observed. Reverse transcriptase–polymerase chain reaction (RT-PCR) analysis of human brain mRNA demonstrated expression of both alleles with different lengths of CAG repeats, suggesting that each allele had lengths of CAG repeats that were polymorphic in nature. In addition, the originally identified stop codon was later found to be polymorphically changed to a tyrosine-coding TAC in about 60% of the alleles in Japanese (Fig. 1B) (Goto et al., 1997). A splice variant with a different *C*-terminal end (Fig. 1B) was also identified (Nakamoto et al., 1998).

Despite our success with RT-PCR, we were unable to amplify genomic fragments containing the CAG repeats by PCR using any PCR primers that were constructed from the coding sequence. We assumed the presence of an intron close to the CAG repeat and thus screened a human genomic library to isolate the corresponding genomic clones; in the process, we obtained four related genomic fragments. We gave the symbol MJD1 to the gene corresponding to the cDNA and the symbols MJD2, MJD3, and MJD4 to the other three putatively related genes (Kawaguchi et al., 1994). MJD1 and the other three genomic fragments were mapped at 14q32.1, 8q23, 14q21, and Xp22.1, respectively (Kawaguchi et al., 1994). The physiological functions of MJD1 remain to be clarified. However, MJD1 has two ubiquitin-interacting motifs in the *N*-terminal side of polyQ and carries another motif, named "Josephin," at the *N*-terminus (Fig. 1C). A recent report shows that MJD1 is able to bind to HHR23A and HHR23B (the human homologs of the yeast DNA repair protein RAD23) through their ubiquitin-like (UBL) domains (Wang et al., 2000).

As predicted, we identified an intron just upstream of the CAG repeat in MJD1. This intron sequence allowed us to amplify by PCR the MJD1 genomic sequence surrounding the CAG repeat. Then, we compared the CAG repeat numbers in MJD1 of healthy individuals with those of clinically diagnosed MJD patients (Fig. 2A) (Kawaguchi et al., 1994). Consistent with our RT-PCR analysis on human brain mRNA, CAG repeat numbers in MJD1 were found to be polymorphic in healthy individuals, from 13 to 36 repeats. The most frequent allele was found to contain 14 repeats, representing about 30% of the total, followed by those with 27, 28, and 21 repeats, representing approximately 15% each. In contrast, MJD patients were found to have longer CAG repeats, from 62 to 84 repeats, and a clear reverse correlation was observed between the repeat numbers and the age of disease onset (Fig. 2B). One repeat elongation was estimated to accelerate onset by approximately 2.6 yr. A recent report shows that rare intermediate repeat lengths, such as 53 and 54 repeats, are also pathogenic (van Alfen et al., 2001). In patients with CAG expansions in both MJD1 genes, onset occurred much earlier than predicted from the repeat numbers, indicating a gene-dosage effect.

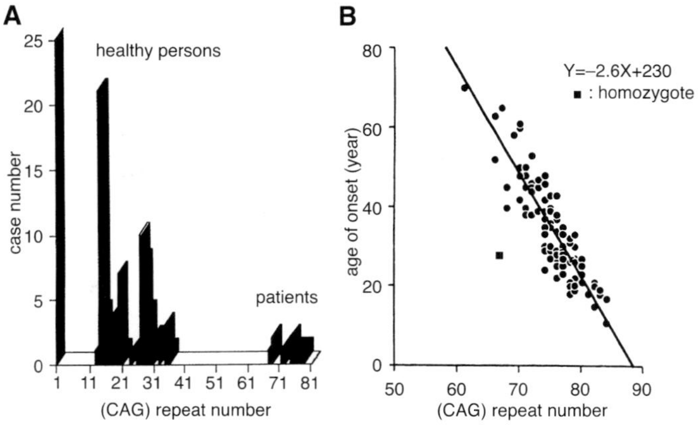

Fig. 2. Clinical features of the CAG repeat numbers of the *MJD1* gene. (**A**) Initial estimation of the CAG repeat numbers in 72 normal alleles and 11 MJD alleles, which were subsequently found to contain 13–36 and 68–79 CAG repeats, respectively.(**B**) Correlation between the number of CAG repeats and age of disease onset. Solid circles: data from 108 heterozygous MJD patients; solid box: number of CAG repeats and age of onset in one homozygous MJD patient. See text for detailed explanation.

Neuronal degeneration by ex-polyQs in the MJD protein

At almost the same time, an additional two inherited neurodegenerative diseases, spinocerebellar ataxia type 1 (SCA1) and DRPLA, were found to be caused by the expansion of CAG repeats in different responsible genes (Banfi et al., 1993; Koide et al., 1994; Nagafuchi et al., 1994). It was predicted that all of the CAG repeats that had been identified as causing these five inherited neurodegenerations by expansion were transcribed and translated into polyQ stretches as portions of apparent intracellular proteins. Thus, the simplest and easiest idea to test was that expanded polyQ (ex-polyQ) itself plays a toxic function that is causally related to the neurodegeneration.

A cultured cell model of the polyQ diseases

To test this idea, expression vectors for MJD cDNAs with various CAG repeat lengths were transfected in cultured COS7 cells and neuronal PC12 cells, and their phenotypes were examined (Fig. 3A) (Ikeda et al., 1996; Yasuda et al., 1999). All of these proteins were expressed with a hemagglutinin epitope (HA) being tagged in their *N*-terminals. No obvious morphological changes were observed by the expression of full-length MJD proteins with normal (MJD27 and MJD35) and even diseased (MJD79) lengths of polyQs, nor were they observed in a *C*-terminal portion of the MJD protein with a normal polyQ length (Q27C). Conversely, cell death was induced when the C-terminal portion with ex-polyQ (Q79C) and an ex-polyQ itself (Q79) were expressed (Fig. 3B) (Yasuda et al., 1999). An immunocytochemical analysis using an anti-HA antibody showed that Q79C and Q79 had stained punctate patterns, suggesting that they aggregated in the cells (Fig. 3C) (Yasuda et al., 1999). In cells with such ex-polyQ aggregates, nuclear condensation and fragmentation of DNA—phenotypes related to apoptosis—were observed. The strength of the aggregates and cell death were parallel, both depending on the length and amount of ex-polyQ. Interestingly, as the protein portions surrounding polyQ became shorter, the strength of the aggregates and the amount of induced cell death increased (see below).

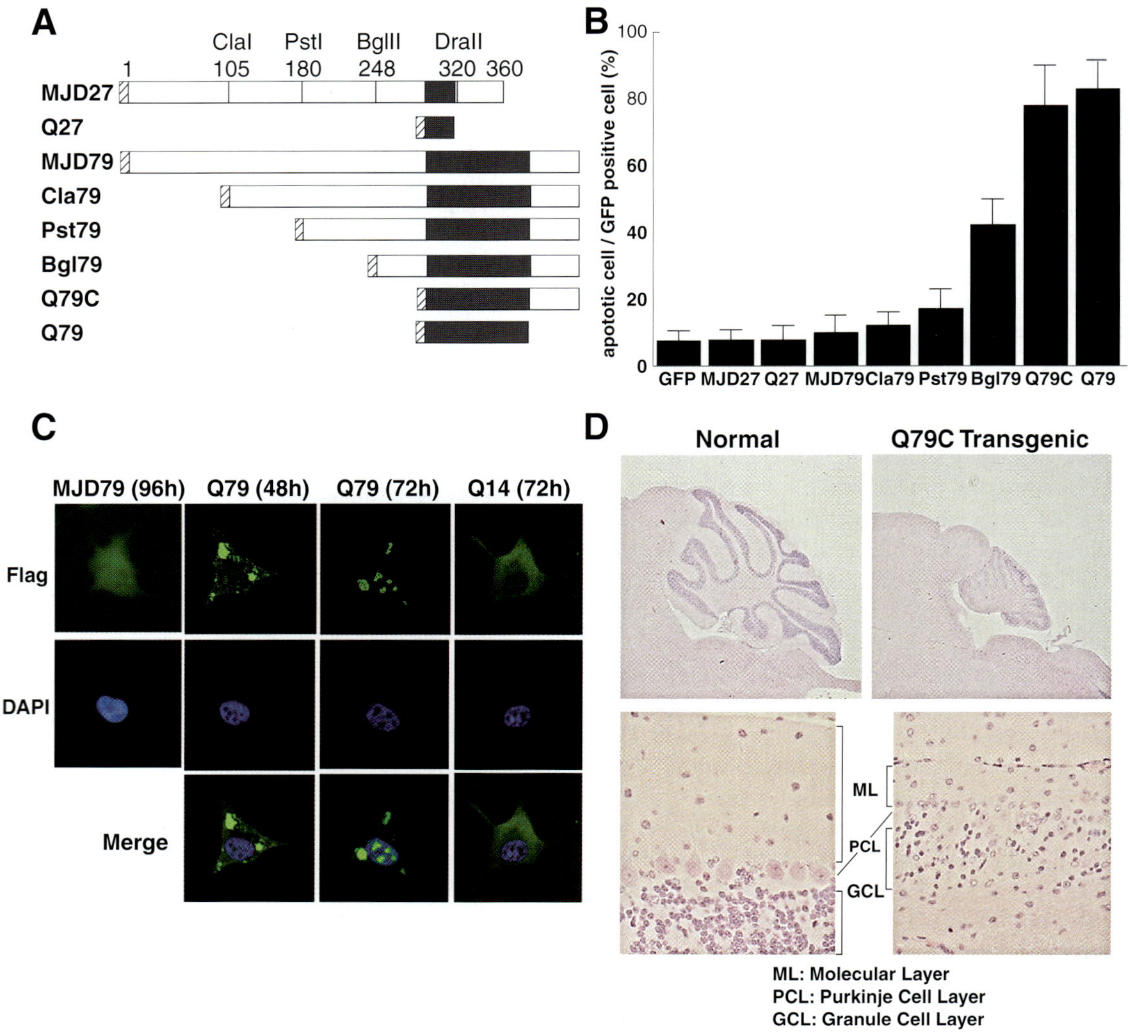

Fig. 3. Cell culture and mouse models of the PolyQ disease. (**A**) Schematic structures of various MJD proteins whose ability to induce cell death was examined in cultured PC12 cells via the transient transfection assay. Cross-hatched and solid areas indicate the flag epitopes and polyQ regions, respectively. Based on Yasuda et al. (1999), with modifications. (**B**) PC12 cells were transfected with expression vectors for GFP and one of the MJD proteins shown in the first panel. Nerve growth factor (NGF) was supplemented 24 h after the transfection. The percentage of apoptotic cells among GFP-positive cells was determined 72 h after the addition of NGF, and the mean values of triplicate experiments are presented, and standard deviations are included as bars. Based on Yasuda et al. (1999), with modifications. (**C**) Immunocytochemical views of PC12 cells expressing MJD79, Q79, or Q14. PC12 cells expressing MJD79, Q79, or Q14 under tet-off promoter were constructed. At indicated time points after tet-off, the expressed proteins were visualized as green signals by treatment with anti-flag antibody followed by treatment with FITC-conjugated secondary antibody (upper panels), and the nuclei were visualized as blue signals with TOTO-3 staining (middle panels). Superimposed views are also presented (lower panels). Based on Yasuda et al. (1999), with modifications. (**D**) Histological analysis of cerebellums of normal and ex-polyQ transgenic mice. Brain sections of an 8-week-old Q79C transgenic mouse and an age-matched littermate were examined. Hematoxylin and eosin staining at low magnification (upper panels) or at high magnification (lower panels) is presented. The cerebellums of the transgenic mouse were extremely atrophic. ML = molecular layer; PCL = Purkinje cell layer; GCL = granule cell layer. Based on Ikeda et al. (1996), with modifications.

A transgenic mouse model of the polyQ diseases

To investigate whether ex-polyQ could also induce neurodegeneration or neuronal cell death in vivo, we created transgenic mice in which a similar set of MJD proteins (Q79C, Q79, Q35C, and MJD79) were expressed specifically in mouse Purkinje cells using the L7 gene cassette (Ikeda et al., 1996). All of the Q79C transgenic mice, their F_1 offspring, and two of six Q79 transgenic mice were found to have an ataxic posture and gait disturbance. The affected cerebellums were extremely atrophic, which could be seen at a glance. Their volumes were approximately one eighth those of the cerebellums of their non-transgenic littermates (Fig. 3D) (Ikeda et al., 1996). The target Purkinje cells were scarce. However, we obtained clear evidence of cell death. In contrast, none of the Q35C and MJD79 proteins in the transgenic mice showed any abnormal phenotypes.

These results clearly show that the ex-polyQ is capable of causing cell death in cultured cell and neurodegeneration in transgenic mice. In contrast, the full-length MJD proteins, even those with ex-polyQs, appeared to have a trivial amount of cellular toxicity, if toxicity was present at all, and the toxicity only became evident when certain *C*-terminal portions with ex-

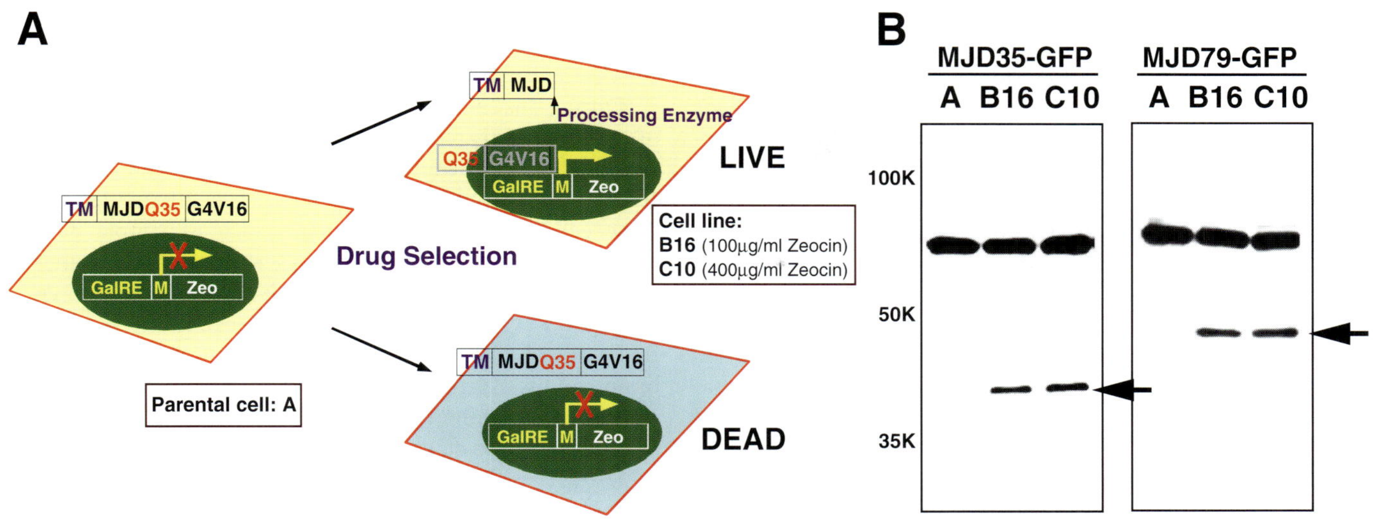

Fig. 4. MJD protein processing in B16 and C10 cells. Based on Yamamoto et al. (2001), with modifications. (**A**) Schematic drawing of the "cleavage-dependent cell-selection system." Live and dead cells are shown in yellow and blue, respectively, and the cell membrane and the nuclei are shown in red and green, respectively. TM-MJDQ35-G4V16 indicates the MJD protein with a 35-glutamine stretch (MJDQ35); with a Fas transmembrane domain (TM) fused to its *N*-terminus; and with Gal4VP16, a strong artificial transcription factor (G4V16), fused to its *C*-terminus. GalRE-M-Zeo indicates a reporter construct containing Gal4-response elements (GalRE) and an MTV minimum promoter (M) upstream of the Zeocin-resistant gene (Zeo). B16 and C10 were selected by the cleavage-dependent cell-selection system via sequential drug selection with 100 µg/ml and 400 µg/ml Zeocin, respectively. A cells, parental PC12 cells. (**B**) The results of Western blot analyses of cell extracts from the parental PC12 cells (A), B16, and C10 cells 6 days after transfection with MJD35-GFP and MJD79-GFP are presented. Proteins were detected using an anti-GFP antibody. Arrows indicate bands representing cleaved proteins.

polyQs were cut off and released in the cells. Based on these results, we suggested a "processing model" that proposes that cleavage or processing is a crucial step in the pathogenesis of the polyQ diseases, namely, that the brain region-specific cleavage of the diseased proteins with ex-polyQs leads to the degeneration of specific brain regions for each disease. This model is able to account for the apparent discrepancy between the affected brain regions and expression patterns of the genes responsible for the polyQ diseases.

Protein processing and seed formation for ex-polyQ aggregates

To find further evidence to substantiate the processing model, we attempted to identify cultured cells in which there is MJD protein processing activity. For this purpose, we expressed MJD79 in a variety of cell lines and examined these MJD79 proteins immunocytochemically. Initially, we were unable to find any evidence suggesting cleavage of the MJD protein in any of the cultured cells examined, which were all dividing cells. However, when neuronally differentiated postmitotic PC12 cells were examined several days after transfection of MJD79, we noticed that ex-polyQ–mediated aggregates were made in approximately one in a few hundred of the transfected PC12 cells; these aggregates could be detected only with an antibody that recognized an amino acid stretch just downstream of the polyglutamine portion of the MJD protein, and not with an antibody to a tag attached to the *N*-terminus of the MJD protein (Yasuda et al., 1999).

We then designed a "cleavage-dependent cell selection system" in which weak protein processing activity is converted into a strong transcriptional activity, thus allowing the expression of drug-resistant proteins required for selection. This system will provide a novel cell selection system for isolating cells containing processing activities for any protein concerned (Fig. 4A) (Yamamoto et al., 2001). The full-length MJD35 protein fused with Ga14VP16, a strong transcriptional activator, at the *C*-terminus and with the transmembrane region of the Fas receptor at the *N*-terminus, is expressed together with a Gal4 reporter plasmid, which drives the expression of a drug-resistant marker protein, in this case for Zeocin resistance. More specifically, this system should work in the following way: when the MJD protein, in this case MJD35, is cleaved by its processing enzyme, the cleaved portion with Gal4VP16 will translocate to the nucleus and transactivate the reporter gene, thereby resulting in Zeocin resistance in dividing pre-differentiated PC12 cells. The Zeocin-resistant cells obtained from this system should contain PC12 subclones with enhanced processing activity of the MJD protein. By repeating this selection process, cells with much higher processing activity for the MJD protein should be obtained. We used MJD35, which contains a normal length of polyglutamine repeats in the selection process, to avoid aggregate formation of the cleaved fragments and consequent cell death. We were indeed able to obtain several PC12 subclones after sequential selection with 100 µg/ml and 400 µg/ml Zeocin. Among these subclones, two representative subclones (B16 and C10) were analyzed by Western blots and were confirmed to have increased MJD processing activity (Fig. 4B) (Yamamoto et al., 2001). A rough estimation of the cleavage site of the MJD protein, based on the sizes of the cleaved products, suggests that the MJD protein is cleaved in these cells around the 250th amino acid residue (Yamamoto et al., 2001).

This processing model appears to be applicable not only for MJD but also for several other polyQ diseases. Apparent processed fragments were found in the nuclear inclusion (NI) in patients' brain samples of HD, MJD, SBMA, and DRPLA as truncated proteins without the epitopes covering the entire proteins (DiFiglia et al., 1997; Li et al., 1998; Schmidt et al., 1998; Schilling et al., 1999). Recently, it has been reported that an aspartic protease inhibitor inhibits the HD protein cleavage in cultured cells, but the responsible protease still remains to be identified (DiFiglia, 2002; Lunkes et al., 2002).

What is the significance of the processing in polyQ diseases? In vitro analysis using GST-fused ex-polyQ and several cellular models clearly demonstrated that polypeptides containing long polyQ aggregate easily when surrounding protein portions become short (Ikeda et al., 1996; Scherzinger et al., 1997). In addition, the truncated fragment of the MJD protein was shown in a cellular model to recruit full-length MJD proteins into insoluble aggregates (Perez et al., 1998). These results suggest that, by steric hindrance, surrounding portions of ex-polyQs inhibit the seed formation in a critical initial step of aggregate formation. In the case of the MJD protein, it possesses a higher-ordered structure in its *N*-terminal portion and a relatively free structure around the *C*-terminus where the polyQ stretch resides (unpublished observations), which fits well with the observation that the *C*-terminal portion of MJD tends to make an aggregate (Yasuda et al., 1999). Another report showed that the overexpression of polyQ-expanded full-length MJD proteins is capable of causing the formation of intranuclear inclusions and cell death, suggesting that a high concentration of the MJD protein may overcome the steric hindrance (Evert et al., 1999).

Formation and toxicity of the protein aggregate

Why does aggregate formation correlate with cell death in neural cells? To analyze cellular response to aggregate formation toward cell death, we constructed PC12 cell lines expressing flag-tagged polyQ (Q14) or ex-polyQ (Q79) under control of the tet-off promoter (Yasuda et al., 1999). These cells were named PC12-Q14 and PC12-Q79, respectively. PC12 cells are a well-characterized cell line with the ability to differentiate into postmitotic neuron-like cells by the simple addition of nerve growth factor (NGF). NGF was added when tetracycline was removed from the medium in this experiment. When Q79 was expressed by the tet-off, the PC12-Q79 cells died gradually, accomplishing differentiation; almost all cells became extinct 96–120 h after tet-off. We analyzed the subcellular localization of ex-polyQ by an immunocytochemical analysis using an anti-flag antibody (Fig. 3C) (Yasuda et al., 1999). Q79 was diffusely distributed in the cytoplasm 24 h after tet-off. At 48 h, most cells contained one or more cytoplasmic ex-polyQ aggregates, together with many small aggregates in the nucleus. At 72 h, large ex-polyQ aggregates were observed in the nucleus, as well as in the cytoplasm. Then the cells started dying, indicating that cytoplasmic and nuclear ex-polyQ aggregate formations are events that precede cell death. Such aggregate formation and cell death were not observed in PC12-Q14 cells.

During this cell-death assay, we occasionally found colonies of cells that appeared to be resistant to ex-polyQ–induced cell death. These cells contained large ex-polyQ aggregates exclusively in the cytoplasm. We unsuccessfully attempted to clone these cells. Any subtle treatment, such as simple passage, caused cell death. These results suggest that nuclear aggregates may cause stronger cell death signals than cytoplasmic aggregates and may correlate with the evidence that NIs are more frequently observed than cytoplasmic aggregates in the affected brain regions of patients with MJD, SCA1, DRPLA, HD, and other polyglutamine diseases (Burright et al., 1995; Davies et al., 1997; Paulson et al., 1997; Shimohata et al., 2000).

Using these cells, cell death-related factors were examined during the course of ex-polyQ–induced cell death. First, we examined the potential involvement of caspases in cell death. Contrary to our expectation, only marginal activation of caspase 3 was observed. In addition, several caspase inhibitors (e.g., DEVD, Z-VAD) were unable to inhibit cell death, although some apoptotic phenotypes, such as chromosomal condensation, were abolished. These results suggest that caspases were involved in the manifestation of certain apoptotic phenotypes, but not in the commitment of cell death. Next, we examined the activation of other apoptosis-related signal transduction factors and found that SEK1-JNK, stress-activated protein kinases (Sánchez et al., 1994), were clearly activated. Twenty-four hours after tet-off, the small population of cells showed small aggregates in the nucleus in which SEK1 was activated. In the cells with a large NI at 72 h, SEK1 was activated only in the place that was in agreement with the inclusion. Moreover, we found that c-Jun, the target protein of JNK, had been activated in the same place. Consistently, the dominant-negative SEK1 mutant inhibited ex-polyQ–induced cell death. These results suggest the possibility that the accumulation of ex-polyQ aggregates in the nucleus provides a platform for the death signal kinase cascade. Then, what activates SEK1? Our recent analysis found that apoptosis-regulating signal kinase 1 (ASK1), an MAPKKK (mitogen-activated protein kinase kinase kinase), is responsible for SEK1 activation in ex-polyQ–mediated cell death (Nishitoh et al., 2002).

Several reports published at around the same time have shown that many cellular factors, including transcription factors, were incorporated into the ex-polyQ aggregates, thereby leading to the hypothesis that NIs may sequester important nuclear factors, which, in turn, alter gene expression and then induce cellular abnormality, most typically cell death. Indeed, it has been shown that ex-polyQ aggregates can trap transcriptional factors, such as the TATA-binding protein (TBP), CREB binding protein (CBP), p53, mSin3A, and TAFII130 (Shimohata et al., 2000). In addition, a different transcriptional repression mechanism has been proposed as being a potential pathological mechanism of polyQ disease, in which the histone acetyltransferase activity of coactivators, such as CBP/p300, are inhibited by direct interaction with diseased proteins (Steffan et al., 2001; Li et al., 2002).

There is another argument, this one proposing that NIs do not cause cell death but, rather, function in a protective manner (Saudou et al., 1998). In the case of Alzheimer's disease, small soluble Aβ fibril, rather than aggregated Aβ fiber itself, has been argued to be toxic to the cell (Ellis and Pinheiro, 2002; Walsh et al., 2002). However, it is not yet known whether this scenario is applicable to ex-polyQ or other disease-related aggregates.

Approaches to elucidate the molecule mechanism of ex-polyQ toxicity

Ex-polyQ NIs are highly ubiquitinated in the brains of MJD patients and in animal and cellular models of MJD. Furthermore, it has recently been reported that several forms of proteasome complexes, as well as Hsp70 and Hsp40 (Hdj1 and Hdj2), are co-immunostained on NIs in the brains of MJD patients (Chai et al., 1999a, b). Ex-polyQ aggregates are a common feature of polyQ disease. These lines of evidence have raised another possibility that ex-polyQ aggregates sequester cellular factors that are involved in protein quality control and inhibit their functions. In cultured cells, ex-polyQ aggregates have indeed been shown to decrease proteasome activity, which, in turn, is predicted to result in further promotion of ex-polyQ aggregates (Bence et al., 2001; Nishitoh et al., 2002). Similar weakened protein degradation has been also reported in other neurodegenerative disorders with cytoplasmic protein aggregates (Leroy et al., 1998; Lam et al., 2000; Shimura et al., 2000). These observations suggest a potential link between neuronal degeneration and dysfunction of protein degradation pathways, most likely via the ubiquitin-proteasome system.

Identification of a putative sensor protein for abnormal protein accumulation

Since the accumulation of abnormal proteins definitely puts stress on cells, especially on non-dividing neurons, we assumed the existence of a cellular factor, namely, a sensor protein, which recognizes and monitors the accumulation of abnormal proteins in the cells. To identify such a potential sensor protein, we performed a pull-down experiment using GST-fused MJD79 protein as a representative of an abnormal protein. All mammalian cultured cells subsequently examined were found to contain a protein with a molecular weight of approximately 100 kDa that interacted with GST-MJD79 much more strongly than GST-MJD35 (Hirabayashi et al., 2001). This protein was then purified from HeLa cells and COS cells by affinity purification using GST-MJD79. Mass analysis and protein sequencing revealed that the 100-kDa proteins purified from both cells were identical and were a previously reported protein named p97/VCP, a member of the AAA ATPase family. Northern blot analysis on human RNA showed that p97/VCP is expressed not only throughout brain but all other tissues as well. p97/VCP is reportedly one of the most abundant intracellular proteins and has been shown to be involved in postmitotic membrane fusion processes in endoplasmic reticulum, the Golgi apparatus, and nuclear envelopes (Latterich et al., 1995; Rabouille et al., 1995; Kondo et al., 1997; Hetzer et al., 2001). In these processes, p97/VCP differentially utilizes its partners, such as p47 and the Ufd1/Npl4 complex.

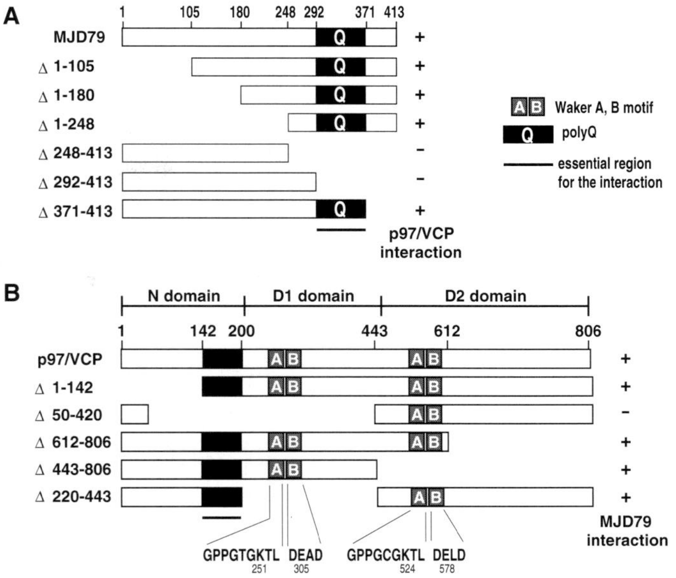

Fig. 5. Schematic drawings of the MJD and p97/VCP proteins. Regions essential for the physical interaction between MJD79 and p97/VCP are underlined. See text for detailed explanation. Based on Hirabayashi et al. (2001), with modifications.

Next, using deletion mutants of MJD79, we investigated which portion of MJD79 is recognized by p97/VCP. As expected, the ex-polyQ portion was found to be indispensable for binding with p97/VCP (Fig. 5A) (Hirabayashi et al., 2001). Moreover, deletion analysis of p97/VCP revealed that the portion containing amino acid residues 143–199, located near the *N*-terminus, is responsible for binding to MJD79 (Fig. 5B) (Hirabayashi et al., 2001). p97/VCP consists of three domains: N, D1, and D2 (Zhang et al., 2000). The D1 and D2 domains are ATPase domains, whereas the N domain is flexible and moves dynamically via nucleotide binding (Rouiller et al., 2000). Interestingly, the N domain is predicted to be the binding domain to its cofactor p47 and multi-ubiquitin (Rouiller et al., 2000). We then tried to determine whether p97/VCP acts as a chaperon toward ex-polyQs as observed by Hsp104, another member of the AAA family (Krobitsch and Lindquist, 2000; Kimura et al., 2001). Although we have not yet succeeded in using mammalian cells to obtain results supporting this idea, p97/VCP can indeed function to enhance relatively short polyQs to aggregate in yeast cells (Kimura et al., 2002; Y. Kimura, personal communication).

p97/VCP was observed in normal condition throughout mammalian cells, such as neuronally differentiated PC12 cells. Interestingly, p97/VCP was found to co-localize not only with ex-polyQ aggregates but also with several other protein aggregates in neuronal cells. We observed p97/VCP immunostaining in ex-polyQ–mediated nuclear and cytoplasmic inclusions in cell culture models (Fig. 6A) (Hirabayashi et al., 2001). p97/VCP co-staining with nuclear inclusions was also confirmed in our transgenic rat model and in brain samples from HD

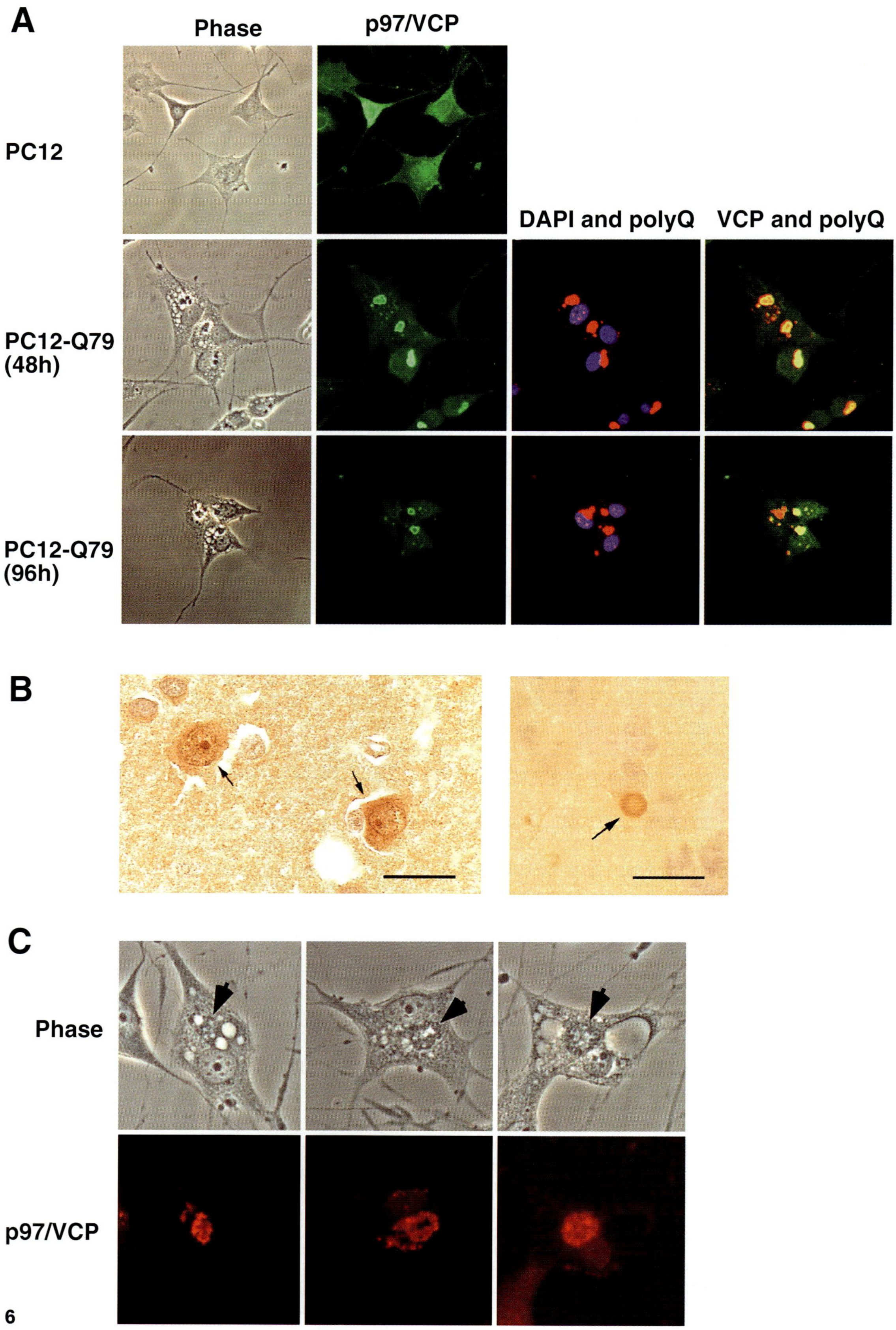
A
Phase
p97/VCP
PC12
DAPI and polyQ
VCP and polyQ
PC12-Q79 (48h)
PC12-Q79 (96h)
B
C
Phase
p97/VCP
6

patients (Fig. 6B) (Hirabayashi et al., 2001). Consistent with the idea that p97/VCP can recognize a broad range of abnormal proteins, p97/VCP was found to bind to α-synuclein and the *C*-terminal portion of β-cleaved APP containing Aβ (unpublished observations). In addition, p97/VCP was found to co-localize with Lewy bodies in brain samples of patients with sporadic Parkinson's disease or dementia with Lewy bodies (Fig. 6B) (Hirabayashi et al., 2001). p97/VCP also co-localized with aggresomes, which were induced at a perinuclear position by proteasome inhibitor treatments in PC12 cells (Fig. 6C) (Hirabayashi et al., 2001). It is notable that in these cells with aggresome or ex-polyQ aggregates, many large cytoplasmic vacuoles were concomitantly observed, followed by cell death.

Identification of ter94/VCP as a modulator of ex-polyQ-induced cell death via genetic screening in Drosophila

At the same time, we took another approach, this time using *Drosophila* genetics for identifying genes involved in ex-polyQ–mediated neurodegeneration. We first created a *Drosophila* model of polyQ diseases (Higashiyama et al., 2002). Transgenes encoding a flag epitope-tagged Q22, Q79, and Q92 were expressed in the compound eyes of *Drosophila* under the control of an eye-specific glass (GMR) promoter. Only pathogenic lengths of polyQs (Q79 and Q92) induced cell death, showing eye-degeneration phenotypes with loss of red pigmentation, loss of internal eye integrity, and severe degeneration of the photoreceptor neurons. The degenerative phenotypes were dependent on ex-polyQ lengths and transgene dosages and were progressive (Fig. 7) (Higashiyama et al., 2002).

We then performed genetic screening to search for dominant suppressors and enhancers of the eye degeneration. Q79 flies were crossed with about 180 fly lines from a stock that had chromosomal deficiencies. Our survey revealed that one chromosomally deficient fly line, Df(2R)X1, acts as a dominant suppressor. To identify the gene responsible for this suppression, available P-element-inserted mutants located within the deleted region were tested, and two weak dominant suppressors were found, the independent P-element-inserted mutants I(2)k15502 and I(2)03775, both having slightly suppressed eye abnormality (Fig. 8A) (Higashiyama et al., 2002). Then, using a plasmid rescue experiment, the locations of both P-elements were identified to be within the 5′-noncoding region of the

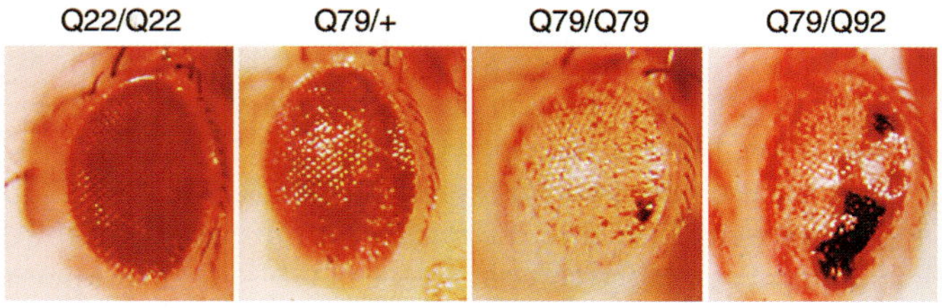

Fig. 7. PolyQ length-dependent eye-degeneration in ex-polyQ flies. Light photomicrographs of compound eyes from flies with the following genotypes: *GMR-Q22/GMR-Q22, GMR-Q79/+, GMR-Q79/GMR-Q79,* and *GMR-Q79/GMR-Q92.* Based on Higashiyama et al. (2002), with modifications.

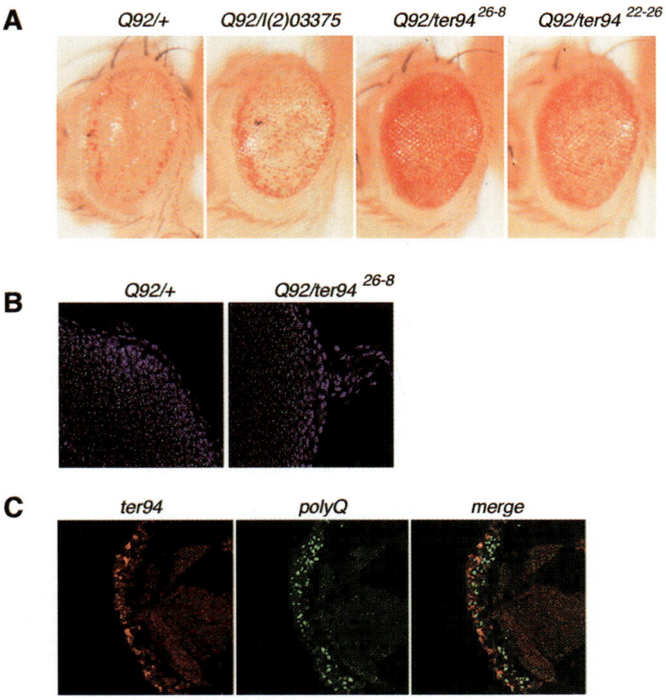

Fig. 8. Suppression of ex-polyQ-induced eye degeneration in *ter94* mutants. Based on Higashiyama et al. (2002), with modifications. (**A**) Light photomicrographs of the compound eye of *GMR-Q92/+*, *GMR-Q92/ter94$^{l(2)03775}$*, *GMR-Q92/ter94$^{26\text{-}8}$*, and *GMR-Q92/ter94$^{22\text{-}26}$* flies are shown. (**B**) Suppression of ex-polyQ-induced eye degeneration in *ter94$^{26\text{-}8}$* without the inhibition of ex-polyQ-aggregate formation. Confocal microscopy images of the eye imaginal discs from third instar larvae with the following genotypes are shown: *GMR-Q92/+* (left panel) and GMR-Q92/ter94$^{26\text{-}8}$ (right panel). Ex-polyQ was detected with anti-flag antibody (green). The nuclei were stained with DAPI (blue). There appeared to be no difference in the amount and localization of ex-polyQ aggregates between the wild-type and the *ter94$^{26\text{-}8}$* background. (**C**) Horizontal sections of adult *GMR-Q92/+* fly heads were stained with anti-ter94 polyclonal antibody (red), anti-flag antibody (green) for detecting ex-polyQ (Q92), and DAPI (blue) for visualizing cell nuclei.

Fig. 6. Co-localization of endogenous p97/VCP with protein aggregates in neuronal cells. Based on Hirabayashi et al. (2001), with modifications. (**A**) Endogenous p97/VCP protein in PC12 cells (upper panels) and PC12-Q79 cells (middle and lower panels) was stained with an anti-VCP antibody and visualized by FITC signals (green). Expressed Q79 was stained with an anti-flag antibody and visualized by Texas Red signals (red) in PC12-Q79 cells, 48 h (upper panels) and 96 h (lower panels) after the induction of Q79. The yellow colors in the superimposed panels show the co-localization of endogenous p97/VCP and Q79. The nuclei were stained with DAPI (blue). (**B**) Identification of endogenous p97/VCP protein in nuclear inclusions from HD patients (left panel) and in Lewy bodies from DLB patients (right panel). The signals were visualized by the ABC method (brown signals). Bar = 20 μm. (**C**) Proteasome inhibitor treatment induces p97/VCP-positive protein aggregates and large cytoplasmic vacuoles in PC12 cells. Phase-contrast (upper panels) and immunohistochemical (lower panels) photomicrographs of PC12 cells treated with 0.1 μM PSI for 48 h. Endogenous p97/VCP protein was stained with an anti-VCP antibody and visualized by Texas Red signals (lower panels). Arrows indicate the protein aggregates observed in cells treated with proteasome inhibitors (upper panels).

ter94 gene. Surprisingly, *Ter94* is the *Drosophila* homolog of p97/VCP. To determine if the *ter94* mutation is responsible for the recovery of ex-polyQ-induced eye degeneration, we crossed GMR-Q92 flies with several *ter94* mutants, such as the reportedly strong ($ter94^{26\text{-}8}$) and weak ($ter94^{22\text{-}26}$) loss of function mutants (Ruden et al., 2000). In flies having $ter94^{26\text{-}8}$ in one allele ($ter94^{26\text{-}8}$/+), eye degeneration was dramatically suppressed (Fig. 8A) (Higashiyama et al., 2002). In flies having $ter94^{22\text{-}26}$ ($ter94^{22\text{-}26}$/+), degeneration was also suppressed, but much less so than in the observed $ter94^{26\text{-}8}$/+ flies. These results indicate that the ter94 protein plays a necessary function in eye degeneration, namely, in ex-polyQ–induced cell death.

Examination of ex-polyQ expression revealed that ex-polyQ aggregates were formed in the nucleus of the eye disc of third instar larvae and in the compound eye in GMR-Q92 fly pupae. Similar nuclear aggregates were also formed in $ter94^{26\text{-}8}$/+ flies (Higashiyama et al., 2002) (Fig. 8B). The total amounts and sizes of ex-polyQ aggregates of both fly lines were apparently indistinguishable. Some aggregates were co-immunostained with ter94 (Fig. 8C) (Higashiyama et al., 2002). Co-expression of ter94 with Q92 or Q79 led to enhancement of rough eye phenotypes in these flies, but the amounts and sizes of ex-polyQ aggregates did not recognizably change upon ter94 co-expression. These results indicated that the ter94 protein positively mediates ex-polyQ–induced death signals.

Recent genetic studies by other researchers using similar *Drosophila* polyQ disease models have shown that overexpression of molecular chaperones, such as Hsp70 and Hsp40 (dHdj1 and dHdj2), is able to suppress ex-polyQ toxicity in the compound eyes, although the suppression levels differ among them (Warrick et al., 1999; Chan et al., 2000; Kazemi-Esfarjani and Benzer, 2000). In these cases, NIs were also formed upon chaperone expression. These observations suggest that the targets of these chaperones in mitigating eye phenotypes are different from those of ex-polyQs. Ter94 protein might be the target. This possibility remains to be tested.

Characterization of p97/VCP functions

p97/VCP was biochemically purified as a factor that physically interacted with the diseased MJD protein and was found to bind to an Aβ-containing peptide and α-synuclein (unpublished observations). In addition, p97/VCP immunoreactivities were detected not only in ex-polyQ NIs but also in Lewy bodies in Parkinson's disease and DLB. Furthermore, mutants of *ter94* (*Drosophila* p97/VCP) were genetically identified as dominant suppressors for ex-polyQ-induced eye degeneration. These results clearly imply that p97/VCP is a key molecule in neuronal cell death by ex-polyQ and probably by other misfolded proteins as well. Thus, we next addressed the pathophysiological roles of p97/VCP from a functional point of view (Hirabayashi et al., 2001; Kobayashi et al., 2002). p97/VCP is a well-conserved ATPase with two ATPase domains (D1 and D2) from yeast to mammals and is essential for growth in yeast (yeast p97/VCP is called CDC48p). We made a series of mutants with single amino acid substitutions of several conserved amino acids (e.g., K251A, E305Q, L306P, Y495A, Y495E, K524A, E578Q, L579P, Y805A, and Y805E) as GFP-fused proteins, expressed these mutants in different cells, and examined the morphology of GFP-positive cells. Among these mutants, only p97[K524A] and p97[E578Q] induced the formation of many large vacuoles in the cytoplasm; the latter produced fewer vacuoles than the former (Hirabayashi et al., 2001; Kobayashi et al., 2002). Amino acid substitutions in both mutants were on those conserved in the Walker A and B motifs, both in the D2 domain (Fig. 5).

We then constructed stable cell lines expressing GFP-fused wild-type p97/VCP or p97[K524A], which were referred to as TV cells and TmV cells, respectively, under the tet-off promoter, and we further characterized the phenotypes of these cells (Fig. 9A) (Kobayashi et al., 2002). Twenty-four hours after tet-off, both proteins became localized throughout the cells. However, we found clear differences between the morphologies of TV and TmV cells. TV cells maintained a healthy appearance and were indistinguishable from the parental PC12 cells with numerous neurite outgrowths, whereas TmV cells appeared sick and lost the ability to extend neuritis, forming many cytoplasmic vacuoles, which was followed by cell death.

A homo-oligomeric ring structure is the most characteristic feature of AAA ATPase. Both exogenous p97/VCPs were similarly incorporated into the hexameric complex together with the endogenous p97/VCP and were also capable of being bound to co-factors, such as p47, Ufd1, and Npl4. It is notable that recombinant p97[K524A] made by the baculovirus system showed much lower ATPase activity (less than 1/20) than wild-type p97/VCP in vitro (Fig. 9B) (Kobayashi et al., 2002). These results clearly showed that alanine substitution of lysine 524 abolished ATPase activity without affecting the oligomeric state or cofactor binding and, further, indicated that the second ATPase domain represents major ATPase activity, loss of which makes many vacuoles in the cytoplasm, followed by cell death. In addition to neuronal cell death, cytoplasmic vacuolization is another hallmark observed in the affected neurons of the polyQ diseases, prion disease, Alzheimer's disease, amyotrophic lateral sclerosis (ALS), and other neurodegenerative diseases (Landis et al., 1981; Wong et al., 1995; Ghoshal et al., 1999; Skinner et al., 2001).

To investigate the origin of these vacuoles, we examined which organelle markers (e.g., mitochondria, Golgi apparatus, lysosomes, endosomes, and endoplasmic reticulum [ER]) could stain them. Among those tested, only ER-CFP (a CFP-fused ER-residing protein that contains the ER-localization signal of calreticulin and the KDEL motif) could do so (Fig. 10A, B) (Kobayashi et al., 2002). Consistently in TmV cells, but not in TV cells, significant up-regulation of GRP78 mRNA (an ER stress marker) was observed (Fig. 11A) (Kobayashi et al., 2002) at a level similar to that induced after tunicamycin treatment, a well-known ER stress inducer. CHOP protein (Zinszner et al., 1998), another ER stress marker, was also clearly up-regulated in TmV cells, but again not in TV cells. These results indicated that the loss or severe decrease of p97/VCP ATPase activity results in ER stress and abnormal ER expansion. Furthermore, proteasome inhibitors in PC12 cells induced indistinguishable cytoplasmic vacuoles stained with CFP-ER as well (Fig. 10C, D) (Kobayashi et al., 2002), suggesting a link among p97/VCP, ER maintenance, and protein degradation in the ubiquitin-proteasome system.

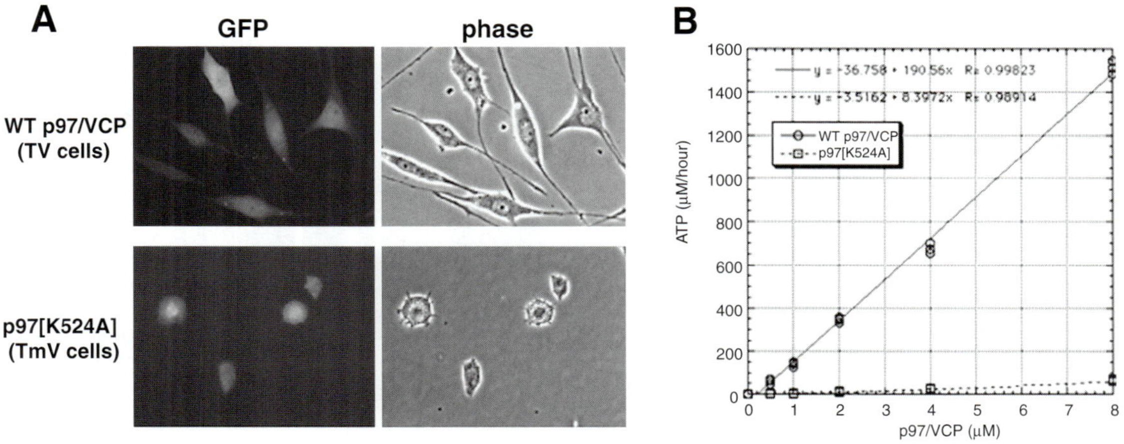

Fig. 9. Characterization of p97[K524A] in neuronal PC12 cells and in vitro. Based on Kobayashi et al. (2002), with modifications. (**A**) Typical GFP fluorescent (left panels) and phase contrast (right panels) images of PC12 cells expressing GFP-fused wild-type p97/VCP (TV cells; upper panels) or p97[K524A] (TmV cells; lower panels) 24 h after removal of tetracycline from the culture medium. NGF (50 ng/ml) was supplemented upon removal of tetracycline. Note that TmV cells lost neurite-like extensions and contained many large vacuoles, followed by cell death. (**B**) Comparison of ATPase activities between recombinant wild-type p97/VCP and p97[K524A]. Recombinant proteins were produced with the baculovirus-mediated expression system. ATPase activities were measured at various concentrations of the recombinant proteins using an enzyme-coupling assay system at 37 °C. Three independent measurements of wild-type p97/VCP (open circles) and p97[K524A] (open square) are shown.

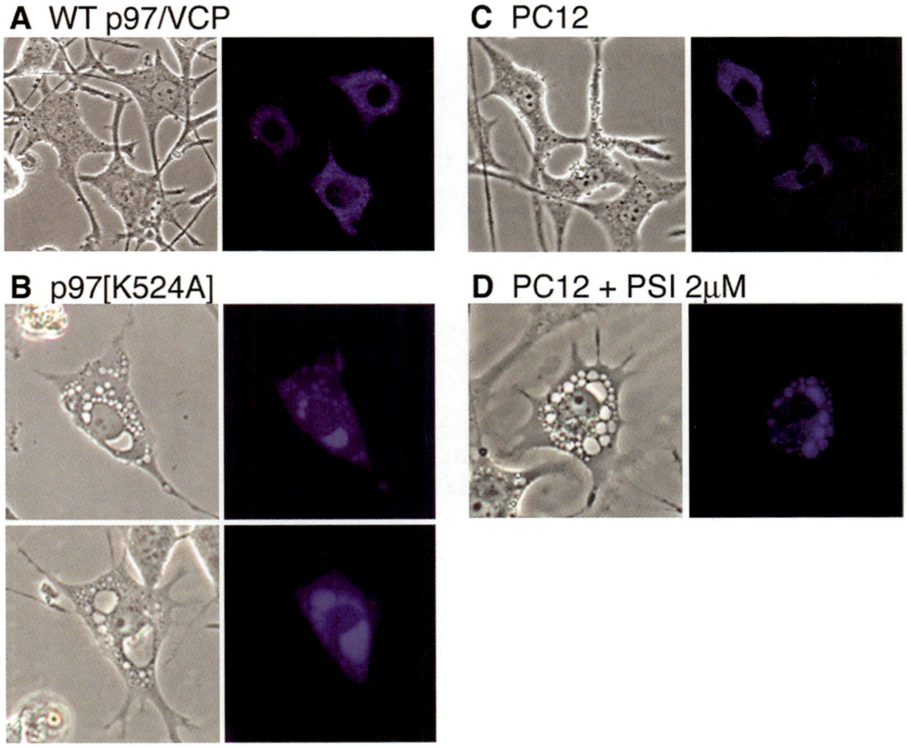

Fig. 10. Endoplasmic reticulum (ER) expansion by p97[K524A] or PSI, a proteasome inhibitor, in PC12 cells. An expression plasmid for ER-CFP, an ER marker protein, was transiently expressed in PC12 cells with that for wild-type p97/VCP (**A**) or p97[K525A] mutant (**B**), or in non-treated PC12 cells (**C**) and those treated with PSI (2 μM) for 24 h (**D**). Localization of ER-CFP proteins (right panels) coincided well with cytoplasmic vacuoles in phase-contrast views (left panels) (**B**, **D**). Based on Kobayashi et al. (2002), with modifications.

Indeed, we found an accumulation of ubiquitinated proteins, which are likely misfolded proteins in the membrane and nuclear fractions, but not in the cytoplasmic fraction, of TmV cells. In the membrane fraction, such ubiquitinated proteins were bound to p97[K524A] (Fig. 11B) (Kobayashi et al., 2002). It is noteworthy that approximately one third of newly synthesized proteins in the cells are not appropriately folded, and these misfolded proteins, especially those originating in ER, are degraded by the ER-associated degradation (ERAD) system, in which misfolded proteins are translocated into the cytoplasm and then degraded by the proteasome (Kopito, 1997). Thus, our observations indicated that in the ERAD system, the ATPase activity of p97/VCP is mainly used for pulling misfolded membrane proteins out into the cytoplasm (Fig. 11C)

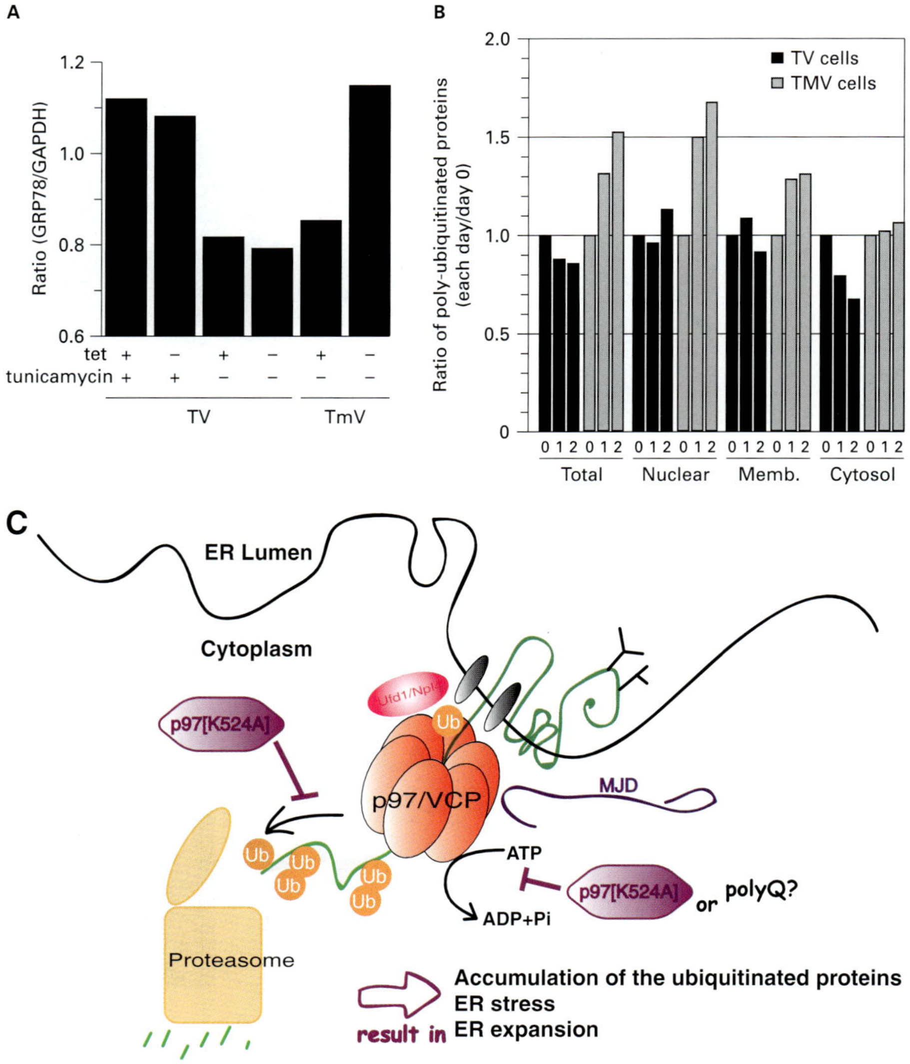

Fig. 11. Cellular phenotypes induced by p97[K524A] expression. (**A**) Up-regulation of GRP78 mRNA, an ER stress marker in TmV cells. Based on Kobayashi et al. (2002), with modifications. (**B**) Accumulation of ubiquitinated proteins in nuclear and membrane fractions in TmV cells. Based on Kobayashi et al. (2002), with modifications. (**C**) A schematic model of p97/VCP functions. See text for detailed explanation.

(Bays et al., 2001; Ye et al., 2001; Rabinovich et al., 2002). In fact, CFTR[ΔF508], a well-known ERAD substrate (Ward et al., 1995), was found to accumulate more in cells expressing p97[K524A] than in cells expressing wild-type p97/VCP or p97[K251A]. These results collectively indicated that the ATPase activity of p97/VCP is an essential factor in the ER quality control system and that its severe dysfunction leads to ER stress and cell death.

Discussion

Many neurodegenerative disorders manifest disease-specific phenotypes (e.g., dementias, ataxias, movement disorders, and spinal cord diseases), and thus it was thought impossible to deduce a common molecular mechanism underlying many, if not all, neurodegenerative disorders. However, neurodegenerative disorders share several common characteristics, such as autosomal dominant inheritance (if inherited), late onset, progressiveness, and neuronal cell loss and degeneration in the affected regions of the central nervous system. These characteristics suggest the existence of similar molecular processes during neurodegeneration and neuronal cell death. Recently, many researchers in this field have gradually accepted this idea, and a yet unknown toxic gain-of-function has been assumed to be rendered in certain disorder-causing proteins, which should typically be observed in inherited neurodegeneration diseases with an autosomal dominant pattern of inheritance.

In 1994, we identified the gene responsible for MJD (Kawaguchi et al., 1994), and, as a result, this disease has been found to be the most common inherited spinocerebellar ataxia throughout the world, regardless of ethnic group. Since then, we have aimed, via molecular analysis of MJD, to elucidate potential common molecular mechanisms underlying neurodegeneration. Not only MJD but also another eight inherited neurodegenerative diseases (SBMA, HD, DRPLA, SCA1, SCA2, SCA6, etc.) have been shown to be caused by the expansion of CAG repeats that encode polyglutamines (Kakizuka, 1998). Indeed, we have shown that polyglutamines have the ability to self-aggregate and induce neurodegeneration and cell death in transgenic animals (e.g., mice, rats, and flies) and cultured cells, leading to the proposal of a "polyglutamine disease" (Ikeda et al., 1996; Yasuda et al., 1999; Higashiyama et al., 2002). These characteristics of polyglutamines closely resemble those of Aβ, prion protein, and α-synuclein, which respectively cause Alzheimer's disease, Creutzfeldt Jacob disease, and Parkinson's disease. Thus, the creation of aggregate-prone proteins is most likely to be a core event underlying toxic gain of function in neurodegenerations. Then, what processes will occur after the creation of such aggregate-prone proteins?

We have identified ter94/p97/VCP, a member of the AAA ATPase, not only as a polyglutamine-interacting protein via biochemical purification but also as a cell death mediator via genetic screening with our fly model of polyglutamine disease (Hirabayashi et al., 2001; Higashiyama et al., 2002). p97/VCP colocalizes not only with polyglutamine aggregates but also other protein aggregates and Lewy bodies, which are neuronal inclusions typically found in Parkinson's disease and a type of dementia (Hirabayashi et al., 2001). Moreover, profound deficits in its ATPase activity are found to severely affect ER quality control, leading to abnormal ER expansion and cell death, phenotypes frequently observed in many neurodegenerative diseases (Kobayashi et al., 2002). These lines of evidence indicate that p97/VCP functions not only as a common sensor for abnormal protein accumulations but also as a mediator of neurodegenerative phenotypes; excessive accumulation of abnormal proteins may inactivate p97/VCP ATPase in several neurodegenerative disorders, eventually leading to neurodegeneration.

p97/VCP is now known to function in the ERAD system, in which its ATPase activity is essential for pulling polyubiquitinated proteins from the ER (Bays and Hampton, 2002; Kobayashi et al., 2002; Lord et al., 2002; Wojcik, 2002). Even when present in an approximately equal amount to endogenous p97/VCP, the expression of ATPase-deficient mutants of p97/VCP, such as p97[K524A], lead to accumulation of polyubiquitinated proteins that are most likely misfolded proteins in the ER and nucleus (Kobayashi et al., 2002). Since p97/VCP functions as a hexamer, seven different ATPase activities could be created even in one complex containing from zero to six ATPase-deficient p97/VCPs, which could, in turn, create varying degrees of ERAD inhibition. Furthermore, considering the high level of expression of endogenous p97/VCP, the overall ATPase activities of p97/VCP in the cells may be expected to change dramatically in relation to the amount of ATPase-deficient p97/VCPs.

In recent studies, the accumulation of misfolded proteins has been shown to impair proteasome activity in several experimental models of neurodegenerative disorders, including polyglutamine diseases (Bence et al., 2001; Nishitoh et al., 2002). Our studies have shown that mild inhibition of proteasomes by inhibitory drugs gives rise to phenotypes resembling those observed in cells expressing p97[K524A] or those observed via pathological and biochemical analyses of several neurodegenerative disorders (Hirabayashi et al., 2001). These results suggest that accumulation of misfolded proteins inhibits not only proteasome activity, but also ATPase activity of p97/VCP to different degrees, which in turn contributes to the inhibition of the ERAD system, also to different degrees.

Approximately 30% of the newly synthesized proteins are reportedly misfolded, and therefore cells are continuously burdened by a large quantity of misfolded proteins that should be degraded (Schubert et al., 2000), indicating that cells need to monitor the amount of misfolded proteins present within them and respond to maintain this amount at a constant level. In regard to this mechanism, we propose the following model for physiological functions of p97/VCP, as well as for those related to neurodegenerative disorders. First, accumulation of misfolded proteins in the cytoplasm is sensed by p97/VCP via either direct binding or ubiquitin-mediated binding. Second, ATPase activity is reduced in proportion to the amount of p97/VCP bound by misfolded proteins; the more p97/VCP units that are occupied by misfolded proteins in the hexametric complex, which reflects the increase in misfolded protein concentration in the cells, the lower its ATPase activity is in the complex. Third, following the degree of ATPase decrease, the ERAD system is proportionally inhibited, which, in turn, inhibits or decreases a major supply of misfolded proteins and hides them in different compartments, such as the ER. As a result, the burden on the proteasome is lowered, accumulated misfolded proteins are degraded, and their levels are reduced in the cytoplasm, which then recovers p97/VCP ATPase activity because of the decrease in p97/VCP bound by misfolded proteins. As a result, the ERAD system is rebooted and accumulated membrane-bound misfolded proteins are transported into the cytoplasm and degraded by the proteasome. This scenario appears to work well when misfolded protein accumulation is transient. However, when misfolded proteins are continuously provided, as in the case of neurodegenerative disorders, this continuous stress on the ER may lead to abnormal ER expansion and cell death. During these processes, ASK1 is activated by ER stress, leading to subsequent activation of SEK1-JNK, followed by cell death (Nishitoh et al., 2002).

This model is still hypothetical and does not exclude other possibilities. It is notable that p97/VCP is one of the most abundant proteins in cells (estimated by Dalal and Hanson [2001] to comprise 1% of all cellular proteins) and is involved in many cellular activities. Thus, it is easy to imagine that not only the misfolded protein levels but also other mechanisms (e.g., classic protein phosphorylation) potentially regulate its ATPase activity. Unraveling the modifications of p97/VCP and correlating them with the regulation of its ATPase activity and interaction with its partners would elucidate the molecular mechanisms underlying the physiological role(s) of p97/VCP in several neurodegenerative disorders and may provide future strategies for the treatment and prevention of such disorders.

Reference

Alfen N van, Sinke RJ, Zwarts MJ, Gabreels-Festen A, Praamstra P, Kremer BP, Horstink MW: Intermediate CAG repeat lengths (53, 54) for MJD/SCA3 are associated with an abnormal phenotype. Ann Neurol 49:805–807 (2001).

Banfi S, Servadio A, Chung MY, Kwiatkowski TJJ, McCall AE, Duvick LA, Shen Y, Roth EJ, Orr HT, Zoghbi HY: Identification and characterization of the gene causing type 1 spinocerebellar ataxia. Nature Genet 7:513–520 (1993).

Bays NW, Hampton RY: Cdc48-Ufd1-Npl4: stuck in the middle with Ub. Curr Biol 12:R366–371 (2002).

Bays NW, Wilhovsky SK, Goradia A, Hodgkiss-Harlow K, Hampton RY: *HRD4/NPL4* is required for the proteasomal processing of ubiquitinated ER proteins. Mol Biol Cell 12:4114–4128 (2001).

Bence NF, Sampat RM, Kopito RR: Impairment of the ubiquitin-proteasome system by protein aggregation. Science 292:1552–1555 (2001).

Burright EN, Clark HB, Servadio A, Matilla T, Feddersen RM, Yunis WS, Duvik LA, Zoghbi HY, Orr HT: SCA1 transgenic mice: a model for neurodegeneration caused by an expanded CAG trinucleotide repeat. Cell 82:937–948 (1995).

Chai Y, Koppenhafer SL, Bonini NM, Paulson HL: Analysis of the role of heat shock protein (Hsp) molecular chaperones in polyglutamine disease. J Neurosci 19:10338–10347 (1999a).

Chai Y, Koppenhafer SL, Shoesmith SJ, Perez MK, Paulson HL: Evidence for proteasome involvement in polyglutamine disease: localization to nuclear inclusions in SCA3/MJD and suppression of polyglutamine aggregation in vitro. Hum molec Genet 8:673–682 (1999b).

Chan HY, Warrick JM, Gray-Board GL, Paulson HL, Bonini NM: Mechanisms of chaperone suppression of polyglutamine disease: selectivity, synergy and modulation of protein solubility in *Drosophila*. Hum molec Genet 9:2811–2820 (2000).

Dalal S, Hanson PI: Membrane traffic: what drives the AAA motor? Cell 104:5–8 (2001).

Davies SW, Turmaine M, Cozens BA, DiFiglia M, Sharp AH, Ross CA, Scherzinger E, Wanker EE, Mangiarini L, Bates GP: Formation of neuronal intranuclear inclusions underlies the neurological dysfunction in mice transgenic for the HD mutation. Cell 90:537–548 (1997).

DiFiglia M: Huntingtin fragments that aggregate go their separate ways. Mol Cell 10:224–225 (2002).

DiFiglia M, Sapp E, Chase KO, Davies SW, Bates GP, Vonsattel JP, Aronin N: Aggregation of huntingtin in neuronal intranuclear inclusions and dystrophic neurites in brain. Science 277:1990–1993 (1997).

Duyao M, Ambrose C, Myers R, Novelletto A, Persichetti F, Frontali M, Folstein S, Rosss C, Franz M, Abbott M, Gray J, Conneally P, Young A, Penney J, Hollingsworth Z, Shoulson I, Lazzarini A, Falek A, Koroshetz W, Sax D, Bird E, Vonsattel J, Bonilla E, Alvir J, Conde JB, Cha J-H, Dure L, Gomez F, Ramos M, Sanchez-Ramos J, Snodgrass S, de Young M, Wexler N, Moscowitz C, Penchaszadeh G, MacFarlane H, Anderson M, Jenkins B, Srinidhi J, Barnes G, Gusella J, MacDonald M: Trinucleotide repeat length instability and age of onset in Huntington's disease. Nature Genet 4:387–392 (1993).

Ellis RJ, Pinheiro TJT: Danger-misfolding proteins. Nature 416:483–484 (2002).

Evert BO, Wüllner U, Schulz JB, Weller M, Groscurth P, Trottier Y, Brice A, Klockgether T: High level expression of expanded full-length ataxin-3 in vitro causes cell death and formation of intranuclear inclusions in neuronal cells. Hum molec Genet 8:1169–1176 (1999).

Ghoshal N, Smiley JF, DeMaggio AJ, Hoekstra MF, Cochran EJ, Binder LI, Kuret J: A new molecular link between the fibrillar and granulovacuolar lesions of Alzheimer's disease. Am J Pathol 155: 1163–1172 (1999).

Goto J, Watanabe M, Ichikawa Y, Yee SB, Ihara N, Endo K, Igarashi S, Takiyama Y, Gaspar C, Maciel P, Tsuji S, Rouleau GA, Kanazawa I: Machado-Joseph disease gene products carrying different carboxyl termini. Neurosci Res 28:373–377 (1997).

Healton EB, Brust JC, Kerr DL, Resor S, Penn A: Presumably Azorean disease in a presumably non-Portuguese family. Neurology 30:1084–1089 (1980).

Hetzer M, Meyer HH, Walther TC, Bilbao-Cortes D, Warren G, Mattaj IW: Distinct AAA-ATPase p97 complexes function in discrete steps of nuclear assembly. Nature Cell Biol 3:1086–1091 (2001).

Higashiyama H, Hirose F, Yamaguchi M, Inoue YH, Fujikake N, Matsukage A, Kakizuka A: Identification of *ter94, Drosophila VCP*, as a modulator of polyglutamine-induced eye degenerations in *Drosophila*. Cell Death Differ 9:264–273 (2002).

Hirabayashi M, Inoue K, Tanaka K, Nakadate K, Ohsawa Y, Kamei Y, Popiel AH, Sinohara A, Iwamatsu A, Kimura Y, Uchiyama Y, Hori S, Kakizuka A: VCP/97 in abnormal protein aggregates, cytoplasmic vacuoles, and cell death, phenotypes relevant to neurodegeneration. Cell Death Differ 8:977–984 (2001).

Ikeda H, Yamaguchi M, Sugai S, Aze Y, Narumiya S, Kakizuka A: Expanded poly-glutamine in the Machado-Joseph disease protein induces cell death in vitro and in vivo. Nature Genet 13:196–202 (1996).

Kakizuka A: Protein precipitation: a common etiology in neurodegenerative disorders? Trends Genet 14:396–402 (1998).

Kawaguchi Y, Okamoto T, Taniwaki M, Aizawa M, Inoue M, Katayama S, Kawakami H, Nakamura S, Nishimura M, Akiguchi I, Kimura J, Narumiya S, Kakizuka A: CAG expansions in a novel gene for Machado-Joseph disease at chromosome 14q32.1. Nature Genet 8:221–228 (1994).

Kazemi-Esfarjani P, Benzer S: Genetic suppression of polyglutamine toxicity in *Drosophila*. Science 287:1837–1840 (2000).

Kimura Y, Koitabashi S, Kakizuka A, Fujita T: Initial process of polyglutamine aggregate formation in vivo. Genes Cells 6:887–897 (2001).

Kimura Y, Koitabashi S, Kakizuka A, Fujita T: Circumvention of chaperone requirement for aggregate formation of a short polyglutamine tract by the co-expression of a long polyglutamine tract. J biol Chem 277:37536–37541 (2002).

Kobayashi T, Tanaka K, Inoue K, Kakizuka A: Functional ATPase activity of p97/VCP is required for the quality control of endoplasmic reticulum in neuronally differentiated mammalian PC12 cells. J biol Chem 277:47358–47365 (2002).

Koide R, Ikeuchi T, Onodera O, Tanaka H, Igarashi S, Endo K, Takahashi H, Kondo R, Ishikawa A, Hayashi T, Saito M, Tomoda A, Miike T, Naito H, Ikuta F, Tsuji S: Unstable expansion of CAG repeat in hereditary dentatorubral-pallidoluysian atrophy (DRPLA). Nature Genet 6:9–13 (1994).

Kondo H, Rabouille C, Newman R, Levine TP, Pappin D, Fremont P, Warren G: p47 is a cofactor for p97-mediated membrane fusion. Nature 388:75–78 (1997).

Kopito RR: ER quality control: the cytoplasmic connection. Cell 88:427–430 (1997).

Krobitsch S, Lindquist S: Aggregation of huntingtin in yeast varies with the length of the polyglutamine expansion and the expression of chaperone proteins. Proc natl Acad Sci, USA 97:1589–1594 (2000).

La Spada AR, Wilson EM, Lubahn DB, Harding AE, Fischbeck KH: Androgen receptor gene mutations in X-linked spinal and bulbar muscular atrophy. Nature 352:77–79 (1991).

Lam YA, Pickart CM, Alban A, Landon M, Jamieson C, Ramage R, Mayer RJ, Layfield R: Inhibition of the ubiquitin-proteasome system in Alzheimer's disease. Proc natl Acad Sci, USA 97:9902–9906 (2000).

Landis DMD, Williams RS, Masters CL: Golgi and electronmicroscopic studies of spongiform encephalopathy. Neurology 31:538–549 (1981).

Latterich M, Fröhlich K-U, Schekman R: Membrane fusion and the cell cycle: cdc48p participates in the fusion of ER membranes. Cell 82:885–893 (1995).

Leroy E, Boyer R, Auburger G, Leube B, Ulm G, Mezey E, Harta G, Brownstein MJ, Jonnalagada S, Chernova T, Dehejia A, Lavedan C, Gasser T, Steinbach PJ, Wilkinson KD, Polymeropoulos MH: The ubiquitin pathway in Parkinson's disease. Nature 395:451–452 (1998).

Li F, Macfarlan T, Pittman RN, Chakravarti D: Ataxin-3 is a histone binding protein with two independent transcriptional corepressor activities. J biol Chem 277:45004–45012 (2002).

Li M, Miwa S, Kobayashi Y, Merry DE, Yamamoto M, Tanaka F, Doyu M, Hashizume Y, Fischbeck KH, Sobue G: Nuclear inclusions of the androgen receptor protein in spinal and bulbar muscular atrophy. Ann Neurol 44:249–254 (1998).

Lima L, Coutinho P: Clinical criteria for diagnosis of Machado-Joseph disease: report of a non-Azorean Portuguese family. Neurology 30:319–322 (1980).

Lord JM, Ceriotti A, Roberts LM: ER dislocation: Cdc48/p97 get into the AAAct. Curr Biol 12: R182–184 (2002).

Lunkes A, Lindenberg KS, Ben-Haïem L, Weber C, Devys D, Landwehrmeyer GB, Mandel J-L, Trottier Y: Proteases acting on mutant huntingtin generate cleaved products that differentially build up cytoplasmic and nuclear inclusions. Mol Cell 10:259–269 (2002).

Nagafuchi S, Yanagisawa H, Sato K, Shirayama T, Ohsaki E, Bundo M, Takeda T, Tadokoro K, Kondo I, Murayama N, Tanaka Y, Kikushima H, Umino K, Kurosawa H, Furukawa T, Nihei K, Inoue T, Sano A, Komure O, Takahashi M, Yoshizawa T, Kanazawa I, Yamada M: Dentatorubral and pallidoluysian atrophy expansion of an unstable CAG trinucleotide on chromosome 12p. Nature Genet 6:14–18 (1994).

Nakamoto M, Ikeda H, Kakizuka A: Genetics and molecular studies of Machado-Joseph disease, in Wells, RD, Warren ST, Sarmiento M (eds): Genetic Instabilities and Hereditary Neurological Disease, pp 283–297 (Academic Press, New York 1998).

Nakano KK, Dawson DM, Spence A: Machado disease. A hereditary ataxia in Portuguese emigrants to Massachusetts. Neurology 22:49–55 (1972).

Nishitoh H, Matsuzawa A, Tobiume K, Saegusa K, Takeda K, Inoue K, Hori S, Kakizuka A, Ichijo H: ASK1 is essential for endoplasmic reticulum stress-induced neuronal cell death triggered by expanded polyglutamine repeats. Genes Dev 16:1345–1355 (2002).

Paulson HL, Perez MK, Trottier Y, Trojanowski JQ, Subramony SH, Das SS, Vig P, Mandel J-L, Fischbeck KH, Pittman RN: Intranuclear inclusions of expanded polyglutamine protein in spinocerebellar ataxia type3. Neuron 19:333–344 (1997).

Perez MK, Paulson HL, Pendse SJ, Saionz SJ, Bonini NM, Pittman RN: Recruitment and the role of nuclear localization in polyglutamine-mediated aggregation. J Cell Biol 143:1457–1470 (1998).

Rabinovich E, Kerem A, Fröhlich K-U, Diamant N, Bar-Nun S: AAA-ATPase p97/Cdc48, a cytosolic chaperone required for endoplasmic reticulum-associated protein degradation. Mol Cell Biol 22:626–634 (2002).

Rabouille C, Levine TP, Peters JM, Warren G: An NSF-like ATPase, p97 and NSF mediate cisternal regrowth from mitotic Golgi fragments. Cell 82: 905–914 (1995).

Romanul FC, Fowler HL, Radvany J, Feldman RG, Feingold M: Azorean disease of the nervous system. N Engl J Med 296:1505–1508 (1997).

Rosenberg RN, Nyhan WL, Bay C, Shore P: Autosomal dominant striatonigral degeneration: a clinical, pathologic, and biochemical study of a new genetic disorder. Neurology 26:703–714 (1976).

Rouiller I, Butel VM, Latterich M, Milligan RA, Wilson-Kubalek EM: A major conformational change in p97 AAA ATPase upon ATP binding. Mol Cell 6:1485–1490 (2000).

Ruden DM, Sollars V, Wang X, Mori D, Alterman M, Lu X: Membrane fusion proteins are required for *osker* mRNA localization in the *Drosophila* egg chamber. Dev Biol 218:314–325 (2000).

Sakai T, Ohta M, Ishino H: Joseph disease in a non-Portuguese family. Neurology 33:74–80 (1983).

Sánchez I, Hughes RT, Mayer BJ, Yee K, Woodgett JR, Avruch J, Kyriakis JM, Zon LI: Role of SAPK/ERK kinase-1 in the stress-activated pathway regulating transcription factor c-Jun. Nature 372:794–798 (1994).

Saudou F, Finkbeiner S, Devys D, Greenberg ME: Huntingtin acts in the nucleus to induce apoptosis but death does not correlate with the formation of intranuclear inclusions. Cell 95:55–66 (1998).

Scherzinger E, Lurz R, Turmaine M, Mangiarini L, Hollenbach B, Hasenbank R, Bates GP, Davies SW, Lehrach H, Wanker EE: Huntingtin-encoded polyglutamine expansions form amyloid-like protein aggregates in vitro and in vivo. Cell 90:549–558 (1997).

Schilling G, Wood JD, Duan K, Slunt HH, Gonzales V, Yamada M, Cooper JK, Margolis RL, Jenkins NA, Copeland NG, Takahashi H, Tsuji S, Price DL, Borchelt DR, Ross CA: Nuclear accumulation of truncated atrophin-1 fragments in a transgenic mouse model of DRPLA. Neuron 24:275–286 (1999).

Schmidt T, Landwehrmeyer GB, Schmitt I, Trottier Y, Auburger G, Laccone F, Klockgether T, Volpel M, Epplen JT, Schols L, Riess O: An isoform of ataxin-3 accumulates in the nucleus of neuronal cells in affected brain regions of SCA3 patients. Brain Pathol 8:669–679 (1998).

Schubert U, Antón LC, Gibbs J, Norbury CC, Yewdell JW, Bennink JR: Rapid degradation of a large fraction of newly synthesized proteins by proteasomes. Nature 404:770–774 (2000).

Shimohata T, Nakajima T, Yamada M, Uchida C, Onodera O, Naruse S, Kimura T, Koide R, Nozaki K, Sano Y, Ishiguro H, Sakoe K, Ooshima T, Sato A, Ikeuchi T, Oyake M, Sato T, Aoyagi Y, Hozumi I, Nagatsu T, Takiyama Y, Nishizawa M, Goto J, Kanazawa I, Davidson I, Tanese N, Takahashi H, Tsuji S: Expanded polyglutamine stretches interact with TAFII130, interfering with CREB-dependent transcription. Nature Genet 26:29–36 (2000).

Shimura H, Hattori N, Kubo S-i, Mizuno Y, Asakawa S, Minoshima S, Shimizu N, Iwai K, Chiba T, Tanaka K, Suzuki T: Familial Parkinson disease gene product, parkin, is a ubiquitin-protein ligase. Nature Genet 25:302–305 (2000).

Skinner PJ, Vierra-Green CA, Clark HB, Zoghbi HY, Orr HT: Altered trafficking of membrane proteins in Purkinje cells of *SCA1* transgenic mice. Am J Pathol 159:905–913 (2001).

Steffan JS, Bodai L, Pallos J, Poelman M, Mccampbell A, Apostol BL, Kazantsev A, Schmidt E, Zhu Y-Z, Greenwald M, Kurokawa R, Housman DE, Jackson GR, Marsh JL, Thompson LM: Histone deacetylase inhibitors arrest polyglutamine-dependent neurodegeneration in *Drosophila*. Nature 413:739–743 (2001).

Stevanin G, Le Guern E, Ravise N, Chneiweiss H, Durr A, Cancel G, Vignal A, Boch AL, Ruberg M, Penet C, et al.: A third locus for autosomal dominant cerebellar ataxia type I maps to chromosome 14q24.3-qter: evidence for the existence of a fourth locus. Am J hum Genet 54:11–20 (1994).

Suite ND, Sequeiros J, McKhann GM: Machado-Joseph disease in a Sicilian-American family. J Neurogenet 3:177–182 (1986).

Takiyama Y, Ikemoto S, Tanaka Y, Mizuno Y, Yoshida M, Yasuda N: A large Japanese family with Machado-Joseph disease: clinical and genetic studies. Acta neurol Scand 79:214–222 (1989).

Takiyama Y, Nishizawa M, Tanaka H, Kawashima S, Sakamoto H, Karube Y, Shimazaki H, Soutome M, Endo K, Ohta S, Kagawa Y, Kanazawa I, Mizuno Y, Yoshida M, Yuasa T, Horikawa Y, Oyanagi K, Nagai H, Kondo T, Inuzuka T, Onodera O, Tsuji S: The gene for Machado-Joseph disease maps to human chromosome 14q. Nature Genet 4:300–304 (1993).

The Huntington's Disease Collaborative Research Group: A novel gene containing a trinucleotide repeat that is expanded and unstable on Huntington's disease chromosomes. Cell 72:971–983 (1993).

Walsh DM, Klyubin I, Fadeeva JV, Cullen WK, Anwyl R, Wolfe MS, Rowan MJ, Selkoe DJ: Naturally secreted oligomers of amyloid β protein potently inhibit hippocampal long-term potentiation in vivo. Nature 416:535–539 (2002).

Wang G-H, Sawai N, Kotliarova S, Kanazawa I, Nukina N: Ataxin-3, the *MJD1* gene product, interacts with the two human homologs of yeast DNA repair protein RAD23, HHRA23A and HHR23B. Hum molec Genet 9:1795–1803 (2000).

Ward CL, Omura S, Kopito RR: Degradation of CFTR by the ubiquitin-proteasome pathway. Cell 83: 121–127 (1995).

Warrick JM, Chan HYE, Gray-Board GL, Chai Y, Paulson HL, Bonini NM: Suppression of polyglutamine-mediated neurodegeneration in *Drosophila* by the molecular chaperone Hsp70. Nature Genet 23:425–428 (1999).

Wojcik C: VCP–the missing link in protein degradation? Trends Cell Biol 12:212 (2002).

Wong PC, Pardo CA, Borchelt DR, Lee MK, Copeland NG, Jenkins NA, Sisodia SS, Cleveland DW, Price DL: An adverse property of a familial ALS-linked SOD1 mutation causes motor neuron disease characterized by vacuolar degeneration of mitochondria. Neuron 14:1105–1116 (1995).

Yamamoto Y, Hasegawa H, Tanaka K, Kakizuka A: Isolation of neuronal cells with high processing activity for the Machado-Joseph disease protein. Cell Death Differ 8:871–873 (2001).

Yasuda S, Inoue K, Hirabayashi M, Higashiyama H, Yamamoto Y, Fuyuhiro H, Komure O, Tanaka F, Sobue G, Tsuchiya K, Hamada K, Sasaki H, Takeda K, Ichijo H, Kakizuka A: Triggering of neuronal cell death by accumulation of activated SEK1 on nuclear polyglutamine aggregations in PML bodies. Genes Cells 4:743–756 (1999).

Ye Y, Meyer HH, Rapoport TA: The AAA ATPase Cdc48/p97 and its partners transport proteins from the ER into the cytosol. Nature 414:652–656 (2001).

Yuasa T, Ohama E, Harayama H, Yamada M, Kawase Y, Wakabayashi M, Atsumi T, Miyatake T: Joseph's disease: clinical and pathological studies in a Japanese family. Ann Neurol 19:152–157 (1986).

Zhang X, Shaw A, Bates PA, Newman RH, Gowen B, Orlova E, Gorman MA, Kondo H, Dokurno P, Lally J, Leonard G, Meyer H, van Heel M, Freemont PS: Structure of the AAA ATPase p97. Mol Cell 6:1473–1484 (2000).

Zinszner H, Kuroda M, Wang X, Batchvarova N, Lightfoot RT, Remotti H, Stevens JL, Ron D: CHOP is implicated in programmed cell death in response to impaired function of the endoplasmic reticulum. Genes Dev 12:982–995 (1998).

Cytogenet Genome Res 100:276–286 (2003)
DOI: 10.1159/000072863

The use of transgenic and knock-in mice to study Huntington's disease

M.A. Hickey and M.-F. Chesselet

Department of Neurology, Reed Neurological Research Centre, The David Geffen School of Medicine at UCLA, Los Angeles, CA (USA)

Abstract. The trinucleotide repeat disorders comprise an ever expanding list of diseases, all of which are caused by an unstable expanded trinucleotide repeat tract. Huntington's disease (HD) is a member of this family of diseases and more specifically, is a Type II trinucleotide repeat disorder. This means that the mutation in HD is an unstable expanded polyglutamine repeat tract, which is expressed at protein level. There is no cure or beneficial treatment for this fatal neurodegenerative disorder, and patients suffer from progressive motor, cognitive and psychiatric dysfunction. Recent years has seen the development of many genetic models of HD, which allow study of the early phases of disease process, at several different levels of cell function. In addition, these models are being used to investigate the potential of a variety of therapeutic agents for clinical use. Here we review these findings, and their implication for HD pathogenesis.

Huntington's disease (HD) is a member of the trinucleotide-repeat disorder family (Table 1). These diseases are caused by an unstable expansion in a trinucleotide repeat tract and they are divided into two groups based on whether the mutation is expressed in protein form or not (Cummings and Zoghbi, 2000). The mutation underlying the disease was characterized in 1993 (Huntington's Disease Collaborative Research Group, 1993). In the diseased state, the number of CAG repeats in the polyglutamine tract near the N-terminus of the huntingtin gene (HD alias IT15) is expanded to beyond 36 repeats. Normally, this tract contains between 19 and 35 CAG repeats. Complete penetrance of the disease is seen with repeat lengths of 42 or more, indicating that onset of the disease in patients carrying between 36 and 41 repeats may be influenced by other factors (Brinkman et al., 1997). HD is transmitted in an autosomal dominant manner and symptoms manifest in midlife (adult-onset), or in rarer circumstances, before the age of 20 (juvenile onset). In adult-onset HD, symptoms gradually worsen over a period of 10–15 years and patients eventually become totally dependent and bedridden (Harper, 1996). The disease is fatal with death usually caused by disease related complications. The juvenile form of the disease is more severe in progression and presentation and is caused by CAG repeat lengths exceeding 60 (Harper, 1996; Foroud et al., 1999).

The gene, symptoms and pathology associated with each Type II triplet repeat disorder are disease-specific, but as the repeat length increases, the areas of the brain and nervous system affected begin to overlap between family members (Cummings and Zoghbi, 2000). At shorter CAG lengths, the protein context may be important for disease specificity since all are caused by an expanded polyglutamine tract (Table 1).

Extensive neuropathological assessment of HD patient brains has revealed that the primary target of HD degeneration is the population of striatal γ-aminobutryic acid projection neurons (Fig. 1). Other areas of the brain are also affected at later stages including the cortex, hippocampus, globus pallidus, substantia nigra, subthalamic nucleus, and cerebellum. The extent to which patient brains are affected is reflected in a grading scale, where brains are divided into five grades depending on severity of pathology (Vonsattel et al., 1985).

Supported by Hereditary Disease Foundation Cure HD Initiative.

Received 15 October 2002; manuscript accepted 27 November 2002.

Request reprints from Marie-Françoise Chesselet, M.D., Ph.D.
Department of Neurology, Reed Neurological Research Center, B114
The Geffen School of Medicine at UCLA, 710 Westwood Plaza
Los Angeles, CA 90095 (USA); telephone: 310-267-1781
fax: 310-267-1786; e-mail: mchesselet@mednet.ucla.edu

KARGER Fax + 41 61 306 12 34
E-mail karger@karger.ch
www.karger.com

0301–0171/03/1004–0276$19.50/0

Accessible online at:
www.karger.com/cgr

Table 1. Trinucleotide repeat diseases[a]

Type I diseases	Fragile X syndrome	Fragile XE syndrome	Freidrich's Ataxia	Myotonic dystrophy	SCA8	SCA12	Jacobsen syndrome	
Associated gene	FMR1	FMR2	X25	DMPK	SCA8	SCA12	FRA11B[b]	
Trinucleotide repeat	CGG	GCC	GAA	CTG	CTG	CAG	CCG	
Type II diseases	HD	SCA1	SCA2	SCA3	SCA6	SCA7	SBMA	DRPLA
Associated protein	Huntingtin	Ataxin-1	Ataxin-2	Ataxin-3	α_{1a}-voltage dependent calcium channel	Ataxin-7	Androgen receptor	Atrophin-1
Trinucleotide repeat	CAG	CAG	CAG	CAG	CAG	CAG	CAG	CAG

[a] Diseases caused by an unstable expansion in a trinucleotide repeat tract. Type I disease mutations are not expressed in protein form, Type II mutations are expressed in protein. In general the expansions observed in Type I diseases far exceed those in the Type II group (Cummings and Zoghbi, 2000). Data from Cummings and Zoghbi (2000) and Jones et al. (2000). FMR1, fragile X MR gene; FMR2, fragile XE MR gene; DMPK, myotonic dystrophy protein kinase; SCA, spinocerebellar ataxia; HD, Huntington's disease; SBMA, Spinobulbar muscular atrophy; DRPLA, Dentatorubral-pallidoluysian atrophy.

[b] Fragile sites also found in 11q23.3→q24 (Jones et al., 2000).

Modeling HD

Animal models of disease are invaluable for research, however as a matter of course, rodent, invertebrate and non-human primate models can never exactly replicate a human disease. With this caveat, the best models are those that show relevant behavioral and neuropathological changes. The earliest models of HD utilized glutamate neurotoxins, including glutamate itself, kainic acid, ibotenic acid and quinolinic acid (respectively, Olney and de Gubareff, 1978; Coyle and Schwarz, 1976; Isacson et al., 1989; Beal et al., 1986). The use of these toxins in the generation of models of HD pathology led to the excitotoxic theory of cell death (DiFiglia, 1990). This was later found to be more successful as an explanation for acute brain damage (Albin and Greenamyre, 1992; Greene and Greenamyre, 1996). Degeneration in HD progresses over a long period of time and it was proposed that indirect excitotoxicity or secondary excitotoxicity could lead to degeneration in HD. Secondary excitotoxicity is caused by impaired cellular metabolism, leading to a reduction in energy levels and consequent membrane depolarization, which in turn increases the likelihood of NMDA receptor activation and slow cell death (Albin and Greenamyre, 1992; Greene and Greenamyre, 1996). 3-Nitropropionic acid and also malonate inhibit complex II/III of the mitochondrial transport chain and administration of these toxins to humans and animals causes degeneration of striatal GABAergic neurons (Brouillet et al., 1999). However, recently, the genetic mouse models of HD (see below) have been shown to be either resistant or at least show no increased sensitivity to excitotoxins and secondary excitotoxins (Hansson et al., 1999, 2001a, 2001b; Hickey and Morton, 2000; Petersén et al., 2002). In addition, direct measurement of mitochondrial complex activity has shown no dysfunction in transgenic mice and in grade 1 HD brains (Guidetti et al., 2001). Furthermore, changes in transcription have been noted in cellular models of HD in the absence of mitochondrial impairment (Kita et al., 2002). Interestingly, a recent paper showed the direct association of mutant huntingtin with mitochondrial membranes (Panov et al., 2002). These authors showed that mitochondria from HD patient lymphoblasts had reduced membrane potentials which they could recreate with long polyglutamine repeats. Anomalies in lymphoblast mitochondria had already been reported by Sawa et al. (1999), however others have found no evidence for mitochondrial impairment in peripheral cells (Gu et al., 1996; Tabrizi et al., 1999). The susceptibility to depolarization appears to be proportional to CAG repeat length (Sawa et al., 1999; Panov et al., 2002) and expression levels (Panov et al., 2002). Sawa and colleagues found a threshold at 60 CAGs, and Panov et al. showed that lymphoblasts carrying juvenile-onset repeat lengths were more sensitive to depolarization than those carrying shorter repeats. Thus, mitochondrial impairment may be a secondary impairment in HD and not the primary cause of cell death. In addition, neurochemically-induced pathology does not progress and is relatively acute. The most recent crop of models for HD are the genetic models, and include the most accurate to date. These genetic models include viral models, transgenic mouse models and knock-in mouse models (Fig. 2), the latter faithfully replicating the exact human mutation found in HD patients.

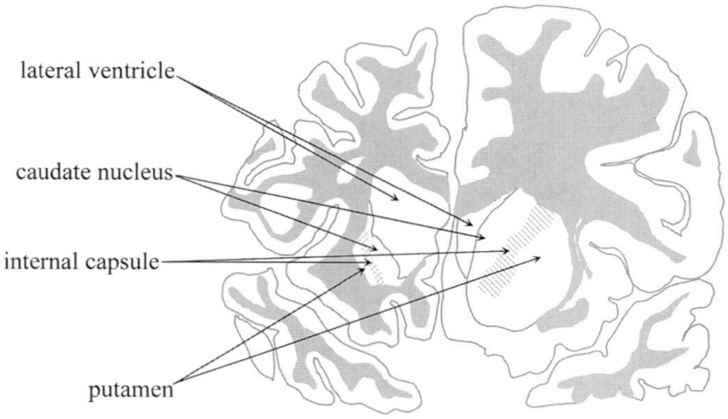

Fig. 1. Schematic representation of the basal ganglia in the human. Line drawing of coronal sections at the level of the frontal lobe, of a brain with HD (left) and a normal brain (right). The cerebral cortex, caudate nucleus and putamen all show atrophy, while the lateral ventricle is enlarged. Adapted from Harper (1996).

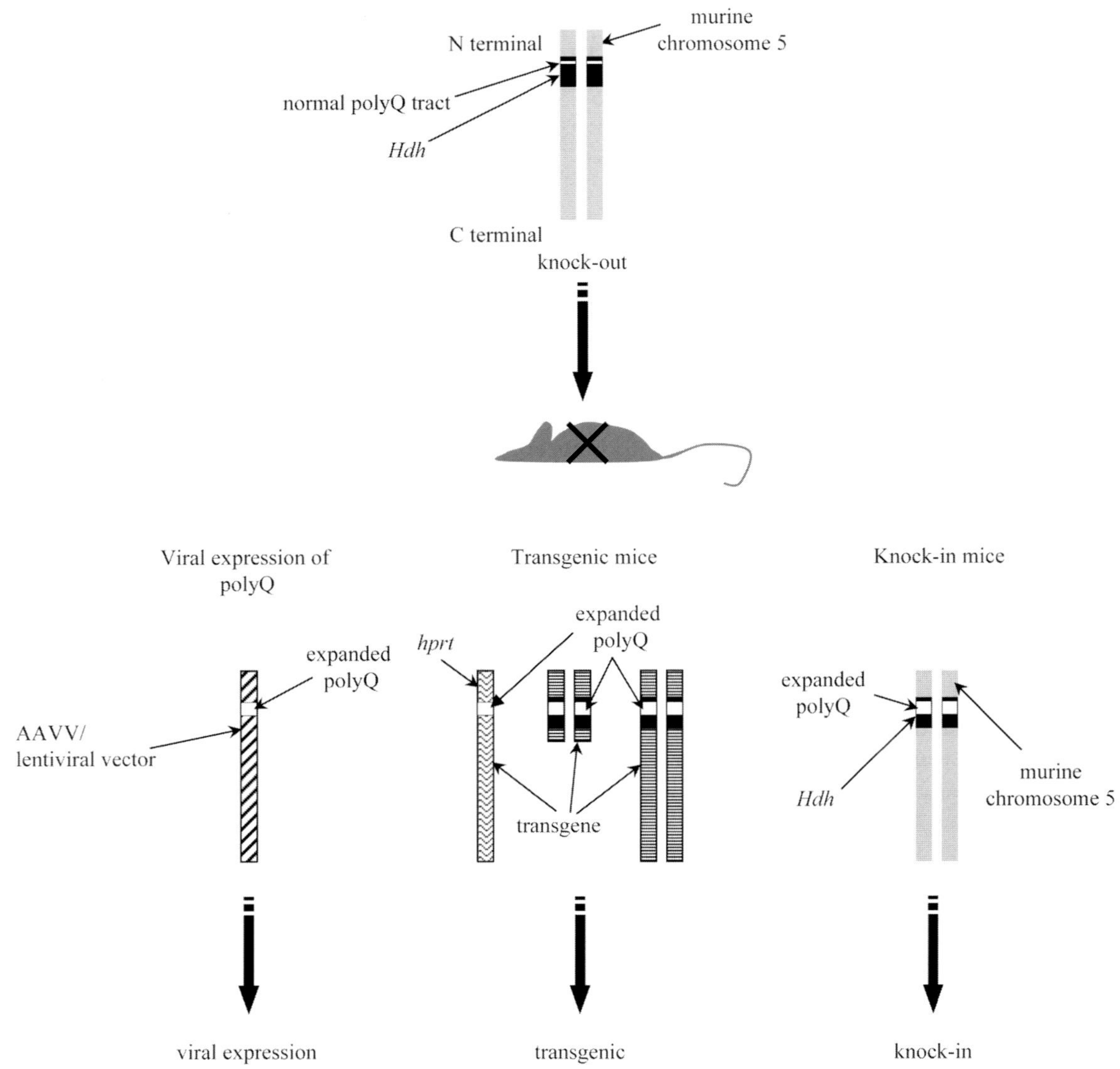

Fig. 2. Schematic diagram of the manipulations used to generate mouse models of HD and polyglutamine repeat disorders AAVV, adeno-associated viral vector; polyQ, polyglutamine; *Hdh*, murine homologue of the human HD gene (encodes huntingtin).

Genetic models of HD

Knock-out mice

At first, the HD mutation was thought to cause a loss in function of the normal protein. However, homozygous deletion of the HD murine analogue *Hdh* (Barnes et al., 1994; Lin et al., 1994) is lethal, suggesting that expression of huntingtin was necessary for normal development (Duyao et al., 1995; Nasir et al., 1995; Zeitlin et al., 1995). Even the expression of one allele of mutant huntingtin rescues the lethal knock-out effect (White et al., 1997). Furthermore, hemizygosity for the HD locus (Ambrose et al., 1994; Estabrooks et al., 1994) does not manifest in HD. It may be that both a gain and loss in huntingtin function contributes to HD (Cattaneo et al., 2001). Mutant heterozygous $Hdh^{+/-}$ mice display behavioral abnormalities and basal ganglia cell loss (Nasir et al., 1995). Late disruption of *Hdh* (in late gestation or up to seven days post partum) also causes behavioral anomalies and pathology mirrors that seen in transgenic models of HD (Dragatsis et al., 2000). This combination of gain and loss in function is also observed in Spinal and Bulbar Muscular Atrophy (SBMA), another Type II disease, which manifests as decreased androgen sensitivity (La Spada et al., 1992). However, complete lack of androgen receptors causes a phenotype distinct from that of SBMA (La Spada et al., 1991), indicating that loss of function alone does not explain the effect of the mutation.

Virally-inserted mutated polyQ tracts

Viral vector aided insertion of genes or parts of genes, has the advantage of precise localization of injection and gene expression. There are two types of viral vector used currently; adeno-associated and lentiviral, both of which integrate into the host genome (Pfeifer and Verma, 2001). Expression is not achieved at the level of the organism, and as models, they are more labor intensive than transgenic or knock-in mouse models. However, this technology is applicable to non-rodent species in a way that is not currently possible with transgene or knock-in technology.

These models can be used to study the effects of discrete amounts of the mutant protein and can also be used to study load, temporal and spatial effects of the mutation, in more controlled circumstances than in transgenic or knock-in animals. In addition the effect of different viral vector inserts can be compared, potentially in the same animal. This may be of importance since genetically identical animals with the same mutant polyglutamine transgene developed very different amounts of aggregates of the mutant huntingtin protein (Morley et al., 2002). This has also been noted in knock-in mouse models of HD (see below, Menalled et al., 2002). Aggregates of the mutant huntingtin protein are a major area of investigation in HD and are discussed in more detail below. They are observed in many different genetic models of HD and are also seen in other polyglutamine diseases (Menalled and Chesselet, 2002).

Thus far, two groups have investigated viral vector delivery of a mutant polyglutamine tract into rat striata. Senut and colleagues (2000) examined the effect of expression of a polyglutamine (97Q) repeat in the context of an adeno-associated viral vector (AAVV). High levels of expression of the polyQ tract led to development of nuclear and cytoplasmic (in the cell soma and axon) aggregates as early as five days post infection. This indicates that expression of the polyglutamine alone is highly associated with development of inclusion bodies. This had previously been noted by Ordway et al. (1997), who developed a model in which ectopic expression of a polyglutamine tract in the hypoxanthine-guanine phosphoribosyltransferase *(Hprt)* gene resulted in aggregate formation in mice. Senut and colleagues (2000) found that when a polyglutamine was overexpressed by way of gene delivery to the striatum, there was a loss in signal at five weeks post infection. This was attributed to apoptotic cell death with the caveat that cell death was not directly associated with cells expressing the GFP-conjugated mutant protein. Aggregates were also observed at a distance from the injection site, indicating that anterograde and retrograde transport of the vector had taken place.

Lentiviral vectors are modified forms of the human HIV retrovirus, and are capable of carrying larger transgenes than AAVVs, in theory, for indefinite periods of time (Pfeifer and Verma, 2001). de Almeida and collegues (2002), with this vector, also found that expression levels of the transgene were proportional to severity of neuropathology. In addition, increased expression led to production of nuclear inclusions, whereas neuritic and nuclear aggregates were observed when expression was driven by a weaker promoter. Longer CAG repeat lengths correlated with increased aggregate formation. Loss of staining for aggregated mutant huntingtin near the injection site was attributed to cell death and proportional to high expression levels. Importantly, ChAT and NADPH-d interneurones were relatively preserved; a feature of HD in humans.

Thus, these models show that the polyglutamine tract is itself inherently toxic, however, the genomic context of the mutation is lost since the viral vector is inserted at random. Were the surrounding protein tract actually unimportant, then the type II diseases would presumably present as similar diseases. Thus both the genomic context and protein context are extremely important for accurate modeling of the disease and providing accurate information for assessing efficacy of therapies. This is also relevant for transgenic models (see below). However, the information that these models provide is valuable in elucidating the role of aggregate formation in HD and the early effects of the mutation.

Transgenic mice

Several transgenic mouse models of HD have been generated (Table 2). These mice also show the importance of the expression levels of the transgene and the length of the transgene and CAG repeat in relation to neuropathology and symptom presentation. In general, it seems that the shorter the transgene, the longer the CAG repeat, and the greater the expression levels, the more severe the phenotype. These findings are important since it is known that huntingtin is cleaved by caspases and calpains thus generating small N-terminal fragments containing the mutant polyglutamine tract (Goldberg et al., 1996b; Wellington et al., 2000, 2002; Mende-Mueller et al., 2001; Gafni and Ellerby, 2002; Toneff et al., 2002). Further, CAG repeat length increases with successive generations in human patients (Telenius et al., 1993; Zühlke et al., 1993). With regard to expression levels, there is no evidence to date to suggest that homozygote human patients manifest with a more severe form of the disease than heterozygotes (Wexler et al., 1987), however mouse studies show a clear gene dosage effect (Mangiarini et al., 1996; Lin et al., 2001). It is possible that with time, a dosage effect will become evident in humans.

The first transgenic mouse models generated were those of Goldberg et al. (1996a) and Mangiarini et al. (1996). Goldberg and colleagues (1996a) generated transgenic mice that carried full-length human huntingtin with 44 CAG repeats accompanied by the 3′ and 5′ untranslated regions. No protein expression was detected, and there was no indication of neuronal loss, neurodegeneration or behavioral abnormalities. Mangiarini et al. (1996) created the R6 lines of HD transgenic mice. These mice carry a short N-terminal region of a mutant human huntingtin gene that includes the promoter, exon 1 and part of intron 1. The authors noted the development of a neurological phenotype (see below) in the mice, with a variable age of onset of symptoms depending on the CAG repeat length and transcript copy number. The most studied of these lines is the R6/2.

One of the most important points to come from examination of the R6/2 mice is that a severe neurological phenotype develops in the absence of overt neuronal loss (Turmaine et al., 2000). DNA damage (Keene et al., 2002; but see Turmaine et al., 2000) and non-apoptotic cell death (Turmaine et al., 2000) has been noted in mice over 13 weeks of age and medium spiny GABAergic neuron dendrite fields are reduced in mice over 12 weeks of age (Klapstein et al., 2001). Despite this late pathology, overt symptoms are generally observed at eight weeks of age in R6/2 mice and closer examination reveals deficits starting as early as 3.5 weeks (Lione et al., 1999). Thus, functional deficits precede cell loss in these mice.

Table 2. Transgenic mouse models of HD carrying a fragment of the human huntingtin gene

Name	Promoter[a]	Background	Polyglutamine repeat length	Expression levels in comparison to endogenous *Hdh* levels	Phenotype[b]: Tremor	hlc	Weight loss	Hyper/hypoactivity	Rotarod impairment	Early death	Neuropathology[b]: Neurodegeneration	NIIs or aggregates	Astrocytosis/gliosis	References[c]
R6/0	HD	CBA × C57BL/6J	142 CAGs	0	–	–	–	–	–	–	–	–	–	1
R6/1			115 CAGs	R6/1, R6/2, R6/5: mRNA < endogenous; R6/2, R6/5: protein > endogenous	√	√	√	–/√	√	√	√	√	–	1, 2, 3, 4
R6/2			145 CAGs		√	√	√	√/√	√	√	√	√	–	1, 5, 6, 7, 8, 12
R6/5			128-156 CAGs		√	√	√	n.r.	n.r.	√	√	√	–	1
N171	*Prp*	C3H/HEJ × C57BL/6J	18 CAGs, line 8	mRNA >> endogenous protein < endogenous	–	–	–	–	–	–	n.r.	–	–	10
			44 CAGs, line 9		–	–	–	–	n.r.	–	–	–	–	10, 11
			82 CAGs, line 6		√	√	√	–/√	n.r.	√	–/?	√	–	10, 11
			82 CAGs, line 77 (died out)		n.r.	n.r.	n.r.	n.r.	n.r.	√	–/?	n.r.	–	10, 11
			82 CAGs, line 81		√	√	√	–/√	√	√	–/?	√	–	10, 11
			82 CAGs, line 100		√	√	√	–/√	√	√	–/?	n.r.	–	10, 11
CT18	*Nse*	C57BL/6 × SJL	18 CAGs	protein < endogenous, varied in expression levels between mice	–	–	–	–	–	–	–	–	–	12
HD46			46 CAGs		n.r.	√	n.r.	*	√	n.r.	√	√	–	12
HD100			100 CAGs		n.r.	√	n.r.	*	√	n.r.	√	√	–	12
HD94-Tet regulated	tetO responsive operator, miminal CMV promoter	CBA × C57BL/6J	94 CAGs	? gene on	√	√	–	–/√	√	√	–	√	√	13, 14
				gene off	n.r.	↓	n.r.	n.r.	↓	↓	–	↓	↓	

[a] Prp, prion protein promoter; CMV, cytomegalovirus promoter; tetO, operator for the repressor of the tetracycline resistance operon, *Nse,* neuron-specific enolase.

[b] hlc, hindlimb clasping; √ present; - not present; n.r. not reported; ↓ reduced occurrence; -/? some neurodegeneration may be present, see note at end of paper by Schilling et al. (1999); ? protein levels not reported; * differences in activity noted, whether hyperactivity or hypoactivity occurred was not reported.

[c] 1, Mangiarini et al. (1996); 2, Iannicola et al. (2000); 3, Hansson et al. (2001b); 4, Clifford et al. (2002); 5, Davies et al. (1997); 6, Dunnett et al. (1998); 7, Carter et al. (1999); 8, Turmaine et al. (2000); 9, Lüesse et al. (2002); 10, Schilling et al. (1999); 11, Schilling et al. (2001); 12, Laforet et al. (2001); 13, Yamamoto et al. (2000); 14, Martin-Aparicio et al. (2001).

Study of the R6/2 mice also led to the rediscovery of protein aggregates of the mutant huntingtin protein, originally described by Roizin et al. (1979). These aggregates have now been identified in HD brain tissue, several HD mouse models and in tissue from the other Type II trinucleotide repeat diseases (Menalled and Chesselet, 2002). R6/2 mice display intranuclear and cytoplasmic inclusions (Davies et al., 1997; Li et al., 1999; Morton et al., 2000), mimicking the human disease where inclusion bodies have been found in both the nucleus and cytoplasm (Gutekunst et al., 1999; Gomez-Tortosa et al., 2001). The R6/2 phenotype is complex and includes biphasic changes in activity (Lüesse et al., 2001) and progressive reduction in body weight, motor and cognitive performance and early death (Mangiarini et al., 1996; Dunnett et al., 1998; Carter et al., 1999; Lione et al., 1999). Decreased protein and mRNA levels of various receptors, enzymes and signalling molecules is also reported (Cha et al., 1998, 1999; Bibb et al., 2000; Luthi-Carter et al., 2000, 2002a, 2002b). Reduced body weight, early death, neuropathological change and progressive motor and cognitive decline have also all been reported in HD patients (Furtado and Suchowersky, 1995; Antonini et al., 1996; Vonsattel and DiFiglia, 1998; Andrews et al., 1999). Diabetes has been detected in R6/2 mice (Hurlbert et al., 1999), and although it is not found in other models (Schilling et al., 1999), it has been reported in some studies of HD patients (Podolsky et al., 1972; Podolsky and Leopold, 1977; Farrer, 1985).

One of the most interesting features of these transgenic models is that despite the differences in genetic makeup and background strain, there are striking similarities in phenotype and neuropathological presentation despite different ages at onset. Hindlimb grooming, resting tremor, weight loss, hypoactivity, early death, neuronal intranuclear inclusion formation and neurodegeneration are displayed by transgenic mouse models of HD (see Table 2). These models include mice that carry the full-length huntingtin (Reddy et al., 1998; Hodgson et al., 1999) and mice that carry a truncated version of the gene (Schilling et al., 1999; Laforet et al., 2001). These models have confirmed the effect of expression levels on phenotype expression and severity of phenotype and neuropathology. In particular the full-length models have shown that use of different promoter constructs leads to very different age of onset even in mice from the same background (FVB/N, see Table 3). Similarly, mice carrying similar CAG repeat tracts can show different phenotypes. As for transgenic mice that express a truncated protein, neurodegeneration was observed after onset of phenotype (Turmaine et al., 2000). However, very few inclusions were found (Reddy et al., 1998), if any (Hodgson et al., 1999) in transgenic mice carrying the full-length protein. It is notewor-

Table 3. Transgenic mouse models of HD carrying the full length human huntingtin gene

Name	Promoter[a]	Background	Genetic insertion	Expression levels in comparison to endogenous *Hdh* levels	Phenotype[b]						Neuropathology[b]			References[c]
					Tremor	hlc	Weight loss	Hyper/hypoactivity	Rotarod impairment	Early death	Neurodegeneration	NIIs or aggregates	Astrocytosis/gliosis	
HD16	CMV and SV40 enhancer	FVB/N	16 CAGs, lines A–D	protein >> → ≈ endogenous	n.r.	–	–	–/–	n.r.	–	–	–	–	1
HD48			48 CAGs, line C	protein ≈ endogenous	n.r.	√	–	√?	n.r	–	–	–	–	1
			48 CAGs, line D	protein > endogenous	n.r.	√	n.r.	√*/√	n.r.	√	n.r.	√	n.r.	1
			48 CAGs, line B	protein >> endogenous	n.r.	√	√^	√*/√	n.r	√	√	√	√	1
HD89			89 CAGs, line A	protein: HD89A, B, C > endogenous	n.r.	√	√^	√*/√	n.r.	√	√	√	√	1
			89 CAGs, line B		n.r.	√	√^	√*/√	n.r.	√	√	√	√	1
			89 CAGs, line C		n.r.	√	√^	√*/√	n.r.	√	√	√	√	1
YAC18	HD/*YAC*	FVB/N	18 CAGs, line 30	protein < endogenous	–	–	–	–	n.r.	–	–	–	n.r.	2
YAC46			46 CAGs, line 668, 1747		–	–	–	–	n.r.	–	–	–	n.r.	2
YAC72			72 CAGs line 2511, 2498	2511 protein < endogenous 2498 protein > endogenous	n.r.	√	√/–~	√/–#	n.r.	√	√	√	n.r.	2

[a] HD, human huntingtin promoter; CMV, cytomegalovirus promoter; SV40, simian virus 40; YAC, yeast artificial chromosome.
[b] hlc, hindlimb clasping; √, present; -, not present; n.r., not reported; √?, one animal showed hyperactivity; √*, 37% of these animals showed hyperactivity; √^ these animals did not gain weight while hyperactive; √/-~, mouse from 2498 lost weigh, mice from line 2511 did not; √/- # both lines exhibited hyperactivity at different times.
[c] 1, Reddy et al. (1998); 2, Hodgson et al. (1999).

thy that even very low levels of the truncated mutant protein also led to phenotype development and aggregate formation (Schilling et al., 1999), showing the low threshold for disease.

Another transgenic model of HD that has been important in understanding disease process was generated by Yamamoto and colleagues (2000). Using a tetracycline-responsive gene system, gene expression was analysed by detection of the linked β-galactosidase gene and could be "turned off" by oral administration of tetracycline derivatives. When the gene was turned on from birth, all transgenic mice showed development of a progressive phenotype, starting at eight weeks of age. Intranuclear and cytoplasmic inclusions were found in both the striatum and cortex, with later loss in striatal volume, and astrocytosis. When the transgene was turned off, nuclear staining for the transgene in striatal and cortical neurons, and neuronal intranuclear inclusions and extranuclear inclusions disappeared. Astrocytosis was reduced and duration of hindlimb clasping reverted to levels of control mice. Therefore, the progressive changes brought about by mutant huntingtin can be arrested and even reversed, if treated in time. Subsequent study has now also shown recovery of rotorod performance following shut down of the gene (Martín-Aparicio et al., 2001). These observations bring new hope for effective treatment of this devastating disorder.

Knock-in mice

Knock-in mice represent faithfully the HD mutation carried by patients, since elongated polyglutamine tracts are placed within the correct context of the murine *Hdh* gene (see Table 4). Initially, behavioral deficits were reported as absent, except for some changes in aggression (Shelbourne et al., 1999). However, homozygous knock-in mice have now been shown to develop very early behavioral anomalies, prior to any detectable neuropathology (Menalled et al., 2002). These mice carry 94 CAG repeats and behavioral data suggest that cellular dysfunction (prior to aggregation of mutant huntingtin) is responsible for the initial symptoms of HD. These mice show abnormal activity levels that are biphasic and mirror the progression of motor symptoms in HD (Harper, 1996; Kirkwood et al., 2001). When neuropathology did develop (in the form of microaggregates) it was preferentially located in the striosomes (Menalled et al., 2002), an area initially targeted in HD patient brains (Hedreen and Folstein, 1995). Lin et al. (2001) also described a knock-in line where old heterozygous and younger homozygous mice carrying 150 CAG repeats displayed progressive hypoactivity, hindlimb clasping and gait abnormalities. Homozygous mice carrying 150 repeats also displayed striatal gliosis and neuronal intranuclear inclusions. A further knock-in mouse model has been shown to develop gait deficits after onset of neurodegeneration (Wheeler et al., 2002). However, we have found that gait changes are one of the later behavioral changes that occur in knock-in mice (Menalled et al., 2003), and as such, it is probable that initial behavioral symptoms and cell dysfunction are followed by nuclear accumulation of mutant huntingtin. Therefore knock-in mouse models of HD represent a valid model of HD, with the added advantage of a slower progression of phenotype and pathology, thus allowing more detailed analysis. Indeed, analysis of the knock-in mouse brains shows early development of nuclear staining and then microaggregates, prior to the development of inclusions (Wheeler et al., 2000; Li et al., 2001; Menalled et al., 2002). These microaggregates may be a precursor to inclusion formation (Menalled et al., 2002). In

Table 4. Knock-in mouse models of HD

Name	Promoter[a]	Background	Genetic insertion	Expression levels in comparison to endogenous *Hdh* levels	Phenotype[b] Tremor	hlc	Weight loss	Hyper/hypoactivity	Rotarod impairment	Early death	Neuropathology[b] Neurodegeneration	NIIs or aggregates	Astrocytosis/gliosis	References[c]
Hdh6/72	*Hdh*	C57BL/6J, FVB/N	72 CAGs	protein < endogenous	–	–	–	–	n.r.	–	√	√	–	1, 2
Hdh4/80			80 CAGs	protein < endogenous	–	–	–	–	n.r.	–	√	√	–	1, 2
Hdh71	*Hdh*	129Sv × C57BL/6J	*Hdh* exon 1 replaced with chimeric HD/*Hdh* exon 1 71 CAGs	protein ≈ endogenous	–	–	–	–	n.r.	–	–	–	n.r.	3, 4
Hdh94			*Hdh* exon 1 replaced with chimeric HD/*Hdh* exon 1 94 CAGs	protein ≈ endogenous	–	–	–	√/√	n.r.	–	–	√	n.r.	3, 4
Hdh140			*Hdh* exon 1 replaced with chimeric HD/*Hdh* exon 1 140 CAGs	protein ≈ endogenous	–	–	–	√/√	n.r.	–	–	√	n.r.	5
Hdh50	*Hdh*	CD1 × 129Sv	48 CAGs	protein ≈ endogenous	–	–	–	–	n.r.	n.r.	–	n.r.	–	6, 7, 8
Hdh92			90 CAGs	n.r.	n.r.	n.r.	n.r.	n.r.	n.r.	n.r.	n.r.	√	n.r.	6, 7, 8
Hdh111			109 CAGs	n.r	n.r.	n.r.	n.r.	n.r.	n.r.	n.r.	n.r.	√	n.r.	6, 7, 8
Hdh80	*Hdh*	129/Ola × C57BL/6J	chimeric HD/*Hdh* exon 1 knock-in 80 CAGs	protein ≈ endogenous	n.r.	√/–	√/–	–/√	n.r.	n.r.	–	√	–	9
Hdh150			chimeric HD/*Hdh* exon 1 knock-in 150 CAGs	protein ≈ endogenous	n.r.	√	√/–	–/√	√	√/?	–	√	√	9

[a] *Hdh*, mouse huntingtin homologue.
[b] hlc, hindlimb clasping; √, present; -, not present; n.r., not reported; √/?, early death shown by some mice, although numbers were small; √/- some mice exhibited hlc, weight loss.
[c] 1, Shelbourne et al. (1999); 2, Li et al. (2001); 3, Levine et al. (1999); 4, Menalled et al. (2002); 5, Menalled et al. (2003); 6, White et al. (1997); 7, Wheeler et al. (2000), 8, Wheeler et al. (2002); 9, Lin et al. (2001).

these knock-in mice, pathology is clearly dependent on CAG repeat length. With shorter (94 CAG) repeat lengths, nuclear staining and aggregates are strikingly restricted to the striatum, whereas they are more widespread, and accompanied by cytoplasmic aggregates in mice with longer repeats (140 CAGs, Menalled et al., 2003). The role of the cytoplasmic aggregates is uncertain, but they have been shown to be present in terminals and in striatal output nuclei (Li et al., 2001; Menalled et al., 2002, 2003). This may be of importance in view of the fact that dysfunctional synaptic release has been reported in several mouse models of HD (Klapstein et al., 2001; NicNiocaill et al., 2001). These progressive changes in aggregate development may be very important since it shows that they do not contribute to initial behavioral deficits, thus may not be the most efficient target for early therapy. However, they may contribute to the functional deficits that develop at later stages of the disease.

The fact that behavioral deficits occur prior to cell loss and prior to nuclear accumulation of huntingtin indicates that cell dysfunction at some other level is occurring. This could entail transcriptional changes, as first proposed by Cha et al. (1998, 1999; Cha, 2000). Dysregulation of gene expression has been noted in several models of HD and in HD itself (Richfield et al., 1995; Augood et al., 1996, 1997; Cha et al., 1998, 1999; Bibb et al., 2000; Iannicola et al., 2000; Luthi-Carter et al., 2000, 2002a, 2002b; Menalled et al., 2000; Chan et al., 2002). p53 and cyclic adenosine monophosphate responsive element binding protein (CREB) binding protein- (CBP) transcription is reduced by mutant huntingtin (Steffan et al., 2000) and if CREB function is disrupted, the hippocampus and striatum degenerate (Mantamadiotis et al., 2002). Interestingly, soluble, but not aggregated, huntingtin bound Sp1 and inhibited its interaction with DNA (Dunah et al., 2002; Li et al., 2002). Sp1 is an activator of TFIID-driven transcription (Pugh and Tjian, 1990). Therefore, disruption of transcription factors could occur at early timepoints in HD. Recent data showed that transcriptional changes occurred in PC12 cells within hours of turning on expression of an exon 1 fragment of mutant huntingtin (Kita et al., 2002). The down-regulated genes included those involved in glucose metabolism and oxidative stress. Interestingly, some of these downregulated genes are regulated by Sp1 (GSTM2, Kumar and Reddy, 2001; SLC2A1, Okamoto et al., 2001). Mouse studies have shown that the length of the transgene correlates negatively with transcriptional changes (Chan et al., 2002). This is important since in general, it appears that the shorter the mutant protein, the more severe the effects of the protein (see Tables 2–4). Truncated huntingtin protein has been detected in HD patient brains and in HD mouse brains prior to degeneration (Wellington et al., 2002). In view of these observations, treatments aimed at preventing transcription disruption are currently being developed, including treatment with histone deacetylase inhibitors (Hughes et al., 2001; Steffan et al., 2001).

Activation of apoptotic pathways may also be important in their own right. Studies of denatured proteins have shown that mutant huntingtin is cleaved in patient brains (Mende-Mueller et al., 2001), and as already discussed, cleavage occurs earlier than neurodegeneration (Wellington et al., 2002). Indeed, caspase inhibition is effective in preventing onset of phenotype in R6/2 mice (Ona et al., 1999; Chen et al., 2000) and in human

HD patients (Denovan-Wright et al., 2001). This suggests that the specific proteases that generate small, and more toxic, fragments of mutated huntingtin could provide a useful therapeutic target. In addition, the proteasome may prove to be vitally important in HD, since clearance of the aggregates in the inducible model of HD depended in part on proteasome function (Martín-Aparicio et al., 2001). Mutant ataxin-1, the affected protein in SCA1, a related polyglutamine disorder, inhibits the proteasome and this may contribute to the development of aggregates (Cummings et al., 1999).

Mouse models are particularly useful to elucidate aspects of the disease that are related to neuronal circuitry, a feature not reproduced in in vitro models or lower animals. The activity of the NMDA 2b receptor subunit has been shown to be increased from an early stage in HD models (Zeron et al., 2002). Thus it is possible that corticostriatal projections to the striatum could mediate an excitotoxic insult on striatal neurons from an early age. Increased sensitivity to NMDA receptor agonists has been noted in several transgenic mouse models of HD (Levine et al., 1999; Laforet et al., 2001). Furthermore, mutant huntingtin interferes with the ability of wildtype huntingtin to bind to NMDA receptors via post-synaptic density 95 protein: this effect reduces the anti-excitotoxic effect of wildtype huntingtin (Sun et al., 2001). Administration of glutamatergic antagonists have not been very successful in HD patients (Murman et al., 1997; Kremer et al., 1999; Rosas et al., 1999; Verhagen Metman et al., 2002), but the new data emerging from the study of mouse models suggest that these should be administered very early in the disease to be of benefit.

Finally, environmental stimulation may be very important in HD, since recent studies have shown progressive reduction in activity in HD patients (van Vugt et al., 2001), as seen in HD mouse models (Dunnett et al., 1998; Menalled et al., 2002). Pathological and behavioral deficits are increased in neurochemical animal models of Parkinson's disease when activity is reduced (Tillerson et al., 2002) and increasing activity reduces these deficits (Tillerson et al., 2001). Several studies have now shown that environmental stimulation is beneficial in HD mice (Carter et al., 2000; van Dellen et al., 2000; Hockly et al., 2002). Thus, more effective methods of controlling chorea may allow patients to produce more purposeful movement, with consequent neuroprotection.

Several other targets are currently under investigation as potential therapeutics, including administration of highly unsaturated fatty acids in human HD patients (Puri et al., 2002; Vaddadi et al., 2002; but see Rosser and Dunnett, 2002). There is also renewed interest in inhibition of tissue transglutaminase, but the effects on neuropathology are inconsistent between groups (Karpuj et al., 2002; Mastroberardino et al., 2002). How these latter treatments translate to actual meaningful human therapies remains to be seen.

Conclusion

The major goal of scientific research in HD is of course to find beneficial agents for treatment of HD patients and ultimately to find a cure. Characterization of multiple animal models is necessary to understand the effects of potential therapies. Several models have now been described in detail and experimenters have the choice to use animal models with phenotypes ranging from severe and short to subtle and slow, the latter mirroring early HD development. All of these models have their own advantages but an added advantage of the models with slower progression is that analysis of early pathogenesis will hopefully shed light on the mechanisms underlying the disease process itself, thus allowing optimal drug design for HD patients.

References

Albin RL, Greenamyre JT: Alternative excitotoxic hypotheses. Neurol 42:733–738 (1992).

Ambrose CM, Duyao MP, Barnes G, Bates GP, Lin CS, Srinidhi J, Baxendale S, Hummerich H, Lehrach H, Altherr M, Wasmuth J, Buckler A, Church D, Housman D, Berks M, Micklem G, Durbin R, Dodge A, Read A, Gusella J, MacDonald ME: Structure and expression of the Huntington's disease gene: evidence against simple inactivation due to an expanded CAG repeat. Somat Cell molec Genet 20:27–38 (1994).

Andrews TC, Weeks RA, Turjanski N, Gunn RN, Watkins LHA, Sahakian B, Hodges JR, Rosser AE, Wood NW, Brooks DJ: Huntington's disease progression PET and clinical observations. Brain 122:2353–2363 (1999).

Antonini A, Leenders KL, Spiegel R, Meier D, Vontobel P, Weigell-Weber M, Sanchez-Pernaute R, de Yebenez JG, Boesiger P, Weindl A, Maguire RP: Striatal glucose metabolism and dopamine D2 receptor binding in asymptomatic gene carriers and patients with Huntington's disease. Brain 119: 2085–2095 (1996).

Augood SJ, Faull RLM, Love DR, Emson PC: Reduction in enkephalin and substance P messenger RNA in the striatum of early grade Huntington's disease: a detailed cellular in situ hybridization study. Neurosci 72:1023–1036 (1996).

Augood SJ, Faull RLM, Emson PC: Dopamine D1 and D2 receptor gene expression in the striatum in Huntington's disease. Ann Neurol 42:215–221 (1997).

Barnes GT, Duyao MP, Ambrose CM, McNeil S, Persichetti F, Srinidhi J, Gusella JF, MacDonald ME: Mouse Huntington's disease gene homolog *(Hdh)*. Somat Cell molec Genet 20:87–97 (1994).

Beal MF, Kowall NW, Ellison DW, Mazurek MF, Swartz KJ, Martin JB: Replication of the neurochemical characteristics of Huntington's disease by quinolinic acid. Nature 321:168–171 (1986)

Bibb JA, Yan Z, Svenningsson P, Snyder GL, Pieribone VA, Horiuchi A, Nairn AC, Messer A, Greengard P: Severe deficiences in dopamine signaling in presymptomatic Huntington's disease mice. Proc natl Acad Sci, USA 97:6809–6814 (2000).

Brinkman RR, Mezei MM, Theilmann J, Almqvist E, Hayden MR: The likelihood of being affected with Huntington disease by a particular age for a specific CAG size. Am J hum Genet 60:1202–1210 (1997).

Brouillet E, Condé F, Beal MF, Hantraye P: Replicating Huntington's disease phenotype in experimental animals. Prog Neurobiol 59:427–468 (1999).

Carter RJ, Lione LA, Humby T, Mangiarini L, Mahal A, Bates GP, Dunnett SB, Morton AJ: Characterization of progressive motor deficits in mice transgenic for the human Huntington's disease mutation. J Neurosci 19:3248–3257 (1999).

Carter RJ, Hunt MJ, Morton AJ: Environmental stimulation increases survival in mice transgenic for exon 1 of the Huntington's disease gene. Mov Disord 15:925–937 (2000).

Cattaneo E, Rigamonti D, Goffredo D, Zuccato C, Squitieri F, Sipione S: Loss of normal huntingtin function: new developments in Huntington's disease research. Trends Neurosci 24:182–188 (2001).

Cha J-HJ: Transcriptional dysregulation in Huntington's disease. Trends Neurosci 23:387–392 (2000).

Cha J-HJ, Kosinski CM, Kerner JA, Alsdorf SA, Mangiarini L, Davies SW, Penney JB, Bates GP, Young AB: Altered brain neurotransmitter receptors in transgenic mice expressing a portion of an abnormal human Huntington disease gene. Proc natl Acad Sci, USA 95:6480–6485 (1998).

Cha J-HJ, Frey AS, Alsdorf SA, Kerner JA, Kosinski CM, Mangiarini L, Penney JB Jr, Davies SW, Bates GP, Young AB: Altered neurotransmitter receptor expression in transgenic mouse models of Huntington's disease. Philos Trans R Soc Lond B Biol Sci 354:981–989 (1999).

Chan EY, Luthi-Carter R, Strand A, Solano SM, Hanson SA, DeJohn MM, Kooperberg C, Chase KO, DiFiglia M, Young AB, Leavitt BR, Cha J-HJ, Aronin N, Hayden MR, Olson JM: Increased huntingtin protein length reduces the number of polyglutamine-induced gene expression changes in mouse models of Huntington's disease. Hum molec Genet 11:1939–1951 (2002).

Chen M, Ona VO, Li M, Ferrante RJ, Fink KB, Zhu S, Bian J, Guo L, Farrell LA, Hersch SM, Hobbs W, Vonsattel J-P, Cha J-HJ, Friedlander RM: Minocycline inhibits caspase-1 and caspase-3 expression and delays mortality in a transgenic mouse model of Huntington disease. Nature Med 6:797–801 (2000).

Clifford JJ, Drago J, Natoli AL, Wong JY, Kinsella A, Waddington JL, Vaddadi KS: Essential fatty acids given from conception prevent topographies of motor deficit in a transgenic model of Huntington's disease. Neurosci 109:81–88 (2002).

Coyle JT, Schwarcz R: Lesion of striatal neurones with kainic acid provides a model for Huntington's chorea. Nature 263:244–246 (1976).

Cummings CJ, Reinstein E, Sun Y, Antalffy B, Jiang Y-H, Ciechanover A, Orr HT, Beaudet AL, Zoghbi HY: Mutation of the E6-AP ubiquitin ligase reduces nuclear inclusion frequency while accelerating polyglutamine-induced pathology in SCA1 mice. Neuron 24:879–892 (1999).

Cummings CJ, Zoghbi HY: Trinucleotide repeats: mechanisms and pathophysiology. Ann Rev Genomics Hum Genet 1:281–328 (2000).

Davies SW, Turmaine M, Cozens BA, DiFiglia M, Sharp AH, Ross CA Scherzinger E, Wanker EE, Mangiarini L, Bates GP: Formation of neuronal intranuclear inclusions underlies the neurological dysfunction in mice transgenic for the HD mutation. Cell 90:537–548 (1997).

de Almeida LP, Ross CA, Zala D, Aebischer P, Deglon N: Lentiviral-mediated delivery of mutant huntingtin in the striatum of rats induces a selective neuropathology modulated by polyglutamine repeat size huntingtin expression levels and protein length. J Neurosci 22:3473–3483 (2002).

DiFiglia M: Excitotoxic injury of the neostriatum: a model for Huntington's disease. Trends Neurosci 13:286–289 (1990).

Denovan-Wright EM, Broster B, Geyer L, Dursun S, Devarajan S, Fisk J, Darvesh S, Cook A, Robertson HA: Minocycline treatment for presymptomatic and symptomatic Huntington's disease patients. Soc Neurosci Abst 968.7 (2001).

Dragatsis I, Levine MS, Zeitlin S: Inactivation of *Hdh* in the brain and testis results in progressive neurodegeneration and sterility in mice Nature Genet 26:300–306 (2000).

Dunah AW, Jeong H, Griffin A, Kim YM, Standaert DG, Hersch SM, Mouradian MM, Young AB, Tanese N, Krainc D: Sp1 and TAFII130 transcriptional activity disrupted in early Huntington's disease. Science 296:2238–2243 (2002).

Dunnett SB, Carter RJ, Watts C, Torres EM, Mahal A, Mangiarini L, Bates G, Morton AJ: Striatal transplantation in a transgenic mouse model of Huntington's disease. Exp Neurol 154:31–40 (1998)

Duyao MP, Auerbach AB, Ryan A, Persichetti F, Barnes GT, McNeil SM, Ge P, Vonsattel J-P, Gusella JF, Joyner AL, MacDonald ME: Inactivation of the mouse Huntington's disease gene homolog *Hdh*. Science 269:407–410 (1995).

Estabrooks LL, Breg WR, Hayden MR, Ledbetter DH, Myers RM, Wyandt HE, Yang-Feng TL, Hirschhorn K: Summary of 1993 ASHG Ancilliary Meeting: Recent research on chromosome 4p syndromes and genes. Am J med Genet 55:453–458 (1994).

Farrer LA: Diabetes mellitus in Huntington disease. Clin Genet 27:62–67 (1985).

Foroud T, Gray J, Ivashina J, Conneally PM: Differences in duration of Huntington's disease based on age at onset. J Neurol Neurosurg Psychiatry 66:52–56 (1999).

Furtado S, Suchowersky O: Huntington's disease: recent advances in diagnosis and management. Can J Neurol Sci 22:5–12 (1995).

Gafni J, Ellerby LM: Calpain activation in Huntington's disease. J Neurosci 22:4842–4849 (2002).

Goldberg YP, Kalchman MA, Metzler M, Nasir J, Zeisler J, Graham R, Koide HB, O'Kusky J, Sharp AH, Ross CA, Jirik F, Hayden MR: Absence of disease phenotype and intergenerational stability of the CAG repeat in transgenic mice expressing the human Huntington disease transcript. Hum molec Genet 5:177–185 (1996a).

Goldberg YP, Nicholson DW, Rasper DM, Kalchman MA, Koide HB, Graham RK, Bromm M, Kazemi-Esfarjani P, Thornberry NA, Vaillancourt JP, Hayden MR: Cleavage of huntingtin by apopain a proapoptotic cysteine protease is modulated by the polyglutamine tract. Nature Genet 13:442–449 (1996b).

Gomez-Tortosa E, MacDonald ME, Friend JC, Taylor SA, Weiler LJ, Cupples LA, Srinidhi J, Gusella JF, Bird ED, Vonsattel J-P, Myers RH: Quantitative neuropathological changes in presymptomatic Huntington's disease. Ann Neurol 49:29–34 (2001).

Greene JG, Greenamyre JT: Bioenergetics and glutamate excitotoxicity. Prog Neurobiol 48:613–634 (1996).

Gu M, Gash MT, Mann VM, Javoy-Agid F, Cooper JM, Schapira AH: Mitochondrial defect in Huntington's disease caudate nucleus. Ann Neurol 39:385–389 (1996).

Guidetti P, Charles V, Chen EY, Reddy PH, Kordower JH, Whetsell WO Jr, Schwarcz R, Tagle DA: Early degenerative changes in transgenic mice expressing mutant huntingtin involve dendritic abnormalities but no impairment of mitochondrial energy production. Exp Neurol 169:340–350 (2001).

Gutekunst C-A, Li S-H, Yi H, Mulroy JS, Kuemmerle S, Jones R, Rye D, Ferrante RJ, Hersch SM, Li X-J: Nuclear and neuropil aggregates in Huntington's disease: Relationship to neuropathology. J Neurosci 19:2522–2534 (1999).

Hansson O, Petersén Å, Leist M, Nicotera P, Castilho RF, Brundin P: Transgenic mice expressing a Huntington's disease mutation are resistant to quinolinic acid-induced striatal excitotoxicity. Proc natl Acad Sci, USA 96:8727–8732 (1999).

Hansson O, Castilho RF, Korhonen L, Lindholm D, Bates GP, Brundin P: Partial resistance to malonate-induced striatal cell death in transgenic mouse models of Huntington's disease is dependent on age and CAG repeat length. J Neurochem 78:694–703 (2001a).

Hansson O, Guatteo E, Mercuri NB, Bernardi G, Li X-J, Castilho RF, Brundin P: Resistance to NMDA toxicity correlates with appearance of nuclear inclusions behavioral deficits and changes in calcium homeostasis in mice transgenic for exon 1 of the huntington gene. Eur J Neurosci 14:1492–1504 (2001b),

Harper PS: Huntington's Disease (Saunders, London 1996).

Hedreen JC, Folstein SE: Early loss of neostriatal striosome neurons in Huntington's disease. J Neuropathol Exp Neurol 54:105–120 (1995).

Hickey MA, Morton AJ: Mice transgenic for the Huntington's disease mutation are resistant to chronic 3-nitropropionic acid-induced striatal toxicity. J Neurochem 75:2163–2171 (2000).

Hockly E, Cordery PM, Woodman B, Mahal A, van Dellen A, Blakemore C, Lewis CM, Hannan AJ, Bates GP: Environmental enrichment slows disease progression in R6/2 Huntington's disease mice. Ann Neurol 51:235–242 (2002).

Hodgson JG, Agopyan N, Gutekunst C-A, Leavitt BR, LePiane F, Singaraja R, Smith DJ, Bissada N, McCutcheon K, Nasir J, Jamot L, Li X-J, Stevens ME, Rosemond E, Roder JC, Phillips AG, Rubin EM, Hersch SM, Hayden MR: A YAC mouse model for Huntington's disease with full-length mutant huntingtin cytoplasmic toxicity and selective striatal neurodegeneration. Neuron 23:181–192 (1999).

Hughes RE, Lo RS, Davis C, Strand AD, Neal CL, Olson JM, Fields S: Altered transcription in yeast expressing expanded polyglutamine. Proc natl Acad Sci, USA 98:13201–13206 (2001).

The Huntington's Disease Collaborative Research Group: A novel gene containing a trinucleotide repeat that is expanded and unstable on Huntington's disease chromosomes. Cell 72:971–983 (1993).

Hurlbert MS, Zhou W, Wasmeier C, Kaddis FG, Hutton JC, Freed CR: Mice transgenic for an expanded CAG repeat in the Huntington's disease gene develop diabetes. Diabetes 48:649–651 (1999).

Iannicola C, Moreno S, Oliverio S, Nardacci R, Ciofi-Luzzatto A, Piacentini M: Early alterations in gene expression and cell morphology in a mouse model of Huntington's disease. J Neurochem 75:830–839 (2000).

Isacson O, Riche D, Hantraye P, Sofroniew MV, Maziere M: A primate model of Huntington's disease: cross-species implantation of striatal precursor cells to the excitotoxically lesioned baboon caudate-putamen. Exp Brain Res 75:213–220 (1989).

Jones C, Mullenbach R, Grossfeld P, Auer R, Favier R, Chien K, James M, Tunnacliffe A, Cotter F: Co-localisation of CCG repeats and chromosome deletion breakpoints in Jacobsen syndrome: evidence for a common mechanism of chromosome breakage. Hum molec Genet 9:1201–1208 (2000).

Karpuj MV, Becher MW, Springer JE, Chabas D, Youssef S, Pedotti R, Mitchell D, Steinman L: Prolonged survival and decreased abnormal movements in transgenic model of Huntington disease with administration of the transglutaminase inhibitor cystamine. Nature Med 8:143–149 (2002).

Keene CD, Rodrigues CMP, Eich T, Chhabra MS, Steer CJ, Low WC: Tauroursodeoxycholic acid a bile acid is neuroprotective in a transgenic animal model of Huntington's disease. Proc natl Acad Sci, USA 99:10671–10676 (2002).

Kirkwood SC, Su JL, Conneally P, Foroud T: Progression of symptoms in the early and middle stages of Huntington disease. Arch Neurol 58:273–278 (2001).

Kita H, Carmichael J, Swartz J, Muro S, Wyttenbach A, Matsubara K, Rubinsztein DC, Kato K: Modulation of polyglutamine-induced cell death by genes identified by expression profiling. Hum molec Genet 11:2279–2287 (2002).

Klapstein GJ, Fisher RS, Zanjani H, Cepeda C, Jokel ES, Chesselet MF, Levine MS: Electrophysiological and morphological changes in striatal spiny neurons in R6/2 Huntington's disease transgenic mice. J Neurophysiol 86:2667–2677 (2001).

Kremer B, Clark CM, Almqvist EW, Raymond LA, Graf P, Jacova C, Mezei M, Hardy MA, Snow B Martin W, Hayden MR: Influence of lamotrigine on progression of early Huntington disease: a randomized clinical trial. Neurol 53:1000–1011 (1999).

Kumar A, Reddy EP: Genomic organization and characterization of the promoter region of murine GSTM2 gene. Gene 270:221–229 (2001).

La Spada AR, Wilson EM, Lubahn DB, Harding AE, Fischbeck KH: Androgen receptor gene mutations in X-linked spinal and bulbar muscular atrophy. Nature 352:77–79 (1991).

La Spada AR, Roling DB, Harding AE, Warner CL, Spiegel R, Hausmanowa-Petrusewicz I, Yee WC, Fischbeck KH: Meiotic stability and genotype-phenotype correlation of the trinucleotide repeat in X-linked spinal and bulbar muscular atrophy. Nature Genet 2:301–304 (1992).

Laforet GA, Sapp E, Chase K, McIntyre C, Boyce FM, Campbell M, Cadigan BA, Warzecki L, Tagle DA, Reddy PH, Cepeda C, Calvert CR, Jokel ES, Klapstein GJ, Ariano MA, Levine MS, DiFiglia M, Aronin N: Changes in cortical and striatal neurons predict behavioral and electrophysiological abnormalities in a transgenic murine model of Huntington's disease. J Neurosci 21:9112–9123 (2001).

Levine MS, Klapstein GJ, Koppel A, Gruen E, Cepeda C, Vargas ME, Jokel ES, Carpenter EM, Zanjani H, Hurst RS, Efstratiadis A, Zeitlin S, Chesselet M-F: Enhanced sensitivity to N-methyl-D-aspartate receptor activation in transgenic and knockin mouse models of Huntington's disease. J Neurosci Res 58:515–532 (1999).

Li H, Li S-H, Cheng AL, Mangiarini L, Bates GP, Li X-J: Ultrastructural localization and progressive formation of neuropil aggregates in Huntington's disease transgenic mice. Hum molec Genet 8:1227–1236 (1999).

Li H, Li S-H, Yu Z-X, Shelbourne P, Li X-J: Huntingtin aggregate-associated axonal degeneration is an early pathological event in Huntington's disease mice. J Neurosci 21:8473–8481 (2001).

Li SH, Cheng AL, Zhou H, Lam S, Rao M, Li H, Li X J: Interaction of Huntington disease protein with transcriptional activator Sp1. Mol Cell Biol 22:1277–1287 (2002).

Lin B, Nasir J, MacDonald H, Hutchinson G, Graham RK, Rommens JM, Hayden MR: Sequence of the murine Huntington disease gene: evidence for conservation alternate splicing and polymorphism in a triplet (CCG) repeat. Hum molec Genet 3:85–92 (1994) Erratum 3:530 (1994).

Lin CH, Tallaksen-Greene S, Chien W-M, Cearley JA, Jackson WS, Crouse AB, Ren S, Li X-J, Albin RL, Detloff PJ: Neurological abnormalities in a knock-in mouse model of Huntington's disease. Hum molec Genet 10:137–144 (2001).

Lione LA, Carter RJ, Hunt MJ, Bates GP, Morton AJ, Dunnett SB: Selective discrimination learning impairments in mice expressing the human Huntington's disease mutation. J Neurosci 19:10428–10437 (1999).

Luesse HG, Schiefer J, Spruenken A, Puls C, Block F, Kosinski CM: Evaluation of R6/2 HD transgenic mice for therapeutic studies in Huntington's disease: behavioral testing and impact of diabetes mellitus. Behav Brain Res 126:185–195 (2001).

Luthi-Carter R, Strand A, Peters NL, Solano SM, Hollingsworth ZR, Menon AS, Frey AS, Spektor BS, Penney EB, Schilling G, Ross CA, Borchelt DR, Tapscott SJ, Young AB, Cha J-HJ, Olson JM: Decreased expression of striatal signaling genes in a mouse model of Huntington's disease. Hum molec Genet 9:1259–1271 (2000).

Luthi-Carter R, Hanson SA, Strand AD, Bergstrom DA, Chun W, Peters NL, Woods AM, Chan EY, Kooperberg C, Krainc D, Young AB, Tapscott SJ, Olson JM: Dysregulation of gene expression in the R6/2 model of polyglutamine disease: parallel changes in muscle and brain. Hum molec Genet 11:1911–1926 (2002a)

Luthi-Carter R, Strand AD, Hanson SA, Kooperberg C, Schilling G, La Spada AR, Merry DE, Young AB, Ross CA, Borchelt DR, Olson JM: Polyglutamine and transcription: gene expression changes shared by DRPLA and Huntington's disease mouse models reveal context-independent effects. Hum molec Genet 11:1927–1937 (2002b).

Mangiarini L, Sathasivam K, Seller M, Cozens B, Harper A, Hetherington C, Lawton M, Trottier Y, Lehrach H, Davies SW, Bates GP: Exon 1 of the HD gene with an expanded CAG repeat is sufficient to cause a progressive neurological phenotype in transgenic mice. Cell 87:493–506 (1996).

Mantamadiotis T, Lemberger T, Bleckmann SC, Kern H, Kretz O, Martin Villalba A, Tronche F, Kellendonk C, Gau D, Kapfhammer J, Otto C, Schmid W, Schutz G: Disruption of CREB function in brain leads to neurodegeneration. Nature Genet 31:47–54 (2002).

Martín-Aparicio E, Yamamoto A, Hernandez F, Hen R, Avila J, Lucas JJ: Proteasomal-dependent aggregate reversal and absence of cell death in a conditional mouse model of Huntington's disease, J Neurosci 21:8772–8781 (2001),

Mastroberardino PG, Iannicola C, Nardacci R, Bernassola F, De Laurenzi V, Melino G, Moreno S, Pavone F, Oliverio S, Fesus L, Piacentini M: 'Tissue' transglutaminase ablation reduces neuronal death and prolongs survival in a mouse model of Huntington's disease. Cell Death Differ 9:873–880 (2002).

Menalled LB, Chesselet MF: Mouse models of Huntington's disease. Trends Pharmacol Sci 23:32–39 (2002).

Menalled L, Zanjani H, MacKenzie L, Koppel A, Carpenter E, Zeitlin S, Chesselet MF: Decrease in striatal enkephalin mRNA in mouse models of Huntington's disease. Exp Neurol 162:328–342 (2000).

Menalled LB, Sison JD, Wu Y, Olivieri M, Li X-J, Li H, Zeitlin S, Chesselet MF: Early motor dysfunction and striosomal distribution of huntingtin microaggregates in Huntington's disease knock-in mice. J Neurosci 22:8266–8276 (2002).

Menalled LB, Sison JD, Dragatsis I, Zeitlin S, Chesselet MF: Time course of early motor and neuropathological anomalies in a knock-in mouse model of Huntington's disease with 140 CAG repeats. J Comp Neurol in press (2003).

Mende-Mueller LM, Toneff T, Hwang S-R, Chesselet M-F, Hook VY: Tissue-specific proteolysis of Huntingtin (htt) in human brain: evidence of enhanced levels of N- and C-terminal htt fragments in Huntington's disease striatum. J Neurosci 21:1830–1837 (2001).

Morley JF, Brignull HR, Weyers JJ, Morimoto RI: The threshold for polyglutamine-expansion protein aggregation and cellular toxicity is dynamic and influenced by aging in *Caenorhabditis elegans*. Proc natl Acad Sci, USA 99:10417–10422 (2002).

Morton AJ, Lagan MA, Skepper JN, Dunnett SB: Progressive formation of inclusions in the striatum and hippocampus of mice transgenic for the human Huntington's disease mutation. J Neurocytol 29:679–702 (2000).

Murman DL, Giordani B, Mellow AM, Johanns JR, Little RJ, Hariharan M, Foster NL: Cognitive behavioral and motor effects of the NMDA antagonist ketamine in Huntington's disease. Neurol 49:153–161 (1997).

Nasir J, Floresco SB, O'Kusky JR, Diewert VM, Richman JM, Zeisler J, Borowski A, Marth JD, Phillips AG, Hayden MR: Targeted disruption of the Huntington's disease gene results in embryonic lethality and behavioral and morphological changes in heterozygotes. Cell 81:811–823 (1995).

NicNiocaill B, Haraldsson B, Hansson O, O'Connor WT, Brundin P: Altered striatal amino acid neurotransmitter release monitored using microdialysis in R6/1 Huntington transgenic mice. Eur J Neurosci 13:206–210 (2001).

Okamoto Y, Sakata M, Yamamoto T, Nishio Y, Adachi K, Ogura K, Yamaguchi M, Takeda T, Tasaka K, Murata Y: Involvement of nuclear transcription factor Sp1 in regulating glucose transporter-1 gene expression during rat trophoblast differentiation. Biochem biophys Res Commun 288:940–948 (2001).

Olney JW, de Gubareff T: Glutamate neurotoxicity and Huntington's chorea. Nature 271:557–559 (1978).

Ona VO, Li M, Vonsattel J-P, Andrews LJ, Khan SQ, Chung WM, Frey AS, Menon AS, Li X-J, Stieg PE, Yuan J, Penney JB, Young AB, Cha J-HJ, Friedlander RM: Inhibition of caspase-1 slows disease progression in a mouse model of Huntington's disease. Nature 399:263–267 (1999).

Ordway JM, Tallaksen-Greene S, Gutekunst C-A, Bernstein EM, Cearley JA, Wiener HW, Dure LS IV, Lindsey R, Hersch SM, Jope RS, Albin RL, Detloff PJ: Ectopically expressed CAG repeats cause intranuclear inclusions and a progressive late onset neurological phenotype in the mouse. Cell 91:753–763 (1997).

Panov AV, Gutekunst C-A, Leavitt BR, Hayden MR, Burke JR, Strittmatter WJ, Greenamyre JT: Early mitochondrial calcium defects in Huntington's disease are a direct effect of polyglutamines. Nature Neurosci 5:731–736 (2002).

Petersén Å, Chase K, Puschban Z, DiFiglia M, Brundin P, Aronin N: Maintenance of susceptibility to neurodegeneration following intrastriatal injections of quinolinic acid in a new transgenic mouse model of Huntington's disease. Exp Neurol 175:297–300 (2002).

Pfeifer A, Verma IM: Gene therapy: promises and problems. Ann Rev Genomics Hum Genet 2:177–211 (2001).

Podolsky S, Leopold NA: Abnormal glucose tolerance and arginine tolerance tests in Huntington's disease. Gerontol 23:55–63 (1977).

Podolsky S, Leopold NA, Sax DS: Increased frequency of diabetes mellitus in patients with Huntington's chorea. Lancet 1:1356–1358 (1972).

Pugh BF, Tjian R: Mechanism of transcriptional activation by Sp1: evidence for coactivators. Cell 61:1187–1197 (1990).

Puri BK, Bydder GM, Counsell SJ, Corridan BJ, Richardson AJ, Hajnal JV, Appel C, McKee HM, Vaddadi KS, Horrobin DF: MRI and neuropsychological improvement in Huntington disease following ethyl-EPA treatment. Neuroreport 13:123–126 (2002).

Reddy PH, Williams M, Charles V, Garrett L, Pike-Buchanan Whetsell WO Jr, Miller G, Tagle DA: Behavioral abnormalities and selective neuronal loss in HD transgenic mice expressing mutated full-length HD cDNA. Nature Genet 20:198–202 (1998).

Richfield EK, Maguire-Zeiss KA, Vonkeman HE, Voorn P: Preferential loss of preproenkephalin versus preprotachykinin neurons from the striatum of Huntington's disease patients. Ann Neurol 38:852–861 (1995).

Roizin L, Stellar S, Liu JC: Neuronal nuclear-cytoplasmic changes in Huntington's chorea: Electron microscope investigations, in Chase TN, Wexler NS, Barbeau A (eds): Advances in Neurology, pp 95–122 (Raven Press New York 1979).

Rosas HD, Koroshetz WJ, Jenkins BG, Chen YI, Hayden DL, Beal MF, Cudkowicz ME: Riluzole therapy in Huntington's disease (HD). Mov Disord 14:326–330 (1999).

Rosser AE, Dunnett SB: New drugs for Huntington's disease, NeuroReport 13:A21–A22 (2002),

Sawa A, Wiegand GW, Cooper J, Margolis RL, Sharp AH, Lawler JF Jr, Greenamyre JT, Snyder SH, Ross CA: Increased apoptosis of Huntington disease lymphoblasts associated with repeat length-dependent mitochondrial depolarization. Nature Med 5:1194–1198 (1999).

Schilling G, Becher MW, Sharp AH, Jinnah HA, Duan K, Kotzuk JA, Slunt HH, Ratovitski T, Cooper JK, Jenkins NA, Copeland NG, Price DL, Ross CA, Borchelt DR: Intranuclear inclusions and neuritic aggregates in transgenic mice expressing a mutant N-terminal fragment of huntingtin. Hum molec Genet 8:397–407 (1999).

Schilling G, Jinnah HA, Gonzales V, Coonfield ML, Kim Y, Wood JD, Price DL, Li X-J, Jenkins N, Copeland N, Moran T, Ross CA, Borchelt DR: Distinct behavioral and neuropathological abnormalities in transgenic mouse models of HD and DRPLA. Neurobiol Dis 8:405–418 (2001).

Senut M-C, Suhr ST, Kaspar B, Gage FH: Intraneuronal aggregate formation and cell death after viral expression of expanded polyglutamine tracts in the adult rat brain. J Neurosci 20:219–229 (2000).

Shelbourne PF, Killeen N, Hevner RF, Johnston HM, Tecott L, Lewandoski M, Ennis M, Ramirez L, Li Z, Iannicola C, Littman DR, Myers RM: A Huntington's disease CAG expansion at the murine *Hdh* locus is unstable and associated with behavioral abnormalities in mice. Hum molec Genet 8:763–774 (1999).

Steffan JS, Kazantsev A, Spasic-Boskovic O, Greenwald M, Zhu Y-Z, Gohler H, Wanker EE, Bates GP, Housman DE, Thompson LM: The Huntington's disease protein interacts with p53 and CREB-binding protein and represses transcription. Proc natl Acad Sci, USA 97:6763–6768 (2000).

Steffan JS, Bodai L, Pallos J, Poelman M, McCampbell A, Apostol BL, Kazantsev A, Schmidt E, Zhu YZ, Greenwald M, Kurokawa R, Housman DE, Jackson GR, Marsh JL, Thompson LM: Histone deacetylase inhibitors arrest polyglutamine-dependent neurodegeneration in Drosophila. Nature 413: 739–743 (2001).

Sun Y, Savanenin A, Reddy PH, Liu YF: Polyglutamine-expanded huntingtin promotes sensitization of N-methyl-D-aspartate receptors via post-synaptic density 95. J biol Chem 276:24713–24718 (2001).

Tabrizi SJ, Cleeter MW, Xuereb J, Taanman J-W, Cooper JM, Schapira AH: Biochemical abnormalities and excitotoxicity in Huntington's disease brain. Ann Neurol 45:25–32 (1999).

Telenius H, Kremer HPH, Theilmann J, Andrew SE, Almqvist E, Anvret M, Greenberg C, Greenberg J, Lucotte G, Squiteri G, Starr E, Goldberg YP, Hayen MR: Molecular analysis of juvenile Huntington's disease: The major influence on (CAG) repeat length is the sex of the affected parent. Hum molec Genet 2:1535–1540 (1993).

Tillerson JL, Cohen AD, Philhower J, Miller GW, Zigmond MJ, Schallert T: Forced limb-use effects on the behavioral and neurochemical effects of 6-hydroxydopamine. J Neurosci 21:4427–4435 (2001).

Tillerson JL, Cohen AD, Caudle WM, Zigmond MJ, Schallert T, Miller GW: Forced nonuse in unilateral parkinsonian rats exacerbates injury. J Neurosci 22:6790–6799 (2002).

Toneff T, Mende-Mueller L, Wu Y, Hwang SR, Bundey R, Thompson LM, Chesselet MF, Hook V: Comparison of huntingtin proteolytic fragments in human lymphoblast cell lines and human brain. J Neurochem 82:84–92 (2002).

Turmaine M, Raza A, Mahal A, Mangiarini L, Bates GP, Davies SW: Nonapoptotic neurodegeneration in a transgenic mouse model of Huntington's disease. Proc natl Acad Sci, USA 97:8093–8097 (2000).

Vaddadi KS, Soosai E, Chiu E, Dingjan P: A randomised placebo-controlled double blind study of treatment of Huntington's disease with unsaturated fatty acids. Neuroreport 13:29–33 (2002).

van Dellen A, Blakemore C, Deacon R, York D, Hannon AJ: Delaying the onset of Huntington's disease in mice. Nature 404:721–722 (2000).

van Vugt JP, Siesling S, Piet KK, Zwinderman AH, Middelkoop HA, van Hilten JJ, Roos RA: Quantitative assessment of daytime motor activity provides a responsive measure of functional decline in patients with Huntington's disease. Mov Disord 16:481–488 (2001).

Verhagen Metman L, Morris MJ, Farmer C, Gillespie M, Mosby K, Wuu J, Chase TN: Huntington's disease: A randomised controlled trial using the NMDA-antagonist amantadine. Neurology 59: 694–699 (2002).

Vonsattel J-P, Myers RH, Stevens TJ, Ferrante RJ, Bird ED, Richardson EP Jr: Neuropathological classification of Huntington's disease. J Neuropathol Exp Neurol 44:559–577 (1985).

Vonsattel J-P, DiFiglia M: Huntington disease. J Neuropathol Exp Neurol 57:369–384 (1998).

Wellington CL, Singaraja R, Ellerby L, Savill J, Roy S, Leavitt B, Cattaneo E, Hackam A, Sharp A, Thornberry N, Nicholson DW, Bredesen DE, Hayden MR: Inhibiting caspase cleavage of huntingtin reduces toxicity and aggregate formation in neuronal and nonneuronal cells. J biol Chem 275:19831–19838 (2000).

Wellington CL, Ellerby LM, Gutekunst C-A, Rogers D, Warby S, Graham RK, Loubser O, van Raamsdonk J, Singaraja R, Yang YZ, Gafni J, Bredesen D, Hersch SM, Leavitt BR, Roy S, Nicholson DW, Hayden MR: Caspase cleavage of mutant huntingtin precedes neurodegeneration in Huntington's disease J Neurosci 22:7862–7872 (2002).

Wexler NS, Young AB, Tanzi RE Travers H, Starosta-Rubenstein S, Penny JB, Snodgrass SR, Shoulson I, Gomez F, Ramos Arroyo MA, Penchaszadeh GK, Moreno H, Gibbons K, Faryniarz A, Hobbs W, Anderson MA, Bonilla E, Conneally PM, Gusella JF: Homozygotes for Huntington's disease. Nature 326:194–197 (1987).

Wheeler VC, White JK, Gutekunst C-A, Vrbanac V, Weaver M, Li X-J, Li S-H, Yi H, Vonsattel J-P, Gusella JF, Hersch SM, Auerbach W, Joyner AL, MacDonald ME: Long glutamine tracts cause nuclear localization of a novel form of huntingtin in medium spiny striatal neurons in HdhQ92 and HdhQ111 knock-in mice. Hum molec Genet 9:503–513 (2000).

Wheeler VC, Gutekunst CA, Vrbanac V, Lebel LA, Schilling G, Hersch S, Friedlander RM, Gusella JF, Vonsattel J-P, Borchelt DR, MacDonald ME: Early phenotypes that presage late-onset neurodegenerative disease allow testing of modifiers in *Hdh* CAG knock-in mice. Hum molec Genet 11:633–640 (2002).

White JK, Auerbach W, Duyao MP, Vonsattel J-P, Gusella JF, Joyner AL, MacDonald ME: Huntingtin is required for neurogenesis and is not impaired by the Huntington's disease CAG expansion. Nature Genet 17:404–410 (1997).

Yamamoto A, Lucas JJ, Hen R: Reversal of neuropathology and motor dysfunction in a conditional model of Huntington's disease. Cell 101:57–66 (2000).

Zeitlin S, Liu J-P, Chapman DL, Papaioannou VE, Efstratiadis A: Increased apoptosis and early embryonic lethality in mice nullizygous for the Huntington's disease gene homologue. Nature Genet 11:155–163 (1995).

Zeron MM, Hansson O, Chen N, Wellington CL, Leavitt BR, Brundin P, Hayden MR, Raymond LA: Increased sensitivity to N-methyl-D-aspartate receptor-mediated excitotoxicity in a mouse model of Huntington's disease. Neuron 33:849–860 (2002).

Zühlke C, Riess O, Bockel B, Lange H, Thies U: Mitotic stability and meiotic variability of the (CAG)n repeat in the Huntington disease gene. Hum molec Genet 2:2063–2067 (1993).

Cytogenet Genome Res 100:287–295 (2003)
DOI: 10.1159/000072864

Mechanisms of neuronal cell death in Huntington's disease

A. Sawa,[a,b] T. Tomoda,[c] and B.-I. Bae[b]

[a] Department of Psychiatry and Behavioral Sciences, Johns Hopkins University School of Medicine, Baltimore, MD;
[b] Department of Neuroscience, Johns Hopkins University School of Medicine, Baltimore, MD; and
[c] Laboratory of Developmental Neurobiology, The Rockefeller University, New York, NY (USA)

Abstract. Huntington's disease (HD) is a genetically dominant neurodegenerative condition caused by an unique mutation in the disease gene *huntingtin*. Although the Huntington protein (Htt) is ubiquitously expressed, expansion of the polyglutamine tract in Htt leads to the progressive loss of specific neuronal subpopulations in HD brains. In this article, we will summarize the current understanding on mechanisms of how mutant Htt can elicit cytotoxicity, as well as how the selective sets of neuronal cell death occur in HD brains.

Huntington's disease (HD) is a genetically dominant neurodegenerative condition characterized by abnormal involuntary movements, including chorea and dystonia, and cognitive impairment (Huntington, 1872). HD is one of a number of diseases caused by an expansion of CAG repeats in the open reading frame of a disease gene, which in turn translates into a protein with an expanded polyglutamine tract (Ross, 2002). The disease gene for HD, *huntingtin*, was identified in 1993, and it encodes a large protein (350 kDa) with a polyglutamine stretch named Huntintin (Htt) (The Huntington's Disease Collaborative Research Group, 1993). Once the polyglutamine repeat becomes expanded over approximately 40, HD occurs. Htt is widely expressed in the whole body, but robust pathology occurs only in restricted areas, in particular, the caudate (Gutekunst et al., 1995; Sharp et al., 1995). In HD brains, medium spiny neurons in the caudate are selectively impaired. The extent of the polyglutamine expansion is correlated with the severity of symptoms, such as age of onset. The causal relationship between neuronal cell death and the onset of neurological dysfunction in HD is unclear; however, the progressive loss of specific neuronal subpopulations in HD coincides with disease progression.

Ten years have already passed since *huntingtin* and its unique mutation were identified. However, we still do not know the mechanisms of how the selective sets of neuronal cell death occur in HD brains.

In the first section, we will overview general understanding of neuronal cell death. In the second section, we will summarize current ideas on how mutant Htt can induce cytotoxicity. Finally, we will discuss possible mechanisms that may account for the selective neuronal cell death of HD brains.

Neuronal cell death

The classic definitions of necrosis and apoptosis were first proposed by intensive morphological examination using cellular systems outside the brain (Kerr et al., 1972; Kerr and Harmon, 1991) (Fig. 1). Apoptosis displays distinctly different morphological features from necrosis, including nuclear condensation and fragmentation, as well as convulsion of membrane. Apoptosis is also characterized by biochemical markers, such as internucleosomal DNA cleavage or DNA ladder formation. Apoptosis is a type of programmed cell death, in which cells participate actively toward their own demise. In contrast, necrotic cells dis-

Supported by grants from the Huntington's Disease Society of America, the Stanley Medical Research Institute, NARSAD, and the S-R Foundation (Washington, DC) (to A.S.).

Received 19 December 2002; revision accepted 21 January 2003.

Request reprints from Dr. Akira Sawa
Departments of Psychiatry and Neuroscience
Johns Hopkins University School of Medicine
600 N. Wolfe Street, Meyer 2-181
Baltimore, MD 21287 (USA)
telephone: 410-955-4726; fax: 410-955-8233
e-mail: asawa1@jhmi.edu

KARGER
Fax + 41 61 306 12 34
E-mail karger@karger.ch
www.karger.com

0301–0171/03/1004–0287$19.50/0

Accessible online at:
www.karger.com/cgr

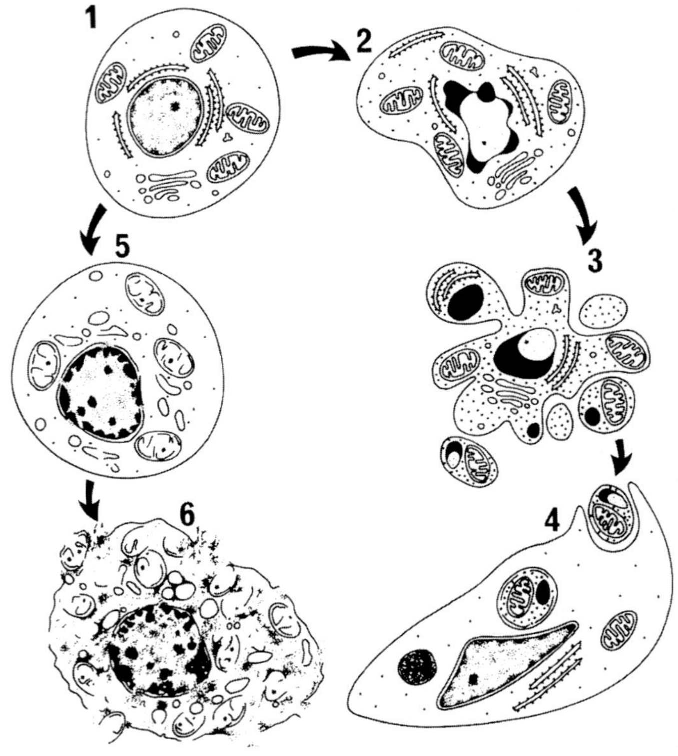

Fig. 1. Schematic description of necrosis and apoptosis. Modified, with permission, from Kerr and Harmon (1991).

play organelle swelling and subsequent cellular disintegration. Necrosis is frequently associated with cellular energy deficit and is regarded as passive cell damage and death.

Robust neuronal cell death is observed in both physiological brain development and in pathological conditions of adult brains, such as neurodegeneration (Server and Mobley, 1991). Although the simple classification of cell death into apoptosis and necrosis is widely accepted in most biological contexts, it has become clearer that neuronal cell death is not easily classified into these two categories, but has more variety of cell death modes (Fig. 2).

Clarke (1990) extensively reviewed the morphological characteristics of neurons dying during development. He classified dying neurons based on morphological aspects into three types. Type 1 cell death is observed among developing thoracic spinal neurons, cervical visceromotor neurons, and lateral motor neurons of the lumbar spinal cord of the chick embryo. In mammals, this type of neuronal cell death was reported in the postnatal rat superior colliculus and retina. Type 1 cell death is characteristic of nuclear condensation and fragmentation, as well as convulsion of membrane. Cell shrinkage is seen during this type of cell death. These features are compatible with the classical definition of apoptosis. Type 2 cell death is characterized mainly by the existence of numerous autophagic vacuoles. Autophagy is a process whereby cells remove cytosolic proteins and organelles and degrade themselves from within. Type 3 cell death is seen in several developmental contexts, including embryonic chick ciliary ganglion cells and lumber motor neurons. This type of cell death, also called cytoplasmic cell death, is characterized by initial dilation of the endoplasmic reticulum (ER), as well as the nuclear envelope and Golgi apparatus. In some cases, mitochondrial dilation also occurs.

In the past decade, detailed molecular cascades for apoptosis, or type 1 cell death by Clarke, have been intensively clarified. Most of apoptosis includes a chain of activation of cysteinyl aspartic-specific proteases, called caspases (Thornberry and Lazebnik, 1998). Two major pathways for apoptosis via caspases exist: the intrinsic pathway, originating from mitochondrial release of cytochrome C and associated activation of caspase-9, and the extrinsic pathway, originating from the activation of cell-surface death receptors, such as Fas, and resulting in the activation of caspase-8 (Ashkenazi and Dixit, 1998; Green and Reed, 1998). Damages to genome DNA activate p53, which activates the intrinsic pathway via mitochondria (Vogelstein et al., 2000). Caspase-independent apoptosis also exists. Apoptosis-inducing factor (AIF), a mitochondrial intermembrane flavoprotein, is released from the mitochondria to the nucleus in response to specific death signals (Susin et al., 1999). AIF causes high-molecular-weight DNA fragmentation and chromatin condensation in cells and isolated nuclei in a caspase-independent manner. Neuronal cell death mediated by p53 reportedly occurs in both a caspase-dependent and an AIF-dependent manner, with different kinetics (Cregan et al., 2002). PML, a main component of a nuclear matrix-associated structure, is another factor that mediates caspase-independent apoptosis (Quignon et al., 1998). Roles of PML in neuronal apoptosis remain elusive; however, PML is necessary for toxic actions of p53 (Zhong et al., 2000).

Molecular mechanisms for autophagic cell death (type 2 cell death) still remain elusive. Recently, *beclin 1*, the mammalian ortholog of a yeast autophagy gene *(Apg6/Vps30)*, was identified (Liang et al., 1999). Interestingly, Beclin 1 protein binds to Bcl-2, a key pro-apoptotic gene product, and inhibits tumorigenesis. Moreover, in Purkinje cell death of *lurcher* mutant mice, the causative mutation in GluRδ2 can activate protein signaling to Beclin 1 (Yue et al., 2002). These findings suggest that autophagic cell death, like apoptosis, is programmed via intracellular molecular signaling.

Cytoplasmic cell death (type 3 cell death) has not yet been characterized at the molecular level. Bredesen and colleagues (Sperandio et al., 2000) recently reported a unique programmed cell death (paraptosis) distinct from classic apoptosis. In 293T cells and fibroblasts, expression of insulin-like growth factor I receptor intracellular domain (IGFIR-IC) can induce cell death characterized by cytoplasmic vacuolation and resistance to well-known apoptosis inhibitors. In this setting, caspase-9 is activated independently of Apaf-1. Valosin containing protein (VCP)/p97 is a member of the AAA+ family of ATPase proteins. Expression of VCP/p97 with a mutation in the second ATP binding domain causes cytoplasmic vacuolation followed by cell death (Hirabayashi et al., 2001). These two observations may not be directly linked to the molecular mechanisms for cytoplasmic cell death (type 3 cell death); their existence, however, suggests interesting connotations.

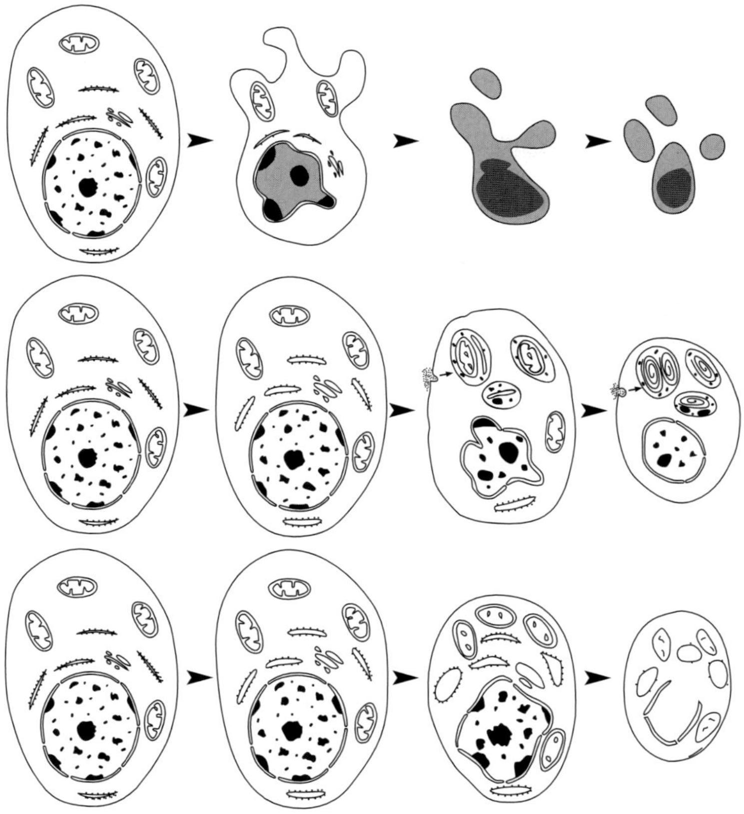

Fig. 2. Schematic description of Clarke's classification for neuronal cell death. Modified, with permission, from Server and Mobley (1991) and Clarke (1990).

Question 1: How can mutant Htt induce cytotoxicity? (Fig. 3)

HD occurs in an autosomal dominant fashion. The dominant model can include both "gain of function" and "loss of function" (haploinsufficiency). After the identification of Htt, its toxic mechanisms were initially addressed as "gain of function" for the following reasons:

First, the overexpression of mutant Htt, especially the *N*-terminal short fragment including the polyglutamine stretch, causes cytotoxicity as well as inclusion formation that is characteristic of HD brains (Cooper et al., 1998; Hackam et al., 1998; Wellington et al., 2000). Analogously, transgenic mice with overexpression of the *N*-terminal fragment with expanded polyglutamine display neurological deficits similar to HD, as well as early death and the inclusion formation (Mangiarini et al., 1996; Schilling et al., 1999).

Second, several disease genes for other types of neurodegenerative conditions, such as spinocerebellar ataxia (SCA) and dentatorubral and pallidoluysian atrophy (DRPLA), also possess a similar type of mutation, expansion of polyglutamine in the disease gene products. The genes that cause these disorders have no homology with each other except for the polyglutamine stretch itself. The other portions of the disease proteins may serve as regulatory units. Therefore, people have reasonably grouped these diseases as polyglutamine diseases (Ross, 2002). Transgenic mice expressing expanded polyglutamine in an unrelated gene also display neurological deficit (Ordway et al., 1997). This reinforces the notion of the polyglutamine disease.

In addition to the "gain of function" theory, studies of functions of normal (wild-type) Htt have disclosed the importance of Htt for cell survival and implicated its "loss of function" in HD pathology (Cattaneo et al., 2001). The normal form of full-length Htt has several important roles, such as endocytosis and up-regulation of brain-derived neurotrophic factor (BDNF), a pro-survival factor (Zuccato et al., 2001). In cell models, normal Htt protects from apoptosis (Rigamonti et al., 2000). This observation was confirmed by animal models in which inactivation of the mouse *huntingtin* gene in the brain and testis results in progressive neurodegeneration and sterility (Dragatsis et al., 2000). In addition to simple loss of normal beneficial functions of Htt by expanded polyglutamine, mutant Htt may recruit wild-type Htt into insoluble aggregates, which results in an accelerated depletion of wild-type Htt.

Cell death in HD cellular models

The condition of "gain of function" can be mimicked in cellular models by overexpressing mutant Htt. Overexpression of Htt, especially the *N*-terminal fragment of Htt, with expanded polyglutamine kills many types of cell lines and increases cellular susceptibility to death (Cooper et al., 1998; Hackam et al., 1998; Wellington et al., 2000). Under these conditions, robust

Fig. 3. Schematic description of possible pathological mechanisms induced by mutant Htt. Mutant Htt seems to display cytotoxicity via multiple pathways. Mutant Htt can induce formation of autophagic vacuoles and stress on the endoplasmic reticulum. Mutant Htt influences mitochondrial functions by several mechanisms, inducing oxidative and metabolic stress. Increased susceptibility to mitochondrial membrane depolarization can also occur in this setting, followed by increased levels of cytosolic calcium and a caspase activation chain. Increased level of calcium can be associated with vulnerability to excitotoxicity. The activated caspase can cleave Htt. Mutant Htt, especially cleaved Htt, can translocate to the nucleus and disturb gene transcription via interaction with CREB binding protein (CBP), Sp1, and p53, while p53 regulates cell death by modulating many mitochondrial proteins and proteins associated with oxidative stress. In the nucleus, nuclear inclusion stained with Htt also occurs, but its pathophysiological significance is not clear. Dystrophic neurites occur in the presence of mutant Htt.

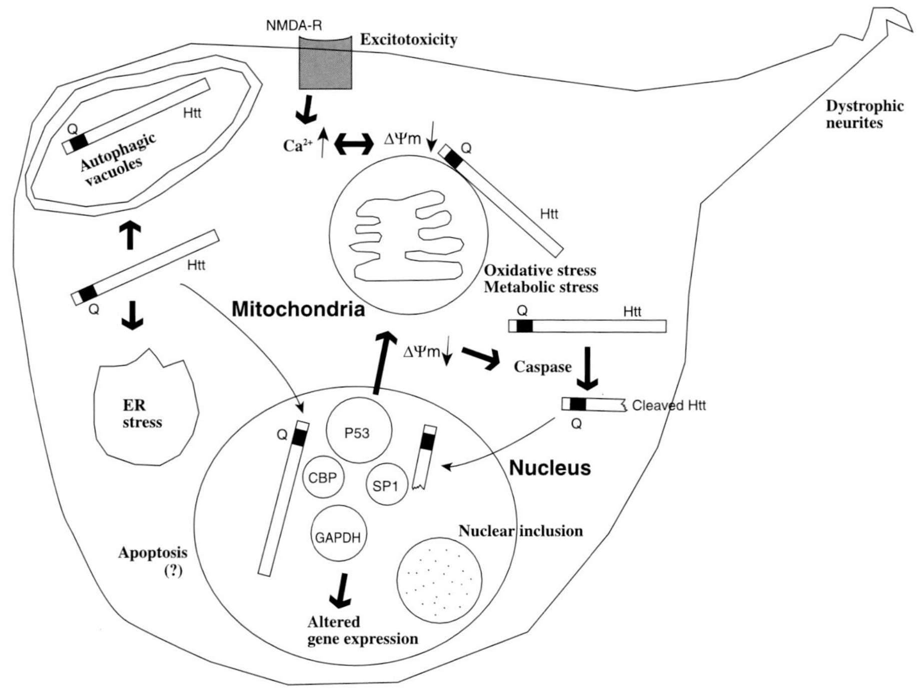

inclusion body formation inside both the nucleus and cytoplasm, followed by cell death, is observed. The pathological significance of inclusion body formation for cell death is unclear. The cells dying from mutant Htt frequently manifest several apoptotic features, including nuclear changes and caspase-3 activation.

In cultured striatal neurons, the *N*-terminal Htt fragments with expanded polyglutamine induce apoptosis with the formation of intracellular inclusions (Saudou et al., 1998). Apoptotic cell death was characterized by nuclear condensation in this study. Anti-apoptotic compounds and neurotrophic factors such as CNTF and BDNF can prevent the apoptosis induced by mutant Htt. The Htt-induced neural degeneration was not correlated with the formation of intranuclear inclusions in this study.

Mutant Htt can augment susceptibility to apoptotic stress. Lymphoblasts from HD patients, but not control subjects display susceptibility to mitochondrial membrane depolarization, an initial event for cell death, upon receiving low doses of cyanide or staurosporine (STS) (Sawa et al., 1999). After STS stress, caspase-9 and caspase-3, downstream components of the apoptotic cascade from mitochondria, are sequentially overactivated in HD lymphoblasts compared to normal lymphoblasts. This leads to accelerated apoptosis in HD cells. Apoptosis in this cell system is characterized by nuclear fragmentation and condensation, as well as DNA ladder formation.

In contrast to frequent reports of apoptotic death by mutant Htt in HD cell models, Johnson and colleagues (Moulder et al., 1999) reported an instance of expanded polyglutamine-induced cell death resistant to z-VAD-fmk, a potent general apoptosis inhibitor. As the polyglutamine-induced cell death shares, at least in part, the molecular mechanism with mutant-Htt-induced cell death, this report may shed some light on our overall understanding of Htt-induced cell death.

In some settings, vacuole formation is also observed in Htt-induced cell death. Most of the vacuoles in these systems have features of autophagy.

Exogenous overexpression of mutant Htt in clonal striatal neurons leads to Htt accumulation in cytoplasmic vacuoles with cell shrinkage (Kegel et al., 2000; Petersen et al., 2001). The Htt-containing vacuoles look similar to autophagosomes. Autophagic mechanisms are believed to play a role in the degradation of Htt, especially the *N*-terminal cleaved fragments with expanded polyglutamine, for the following reasons (Ravikumar et al., 2002): Htt tends to accumulate more when cells are treated with inhibitors to the autophagy-lysosome pathway. In contrast, rapamycin, which stimulates autophagy, enhances the clearance of aggregate prone mutant Htt. It is not yet clear what triggers the autophagic cell response and what signaling bridges from mutant Htt to autophagosome formation. Moreover, we observe the existence of large vacuoles in HD lymphoblasts, but not in control subjects (Nagata et al., manuscript submitted for publication). The vacuoles are stained with anti-Htt antibody, as well as with an antibody against cathepsin B, a marker for lysosome. The vacuoles in HD lymphoblasts are even more enlarged upon an apoptotic stress, such as STS addition. It remains elusive whether the autophagic vacuoles seen in HD cell models and brains are cytotoxic or cytoprotective.

As described above, a mutation in VCP/p97 causes cell death with associated cytoplasmic vacuoles (Hirabayashi et al., 2001). VCP/p97 binds and co-localizes to polyglutamine tracts. Interestingly, *Drosophila* VCP is reported as a modulator of polyglutamine-induced neurodegeneration (Higashiyama et al., 2002). The relationship between VCP/p97-induced vacuoles and autophagic vacuoles is still unclear.

Frequent observation of apoptosis and autophagy in HD cell models may imply certain molecular mechanisms by which mutant Htt can induce both apoptosis and an autophagic cell response. However, Htt-induced cell death in previous studies usually occurred by overexpression of the *N*-terminal fragments of Htt or upon several apoptotic stimuli, so caution when interpreting the data is necessary.

Neuronal cell death in HD animal models

Both drug-induced and genetically engineered animals are currently available for HD models.

Chronic, systemic administration of 3-nitropropionic acid (3-NPA), a mitochondrial toxin, to rats (Beal et al., 1993) or primates (Brouillet et al., 1995) can cause a selective loss of medium spiny neurons in the striatum. Their pathology is similar to HD neuropathology, such as selective loss of medium spiny neurons and some sparing of large neurons. Both freeze-clamp measurements and chemical shift magnetic resonance spectroscopy show that 3-NPA impairs energy metabolism (Beal et al., 1993). Deficit of energy metabolism in neurons is frequently associated with necrotic cell death. Dysfunction of mitochondria is, in some contexts, associated with apoptotic cell death.

Transgenic mice overexpressing exon 1 of the *huntingtin* gene are widely used as HD animal models, with HD-relevant manifestations in neurological phenotype as well as neuropathology including neuronal intranuclear inclusions and dystrophic neurites (Mangiarini et al., 1996; Schilling et al., 1999). In more detailed examinations, dying neurons with neuronal intranuclear inclusions display condensation of both the cytoplasm and nucleus, as well as ruffling of the plasma membrane, while maintaining ultrastructural preservation of cellular organelles. These cells do not develop blebbing of the nucleus or cytoplasm, apoptotic bodies, or fragmentation of DNA. Taking these observations together, Davies and colleagues (Turmaine et al., 2000) propose that neuronal cell death in the mice model might be neither apoptosis not necrosis. In contrast, Tagle's group (Reddy et al., 1998) reports TdT-mediated dUTP-biotin nick end labeling (TUNEL) positive dying neurons in mice expressing the full length of mutant Htt, suggesting the existence of apoptotic cell death. Mitochondrial abnormalities, especially a deficit of energy metabolism in brains, have been reported in Htt transgenic mice (Tabrizi et al., 2000).

Although it is still controversial in the existence of apoptosis in HD animal brains by neuropathological access, genetic intervention of an apoptotic cascade can ameliorate the pathology and phenotype of HD mice (Ona et al., 1999). Cross-breeding of the R6/2 line of Htt transgenic with mice expressing a dominant negative form of caspase-1 leads to extended survival, delay in appearance of neuronal inclusions, and neurotransmitter receptor alternations, as well as the onset of symptoms. Involvement of caspases in HD pathology is also suggested by a YAC transgenic mice expressing full-length Htt (Hodgson et al., 1999, Wellington et al., 2002). The YAC mice expressing mutant Htt, but not wild-type Htt, show neuronal degeneration and accumulation of *N*-terminal fragments of Htt that have occurred by caspase cleavage. This observation in YAC mice fits well with the cytotoxicity related to overexpression of the *N*-terminal fragment of mutant Htt in HD cell models described above (Cooper et al., 1998; Hackam et al., 1998; Wellington et al., 2000).

Dystrophic neurites are remarkable features of dying neurons in HD animal models. Li and associates (Li et al., 2001) report neuropil aggregates and dystrophic neurites prior to intranuclear accumulation of Htt, suggesting that such axonal degeneration is an early pathological event in HD mice.

Neuronal cell death in HD patient brains

Gross examination of coronal sections reveals bilateral, symmetric atrophy of the striatum in 95% of HD brains. In later stages of the disease, atrophy of the cerebral cortex becomes pronounced (Vonsattel, 2000).

Immunohistochemical analysis using antibodies against the *N*-terminal portions of Htt clarified the localization of Htt to neuronal intranuclear inclusions and dystrophic neurites in the HD striatum and cortex (DiFiglia et al., 1997). The prevalence of dystrophic neurites in deep layers of the cortex correlates with greater neurodegeneration in these layers, and their appearance in a presymptomatic adult suggests that they precede clinical onset. Dystrophic neurites are associated with neurofilament-positive axonal fibers, which agrees with evidence that dystrophic neurites are distended axon terminals. In addition, perinuclear and nuclear staining of Htt is observed with an antibody against the *N*-terminal Htt (Sapp et al., 1997).

There is some evidence using TUNEL assays for detecting apoptotic neurons to support the existence of apoptosis in HD brains (Dragunow et al., 1995; Portera-Cailliau et al., 1995; Thomas et al., 1995). However, it is still controversial whether the major form of neuronal cell death in HD patient brains is apoptotic.

Fitting with the observation of autophagic vacuoles in some HD cell models, endosomal- and lysosomal-like organelles and tubulovesicular structures with Htt immunoreactivity are more prominent in HD brains than in brains of control subjects (Sapp et al., 1997).

Biochemical analysis of HD brains has characterized mitochondrial abnormalities (Browne et al., 1997; Gu et al., 1996; Schapira, 1997; Tabrizi et al., 1999). The enzymatic activities of several mitochondrial proteins involved in oxidative phosphorylation, such as complex II, III, and IV, are decreased in HD brains. This type of deficit can link to necrosis, as well as apoptotic cell death. These enzymatic abnormalities are consistent with the findings in drug-induced and Htt transgenic mouse models.

Intracellular signaling

Modes of cell death by mutant Htt are, as described, very complex. However, at least two organelles play important roles for its cytotoxicity: the mitochondria and the nucleus (Sawa, 2001).

Involvement of mitochondria in HD pathology has been originally supported by biochemical analyses of patient tissues (Browne et al., 1997; Gu et al., 1996; Schapira, 1997) and brains from HD animal models (Beal et al., 1993; Brouillet et al., 1995; Tabrizi et al., 1999). Other classic hypotheses for explaining HD pathogenesis, such as excitotoxicity, metabolic stress, and the involvement of free radicals, are also associated with mitochondrial dysfunction (Sawa, 2001). Increased levels of free radicals impair mitochondrial functions and energy production, and metabolic inhibition predisposes neurons to excitotoxic damage.

In addition, our group (Sawa et al., 1999) reported increased vulnerability to mitochondrial membrane depolarization and apoptotic cell death in HD lymphoblasts. Greenamyre and associates (Panov et al., 2002) also observed abnormalities of mitochondrial membrane potentials. In their study, lymphoblast mitochondria from HD patients had a lower membrane potential and depolarized at lower calcium loads than did mitochondria from controls. In addition, transgenic mice expressing full-length mutant Htt display similar mitochondrial abnormalities that precede the onset of pathological or behavioral abnormalities.

The mitochondrial abnormalities found in HD patient tissues and model systems may underlie cytotoxic molecular signaling in HD. First, such susceptibility to mitochondrial depolarization can be directly associated with apoptotic signaling, which is initiated by release of cytochrome C or AIF (Green and Reed, 1998; van Loo et al., 2002). Released cytochrome C activates a chain of caspases via activation of Apaf-1. Secondly, such abnormalities impair mitochondrial buffering capacity of cytosolic calcium, which increases susceptibility to excitotoxicity. As a result, intracellular calcium levels are reportedly increased in the YAC-Htt transgenic mice (Hodgson et al., 1999) and PC12 cells expressing the *N*-terminal portion of mutant Htt (Bae, Snyder, and Sawa; unpublished observation). The elevated levels of cytosolic calcium can lead to several cellular dysfunctions, including over-activation of a calcium-dependent protease, calpain. Interestingly, cleavage of Htt by calpain has been reported in HD brains (Kim et al., 2001).

People have paid particular attention to the nucleus in HD pathology, since intranuclear inclusions with Htt immunoreactivity were found in brains of both HD patients and animal models of HD (Davies et al., 1997). Currently, the pathological significance of nuclear inclusions is controversial (Sisodia, 1998). However, the accumulation of Htt in the nucleus may play a role in the pathogenesis. In the YAC-Htt transgenic mice with expanded polyglutamine repeat, nuclear accumulation of Htt, particularly the *N*-terminal truncated form of Htt, was observed from earlier pathological stages (Hodgson et al., 1999). In Htt knock-in mice with the expanded polyglutamine tract, the nuclear accumulation of full-length Htt occurs prior to cleaved Htt and neuronal atrophy in striatal neurons (Wheeler et al., 2000).

Cytotoxicity by mutant Htt in the nucleus may be mediated via interactions with Htt-binding partners, such as CREB-binding protein (CBP) (Steffan et al., 2000), Sp1 (Dunah et al., 2002), TAFII130 (Dunah et al., 2002), glyceraldehydes-3-phosphate dehydrogenase (GAPDH) (Burke et al., 1996), and p53 (Steffan et al., 2000). CBP is depleted from its normal nuclear localization and is present in intranuclear inclusions in HD cell models, Htt transgenic mice brains, and HD autopsy brains (Nucifora et al., 2001). Such recruitment of CBP by mutant Htt decreases CBP-activated gene transcription, and overexpression of CBP rescues mutant-Htt-induced neuronal toxicity.

Htt interacts with the transcriptional activator Sp1 and coactivator TAFII130 (Dunah et al., 2002). Coexpression of Sp1 and TAFII130 in cultured striatal cells from Htt transgenic mice normalizes transcriptional inhibition and cytotoxicity caused by mutant Htt. The blockade of Sp1 function by mutant Htt may be due to a direct interaction of soluble mutant Htt with Sp1, which results in inhibition of Sp1 binding to DNA.

Htt binds to GAPDH, but the pathological significance of this is unclear (Burke et al., 1996). GAPDH translocates to the nucleus under oxidative stress, and nuclear GAPDH displays a gain-of-toxic function (Sawa et al., 1997). Nuclear translocation of Htt seems to be facilitated by GAPDH in some cellular models (Sawa and Snyder, unpublished observations). Thus, GAPDH together with Htt in the nucleus may, at least in part, mediate cytotoxicity of mutant Htt.

The protein p53 may have a particular importance in linking nuclear pathology and mitochondrial pathology. Many genes under the transcriptional control of p53 encode mitochondrial proteins that are directly associated with apoptosis and oxidative stress and include Bax, PUMA, proline oxidase, and PIG3 (Vogelstein et al., 2000). Mitochondrial dysfunction frequently occurs by oxidative stress. Since mutant Htt augments p53, it is conceivable that augmented p53 in the nucleus modifies mitochondrial functions, which may be associated with cell death signaling described above (Bae et al., manuscript submitted for publication).

Taken together, mutant Htt is expected to modulate functions of nuclear proteins, including important transcription factors. Thus, several groups have studied gene-expression changes mediated by mutant Htt, and one representative study reported diminished levels of mRNAs encoding components of the neurotransmitter, calcium, and retinoid signaling pathways in HD animal models (Luthi-Carter et al., 2000).

In addition to signaling cascades associated with the mitochondria and the nucleus, there are several other important death cascades that may play roles in HD pathology.

In neurons, the PI3K and Akt pathway functions as cytoprotective. In contrast, JNK and p38 play cytotoxic roles. Htt is reportedly phosphorylated directly by Akt, and the phosphorylation seems to be cytoprotective in striatal neurons with mutant Htt (Humbert et al., 2002). Apoptosis signal-regulating kinase 1 (ASK1) is activated under oxidative stress and ER stress that occurs in association with unfolded proteins (Matsuzawa and Ichijo, 2001). ASK1 is located upstream of JNK signaling. Essential roles of ASK1 in ER stress-induced neuronal cell death triggered by the expanded polyglutamine tract have been reported (Nishitoh et al., 2002).

Coexpression of human Hsp70 ameliorates polyglutamine toxicity, suppressing degeneration (Warrick et al., 1999). Despite the controversial role of Htt aggregate formations, unfolded and mis-folded proteins occurring in HD brains, as intermediate materials before forming aggregates, may have toxic properties. Molecular chaperones and heat-shock proteins in HD should have an important role in preventing these toxic cascades.

Question 2: How can mutant Htt cause selective neuronal cell death of HD brains?

Mutant Htt protein is widely expressed in both the central nervous system (CNS) and non-CNS tissues (Gutekunst et al., 1995; Sharp et al., 1995). Thus, it is still unclear as to why medium spiny striatal neurons selectively die in HD. One possible explanation attributes cell-type-specific interactions of Htt with other proteins that are associated with cell-death signaling. Htt interactors that are expressed exclusively in the striatum may play roles in HD pathology. As the cleavage of Htt by caspases and other enzymes is believed to mediate, or at least accelerate, HD pathology, cell-type preferential Htt cleavage in medium spiny striatal neuron can also be postulated.

An alternative possibility is an increase in the copy number of CAG repeats present in the mutant gene. Kennedy and Shelbourne (2000) investigated the mutation profile in somatic tissues derived from an accurate knock-in mouse model of HD. These mice, generated by inserting a 72–80 CAG repeat into the mouse counterpart of the human Htt gene, express full-length mutant Htt appropriately and display behavioral and pathological features reminiscent of early HD. These workers observed unexpectedly high levels of CAG repeat length variation in many tissues of the HD mice. The size of the mutations, particularly in some striatal cells, was much greater than previously seen in corresponding human tissue. Age-dependent and tissue-specific CAG repeat instability in Htt knock-in mice was reported by another group (Ishiguro et al., 2001).

As described above, much evidence suggests the involvement of excitotoxicity in HD (Beal, 1994). In medium spiny neurons, a selective combination of NMDA receptor subtypes, NR1A/NR2B is expressed. In HEK293 cells expressing mutant Htt, the co-expression of NR1A/NR2B-type NMDA receptors endows susceptibility to apoptotic cell death, but the co-expression of NR1A/NR2A does not (Zeron et al., 2001). Moreover, striatal neurons from YAC transgenic mice expressing full-length mutant Htt display a greater vulnerability to excitotoxic death, compared with striatal neurons from wild-type mice, and the NMDA-induced cell death is abolished by an NR2B subtype-specific antagonist (Zeron et al., 2002). These results suggest that the unique expression of a specific NMDA receptor subtypes in medium spiny neurons may, at least partly, contribute to their specific neuronal death in HD.

Conclusion

Mutant Htt seems to activate many death cascades. We should be cautious in considering their pathophysiological relevance to HD pathology. We will need efforts to integrate the cascades with hierarchy. Such organization may help to narrow down future therapeutic targets for HD.

Acknowledgements

We thank Ms. Y. Lema for preparation of the figures and the references. We thank Ms. J. Cheah and N. Zaller for their critical reading of the manuscript.

References

Ashkenazi A, Dixit VM: Death receptors: signaling and modulation. Science 281:1305–1308 (1998).

Beal MF: Huntington's disease, energy, and excitotoxicity. Neurobiol Aging 15:275–276 (1994).

Beal MF, Brouillet E, Jenkins BG, Ferrante RJ, Kowall NW, Miller JM, Storey E, Srivastava R, Rosen BR, Hyman BT: Neurochemical and histologic characterization of striatal excitotoxic lesions produced by the mitochondrial toxin 3-nitropropionic acid. J Neurosci 13:4181–4192 (1993).

Brouillet E, Hantraye P, Ferrante RJ, Dolan R, Leroy-Willig A, Kowall NW, Beal MF: Chronic mitochondrial energy impairment produces selective striatal degeneration and abnormal choreiform movements in primates. Proc natl Acad Sci, USA 92:7105–7109 (1995).

Browne SE, Bowling AC, MacGarvey U, Baik MJ, Berger SC, Muqit MM, Bird ED Beal MF: Oxidative damage and metabolic dysfunction in Huntington's disease: selective vulnerability of the basal ganglia. Ann Neurol 41:646–653 (1997).

Burke JR, Enghild JJ, Martin ME, Jou YS, Myers RM, Roses AD, Vance JM, Strittmatter WJ: Huntingtin and DRPLA proteins selectively interact with the enzyme GAPDH. Nature Med 2:347–350 (1996).

Cattaneo E, Rigamonti D, Goffredo D, Zuccato C, Squitieri F, Sipione S: Loss of normal huntingtin function: new developments in Huntington's disease research. Trends Neurosci 24:182–188 (2001).

Clarke PGH: Developmental cell death: morphological diversity and multiple mechanisms. Anat Embryol 181:195–213 (1990).

Cooper JK, Schilling G, Peters MF, Herring WJ, Sharp AH, Kaminsky Z, Masone J, Khan FA, Delanoy M, Borchelt DR, Dawson VL, Dawson TM, Ross CA: Truncated N-terminal fragments of huntingtin with expanded glutamine repeats form nuclear and cytoplasmic aggregates in cell culture. Hum molec Genet 7:783–790 (1998).

Cregan SP, Fortin A, MacLaurin JG, Callaghan SM, Cecconi F, Yu SW, Dawson TM, Dawson VL, Park DS, Kroemer G, Slack RS: Apoptosis-inducing factor is involved in the regulation of caspase-independent neuronal cell death. J Cell Biol 158:507–517 (2002).

Davies SW, Turmaine M, Cozens BA, DiFiglia M, Sharp AH, Ross CA, Scherzinger E, Wanker EE, Mangiarini L, Bates GP: Formation of neuronal intranuclear inclusions underlies the neurological dysfunction in mice transgenic for the HD mutation. Cell 90:537–548 (1997).

DiFiglia M, Sapp E, Chase KO, Davies SW, Bates GP, Vonsattel JP, Aronin N: Aggregation of huntingtin in neuronal intranuclear inclusions and dystrophic neurites in brain. Science 277:1990–1993 (1997).

Dragatsis I, Levine MS, Zeitlin S: Inactivation of Hdh in the brain and testis results in progressive neurodegeneration and sterility in mice. Nature Genet 26:300–306 (2000).

Dragunow M, Faull RL, Lawlor P, Beilharz EJ, Singleton K, Walker EB, Mee E: In situ evidence for DNA fragmentation in Huntington's disease striatum and Alzheimer's disease temporal lobes. Neuroreport 6:1053–1057 (1995).

Dunah AW, Jeong H, Griffin A, Kim YM, Standaert DG, Hersch SM, Mouradian MM, Young AB, Tanese N, Krainc D: Sp1 and TAFII130 transcriptional activity disrupted in early Huntington's disease. Science 296:2238–2243 (2002).

Green DR, Reed JC: Mitochondria and apoptosis. Science 281:1309–1312 (1998).

Gu M, Gash MT, Mann VM, Javoy-Agid F, Cooper JM, Schapira AH: Mitochondrial defect in Huntington's disease caudate nucleus. Ann Neurol 39:385–389 (1996).

Gutekunst CA, Levey AI, Heilman CJ, Whaley WL, Yi H, Nash NR, Rees HD, Madden JJ, Hersch SM: Identification and localization of huntingtin in brain and human lymphoblastoid cell lines with anti-fusion protein antibodies. Proc natl Acad Sci, USA 92:8710–8714 (1995).

Hackam AS, Singaraja R, Wellington CL, Metzler M, McCutcheon K, Zhang T, Kalchman M, Hayden MR: The influence of huntingtin protein size on nuclear localization and cellular toxicity. J Cell Biol 141:1097–1105 (1998).

Higashiyama H, Hirose F, Yamaguchi M, Inoue YH, Fujikake N, Matsukage A, Kakizuka A: Identification of ter94, Drosophila VCP, as a modulator of polyglutamine-induced neurodegeneration. Cell Death Differ 9:264–273 (2002).

Hirabayashi M, Inoue K, Tanaka K, Nakadate K, Ohsawa Y, Kamei Y, Popiel AH, Sinohara A Iwamatsu A, Kimura Y, Uchiyama Y, Hori S, Kakizuka A: VCP/p97 in abnormal protein aggregates, cytoplasmic vacuoles, and cell death, phenotypes relevant to neurodegeneration. Cell Death Differ 8:977–984 (2001).

Hodgson JG, Agopyan N, Gutekunst CA, Leavitt BR, LePiane F, Singaraja R, Smith DJ, Bissada N, McCutcheon K, Nasir J, Jamot L, Li XJ, Stevens ME, Rosemond E, Roder JC, Phillips AG, Rubin EM, Hersch SM, Hayden MR: A YAC mouse model for Huntington's disease with full-length mutant huntingtin, cytoplasmic toxicity, and selective striatal neurodegeneration. Neuron 23:181–192 (1999).

Humbert S, Bryson EA, Cordelieres FP, Connors NC, Datta SR, Finkbeiner S, Greenberg ME, Saudou F: The IGF-1/Akt pathway is neuroprotective in Huntington's disease and involves Huntingtin phosphorylation by Akt. Dev Cell 2:831–837 (2002).

The Huntington's Disease Collaborative Research Group: A novel gene containing a trinucleotide repeat that is expanded and unstable on Huntington's disease chromosomes. Cell 72:971–983 (1993).

Ishiguro H, Yamada K, Sawada H, Nishii K, Ichino N, Sawada M, Kurosawa Y, Matsushita N, Kobayashi K, Goto J, Hashida H, Masuda N, Kanazawa I, Nagatsu T: Age-dependent and tissue-specific CAG repeat instability occurs in mouse knock-in for a mutant Huntington's disease gene. J neurosci Res 65:289–297 (2001).

Kegel KB, Kim M, Sapp E, McIntyre C, Castano JG, Aronin N, DiFiglia M: Huntingtin expression stimulates endosomal-lysosomal activity, endosome tubulation, and autophagy. J Neurosci 20:7268–7278 (2000).

Kennedy L, Shelbourne PF: Dramatic mutation instability in HD mouse striatum: does polyglutamine load contribute to cell-specific vulnerability in Huntington's disease? Hum molec Genet 9:2539–2544 (2000).

Kerr JF, Wyllie AH, Currie AR: Apoptosis: a basic biological phenomenon with wide-ranging implications in tissue kinetics. Br J Cancer 26:239–257 (1972).

Kerr JFR, Harmon BV: Definition and incidence of apoptosis: an historical perspective, in: Apoptosis: The Molecular Basis of Cell Death, pp 5–29 (Cold Spring Harbor Laboratory Press, Cold Spring Harbor 1991).

Kim YJ, Yi Y, Sapp E, Wang Y, Cuiffo B, Kegel KB, Qin ZH, Aronin N, DiFiglia M: Caspase 3-cleaved N-terminal fragments of wild-type and mutant huntingtin are present in normal and Huntington's disease brains, associate with membranes, and undergo calpain-dependent proteolysis. Proc natl Acad Sci, USA 98:12784–12789 (2001).

Li H, Li SH, Yu ZX, Shelbourne P, Li XJ: Huntingtin aggregate-associated axonal degeneration is an early pathological event in Huntington's disease mice. J Neurosci 21:8473–8481 (2001).

Liang XH, Jackson S, Seaman M, Brown K, Kempkes B, Hibshoosh H, Levine B: Induction of autophagy and inhibition of tumorigenesis by beclin 1. Nature 402:672–676 (1999).

Loo G van, Saelens X, van Gurp M, MacFarlane M, Martin SJ, Vandenabeele P: The role of mitochondrial factors in apoptosis: a Russian roulette with more than one bullet. Cell Death Differ 9:1031–1042 (2002).

Luthi-Carter R, Strand A, Peters NL, Solano SM, Hollingsworth ZR, Menon AS, Frey AS, Spektor BS, Penney EB, Schilling G, Ross CA, Borchelt DR, Tapscott SJ, Young AB, Cha JH, Olson JM: Decreased expression of striatal signaling genes in a mouse model of Huntington's disease. Hum molec Genet 9:1259–1271 (2000).

Mangiarini L, Sathasivam K, Seller M, Cozens B, Harper A, Hetherington C, Lawton M, Trottier Y, Lehrach H, Davies SW, Bates GP: Exon 1 of the HD gene with an expanded CAG repeat is sufficient to cause a progressive neurological phenotype in transgenic mice. Cell 87:493–506 (1996).

Matsuzawa A, Ichijo, H: Molecular mechanisms of the decision between life and death: regulation of apoptosis by apoptosis signal-regulating kinase 1. J Biochem, Tokyo 130:1–8 (2001).

Moulder KL, Onodera O, Burke JR, Strittmatter WJ, Johnson Jr EM: Generation of neuronal intranuclear inclusions by polyglutamine-GFP: analysis of inclusion clearance and toxicity as a function of polyglutamine length. J Neurosci 19:705–715 (1999).

Nishitoh H, Matsuzawa A, Tobiume K, Saegusa K, Takeda K, Inoue K, Hor S, Kakizuka A, Ichijo H: ASK1 is essential for endoplasmic reticulum stress-induced neuronal cell death triggered by expanded polyglutamine repeats. Genes Dev 16:1345–1355 (2002).

Nucifora Jr FC, Sasaki M, Peters MF, Huan H, Cooper JK, Yamad M, Takahashi H, Tsuji S, Troncoso J, Dawson VL, Dawson TM, Ross CA: Interference by huntingtin and atrophin-1 with cbp-mediated transcription leading to cellular toxicity. Science 291:2423–2428 (2001).

Ona VO, Li M, Vonsattel JP, Andrews LJ, Khan SQ, Chung WM, Frey AS, Menon AS, Li XJ, Stieg PE, Yuan J, Penney JB, Young AB, Cha JH, Friedlander RM: Inhibition of caspase-1 slows disease progression in a mouse model of Huntington's disease. Nature 399:263–267 (1999).

Ordway JM, Tallaksen-Greene S, Gutekunst CA, Bernstein EM, Cearley JA, Wiener HW, Dure LSt, Lindsey R, Hersch SM, Jope RS, Albin RL, Detloff PJ: Ectopically expressed CAG repeats cause intranuclear inclusions and a progressive late onset neurological phenotype in the mouse. Cell 91:753–763 (1997).

Panov AV, Gutekunst CA, Leavitt BR, Hayden MR, Burke JR, Strittmatter WJ, Greenamyre JT: Early mitochondrial calcium defects in Huntington's disease are a direct effect of polyglutamines. Nature Neurosci 5:731–736 (2002).

Petersen A, Larsen KE, Behr GG, Romero N, Przedborski S, Brundin P, Sulzer D: Expanded CAG repeats in exon 1 of the Huntington's disease gene stimulate dopamine-mediated striatal neuron autophagy and degeneration. Hum molec Genet 10:1243–1254 (2001).

Portera-Cailliau C, Hedreen JC, Price DL, Koliatsos VE: Evidence for apoptotic cell death in Huntington disease and excitotoxic animal models. J Neurosci 15:3775–3787 (1995).

Quignon F, De Bels F, Koken M, Feunteun J, Ameisen JC, The H de: PML induces a novel caspase-independent death process. Nature Genet 20:259–265 (1998).

Ravikumar B, Duden R, Rubinsztein DC: Aggregate-prone proteins with polyglutamine and polyalanine expansions are degraded by autophagy. Hum molec Genet 11:1107–1117 (2002).

Reddy PH, Williams M, Charles V, Garrett L, Pike-Buchanan L, Whetsell Jr WO, Miller G, Tagle DA: Behavioural abnormalities and selective neuronal loss in HD transgenic mice expressing mutated full-length HD cDNA. Nature Genet 20:198–202 (1998).

Rigamonti D, Bauer JH, De-Fraja C, Conti L, Sipione S, Sciorati C, Clementi E, Hackam A, Hayden MR, Li Y, Cooper JK, Ross CA, Govoni S, Vincenz C, Cattaneo E: Wild-type huntingtin protects from apoptosis upstream of caspase-3. J Neurosci 15:3705–3713 (2000).

Ross CA: Polyglutamine pathogenesis: emergence of unifying mechanisms for Huntington's disease and related disorders. Neuron 35:819–822 (2002).

Sapp E, Schwarz C, Chase K, Bhide PG, Young AB, Penney J, Vonsattel JP, Aronin N, DiFiglia M: Huntingtin localization in brains of normal and Huntington's disease patients. Ann Neurol 42:604–612 (1997).

Saudou F, Finkbeiner S, Devys D, Greenberg ME: Huntingtin acts in the nucleus to induce apoptosis but death does not correlate with the formation of intranuclear inclusions. Cell 95:55–66 (1998).

Sawa A: Mechanisms for neuronal cell death and dysfunction in Huntington's disease: pathological cross-talk between the nucleus and the mitochondria? J molec Med 79:375–381 (2001).

Sawa A, Khan AA, Hester LD, Snyder SH: Glyceraldehyde-3-phosphate dehydrogenase: nuclear translocation participates in neuronal and nonneuronal cell death. Proc natl Acad Sci, USA 94:11669–11674 (1997).

Sawa A, Wiegand GW, Cooper J, Margolis RL, Sharp AH, Lawler Jr JF, Greenamyre JT, Snyder SH, Ross CA: Increased apoptosis of Huntington disease lymphoblasts associated with repeat length-dependent mitochondrial depolarization. Nature Med 5:1194–1198 (1999).

Schapira AH: Mitochondrial function in Huntington's disease: clues for pathogenesis and prospects for treatment. Ann Neurol 41:141–142 (1997).

Schilling G, Becher MW, Sharp AH, Jinnah HA, Duan K, Kotzuk JA, Slunt HH, Ratovitski T, Cooper JK, Jenkins NA, Copeland NG, Price DL, Ross CA, Borchelt DR: Intranuclear inclusions and neuritic aggregates in transgenic mice expressing a mutant N-terminal fragment of huntingtin. Hum molec Genet 8:397–407 (1999).

Server AC, Mobley WC: Neuronal cell death and the role of apoptosis, in: The Molecular Basis of Cell Death, pp 263–278 (Cold Spring Harbor Laboratory Press, Cold Spring Harbor 1991).

Sharp AH, Loev SJ, Schilling G, Li SH, Li XJ, Bao J, Wagster MV, Kotzuk JA, Steiner JP, Lo A, et al.: Widespread expression of Huntington's disease gene (IT15) protein product. Neuron. 14:1065–1074 (1995).

Sisodia SS: Nuclear inclusions in glutamine repeat disorders: are they pernicious, coincidental, or beneficial? Cell 95:1–4 (1998).

Sperandio S, Belle I de, Bredesen DE: An alternative, nonapoptotic form of programmed cell death. Proc natl Acad Sci, USA 97:14376–14381 (2000).

Steffan JS, Kazantsev A, Spasic-Boskovic O, Greenwald M, Zhu YZ, Gohler H, Wanker EE, Bates GP, Housman DE, Thompson LM: The Huntington's disease protein interacts with p53 and CREB-binding protein and represses transcription. Proc natl Acad Sci, USA 97:6763–6768 (2000).

Susin SA, Lorenzo HK, Zamzami N, Marzo I, Snow BE, Brothers GM, Mangion J, Jacotot E, Costantini P, Loeffler M, Larochette N, Goodlett DR, Aebersold R, Siderovski DP, Penninger JM, Kroemer G: Molecular characterization of mitochondrial apoptosis-inducing factor. Nature 397:441–446 (1999).

Tabrizi SJ, Cleeter MW, Xuereb J, Taanman JW, Cooper JM, Schapira AH: Biochemical abnormalities and excitotoxicity in Huntington's disease brain. Ann Neurol 45:25–32 (1999).

Tabrizi SJ, Workman J, Hart PE, Mangiarini L, Mahal A, Bates G, Cooper JM, Schapira AH: Mitochondrial dysfunction and free radical damage in the Huntington R6/2 transgenic mouse. Ann Neurol 47:80–86 (2000).

Thomas LB, Gates DJ, Richfield EK, O'Brien TF, Schweitzer JB, Steindler DA: DNA end labeling (TUNEL) in Huntington's disease and other neuropathological conditions. Expl Neurol 133:265–272 (1995).

Thornberry NA, Lazebnik Y: Caspases: enemies within. Science 281:1312–1316 (1998).

Turmaine M, Raza A, Mahal A, Mangiar ini L, Bates GP, Davies SW: Nonapoptotic neurodegeneration in a transgenic mouse model of Huntington's disease. Proc natl Acad Sci, USA 97:8093–8097 (2000).

Vogelstein B, Lane D, Levine AJ: Surfing the p53 network. Nature 408:307–310 (2000).

Vonsattel JPG: Neuropathology of Huntington's disease. Neurosci News 3:45–53 (2000).

Warrick JM, Chan HY, Gray-Board GL, Chai Y, Paulson HL, Bonini NM: Suppression of polyglutamine-mediated neurodegeneration in *Drosophila* by the molecular chaperone HSP70. Nature Genet 23:425–428 (1999).

Wellington CL, Ellerby LM, Gutekunst CA, Rogers D, Warby S, Graham RK, Loubser O, Raamsdonk JV, Singaraja R, Yang YZ, Gafni J, Bredesen D, Hersch SM, Leavitt BR, Roy S, Nicholson DW, Hayden MR: Caspase cleavage of mutant Huntingtin precedes neurodegeneration in Huntington's disease. J Neurosci 22:7862–7872 (2000).

Wellington CL, Singaraja R, Ellerby L, Savill J, Roy S, Leavitt B, Cattaneo E, Hackam A, Sharp A, Thornberry N, Nicholson DW, Bredesen DE, Hayden MR: Inhibiting caspase cleavage of huntingtin reduces toxicity and aggregate formation in neuronal and nonneuronal cells. J biol Chem 275:19831–19838 (2000).

Wheeler VC, White JK, Gutekunst CA, Vrbanac V, Weaver M, Li XJ, Li SH, Yi H, Vonsattel JP, Gusella JF, Hersch S, Auerbach W, Joyner AL, MacDonald ME: Long glutamine tracts cause nuclear localization of a novel form of huntingtin in medium spiny striatal neurons in HdhQ92 and HdhQ111 knock-in mice. Hum molec Genet 9:503–513 (2000).

Yue Z, Horton A, Bravin M, DeJager PL, Selimi F, Heintz N: A novel protein complex linking the delta 2 glutamate receptor and autophagy: implications for neurodegeneration in *lurcher* mice. Neuron 35:921–933 (2002).

Zeron MM, Chen N, Moshaver A, Lee AT, Wellington CL, Hayden MR, Raymond LA: Mutant huntingtin enhances excitotoxic cell death. Mol Cell Neurosci 17:41–53 (2001).

Zeron MM, Hansson O, Chen N, Wellington CL, Leavitt BR, Brundin P, Hayden MR, Raymond LA: Increased sensitivity to N-methyl-D-aspartate receptor-mediated excitotoxicity in a mouse model of Huntington's disease. Neuron 33:849–860 (2002).

Zhong S, Salomoni P, Pandolfi PP: The transcriptional role of PML and the nuclear body. Nature Cell Biol 2:E85–90 (2000).

Zuccato C, Ciammola A, Rigamonti D, Leavitt BR, Goffredo D, Conti L, MacDonald ME, Friedlander RM, Silani V, Hayden MR, Timmusk T, Sipione S, Cattaneo E: Loss of huntingtin-mediated BDNF gene transcription in Huntington's disease. Science 293:493–498 (2001).

Author Index

KARGER

Fax + 41 61 306 12 34
E-Mail karger@karger.ch
www.karger.com

Subject Index

Fax + 41 61 306 12 34
E-mail karger@karger.ch
www.karger.com